An Overview of Complex Systems

Edited by Paul F. Kisak

Contents

1 **Complex system** 1
   1.1 Overview 1
   1.2 Key concepts 1
      1.2.1 Systems 1
      1.2.2 Complexity 2
      1.2.3 Networks 2
      1.2.4 Nonlinearity 2
      1.2.5 Emergence 3
      1.2.6 Adaptation 3
   1.3 Features 3
   1.4 History 4
   1.5 Applications of complex systems 5
      1.5.1 Complexity in practice 5
      1.5.2 Complexity management 5
      1.5.3 Complexity economics 5
      1.5.4 Complexity and education 5
      1.5.5 Complexity and modeling 5
      1.5.6 Complexity and chaos theory 6
      1.5.7 Complexity and network science 6
      1.5.8 General form of complexity computation 6
   1.6 Notable figures 7
   1.7 See also 8
   1.8 References 8
   1.9 Further reading 9
   1.10 External links 9

2 **Emergence** 10
   2.1 In philosophy 10
      2.1.1 Definitions 11
      2.1.2 Strong and weak emergence 11

		2.1.3 Objective or subjective quality	13
	2.2	In religion, art and humanities	13
	2.3	Emergent properties and processes	14
	2.4	Emergent structures in nature	14
		2.4.1 Non-living, physical systems	15
		2.4.2 Living, biological systems	16
	2.5	In humanity	17
		2.5.1 Spontaneous order	17
		2.5.2 Computer AI	19
		2.5.3 Language	19
		2.5.4 Emergent change processes	19
	2.6	See also	19
	2.7	References	20
	2.8	Bibliography	21
	2.9	Further reading	21
	2.10	External links	23
3	**Self-organization**		**24**
	3.1	Overview	24
	3.2	Principles	24
	3.3	History	25
	3.4	By field	25
		3.4.1 Physics	25
		3.4.2 Chemistry	25
		3.4.3 Biology	25
		3.4.4 Computer science	26
		3.4.5 Cybernetics	26
		3.4.6 Human society	27
		3.4.7 In learning	27
		3.4.8 Traffic flow	28
		3.4.9 In linguistics	28
	3.5	Criticism	28
	3.6	See also	28
	3.7	Notes	28
	3.8	References	28
	3.9	Further reading	31
	3.10	External links	32
4	**Collective consciousness**		**34**

4.1	In Durkheimian social theory		34
4.2	Other uses of the term		34
4.3	See also		35
4.4	Notes		35
4.5	References		36

5 Collective behavior — 37

- 5.1 Defining the field — 37
- 5.2 Examples — 37
  - 5.2.1 Four forms — 37
- 5.3 Theories developed to explain — 38
- 5.4 See also — 39
- 5.5 Bibliography — 40
- 5.6 External links — 40
- 5.7 References — 40

6 Social dynamics — 41

- 6.1 Topics — 41
- 6.2 See also — 41
- 6.3 References — 41
- 6.4 Further reading — 42
- 6.5 External links — 42

7 Collective intelligence — 43

- 7.1 History — 44
- 7.2 Dimensions — 44
- 7.3 Collective intelligence factor c — 46
  - 7.3.1 Causes — 46
  - 7.3.2 Processes — 47
  - 7.3.3 Evidence — 48
  - 7.3.4 Predictive validity — 49
  - 7.3.5 Potential connections to individual intelligence — 49
  - 7.3.6 Controversies — 49
- 7.4 Alternative mathematical techniques — 50
  - 7.4.1 Computational collective intelligence — 50
  - 7.4.2 Collective intelligence quotient — 50
- 7.5 Applications — 51
  - 7.5.1 Cognition — 51
  - 7.5.2 Cooperation — 51

		7.5.3 Coordination	54
7.6	Alternative views		55
	7.6.1	A tool for combating self-preservation	55
	7.6.2	Separation from IQism	55
	7.6.3	Artificial intelligence views	55
7.7	See also		56
	7.7.1	Similar concepts and applications	56
	7.7.2	Computation and computer science	56
	7.7.3	Others	56
7.8	Notes and references		56
7.9	Bibliography		62
7.10	External links		63

8 Collective action — 64

8.1	The social identity model		64
	8.1.1	Perceived injustice	64
	8.1.2	Perceived efficacy	64
	8.1.3	Social identity	64
	8.1.4	Model refinement	64
8.2	Public good		65
	8.2.1	Collective action problem	65
	8.2.2	Exploitation of the great by the small	65
	8.2.3	Institutional design	65
8.3	In philosophy		66
8.4	Spontaneous consensus		66
	8.4.1	Dimensions	66
	8.4.2	Equilibrium mechanisms	67
	8.4.3	Methods and techniques	67
8.5	See also		67
8.6	Footnotes		68
8.7	Bibliography		69

9 Self-organized criticality — 70

9.1	Overview	70
9.2	Examples of self-organized critical dynamics	71
9.3	See also	71
9.4	References	72
9.5	Further reading	72

10 Herd mentality — 74

- 10.1 History — 74
- 10.2 Evidence based examples — 74
- 10.3 See also — 74
- 10.4 References — 75
  - 10.4.1 Further reading — 75
- 10.5 External links — 75

11 Phase transition — 76

- 11.1 Types of phase transition — 76
- 11.2 Classifications — 78
  - 11.2.1 Ehrenfest classification — 78
  - 11.2.2 Modern classifications — 78
- 11.3 Characteristic properties — 78
  - 11.3.1 Phase coexistence — 79
  - 11.3.2 Critical points — 79
  - 11.3.3 Symmetry — 79
  - 11.3.4 Order parameters — 79
  - 11.3.5 Relevance in cosmology — 79
  - 11.3.6 Critical exponents and universality classes — 80
  - 11.3.7 Critical slowing down and other phenomena — 81
  - 11.3.8 Percolation theory — 81
  - 11.3.9 Phase transitions in biological systems — 81
- 11.4 See also — 81
- 11.5 References — 81
- 11.6 Further reading — 83
- 11.7 External links — 84

12 Agent-based model — 85

- 12.1 History — 85
  - 12.1.1 Early developments — 85
  - 12.1.2 1970s and 1980s: the first models — 85
  - 12.1.3 1990s: expansion — 86
  - 12.1.4 2000s and later — 86
- 12.2 Theory — 87
  - 12.2.1 Framework — 87
- 12.3 Applications — 87
  - 12.3.1 In biology — 87
  - 12.3.2 In business, technology and network theory — 88

12.3.3	In economics and social sciences	88
12.3.4	Organizational ABM: agent-directed simulation	89

12.4 Implementation 89
12.5 Verification and validation 89
    12.5.1 Complex systems modelling 89
12.6 See also 90
12.7 References 91
    12.7.1 Inline 91
    12.7.2 General 94
12.8 External links 95
    12.8.1 Articles/general Information 95
    12.8.2 Simulation models 95

13 Synchronization   96

13.1 Transport 96
13.2 Communication 96
13.3 Dynamical systems 97
13.4 Human movement 97
13.5 Uses 97
13.6 See also 98
13.7 References 99
13.8 External links 99

14 Ant colony optimization algorithms   100

14.1 Overview 100
14.2 Common extensions 100
    14.2.1 Elitist ant system 100
    14.2.2 Max-min ant system (MMAS) 101
    14.2.3 Ant colony system 101
    14.2.4 Rank-based ant system (ASrank) 101
    14.2.5 Continuous orthogonal ant colony (COAC) 101
    14.2.6 Recursive ant colony optimization 101
14.3 Convergence 101
14.4 Example pseudo-code and formula 101
    14.4.1 Edge selection 101
    14.4.2 Pheromone update 102
14.5 Applications 102
    14.5.1 Scheduling problem 102
    14.5.2 Vehicle routing problem 103

- 14.5.3 Assignment problem 103
- 14.5.4 Set problem 103
- 14.5.5 Device sizing problem in nanoelectronics physical design 103
- 14.5.6 Antennas Optimization and Synthesis 103
- 14.5.7 Image processing 103
- 14.5.8 Others 104
- 14.6 Definition difficulty 104
- 14.7 Stigmergy algorithms 105
- 14.8 Related methods 105
- 14.9 History 106
- 14.10 References 107
- 14.11 Publications (selected) 111
- 14.12 External links 111

15 Particle swarm optimization   112

- 15.1 Algorithm 112
- 15.2 Parameter selection 113
- 15.3 Neighbourhoods and topologies 113
- 15.4 Inner workings 113
  - 15.4.1 Convergence 114
  - 15.4.2 Biases 114
- 15.5 Variants 114
  - 15.5.1 Hybridization 114
  - 15.5.2 Alleviate premature 114
  - 15.5.3 Simplifications 115
  - 15.5.4 Multi-objective optimization 115
  - 15.5.5 Binary, discrete, and combinatorial 115
- 15.6 See also 115
- 15.7 References 115
- 15.8 External links 118

16 Swarm behaviour   119

- 16.1 Models 119
  - 16.1.1 Mathematical models 119
  - 16.1.2 Evolutionary models 120
  - 16.1.3 Agents 120
  - 16.1.4 Self-organization 120
  - 16.1.5 Emergence 120
  - 16.1.6 Stigmergy 121

- 16.1.7 Swarm intelligence ... 121
- 16.1.8 Algorithms ... 121
- 16.2 Biological swarming ... 122
  - 16.2.1 Insects ... 123
  - 16.2.2 Birds ... 125
  - 16.2.3 Marine life ... 126
  - 16.2.4 Plants ... 128
  - 16.2.5 Other organisms ... 129
- 16.3 People ... 129
- 16.4 Robotics ... 130
- 16.5 Military ... 131
- 16.6 Gallery ... 131
- 16.7 Myths ... 132
- 16.8 See also ... 132
- 16.9 References ... 132
- 16.10 Sources ... 137
- 16.11 External links ... 138

17 Network science — 139

- 17.1 Background and history ... 139
  - 17.1.1 Department of Defense initiatives ... 140
- 17.2 Network properties ... 140
  - 17.2.1 Size ... 141
  - 17.2.2 Density ... 141
  - 17.2.3 Planar Network Density ... 141
  - 17.2.4 Average degree ... 141
  - 17.2.5 Average path length (or characteristic path length) ... 141
  - 17.2.6 Diameter of a network ... 141
  - 17.2.7 Clustering coefficient ... 141
  - 17.2.8 Connectedness ... 141
  - 17.2.9 Node centrality ... 142
  - 17.2.10 Node influence ... 142
- 17.3 Network models ... 142
  - 17.3.1 Erdős–Rényi random graph model ... 142
  - 17.3.2 Watts–Strogatz small world model ... 142
  - 17.3.3 Barabási–Albert (BA) preferential attachment model ... 143
  - 17.3.4 Fitness model ... 144
- 17.4 Network analysis ... 144
  - 17.4.1 Social network analysis ... 144

- 17.4.2 Dynamic network analysis 144
- 17.4.3 Biological network analysis 144
- 17.4.4 Link analysis 145
- 17.4.5 Centrality measures 146
- 17.5 Spread of content in networks 146
  - 17.5.1 The SIR model 146
  - 17.5.2 The master equation approach 147
- 17.6 Interdependent networks 147
- 17.7 Multilayer networks 148
- 17.8 Network optimization 148
- 17.9 See also 148
- 17.10 Further reading 148
- 17.11 Notes 149

18 Scale-free network — 152
- 18.1 History 152
- 18.2 Characteristics 153
- 18.3 Examples 154
- 18.4 Generative models 154
- 18.5 Generalized scale-free model 155
  - 18.5.1 Features 155
  - 18.5.2 Examples 155
- 18.6 Scale-free ideal network 157
- 18.7 See also 157
- 18.8 References 157
- 18.9 Further reading 158

19 Social network analysis — 161
- 19.1 History 161
- 19.2 Metrics 162
  - 19.2.1 Connections 162
  - 19.2.2 Distributions 162
  - 19.2.3 Segmentation 162
- 19.3 Modelling and visualization of networks 162
  - 19.3.1 Social networking potential 163
- 19.4 Practical applications 163
  - 19.4.1 In computer-supported collaborative learning 164
- 19.5 See also 165
- 19.6 References 166

19.7 External links . . . 168
    19.7.1 Further reading . . . 168
    19.7.2 Organizations . . . 168
    19.7.3 Peer-reviewed journals . . . 168
    19.7.4 Textbooks and educational resources . . . 169

20 Small-world network — 170

20.1 Properties of small-world networks . . . 171
20.2 Examples of small-world networks . . . 171
20.3 Examples of non-small-world networks . . . 171
20.4 Network robustness . . . 172
20.5 Construction of small-world networks . . . 172
20.6 Applications . . . 172
    20.6.1 Applications to sociology . . . 172
    20.6.2 Applications to earth sciences . . . 173
    20.6.3 Applications to computing . . . 173
    20.6.4 Small-world neural networks in the brain . . . 173
20.7 Small world with a distribution of link length . . . 173
20.8 See also . . . 173
20.9 References . . . 174
    20.9.1 Books . . . 175
    20.9.2 Journal articles . . . 175
20.10 External links . . . 175

21 Centrality — 176

21.1 Definition and characterization of centrality indices . . . 176
    21.1.1 Characterization by network flows . . . 177
    21.1.2 Characterization by walk structure . . . 177
    21.1.3 Radial-volume centralities exist on a spectrum . . . 177
21.2 Important limitations . . . 178
21.3 Degree centrality . . . 178
21.4 Closeness centrality . . . 179
    21.4.1 Harmonic centrality . . . 179
21.5 Betweenness centrality . . . 179
21.6 Eigenvector centrality . . . 180
    21.6.1 Using the adjacency matrix to find eigenvector centrality . . . 180
21.7 Katz centrality . . . 180
21.8 PageRank centrality . . . 180
21.9 Percolation centrality . . . 181

21.10 Cross-clique centrality 181

21.11 Freeman Centralization 181

21.12 Dissimilarity based centrality measures 182

21.13 Extensions 182

21.14 See also 182

21.15 Notes and references 182

21.16 Further reading 184

22 Network motif   185

22.1 Definition 185

22.2 History 186

22.3 Motif Discovery Algorithms 186

    22.3.1 mfinder 186

    22.3.2 FPF (Mavisto) 187

    22.3.3 ESU (FANMOD) 187

    22.3.4 NeMoFinder 189

    22.3.5 Grochow-Kellis 189

    22.3.6 Color-Coding Approach 190

    22.3.7 MODA 190

    22.3.8 Kavosh 191

    22.3.9 G-Tries 192

    22.3.10 Comparison 192

    22.3.11 Classification of Algorithms 192

22.4 Well-Established Motifs and Their Functions 193

    22.4.1 Negative auto-regulation (NAR) 193

    22.4.2 Positive auto-regulation (PAR) 193

    22.4.3 Feed-forward loops (FFL) 193

    22.4.4 Coherent type 1 FFL (C1-FFL) 194

    22.4.5 Incoherent type 1 FFL (I1-FFL) 194

    22.4.6 Multi-output FFLs 194

    22.4.7 Single-input modules (SIM) 194

    22.4.8 Dense overlapping regulons (DOR) 194

22.5 Activity motifs 194

22.6 Criticism 195

22.7 See also 195

22.8 References 195

22.9 External links 198

23 Graph theory   199

23.1 Definitions 199
    23.1.1 Graph 199
23.2 Applications 199
23.3 History 201
23.4 Graph drawing 202
23.5 Graph-theoretic data structures 202
23.6 Problems 203
    23.6.1 Enumeration 203
    23.6.2 Subgraphs, induced subgraphs, and minors 203
    23.6.3 Graph coloring 203
    23.6.4 Subsumption and unification 203
    23.6.5 Route problems 204
    23.6.6 Network flow 204
    23.6.7 Visibility problems 204
    23.6.8 Covering problems 204
    23.6.9 Decomposition problems 204
    23.6.10 Graph classes 204
23.7 See also 204
    23.7.1 Related topics 205
    23.7.2 Algorithms 205
    23.7.3 Subareas 205
    23.7.4 Related areas of mathematics 205
    23.7.5 Generalizations 205
    23.7.6 Prominent graph theorists 205
23.8 Notes 206
23.9 References 207
23.10 External links 207
    23.10.1 Online textbooks 207

24 Scalability   208

24.1 Measures 208
24.2 Examples 208
24.3 Horizontal and vertical scaling 209
24.4 Database scalability 209
24.5 Strong versus eventual consistency (storage) 210
24.6 Performance tuning versus hardware scalability 210
24.7 Weak versus strong scaling 211
24.8 See also 211
24.9 References 211

24.10 External links 211

25 Robustness (computer science) — 212
- 25.1 Introduction 212
- 25.2 Challenges 212
- 25.3 Areas 212
  - 25.3.1 Robust programming 213
  - 25.3.2 Robust machine learning 213
  - 25.3.3 Robust network design 213
- 25.4 Examples 213
- 25.5 See also 213
- 25.6 References 213

26 Systems biology — 214
- 26.1 Overview 214
- 26.2 History 215
- 26.3 Associated disciplines 216
- 26.4 Bioinformatics and data analysis 217
- 26.5 See also 218
- 26.6 References 218
- 26.7 Further reading 220
- 26.8 External links 220

27 Dynamic network analysis — 221
- 27.1 Meta-network 221
- 27.2 Illustrative problems that people in the DNA area work on 222
- 27.3 See also 222
- 27.4 References 222
- 27.5 Further reading 223
- 27.6 External links 224

28 Complex adaptive system — 225
- 28.1 Overview 225
  - 28.1.1 General properties 225
  - 28.1.2 Characteristics 225
- 28.2 Modeling and simulation 226
- 28.3 Evolution of complexity 226
- 28.4 See also 227
- 28.5 References 227
- 28.6 Literature 228

29 Evolution — 230

- 29.1 History of evolutionary thought . . . 231
  - 29.1.1 Classical times . . . 231
  - 29.1.2 Medieval . . . 231
  - 29.1.3 Pre-Darwinian . . . 232
  - 29.1.4 Darwinian revolution . . . 232
  - 29.1.5 Pangenesis and heredity . . . 233
  - 29.1.6 The 'modern synthesis' . . . 233
  - 29.1.7 Further syntheses . . . 233
- 29.2 Heredity . . . 233
- 29.3 Variation . . . 235
  - 29.3.1 Mutation . . . 235
  - 29.3.2 Sex and recombination . . . 235
  - 29.3.3 Gene flow . . . 236
- 29.4 Mechanisms . . . 237
  - 29.4.1 Natural selection . . . 237
  - 29.4.2 Biased mutation . . . 238
  - 29.4.3 Genetic drift . . . 239
  - 29.4.4 Genetic hitchhiking . . . 239
  - 29.4.5 Gene flow . . . 240
- 29.5 Outcomes . . . 240
  - 29.5.1 Adaptation . . . 241
  - 29.5.2 Coevolution . . . 242
  - 29.5.3 Cooperation . . . 242
  - 29.5.4 Speciation . . . 243
  - 29.5.5 Extinction . . . 244
- 29.6 Evolutionary history of life . . . 245
  - 29.6.1 Origin of life . . . 246
  - 29.6.2 Common descent . . . 246
  - 29.6.3 Evolution of life . . . 247
- 29.7 Applications . . . 247
- 29.8 Social and cultural responses . . . 248
- 29.9 See also . . . 249
- 29.10 References . . . 249
- 29.11 Bibliography . . . 265
- 29.12 Further reading . . . 269
- 29.13 External links . . . 270

28.7 External links . . . 229

30 Adaptation — 271

- 30.1 History — 271
- 30.2 General principles — 271
  - 30.2.1 What adaptation is — 271
  - 30.2.2 What adaptation is not — 272
  - 30.2.3 Adaptedness and fitness — 273
  - 30.2.4 Genetic basis — 273
- 30.3 Types — 273
  - 30.3.1 Changes in habitat — 274
  - 30.3.2 Genetic change — 274
  - 30.3.3 Co-adaptation — 274
  - 30.3.4 Mimicry — 274
  - 30.3.5 Internal adaptations — 275
  - 30.3.6 Trade-offs — 275
- 30.4 Shifts in function — 276
  - 30.4.1 Pre-adaptations — 276
  - 30.4.2 Co-option of existing traits: exaptation — 276
- 30.5 Non-adaptive traits — 277
- 30.6 Extinction and coextinction — 277
- 30.7 Philosophical issues — 277
- 30.8 See also — 277
- 30.9 References — 277
- 30.10 Sources — 280

31 Artificial neural network — 283

- 31.1 History — 283
  - 31.1.1 Hebbian learning — 284
  - 31.1.2 Backpropagation — 284
  - 31.1.3 Hardware-based designs — 285
  - 31.1.4 Contests — 285
  - 31.1.5 Convolutional networks — 285
- 31.2 Models — 285
  - 31.2.1 Components of an artificial neural network — 285
  - 31.2.2 Neural networks as functions — 286
  - 31.2.3 Learning — 287
  - 31.2.4 Learning paradigms — 288
  - 31.2.5 Learning algorithms — 289
- 31.3 Variants — 289
  - 31.3.1 Group method of data handling — 289

- 31.3.2 Convolutional neural networks ... 289
- 31.3.3 Long short-term memory ... 290
- 31.3.4 Deep reservoir computing ... 290
- 31.3.5 Deep belief networks ... 290
- 31.3.6 Large memory storage and retrieval neural networks ... 291
- 31.3.7 Stacked (de-noising) auto-encoders ... 291
- 31.3.8 Deep stacking networks ... 291
- 31.3.9 Tensor deep stacking networks ... 292
- 31.3.10 Spike-and-slab RBMs ... 292
- 31.3.11 Compound hierarchical-deep models ... 292
- 31.3.12 Deep predictive coding networks ... 293
- 31.3.13 Networks with separate memory structures ... 293
- 31.4 Multilayer kernel machine ... 294
- 31.5 Use ... 294
- 31.6 Applications ... 295
  - 31.6.1 Neuroscience ... 295
- 31.7 Theoretical properties ... 296
  - 31.7.1 Computational power ... 296
  - 31.7.2 Capacity ... 296
  - 31.7.3 Convergence ... 296
  - 31.7.4 Generalization and statistics ... 296
- 31.8 Criticism ... 296
  - 31.8.1 Training issues ... 297
  - 31.8.2 Theoretical issues ... 297
  - 31.8.3 Hardware issues ... 297
  - 31.8.4 Practical counterexamples to criticisms ... 297
  - 31.8.5 Hybrid approaches ... 298
- 31.9 Types ... 298
- 31.10 Gallery ... 298
- 31.11 See also ... 299
- 31.12 References ... 299
- 31.13 Bibliography ... 308
- 31.14 External links ... 309

32 Evolutionary computation — 310

- 32.1 History ... 310
- 32.2 Techniques ... 310
- 32.3 Evolutionary algorithms ... 311
- 32.4 Practitioners ... 311

	32.5	See also	312
	32.6	Bibliography	312
	32.7	References	313

33 Genetic algorithm — 314

- 33.1 Methodology — 314
  - 33.1.1 Optimization problems — 314
- 33.2 The building block hypothesis — 316
- 33.3 Limitations — 316
- 33.4 Variants — 317
  - 33.4.1 Chromosome representation — 317
  - 33.4.2 Elitism — 318
  - 33.4.3 Parallel implementations — 318
  - 33.4.4 Adaptive GAs — 318
- 33.5 Problem domains — 319
- 33.6 History — 319
  - 33.6.1 Commercial products — 320
- 33.7 Related techniques — 320
  - 33.7.1 Parent fields — 320
  - 33.7.2 Related fields — 320
- 33.8 See also — 322
- 33.9 References — 322
- 33.10 Bibliography — 324
- 33.11 External links — 325
  - 33.11.1 Resources — 325
  - 33.11.2 Tutorials — 325

34 Genetic programming — 326

- 34.1 History — 326
- 34.2 Program representation — 326
- 34.3 Other approaches — 327
- 34.4 Meta-genetic programming — 327
- 34.5 See also — 327
- 34.6 References — 328
- 34.7 External links — 328

35 Artificial life — 329

- 35.1 Overview — 329
- 35.2 Philosophy — 329

- 35.3 Organizations 330
- 35.4 Software-based ("soft") 330
  - 35.4.1 Techniques 330
  - 35.4.2 Notable simulators 330
  - 35.4.3 Complex systems modelling 330
- 35.5 Hardware-based ("hard") 331
- 35.6 Biochemical-based ("wet") 331
- 35.7 Open problems 331
- 35.8 Related subjects 332
- 35.9 History 332
- 35.10 Criticism 332
- 35.11 See also 332
- 35.12 References 333
- 35.13 External links 333

36 Machine learning   334

- 36.1 Overview 334
  - 36.1.1 Types of problems and tasks 334
- 36.2 History and relationships to other fields 335
  - 36.2.1 Relation to statistics 336
- 36.3 Theory 336
- 36.4 Approaches 337
  - 36.4.1 Decision tree learning 337
  - 36.4.2 Association rule learning 337
  - 36.4.3 Artificial neural networks 337
  - 36.4.4 Deep learning 337
  - 36.4.5 Inductive logic programming 337
  - 36.4.6 Support vector machines 337
  - 36.4.7 Clustering 337
  - 36.4.8 Bayesian networks 338
  - 36.4.9 Reinforcement learning 338
  - 36.4.10 Representation learning 338
  - 36.4.11 Similarity and metric learning 338
  - 36.4.12 Sparse dictionary learning 338
  - 36.4.13 Genetic algorithms 339
  - 36.4.14 Rule-based machine learning 339
- 36.5 Applications 339
- 36.6 Model assessments 340
- 36.7 Ethics 340

36.8 Software 340
    36.8.1 Free and open-source software 340
    36.8.2 Proprietary software with free and open-source editions 341
    36.8.3 Proprietary software 341
36.9 Journals 341
36.10 Conferences 341
36.11 See also 341
36.12 References 342
36.13 Further reading 343
36.14 External links 344

37 Evolutionary developmental biology   345

37.1 History 345
    37.1.1 Recapitulation 345
    37.1.2 Evolutionary morphology 346
    37.1.3 The modern synthesis of the early 20th century 347
    37.1.4 The lac operon 347
    37.1.5 The birth of evo-devo and a second synthesis 347
37.2 The control of body structure 347
    37.2.1 Deep homology 347
    37.2.2 Gene toolkit 348
    37.2.3 The embryo's regulatory networks 348
37.3 The origins of novelty 350
    37.3.1 Variations in the toolkit 350
    37.3.2 Consolidation of epigenetic changes 350
37.4 Eco-evo-devo 350
37.5 See also 350
37.6 Notes 351
37.7 References 353

38 Evolutionary robotics   354

38.1 History 354
38.2 Objectives 354
38.3 Motivation 355
38.4 Conferences and institutes 355
    38.4.1 Main conferences 355
    38.4.2 Academic institutes and researchers 355
38.5 See also 356
38.6 References 356

39 Pattern formation — 357
- 38.7 External links . . . 356

39 Pattern formation — 357
- 39.1 Examples . . . 357
  - 39.1.1 Biology . . . 357
  - 39.1.2 Chemistry . . . 358
  - 39.1.3 Physics . . . 358
  - 39.1.4 Mathematics . . . 358
  - 39.1.5 Computer graphics . . . 358
- 39.2 References . . . 358
- 39.3 Bibliography . . . 359
- 39.4 External links . . . 359

40 Reaction–diffusion system — 360
- 40.1 One-component reaction–diffusion equations . . . 360
- 40.2 Two-component reaction–diffusion equations . . . 361
- 40.3 Three- and more-component reaction–diffusion equations . . . 363
- 40.4 Applications and universality . . . 363
- 40.5 Experiments . . . 363
- 40.6 Numerical treatments . . . 364
- 40.7 See also . . . 364
- 40.8 Some examples of reaction-diffusion equations . . . 364
- 40.9 References . . . 364
- 40.10 External links . . . 365

41 Partial differential equation — 366
- 41.1 Introduction . . . 366
- 41.2 Existence and uniqueness . . . 367
- 41.3 Notation . . . 367
- 41.4 Classification . . . 367
  - 41.4.1 Equations of first order . . . 367
  - 41.4.2 Linear equations of second order . . . 368
  - 41.4.3 Systems of first-order equations and characteristic surfaces . . . 368
  - 41.4.4 Equations of mixed type . . . 369
  - 41.4.5 Infinite-order PDEs in quantum mechanics . . . 369
- 41.5 Analytical solutions . . . 369
  - 41.5.1 Separation of variables . . . 369
  - 41.5.2 Method of characteristics . . . 369
  - 41.5.3 Integral transform . . . 370

- 41.5.4 Change of variables 370
- 41.5.5 Fundamental solution 370
- 41.5.6 Superposition principle 370
- 41.5.7 Methods for non-linear equations 370
- 41.5.8 Lie group method 370
- 41.5.9 Semianalytical methods 371
- 41.6 Numerical solutions 371
  - 41.6.1 Finite element method 371
  - 41.6.2 Finite difference method 371
  - 41.6.3 Finite volume method 371
- 41.7 See also 371
- 41.8 Notes 372
- 41.9 References 372
- 41.10 Further reading 373
- 41.11 External links 373

42 Dissipative system 374

- 42.1 Overview 374
- 42.2 Dissipative structures in thermodynamics 374
- 42.3 Dissipative systems in control theory 375
- 42.4 Quantum dissipative systems 375
- 42.5 See also 375
- 42.6 Notes 375
- 42.7 References 376
- 42.8 External links 376

43 Percolation 377

- 43.1 Background 377
- 43.2 Examples 377
- 43.3 See also 378
- 43.4 References 378
- 43.5 Further reading 378
- 43.6 External links 378

44 Cellular automaton 379

- 44.1 Overview 380
- 44.2 History 381
- 44.3 Classification 382
  - 44.3.1 Reversible 383

- 44.3.2 Totalistic 383
- 44.3.3 Related automata 383
- 44.4 Elementary cellular automata 384
- 44.5 Rule space 385
- 44.6 Biology 385
- 44.7 Chemical types 386
- 44.8 Applications 386
  - 44.8.1 Computer processors 386
  - 44.8.2 Cryptography 386
  - 44.8.3 Error correction coding 386
- 44.9 Modeling physical reality 386
- 44.10 Specific rules 387
- 44.11 Problems solved 387
- 44.12 See also 387
- 44.13 Reference notes 387
- 44.14 References 390
- 44.15 External links 390

45 Spatial ecology 392
- 45.1 Overview 392
- 45.2 History 392
- 45.3 Concepts 392
  - 45.3.1 Scale 392
  - 45.3.2 Spatial autocorrelation 393
  - 45.3.3 Pattern 393
- 45.4 Applications 393
  - 45.4.1 Research 393
  - 45.4.2 Interdisciplinary 394
- 45.5 Statistical tests 394
  - 45.5.1 Tests based on distance 394
- 45.6 See also 394
- 45.7 References 394
- 45.8 External links 395

46 Self-replication 396
- 46.1 Overview 396
  - 46.1.1 Theory 396
  - 46.1.2 Classes of self-replication 396
  - 46.1.3 A self-replicating computer program 397

        46.1.4 Self-replicating tiling 397
        46.1.5 Applications 397
   46.2 Mechanical self-replication 398
   46.3 Fields 398
   46.4 In industry 399
        46.4.1 Space exploration and manufacturing 399
        46.4.2 Molecular manufacturing 399
   46.5 See also 399
   46.6 References 400

47 Geomorphology **401**

   47.1 Overview 401
   47.2 History 402
        47.2.1 Ancient geomorphology 403
        47.2.2 Early modern geomorphology 403
        47.2.3 Climatic geomorphology 404
        47.2.4 Quantitative and process geomorphology 404
        47.2.5 Contemporary geomorphology 405
   47.3 Processes 405
        47.3.1 Aeolian processes 405
        47.3.2 Biological processes 406
        47.3.3 Fluvial processes 406
        47.3.4 Glacial processes 406
        47.3.5 Hillslope processes 407
        47.3.6 Igneous processes 407
        47.3.7 Tectonic processes 407
        47.3.8 Marine processes 408
   47.4 Scales in geomorphology 408
   47.5 Overlap with other fields 408
   47.6 See also 408
   47.7 References 409
   47.8 Further reading 410
   47.9 External links 411
   47.10 Text and image sources, contributors, and licenses 412
        47.10.1 Text 412
        47.10.2 Images 428
        47.10.3 Content license 438

Chapter 1

Complex system

"Complex systems" redirects here. For the journal, see Complex Systems (journal).

A **complex system** is a system composed of many components which may interact with each other. In many cases it is useful to represent such a system as a network where the nodes represent the components and the links their interactions. Examples of complex systems are Earth's global climate, organisms, the human brain, social and economic organizations (like cities), an ecosystem, a living cell, and ultimately the entire universe.

Complex systems are systems whose behavior is intrinsically difficult to model due to the dependencies, relationships, or interactions between their parts or between a given system and its environment. Systems that are "complex" have distinct properties that arise from these relationships, such as nonlinearity, emergence, spontaneous order, adaptation, and feedback loops, among others. Because such systems appear in a wide variety of fields, the commonalities among them have become the topic of their own independent area of research.

1.1 Overview

The term *complex systems* often refers to the study of complex systems, which is an approach to science that investigates how relationships between a system's parts give rise to its collective behaviors and how the system interacts and forms relationships with its environment.[1] The study of complex systems regards collective, or system-wide, behaviors as the fundamental object of study; for this reason, complex systems can be understood as an alternative paradigm to reductionism, which attempts to explain systems in terms of their constituent parts and the individual interactions between them.

As an interdisciplinary domain, complex systems draws contributions from many different fields, such as the study of self-organization from physics, that of spontaneous order from the social sciences, chaos from mathematics, adaptation from biology, and many others. *Complex systems* is therefore often used as a broad term encompassing a research approach to problems in many diverse disciplines, including statistical physics, information theory, nonlinear dynamics, anthropology, computer science, meteorology, sociology, economics, psychology, and biology.

1.2 Key concepts

1.2.1 Systems

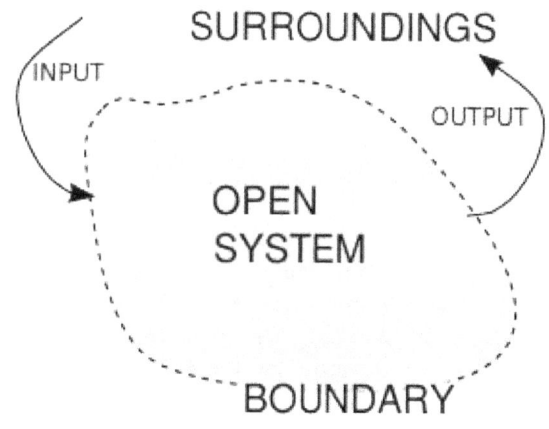

Open systems *have input and output flows, representing exchanges of matter, energy or information with their surroundings.*

Complex systems is chiefly concerned with the behaviors and properties of *systems*. A system, broadly defined, is a set of entities that, through their interactions, relationships, or dependencies, form a unified whole. It is always defined in terms of its *boundary*, which determines the entities that are or are not part of the system. Entities lying outside the system then become part of the system's *environment*.

A system can exhibit *properties* that produce *behaviors*

which are distinct from the properties and behaviors of its parts: these system-wide or *global* properties and behaviors are characteristics of how the system interacts with or appears to its environment, or of how its parts behave (say, in response to external stimuli) by virtue of being within the system. The notion of *behavior* implies that the study of systems is also concerned with processes that take place over time (or, in mathematics, some other phase space parameterization). Because of their broad, interdisciplinary applicability, systems concepts play a central role in complex systems.

As a field of study, complex systems is a subset of systems theory. General systems theory focuses similarly on the collective behaviors of interacting entities, but it studies a much broader class of systems, including non-complex systems where traditional reductionist approaches may remain viable. Indeed, systems theory seeks to explore and describe *all* classes of systems, and the invention of categories that are useful to researchers across widely varying fields is one of systems theory's main objectives.

As it relates to complex systems, systems theory contributes an emphasis on the way relationships and dependencies between a system's parts can determine system-wide properties. It also contributes the interdisciplinary perspective of the study of complex systems: the notion that shared properties link systems across disciplines, justifying the pursuit of modeling approaches applicable to complex systems wherever they appear. Specific concepts important to complex systems, such as emergence, feedback loops, and adaptation, also originate in systems theory.

1.2.2 Complexity

Systems exhibit complexity when difficulties with modeling them are endemic. This means their behaviors cannot be understood apart from the very properties that make them difficult to model, and they are governed entirely, or almost entirely, by the behaviors those properties produce. Any modeling approach that ignores such difficulties or characterizes them as noise, then, will necessarily produce models that are neither accurate nor useful. As yet no fully general theory of complex systems has emerged for addressing these problems, so researchers must solve them in domain-specific contexts. Researchers in complex systems address these problems by viewing the chief task of modeling to be capturing, rather than reducing, the complexity of their respective systems of interest.

While no generally accepted exact definition of complexity exists yet, there are many archetypal examples of complexity. Systems can be complex if, for instance, they have chaotic behavior (behavior that exhibits extreme sensitivity to initial conditions), or if they have emergent properties (properties that are not apparent from their components in isolation but which result from the relationships and dependencies they form when placed together in a system), or if they are computationally intractable to model (if they depend on a number of parameters that grows too rapidly with respect to the size of the system).

1.2.3 Networks

The interacting components of a complex system form a network, which is a collection of discrete objects and relationships between them, usually depicted as a graph of vertices connected by edges. Networks can describe the relationships between individuals within an organization, between logic gates in a circuit, between genes in gene regulatory networks, or between any other set of related entities.

Networks often describe the sources of complexity in complex systems. Studying complex systems as networks therefore enables many useful applications of graph theory and network science. Some complex systems, for example, are also complex networks, which have properties such as power-law degree distributions that readily lend themselves to emergent or chaotic behavior. The fact that the number of edges in a complete graph grows quadratically in the number of vertices sheds additional light on the source of complexity in large networks: as a network grows, the number of relationships between entities quickly dwarfs the number of entities in the network.

1.2.4 Nonlinearity

Complex systems often have nonlinear behavior, meaning they may respond in different ways to the same input depending on their state or context. In mathematics and physics, nonlinearity describes systems in which a change in the size of the input does not produce a proportional change in the size of the output. For a given change in input, such systems may yield significantly greater than or less than proportional changes in output, or even no output at all, depending on the current state of the system or its parameter values.

Of particular interest to complex systems are nonlinear dynamical systems, which are systems of differential equations that have one or more nonlinear terms. Some nonlinear dynamical systems, such as the Lorenz system, can produce a mathematical phenomenon known as chaos. Chaos as it applies to complex systems refers to the sensitive dependence on initial conditions, or "butterfly effect," that a complex system can exhibit. In such a system, small changes to initial conditions can lead to dramatically different outcomes. Chaotic behavior can therefore be extremely hard to model numerically, because small rounding errors at an interme-

A sample solution in the Lorenz attractor when $\varrho = 28$, $\sigma = 10$, and $\beta = 8/3$

diate stage of computation can cause the model to generate completely inaccurate output. Furthermore, if a complex system returns to a state similar to one it held previously, it may behave completely differently in response to exactly the same stimuli, so chaos also poses challenges for extrapolating from past experience.

1.2.5 Emergence

Gosper's Glider Gun creating "gliders" in the cellular automaton Conway's Game of Life[2]

Another common feature of complex systems is the presence of emergent behaviors and properties: these are traits of a system which are not apparent from its components in isolation but which result from the interactions, dependencies, or relationships they form when placed together in a system. Emergence broadly describes the appearance of such behaviors and properties, and has applications to systems studied in both the social and physical sciences. While emergence is often used to refer only to the appearance of unplanned organized behavior in a complex system, emergence can also refer to the breakdown of organization; it describes any phenomena which are difficult or even impossible to predict from the smaller entities that make up the system.

One example of complex system whose emergent properties have been studied extensively is cellular automata. In a cellular automaton, a grid of cells, each having one of finitely many states, evolves over time according to a simple set of rules. These rules guide the "interactions" of each cell with its neighbors. Although the rules are only defined locally, they have been shown capable of producing globally interesting behavior, for example in Conway's Game of Life.

Spontaneous order and self-organization

When emergence describes the appearance of unplanned order, it is spontaneous order (in the social sciences) or self-organization (in physical sciences). Spontaneous order can be seen in herd behavior, whereby a group of individuals coordinates their actions without centralized planning. Self-organization can be seen in the global symmetry of certain crystals, for instance the apparent radial symmetry of snowflakes, which arises from purely local attractive and repulsive forces both between water molecules and between water molecules and their surrounding environment.

1.2.6 Adaptation

Complex adaptive systems are special cases of complex systems that are adaptive in that they have the capacity to change and learn from experience. Examples of complex adaptive systems include the stock market, social insect and ant colonies, the biosphere and the ecosystem, the brain and the immune system, the cell and the developing embryo, manufacturing businesses and any human social group-based endeavor in a cultural and social system such as political parties or communities.

1.3 Features

Complex systems may have the following features:[3]

Cascading failures Due to the strong coupling between components in complex systems, a failure in one or more components can lead to cascading failures which may have catastrophic consequences on the functioning of the system.[4]

Localized attack may lead to cascading failures in spatial networks.[5]

Complex systems may be open Complex systems are usually open systems — that is, they exist in a thermodynamic gradient and dissipate energy. In other words, complex systems are frequently far from energetic equilibrium: but despite this flux, there may be pattern stability, see synergetics.

Complex systems may have a memory The history of a complex system may be important. Because complex systems are dynamical systems they change over time, and prior states may have an influence on present states. More formally, complex systems often exhibit spontaneous failures and recovery as well as hysteresis.[6]

Interacting systems may have complex hysteresis of many transitions.[7]

Complex systems may be nested The components of a complex system may themselves be complex systems. For example, an economy is made up of organisations, which are made up of people, which are made up of cells - all of which are complex systems.

Dynamic network of multiplicity As well as coupling rules, the dynamic network of a complex system is important. Small-world or scale-free networks[8][9][10] which have many local interactions and a smaller number of inter-area connections are often employed. Natural complex systems often exhibit such topologies. In the human cortex for example, we see dense local connectivity and a few very long axon projections between regions inside the cortex and to other brain regions.

May produce emergent phenomena Complex systems may exhibit behaviors that are emergent, which is to say that while the results may be sufficiently determined by the activity of the systems' basic constituents, they may have properties that can only be studied at a higher level. For example, the termites in a mound have physiology, biochemistry and biological development that are at one level of analysis, but their social behavior and mound building is a property that emerges from the collection of termites and needs to be analysed at a different level.

Relationships are non-linear In practical terms, this means a small perturbation may cause a large effect (see butterfly effect), a proportional effect, or even no effect at all. In linear systems, effect is *always* directly proportional to cause. See nonlinearity.

Relationships contain feedback loops Both negative (damping) and positive (amplifying) feedback are always found in complex systems. The effects of an element's behaviour are fed back to in such a way that the element itself is altered.

1.4 History

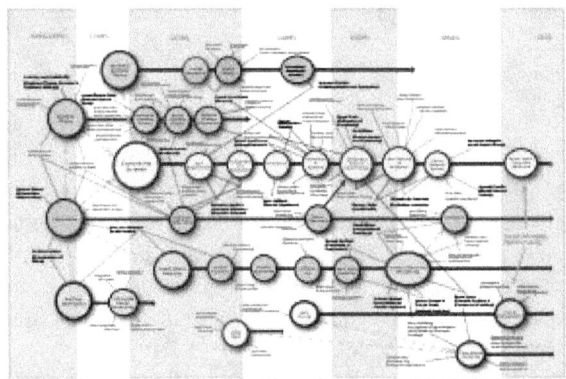

A perspective on the development of complexity science: http://www.art-sciencefactory.com/complexity-map_feb09.html

Although it is arguable that humans have been studying complex systems for thousands of years, the modern scientific study of complex systems is relatively young in comparison to established fields of science such as physics and chemistry. The history of the scientific study of these systems follows several different research trends.

In the area of mathematics, arguably the largest contribution to the study of complex systems was the discovery of chaos in deterministic systems, a feature of certain dynamical systems that is strongly related to nonlinearity.[11] The study of neural networks was also integral in advancing the mathematics needed to study complex systems.

The notion of self-organizing systems is tied up to work in nonequilibrium thermodynamics, including that pioneered by chemist and Nobel laureate Ilya Prigogine in his study of dissipative structures. Even older is the work by Hartree-Fock c.s. on the quantum-chemistry equations and later calculations of the structure of molecules which can be regarded as one of the earliest examples of emergence and emergent wholes in science.

The earliest precursor to modern complex systems theory can be found in the classical political economy of the Scottish Enlightenment, later developed by the Austrian school of economics, which argues that order in market systems is spontaneous (or emergent) in that it is the result of human action, but not the execution of any human design.[12][13]

Upon this the Austrian school developed from the 19th to the early 20th century the economic calculation problem, along with the concept of dispersed knowledge, which were to fuel debates against the then-dominant Keynesian economics. This debate would notably lead economists, politicians and other parties to explore the question of computational complexity.

A pioneer in the field, and inspired by Karl Popper's and Warren Weaver's works, Nobel prize economist and philosopher Friedrich Hayek dedicated much of his work, from early to the late 20th century, to the study of complex phenomena,[14] not constraining his work to human economies but venturing into other fields such as psychology,[15] biology and cybernetics. Gregory Bateson played a key role in establishing the connection between anthropology and systems theory; he recognized that the interactive parts of cultures function much like ecosystems.

In mathematics, arguably the largest contribution to the study of complex systems was the discovery of chaos in deterministic systems, a feature of certain dynamical systems that is strongly related to nonlinearity.[16]

The notion of self-organizing systems is tied to work in nonequilibrium thermodynamics, including that pioneered by chemist and Nobel laureate Ilya Prigogine in his study of dissipative structures. Even older is the work by Hartree-Fock c.s. on the quantum-chemistry equations and later calculations of the structure of molecules which can be regarded as one of the earliest examples of emergence and emergent wholes in science.

The first research institute focused on complex systems, the Santa Fe Institute, was founded in 1984.[17] Early Santa Fe Institute participants included physics Nobel laureates Murray Gell-Mann and Philip Anderson, economics Nobel laureate Kenneth Arrow, and Manhattan Project scientists George Cowan and Herb Anderson.[18] Today, there are over 50 institutes and research centers focusing on complex systems.

1.5 Applications of complex systems

1.5.1 Complexity in practice

The traditional approach to dealing with complexity is to reduce or constrain it. Typically, this involves compartmentalisation: dividing a large system into separate parts. Organizations, for instance, divide their work into departments that each deal with separate issues. Engineering systems are often designed using modular components. However, modular designs become susceptible to failure when issues arise that bridge the divisions.

1.5.2 Complexity management

As projects and acquisitions become increasingly complex, companies and governments are challenged to find effective ways to manage mega-acquisitions such as the Army Future Combat Systems. Acquisitions such as the FCS rely on a web of interrelated parts which interact unpredictably. As acquisitions become more network-centric and complex, businesses will be forced to find ways to manage complexity while governments will be challenged to provide effective governance to ensure flexibility and resiliency.[19]

1.5.3 Complexity economics

Over the last decades, within the emerging field of complexity economics new predictive tools have been developed to explain economic growth. Such is the case with the models built by the Santa Fe Institute in 1989 and the more recent economic complexity index (ECI), introduced by the MIT physicist Cesar A. Hidalgo and the Harvard economist Ricardo Hausmann. Based on the ECI, Hausmann, Hidalgo and their team of The Observatory of Economic Complexity have produced GDP forecasts for the year 2020.

1.5.4 Complexity and education

Focusing on issues of student persistence with their studies, Forsman, Moll and Linder explore the "viability of using complexity science as a frame to extend methodological applications for physics education research," finding that "framing a social network analysis within a complexity science perspective offers a new and powerful applicability across a broad range of PER topics."[20]

1.5.5 Complexity and modeling

One of Friedrich Hayek's main contributions to early complexity theory is his distinction between the human capacity to predict the behaviour of simple systems and its capacity to predict the behaviour of complex systems through modeling. He believed that economics and the sciences of complex phenomena in general, which in his view included biology, psychology, and so on, could not be modeled after the sciences that deal with essentially simple phenomena like physics.[21] Hayek would notably explain that complex phenomena, through modeling, can only allow pattern predictions, compared with the precise predictions that can be made out of non-complex phenomena.[22]

1.5.6 Complexity and chaos theory

Complexity theory is rooted in chaos theory, which in turn has its origins more than a century ago in the work of the French mathematician Henri Poincaré. Chaos is sometimes viewed as extremely complicated information, rather than as an absence of order.[23] Chaotic systems remain deterministic, though their long-term behavior can be difficult to predict with any accuracy. With perfect knowledge of the initial conditions and of the relevant equations describing the chaotic system's behavior, one can theoretically make perfectly accurate predictions about the future of the system, though in practice this is impossible to do with arbitrary accuracy. Ilya Prigogine argued[24] that complexity is non-deterministic, and gives no way whatsoever to precisely predict the future.[25]

The emergence of complexity theory shows a domain between deterministic order and randomness which is complex.[26] This is referred as the "edge of chaos".[27]

A plot of the Lorenz attractor.

When one analyzes complex systems, sensitivity to initial conditions, for example, is not an issue as important as it is within chaos theory, in which it prevails. As stated by Colander,[28] the study of complexity is the opposite of the study of chaos. Complexity is about how a huge number of extremely complicated and dynamic sets of relationships can generate some simple behavioral patterns, whereas chaotic behavior, in the sense of deterministic chaos, is the result of a relatively small number of non-linear interactions.[26]

Therefore, the main difference between chaotic systems and complex systems is their history.[29] Chaotic systems do not rely on their history as complex ones do. Chaotic behaviour pushes a system in equilibrium into chaotic order, which means, in other words, out of what we traditionally define as 'order'. On the other hand, complex systems evolve far from equilibrium at the edge of chaos. They evolve at a critical state built up by a history of irreversible and unexpected events, which physicist Murray Gell-Mann called "an accumulation of frozen accidents."[30] In a sense chaotic systems can be regarded as a subset of complex systems distinguished precisely by this absence of historical dependence. Many real complex systems are, in practice and over long but finite time periods, robust. However, they do possess the potential for radical qualitative change of kind whilst retaining systemic integrity. Metamorphosis serves as perhaps more than a metaphor for such transformations.

1.5.7 Complexity and network science

A complex system is usually composed of many components and their interactions. Such a system can be represented by a network where nodes represent the components and links represent their interactions.[31][32][33] for example, the INTERNET can be represented as a network composed of nodes (computers) and links (direct connections between computers). Its resilience to failures was studied using percolation theory in.[34] Other examples are social networks, airline networks,[35] biological networks and climate networks.[36] Networks can also fail and recover spontaneously. For modeling this phenomenon see ref.[6] Interacting complex systems can be modeled as networks of networks. For their breakdown and recovery properties see [37] [7]

1.5.8 General form of complexity computation

The computational law of reachable optimality[38] is established as a general form of computation for ordered system and it reveals complexity computation is a compound computation of optimal choice and optimality driven reaching pattern overtime underlying a specific and any experience path of ordered system within the general limitation of system integrity.

The computational law of reachable optimality has four key components as described below.

1. **Reachability of Optimality**: Any intended optimality shall be reachable. Unreachable optimality has no meaning for a member in the ordered system and even for the ordered system itself.

2. **Prevailing and Consistency**: Maximizing reachability to explore best available optimality is the prevailing computation logic for all members in the ordered system and is accommodated by the ordered system.

3. **Conditionality**: Realizable tradeoff between reachability and optimality depends primarily upon the initial bet capacity and how the bet capacity evolves along with the payoff table update path triggered by bet behavior and empowered by the underlying law of reward and punishment. Precisely, it is a sequence of conditional events where the next event happens upon reached status quo from experience path.

4. **Robustness**: The more challenge a reachable optimality can accommodate, the more robust it is in term of path integrity.

There are also four computation features in the law of reachable optimality.

1. **Optimal Choice**: Computation in realizing Optimal Choice can be very simple or very complex. A simple rule in Optimal Choice is to accept whatever is reached, Reward As You Go (RAYG). A Reachable Optimality computation reduces into optimizing reachability when RAYG is adopted. The Optimal Choice computation can be more complex when multiple NE strategies present in a reached game.

2. **Initial Status**: Computation is assumed to start at an interested beginning even the absolute beginning of an ordered system in nature may not and need not present. An assumed neutral Initial Status facilitates an artificial or a simulating computation and is not expected to change the prevalence of any findings.

3. **Territory**: An ordered system shall have a territory where the universal computation sponsored by the system will produce an optimal solution still within the territory.

4. **Reaching Pattern**: The forms of Reaching Pattern in the computation space, or the Optimality Driven Reaching Pattern in the computation space, primarily depend upon the nature and dimensions of measure space underlying a computation space and the law of punishment and reward underlying the realized experience path of reaching. There are five basic forms of experience path we are interested in, persistently positive reinforcement experience path, persistently negative reinforcement experience path, mixed persistent pattern experience path, decaying scale experience path and selection experience path.

The compound computation in selection experience path includes current and lagging interaction, dynamic topological transformation and implies both invariance and variance characteristics in an ordered system's experience path.

In addition, the computation law of reachable optimality gives out the boundary between complexity model, chaotic model and determination model. When RAYG is the Optimal Choice computation, and the reaching pattern is a persistently positive experience path, persistently negative experience path, or mixed persistent pattern experience path, the underlying computation shall be a simple system computation adopting determination rules. If the reaching pattern has no persistent pattern experienced in RAYG regime, the underlying computation hints there is a chaotic system. When the optimal choice computation involves non-RAYG computation, it's a complexity computation driving the compound effect.

1.6 Notable figures

- Christopher Alexander
- Gregory Bateson
- Ludwig von Bertalanffy
- Samuel Bowles
- Paul Cilliers
- Murray Gell-Mann
- Arthur Iberall
- Stuart Kauffman
- Cris Moore
- Bill McKelvey
- Jerry Sabloff
- Geoffrey West
- Yaneer Bar-Yam
- Walter Clemens, Jr.
- Edgar Morin

1.7 See also

1.8 References

[1] Bar-Yam, Yaneer (2002). "General Features of Complex Systems" (PDF). *Encyclopedia of Life Support Systems*. EOLSS UNESCO Publishers, Oxford, UK. Retrieved 16 September 2014.

[2] Daniel Dennett (1995), *Darwin's Dangerous Idea*, Penguin Books, London, ISBN 978-0-14-016734-4, ISBN 0-14-016734-X

[3] Alan Randall (2011). *Risk and Precaution*. Cambridge University Press. ISBN 9781139494793.

[4] S. V. Buldyrev; R. Parshani; G. Paul; H. E. Stanley; S. Havlin (2010). "Catastrophic cascade of failures in interdependent networks". *Nature*. **464** (7291): 08932. Bibcode:2010Natur.464.1025B. PMID 20393559. arXiv:0907.1182. doi:10.1038/nature08932.

[5] Berezin, Yehiel; Bashan, Amir; Danziger, Michael M.; Li, Daqing; Havlin, Shlomo (2015). "Localized attacks on spatially embedded networks with dependencies". *Scientific Reports*. **5** (1). ISSN 2045-2322. doi:10.1038/srep08934.

[6] Majdandzic, Antonio; Podobnik, Boris; Buldyrev, Sergey V.; Kenett, Dror Y.; Havlin, Shlomo; Eugene Stanley, H. (2013). "Spontaneous recovery in dynamical networks". *Nature Physics*. **10** (1): 34–38. ISSN 1745-2473. doi:10.1038/nphys2819.

[7] Majdandzic, Antonio; Braunstein, Lidia A.; Curme, Chester; Vodenska, Irena; Levy-Carciente, Sary; Eugene Stanley, H.; Havlin, Shlomo (2016). "Multiple tipping points and optimal repairing in interacting networks". *Nature Communications*. **7**: 10850. ISSN 2041-1723. doi:10.1038/ncomms10850.

[8] A. L. Barabási, R. Albert (2002). "Statistical mechanics of complex networks". *Reviews of Modern Physics*. **74**: 47–94. Bibcode:2002RvMP...74...47A. arXiv:cond-mat/0106096. doi:10.1103/RevModPhys.74.47.

[9] M. Newman (2010). *Networks: An Introduction*. Oxford University Press. ISBN 978-0-19-920665-0.

[10] Reuven Cohen, Shlomo Havlin (2010). *Complex Networks: Structure, Robustness and Function*. Cambridge University Press. ISBN 978-0-521-84156-6.

[11] History of Complex Systems

[12] Ferguson, Adam (1767). *An Essay on the History of Civil Society*. London: T. Cadell. Part the Third, Section II, p. 205.

[13] Friedrich Hayek, "The Results of Human Action but Not of Human Design" in *New Studies in Philosophy, Politics, Economics*, Chicago: University of Chicago Press, 1978, pp. 96–105.

[14] Bruce J. Caldwell, Popper and Hayek: Who influenced whom?, Karl Popper 2002 Centenary Congress, 2002.

[15] Friedrich von Hayek, *The Sensory Order: An Inquiry into the Foundations of Theoretical Psychology*, The University of Chicago Press, 1952.

[16] History of Complex Systems

[17] Ledford, H (2015). "How to solve the world's biggest problems". *Nature*. **525** (7569): 308–311. doi:10.1038/525308a.

[18] Waldrop, M. M. (1993). Complexity: The emerging science at the edge of order and chaos. Simon and Schuster.

[19] CSIS paper: "Organizing for a Complex World: The Way Ahead

[20] Forsman, Jonas; Moll, Rachel; Linder, Cedric (2014). "Extending the theoretical framing for physics education research: An illustrative application of complexity science". *Physical Review Special Topics: Physics Education Research*. **10** (2). doi:10.1103/PhysRevSTPER.10.020122. http://hdl.handle.net/10613/2583.

[21] Reason Magazine - The Road from Serfdom

[22] Friedrich August von Hayek - Prize Lecture

[23] Hayles, N. K. (1991). *Chaos Bound: Orderly Disorder in Contemporary Literature and Science*. Cornell University Press, Ithaca, NY.

[24] Prigogine, I. (1997). *The End of Certainty*, The Free Press, New York.

[25] See also D. Carfì (2008). "Superpositions in Prigogine approach to irreversibility". *AAPP: Physical, Mathematical, and Natural Sciences*. **86** (1): 1–13..

[26] Cilliers, P. (1998). *Complexity and Postmodernism: Understanding Complex Systems*, Routledge, London.

[27] Per Bak (1996). *How Nature Works: The Science of Self-Organized Criticality*, Copernicus, New York, U.S.

[28] Colander, D. (2000). *The Complexity Vision and the Teaching of Economics*, E. Elgar, Northampton, Massachusetts.

[29] Buchanan, M. (2000). *Ubiquity : Why catastrophes happen*. three river press, New-York.

[30] Gell-Mann, M. (1995). What is Complexity? Complexity 1/1, 16-19

[31] Dorogovtsev, S.N.; Mendes, J.F.F. (2003). "Evolution of Networks". doi:10.1093/acprof:oso/9780198515906.001.0001.

[32] Fortunato, Santo (2011). "Reuven Cohen and Shlomo Havlin: Complex Networks". *Journal of Statistical Physics*. **142** (3): 640–641. ISSN 0022-4715. doi:10.1007/s10955-011-0129-7.

[33] Newman, Mark (2010). "Networks". doi:10.1093/acprof:oso/9780199206650.001.0001.

[34] Cohen, Reuven; Erez, Keren; ben-Avraham, Daniel; Havlin, Shlomo (2001). "Cohen, Erez, ben-Avraham, and Havlin Reply:". *Physical Review Letters.* **87** (21). Bibcode:2001PhRvL..87u9802C. ISSN 0031-9007. doi:10.1103/PhysRevLett.87.219802.

[35] Barrat, A.; Barthelemy, M.; Pastor-Satorras, R.; Vespignani, A. (2004). "The architecture of complex weighted networks". *Proceedings of the National Academy of Sciences.* **101** (11): 3747–3752. ISSN 0027-8424. PMC 374315. PMID 15007165. doi:10.1073/pnas.0400087101.

[36] Yamasaki, K.; Gozolchiani, A.; Havlin, S. (2008). "Climate Networks around the Globe are Significantly Affected by El Niño". *Physical Review Letters.* **100** (22): 228501. ISSN 0031-9007. PMID 18643467. doi:10.1103/PhysRevLett.100.228501.

[37] Gao, Jianxi; Buldyrev, Sergey V.; Stanley, H. Eugene; Havlin, Shlomo (2011). "Networks formed from interdependent networks". *Nature Physics.* **8** (1): 40–48. Bibcode:2012NatPh...8...40G. ISSN 1745-2473. doi:10.1038/nphys2180.

[38] Wenliang Wang (2015). Pooling Game Theory and Public Pension Plan. ISBN 978-1507658246. Chapter 4.

1.9 Further reading

- Bazin, A. (2014). Defeating ISIS and Their Complex Way of War Small Wars Journal.
- Syed M. Mehmud (2011), *A Healthcare Exchange Complexity Model*
- Chu, D.; Strand, R.; Fjelland, R. (2003). "Theories of complexity". *Complexity.* **8** (3): 19–30. doi:10.1002/cplx.10059.
- L.A.N. Amaral and J.M. Ottino, *Complex networks — augmenting the framework for the study of complex system*, 2004.
- Gell-Mann, Murray (1995). "Let's Call It Plectics" (PDF). *Complexity.* **1** (5).
- Nigel Goldenfeld and Leo P. Kadanoff, *Simple Lessons from Complexity*, 1999
- A. Gogolin, A. Nersesyan and A. Tsvelik, *Theory of strongly correlated systems*, Cambridge University Press, 1999.
- Kelly, K. (1995). *Out of Control*, Perseus Books Group.

- Donald Snooks, Graeme (2008). "A general theory of complex living systems: Exploring the demand side of dynamics". *Complexity.* **13** (6): 12–20. doi:10.1002/cplx.20225.
- Sorin Solomon and Eran Shir, *Complexity; a science at 30*, 2003.
- Preiser-Kapeller, Johannes, "Calculating Byzantium. Social Network Analysis and Complexity Sciences as tools for the exploration of medieval social dynamics". August 2010
- Walter Clemens, Jr., *Complexity Science and World Affairs*, SUNY Press, 2013.

1.10 External links

- "The Open Agent-Based Modeling Consortium".
- "Complexity Science Focus".
- "Santa Fe Institute".
- "The Center for the Study of Complex Systems, Univ. of Michigan Ann Arbor".
- "INDECS". (Interdisciplinary Description of Complex Systems)
- "Introduction to complex systems - Short course by Shlomo Havlin".
- Jessie Henshaw (October 24, 2013). "Complex Systems". Encyclopedia of Earth.
- Introduction to complex systems-short course by Shlomo Havlin
- Complex systems in scholarpedia.
- (European) Complex Systems Society
- (Australian) Complex systems research network.
- Complex Systems Modeling based on Luis M. Rocha, 1999.
- CRM Complex systems research group
- The Center for Complex Systems Research, Univ. of Illinois at Urbana-Champaign
- FuturICT - Exploring and Managing our Future

Chapter 2

Emergence

For other uses, see Emergence (disambiguation).
See also: Emergent (disambiguation), Spontaneous order, and Self-organization

In philosophy, systems theory, science, and art, **emergence**

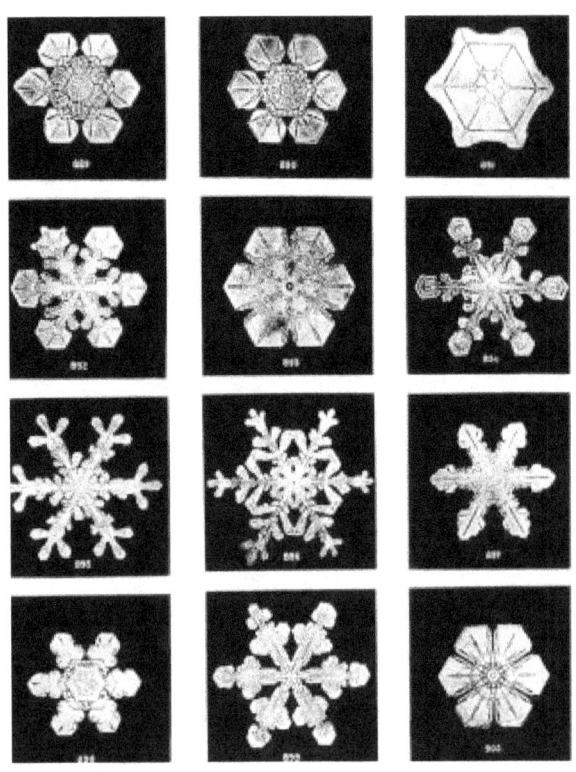

The formation of complex symmetrical and fractal patterns in snowflakes exemplifies emergence in a physical system.

is a phenomenon whereby larger entities arise through interactions among smaller or simpler entities such that the larger entities exhibit properties the smaller/simpler entities do not exhibit.

Emergence is central in theories of integrative levels and of complex systems. For instance, the phenomenon of *life* as studied in biology is an emergent property of chemistry, and psychological phenomena emerge from the neurobiological

A termite "cathedral" mound produced by a termite colony offers a classic example of emergence in nature.

phenomena of living things.

In philosophy, theories that emphasize emergent properties have been called emergentism. Almost all accounts of emergentism include a form of epistemic or ontological irreducibility to the lower levels.[1]

2.1 In philosophy

Main article: Emergentism

In philosophy, emergence is often understood to be a claim about the etiology of a system's properties. An emergent property of a system, in this context, is one that is not a property of any component of that system, but is still a feature of the system as a whole. Nicolai Hartmann, one of the first modern philosophers to write on emergence, termed this *categorial novum* (new category).

2.1.1 Definitions

This idea of emergence has been around since at least the time of Aristotle.[2] John Stuart Mill[3] and Julian Huxley[4] are two of many scientists and philosophers who have written on the concept.

The term "emergent" was coined by philosopher G. H. Lewes, who wrote:

> Every resultant is either a sum or a difference of the co-operant forces; their sum, when their directions are the same – their difference, when their directions are contrary. Further, every resultant is clearly traceable in its components, because these are homogeneous and commensurable. It is otherwise with emergents, when, instead of adding measurable motion to measurable motion, or things of one kind to other individuals of their kind, there is a co-operation of things of unlike kinds. The emergent is unlike its components insofar as these are incommensurable, and it cannot be reduced to their sum or their difference.[5][6]

Economist Jeffrey Goldstein provided a current definition of emergence in the journal *Emergence*.[7] Goldstein initially defined emergence as: "the arising of novel and coherent structures, patterns and properties during the process of self-organization in complex systems".

Goldstein's definition can be further elaborated to describe the qualities of this definition in more detail:

> The common characteristics are: (1) radical novelty (features not previously observed in systems); (2) coherence or correlation (meaning integrated wholes that maintain themselves over some period of time); (3) A global or macro "level" (i.e. there is some property of "wholeness"); (4) it is the product of a dynamical process (it evolves); and (5) it is "ostensive" (it can be perceived).[8]

Systems scientist Peter Corning also says that living systems cannot be reduced to underlying laws of physics:

> Rules, or laws, have no causal efficacy; they do not in fact "generate" anything. They serve merely to describe regularities and consistent relationships in nature. These patterns may be very illuminating and important, but the underlying causal agencies must be separately specified (though often they are not). But that aside, the game of chess illustrates ... why any laws or rules of emergence and evolution are insufficient. Even in a chess game, you cannot use the rules to predict "history" – i.e., the course of any given game. Indeed, you cannot even reliably predict the next move in a chess game. Why? Because the "system" involves more than the rules of the game. It also includes the players and their unfolding, moment-by-moment decisions among a very large number of available options at each choice point. The game of chess is inescapably historical, even though it is also constrained and shaped by a set of rules, not to mention the laws of physics. Moreover, and this is a key point, the game of chess is also shaped by teleonomic, cybernetic, feedback-driven influences. It is not simply a self-ordered process; it involves an organized, "purposeful" activity.[8]

2.1.2 Strong and weak emergence

Usage of the notion "emergence" may generally be subdivided into two perspectives, that of "weak emergence" and "strong emergence". In terms of physical systems, weak emergence is a type of emergence in which the emergent property is amenable to computer simulation. This is opposed to the older notion of strong emergence, in which the emergent property cannot be simulated by a computer.

Some common points between the two notions are that emergence concerns new properties produced as the system grows, which is to say ones which are not shared with its components or prior states. Also, it is assumed that the properties are supervenient rather than metaphysically primitive (Bedau 1997).

Weak emergence describes new properties arising in systems as a result of the interactions at an elemental level. However, it is stipulated that the properties can be determined by observing or simulating the system, and not by any process of a priori analysis.

Bedau notes that weak emergence is not a universal metaphysical solvent, as weak emergence leads to the conclusion that matter itself contains elements of awareness to it. However, Bedau concludes that adopting this view would provide a precise notion that emergence is involved in consciousness, and second, the notion of weak emergence is metaphysically benign.(Bedau 1997)

Strong emergence describes the direct causal action of a high-level system upon its components; qualities produced this way are irreducible to the system's constituent parts (Laughlin 2005). The whole is other than the sum of its parts. An example from physics of such emergence is water, being seemingly unpredictable even after an exhaustive study of the properties of its constituent atoms of hydrogen and oxygen.[9] It follows then that no simulation of the system can exist, for such a simulation would itself constitute a reduction of the system to its constituent parts.(Bedau 1997)

However, "the debate about whether or not the whole can be predicted from the properties of the parts misses the point. Wholes produce unique combined effects, but many of these effects may be co-determined by the context and the interactions between the whole and its environment(s)" (Corning 2002). In accordance with his Synergism Hypothesis, (Corning 1983 2005) Corning also stated, "It is the synergistic effects produced by wholes that are the very cause of the evolution of complexity in nature." Novelist Arthur Koestler used the metaphor of Janus (a symbol of the unity underlying complements like open/shut, peace/war) to illustrate how the two perspectives (strong vs. weak or holistic vs. reductionistic) should be treated as non-exclusive, and should work together to address the issues of emergence.(Koestler 1969) Further,

> The ability to reduce everything to simple fundamental laws does not imply the ability to start from those laws and reconstruct the universe. The constructionist hypothesis breaks down when confronted with the twin difficulties of scale and complexity. At each level of complexity entirely new properties appear. Psychology is not applied biology, nor is biology applied chemistry. We can now see that the whole becomes not merely more, but very different from the sum of its parts.(Anderson 1972)

The plausibility of strong emergence is questioned by some as contravening our usual understanding of physics. Mark A. Bedau observes:

> Although strong emergence is logically possible, it is uncomfortably like magic. How does an irreducible but supervenient downward causal power arise, since by definition it cannot be due to the aggregation of the micro-level potentialities? Such causal powers would be quite unlike anything within our scientific ken. This not only indicates how they will discomfort reasonable forms of materialism. Their mysteriousness will only heighten the traditional worry that emergence entails illegitimately getting something from nothing.[10]

Strong emergence can be criticized for being causally overdetermined. The canonical example concerns emergent mental states (M and M∗) that supervene on physical states (P and P∗) respectively. Let M and M∗ be emergent properties. Let M∗ supervene on base property P∗. What happens when M causes M∗? Jaegwon Kim says:

> In our schematic example above, we concluded that M causes M∗ by causing P∗. So M causes P∗. Now, M, as an emergent, must itself have an emergence base property, say P. Now we face a critical question: if an emergent, M, emerges from basal condition P, why cannot P displace M as a cause of any putative effect of M? Why cannot P do all the work in explaining why any alleged effect of M occurred? If causation is understood as nomological (law-based) sufficiency, P, as M's emergence base, is nomologically sufficient for it, and M, as P∗'s cause, is nomologically sufficient for P∗. It follows that P is nomologically sufficient for P∗ and hence qualifies as its cause...If M is somehow retained as a cause, we are faced with the highly implausible consequence that every case of downward causation involves overdetermination (since P remains a cause of P∗ as well). Moreover, this goes against the spirit of emergentism in any case: emergents are supposed to make distinctive and novel causal contributions.[11]

If M is the cause of M∗, then M∗ is overdetermined because M∗ can also be thought of as being determined by P. One escape route that a strong emergentist could take would be to deny downward causation. However, this would deny that emergent mental states must supervene on physical states, which in turn would deny physicalism, and thus be unpalatable for some philosophers and physicists.

Meanwhile, others have worked towards developing analytical evidence of strong emergence. In 2009, Gu et al. presented a class of physical systems that exhibits non-computable macroscopic properties.[12][13] More precisely, if one could compute certain macroscopic properties of these systems from the microscopic description of these systems, then one would be able to solve computational problems known to be undecidable in computer science. They concluded that

> Although macroscopic concepts are essential for understanding our world, much of fundamental physics has been devoted to the search for

a 'theory of everything', a set of equations that perfectly describe the behavior of all fundamental particles. The view that this is the goal of science rests in part on the rationale that such a theory would allow us to derive the behavior of all macroscopic concepts, at least in principle. The evidence we have presented suggests that this view may be overly optimistic. A 'theory of everything' is one of many components necessary for complete understanding of the universe, but is not necessarily the only one. The development of macroscopic laws from first principles may involve more than just systematic logic, and could require conjectures suggested by experiments, simulations or insight.[12]

Emergent structures are patterns that emerge via collective actions of many individual entities. To explain such patterns, one might conclude, per Aristotle,[2] that emergent structures are other than the sum of their parts on the assumption that the emergent order will not arise if the various parts simply interact independently of one another. However, there are those who disagree.[14] According to this argument, the interaction of each part with its immediate surroundings causes a complex chain of processes that can lead to order in some form. In fact, some systems in nature are observed to exhibit emergence based upon the interactions of autonomous parts, and some others exhibit emergence that at least at present cannot be reduced in this way. In particular renormalization are methods in theoretical physics which enables scientists to study systems that are not tractable as the combination of their parts.[15]

2.1.3 Objective or subjective quality

The properties of complexity and organization of any system are considered by Crutchfield to be subjective qualities determined by the observer.

> Defining structure and detecting the emergence of complexity in nature are inherently subjective, though essential, scientific activities. Despite the difficulties, these problems can be analysed in terms of how model-building observers infer from measurements the computational capabilities embedded in non-linear processes. An observer's notion of what is ordered, what is random, and what is complex in its environment depends directly on its computational resources: the amount of raw measurement data, of memory, and of time available for estimation and inference. The discovery of structure in an environment depends more critically and subtly, though, on how those resources are organized. The descriptive power of the observer's chosen (or implicit) computational model class, for example, can be an overwhelming determinant in finding regularity in data.(Crutchfield 1994)

On the other hand, Peter Corning argues "Must the synergies be perceived/observed in order to qualify as emergent effects, as some theorists claim? Most emphatically not. The synergies associated with emergence are real and measurable, even if nobody is there to observe them."(Corning 2002)

2.2 In religion, art and humanities

In religion, emergence grounds expressions of religious naturalism and syntheism in which a sense of the sacred is perceived in the workings of entirely naturalistic processes by which more complex forms arise or evolve from simpler forms. Examples are detailed in *The Sacred Emergence of Nature* by Ursula Goodenough & Terrence Deacon and *Beyond Reductionism: Reinventing the Sacred* by Stuart Kauffman, both from 2006, and in *Syntheism – Creating God in The Internet Age* by Alexander Bard & Jan Söderqvist from 2014. An early argument (1904–05) for the emergence of social formations, in part stemming from religion, can be found in Max Weber's most famous work, *The Protestant Ethic and the Spirit of Capitalism*.[16]

In art, emergence is used to explore the origins of novelty, creativity, and authorship. Some art/literary theorists (Wheeler, 2006;[17] Alexander, 2011[18]) have proposed alternatives to postmodern understandings of "authorship" using the complexity sciences and emergence theory. They contend that artistic selfhood and meaning are emergent, relatively objective phenomena. Michael J. Pearce has used emergence to describe the experience of works of art in relation to contemporary neuroscience.[19])

In international development, concepts of emergence have been used within a theory of social change termed SEED-SCALE to show how standard principles interact to bring forward socio-economic development fitted to cultural values, community economics, and natural environment (local solutions emerging from the larger socio-econo-biosphere). These principles can be implemented utilizing a sequence of standardized tasks that self-assemble in individually specific ways utilizing recursive evaluative criteria.[20]

In postcolonial studies, the term "Emerging Literature" refers to a contemporary body of texts that is gaining momentum in the global literary landscape (v. esp.: J.M. Grassin, ed. *Emerging Literatures*, Bern, Berlin, etc. : Peter Lang, 1996). By opposition, "emergent literature" is rather a concept used in the theory of literature.

2.3 Emergent properties and processes

An emergent behavior or emergent property can appear when a number of simple entities (agents) operate in an environment, forming more complex behaviors as a collective. If emergence happens over disparate size scales, then the reason is usually a causal relation across different scales. In other words, there is often a form of top-down feedback in systems with emergent properties.[21] The processes from which emergent properties result may occur in either the observed or observing system, and can commonly be identified by their patterns of accumulating change, most generally called 'growth'. Emergent behaviours can occur because of intricate causal relations across different scales and feedback, known as interconnectivity. The emergent property itself may be either very predictable or unpredictable and unprecedented, and represent a new level of the system's evolution. The complex behaviour or properties are not a property of any single such entity, nor can they easily be predicted or deduced from behaviour in the lower-level entities, and might in fact be irreducible to such behavior. The shape and behaviour of a flock of birds or school of fish are good examples of emergent properties.

One reason why emergent behaviour is hard to predict is that the number of interactions between components of a system increases exponentially with the number of components, thus potentially allowing for many new and subtle types of behaviour to emerge. Emergence is often a product of particular patterns of interaction. Negative feedback introduces constraints that serve to fix structures or behaviours. In contrast, positive feedback promotes change, allowing local variations to grow into global patterns. Another way in which interactions leads to emergent properties is dual-phase evolution. This occurs where interactions are applied intermittently, leading to two phases: one in which patterns form or grow, the other in which they are refined or removed.

On the other hand, merely having a large number of interactions is not enough by itself to guarantee emergent behaviour; many of the interactions may be negligible or irrelevant, or may cancel each other out. In some cases, a large number of interactions can in fact work against the emergence of interesting behaviour, by creating a lot of "noise" to drown out any emerging "signal"; the emergent behaviour may need to be temporarily isolated from other interactions before it reaches enough critical mass to be self-supporting. Thus it is not just the sheer number of connections between components which encourages emergence; it is also how these connections are organised. A hierarchical organisation is one example that can generate emergent behaviour (a bureaucracy may behave in a way quite different from that of the individual humans in that bureaucracy); but perhaps more interestingly, emergent behaviour can also arise from more decentralized organisational structures, such as a marketplace. In some cases, the system has to reach a combined threshold of diversity, organisation, and connectivity before emergent behaviour appears.

Unintended consequences and side effects are closely related to emergent properties. Luc Steels writes: "A component has a particular functionality but this is not recognizable as a subfunction of the global functionality. Instead a component implements a behaviour whose side effect contributes to the global functionality [...] Each behaviour has a side effect and the sum of the side effects gives the desired functionality".(Steels 1990) In other words, the global or macroscopic functionality of a system with "emergent functionality" is the sum of all "side effects", of all emergent properties and functionalities.

Systems with emergent properties or emergent structures may appear to defy entropic principles and the second law of thermodynamics, because they form and increase order despite the lack of command and central control. This is possible because open systems can extract information and order out of the environment.

Emergence helps to explain why the fallacy of division is a fallacy.

2.4 Emergent structures in nature

Main article: Patterns in nature

Emergent structures can be found in many natural phenom-

Ripple patterns in a sand dune created by wind or water is an example of an emergent structure in nature.

2.4. EMERGENT STRUCTURES IN NATURE

Giant's Causeway in Northern Ireland is an example of a complex emergent structure created by natural processes.

ena, from the physical to the biological domain. For example, the shape of weather phenomena such as hurricanes are emergent structures. The development and growth of complex, orderly crystals, as driven by the random motion of water molecules within a conducive natural environment, is another example of an emergent process, where randomness can give rise to complex and deeply attractive, orderly structures.

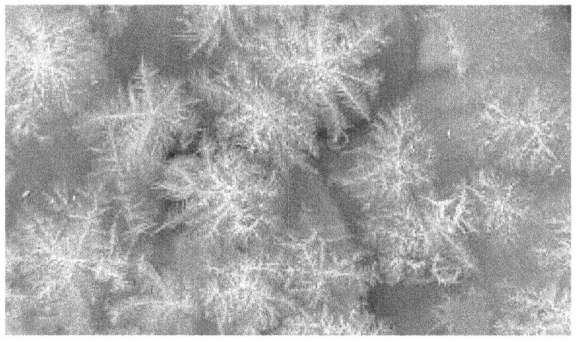

Water crystals forming on glass demonstrate an emergent, fractal natural process occurring under appropriate conditions of temperature and humidity.

However, crystalline structure and hurricanes are said to have a self-organizing phase.

It is useful to distinguish three forms of emergent structures. A *first-order* emergent structure occurs as a result of shape interactions (for example, hydrogen bonds in water molecules lead to surface tension). A *second-order* emergent structure involves shape interactions played out sequentially over time (for example, changing atmospheric conditions as a snowflake falls to the ground build upon and alter its form). Finally, a *third-order* emergent structure is a consequence of shape, time, and heritable instructions. For example, an organism's genetic code sets boundary conditions on the interaction of biological systems in space and time.

2.4.1 Non-living, physical systems

In physics, emergence is used to describe a property, law, or phenomenon which occurs at macroscopic scales (in space or time) but not at microscopic scales, despite the fact that a macroscopic system can be viewed as a very large ensemble of microscopic systems.

An emergent property need not be more complicated than the underlying non-emergent properties which generate it. For instance, the laws of thermodynamics are remarkably simple, even if the laws which govern the interactions between component particles are complex. The term emergence in physics is thus used not to signify complexity, but rather to distinguish which laws and concepts apply to macroscopic scales, and which ones apply to microscopic scales.

Some examples include:

- Classical mechanics: The laws of classical mechanics can be said to emerge as a limiting case from the rules of quantum mechanics applied to large enough masses. This is particularly strange since quantum mechanics is generally thought of as *more* complicated than classical mechanics.

- Friction: Forces between elementary particles are conservative. However, friction emerges when considering more complex structures of matter, whose surfaces can convert mechanical energy into heat energy when rubbed against each other. Similar considerations apply to other emergent concepts in continuum mechanics such as viscosity, elasticity, tensile strength, etc.

- Patterned ground: the distinct, and often symmetrical geometric shapes formed by ground material in periglacial regions.

- Statistical mechanics was initially derived using the concept of a large enough ensemble that fluctuations about the most likely distribution can be all but ignored. However, small clusters do not exhibit sharp first order phase transitions such as melting, and at the boundary it is not possible to completely categorize the cluster as a liquid or solid, since these concepts are (without extra definitions) only applicable to macroscopic systems. Describing a system using statistical mechanics methods is much simpler than using a low-level atomistic approach.

- Electrical networks: The bulk conductive response of binary (RC) electrical networks with random arrangements can be seen as emergent properties of such physical systems. Such arrangements can be used as simple physical prototypes for deriving mathematical formulae for the emergent responses of complex systems.[22]

- Weather

Temperature is sometimes used as an example of an emergent macroscopic behaviour. In classical dynamics, a *snapshot* of the instantaneous momenta of a large number of particles at equilibrium is sufficient to find the average kinetic energy per degree of freedom which is proportional to the temperature. For a small number of particles the instantaneous momenta at a given time are not statistically sufficient to determine the temperature of the system. However, using the ergodic hypothesis, the temperature can still be obtained to arbitrary precision by further averaging the momenta over a long enough time.

Convection in a liquid or gas is another example of emergent macroscopic behaviour that makes sense only when considering differentials of temperature. Convection cells, particularly Bénard cells, are an example of a self-organizing system (more specifically, a dissipative system) whose structure is determined both by the constraints of the system and by random perturbations: the possible realizations of the shape and size of the cells depends on the temperature gradient as well as the nature of the fluid and shape of the container, but which configurations are actually realized is due to random perturbations (thus these systems exhibit a form of symmetry breaking).

In some theories of particle physics, even such basic structures as mass, space, and time are viewed as emergent phenomena, arising from more fundamental concepts such as the Higgs boson or strings. In some interpretations of quantum mechanics, the perception of a deterministic reality, in which all objects have a definite position, momentum, and so forth, is actually an emergent phenomenon, with the true state of matter being described instead by a wavefunction which need not have a single position or momentum. Most of the laws of physics themselves as we experience them today appear to have emerged during the course of time making emergence the most fundamental principle in the universe and raising the question of what might be the most fundamental law of physics from which all others emerged. Chemistry can in turn be viewed as an emergent property of the laws of physics. Biology (including biological evolution) can be viewed as an emergent property of the laws of chemistry. Similarly, psychology could be understood as an emergent property of neurobiological laws. Finally, free-market theories understand economy as an emergent feature of psychology.

According to Laughlin (2005), for many particle systems, nothing can be calculated exactly from the microscopic equations, and macroscopic systems are characterised by broken symmetry: the symmetry present in the microscopic equations is not present in the macroscopic system, due to phase transitions. As a result, these macroscopic systems are described in their own terminology, and have properties that do not depend on many microscopic details. This does not mean that the microscopic interactions are irrelevant, but simply that you do not see them anymore — you only see a renormalized effect of them. Laughlin is a pragmatic theoretical physicist: if you cannot, possibly ever, calculate the broken symmetry macroscopic properties from the microscopic equations, then what is the point of talking about reducibility?

2.4.2 Living, biological systems

Emergence and evolution

See also: Abiogenesis

Life is a major source of complexity, and evolution is the major process behind the varying forms of life. In this view, evolution is the process describing the growth of complexity in the natural world and in speaking of the emergence of complex living beings and life-forms, this view refers therefore to processes of sudden changes in evolution.

Life is thought to have emerged in the early RNA world when RNA chains began to express the basic conditions necessary for natural selection to operate as conceived by Darwin: heritability, variation of type, and competition for limited resources. Fitness of an RNA replicator (its per capita rate of increase) would likely be a function of adaptive capacities that were intrinsic (in the sense that they were determined by the nucleotide sequence) and the availability of resources.[23][24] The three primary adaptive capacities may have been (1) the capacity to replicate with moderate fidelity (giving rise to both heritability and variation of type); (2) the capacity to avoid decay; and (3) the capacity to acquire and process resources.[23][24] These capacities would have been determined initially by the folded configurations of the RNA replicators (see "Ribozyme") that, in turn, would be encoded in their individual nucleotide sequences. Competitive success among different replicators would have depended on the relative values of these adaptive capacities.

Regarding causality in evolution Peter Corning observes:

> Synergistic effects of various kinds have played a major causal role in the evolutionary process generally and in the evolution of cooperation and complexity in particular... Natural selection is often portrayed as a "mechanism",

or is personified as a causal agency... In reality, the differential "selection" of a trait, or an adaptation, is a consequence of the functional effects it produces in relation to the survival and reproductive success of a given organism in a given environment. It is these functional effects that are ultimately responsible for the trans-generational continuities and changes in nature.(Corning 2002)

Per his definition of emergence, Corning also addresses emergence and evolution:

> [In] evolutionary processes, causation is iterative; effects are also causes. And this is equally true of the synergistic effects produced by emergent systems. In other words, emergence itself... has been the underlying cause of the evolution of emergent phenomena in biological evolution; it is the synergies produced by organized systems that are the key.(Corning 2002)

Swarming is a well-known behaviour in many animal species from marching locusts to schooling fish to flocking birds. Emergent structures are a common strategy found in many animal groups: colonies of ants, mounds built by termites, swarms of bees, shoals/schools of fish, flocks of birds, and herds/packs of mammals.

An example to consider in detail is an ant colony. The queen does not give direct orders and does not tell the ants what to do. Instead, each ant reacts to stimuli in the form of chemical scent from larvae, other ants, intruders, food and buildup of waste, and leaves behind a chemical trail, which, in turn, provides a stimulus to other ants. Here each ant is an autonomous unit that reacts depending only on its local environment and the genetically encoded rules for its variety of ant. Despite the lack of centralized decision making, ant colonies exhibit complex behavior and have even demonstrated the ability to solve geometric problems. For example, colonies routinely find the maximum distance from all colony entrances to dispose of dead bodies.[25]

It appears that environmental factors may play a role in influencing emergence. Research suggests induced emergence of the bee species Macrotera portalis. In this species, the bees emerge in a pattern consistent with rainfall. Specifically, the pattern of emergence is consistent with southwestern deserts' late summer rains and lack of activity in the spring.[26]

Organization of life

A broader example of emergent properties in biology is viewed in the biological organisation of life, ranging from the subatomic level to the entire biosphere. For example, individual atoms can be combined to form molecules such as polypeptide chains, which in turn fold and refold to form proteins, which in turn create even more complex structures. These proteins, assuming their functional status from their spatial conformation, interact together and with other molecules to achieve higher biological functions and eventually create an organism. Another example is how cascade phenotype reactions, as detailed in chaos theory, arise from individual genes mutating respective positioning.[27] At the highest level, all the biological communities in the world form the biosphere, where its human participants form societies, and the complex interactions of meta-social systems such as the stock market.

Emergence of mind

Among the considered phenomena in the evolutionary account of life, as a continuous history, marked by stages at which fundamentally new forms have appeared - the origin of sapiens intelligence.[28] The emergence of mind and its evolution is researched and considered as a separate phenomenon in a special system knowledge noogenesis[29]

2.5 In humanity

2.5.1 Spontaneous order

See also: Spontaneous order and Self-organization

Groups of human beings, left free to each regulate themselves, tend to produce spontaneous order, rather than the meaningless chaos often feared. This has been observed in society at least since Chuang Tzu in ancient China. A classic traffic roundabout is a good example, with cars moving in and out with such effective organization that some modern cities have begun replacing stoplights at problem intersections with traffic circles, and getting better results. Open-source software and Wiki projects form an even more compelling illustration.

Emergent processes or behaviours can be seen in many other places, such as cities, cabal and market-dominant minority phenomena in economics, organizational phenomena in computer simulations and cellular automata. Whenever you have a multitude of individuals interacting with one another, there often comes a moment when disorder gives way to order and something new emerges: a pattern, a decision, a structure, or a change in direction (Miller 2010, 29).[30]

Economics

The stock market (or any market for that matter) is an example of emergence on a grand scale. As a whole it precisely regulates the relative security prices of companies across the world, yet it has no leader; when no central planning is in place, there is no one entity which controls the workings of the entire market. Agents, or investors, have knowledge of only a limited number of companies within their portfolio, and must follow the regulatory rules of the market and analyse the transactions individually or in large groupings. Trends and patterns emerge which are studied intensively by technical analysts.. Emergence has been shown to occur in certain econophysics models of economics systems.[31]

World Wide Web and the Internet

The World Wide Web is a popular example of a decentralized system exhibiting emergent properties. There is no central organization rationing the number of links, yet the number of links pointing to each page follows a power law in which a few pages are linked to many times and most pages are seldom linked to. A related property of the network of links in the World Wide Web is that almost any pair of pages can be connected to each other through a relatively short chain of links. Although relatively well known now, this property was initially unexpected in an unregulated network. It is shared with many other types of networks called small-world networks.(Barabasi, Jeong, & Albert 1999, pp. 130–31)

Internet traffic can also exhibit some seemingly emergent properties. In the congestion control mechanism, TCP flows can become globally synchronized at bottlenecks, simultaneously increasing and then decreasing throughput in coordination. Congestion, widely regarded as a nuisance, is possibly an emergent property of the spreading of bottlenecks across a network in high traffic flows which can be considered as a phase transition [see review of related research in (Smith 2008, pp. 1–31)].

Another important example of emergence in web-based systems is social bookmarking (also called collaborative tagging). In social bookmarking systems, users assign tags to resources shared with other users, which gives rise to a type of information organisation that emerges from this crowdsourcing process. Recent research which analyzes empirically the complex dynamics of such systems[32] has shown that consensus on stable distributions and a simple form of shared vocabularies does indeed emerge, even in the absence of a central controlled vocabulary. Some believe that this could be because users who contribute tags all use the same language, and they share similar semantic structures underlying the choice of words. The convergence in social tags may therefore be interpreted as the emergence of structures as people who have similar semantic interpretation collaboratively index online information, a process called semantic imitation.[33] [34]

Architecture and cities

Traffic patterns in cities can be seen as an example of spontaneous order

Emergent structures appear at many different levels of organization or as spontaneous order. Emergent self-organization appears frequently in cities where no planning or zoning entity predetermines the layout of the city.(Krugman 1996, pp. 9–29) The interdisciplinary study of emergent behaviors is not generally considered a homogeneous field, but divided across its application or problem domains.

Architects and Landscape Architects may not design all the pathways of a complex of buildings. Instead they might let usage patterns emerge and then place pavement where pathways have become worn, such as a desire path.

The on-course action and vehicle progression of the 2007 Urban Challenge could possibly be regarded as an example of cybernetic emergence. Patterns of road use, indeterministic obstacle clearance times, etc. will work together to form a complex emergent pattern that can not be deterministically planned in advance.

The architectural school of Christopher Alexander takes a deeper approach to emergence attempting to rewrite the process of urban growth itself in order to affect form, establishing a new methodology of planning and design tied to traditional practices, an Emergent Urbanism. Urban emergence has also been linked to theories of urban complexity (Batty 2005) and urban evolution.(Marshall 2009)

Building ecology is a conceptual framework for understanding architecture and the built environment as the interface between the dynamically interdependent elements of buildings, their occupants, and the larger environment. Rather than viewing buildings as inanimate or static objects, building ecologist Hal Levin views them as interfaces or intersecting domains of living and non-living systems.[35] The microbial ecology of the indoor environment is strongly dependent on the building materials, occupants, contents, environmental context and the indoor and outdoor climate. The strong relationship between atmospheric chemistry and indoor air quality and the chemical reactions occurring indoors. The chemicals may be nutrients, neutral or biocides for the microbial organisms. The microbes produce chemicals that affect the building materials and occupant health and well being. Humans manipulate the ventilation, temperature and humidity to achieve comfort with the concomitant effects on the microbes that populate and evolve.[35][36][37]

Eric Bonabeau's attempt to define emergent phenomena is through traffic: "traffic jams are actually very complicated and mysterious. On an individual level, each driver is trying to get somewhere and is following (or breaking) certain rules, some legal (the speed limit) and others societal or personal (slow down to let another driver change into your lane). But a traffic jam is a separate and distinct entity that emerges from those individual behaviors. Gridlock on a highway, for example, can travel backward for no apparent reason, even as the cars are moving forward." He has also likened emergent phenomena to the analysis of market trends and employee behavior.[38]

Computational emergent phenomena have also been utilized in architectural design processes, for example for formal explorations and experiments in digital materiality.[39]

2.5.2 Computer AI

Some artificially intelligent (AI) computer applications utilize emergent behavior for animation. One example is Boids, which mimics the swarming behavior of birds.

2.5.3 Language

It has been argued that the structure and regularity of language grammar, or at least language change, is an emergent phenomenon (Hopper 1998). While each speaker merely tries to reach his or her own communicative goals, he or she uses language in a particular way. If enough speakers behave in that way, language is changed (Keller 1994). In a wider sense, the norms of a language, i.e. the linguistic conventions of its speech society, can be seen as a system emerging from long-time participation in communicative problem-solving in various social circumstances (Määttä 2000).

2.5.4 Emergent change processes

Within the field of group facilitation and organization development, there have been a number of new group processes that are designed to maximize emergence and self-organization, by offering a minimal set of effective initial conditions. Examples of these processes include SEED-SCALE, Appreciative Inquiry, Future Search, the World Cafe or Knowledge Cafe, Open Space Technology, and others (Holman, 2010[40]).

2.6 See also

- Abstraction
- Abiogenesis
- Agent-based model
- Anthropic principle
- Big History
- Connectionism
- Consilience
- Constructal theory
- Dynamical system
- Deus ex machina
- Dual-phase evolution
- Emergenesis
- Emergent algorithm
- Emergent evolution
- Emergent gameplay
- Emergent organization
- Emergentism
- Epiphenomenon
- Externality
- Free Will
- Generative sciences
- Innovation butterfly
- Interconnectedness
- Irreducible complexity

- Langton's ant
- Law of Complexity-Consciousness
- Libertarianism (metaphysics)
- Mass action (sociology)
- Neural networks
- Noogenesis
- Organic Wholes of G.E. Moore
- Polytely
- Society of Mind theory
- Structuralism
- Superorganism
- Swarm intelligence
- System of systems
- Teleology
- Spontaneous order
- Synergetics (Fuller)
- Synergetics (Haken)

2.7 References

[1] O'Connor, Timothy; Wong, Hong Yu (February 28, 2012). Edward N. Zalta, ed. "Emergent Properties". *The Stanford Encyclopedia of Philosophy (Spring 2012 Edition)*.

[2] Aristotle, *Metaphysics*, Book H 1045a 8–10: "... the totality is not, as it were, a mere heap, but the whole is something besides the parts ...", i.e., the whole is other than the sum of the parts.

[3] "The chemical combination of two substances produces, as is well known, a third substance with properties different from those of either of the two substances separately, or of both of them taken together" (Mill 1843)

[4] Julian Huxley: "now and again there is a sudden rapid passage to a totally new and more comprehensive type of order or organization, with quite new emergent properties, and involving quite new methods of further evolution" (Huxley & Huxley 1947)

[5] (Lewes 1875, p. 412)

[6] (Blitz 1992)

[7] (Goldstein 1999)

[8] Corning, Peter A. (2002), "The Re-Emergence of "Emergence": A Venerable Concept in Search of a Theory", *Complexity*, **7** (6): 18–30, Bibcode:2002Cmplx...7f..18C, doi:10.1002/cplx.10043

[9] Luisi, Pier L. (2006). *The Emergence of Life: From Chemical Origins to Synthetic Biology*. Cambridge, England: Cambridge University Press. p. 119. ISBN 0521821177.

[10] (Bedau 1997)

[11] Kim, Jaegwon (2016). "Emergence: Core ideas and issues". *Synthese*. **151**: 547–59. doi:10.1007/s11229-006-9025-0.

[12] Gu, Mile; et al. (2009). "More really is different". *Physica D: Nonlinear Phenomena*. **238** (9): 835–39. Bibcode:2009PhyD..238..835G. arXiv:0809.0151. doi:10.1016/j.physd.2008.12.016.

[13] Binder, P-M (2009). "Computation: The edge of reductionism". *Nature*. **459** (7245): 332–34. Bibcode:2009Natur.459..332B. doi:10.1038/459332a.

[14] Steven Weinberg. "A Designer Universe?". Retrieved 2008-07-14. A version of the original quote from address at the Conference on Cosmic Design, American Association for the Advancement of Science, Washington, D.C. in April 1999

[15] Longo, Giuseppe; Montévil, Maël; Pocheville, Arnaud (2012-01-01). "From bottom-up approaches to levels of organization and extended critical transitions". *Fractal Physiology*. **3**: 232. PMC 3429021. PMID 22934001. doi:10.3389/fphys.2012.00232.

[16] McKinnon, AM. (2010). 'Elective affinities of the Protestant ethic: Weber and the chemistry of capitalism'. Sociological Theory, vol 28, no. 1, pp. 108–26.

[17] Wheeler, Wendy (2006). *The Whole Creature: Complexity, Biosemiotics and the Evolution of Culture*. London: Lawrence & Wishart. p. 192. ISBN 1-905007-30-2.

[18] Alexander, Victoria N. (2011). *The Biologist's Mistress: Rethinking Self-Organization in Art, Literature, and Nature*. Litchfield Park, AZ: Emergent Publications. ISBN 0-9842165-5-3.

[19] Pearce, Michael J. (2015). *Art in the Age of Emergence*. Manchester, England: Cambridge Scholars Publishing. ISBN 1443870579.

[20] Daniel C. Taylor, Carl E. Taylor, Jesse O. Taylor, *Empowerment on an Unstable Planet: From Seeds of Human Energy to a Scale of Global Change* (New York: Oxford University Press, 2012)

[21] See, e.g., Korotayev, A.; Malkov, A.; Khaltourina, D. (2006). *Introduction to Social Macrodynamics: Compact Macromodels of the World System Growth*. Moscow: URSS. ISBN 5-484-00414-4

[22] Almond, D.P.; Budd, C.J.; Freitag, M.A.; Hunt, G.W.; McCullen, N.J.; Smith, N.D. (2013). "The origin of power-law emergent scaling in large binary networks". *Physica A: Statistical Mechanics and its Applications.* **392**: 1004–1027. Bibcode:2013PhyA..392.1004A. arXiv:1204.5601 ⊙. doi:10.1016/j.physa.2012.10.035.

[23] Bernstein, H; Byerly, HC; Hopf, FA; Michod, RA; Vemulapalli, GK (1983). "The Darwinian Dynamic". *Quarterly Review of Biology.* **58**: 185–207. doi:10.1086/413216.

[24] Michod RE. (2000) Darwinian Dynamics: Evolutionary Transitions in Fitness and Individuality. Princeton University Press, Princeton, New Jersey ISBN 0691050112 ISBN 978-0691050119

[25] Steven Johnson. 2001. Emergence: The Connected Lives of Ants, Brains, Cities, and Software

[26] Danforth, Bryan (1991). "Female Foraging and Intranest Behavior of a Communal Bee, Perdita portalis (Hymenoptera: Andrenidae)". *Annals of the Entomological Society of America.* **84** (5): 537–48. doi:10.1093/aesa/84.5.537.

[27] Campbell, Neil A., and Jane B. Reece. *Biology.* 6th ed. San Francisco: Benjamin Cummings, 2002.

[28] Emergence // Encyclopædia Britannica, 2017

[29] Eryomin A.L. **Noogenesis and Theory of Intellect**. Krasnodar, 2005. 356 pp.

[30] Miller, Peter. 2010. The Smart Swarm: How understanding flocks, schools, and colonies can make us better at communicating, decision making, and getting things done. New York: Avery.

[31] Campbell, Michael J.; Carfi, David (2017). "Bounded Rational Speculative and Hedging Interaction Model in Oil and U.S. Dollar Markets III - Phase Transition". *preprint.*

[32] Valentin Robu, Harry Halpin, Hana Shepherd Emergence of consensus and shared vocabularies in collaborative tagging systems, ACM Transactions on the Web (TWEB), Vol. 3(4), article 14. ACM Press, September 2009.

[33] Fu, Wai-Tat; Kannampallil, Thomas George; Kang, Ruogu (August 2009). "A Semantic Imitation Model of Social Tagging". *Proceedings of the IEEE conference on Social Computing*: 66–72, ISBN 978-1-4244-5334-4, doi:10.1109/CSE.2009.382

[34] Fu, Wai-Tat; Kannampallil, Thomas; Kang, Ruogu; He, Jibo (2010), "Semantic Imitation in Social Tagging", *ACM Transactions on Computer-Human Interaction*, **17** (3): 1–37, doi:10.1145/1806923.1806926

[35] http://www.microbe.net/fact-sheet-building-ecology/

[36] http://www.microbe.net

[37] http://buildingecology.com

[38] Bonabeau E. Predicting the Unpredictable. Harvard Business Review [serial online]. March 2002. 80(3):109–16. Available from: Business Source Complete, Ipswich, MA. Accessed February 1, 2012.

[39] Roudavski, Stanislav and Gwyllim Jahn (2012). 'Emergent Materiality though an Embedded Multi-Agent System', in 15th Generative Art Conference, ed. by Celestino Soddu (Lucca, Italy: Domus Argenia), pp. 348–63

[40] Holman, Peggy (December 2010 – January 2011). "Engaging Emergence: Turning Upheaval into Opportunity" (PDF). *Pegasus Communication: The Systems Thinker.* **21**.

2.8 Bibliography

- Anderson, P.W. (1972), "More is Different: Broken Symmetry and the Nature of the Hierarchical Structure of Science", *Science,* **177** (4047): 393–96, Bibcode:1972Sci...177..393A, PMID 17796623, doi:10.1126/science.177.4047.393

- Bedau, Mark A. (1997), *Weak Emergence* (PDF)

- Corning, Peter A. (1983), *The Synergism Hypothesis: A Theory of Progressive Evolution,* New York: McGraw-Hill

- Koestler, Arthur (1969), A. Koestler; J. R. Smythies, eds., *Beyond Reductionism: New Perspectives in the Life Sciences,* London: Hutchinson

- Laughlin, Robert (2005), *A Different Universe: Reinventing Physics from the Bottom Down,* Basic Books, ISBN 0-465-03828-X

2.9 Further reading

- Alexander, V. N. (2011). *The Biologist's Mistress: Rethinking Self-Organization in Art, Literature and Nature.* Litchfield Park AZ: Emergent Publications.

- Bateson, Gregory (1972), *Steps to an Ecology of Mind,* Ballantine Books, ISBN 0-226-03905-6

- Batty, Michael (2005), *Cities and Complexity,* MIT Press, ISBN 0-262-52479-1

- Blitz, David. (1992). *Emergent Evolution: Qualitative Novelty and the Levels of Reality.* Dordrecht: Kluwer Academic.

- Bunge, Mario Augusto (2003), *Emergence and Convergence: Qualitative Novelty and the Unity of Knowledge,* Toronto: University of Toronto Press

- Chalmers, David J. (2002). "Strong and Weak Emergence" http://consc.net/papers/emergence.pdf Republished in P. Clayton and P. Davies, eds. (2006) *The Re-Emergence of Emergence*. Oxford: Oxford University Press.

- Philip Clayton & Paul Davies (eds.) (2006). *The Re-Emergence of Emergence: The Emergentist Hypothesis from Science to Religion* Oxford: Oxford University Press.

- Corning, Peter A. (2005). "Holistic Darwinism: Synergy, Cybernetics and the Bioeconomics of Evolution." Chicago: University of Chicago Press.

- Felipe Cucker and Stephen Smale (2007), The Japanese Journal of Mathematics, *The Mathematics of Emergence*

- Delsemme, Armand (1998), *Our Cosmic Origins: From the Big Bang to the Emergence of Life and Intelligence*, Cambridge University Press

- Goodwin, Brian (2001), *How the Leopard Changed Its Spots: The Evolution of Complexity*, Princeton University Press

- Hofstadter, Douglas R. (1979), *Gödel, Escher, Bach: an Eternal Golden Braid*, Harvester Press

- Holland, John H. (1998), *Emergence from Chaos to Order*, Oxford University Press, ISBN 0-7382-0142-1

- Kauffman, Stuart (1993), *The Origins of Order: Self-Organization and Selection in Evolution*, Oxford University Press, ISBN 0-19-507951-5

- Keller, Rudi (1994), *On Language Change: The Invisible Hand in Language*, London/New York: Routledge, ISBN 0-415-07671-4

- Kauffman, Stuart (1995), *At Home in the Universe*, New York: Oxford University Press

- Kelly, Kevin (1994), *Out of Control: The New Biology of Machines, Social Systems, and the Economic World*, Perseus Books, ISBN 0-201-48340-8

- Koestler, Arthur (1969), A. Koestler & J. R. Smythies, ed., *Beyond Reductionism: New Perspectives in the Life Sciences*, London: Hutchinson

-

-

- Krugman, Paul (1996), *The Self-organizing Economy*, Oxford: Blackwell, ISBN 1-55786-698-8, ISBN 0-87609-177-X

- Laughlin, Robert (2005), *A Different Universe: Reinventing Physics from the Bottom Down*, Basic Books, ISBN 0-465-03828-X

- Lewin, Roger (2000), *Complexity - Life at the Edge of Chaos* (second ed.), University of Chicago Press, ISBN 0-226-47654-5, ISBN 0-226-47655-3

- Ignazio Licata & Ammar Sakaji (eds) (2008). *Physics of Emergence and Organization*, ISBN 978-981-277-994-6, World Scientific and Imperial College Press.

- Marshall, Stephen (2009), *Cities Design and Evolution*, Routledge, ISBN 978-0-415-42329-8, ISBN 0-415-42329-5

- Morowitz, Harold J. (2002), *The Emergence of Everything: How the World Became Complex*, Oxford University Press, ISBN 0-19-513513-X

- Pearce, Michael J. (2015), *Art in the Age of Emergence.*, Cambridge Scholars Publishing, ISBN 1-443-87057-9, ISBN 1-443-87057-9

- Schelling, Thomas C. (1978), *Micromotives and Macrobehaviour*, W. W. Norton, ISBN 0-393-05701-1

- Smith, John Maynard; Szathmáry, Eörs (1997), *The Major Transitions in Evolution*, Oxford University Press, ISBN 0-19-850294-X

- Smith, Reginald D. (2008), "The Dynamics of Internet Traffic: Self-Similarity, Self-Organization, and Complex Phenomena", *Advances in Complex Systems*, **14**: 905–949, Bibcode:2008arXiv0807.3374S, arXiv:0807.3374, doi:10.1142/S0219525911003451

- Solé, Ricard and Goodwin, Brian (2000) Signs of life: how complexity pervades biology, Basic Books, New York

- Jakub Tkac & Jiri Kroc (2017), Cellular Automaton Simulation of Dynamic Recrystallization: Introduction into Self-Organization and Emergence (Software) "Video - Simulation of DRX"

- Wan, Poe Yu-ze (2011), "Emergence a la Systems Theory: Epistemological *Totalausschluss* or Ontological Novelty?", *Philosophy of the Social Sciences, 41(2)*, pp. 178–210

- Wan, Poe Yu-ze (2011), *Reframing the Social: Emergentist Systemism and Social Theory*, Ashgate Publishing

- Weinstock, Michael (2010), *The Architecture of Emergence - the evolution of form in Nature and Civilisation*, John Wiley and Sons, ISBN 0-470-06633-4

- Wolfram, Stephen (2002), *A New Kind of Science*, ISBN 1-57955-008-8
- Young, Louise B. (2002), *The Unfinished Universe*, ISBN 0-19-508039-4

2.10 External links

- "Emergence". *Internet Encyclopedia of Philosophy*.
- "Emergent Properties". *Stanford Encyclopedia of Philosophy*.
- Emergence at PhilPapers
- Emergence at the Indiana Philosophy Ontology Project
- The Emergent Universe: An interactive introduction to emergent phenomena, from ant colonies to Alzheimer's.
- Exploring Emergence: An introduction to emergence using CA and Conway's Game of Life from the MIT Media Lab
- ISCE group: Institute for the Study of Coherence and Emergence.
- Towards modeling of emergence: lecture slides from Helsinki University of Technology
- Biomimetic Architecture – Emergence applied to building and construction
- Studies in Emergent Order: Studies in Emergent Order (SIEO) is an open-access journal
- Emergence

Chapter 3

Self-organization

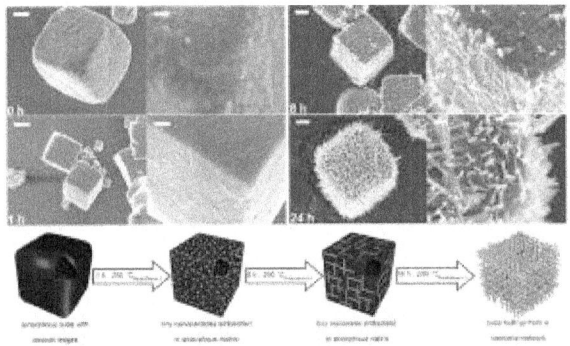

Self-organization in micron-sized $Nb_3O_7(OH)$ cubes during a hydrothermal treatment at 200 °C. Initially amorphous cubes gradually transform into ordered 3D meshes of crystalline nanowires as summarized in the model below.[1]

Self-organization, also called spontaneous order (in the social sciences), is a process where some form of overall order arises from local interactions between parts of an initially disordered system. The process is spontaneous, not needing control by any external agent. It is often triggered by random fluctuations, amplified by positive feedback. The resulting organization is wholly decentralized, distributed over all the components of the system. As such, the organization is typically robust and able to survive or self-repair substantial perturbation. Chaos theory discusses self-organization in terms of islands of predictability in a sea of chaotic unpredictability.

Self-organization occurs in many physical, chemical, biological, robotic, and cognitive systems. Examples can be found in crystallization, thermal convection of fluids, chemical oscillation, animal swarming, and artificial and biological neural networks.

3.1 Overview

Self-organization is realized[2] in the physics of non-equilibrium processes, and in chemical reactions, where it is often described as self-assembly. The concept has proven useful in biology,[3] from molecular to ecosystem level.[4] Cited examples of self-organizing behaviour also appear in the literature of many other disciplines, both in the natural sciences and in the social sciences such as economics or anthropology. Self-organization has also been observed in mathematical systems such as cellular automata.[5] Self-organization is not to be confused with the related concept of emergence.[6]

Self-organization relies on three basic ingredients:[7]

1. strong dynamical non-linearity, often though not necessarily involving positive and negative feedback
2. balance of exploitation and exploration
3. multiple interactions

3.2 Principles

The cybernetician William Ross Ashby formulated the original principle of self-organization in 1947.[8][9] It states that any deterministic dynamic system automatically evolves towards a state of equilibrium that can be described in terms of an attractor in a basin of surrounding states. Once there, the further evolution of the system is constrained to remain in the attractor. This constraint implies a form of mutual dependency or coordination between its constituent components or subsystems. In Ashby's terms, each subsystem has adapted to the environment formed by all other subsystems.[8]

The cybernetician Heinz von Foerster formulated the principle of "order from noise" in 1960.[10] It notes that self-organization is facilitated by random perturbations ("noise") that let the system explore a variety of states in its state space. This increases the chance that the system will arrive into the basin of a "strong" or "deep" attractor, from which it then quickly enters the attractor itself. The thermodynamicist Ilya Prigogine formulated a similar principle as "order through fluctuations"[11] or "order out of chaos".[12] It is applied in the method of simulated annealing for problem solving and machine learning.[13]

3.3 History

Further information: Spontaneous order

The idea that the dynamics of a system can lead to an increase in its organization has a long history. The ancient atomists such as Democritus and Lucretius believed that a designing intelligence is unnecessary to create order in nature, arguing that given enough time and space and matter, order emerges by itself.[14]

The philosopher René Descartes presents it hypothetically in the fifth part of his 1637 *Discourse on Method*. He elaborated on the idea in his unpublished work *The World*.[lower-alpha 1]

The term "self-organizing" was used by Immanuel Kant in his 1790 *Critique of Judgment*, where he argued that teleology is a meaningful concept only if there exists such an entity whose parts or "organs" are simultaneously ends and means. Such a system of organs must be able to behave as if it has a mind of its own, that is, it is capable of governing itself.[15]

Sadi Carnot and Rudolf Clausius discovered the Second Law of Thermodynamics in the 19th century. It states that total entropy, sometimes understood as disorder, will always increase over time in an isolated system. This means that a system cannot spontaneously increase its order, without an external relationship that decreases order elsewhere in the system (e.g. through consuming the low-entropy energy of a battery and diffusing high-entropy heat).[16][17]

18th century thinkers had sought to understand the "universal laws of form" to explain the observed forms of living organisms. This idea became associated with Lamarckism and fell into disrepute until the early 20th century, when D'Arcy Wentworth Thompson attempted to revive it.[18]

The term "self-organizing" was introduced to contemporary science in 1947 by the psychiatrist and engineer W. Ross Ashby.[8] It was taken up by the cyberneticians Heinz von Foerster, Gordon Pask, Stafford Beer, and von Foerster organized a conference on "The Principles of Self-Organization" at the University of Illinois' Allerton Park in June, 1960 which led to a series of conferences on Self-Organizing Systems.[19] Norbert Wiener took up the idea in the second edition of his *Cybernetics: or Control and Communication in the Animal and the Machine* (1961).

Self-organization was associated with general systems theory in the 1960s, but did not become commonplace in the scientific literature until physicists and complex systems researchers adopted it in the 1970s and 1980s.[20] After Ilya Prigogine's 1977 Nobel Prize, the *thermodynamic concept of self-organization* received public attention, and scientists started to migrate from the *cybernetic view* to the *thermodynamic view*.[21]

3.4 By field

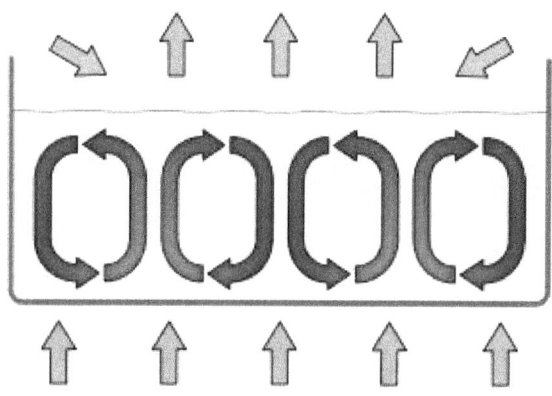

Convection cells in a gravity field

3.4.1 Physics

See also: Self-assembly and Self-assembly of nanoparticles

The many self-organizing phenomena in physics include phase transitions and spontaneous symmetry breaking such as spontaneous magnetization and crystal growth in classical physics, and the laser,[22] superconductivity and Bose–Einstein condensation in quantum physics. It is found in self-organized criticality in dynamical systems, in tribology, in spin foam systems, and in loop quantum gravity.[23]

3.4.2 Chemistry

Self-organization in chemistry includes molecular self-assembly,[25] reaction-diffusion systems and oscillating reactions,[26] autocatalytic networks, liquid crystals,[27] grid complexes, colloidal crystals, self-assembled monolayers,[28][29] micelles, microphase separation of block copolymers, and Langmuir-Blodgett films.[30]

3.4.3 Biology

Further information: Biological organisation

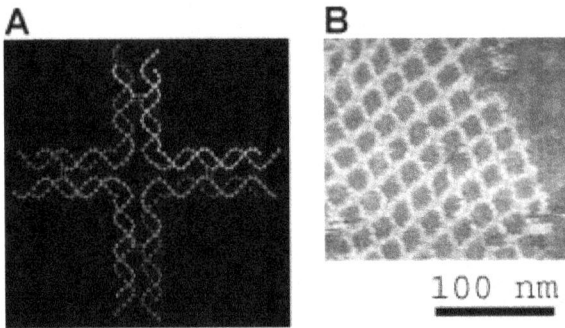

The DNA structure shown schematically at left self-assembles into the structure at right.[24]

Birds flocking, an example of self-organization in biology

Self-organization in biology[3][31] can be observed in spontaneous folding of proteins and other biomacromolecules, formation of lipid bilayer membranes, pattern formation and morphogenesis in developmental biology, the coordination of human movement, social behaviour in insects (bees, ants, termites),[32] and mammals, flocking behaviour in birds and fish.[33]

The mathematical biologist Stuart Kauffman and other structuralists have suggested that self-organization may play roles alongside natural selection in three areas of evolutionary biology, namely population dynamics, molecular evolution, and morphogenesis. However, this does not take into account the essential role of energy in driving biochemical reactions in cells. The systems of reactions in any cell are self-catalyzing but not simply self-organizing as they are thermodynamically open systems relying on a continuous input of energy.[34][35] Self-organization is not an alternative to natural selection, but it constrains what evolution can do and provides mechanisms such as the self-assembly of membranes which evolution then exploits.[36]

3.4.4 Computer science

Phenomena from mathematics and computer science such as cellular automata, random graphs, and some instances of evolutionary computation and artificial life exhibit features of self-organization. In swarm robotics, self-organization is used to produce emergent behavior. In particular the theory of random graphs has been used as a justification for self-organization as a general principle of complex systems. In the field of multi-agent systems, understanding how to engineer systems that are capable of presenting self-organized behavior is an active research area.[37] Optimization algorithms can be considered self-organizing because they aim to find the optimal solution to a problem. If the solution is considered as a state of the iterative system, the optimal solution is the selected, converged structure of the system.[38][39] Self-organizing networks include small-world networks[40] and scale-free networks. These emerge from bottom-up interactions, unlike top-down hierarchical networks within organizations, which are not self-organizing.[41] Cloud computing systems have been argued to be inherently self-organising,[42] but while they have some autonomy, they are not self-managing as they do not have the goal of reducing their own complexity.[43][44]

3.4.5 Cybernetics

Main article: Self-organization in cybernetics

Norbert Wiener regarded the automatic serial identification of a black box and its subsequent reproduction as self-organization in cybernetics.[45] The importance of phase locking or the "attraction of frequencies", as he called it, is discussed in the 2nd edition of his *Cybernetics: Or Control and Communication in the Animal and the Machine*.[46] K. Eric Drexler sees self-replication as a key step in nano and universal assembly. By contrast, the four concurrently connected galvanometers of W. Ross Ashby's Homeostat hunt, when perturbed, to converge on one of many possible stable states.[47] Ashby used his state counting measure of variety[48] to describe stable states and produced the "Good Regulator"[49] theorem which requires internal models for self-organized endurance and stability (e.g. Nyquist stability criterion). Warren McCulloch proposed "Redundancy of Potential Command"[50] as characteristic of the organization of the brain and human nervous system and the necessary condition for self-organization. Heinz von Foerster proposed Redundancy, $R=1 - H/H_{max}$, where H is entropy.[51][52] In essence this states that unused potential communication bandwidth is a measure of self-organization.

In the 1970s Stafford Beer considered self-organization

necessary for autonomy in persisting and living systems. Using Variety analyses he applied his neurophysiologically derived recursive Viable System Model to management. It consists of five parts: the monitoring of performance of the survival processes (1), their management by recursive application of regulation (2), homeostatic operational control (3) and development (4) which produce maintenance of identity (5) under environmental perturbation. Focus is prioritized by an alerting "algedonic loop" feedback: a sensitivity to both pain and pleasure produced from under-performance or over-performance relative to a standard capability.[53]

In the 1990s Gordon Pask argued that von Foerster's H and Hmax were not independent, but interacted via countably infinite recursive concurrent spin processes[54] which he called concepts. His strict definition of concept "a procedure to bring about a relation"[55] permitted his theorem "Like concepts repel, unlike concepts attract"[56] to state a general spin-based principle of self-organization. His edict, an exclusion principle, "There are No Doppelgangers" means no two concepts can be the same. After sufficient time, all concepts attract and coalesce as pink noise. The theory applies to all organizationally closed or homeostatic processes that produce enduring and coherent products which evolve, learn and adapt.[57][54]

3.4.6 Human society

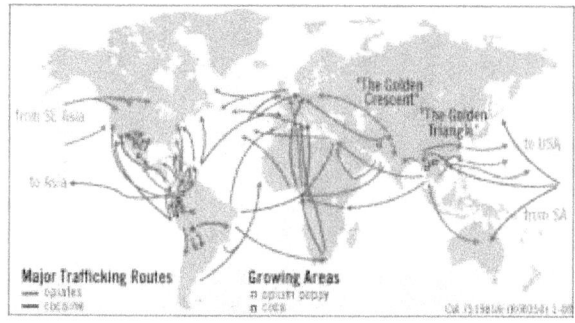

Social self-organization in international drug routes

Main article: Spontaneous order

The self-organizing behaviour of social animals and the self-organization of simple mathematical structures both suggest that self-organization should be expected in human society. Tell-tale signs of self-organization are usually statistical properties shared with self-organizing physical systems. Examples such as critical mass, herd behaviour, groupthink and others, abound in sociology, economics, behavioral finance and anthropology.[58]

In social theory, the concept of self-referentiality has been introduced as a sociological application of self-organization theory by Niklas Luhmann (1984). For Luhmann the elements of a social system are self-producing communications, i.e. a communication produces further communications and hence a social system can reproduce itself as long as there is dynamic communication. For Luhmann human beings are sensors in the environment of the system. Luhmann developed an evolutionary theory of Society and its subsystems, using functional *analyses* and systems *theory*.[59]

In economics, a market economy is sometimes said to be self-organizing. Paul Krugman has written on the role that market self-organization plays in the business cycle in his book "The Self Organizing Economy".[60] Friedrich Hayek coined the term *catallaxy*[61] to describe a "self-organizing system of voluntary co-operation", in regards to the spontaneous order of the free market economy. Neo-classical economists hold that imposing central planning usually makes the self-organized economic system less efficient. On the other end of the spectrum, economists consider that market failures are so significant that self-organization produces bad results and that the state should direct production and pricing. Most economists adopt an intermediate position and recommend a mixture of market economy and command economy characteristics (sometimes called a mixed economy). When applied to economics, the concept of self-organization can quickly become ideologically imbued.[21][62]

3.4.7 In learning

Enabling others to "learn how to learn"[63] is often taken to mean instructing them[64] how to submit to being taught. Self-Organised learning (S.O.L.)[65][66][67] denies that "the expert knows best" or that there is ever "the one best method".[68][69][70] insisting instead on "the construction of personally significant, relevant and viable meaning"[71] to be tested experientially by the learner.[72] This may be collaborative, and more rewarding personally.[73][74] It is seen as a lifelong process, not limited to specific learning environments (home, school, university) or under the control of authorities such as parents and professors.[75] It needs to be tested, and intermittently revised, through the personal experience of the learner.[76] It need not be restricted by either consciousness or language.[77] Fritjof Capra argued that it is poorly recognised within psychology and education.[78] It may be related to cybernetics as it involves a negative feedback control loop,[55] or to systems theory.[79] It can be conducted as a learning conversation or dialogue between learners or within one person.[80][81]

3.4.8 Traffic flow

Main article: Three-phase traffic theory

The self-organizing behavior of drivers in traffic flow determines almost all the spatiotemporal behavior of traffic, such as traffic breakdown at a highway bottleneck, highway capacity, and the emergence of moving traffic jams. In 1996–2002 these complex self-organizing effects were explained by Boris Kerner's three-phase traffic theory.[82]

3.4.9 In linguistics

Order appears spontaneously in the evolution of language as individual and population behaviour interacts with biological evolution.[83]

3.5 Criticism

Heinz Pagels, in a 1985 review of Ilya Prigogine and Isabelle Stengers's book *Order Out of Chaos* in *Physics Today*, appeals to authority:[84]

In theology, Thomas Aquinas (1225–1274) in his *Summa Theologica* assumes a teleological created universe in rejecting the idea that something can be a self-sufficient cause of its own organization:[85]

3.6 See also

- Autopoiesis
- Autowave
- Self-organized criticality control
- Free energy principle
- Information theory

3.7 Notes

[1] For related history, see Aram Vartanian, *Diderot and Descartes*.

3.8 References

[1] Betzler, S. B.; Wisnet, A.; Breitbach, B.; Mitterbauer, C.; Weickert, J.; Schmidt-Mende, L.; Scheu, C. (2014). "Template-free synthesis of novel, highly-ordered 3D hierarchical $Nb_3O_7(OH)$ superstructures with semiconductive and photoactive properties". *Journal of Materials Chemistry A*. **2** (30): 12005. doi:10.1039/C4TA02202E.

[2] Glansdorff, P., Prigogine, I. (1971). *Thermodynamic Theory of Structure, Stability and Fluctuations*, Wiley-Interscience, London. ISBN 0-471-30280-5

[3] Witzany, G (2014). "Biological Self-Organization". *International Journal of Signs and Semiotic Systems*. **3**: 1–11.

[4] Compare: Camazine, Scott (2003). *Self-organization in Biological Systems*. Princeton studies in complexity (reprint ed.). Princeton University Press. ISBN 9780691116242. Retrieved 2016-04-05.

[5] Ilachinski, Andrew (2001). *Cellular Automata: A Discrete Universe*. World Scientific. p. 247. ISBN 9789812381835. We have already seen ample evidence for what is arguably the single most impressive general property of CA, namely their capacity for self-organization

[6] Feltz, Bernard; et al. (2006). *Self-organization and Emergence in Life Sciences*. p. 1. ISBN 978-1-402-03916-4.

[7] Bonabeau, Eric; Dorigo, Marco and Theraulaz, Guy (1999), *Swarm intelligence: from natural to artificial systems*. pp. 9–11. ISBN 0195131592.

[8] Ashby, W. R. (1947). "Principles of the Self-Organizing Dynamic System". *The Journal of General Psychology*. **37** (2): 125–28. PMID 20270223. doi:10.1080/00221309.1947.9918144.

[9] Ashby, W. R. (1962). "Principles of the self-organizing system", pp. 255–78 in *Principles of Self-Organization*. Heinz von Foerster and George W. Zopf, Jr. (eds.) U.S. Office of Naval Research.

[10] Von Foerster, H. (1960). "On self-organizing systems and their environments", pp. 31–50 in *Self-organizing systems*. M.C. Yovits and S. Cameron (eds.), Pergamon Press, London

[11] Nicolis, G. and Prigogine, I. (1977). *Self-organization in nonequilibrium systems: From dissipative structures to order through fluctuations*. Wiley, New York.

[12] Prigogine, I. and Stengers, I. (1984). *Order out of chaos: Man's new dialogue with nature*. Bantam Books.

[13] Ahmed, Furqan; Tirkkonen, Olav (January 2016). "Applied Soft Computing". *Applied Soft Computing*. **38**: 762–70. doi:10.1016/j.asoc.2015.10.028.

[14] Palmer, Ada (October 2014). *Reading Lucretius in the Renaissance*. Harvard University Press. ISBN 978-0-674-72557-7. Ada Palmer explores how Renaissance readers, such as Machiavelli, Pomponio Leto, and Montaigne, actually ingested and disseminated Lucretius, ... and shows how ideas of emergent order and natural selection, so critical to our current thinking, became embedded in Europe's intellectual landscape before the seventeenth century.

3.8. REFERENCES

[15] *German Aesthetic*. CUP Archive. pp. 64–. GGKEY: TFTHBB91ZH2.

[16] Carnot, S. (1824/1986). *Reflections on the motive power of fire*, Manchester University Press, Manchester UK, ISBN 0-7190-1741-6.

[17] Clausius, R. (1850). "Ueber Die Bewegende Kraft Der Wärme Und Die Gesetze, Welche Sich Daraus Für Die Wärmelehre Selbst Ableiten Lassen". *Annalen der Physik*. **79**: 368–97, 500–24. Bibcode:1850AnP...155..500C. doi:10.1002/andp.18501550403. Translated into English: Clausius, R. (July 1851). "On the Moving Force of Heat, and the Laws regarding the Nature of Heat itself which are deducible therefrom". *London, Edinburgh and Dublin Philosophical Magazine and Journal of Science*. 4th. **2** (VIII): 1–21, 102–19. Retrieved 26 June 2012.

[18] Ruse, Michael (2013). "17. From Organicism to Mechanism-and Halfway Back?". In Henning, Brian G.; Scarfe, Adam. *Beyond Mechanism: Putting Life Back Into Biology*. Lexington Books. p. 419.

[19] Asaro, P. (2007). "Heinz von Foerster and the Bio-Computing Movements of the 1960s" in Albert Müller and Karl H. Müller (eds.) *An Unfinished Revolution? Heinz von Foerster and the Biological Computer Laboratory* BCL 1958–1976. Vienna, Austria: Edition Echoraum.

[20] As an indication of the increasing importance of this concept, when queried with the keyword self-organ*, *Dissertation Abstracts* finds nothing before 1954, and only four entries before 1970. There were 17 in the years 1971–1980; 126 in 1981–1990; and 593 in 1991–2000.

[21] Biel, R.; Mu-Jeong Kho (November 2009). "The Issue of Energy within a Dialectical Approach to the Regulationist Problematique" (PDF). *Recherches & Régulation Working Papers*, RR Série ID 2009-1. Association Recherche & Régulation: 1–21. Retrieved 2013-11-09.

[22] Zeiger, H. J. and Kelley, P. L. (1991) "Lasers", pp. 614–19 in *The Encyclopedia of Physics*, Second Edition, edited by Lerner, R. and Trigg, G., VCH Publishers.

[23] Ansari M. H. (2004) Self-organized theory in quantum gravity. arxiv.org

[24] Strong, M. (2004). "Protein Nanomachines". *PLoS Biology*. **2** (3): e73–e74. PMC 368168. PMID 15024422. doi:10.1371/journal.pbio.0020073.

[25] Lehn, J.-M. (1988). "Perspectives in Supramolecular Chemistry-From Molecular Recognition towards Molecular Information Processing and Self-Organization". *Angew. Chem. Int. Ed. Engl.* **27** (11): 89–121. doi:10.1002/anie.198800891.

[26] Bray, William C. (1921). "A periodic reaction in homogeneous solution and its relation to catalysis.". *Journal of the American Chemical Society*. **43** (6): 1262–67. doi:10.1021/ja01439a007.

[27] Rego, J.A.; Harvey, Jamie A.A.; MacKinnon, Andrew L.; Gatdula, Elysse (January 2010). "Asymmetric synthesis of a highly soluble 'trimeric' analogue of the chiral nematic liquid crystal twist agent Merck S1011" (PDF). *Liquid Crystals*. **37** (1): 37–43. doi:10.1080/02678290903359291.

[28] Love; et al. (2005). "Self-Assembled Monolayers of Thiolates on Metals as a Form of Nanotechnology". *Chem. Rev.* **105** (4): 1103–70. PMID 15826011. doi:10.1021/cr0300789.

[29] Barlow, S.M.; Raval R.. (2003). "Complex organic molecules at metal surfaces: bonding, organisation and chirality". *Surface Science report*. **50** (6–8): 201–341. Bibcode:2003SurSR..50..201B. doi:10.1016/S0167-5729(03)00015-3.

[30] Ritu, Harneet (2016). "Large Area Fabrication of Semiconducting Phosphorene by Langmuir-Blodgett Assembly". *Sci. Rep.* **6**: 34095. PMC 5037434. PMID 27671093. doi:10.1038/srep34095.

[31] Camazine, Deneubourg, Franks, Sneyd, Theraulaz, Bonabeau. *Self-Organization in Biological Systems*. Princeton University Press, 2003. ISBN 0-691-11624-5 ISBN 0-691-01211-3 (pbk.) p. 8

[32] Bonabeau, Eric; et al. (May 1997). "Self-organization in social insects". *Trends in Ecology & Evolution*. **12** (5): 188–93. doi:10.1016/S0169-5347(97)01048-3.

[33] Couzin, Iain D.; Krause, Jens (2003). "Self-Organization and Collective Behavior in Vertebrates" (PDF). *Advances in the Study of Behavior*. **32**: 1–75.

[34] Fox, Ronald F. (December 1993). "Review of Stuart Kauffman, The Origins of Order: Self-Organization and Selection in Evolution". *Biophys. J.* **65** (6): 2698–99. PMC 1226010. doi:10.1016/s0006-3495(93)81321-3.

[35] Goodwin, Brian (2009). Ruse, Michael; Travis, Joseph, eds. *Beyond the Darwinian Paradigm: Understanding Biological Forms. Evolution: The First Four Billion Years*. Harvard University Press.

[36] Johnson, Brian R.; Lam, Sheung Kwan (2010). "Self-organization, Natural Selection, and Evolution: Cellular Hardware and Genetic Software". *BioScience*. **60** (11): 879–85. doi:10.1525/bio.2010.60.11.4.

[37] Serugendo, Giovanna Di Marzo; et al. (June 2005). "Self-organization in multi-agent systems". *Knowledge Engineering Review*. **20** (2): 165–89. doi:10.1017/S0269888905000494.

[38] Yang, X. S.; Deb, S.; Loomes, M.; Karamanoglu, M. (2013). "A framework for self-tuning optimization algorithm". *Neural Computing and Applications*. **23** (7–8): 2051–57. doi:10.1007/s00521-013-1498-4.

[39] X. S. Yang (2014) *Nature-Inspired Optimization Algorithms*. Elsevier.

[40] Watts, Duncan J.; Strogatz, Steven H. (June 1998). "Collective dynamics of 'small-world' networks". *Nature*. **393**: 440–42. PMID 9623998. doi:10.1038/30918.

[41] Clauset, Aaron; Cosma Rohilla Shalizi; M. E. J Newman (2007-06-07). "Power-law distributions in empirical data". *SIAM Review*. **51**: 661–703. Bibcode:2009SIAMR..51..661C. arXiv:0706.1062. doi:10.1137/070710111.

[42] Zhang, Q., Cheng, L., and Boutaba, R. (2010). "Cloud computing: state-of-the-art and research challenges" (PDF). *Journal of Internet Services and Applications*. **1** (1): 7–18. doi:10.1007/s13174-010-0007-6.

[43] Marinescu, D. C.; Paya, A.; Morrison, J. P.; Healy, P. (2013). "An auction-driven self-organising cloud delivery model". arXiv:1312.2998 [cs.DC].

[44] Lynn; et al. (2016). "Cloudlightning: A Framework for a Self-organising and Self-managing Heterogeneous Cloud". *Proceedings of the 6th International Conference on Cloud Computing and Services Science*: 333. ISBN 978-989-758-182-3. doi:10.5220/0005921503330338.

[45] Wiener, Norbert (1962) "The mathematics of self-organising systems". *Recent developments in information and decision processes*, Macmillan, N. Y. and Chapter X in *Cybernetics, or control and communication in the animal and the machine*, The MIT Press.

[46] *Cybernetics, or control and communication in the animal and the machine*, The MIT Press, Cambridge, Massachusetts and Wiley, NY, 1948. 2nd Edition 1962 "Chapter X "Brain Waves and Self-Organizing Systems" pp. 201–02.

[47] Ashby, William Ross (1952) *Design for a Brain*, Chapter 5 Chapman & Hall

[48] Ashby, William Ross (1956) *An Introduction to Cybernetics*, Part Two Chapman & Hall

[49] Conant, R. C.; Ashby, W. R. (1970). "Every good regulator of a system must be a model of that system" (PDF). *Int. J. Systems Sci*. **1** (2): 89–97. doi:10.1080/00207727008920220.

[50] *Embodiments of Mind* MIT Press (1965)"

[51] von Foerster, Heinz; Pask, Gordon (1961). "A Predictive Model for Self-Organizing Systems, Part I". *Cybernetica*. **3**: 258–300.

[52] von Foerster, Heinz; Pask, Gordon (1961). "A Predictive Model for Self-Organizing Systems, Part II". *Cybernetica*. **4**: 20–55.

[53] "Brain of the Firm" Alan Lane (1972); see also Viable System Model in "Beyond Dispute", and Stafford Beer (1994) "Redundancy of Potential Command" pp. 157–58.

[54] Pask, Gordon (1996). "Heinz von Foerster's Self-Organisation, the Progenitor of Conversation and Interaction Theories" (PDF). *Systems Research*. **13** (3): 349–62. doi:10.1002/(sici)1099-1735(199609)13:3<349::aid-sres103>3.3.co;2-7.

[55] Pask, G. (1973). *Conversation, Cognition and Learning. A Cybernetic Theory and Methodology*. Elsevier

[56] Green, N. (2001). "On Gordon Pask". *Kybernetes*. **30** (5/6): 673–82. doi:10.1108/03684920110391913.

[57] Pask, Gordon (1993) *Interactions of Actors (IA), Theory and Some Applications*.

[58] *Interactive models for self organization and biological systems* Center for Models of Life, Niels Bohr Institute, Denmark

[59] Luhmann, Niklas (1995) *Social Systems*. Stanford, California: Stanford University Press. ISBN 0804726256. p. 410.

[60] Krugman, P. (1995) *The Self Organizing Economy*. Blackwell Publishers. ISBN 1557866996

[61] Hayek, F. (1976) *Law, Legislation and Liberty, Volume 2: The Mirage of Social Justice*. University of Chicago Press.

[62] Marshall, A. (2002) *The Unity of Nature*. Chapter 5. Imperial College Press. ISBN 1860943306.

[63] Rogers.C. (1969). *Freedom to Learn*. Merrill

[64] Feynman, R. P. (1987) *Elementary Particles and the Laws of Physics*. The Dyrac 1997 Memorial Lecture. Cambridge University Press. ISBN 9780521658621.

[65] Thomas L.F. & Augstein E.S. (1985) "Self-Organised Learning: Foundations of a conversational science for psychology". Routledge (1st Ed.)

[66] Thomas L.F. & Augstein E.S. (1994) "Self-Organised Learning: Foundations of a conversational science for psychology". Routledge (2nd Ed.)

[67] Thomas L.F. & Augstein E.S. (2013) "Self-Organised Learning: Foundations of a conversational science for psychology". Routledge (Psy. Revivals)

[68] Harri-Augstein E. S. and Thomas L. F. (1991) *Learning Conversations: The S-O-L way to personal and organizational growth*. Routledge (1st Ed.)

[69] Harri-Augstein E. S. and Thomas L. F. (2013) *Learning Conversations: The S-O-L way to personal and organizational growth*. Routledge (2nd Ed.)

[70] Harri-Augstein E. S. and Thomas L. F. (2013)*Learning Conversations: The S-O-L way to personal and organizational growth*. BookBaby (eBook)

[71] Illich, I. (1971) *A Celebration of Awareness*. Penguin Books.

[72] Harri-Augstein E. S. (2000) *The University of Learning in transformation*

[73] Schumacher, E. F. (1997) *This I Believe and Other Essays (Resurgence Book)*. ISBN 1870098668.

[74] Revans R. W. (1982) *The Origins and Growth of Action Learning* Chartwell-Bratt, Bromley

[75] Thomas L.F. and Harri-Augstein S. (1993) "On Becoming a Learning Organisation" in *Report of a 7 year Action Research Project with the Royal Mail Business*. CSHL Monograph

[76] Rogers C.R. (1971) *On Becoming a Person*. Constable, London

[77] Prigogyne I. & Sengers I. (1985) *Order out of Chaos* Flamingo Paperbacks, London

[78] Capra F (1989) *Uncommon Wisdom* Flamingo Paperbacks, London

[79] Bohm D. (1994) *Thought as a System*. Routledge.

[80] Maslow, A. H. (1964). *Religions, values, and peak-experiences*, Columbus: Ohio State University Press.

[81] *Conversational Science* Thomas L.F. and Harri-Augstein E.S. (1985)

[82] Kerner, Boris S. (1998). "Experimental Features of Self-Organization in Traffic Flow". *Physical Review Letters*. **81**: 3797–3800. doi:10.1103/physrevlett.81.3797.

[83] De Boer, Bart (2011). Gibson, Kathleen R.; Tallerman, Maggie, eds. *Self-organization and language evolution*. The Oxford Handbook of Language Evolution. Oxford.

[84] Pagels, H. R. (January 1, 1985). "Is the irreversibility we see a fundamental property of nature?" (PDF). *Physics Today*: 97–99.

[85] Article 3. Whether God exists? newadvent.org

3.9 Further reading

- W. Ross Ashby (1966), *Design for a Brain*, Chapman & Hall, 2nd edition.

- Amoroso, Richard (2005) *The Fundamental Limit and Origin of Complexity in Biological Systems* .

- Per Bak (1996), *How Nature Works: The Science of Self-Organized Criticality*, Copernicus Books.

- Philip Ball (1999), *The Self-Made Tapestry: Pattern Formation in Nature*, Oxford University Press.

- Stafford Beer, Self-organization as autonomy: *Brain of the Firm* 2nd edition Wiley 1981 and *Beyond Dispute* Wiley 1994.

- A. Bejan (2000), *Shape and Structure, from Engineering to Nature*, Cambridge University Press, Cambridge, UK, 324 pp.

- Mark Buchanan (2002), *Nexus: Small Worlds and the Groundbreaking Theory of Networks* W. W. Norton & Company.

- Scott Camazine, Jean-Louis Deneubourg, Nigel R. Franks, James Sneyd, Guy Theraulaz, & Eric Bonabeau (2001) *Self-Organization in Biological Systems*, Princeton Univ Press.

- Falko Dressler (2007), *Self-Organization in Sensor and Actor Networks*, Wiley & Sons.

- Manfred Eigen and Peter Schuster (1979), *The Hypercycle: A principle of natural self-organization*, Springer.

- Myrna Estep (2003), *A Theory of Immediate Awareness: Self-Organization and Adaptation in Natural Intelligence*, Kluwer Academic Publishers.

- Myrna L. Estep (2006), *Self-Organizing Natural Intelligence: Issues of Knowing, Meaning, and Complexity*, Springer-Verlag.

- J. Doyne Farmer et al. (editors) (1986), "Evolution, Games, and Learning: Models for Adaptation in Machines and Nature", in: *Physica D*, Vol 22.

- Carlos Gershenson and Francis Heylighen (2003). "When Can we Call a System Self-organizing?" In Banzhaf, W, T. Christaller, P. Dittrich, J. T. Kim, and J. Ziegler, Advances in Artificial Life, 7th European Conference, ECAL 2003, Dortmund, Germany, pp. 606–14. LNAI 2801. Springer.

- Hermann Haken (1983) *Synergetics: An Introduction. Nonequilibrium Phase Transition and Self-Organization in Physics, Chemistry, and Biology*, Third Revised and Enlarged Edition, Springer-Verlag.

- F.A. Hayek *Law, Legislation and Liberty*, RKP, UK.

- Francis Heylighen (2001): "The Science of Self-organization and Adaptivity".

- Arthur Iberall (2016), *Homeokinetics: The Basics*, Strong Voices Publishing, Medfield, Massachusetts.

- Henrik Jeldtoft Jensen (1998), *Self-Organized Criticality: Emergent Complex Behaviour in Physical and Biological Systems*, Cambridge Lecture Notes in Physics 10, Cambridge University Press.

- Steven Berlin Johnson (2001), *Emergence: The Connected Lives of Ants, Brains, Cities, and Software*.

- Stuart Kauffman (1995), *At Home in the Universe*, Oxford University Press.

- Stuart Kauffman (1993), *Origins of Order: Self-Organization and Selection in Evolution* Oxford University Press.

- J. A. Scott Kelso (1995), *Dynamic Patterns: The self-organization of brain and behavior*, The MIT Press, Cambridge, Massachusetts.

- J. A. Scott Kelso & David A Engstrom (2006), "*The Complementary Nature*", The MIT Press, Cambridge, Massachusetts.

- Alex Kentsis (2004), *Self-organization of biological systems: Protein folding and supramolecular assembly*, Ph.D. Thesis, New York University.

- E.V. Krishnamurthy (2009)", Multiset of Agents in a Network for Simulation of Complex Systems", in "Recent advances in Nonlinear Dynamics and synchronization, (NDS-1) – Theory and applications, Springer Verlag, New York,2009, Eds. K.Kyamakya et al.

- Paul Krugman (1996), *The Self-Organizing Economy*, Cambridge, Massachusetts, and Oxford: Blackwell Publishers.

- Elizabeth McMillan (2004) "Complexity, Organizations and Change".

- Marshall, A (2002) The Unity of Nature, Imperial College Press: London (esp. chapter 5)

- Müller, J.-A., Lemke, F. (2000), *Self-Organizing Data Mining*.

- Gregoire Nicolis and Ilya Prigogine (1977) *Self-Organization in Non-Equilibrium Systems*, Wiley.

- Heinz Pagels (1988), *The Dreams of Reason: The Computer and the Rise of the Sciences of Complexity*, Simon & Schuster.

- Gordon Pask (1961), *The cybernetics of evolutionary processes and of self organizing systems*, 3rd. International Congress on Cybernetics, Namur, Association Internationale de Cybernetique.

- Christian Prehofer ea. (2005), "Self-Organization in Communication Networks: Principles and Design Paradigms", in: *IEEE Communications Magazine*, July 2005.

- Mitchell Resnick (1994), *Turtles, Termites and Traffic Jams: Explorations in Massively Parallel Microworlds*, Complex Adaptive Systems series, MIT Press.

- Lee Smolin (1997), *The Life of the Cosmos* Oxford University Press.

- Ricard V. Solé and Brian C. Goodwin (2001), *Signs of Life: How Complexity Pervades Biology*, Basic Books.

- Ricard V. Solé and Jordi Bascompte (2006), *Selforganization in Complex Ecosystems*, Princeton U. Press

- Soodak, Harry; Iberall, Arthur (1978). "Homeokinetics: A Physical Science for Complex Systems". Science. **201**: 579–582. doi:10.1126/science.201.4356.579.

- Steven Strogatz (2004), *Sync: The Emerging Science of Spontaneous Order*, Theia.

- D'Arcy Thompson (1917), *On Growth and Form*, Cambridge University Press, 1992 Dover Publications edition.

- J. Tkac, J Kroc (2017), *Cellular Automaton Simulation of Dynamic Recrystallization: Introduction into Self-Organization and Emergence* "(open source software)" "Video - Simulation of DRX"

- Tom De Wolf, Tom Holvoet (2005), *Emergence Versus Self-Organisation: Different Concepts but Promising When Combined*, In Engineering Self Organising Systems: Methodologies and Applications, Lecture Notes in Computer Science, volume 3464, pp. 1–15.

- K. Yee (2003), "Ownership and Trade from Evolutionary Games", International Review of Law and Economics, 23.2, 183–197.

- Louise B. Young (2002), *The Unfinished Universe*

3.10 External links

- Hermann Haken (ed.). "Self-organization". *Scholarpedia*.

- Max Planck Institute for Dynamics and Self-Organization, Göttingen

- PDF file on self-organized common law with references

- An entry on self-organization at the *Principia Cybernetica* site

- The Science of Self-organization and Adaptivity, a review paper by Francis Heylighen

- The *Self-Organizing Systems (SOS) FAQ* by Chris Lucas, from the USENET newsgroup comp.theory.self-org.sys

- David Griffeath, *Primordial Soup Kitchen* (graphics, papers)

3.10. EXTERNAL LINKS

- nlin.AO, nonlinear preprint archive, (electronic preprints in adaptation and self-organizing systems)
- Structure and Dynamics of Organic Nanostructures
- Metal organic coordination networks of oligopyridines and Cu on graphite
- *Selforganization in complex networks* The Complex Systems Lab, Barcelona
- Computational Mechanics Group at the Santa Fe Institute
- "Organisation must grow" (1939) W. Ross Ashby journal p. 759, from The W. Ross Ashby Digital Archive
- Cosma Shalizi's notebook on self-organization from 2003-06-20, used under the GFDL with permission from author.
- Connectivism:SelfOrganization
- UCLA Human Complex Systems Program
- "Interactions of Actors (IA), Theory and Some Applications" 1993 Gordon Pask's theory of learning, evolution and self-organization (in draft).
- The Cybernetics Society
- Scott Camazine's webpage on self-organization in biological systems
- Mikhail Prokopenko's page on Information-driven Self-organisation (IDSO)
- Lakeside Labs Self-Organizing Networked Systems A platform for science and technology, Klagenfurt, Austria.
- Watch 32 discordant metronomes synch up all by themselves theatlantic.com

Chapter 4

Collective consciousness

Not to be confused with Social consciousness.
For the related philosophical term, see Higher consciousness.

Collective consciousness, **collective conscience**, or **collective conscious** (French: *conscience collective*) is the set of shared beliefs, ideas and moral attitudes which operate as a unifying force within society.[1] The term was introduced by the French sociologist Émile Durkheim in his *Division of Labour in Society* in 1893.

The French word *conscience* generally means "conscience", "consciousness", "awareness",[2] or "perception".[3] Commentators and translators of Durkheim disagree on which is most appropriate, or whether the translation should depend on the context. Some prefer to treat the word 'conscience' as an untranslatable foreign word or technical term, without its normal English meaning.[4] In general, it does not refer to the specifically moral conscience, but to a shared understanding of social norms.[5] To avoid the misleading implications of the English words "conscience" and "collective", it can be translated as "common understanding" or "shared awareness".

As for "collective", Durkheim makes clear that he is not reifying or hypostasizing this concept: for him, it is "collective" simply in the sense that it is common to many individuals;[6] *cf.* social fact.

4.1 In Durkheimian social theory

Durkheim used the term in his books *The Division of Labour in Society* (1893), *Rules of the Sociological Method* (1895), *Suicide* (1897), and *The Elementary Forms of Religious Life* (1912). In *The Division of Labour*, Durkheim argued that in traditional/primitive societies (those based around clan, family or tribal relationships), totemic religion played an important role in uniting members through the creation of a common consciousness (*conscience collective* in the original French). In societies of this type, the contents of an individual's consciousness are largely shared in common with all other members of their society, creating a mechanical solidarity through mutual likeness.

> The totality of beliefs and sentiments common to the average members of a society forms a determinate system with a life of its own. It can be termed the collective or creative consciousness.
> — Emile Durkheim[7]

In *Suicide*, Durkheim developed the concept of anomie to refer to the social rather than individual causes of suicide. This relates to the concept of collective consciousness, as if there is a lack of integration or solidarity in society then suicide rates will be higher.[8]

4.2 Other uses of the term

Various forms of what might be termed "collective consciousness" in modern societies have been identified by other sociologists, such as Mary Kelsey, going from solidarity attitudes and memes to extreme behaviors like group-think, herd behavior, or collectively shared experiences during collective rituals and dance parties.[9][10] Mary Kelsey, sociology lecturer in the University of California, Berkeley, used the term in the early 2000s to describe people within a social group, such as mothers, becoming aware of their shared traits and circumstances, and as a result acting as a community and achieving solidarity. Rather than existing as separate individuals, people come together as dynamic groups to share resources and knowledge. It has also developed as a way of describing how an entire community comes together to share similar values. This has also been termed "hive mind", "group mind", "**mass mind**", and "**social mind**".[11]

According to a theory the character of collective consciousness depends on the type of mnemonic encoding used within a group (Tsoukalas, 2007). The specific type of encoding used has a predictable influence on the groups behavior and collective ideology. Informal groups, that meet infrequently and spontaneously, have a tendency to represent significant aspects of their community as episodic memories. This usually leads to stong social cohesion and solidarity, an indulgent atmosphere, an exclusive ethos and a restriction of social networks. Formal groups, that have scheduled and anonymous meetings, tend to represent significant aspects of their community as semantic memories which usually leads to weak social cohesion and solidarity, a more moderate atmosphere, an inclusive ethos and an expansion of social networks.[12]

Society is made up of various collective groups, such as the family, community, organizations, regions, nations which as Burns and Egdahl state *"can be considered to possess agential capabilities*: to think, judge, decide, act, reform; to conceptualize self and others as well as self's actions and interactions; and to reflect."[13](italics in the original). Burns and Egdahl note that during the Second World War different nations behaved differently towards their Jewish populations.[14] The Jewish populations of Bulgaria and Denmark survived whereas the majority of the Jewish populations in Slovakia and Hungary did not survive the Holocaust. It is suggested that these different national behaviors vary according to the different collective consciousness between nations. This illustrates that differences in collective consciousness can have practical significance.

Edmans, Garcia, and Norlia examined national sporting defeats and correlated them with decreases in the value of stocks. They examined 1,162 football matches in thirty-nine countries and discovered that stock markets of those countries dropped on average forty-nine points after being eliminated from the World Cup, and thirty-one points after being eliminated in other tournaments.[15] Edmans, Garcia, and Norli found similar but smaller effects with international cricket, rugby, ice hockey, and basketball games.

4.3 See also

- Abilene paradox
- Anonymous (group)
- Borg Collective
- Collective effervescence
- Collective identity
- Collective intelligence
- Collective unconscious
- Communal reinforcement
- Crowd psychology
- Deep social mind
- Prána Dharma[16]
- Egregor
- Global brain
- Group behaviour
- Group mind
- Higher consciousness
- Human spirit
- Noogenesis
- Noosphere
- Paradigm, Paradigm shift.
- Peer pressure
- Reality tunnel
- Social Justice
- Social Representations
- Superorganism
- Zeitgeist
- Agenda 21
- Global Goals
- United Nations

4.4 Notes

[1] *Collins Dictionary of Sociology*, p93.

[2] *Collins French-English Dictionary s.v.*

[3] Shaun Best, *A Beginner's Guide to Social Theory*, p. 28

[4] Simpson, George (Trans.) in Durkheim, Emile "The Division of Labour in Society" The Free Press, New York, 1993, pp. ix

[5] Thomas E. Wren, *Conceptions of Culture: What Multicultural Educators Need to Know*, p. 64

[6] Warren Schmaus, *Durkheim's Philosophy of Science and the Sociology of Knowledge: Creating an Intellectual Niche*, 1994, ISBN 0226742512, p. 50-51

[7] Kenneth Allan; Kenneth D. Allan (2 November 2005). *Explorations in Classical Sociological Theory: Seeing the Social World*. Pine Forge Press. p. 108. ISBN 1-4129-0572-9.

[8] Durkheim, E. Suicide, 1897.

[9] Trnka R., Lorencova R. (2016). *Collective consciousness and collective unconscious in anthropology - Chapter 7 in Quantum anthropology: Man, cultures, and groups in a quantum perspective*. Prague: Charles University Karolinum Press. pp. 81–90.

[10] Combs, A., & Krippner, S. (2008). "Collective consciousness and the social brain". *Journal of Consciousness Studies*. 15: 264–276.

[11] John D. Greenwood *The Disappearance of the Social in American Social Psychology* 2004, p. 110

[12] Tsoukalas, I. (2007). Exploring the Microfoundations of Group Consciousness. *Culture and Psychology*, 13(1), 39-81.

[13] Burns, T.R. Engdahl, E. (1998) The Social Construction of Consciousness. Part 1: Collective Consciousness and its Socio-Cultural Foundations, *Journal of Consciousness Studies*, 5 (1) p 72.

[14] Burns, T.R. Engdahl, E. (1998) The Social Construction of Consciousness. Part 1: Collective Consciousness and its Socio-Cultural Foundations, *Journal of Consciousness Studies*, 5 (1) p 77.

[15] Edmans, A. García, D. Norli, O. 2007 Sports Sentiment and Stock Returns. *Journal of Finance* 62 (4) pp. 1967-1998.

[16] Sarkar, Prabhat Ranjan (16 February 1967). "Práńa Dharma-The Sine Qua Non of Human Existence". *PROUT in a Nutshell Part 6*. Ranchi, India: Ánanda Márga Publications. Retrieved 2015-03-02.

4.5 References

Works by Durkheim

- *The Division of Labour in Society* (1893)
- *Rules of the Sociological Method* (1895)
- *Suicide* (1897)
- *The Elementary Forms of Religious Life* (1912)

Works by others

- Gad Barzilai, *Communities and Law: Politics and Cultures of Legal Identities* University of Michigan Press, 2003. ISBN 978-0-472-03079-8

- Jary, David; Julia Jary (1991). *Collins Dictionary of Sociology*. Glasgow: Harper Collins. p. 774. ISBN 0-00-470804-0.

- Williams, Swatos (editor) (1998). "Émile Durkheim" (Web and Print). *Encyclopedia of Religion and Society*. Altamira Press. Retrieved 2007-02-03.

- Lloyd, Annemaree (2007). "Guarding Against Collective Amnesia? Making Significance Problematic: An Exploration of Issues" (Web and Print). *Library Trends*. Retrieved 2008-04-25.

Chapter 5

Collective behavior

The expression **collective behavior** was first used by Franklin Henry Giddings (1908) and employed later by Robert E. Park (1921), Herbert Blumer (1939), Ralph Turner and Lewis Killian (1957), and Neil Smelser (1962) to refer to social processes and events which do not reflect existing social structure (laws, conventions, and institutions), but which emerge in a "spontaneous" way. Use of the term has been expanded to include reference to cells, social animals like birds and fish, and insects including ants.[1] Collective behavior takes many forms but generally violates societal norms (Miller 2000, Locher 2002). Collective behavior can be tremendously destructive, as with riots or mob violence, silly, as with fads, or anywhere in between. Collective behavior is always driven by group dynamics, encouraging people to engage in acts they might consider unthinkable under typical social circumstances (Locher 2002).

5.1 Defining the field

Turner and Killian (1957) were the first sociologists to back their theoretical propositions with visual evidence in the form of photographs and motion pictures of collective behavior in action. Prior to that sociologists relied heavily upon eyewitness accounts, which turned out to be far less reliable than one would hope.

Turner and Killian's approach is based largely upon the arguments of Blumer, who argued that social "forces" are not really forces. The actor is active: He creates an interpretation of the acts of others, and acts on the basis of this interpretation.

5.2 Examples

Here are some instances of collective behavior: the Los Angeles riot of 1992, the hula-hoop fad of 1958, the stock market crashes of 1929, and the "phantom gasser" episodes in Virginia in 1933-34 and Mattoon, IL in 1944 (Locher 2002, Miller 2000). The claim that such diverse episodes all belong to a single field of inquiry is a theoretical assertion, and not all sociologists would agree with it. But Blumer and Neil Smelser did agree, as did others, indicating that the formulation has satisfied some leading sociological thinkers.

5.2.1 Four forms

Although there are several other schema that may be used to classify forms of collective behavior the following four categories from Blumer (1939) are generally considered useful by most sociologists.

The crowd

Scholars differ about what classes of social events fall under the rubric of collective behavior. In fact, the only class of events which all authors include is crowds. Clark McPhail is one of those who treats crowds and collective behavior as synonyms. Although some consider McPhail's work overly simplistic (Locher 2002), his important contribution is to have gone beyond the speculations of others to carry out pioneering empirical studies of crowds. He finds them to form an elaborate set of types.

The classic treatment of crowds is Gustave LeBon, *The Crowd: A Study of the Popular Mind* (1896), in which the author interpreted the crowds of the French Revolution as irrational reversions to animal emotion, and inferred from this that such reversion is characteristic of crowds in general. LeBon believed that crowds somehow induced people to lose their ability to think rationally and to somehow recover this ability once they had left the crowd. He speculated, but could not explain how this might occur. Freud expressed a similar view in *Group Psychology and the Analysis of the Ego* (1922). Such authors have thought that their ideas were confirmed by various kinds of crowds, one of these being the economic bubble. In Holland, during the tulip mania (1637), the prices of tulip bulbs rose to astronomical heights. An array of such crazes and other historical oddities is narrated in Charles MacKay's *Extraordinary Popular Delusions and the Madness of Crowds* (1841).

At the University of Chicago, Robert Park and Herbert Blumer agreed with the speculations of LeBon and other that crowds are indeed emotional. But to them a crowd is capable of any emotion, not only the negative ones of anger and fear.

A number of authors modify the common-sense notion of the crowd to include episodes during which the participants are not assembled in one place but are dispersed over a large area. Turner and Killian refer to such episodes as *diffuse* crowds, examples being Billy Graham's revivals, panics about sexual perils, witch hunts and Red scares. Their expanded definition of the crowd is justified if propositions which hold true among compact crowds do so for diffuse crowds as well.

Some psychologists have claimed that there are three fundamental human emotions: fear, joy, and anger. Neil Smelser, John Lofland, and others have proposed three corresponding forms of the crowd: the panic (an expression of fear), the craze (an expression of joy), and the hostile outburst (an expression of anger). Each of the three emotions can characterize either a *compact* or a *diffuse* crowd, the result being a scheme of six types of crowds. Lofland has offered the most explicit discussion of these types.

The public

Boom distinguishes the **crowd**, which expresses a *common emotion*, from a **public**, which discusses a *single issue*. Thus, a public is not equivalent to all of the members of a society. Obviously, this is not the usual use of the word, "public." To Park and Blumer, there are as many publics as there are issues. A public comes into being when discussion of an issue begins, and ceases to be when it reaches a decision on it.

The mass

To the crowd and the public Blumer adds a third form of collective behavior, the **mass**. It differs from both the crowd and the public in that it is defined not by a form of interaction but by the efforts of those who use the mass media to address an audience. The first mass medium was printing.

The social movement

Main article: Social movement

We change intellectual gears when we confront Blumer's final form of collective behavior, the social movement. He identifies several types of these, among which are *active* social movements such as the French Revolution and *expressive* ones such as Alcoholics Anonymous. An active movement tries to change society; an expressive one tries to change its own members.

The social movement is the form of collective behavior which satisfies least well the first definition of it which was offered at the beginning of this article. These episodes are less fluid than the other forms, and do not change as often as other forms do. Furthermore, as can be seen in the history of the labor movement and many religious sects, a social movement may begin as collective behavior but over time become firmly established as a social institution.

For this reason, social movements are often considered a separate field of sociology. The books and articles about them are far more numerous than the sum of studies of all the other forms of collective behavior put together. Social movements are considered in many Wikipedia articles, and an article on the field of social movements as a whole would be much longer than this essay.

The study of collective behavior spun its wheels for many years, but began to make progress with the appearance of Turner and Killian's "Collective Behavior" (1957) and Smelser's *Theory of Collective Behavior* (1962). Both books pushed the topic of collective behavior back into the consciousness of American sociologists and both theories contributed immensely to our understanding of collective behavior (Locher 2002, Miller 2000). Social disturbances in the U. S. and elsewhere in the late '60s and early '70s inspired another surge of interest in crowds and social movements. These studies presented a number of challenges to the armchair sociology of earlier students of collective behavior.

5.3 Theories developed to explain

Social scientists have developed theories to explain crowd behavior.

1. **Contagion Theory** - the Contagion Theory was formulated by Gustave Le Bon. According to Le Bon crowds exert a hypnotic influence over their members. Shielded by their anonymity, large numbers of people abandon personal responsibility and surrender to the contagious emotions of the crowd. A crowd thus assumes a life of its own, stirring up emotions and driving people toward irrational, even violent action (LeBon 1895). Le Bon's Theory, although one of the earliest explanations of crowd behavior, is still

accepted by many people outside of sociology.[2][3][4] However, critics argue that the "collective mind" has not been documented by systematic studies. Furthermore, although collective behavior may involve strong emotions, such feelings are not necessarily irrational. Turner and Killian (1957) argue convincingly that the "contagion" never actually occurs and participants in collective behavior do not lose their ability to think rationally.

2. **Convergence Theory** - whereas the Contagion Theory states that crowds cause people to act in a certain way, Convergence theory states that people who want to act in a certain way come together to form crowds. Developed by Floyd Allport (1924) and later expanded upon by Neil Miller and John Dollard (1941) as "Learning Theory," the central argument of all convergence theories is that collective behavior reveals the otherwise hidden tendencies of the individuals who take part in the episode. It asserts that people with similar attributes find other like-minded persons with whom they can release these underlying tendencies. People sometimes do things in a crowd that they would not have the courage to do alone because crowds can diffuse responsibility but the behavior itself is claimed to originate within the individuals. Crowds, in addition, can intensify a sentiment simply by creating a critical mass of like-minded people.

3. **Emergent-Norm Theory** - according to Ralph Turner and Lewis Killian (1957), crowds begin as collectivities composed of people with mixed interests and motives. Especially in the case of less stable crowds—expressive, acting and protest crowds—norms may be vague and changing, as when one person decides to break the glass windows of a store and others join in and begin looting merchandise. When people find themselves in a situation that is vague, ambiguous, or confusing new norms "emerge" on the spot and people follow those emergent norms, which may be at odds with normal social behavior. Turner and Killian further argue that there are several different categories of participants, all of whom follow different patterns of behavior due to their differing motivations.

4. **Value-added Theory** - Neil Smelser (1962) argues that collective behavior is actually a sort of release valve for built-up tension ("strain") within the social system, community, or group. If the proper determinants are present then collective behavior becomes inevitable. Conversely, if any of the key determinants are not present no collective behavior will occur unless and until the missing determinants fall into place. These are primarily social, although physical factors such as location and weather may also contribute to or hinder the development of collective behavior.

5. **Complex Adaptive Systems theory** - Dutch scholar Jaap van Ginneken claims that contagion, convergence and emergent norms are just instances of the synergy, emergence and autopoiesis or self-creation of patterns and new entities typical for the newly discovered meta-category of complex adaptive systems. This also helps explain the key role of salient details and path-dependence in rapid shifts.

5.4 See also

- Bandwagon effect
- Bioengineering
- Collective consciousness
- Collective effervescence
- Collective intelligence
- Collective hysteria
- Collective narcissism
- Complex adaptive systems
- Crowd manipulation
- Crowd psychology
- Group behaviour
- Herd behavior
- Herd morality
- Keeping up with the Joneses
- Moral panic
- Penis panic
- Peer pressure
- Sheeple
- Social comparison theory
- Spiral of silence
- Systems science
- Theories of political behavior

5.5 Bibliography

- Charles MacKay, *Extraordinary Popular Delusions and the Madness of Crowds*, 1841.

- Herbert Blumer, "Collective Behavior," in A. M. Lee, ed., *Principles of Sociology*. New York, Barnes & Noble, 1951, pp. 67–121.

- Neil J. Smelser, *Theory of Collective Behavior*, Free Press, Glencoe, Ill., 1962.

- Ralph H. Turner and Lewis M. Killian, *Collective Behavior*, Englewood Cliffs, N. J., Prentice-Hall, 1957 1st ed., 2d ed., 1972; 3d. ed. 1987; 4th ed. 1993.

- Kurt Lang and Gladys Lang, *Collective Dynamics*

- John Lofland, *Protest*...

- James B. Rule, *Theories of Civil Violence*, Berkeley, University of California, 1988.

- Clark McPhail, *The Myth of the Madding Crowd*. New York, Aldine de Gruyter, 1991.

- Basco Michael, *Socio Anthropology " Mendiola Manila*.

- Jaap van Ginneken, *Collective behavior and public opinion – Rapid shifts in opinion and communication*, Mahwah, NJ: Erlbaum, 2003.

- Giovanni Naldi, Lorenzo Pareschi, Giuseppe Toscani, *Mathematical modelling of collective behavior in socio-economic and life sciences*, Birkhauser, (2010).

- Locher, David A., *Collective Behavior*, Upper Saddle River, NJ: Prentice Hall, 2002.

- Miller, David L., *Introduction to Collective Behavior and Collective Action*, Prospect Heights, IL: Waveland Press, 2000 2d ed., 1985.

- Park, Robert E and Ernest W. Burgess. 1921. *Introduction to the Science of Sociology* Chicago, IL: University of Chicago Press.

- Allport, Floyd. 1924. *Social Psychology*. Cambridge: Houghton Mifflin, The Riverside Press.

- Miller, Neil and John Dollard. 1941. *Social Learning and Imitation*. New Haven: Yale University Press.

- LeBon, Gustave. 1895. *The Crowd: A Study of the Popular Mind*. Atlanta: Cherokee Publishing Company.

- Giddings, Franklin Henry. 1908. *Sociology*. New York: Columbia University Press.

5.6 External links

- Group Experiment Environments (GEE) project, sponsored by the Percepts and Concepts Laboratory at Indiana University

5.7 References

[1] Gordon, Deborah M. (11 March 2014). "The Ecology of Collective Behavior". *PLoS Biol.* **12** (3). doi:10.1371/journal.pbio.1001805. Retrieved 28 March 2014.

[2] Castellano, C.; Fortunato, S., & Loreto, V. (2009). Statistical physics of social dynamics. Rev Mod Phys 81(2): 591–646.

[3] Braha, D. (2012). Global Civil Unrest: Contagion, Self-Organization, and Prediction. PLoS ONE 7(10): e48596. https://doi.org/10.1371/journal.pone.0048596

[4] Braha, D., & de Aguiar, M. A. (2017). Voting contagion: Modeling and analysis of a century of U.S. presidential elections. PLoS ONE 12(5): e0177970. https://doi.org/10.1371/journal.pone.0177970

Chapter 6

Social dynamics

Social dynamics can refer to the behavior of groups that results from the interactions of individual group members as well to the study of the relationship between individual interactions and group level behaviors.[1] The field of social dynamics brings together ideas from Economics, Sociology, Social Psychology, and other disciplines, and is a sub-field of complex adaptive systems or complexity science. The fundamental assumption of the field is that individuals are influenced by one another's behavior. The field is closely related to system dynamics. Like system dynamics, social dynamics is concerned with changes over time and emphasizes the role of feedbacks. However, in social dynamics individual choices and interactions are typically viewed as the source of aggregate level behavior, while system dynamics posits that the structure of feedbacks and accumulations are responsible for system level dynamics.[2] Research in the field typically takes a behavioral approach, assuming that individuals are boundedly rational and act on local information. Mathematical and computational modeling are important tools for studying social dynamics. Because social dynamics focuses on individual level behavior, and recognizes the importance of heterogeneity across individuals, strict analytic results are often impossible. Instead, approximation techniques, such as mean field approximations from statistical physics, or computer simulations are used to understand the behaviors of the system. In contrast to more traditional approaches in economics, scholars of social dynamics are often interested in non-equilibrium, or dynamic, behavior.[1][3] That is, behavior that changes over time.

6.1 Topics

- Social networks
- Diffusion of technologies and information
- Cooperation
- Social norms

6.2 See also

- Complex adaptive system
- Complexity science
- Collective intelligence
- Dynamical systems
- Jay Wright Forrester
- Group dynamics
- Operations research
- Population dynamics
- System dynamics
- Social psychology
- Societal collapse
- Sociobiology
- Sociocultural evolution

6.3 References

[1] Durlauf, Steven; Young, Peyton (2001). *Social Dynamics*. Cambridge, MA: MIT Press. ISBN 0-262-04186-3.

[2] Sterman, John (2000). *Business Dynamics*. McGraw Hill. ISBN 0-07-231135-5.

[3] "Brookings Institution, Center for Social Dynamics and Policy". Retrieved 29 Sep 2012.

Weidlich, W. (1997) "Sociodynamics applied to the evolution of urban and regional structures". *Discrete Dynamics in Nature and Society*, Vol. 1, pp. 85–98. Available on line: http://www.hindawi.com/GetArticle.aspx?doi=10.1155/S1026022697000101

6.4 Further reading

- Easley, David; Klienberg, Jon (2010). *Networks, Crowds, and Markets*. New York, NY: Cambridge University Press. ISBN 978-0-521-19533-1.

- Jackson, Matthew O. (2008). *Social and Economic Networks*. Princeton, NJ: Princeton University Press. ISBN 978-0-691-13440-6.

6.5 External links

- Introduction to Social Macrodynamics

- Club of Rome report, quote: "We must also keep in mind the presence of social delays--the delays necessary to allow society to absorb or to prepare for a change. Most delays, physical or social reduce the stability of the world system and increase the likelihood of the overshoot mode"

- Northwestern Institute on Complex Systems—Institute with research focusing on complexity and social dynamics.

- Center for the Study of Complex Systems, University of Michigan—Center with research focusing on complexity and social dynamics.

- Social Dynamics course at the Kellogg School of Management.

- social-dynamics.org—Blog on Social Dynamics from Kellogg School of Management Social Dynamics Scholar

- http://139.142.203.66/pub/www/Journal/vol3/iss2/art4/

- http://www-rcf.usc.edu/~{}read/connectionism_preface2.html

- "Historical Dynamics in a Time of Crisis: Late Byzantium, 1204–1453" (discussion of social dynamics from the point of view of historical studies)

- Watts, D.J.; Strogatz, S.H. (1998). "Collective dynamics of 'small-world' networks". *Nature*. **393** (6684): 440–442. Bibcode:1998Natur.393..440W. PMID 9623998. doi:10.1038/30918.

Chapter 7

Collective intelligence

Types of collective intelligence

Collective intelligence (CI) is shared or group intelligence that emerges from the collaboration, collective efforts, and competition of many individuals and appears in consensus decision making. The term appears in sociobiology, political science and in context of mass peer review and crowdsourcing applications. It may involve consensus, social capital and formalisms such as voting systems, social media and other means of quantifying mass activity. Collective IQ is a measure of collective intelligence, although it is often used interchangeably with the term collective intelligence. Collective intelligence has also been attributed to bacteria[1]:63 and animals.[1]:69

It can be understood as an emergent property from the synergies among: 1) data-information-knowledge; 2) software-hardware; and 3) experts (those with new insights as well as recognized authorities) that continually learns from feedback to produce just-in-time knowledge for better decisions than these three elements acting alone.[2] Or more narrowly as an emergent property between people and ways of processing information.[3] This notion of collective intelligence is referred to as "symbiotic intelligence" by Norman Lee Johnson.[4] The concept is used in sociology, business, computer science and mass communications: it also appears in science fiction. Pierre Lévy defines collective intelligence as, "It is a form of universally distributed intelligence, constantly enhanced, coordinated in real time, and resulting in the effective mobilization of skills. I'll add the following indispensable characteristic to this definition:

The basis and goal of collective intelligence is mutual recognition and enrichment of individuals rather than the cult of fetishized or hypostatized communities."[5] According to researchers Pierre Lévy and Derrick de Kerckhove, it refers to capacity of networked ICTs (Information communication technologies) to enhance the collective pool of social knowledge by simultaneously expanding the extent of human interactions.[6]

Collective intelligence strongly contributes to the shift of knowledge and power from the individual to the collective. According to Eric S. Raymond (1998) and JC Herz (2005), open source intelligence will eventually generate superior outcomes to knowledge generated by proprietary software developed within corporations (Flew 2008). Media theorist Henry Jenkins sees collective intelligence as an 'alternative source of media power', related to convergence culture. He draws attention to education and the way people are learning to participate in knowledge cultures outside formal learning settings. Henry Jenkins criticizes schools which promote 'autonomous problem solvers and self-contained learners' while remaining hostile to learning through the means of collective intelligence.[7] Both Pierre Lévy (2007) and Henry Jenkins (2008) support the claim that collective intelligence is important for democratization, as it is interlinked with knowledge-based culture and sustained by collective idea sharing, and thus contributes to a better understanding of diverse society.

Similar to the g factor (g) for general individual intelligence, a new scientific understanding of collective intelligence aims to extract a general collective intelligence factor **c factor** for groups indicating a group's ability to perform a wide range of tasks.[8] Definition, operationalization and statistical methods are derived from g. Similarly as g is highly interrelated with the concept of IQ,[9][10] this measurement of collective intelligence can be interpreted as intelligence quotient for groups (Group-IQ) even though the score is not a quotient per se. Causes for c and predictive validity are investigated as well.

Writers who have influenced the idea of collective intelligence include Douglas Hofstadter (1979), Peter Russell (1983), Tom Atlee (1993), Pierre Lévy (1994), Howard Bloom (1995), Francis Heylighen (1995), Douglas Engelbart, Louis Rosenberg, Cliff Joslyn, Ron Dembo, Gottfried Mayer-Kress (2003).

7.1 History

H.G. Wells World Brain *(1936-1938)*

The concept (although not so named) originated in 1785 with the Marquis de Condorcet, whose "jury theorem" states that if each member of a voting group is more likely than not to make a correct decision, the probability that the highest vote of the group is the correct decision increases with the number of members of the group (see Condorcet's jury theorem).[11] Many theorists have interpreted Aristotle's statement in the Politics that "a feast to which many contribute is better than a dinner provided out of a single purse" to mean that just as many may bring different dishes to the table, so in a deliberation many may contribute different pieces of information to generate a better decision.[12][13] Recent scholarship,[14] however, suggests that this was probably not what Aristotle meant but is a modern interpretation based on what we now know about team intelligence.[15]

A precursor of the concept is found in entomologist William Morton Wheeler's observation that seemingly independent individuals can cooperate so closely as to become indistinguishable from a single organism (1911).[16] Wheeler saw this collaborative process at work in ants that acted like the cells of a single beast he called a superorganism.

In 1912 Émile Durkheim identified society as the sole source of human logical thought. He argued in "The Elementary Forms of Religious Life" that society constitutes a higher intelligence because it transcends the individual over space and time.[17] Other antecedents are Vladimir Vernadsky's concept of "noosphere" and H.G. Wells's concept of "world brain" (see also the term "global brain"). Peter Russell, Elisabet Sahtouris, and Barbara Marx Hubbard (originator of the term "conscious evolution")[18] are inspired by the visions of a noosphere – a transcendent, rapidly evolving collective intelligence – an informational cortex of the planet. The notion has more recently been examined by the philosopher Pierre Lévy. In a 1962 research report, Douglas Engelbart linked collective intelligence to organizational effectiveness, and predicted that pro-actively 'augmenting human intellect' would yield a multiplier effect in group problem solving: "Three people working together in this augmented mode [would] seem to be more than three times as effective in solving a complex problem as is one augmented person working alone".[19] In 1994, he coined the term 'collective IQ' as a measure of collective intelligence, to focus attention on the opportunity to significantly raise collective IQ in business and society.[20]

The idea of collective intelligence also forms the framework for contemporary democratic theories often referred to as epistemic democracy. Epistemic democratic theories refer to the capacity of the populace, either through deliberation or aggregation of knowledge, to track the truth and relies on mechanisms to synthesize and apply collective intelligence.[21]

7.2 Dimensions

Howard Bloom has discussed mass behavior – collective behavior from the level of quarks to the level of bacterial, plant, animal, and human societies. He stresses the biological adaptations that have turned most of this earth's living beings into components of what he calls "a learning machine". In 1986 Bloom combined the concepts of apoptosis, parallel distributed processing, group selection, and the superorganism to produce a theory of how collective intelligence works.[22] Later he showed how the collective intelli-

7.2. DIMENSIONS

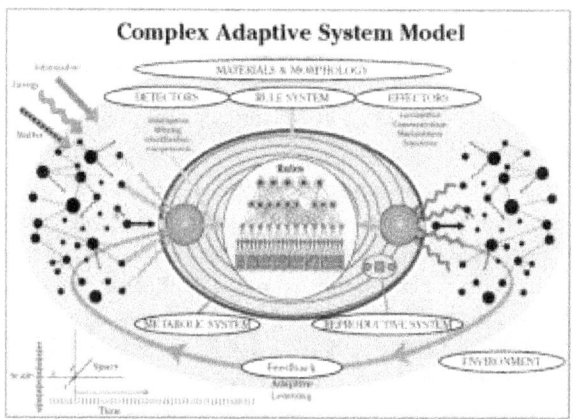

Complex adaptive systems model

gences of competing bacterial colonies and human societies can be explained in terms of computer-generated "complex adaptive systems" and the "genetic algorithms", concepts pioneered by John Holland.[23]

Bloom traced the evolution of collective intelligence to our bacterial ancestors 1 billion years ago and demonstrated how a multi-species intelligence has worked since the beginning of life.[23] Ant societies exhibit more intelligence, in terms of technology, than any other animal except for humans and co-operate in keeping livestock, for example aphids for "milking".[23] Leaf cutters care for fungi and carry leaves to feed the fungi.[23]

David Skrbina[24] cites the concept of a 'group mind' as being derived from Plato's concept of panpsychism (that mind or consciousness is omnipresent and exists in all matter). He develops the concept of a 'group mind' as articulated by Thomas Hobbes in "Leviathan" and Fechner's arguments for a collective consciousness of mankind. He cites Durkheim as the most notable advocate of a "collective consciousness"[25] and Teilhard de Chardin as a thinker who has developed the philosophical implications of the group mind.[26]

Tom Atlee focuses primarily on humans and on work to upgrade what Howard Bloom calls "the group IQ". Atlee feels that collective intelligence can be encouraged "to overcome 'groupthink' and individual cognitive bias in order to allow a collective to cooperate on one process – while achieving enhanced intellectual performance." George Pór defined the collective intelligence phenomenon as "the capacity of human communities to evolve towards higher order complexity and harmony, through such innovation mechanisms as differentiation and integration, competition and collaboration."[27] Atlee and Pór state that "collective intelligence also involves achieving a single focus of attention and standard of metrics which provide an appropriate threshold of action".[28] Their approach is rooted in scientific community metaphor.[28]

The term group intelligence is sometimes used interchangeably with the term collective intelligence. Anita Woolley presents Collective intelligence as a measure of group intelligence and group creativity.[8] The idea is that a measure of collective intelligence covers a broad range of features of the group, mainly group composition and group interaction.[29] The features of composition that lead to increased levels of collective intelligence in groups include criteria such as higher numbers of women in the group as well as increased diversity of the group.[29]

Atlee and Pór suggest that the field of collective intelligence should primarily be seen as a human enterprise in which mind-sets, a willingness to share and an openness to the value of distributed intelligence for the common good are paramount, though group theory and artificial intelligence have something to offer.[28] Individuals who respect collective intelligence are confident of their own abilities and recognize that the whole is indeed greater than the sum of any individual parts.[30] Maximizing collective intelligence relies on the ability of an organization to accept and develop "The Golden Suggestion", which is any potentially useful input from any member.[31] Groupthink often hampers collective intelligence by limiting input to a select few individuals or filtering potential Golden Suggestions without fully developing them to implementation.[28]

Robert David Steele Vivas in *The New Craft of Intelligence* portrayed all citizens as "intelligence minutemen," drawing only on legal and ethical sources of information, able to create a "public intelligence" that keeps public officials and corporate managers honest, turning the concept of "national intelligence" (previously concerned about spies and secrecy) on its head.[32]

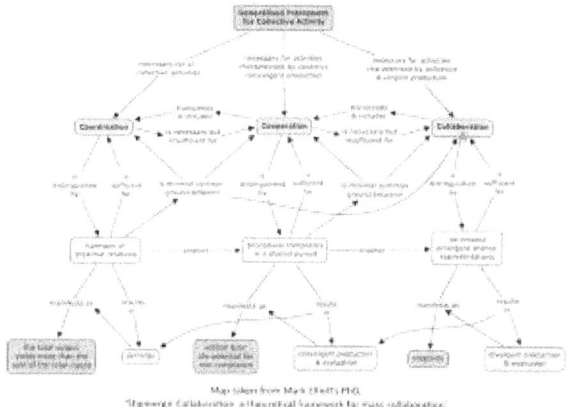

Stigmergic Collaboration: a theoretical framework for mass collaboration

According to Don Tapscott and Anthony D. Williams, collective intelligence is mass collaboration. In order for this concept to happen, four principles need to exist:[33]

Openness Sharing ideas and intellectual property: though these resources provide the edge over competitors more benefits accrue from allowing others to share ideas and gain significant improvement and scrutiny through collaboration.[33]

Peering Horizontal organization as with the 'opening up' of the Linux program where users are free to modify and develop it provided that they make it available for others. Peering succeeds because it encourages self-organization – a style of production that works more effectively than hierarchical management for certain tasks.[33]

Sharing Companies have started to share some ideas while maintaining some degree of control over others, like potential and critical patent rights. Limiting all intellectual property shuts out opportunities, while sharing some expands markets and brings out products faster.[33]

Acting Globally The advancement in communication technology has prompted the rise of global companies at low overhead costs. The internet is widespread, therefore a globally integrated company has no geographical boundaries and may access new markets, ideas and technology.[33]

7.3 Collective intelligence factor c

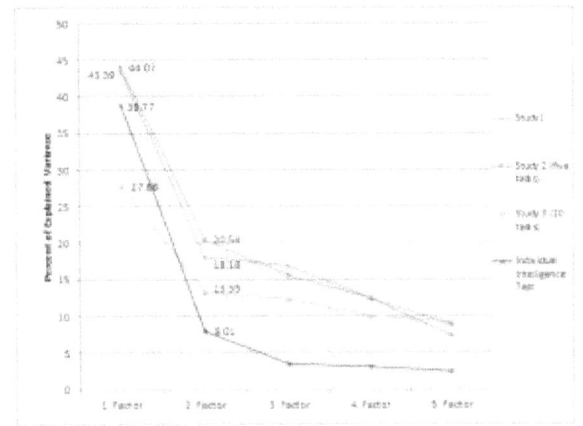

Scree plot showing percent of explained variance for the first factors in Woolley et al.'s (2010) two original studies.

A new scientific understanding of collective intelligence defines it as a group's general ability to perform a wide range of tasks.[8] Definition, operationalization and statistical methods are similar to the psychometric approach of general individual intelligence. Hereby, an individual's performance on a given set of cognitive tasks is used to measure general cognitive ability indicated by the general intelligence factor g extracted via factor analysis.[34] In the same vein as g serves to display between-individual performance differences on cognitive tasks, collective intelligence research aims to find a parallel intelligence factor for groups 'c factor'[8] (also called 'collective intelligence factor' (CI)[35]) displaying between-group differences on task performance. The collective intelligence score then is used to predict how this same group will perform on any other similar task in the future. Yet tasks, hereby, refer to mental or intellectual tasks performed by small groups[8] even though the concept is hoped to be transferrable to other performances and any groups or crowds reaching from families to companies and even whole cities.[36] Since individuals' g factor scores are highly correlated with full-scale IQ scores, which are in turn regarded as good estimates of g,[9][10] this measurement of collective intelligence can also be seen as an intelligence indicator or quotient respectively for a group (Group-IQ) parallel to an individual's intelligence quotient (IQ) even though the score is not a quotient per se.

Mathematically, c and g are both variables summarizing positive correlations among different tasks supposing that performance on one task is comparable with performance on other similar tasks.[37] c thus is a source of variance among groups and can only be considered as a group's standing on the c factor compared to other groups in a given relevant population.[10][38] The concept is in contrast to competing hypotheses including other correlational structures to explain group intelligence,[8] such as a composition out of several equally important but independent factors as found in individual personality research.[39]

Besides, this scientific idea also aims to explore the causes affecting collective intelligence, such as group size, collaboration tools or group members' interpersonal skills.[40] The MIT Center for Collective Intelligence, for instance, announced the detection of *The Genome of Collective Intelligence*[40] as one of its main goals aiming to develop a *taxonomy of organizational building blocks, or genes, that can be combined and recombined to harness the intelligence of crowds*.[40]

7.3.1 Causes

Individual intelligence is shown to be genetically and environmentally influenced.[41][42] Analogously, collective intelligence research aims to explore reasons why certain groups perform more intelligent than other groups given that c is just moderately correlated with the intelligence of individual group members.[8] According to Woolley et al.'s results, neither team cohesion nor motivation or satisfaction correlated with c. However, they claim that three factors were found as significant correlates: the variance in the number of speaking turns, group members' average social sensitivity and the proportion of females. All three had similar predictive power for c, but only social sensitivity was statistically significant (b=0.33, P=0.05).[8]

The number speaking turns indicates that "groups where a few people dominated the conversation were less collectively intelligent than those with a more equal distribution of conversational turn-taking".[35] Hence, providing multiple team members the chance to speak up made a group more intelligent.[8]

Group members' social sensitivity was measured via the Reading the Mind in the Eyes Test[43] (RME) and correlated .26 with c.[8] Hereby, participants are asked to detect thinking or feeling expressed in other peoples' eyes presented on pictures and assessed in a multiple choice format. The test aims to measure peoples' theory of mind (ToM), also called 'mentalizing'[44][45][46][47] or 'mind reading',[48] which refers to the ability to attribute mental states, such as beliefs, desires or intents, to other people and in how far people understand that others have beliefs, desires, intentions or perspectives different from their own ones.[43] RME is a ToM test for adults[43] that shows sufficient test-retest reliability[49] and constantly differentiates control groups from individuals with functional autism or Asperger Syndrome.[43] It is one of the most widely accepted and well-validated tests for ToM within adults.[50] ToM can be regarded as an associated subset of skills and abilities within the broader concept of emotional intelligence.[35][51]

The proportion of females as a predictor of c was **largely mediated by social sensitivity (Sobel z = 1.93, P= 0.03)**[8] which is in vein with previous research showing that women score higher on social sensitivity tests.[43] While a mediation, statistically speaking, clarifies the mechanism underlying the relationship between a dependent and an independent variable,[52] Wolley agreed in an interview with the *Harvard Business Review* that these findings are **saying that groups of women are smarter than groups of men**.[36] However, she relativizes this stating that the actual important thing is the high social sensitivity of group members.[36]

It is theorized that the collective intelligence factor c is an emergent property resulting from bottom-up as well as top-down processes.[29] Hereby, bottom-up processes cover aggregated group-member characteristics. Top-down processes cover group structures and norms that influence a group's way of collaborating and coordinating.[29]

7.3.2 Processes

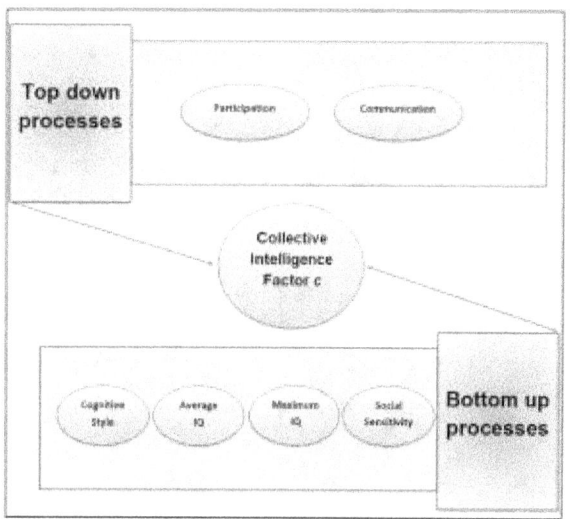

Predictors for the collective intelligence factor c. Suggested by Woolley, Aggarwal & Malone[29] (2015)

Top-down processes

Top-down processes cover group interaction, such as structures, processes, and norms.[53] An example of such top-down processes is conversational turn-taking.[8] Research further suggest that collectively intelligent groups communicate more in general as well as more equally; same applies for participation and is shown for face-to-face as well as online groups communicating only via writing.[35][54]

Bottom-up processes

Bottom-up processes include group composition,[53] namely the characteristics of group members which are aggregated to the team level[29] encompassing. An example of such bottom-up processes is the average social sensitivity or the average and maximum intelligence scores of group members.[8] Furthermore, collective intelligence was found to be related to a group's cognitive diversity[55] including thinking styles and perspectives.[56] Groups that are moderately diverse in cognitive style have higher collective intelligence than those who are very similar in cognitive style or very different. Consequently, groups where members are too similar to each other lack the variety of perspectives and skills needed to perform well. On the other hand, groups whose members are too different seem to have difficulties to communicate and coordinate effectively.[55]

Serial vs Parallel processes

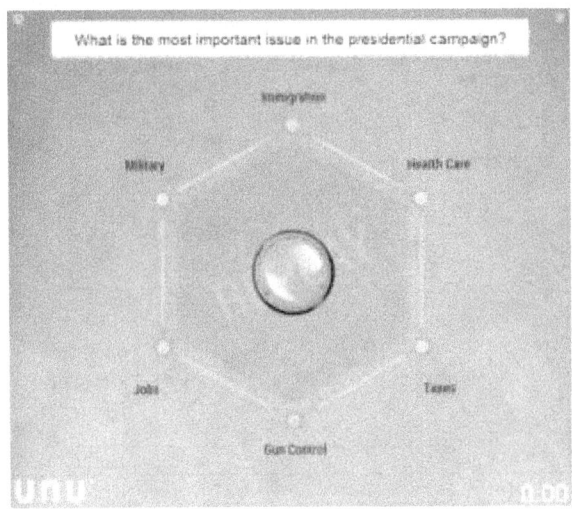

Human Swarm (Rosenberg, 2015) - real-time parallel collective intelligence.

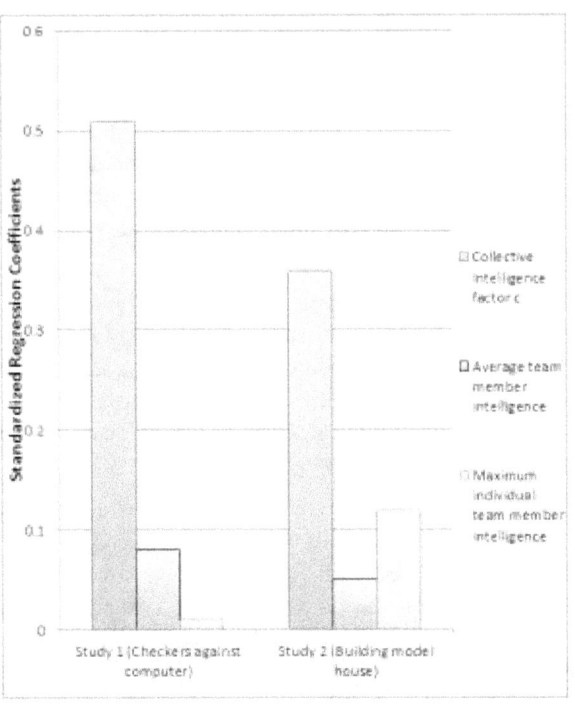

Standardized Regression Coefficients for the collective intelligence factor c as found in Woolley et al.'s[8] (2010) two original studies. c and average (maximum) member intelligence scores are regressed on the criterion tasks.

For most of human history, collective intelligence was confined to small tribal groups in which opinions were aggregated through real-time parallel interactions among members.[57] In modern times, mass communication, mass media, and networking technologies have enabled collective intelligence to span massive groups, distributed across continents and time-zones. To accommodate this shift in scale, collective intelligence in large-scale groups been dominated by serialized polling processes such as aggregating up-votes, likes, and ratings over time. While modern systems benefit from larger group size, the serialized process has been found to introduce substantial noise that distorts the collective output of the group. In one significant study of serialized collective intelligence, it was found that the first vote contributed to a serialized voting system can distort the final result by 34%.[58]

To address the problems of serialized aggregation of input among large-scale groups, recent advancements collective intelligence have worked to replace serialized votes, polls, and markets, with parallel systems such as "human swarms" modeled after synchronous swarms in nature.[59][60] Based on natural process of Swarm Intelligence, these artificial swarms of networked humans enable participants to work together in parallel to answer questions and make predictions as an emergent collective intelligence. In one high profile example, a human swarm challenge by CBS Interactive to predict the Kentucky Derby. The swarm correctly predicted the first four horses, in order, defying 542-1 odds and turning a $20 bet into $10,800.[61]

7.3.3 Evidence

Woolley, Chabris, Pentland, Hashmi, & Malone (2010),[8] the originators of this scientific understanding of collective intelligence, found a single statistical factor for collective intelligence in their research across 192 groups with people randomly recruited from the public. In Woolley et al.'s two initial studies, groups worked together on different tasks from the McGrath Task Circumplex,[62] a well-established taxonomy of group tasks. Tasks were chosen from all four quadrants of the circumplex and included visual puzzles, brainstorming, making collective moral judgments, and negotiating over limited resources. The results in these tasks were taken to conduct a factor analysis. Both studies showed support for a general collective intelligence factor c underlying differences in group performance with an initial eigenvalue accounting for 43% (44% in study 2) of the variance, whereas the next factor accounted for only 18% (20%). That fits the range normally found in research regarding a general individual intelligence factor g typically accounting for 40% to 50% percent of between-individual performance differences on cognitive tests.[37] Afterwards, a more complex criterion task was absolved by each group measuring whether the extracted c factor had predictive power for performance outside the original task batteries. Criterion tasks were playing checkers (draughts) against a standardized computer in the first and a complex architectural design task in the second study. In a regression analysis using both individual intelligence of group members and c to predict performance on the criterion tasks, c had a significant effect, but average and maximum individual intelligence had not. While average (r=0.15, P=0.04) and maximum intelligence (r=0.19, P=0.008) of individual group members were moderately correlated with c, c was still a much better predictor of the criterion tasks. According to Woolley et al., this supports the existence of a collective intelligence factor c, because it demonstrates an effect over and beyond group members' individual intelligence and thus that c is more than just the aggregation of the individual IQs or the influence of the group member with the highest IQ.[8]

Engel et al.[35] (2014) replicated Woolley et al.'s findings applying an accelerated battery of tasks with a first factor in the factor analysis explaining 49% of the between-group variance in performance with the following factors explaining less than half of this amount. Moreover, they found a similar result for groups working together online communicating only via text and confirmed the role of female proportion and social sensitivity in causing collective intelligence in both cases. Similarly to Wolley et al.,[8] they also measured social sensitivity with the RME which is actually meant to measure people's ability to detect mental states in other peoples' eyes. The online collaborating participants, however, did neither know nor see each other at all. The authors conclude that scores on the RME must be related to a broader set of abilities of social reasoning than only drawing inferences from other people's eye expressions.[63]

A collective intelligence factor c in the sense of Woolley et al.[8] was further found in groups of MBA students working together over the course of a semester,[64] in online gaming groups[54] as well as in groups from different cultures[65] and groups in different contexts in terms of short-term versus long-term groups.[65] None of these investigations considered team members' individual intelligence scores as control variables.[54][64][65]

Note as well that the field of collective intelligence research is quite young and published empirical evidence is relatively rare yet. However, various proposals and working papers are in progress or already completed but (supposedly) still in a scholarly peer reviewing publication process.[66][67][68][69]

7.3.4 Predictive validity

Next to predicting a group's performance on more complex criterion tasks as shown in the original experiments,[8] the collective intelligence factor c was also found to predict group performance in diverse tasks in MBA classes lasting over several months.[64] Thereby, highly collectively intelligent groups earned significantly higher scores on their group assignments although their members did not do any better on other individually performed assignments. Moreover, highly collective intelligent teams improved performance over time suggesting that more collectively intelligent teams learn better.[64] This is another potential parallel to individual intelligence where more intelligent people are found to acquire new material quicker.[10][70]

Individual intelligence can be used to predict plenty of life outcomes from school attainment[71] and career success[72] to health outcomes[73] and even mortality.[73] Whether collective intelligence is able to predict other outcomes besides group performance on mental tasks has still to be investigated.

7.3.5 Potential connections to individual intelligence

Gladwell[74] (2008) showed that the relationship between individual IQ and success works only to a certain point and that additional IQ points over an estimate of IQ 120 do not translate into real life advantages. If a similar border exists for Group-IQ or if advantages are linear and infinite, has still to be explored. Similarly, demand for further research on possible connections of individual and collective intelligence exists within plenty of other potentially transferable logics of individual intelligence, such as, for instance, the development over time[75] or the question of improving intelligence.[76][77] Whereas it is controversial whether human intelligence can be enhanced via training,[76][77] a group's collective intelligence potentially offers simpler opportunities for improvement by exchanging team members or implementing structures and technologies.[36] Moreover, social sensitivity was found to be, at least temporarily, improvable by reading literary fiction[78] as well as watching drama movies.[79] In how far such training ultimately improves collective intelligence through social sensitivity remains an open question.[80]

There are further more advanced concepts and factor models attempting to explain individual cognitive ability including the categorization of intelligence in fluid and crystallized intelligence[81][82] or the hierarchical model of intelligence differences.[83][84] Further supplementing explanations and conceptualizations for the factor structure of the **Genomes'[85] of collective intelligence besides a general** c **factor'**, though, are missing yet.

7.3.6 Controversies

Other scholars explain team performance by aggregating team members' general intelligence to the team level[86][87] instead of building an own overall collective intelligence measure. Devine and Philips[88] (2001) showed in a meta-analysis that mean cognitive ability predicts team performance in laboratory settings (.37) as well as field settings (.14) – note that this is only a small effect. Suggesting a strong dependence on the relevant tasks, other scholars showed that tasks requiring a high degree of communication and cooperation are found to be most influenced by the team member with the lowest cognitive ability.[89] Tasks in which selecting the best team member is the most successful strategy, are shown to be most influenced by the member with the highest cognitive ability.[51]

Since Woolley et al.'s[8] results do not show any influence of group satisfaction, group cohesiveness, or motivation, they, at least implicitly, challenge these concepts regarding the importance for group performance in general and thus contrast meta-analytically proven evidence concerning the positive effects of group cohesion,[90][91][92] motivation[93][94] and satisfaction[95] on group performance.

Noteworthy is also that the involved researchers among the confirming findings widely overlap with each other and with the authors participating in the original first study around Anita Woolley.[8][29][29][35][55][63]

7.4 Alternative mathematical techniques

7.4.1 Computational collective intelligence

In 2001, Tadeusz (Tad) Szuba from the AGH University in Poland proposed a formal model for the phenomenon of collective intelligence. It is assumed to be an unconscious, random, parallel, and distributed computational process, run in mathematical logic by the social structure.[96]

In this model, beings and information are modeled as abstract information molecules carrying expressions of mathematical logic.[96] They are quasi-randomly displacing due to their interaction with their environments with their intended displacements.[96] Their interaction in abstract computational space creates multi-thread inference process which we perceive as collective intelligence.[96] Thus, a non-Turing model of computation is used. This theory allows simple formal definition of collective intelligence as the property of social structure and seems to be working well for a wide spectrum of beings, from bacterial colonies up to human social structures. Collective intelligence considered as a specific computational process is providing a straightforward explanation of several social phenomena. For this model of collective intelligence, the formal definition of IQS (IQ Social) was proposed and was defined as "the probability function over the time and domain of N-element inferences which are reflecting inference activity of the social structure".[96] While IQS seems to be computationally hard, modeling of social structure in terms of a computational process as described above gives a chance for approximation.[96] Prospective applications are optimization of companies through the maximization of their IQS, and the analysis of drug resistance against collective intelligence of bacterial colonies.[96]

Computational Collective Intelligence, by Tadeusz Szuba

7.4.2 Collective intelligence quotient

One measure sometimes applied, especially by more artificial intelligence focused theorists, is a "collective intelligence quotient"[97] (or "cooperation quotient") – which can be normalized from the "individual" intelligence quotient (IQ)[97] – thus making it possible to determine the marginal intelligence added by each new individual participating in the collective action, thus using metrics to avoid the hazards of group think and stupidity.[98]

7.5 Applications

Elicitation of point estimates – Here, we try to get an estimate (in a single value) of something. For example, estimating the weight of an object, or the release date of a product or probability of success of a project etc. as are seen in prediction markets like Intrade, HSX or InklingMarkets and also in several implementations of crowdsourced estimation of a numeric outcome. Essentially, we try to get the average value of the estimates provided by the members in the crowd.

Opinion Aggregation – In this situation, we gather opinions from the crowd regarding some idea, issue or product. For example, trying to get a rating (on some scale) of a product sold online (such as Amazon's star rating system). Here, the emphasis is to collect and simply aggregate the ratings provided by customers/users.

Idea Collection – In these problems, someone solicits ideas for projects, designs or solutions from the crowd. For example, ideas on solving a data science problem (as in Kaggle) or getting a good design for a t-shirt (as in Threadless) or in getting answers to simple problems that only humans can do well (as in Amazon's Mechanical Turk). Here, the objective is to gather the ideas and devise some selection criteria to choose the best ideas.

7.5.1 Cognition

Market judgment

Because of the Internet's ability to rapidly convey large amounts of information throughout the world, the use of collective intelligence to predict stock prices and stock price direction has become increasingly viable.[99] Websites aggregate stock market information that is as current as possible so professional or amateur stock analysts can publish their viewpoints, enabling amateur investors to submit their financial opinions and create an aggregate opinion.[99] The opinion of all investor can be weighed equally so that a pivotal premise of the effective application of collective intelligence can be applied: the masses, including a broad spectrum of stock market expertise, can be utilized to more accurately predict the behavior of financial markets.[100][101]

Collective intelligence underpins the efficient-market hypothesis of Eugene Fama[102] – although the term collective intelligence is not used explicitly in his paper. Fama cites research conducted by Michael Jensen[103] in which 89 out of 115 selected funds underperformed relative to the index during the period from 1955 to 1964. But after removing the loading charge (up-front fee) only 72 underperformed while after removing brokerage costs only 58 underperformed. On the basis of such evidence index funds became popular investment vehicles using the collective intelligence of the market, rather than the judgement of professional fund managers, as an investment strategy.[103]

Predictions in politics and technology

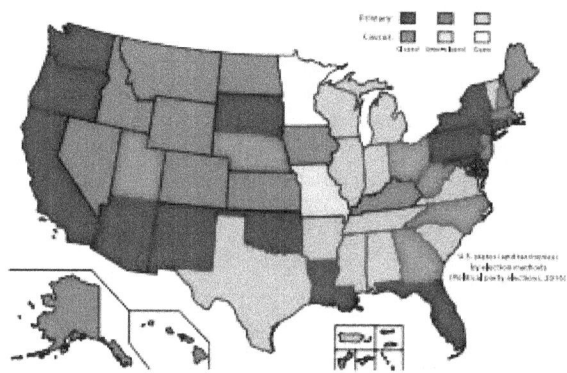

Voting methods used in the United States 2016

Political parties mobilize large numbers of people to form policy, select candidates and finance and run election campaigns.[104] Knowledge focusing through various voting methods allows perspectives to converge through the assumption that uninformed voting is to some degree random and can be filtered from the decision process leaving only a residue of informed consensus.[104] Critics point out that often bad ideas, misunderstandings, and misconceptions are widely held, and that structuring of the decision process must favor experts who are presumably less prone to random or misinformed voting in a given context.[105]

Companies such as Affinnova (acquired by Nielsen), Google, InnoCentive, Marketocracy, and Threadless[106] have successfully employed the concept of collective intelligence in bringing about the next generation of technological changes through their research and development (R&D), customer service, and knowledge management.[106][107] An example of such application is Google's Project Aristotle in 2012, where the effect of collective intelligence on team makeup was examined in hundreds of the company's R&D teams.[108]

7.5.2 Cooperation

Networks of trust

In 2012, the *Global Futures Collective Intelligence System* (GFIS) was created by The Millennium Project,[109] which epitomizes collective intelligence as the synergistic intersection among data/information/knowledge, software/hardware, and expertise/insights that has a recursive

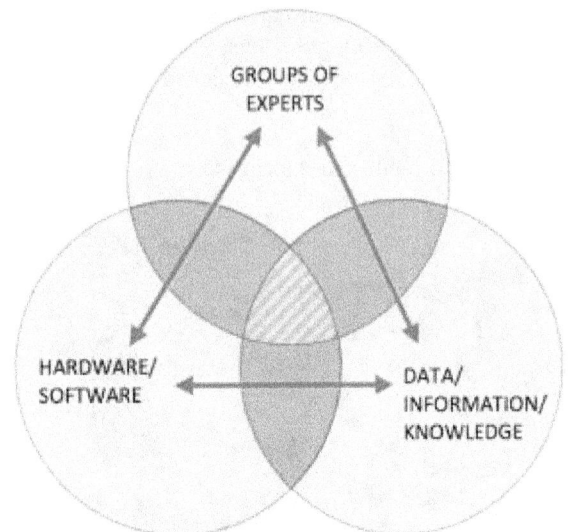

COLLECTIVE INTELLIGENCE

Application of collective intelligence in the Millennium Project

The makeup of a global brain

learning process for better decision-making than the individual players alone.[110]

New media are often associated with the promotion and enhancement of collective intelligence. The ability of new media to easily store and retrieve information, predominantly through databases and the Internet, allows for it to be shared without difficulty. Thus, through interaction with new media, knowledge easily passes between sources (Flew 2008) resulting in a form of collective intelligence. The use of interactive new media, particularly the internet, promotes online interaction and this distribution of knowledge between users.

Francis Heylighen, Valentin Turchin, and Gottfried Mayer-Kress are among those who view collective intelligence through the lens of computer science and cybernetics. In their view, the Internet enables collective intelligence at the widest, planetary scale, thus facilitating the emergence of a global brain.

The developer of the World Wide Web, Tim Berners-Lee, aimed to promote sharing and publishing of information globally. Later his employer opened up the technology for free use. In the early '90s, the Internet's potential was still untapped, until the mid-1990s when 'critical mass', as termed by the head of the Advanced Research Project Agency (ARPA), Dr. J.C.R. Licklider, demanded more accessibility and utility.[111] The driving force of this Internet-based collective intelligence is the digitization of information and communication. Henry Jenkins, a key theorist of new media and media convergence draws on the theory that collective intelligence can be attributed to media convergence and participatory culture (Flew 2008). He criticizes contemporary education for failing to incorporate online trends of collective problem solving into the classroom, stating "whereas a collective intelligence community encourages ownership of work as a group, schools grade individuals". Jenkins argues that interaction within a knowledge community builds vital skills for young people, and teamwork through collective intelligence communities contribute to the development of such skills.[112] Collective intelligence is not merely a quantitative contribution of information from all cultures, it is also qualitative.[112]

Lévy and de Kerckhove consider CI from a mass communications perspective, focusing on the ability of networked information and communication technologies to enhance the community knowledge pool. They suggest that these communications tools enable humans to interact and to share and collaborate with both ease and speed (Flew 2008). With the development of the Internet and its widespread use, the opportunity to contribute to knowledge-building communities, such as Wikipedia, is greater than ever before. These computer networks give participating users the opportunity to store and to retrieve knowledge through the collective access to these databases and allow them to "harness the hive".[113] Researchers at the MIT Center for Collective Intelligence research and explore collective intelligence of groups of people and computers.[114]

In this context collective intelligence is often confused with shared knowledge. The former is the sum total of information held individually by members of a community while the latter is information that is believed to be true and known by all members of the community.[115] Collective intelligence as represented by Web 2.0 has less user engagement than collaborative intelligence. An art project using Web 2.0 platforms is "Shared Galaxy", an experiment developed by an anonymous artist to create a collective identity that shows up as one person on several platforms like MySpace, Facebook, YouTube and Second Life. The password is written in the profiles and the accounts named "Shared Galaxy" are open to be used by anyone. In this way many take part in being one.[116] Another art project using collective intelligence to produce artistic work is Curatron, where a large group of artists together decides on a smaller group that they think would make a good collaborative group. The process is used based on an algorithm computing the collective preferences.[117] In creating what he calls 'CI-Art', Nova Scotia based artist Mathew Aldred follows Pierry Lévy's definition of collective intelligence.[118] Aldred's CI-Art event in March 2016 involved over four hundred people from the community of Oxford, Nova Scotia, and internationally.[119][120] Later work developed by Aldred used the UNU swarm intelligence system to create digital drawings and paintings.[121] The Oxford Riverside Gallery (Nova Scotia) held a public CI-Art event in May 2016, which connected with online participants internationally.[122]

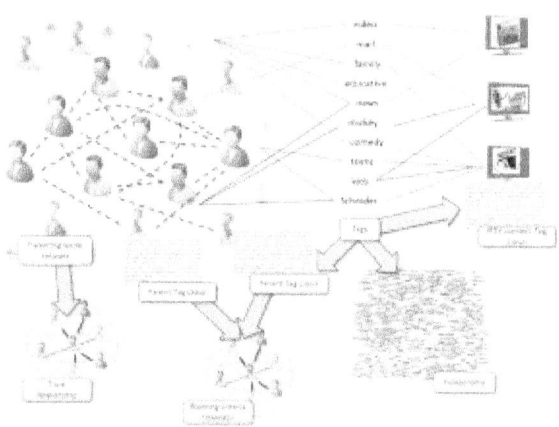

Parenting social network and collaborative tagging as pillars for automatic IPTV content blocking system

In social bookmarking (also called collaborative tagging),[123] users assign tags to resources shared with other users, which gives rise to a type of information organisation that emerges from this crowdsourcing process. The resulting information structure can be seen as reflecting the collective knowledge (or collective intelligence) of a community of users and is commonly called a "Folksonomy", and the process can be captured by models of collaborative tagging.[123]

Recent research using data from the social bookmarking website Delicious, has shown that collaborative tagging systems exhibit a form of complex systems (or self-organizing) dynamics.[124][125][126] Although there is no central controlled vocabulary to constrain the actions of individual users, the distributions of tags that describe different resources has been shown to converge over time to a stable power law distributions.[124] Once such stable distributions form, examining the correlations between different tags can be used to construct simple folksonomy graphs, which can be efficiently partitioned to obtained a form of community or shared vocabularies.[127] Such vocabularies can be seen as a form of collective intelligence, emerging from the decentralised actions of a community of users. The Wall-it Project is also an example of social bookmarking.[128]

P2P business

Research performed by Tapscott and Williams has provided a few examples of the benefits of collective intelligence to business:[33]

Talent utilization At the rate technology is changing, no firm can fully keep up in the innovations needed to compete. Instead, smart firms are drawing on the power of mass collaboration to involve participation of the people they could not employ. This also helps generate continual interest in the firm in the form of those drawn to new idea creation as well as investment opportunities.[33]

Demand creation Firms can create a new market for complementary goods by engaging in open source community. Firms also are able to expand into new fields that they previously would not have been able to without the addition of resources and collaboration from the community. This creates, as mentioned before, a new market for complementary goods for the products in said new fields.[33]

Costs reduction Mass collaboration can help to reduce costs dramatically. Firms can release a specific software or product to be evaluated or debugged by online communities. The results will be more personal, robust and error-free products created in a short amount of time and costs. New ideas can also be generated and explored by collaboration of online communities creating opportunities for free R&D outside the confines of the company.[33]

Open source software

Cultural theorist and online community developer, John Banks considered the contribution of online fan communities in the creation of the Trainz product. He argued that its commercial success was fundamentally dependent upon "the formation and growth of an active and vibrant online fan community that would both actively promote the product and create content- extensions and additions to the game software".[129]

The increase in user created content and interactivity gives rise to issues of control over the game itself and ownership of the player-created content. This gives rise to fundamental legal issues, highlighted by Lessig[130] and Bray and Konsynski,[131] such as intellectual property and property ownership rights.

Gosney extends this issue of Collective Intelligence in videogames one step further in his discussion of alternate reality gaming. This genre, he describes as an "across-media game that deliberately blurs the line between the in-game and out-of-game experiences"[132] as events that happen outside the game reality "reach out" into the player's lives in order to bring them together. Solving the game requires "the collective and collaborative efforts of multiple players"; thus the issue of collective and collaborative team play is essential to ARG. Gosney argues that the Alternate Reality genre of gaming dictates an unprecedented level of collaboration and "collective intelligence" in order to solve the mystery of the game.[132]

7.5.3 Coordination

Ad-hoc communities

Military, trade unions, and corporations satisfy some definitions of CI – the most rigorous definition would require a capacity to respond to very arbitrary conditions without orders or guidance from "law" or "customers" to constrain actions. Online advertising companies are using collective intelligence to bypass traditional marketing and creative agencies.[133]

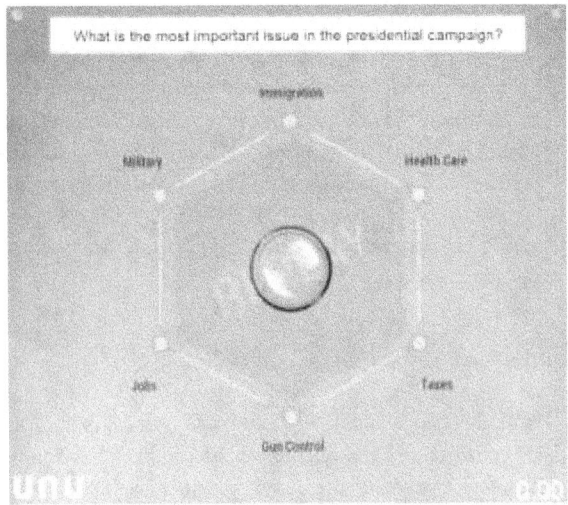

UNU Predicts the most important issue in presidential campaign

The UNU open platform for "human swarming" (or "social swarming") establishes real-time closed-loop systems around groups of networked users molded after biological swarms, enabling human participants to behave as a unified collective intelligence.[134][135] When connected to UNU, groups of distributed users collectively answer questions and make predictions in real-time.[136] Early testing shows that human swarms can out-predict individuals.[134] In 2016, an UNU swarm was challenged by a reporter to predict the winners of the Kentucky Derby, and successfully picked the first four horses, in order, beating 540 to 1 odds.[137][138]

Specialized information sites such as Digital Photography Review[139] or Camera Labs[140] is an example of collective intelligence. Anyone who has an access to the internet can contribute to distributing their knowledge over the world through the specialized information sites.

In learner-generated context a group of users marshal resources to create an ecology that meets their needs often (but not only) in relation to the co-configuration, co-creation and co-design of a particular learning space that allows learners to create their own context.[141][142][143] Learner-generated contexts represent an *ad hoc* community that facilitates coordination of collective action in a network of trust. An example of learner-generated context is found on the Internet when collaborative users pool knowledge in a "shared intelligence space". As the Internet has developed so has the concept of CI as a shared public forum. The global accessibility and availability of the Internet has allowed more people than ever to contribute and access ideas. (Flew 2008)

Games such as *The Sims* Series, and *Second Life* are designed to be non-linear and to depend on collective intelligence for expansion. This way of sharing is gradually evolving and influencing the mindset of the current and future generations.[111] For them, collective intelligence has become a norm. In Terry Flew's discussion of 'interactivity' in the online games environment, the ongoing interactive dialogue between users and game developers,[144] he refers to Pierre Lévy's concept of Collective Intelligence (Lévy 1998) and argues this is active in videogames as clans or guilds in MMORPG constantly work to achieve goals. Henry Jenkins proposes that the participatory cultures emerging between games producers, media companies, and the end-users mark a fundamental shift in the nature of media production and consumption. Jenkins argues that this new participatory culture arises at the intersection of three broad new media trends.[145] Firstly, the development of new media tools/technologies enabling the creation of content. Secondly, the rise of subcultures promoting such creations, and lastly, the growth of value adding media conglomerates, which foster image, idea and narrative flow.

7.6. ALTERNATIVE VIEWS

Coordinating collective actions

The cast of After School Improv learns an important lesson about improvisation and life

Improvisational actors also experience a type of collective intelligence which they term "group mind", as theatrical improvisation relies on mutual cooperation and agreement,[146] leading to the unity of "group mind".[146][147]

Growth of the Internet and mobile telecom has also produced "swarming" or "rendezvous" events that enable meetings or even dates on demand.[148] The full impact has yet to be felt but the anti-globalization movement, for example, relies heavily on e-mail, cell phones, pagers, SMS and other means of organizing.[149] The Indymedia organization does this in a more journalistic way.[150] Such resources could combine into a form of collective intelligence accountable only to the current participants yet with some strong moral or linguistic guidance from generations of contributors – or even take on a more obviously democratic form to advance shared goal.[150]

A further application of collective intelligence is found in the "Community Engineering for Innovations".[151] In such an integrated framework proposed by Ebner et al., idea competitions and virtual communities are combined to better realize the potential of the collective intelligence of the participants, particularly in open-source R&D.[152]

7.6 Alternative views

7.6.1 A tool for combating self-preservation

Tom Atlee reflects that, although humans have an innate ability to gather and analyze data, they are affected by culture, education and social institutions.[153] A single person tends to make decisions motivated by self-preservation. Therefore, without collective intelligence, humans may drive themselves into extinction based on their selfish needs.[31]

7.6.2 Separation from IQism

Phillip Brown and Hugh Lauder quotes Bowles and Gintis (1976) that in order to truly define collective intelligence, it is crucial to separate 'intelligence' from IQism.[154] They go on to argue that intelligence is an achievement and can only be developed if allowed to.[154] For example, earlier on, groups from the lower levels of society are severely restricted from aggregating and pooling their intelligence. This is because the elites fear that the collective intelligence would convince the people to rebel. If there is no such capacity and relations, there would be no infrastructure on which collective intelligence is built (Brown & Lauder 2000, p. 230). This reflects how powerful collective intelligence can be if left to develop.[154]

7.6.3 Artificial intelligence views

Skeptics, especially those critical of artificial intelligence and more inclined to believe that risk of bodily harm and bodily action are the basis of all unity between people, are more likely to emphasize the capacity of a group to take action and withstand harm as one fluid mass mobilization, shrugging off harms the way a body shrugs off the loss of a few cells.[155][156][156] This strain of thought is most obvious in the anti-globalization movement and characterized by the works of John Zerzan, Carol Moore, and Starhawk, who typically shun academics.[155][156][156] These theorists are more likely to refer to ecological and collective wisdom and to the role of consensus process in making ontological distinctions than to any form of "intelligence" as such, which they often argue does not exist, or is mere "cleverness".[155][156][156]

Harsh critics of artificial intelligence on ethical grounds are likely to promote collective wisdom-building methods, such as the new tribalists and the Gaians.[157] Whether these can be said to be collective intelligence systems is an open question. Some, e.g. Bill Joy, simply wish to avoid any form of autonomous artificial intelligence and seem willing to work on rigorous collective intelligence in order to remove any possible niche for AI.[158]

In contrast to these views, Artificial Intelligence companies such as Amazon Mechanical Turk and CrowdFlower are using collective intelligence and crowdsourcing or Consensus-based assessment to collect the enormous amounts of data for machine learning algorithms such as Keras and IBM Watson.

7.7 See also

7.7.1 Similar concepts and applications

- Civic intelligence
- Collaborative filtering
- Collaborative innovation network
- Collective decision-making
- Collective effervescence
- Collective memory
- Collective problem solving
- Crowd psychology
- Global Consciousness Project
- Group behaviour
- Group mind (science fiction)
- Knowledge ecosystem
- Noogenesis
- Open source intelligence
- Recommendation system
- Smart mob
- Social commerce
- Social information processing
- Stigmergy
- Syntality
- *The Wisdom of Crowds*
- Think tank
- Wiki

7.7.2 Computation and computer science

- Bees algorithm
- Cellular automaton
- Collaborative human interpreter
- Collaborative software
- Connectivity (graph theory)
- Enterprise bookmarking
- Human-based computation
- Open-source software
- Organismic computing
- Preference elicitation

7.7.3 Others

- Customer engagement
- Dispersed knowledge
- Distributed cognition
- Facilitation (business)
- Facilitator
- Hundredth monkey effect
- Keeping up with the Joneses
- Library
- Library of Alexandria
- Meme
- Open-space meeting

7.8 Notes and references

[1] Ngoc Thanh Nguyen (25 July 2011). *Transactions on Computational Collective Intelligence III*. Springer. ISBN 978-3-642-19967-7. Retrieved 11 June 2013.

[2] Glenn, Jerome C. Collective Intelligence – One of the Next Big Things. Futura 4/2009, Finnish Society for Futures Studies, Helsinki, Finland

[3] Glenn, Jerome C. Chapter 5, 2008 State of the Future. The Millennium Project, Washington, DC 2008

[4] Norman Lee Johnson, Collective Science site

[5] Pierre Lévy, *Collective Intelligence: Mankind's Emerging World in Cyberspace*, 1994, p. 13

[6] Flew, Terry *New Media: An Introduction*. Oxford University Press, 2007, p. 21

[7] Jenkins, *Henry Convergence Culture: Where old and new media collide*. New York: New York University Press, 2006, p. 259

[8] Woolley, Anita Williams; Chabris, Christopher F.; Pentland, Alex; Hashmi, Nada; Malone, Thomas W. (2010-10-29). "Evidence for a Collective Intelligence Factor in the Performance of Human Groups". *Science*. **330** (6004): 686–688. Bibcode:2010Sci...330..686W. PMID 20929725. doi:10.1126/science.1193147.

[9] Jensen, Arthur, R. (1992). "Understanding g in terms of information processing". *Educational Psychology Review*. **4** (3): 271–308. doi:10.1007/bf01417874.

[10] Jensen, Arthur, R. (1998). *The g factor: The science of mental ability*. Westport, CT: Praeger.

[11] Landemore, Hélène (2012). Landemore, *Democratic Reason: Politics, Collective Intelligence, and the Rule of the Many*. Princeton: Princeton University Press.

[12] Waldron, Jeremy (1995). "The Wisdom of the Multitude: Some Reflections on Book 3, Chapter 11 of Aristotle's Politics". *Political Theory*. **23** (4): 563–584. doi:10.1177/0090591795023004001.

[13] Ober, Josiah (2008). *Democracy and Knowledge*. Princeton, N.J.: Princeton University Press. pp. 110–14.

[14] Cammack, Daniela (2013). "Aristotle and the Virtue of the Multitude". *Political Theory*. **41** (2): 175–202. doi:10.1177/0090591712470423.

[15] Page, Scott (2008). *The Difference: How the Power of Diversity Creates Better Groups, Firms, Schools, and Societies*. Princeton: Princeton University Press.

[16] Its source can be found in this link http://skccblog.tistory.com/716. written by story teller who belongs to SK C&C, story teller explains history of collective intelligence. also, it mentioned principle of collective intelligence. especially, it says William Morton Wheeler studied it at first time. The title is 'collective intelligence that made us who are smarter than me'

[17] Émile Durkheim, *The Elementary Forms of Religious Life*, 1912.

[18] "About the Book - Foundation for Conscious Evolution". *Foundation for Conscious Evolution*. Retrieved 2016-12-04.

[19] Engelbart, Douglas (1962) Augmenting Human Intellect: A Conceptual Framework - section on Team Cooperation

[20] Engelbart, Douglas (1994)Boosting Collective IQ (Slide Handouts) – 'Collective IQ' defined on Slide 4; also (1994) BBN Distinguished Guest Lecture (Video) – 'Collective IQ' defined @16:56 "CoDIAK"

[21] Landemore, Helene (2013). *Democratic Reason: Politics, Collective Intelligence, and the Rule of the Many*. Princeton University Press.

[22] Howard Bloom, *The Lucifer Principle: A Scientific Expedition Into the Forces of History*, 1995

[23] Howard Bloom, *Global Brain: The Evolution of Mass Mind from the Big Bang to the 21st Century*, 2000

[24] Skrbina, D., 2001. Participation, Organization, and Mind: Toward a Participatory Worldview, ch. 8, Doctoral Thesis, Centre for Action Research in Professional Practice, School of Management, University of Bath: England

[25] Levey, Geoffrey Brahm (2015-05-01). *Authenticity, Autonomy and Multiculturalism*. Routledge. ISBN 9781317535928.

[26] Skrbina, David F. (2007-01-26). *Panpsychism in the West* (1 ed.). The MIT Press. ISBN 9780262693516.

[27] George Pór, Blog of Collective Intelligence

[28] Atlee, Tom. "Collective Intelligence as a Field of Multi-disciplinary Study and Practice" (PDF). *Community Intelligence*. CommunityIntelligence. Ltd. Retrieved December 11, 2016.

[29] Woolley, Anita Williams; Aggarwal, Ishani; Malone, Thomas W. (2015-12-01). "Collective Intelligence and Group Performance". *Current Directions in Psychological Science*. **24** (6): 420–424. doi:10.1177/0963721415599543.

[30] Leimeister, Jan Marco (2010-06-24). "Collective Intelligence". *Business & Information Systems Engineering*. **2** (4): 245–248. doi:10.1007/s12599-010-0114-8.

[31] Atlee, T. (2008). Reflections on the evolution of choice and collective intelligence. Retrieved 26 August 2008

[32] Steele, Robert David (2002-04-08). *The New Craft of Intelligence: Personal, Public, & Political--Citizen's Action Handbook for Fighting Terrorism, Genocide, Disease, Toxic Bombs, & Corruption*. Oakton, Va.: Oss Pr. ISBN 9780971566118.

[33] Tapscott, D., & Williams, A. D. (2008). *Wikinomics: How Mass Collaboration Changes Everything*, USA: Penguin Group

[34] Spearman, Charles, E. (1904). ""General intelligence," objectively determined and measured". *American Journal of Psychology*. **15** (2): 201–293. JSTOR 1412107. doi:10.2307/1412107.

[35] Engel, D.; Woolley, A. W.; Jing, L. X.; Chabris, C. F. & Malone, T. W. (2014). "Reading the Mind in the Eyes or reading between the lines? Theory of Mind predicts collective intelligence equally well online and face-to-face". *PLoS ONE*. **9** (12): e115212. Bibcode:2014PLoSO...9k5212E. PMC 4267836. PMID 25514387. doi:10.1371/journal.pone.0115212.

[36] Woolley, A. & Malone, T. (June 2011). "Defend your research: What makes a team smarter? More women". *Harvard Business Review*. **89** (6): 32–33.

[37] Kamphaus, R.W.; Winsor, A.P.; Rowe, E.W. & Kim, S. (2005). *A history of intelligence test interpretation*. In D.P. Flanagan and P.L. Harrison (Eds.), Contemporary intellectual assessment: Theories, tests, and issues (2nd Ed.). New York, NY: Guilford. pp. 23–38.

[38] van der Maas, Han L. J.; Dolan, Conor V.; Grasman, Raoul P. P. P.; Wicherts, Jelte M.; Huizenga, Hilde M.; Raijmakers, Maartje E. J. (2006-10-01). "A dynamical model of general intelligence: the positive manifold of intelligence by mutualism". *Psychological Review*. **113** (4): 842–861. PMID 17014305. doi:10.1037/0033-295X.113.4.842.

[39] McCrae, R. R.; Costa Jr., P. T. (1987). "Validation of the Five-Factor Model of Personality Across Instruments and Observers" (PDF). *Journal of Personality and Social Psychology*. **52** (1): 81–90. PMID 3820081. doi:10.1037/0022-3514.52.1.81.

[40] "MIT Center for Collective Intelligence". *cci.mit.edu*. Retrieved 2016-04-26.

[41] Briley, Daniel A.; Tucker-Drob, Elliot M. (2014-09-01). "Genetic and environmental continuity in personality development: a meta-analysis". *Psychological Bulletin*. **140** (5): 1303–1331. PMC 4152379. PMID 24956122. doi:10.1037/a0037091.

[42] Deary, Ian J.; Spinath, Frank M.; Bates, Timothy C. (2006-01-01). "Genetics of intelligence". *European Journal of Human Genetics*. **14** (6): 690–700. PMID 16721405. doi:10.1038/sj.ejhg.5201588.

[43] Baron-Cohen S, Wheelwright S, Hill J, Raste Y, Plumb I (2001). "The Reading the Mind in the Eyes Test revised version: a study with normal adults, and adults with Asperger syndrome or high-functioning autism". *Journal of Child Psychology and Psychiatry*. **42** (2): 241–251. doi:10.1017/s0021963001006643.

[44] Apperly, Ian A. (2012-05-01). "What is "theory of mind"? Concepts, cognitive processes and individual differences". *The Quarterly Journal of Experimental Psychology*. **65** (5): 825–839. PMID 22533318. doi:10.1080/17470218.2012.676055.

[45] Baron-Cohen, S.; Leslie, A.M.; Frith, U. (1985). "Does the autistic child have a *theory of mind*?". *Cognition*. **21** (1): 37–46. PMID 2934210. doi:10.1016/0010-0277(85)90022-8.

[46] Flavell, J. H. (1999-01-01). "Cognitive development: children's knowledge about the mind". *Annual Review of Psychology*. **50**: 21–45. PMID 10074674. doi:10.1146/annurev.psych.50.1.21.

[47] Premack, David; Woodruff, Guy (1978-12-01). "Does the chimpanzee have a theory of mind?". *Behavioral and Brain Sciences*. **1** (4): 515–526. doi:10.1017/S0140525X00076512.

[48] Heyes, Cecilia M.; Frith, Chris D. (2014-06-20). "The cultural evolution of mind reading". *Science*. **344** (6190): 1243091. PMID 24948740. doi:10.1126/science.1243091.

[49] Hallerbäck, Maria Unenge; Lugnegård, Tove; Hjärthag, Fredrik; Gillberg, Christopher (2009-03-01). "The Reading the Mind in the Eyes Test: Test–retest reliability of a Swedish version". *Cognitive Neuropsychiatry*. **14** (2): 127–143. PMID 19370436. doi:10.1080/13546800902901518.

[50] Pinkham, Amy E.; Penn, David L.; Green, Michael F.; Buck, Benjamin; Healey, Kristin; Harvey, Philip D. (2014-07-01). "The Social Cognition Psychometric Evaluation Study: Results of the Expert Survey and RAND Panel". *Schizophrenia Bulletin*. **40** (4): 813–823. PMC 4059426. PMID 23728248. doi:10.1093/schbul/sbt081.

[51] Yip, Jeremy A.; Côté, Stéphane (2013-01-01). "The Emotionally Intelligent Decision Maker Emotion-Understanding Ability Reduces the Effect of Incidental Anxiety on Risk Taking". *Psychological Science*. **24** (1): 48–55. PMID 23221020. doi:10.1177/0956797612450031.

[52] MacKinnon, D. P. (2008). *Introduction to Statistical Mediation Analysis*. New York, NY: Erlbaum.

[53] Woolley, A. W.; Aggarwal, I.; Malone, T. W. (2015-12-01). "Collective Intelligence and Group Performance". *Current Directions in Psychological Science*. **24** (6): 420–424. doi:10.1177/0963721415599543.

[54] Kim, Y. J.; Engel, D.; Woolley, A. W.; Lin, J.; McArthur, N. & Malone, T. W. (2015). "Work together, play smart: Collective intelligence in League of Legends teams". *Paper presented at the 2015 Collective Intelligence Conference*, Santa Clara, CA.

[55] Aggarwal, I.; Woolley, A. W.; Chabris, C. F. & Malone, T. W. (2015). "Cognitive diversity, collective intelligence, and learning in teams.". *Paper presented at the 2015 Collective Intelligence Conference*, Santa Clara, CA.

[56] Kozhevnikov, M.; Evans, C. & Kosslyn, S. M. (2014). "Cognitive style as environmentally sensitive individual differences in cognition: A modern synthesis and applications in education, business, and management". *Psychological Science in the Public Interest*. **15** (1): 3–33. PMID 26171827. doi:10.1177/1529100614525555.

[57] 1974-, Greene, Joshua David,. *Moral tribes : emotion, reason, and the gap between us and them*. ISBN 0143126059. OCLC 871336785.

[58] Muchnik, Lev; Aral, Sinan; Taylor, Sean J. (2013-08-09). "Social Influence Bias: A Randomized Experiment". *Science*. **341** (6146): 647–651. ISSN 0036-8075. PMID 23929980. doi:10.1126/science.1240466.

[59] Oxenham, Simon. "Why bees could be the secret to superhuman intelligence". Retrieved 2017-05-23.

[60] Rosenberg, L.; Baltaxe, D.; Pescetelli, N. (2016-10-01). "Crowds vs swarms, a comparison of intelligence". *2016 Swarm/Human Blended Intelligence Workshop (SHBI)*: 1–4. doi:10.1109/SHBI.2016.7780278.

[61] "Artificial intelligence turns $20 into $11,000 in Kentucky Derby bet". *Newsweek*. 2016-05-10. Retrieved 2017-05-23.

[62] McGrath, J. E. (1984). *Groups: Interaction and Performance*. Englewood Cliffs, NJ: Prentice-Hall.

[63] Engel, David; Woolley, Anita Williams; Aggarwal, Ishani; Chabris, Christopher F.; Takahashi, Masamichi; Nemoto, Keiichi; Kaiser, Carolin; Kim, Young Ji; Malone, Thomas W. (2015-01-01). "Collective Intelligence in Computer-Mediated Collaboration Emerges in Different Contexts and Cultures". *Proceedings of the 33rd Annual ACM Conference on Human Factors in Computing Systems*. CHI '15. New York, NY, USA: ACM: 3769–3778. ISBN 9781450331456. doi:10.1145/2702123.2702259.

[64] Aggarwal, I. & Woolley, A.W. (2014). "The effects of cognitive diversity on collective intelligence and team learning.". *Symposium presented at the 50th Meeting of the Society of Experimental Social Psychology, Columbus, OH*.

[65] Engel, D.; Woolley, A. W.; Aggarwal, I.; Chabris, C. F.; Takahashi, M.; Nemoto, K.; Malone, T. W. (2015). "Collective intelligence in computer-mediates collaboration emerges in different contexts and cultures.". *In Proceedings of the 33rd Annual ACM Conference on Human Factors in Computing Systems (CHI '15) (pp. 3769–3778). New York, NY: ACM*.

[66] "Collective Intelligence 2016". sites.google.com. Retrieved 2016-04-27.

[67] "Posters | Collective Intelligence 2015". sites.lsa.umich.edu. Retrieved 2016-04-27.

[68] "Proceedings | Collective Intelligence 2014". collective.mech.northwestern.edu. Retrieved 2016-04-27.

[69] Malone, Thomas W.; Luis von Ahn (2012). "Collective Intelligence 2012: Proceedings". arXiv:1204.2991 [cs.SI].

[70] Schmidt, F.L. & Hunter, J.E. (1998). "The validity and utility of selection methods in personnel psychology: Practical and theoretical implications of 85 years of research findings". *Psychological Bulletin*. **124** (2): 262–274. doi:10.1037/0033-2909.124.2.262.

[71] Nathan, B. (1997). "Intelligence, Schooling, and Society". *American Psychologist*. **52** (10): 1046–1050. doi:10.1037/0003-066x.52.10.1046.

[72] Strenze, Tarmo (2007-09-01). "Intelligence and socioeconomic success: A meta-analytic review of longitudinal research". *Intelligence*. **35** (5): 401–426. doi:10.1016/j.intell.2006.09.004.

[73] Deary, I.J.; Weiss, A. & Batty, D.G. (2010). "Intelligence and Personality as Predictors of Illness and Death. How Researchers in Differential Psychology and Chronic Disease Epidemiology Are Collaborating to Understand and Address Health Inequalities". *Psychological Science in the Public Interest*. **11** (2): 53–79. PMID 26168413. doi:10.1177/1529100610387081.

[74] Gladwell, M. (2008). *Outliers. The Story of Success*. New York, NY: Little, Brown and Company. ISBN 978-0-316-01792-3.

[75] Hedden, Trey; Gabrieli, John D. E. (2004-02-01). "Insights into the ageing mind: a view from cognitive neuroscience". *Nature Reviews. Neuroscience*. **5** (2): 87–96. PMID 14735112. doi:10.1038/nrn1323.

[76] Shipstead, Zach; Redick, Thomas S; Engle, Randall W. (2010-10-01). "Does working memory training generalize?". *Psychologica Belgica*. **50** (3–4): 245. doi:10.5334/pb-50-3-4-245.

[77] Buschkuehl, M.; Jaeggi, S.M. (2010). "Improving intelligence a literature review". *Swiss Medical Weekly*. **140** (19): 266–72.

[78] Kidd, David Comer; Castano, Emanuele (2013-10-18). "Reading Literary Fiction Improves Theory of Mind". *Science*. **342** (6156): 377–380. Bibcode:2013Sci...342..377K. PMID 24091705. doi:10.1126/science.1239918.

[79] Black, Jessica; Barnes, Jennifer L. (2015). "Fiction and social cognition: The effect of viewing award-winning television dramas on theory of mind". *Psychology of Aesthetics, Creativity, and the Arts*. **9** (4): 423–429. doi:10.1037/aca0000031.

[80] Malone, T. W. & Bernstein, M.S. (2015). *Handbook of Collective Intelligence*. Cambridge, MA: MIT Press.

[81] Horn, J. (1989). *Models of intelligence. In R.L. Linn (Ed.), Intelligence: Measurement, theory, and public policy (pp. 29–73)*. Urbana, IL: University of Illinois Press.

[82] Cattell, R. B. (1971). *Abilities: Their structure, growth, and action*. Houghton Mifflin: New York, NY.

[83] Carroll, J.B. (1993). *Human cognitive abilities: A survey of factor analytic studies*. Cambridge, England: Cambridge University Press.

[84] Johnson, Wendy; Bouchard Jr., Thomas J. (2005-07-01). "The structure of human intelligence: It is verbal, perceptual, and image rotation (VPR), not fluid and crystallized". *Intelligence*. **33** (4): 393–416. doi:10.1016/j.intell.2004.12.002.

[85] "MIT Center for Collective Intelligence". cci.mit.edu. Retrieved 2016-04-27.

[86] LePine, Jeffery A. (2005). "Adaptation of Teams in Response to Unforeseen Change: Effects of Goal Difficulty and Team Composition in Terms of Cognitive Ability and Goal Orientation". *Journal of Applied Psychology*. **90** (6): 1153–1167. PMID 16316271. doi:10.1037/0021-9010.90.6.1153.

[87] Tziner, Aharon; Eden, Dov (1985). "Effects of crew composition on crew performance: Does the whole equal the sum of its parts?". *Journal of Applied Psychology*. **70** (1): 85–93. doi:10.1037/0021-9010.70.1.85.

[88] Devine, Dennis J.; Philips, Jennifer L. (2001-10-01). "Do Smarter Teams Do Better A Meta-Analysis of Cognitive Ability and Team Performance". *Small Group Research*. **32** (5): 507–532. doi:10.1177/104649640103200501.

[89] O'Brien, G.; Owens, A. (1969). "Effects of organizational structure on correlations between member abilities and group productivity". *Journal of Applied Psychology*. **53** (6): 525–530. doi:10.1037/h0028659.

[90] Evans, Charles R.; Dion, Kenneth L. (1991-05-01). "Group Cohesion and Performance A Meta-Analysis". *Small Group Research*. **22** (2): 175–186. doi:10.1177/1046496491222002.

[91] Gully, Stanley M.; Devine, Dennis J.; Whitney, David J. (2012-12-01). "A Meta-Analysis of Cohesion and Performance Effects of Level of Analysis and Task Interdependence". *Small Group Research*. **43** (6): 702–725. doi:10.1177/1046496412468069.

[92] Beal, Daniel J.; Cohen, Robin R.; Burke, Michael J.; McLendon, Christy L. (December 2003). "Cohesion and Performance in Groups: A Meta-Analytic Clarification of Construct Relations.". *Journal of Applied Psychology*. **88** (6): 989–1004. PMID 14640811. doi:10.1037/0021-9010.88.6.989.

[93] O'leary-kelly, Anne M.; Martocchio, Joseph J.; Frink, Dwight D. (1994-10-01). "A Review of the Influence of Group Goals on Group Performance". *Academy of Management Journal*. **37** (5): 1285–1301. doi:10.2307/256673.

[94] Kleingeld, Ad; Mierlo, Heleen van; Arends, Lidia (2011). "The effect of goal setting on group performance: A meta-analysis". *Journal of Applied Psychology*. **96** (6): 1289–1304. doi:10.1037/a0024315.

[95] Duffy, M. K.; Shaw, J. D. & Stark, E. M. (2000). "Performance and satisfaction in conflicted interdependent groups: When and how does selfesteem make a difference?". *Academy of Management Journal*. **43** (4): 772–782. JSTOR 1556367. doi:10.2307/1556367.

[96] Szuba T., *Computational Collective Intelligence*, 420 pages, Wiley NY, 2001

[97] Kowalczyk, Ryszard (2009-09-23). *Computational Collective Intelligence. Semantic Web, Social Networks and Multi-agent Systems: First International Conference, ICCCI 2009, Wroclaw, Poland, October 5-7, 2009, Proceedings*. Springer Science & Business Media. ISBN 9783642044403.

[98] Administrator. "About Collective IQ - Doug Engelbart Institute". *www.dougengelbart.org*. Retrieved 2016-12-11.

[99] Kaplan, Craig A. (2001). "COLLECTIVE INTELLIGENCE: A NEW APPROACH TO STOCK PRICE FORECASTING" (PDF). *Proceedings of the 2001 IEEE Systems, Man, and Cybernetics Conference*.

[100] Ma, Ying; Li, Guanyi; Dong, Yingsai; Qin, Zengchang (2010). "Minority Game Data Mining for Stock Market Predictions" (PDF). *Agents and Data Mining Interaction, 6th International Workshopon Agents and Data Mining Interaction, ADMI 2010*. Lecture Notes in Computer Science. **5980**. Bibcode:2010LNCS.5980.....C. ISBN 978-3-642-15419-5. doi:10.1007/978-3-642-15420-1.

[101] Yu, Du; Dong, Yingsai; Qin, Zengchang; Wan, Tao (2011). "Exploring Market Behaviors with Evolutionary Mixed-Games Learning Model" (PDF). *Computational Collective Intelligence. Technologies and Applications – Third International Conference, ICCCI 2011*. Lecture Notes in Computer Science. **6922**: 244–253. ISBN 978-3-642-23934-2. doi:10.1007/978-3-642-23935-9_24.

[102] Fama, E.F. (1970). "Efficient Capital Markets: A Review of Theory and Empirical Work". *Journal of Finance*. **25** (2): 383–417. JSTOR 2325486. doi:10.2307/2325486.

[103] Jensen, M.C (1967). "The Performance of Mutual Funds in the Period 1945–1964". *Journal of Finance*. **23** (2): 389–416. doi:10.1111/j.1540-6261.1968.tb00815.x.

[104] "Vote amount according to "intelligence"". *politics.stackexchange.com*. Retrieved 2016-12-12.

[105] "Landemore, H.: Democratic Reason: Politics, Collective Intelligence, and the Rule of the Many. (eBook and Paperback)". *press.princeton.edu*. Retrieved 2016-12-04.

[106] Bonabeau, E (2009). "The power of collective intelligence". *MIT Sloan Management Review*. **50**: 45–52.

[107] Malone, Thomas W.; Laubacher, Robert; Dellarocas, Chrysanthos (2009-02-03). "Harnessing Crowds: Mapping the Genome of Collective Intelligence". Rochester, NY: Social Science Research Network. SSRN 1381502.

[108] Duhigg, Charles (2016-02-25). "What Google Learned From Its Quest to Build the Perfect Team". *The New York Times*. Retrieved 2016-12-11.

[109] "Global Futures Intelligence System". *www.millennium-project.org*. Retrieved 2016-12-07.

[110] "Global Futures Intelligence System". *www.millennium-project.org*. Retrieved 2016-12-11.

[111] Weiss, A. (2005). The Power of Collective Intelligence. Collective Intelligence, pp. 19–23

[112] Henry, Jenkins. "INTERACTIVE AUDIENCES? THE 'COLLECTIVE INTELLIGENCE' OF MEDIA FANS" (PDF). Retrieved December 11, 2016.

[113] Flew, Terry (2008). *New Media: an introduction*. Melbourne: Oxford University Press.

[114] MIT Center for Collective Intelligence Archived 11 June 2010 at the Wayback Machine.. Cci.mit.edu. Retrieved on 2013-07-13.

7.8. NOTES AND REFERENCES

[115] Jenkins, H. 2006. *Convergence Culture*. New York: New York University Press.

[116] Scardamalia, Marlene; Bereiter, Carl (1994-07-01). "Computer Support for Knowledge-Building Communities". *Journal of the Learning Sciences*. **3** (3): 265–283. doi:10.1207/s15327809jls0303_3.

[117] "Math Takes the Guessing Out of Artistic Collaboration". 9 July 2014.

[118] Mathew Aldred, May 2016. ""What is Collective Intelligence Art?"". 2016.

[119] Amherst News Citizen Record, March 17, 2016. ""Community creation taking place in Oxford"". 2016.

[120] Oxford Regional Education Centre""Nexus OREC"". 2016.

[121] UNU Interview with artist, May 23, 2016. ""Artwork from an emergent hive mind"". 2016.

[122] Oxford Riverside Gallery News, May 2016. ""CI-Art event at Oxford Riverside Gallery 'Nexus' opening"". 2016.

[123] Millen, David R.; Feinberg, Jonathan; Kerr, Bernard (2006-01-01). "Dogear: Social Bookmarking in the Enterprise". *Proceedings of the SIGCHI Conference on Human Factors in Computing Systems*. CHI '06. New York, NY, USA: ACM: 111–120. ISBN 1595933727. doi:10.1145/1124772.1124792.

[124] Harry Halpin, Valentin Robu, Hana Shepherd The Complex Dynamics of Collaborative Tagging, Proceedings 6th International Conference on the World Wide Web (WWW'07), Banff, Canada, pp. 211–220, ACM Press, 2007.

[125] Fu, Wai-Tat (2010). "Semantic imitation in social tagging". *ACM Transactions on Computer-Human Interaction*: 229. ISBN 9781605580074. doi:10.1145/1460563.1460600.

[126] Fu, Wai-Tat (August 2009). "A Semantic Imitation Model of Social Tagging.". *Proceedings of the IEEE conference on Social Computing*: 66–72.

[127] Valentin Robu, Harry Halpin, Hana Shepherd Emergence of consensus and shared vocabularies in collaborative tagging systems, ACM Transactions on the Web (TWEB), Vol. 3(4), article 14, ACM Press, September 2009.

[128] Carlos J. Costa, January 2012. "Article on Wall-it project". 2012.

[129] John A.L. Banks. *Negotiating Participatory Culture in the New Media Environment: Auran and the Trainz Online Community – An (Im)possible Relation*, The University of Queensland, School of English, Media Studies and Art History. MelbourneDAC2003

[130] L, Lessig,(2006)Code Version 2.0 (2nd ed.). New York: Basic Books.

[131] Bray, DA & Konsynski, BR. 2007, *Virtual Worlds, Virtual Economies, Virtual Institutions*, viewed 10 October 2008, p. 1-27 <http://ssrn.com/abstract=962501>

[132] Gosney, J.W. 2005, *Beyond Reality: A Guide to Alternate Reality Gaming*, Thomson Course Technology, Boston.

[133] Lee, Sang M., et al. "Success factors of platform leadership in web 2.0 service business." Service Business 4.2 (2010): 89-103.

[134] DNews (3 June 2015). "Swarms of Humans Power A.I. Platform".

[135] Rosenberg, Louis B.; A.I., Unanimous, Francisco, San; California; USA (1 January 2016). "Human Swarms, a real-time method for collective intelligence". pp. 658–659. doi:10.7551/978-0-262-33027-5-ch117.

[136] Rosenberg, L.B., "Human swarming, a real-time method for parallel distributed intelligence," in Swarm/Human Blended Intelligence Workshop (SHBI), 2015, vol., no., pp.1-7, 28-29 Sept. 2015 doi: 10.1109/SHBI.2015.7321685

[137] "Artificial intelligence turns $20 into $11,000 in Kentucky Derby bet". 10 May 2016.

[138] Ohlheiser, Abby (2 June 2016). "What happened when an A.I. hive mind answered Reddit's burning politics questions" – via washingtonpost.com.

[139] "Digital Photography Review". *www.dpreview.com*. Retrieved 2016-12-07.

[140] "Camera reviews, DSLR reviews, lens reviews, photography guides | Cameralabs". *www.cameralabs.com*. Retrieved 2016-12-07.

[141] Luckin, R., du Boulay, B., Smith, H., Underwood, J., Fitzpatrick, G., Holmberg, J., Kerawalla, L., Tunley, H., Brewster, D. and Pearce, D. (2005), 'Using Mobile Technology to Create Flexible Learning Contexts '. Journal of Interactive Media in Education, 22.

[142] Luckin, R. (2006). Understanding Learning Contexts as Ecologies of Resources: From the Zone of Proximal Development to Learner Generated Contexts. Paper presented at the Proceedings of World Conference on Elearning in Corporate, Government, Healthcare, and Higher Education 2006.

[143] Luckin, R., Shurville, S. and Browne, T. (2007), 'Initiating elearning by stealth, participation and consultation in a late majority institution'. Organisational Transformation and Social Change, 3, 4, 317–332.

[144] Flew, Terry and Humphreys, Sal (2005) "Games: Technology, Industry, Culture" in Terry Flew, New Media: An Introduction (2nd edn). Oxford University Press, South Melbourne 101-114.

[145] Henry Jenkins (2002) in Flew, Terry and Humphreys, Sal (2005) *Games: Technology, Industry, Culture* in Terry Flew, New Media: An Introduction (2nd edn), Oxford University Press, South Melbourne 101-114.

[146] Vera, Dusya; Crossan, Mary (2004-06-01). "Theatrical Improvisation: Lessons for Organizations". *Organization Studies*. **25** (5): 727–749. doi:10.1177/0170840604042412.

[147] *. R. Keith Sawyer (2004-06-01). "Improvised lessons: collaborative discussion in the constructivist classroom". *Teaching Education*. **15** (2): 189–201. doi:10.1080/1047621042000213610.

[148] Wolpert, David H.; Tumer, Kagan; Frank, Jeremy (1999-05-10). "Using Collective Intelligence to Route Internet Traffic". *Advances in Information Processing Systems, eds M. Kearns, S. Solla, D. Cohn, MIT Press.* **11** (1999). arXiv:cs/9905004.

[149] Held, David; McGrew, Anthony (2007-11-19). *Globalization / Anti-Globalization: Beyond the Great Divide*. Polity. ISBN 9780745639116.

[150] "'Anti-Globals' Use Internet As Collective Intelligence - UK Indymedia". *www.indymedia.org.uk*. Retrieved 2016-12-11.

[151] Jan Marco Leimeister, Michael Huber, Ulrich Bretschneider, Helmut Krcmar (2009): Leveraging Crowdsourcing: Activation-Supporting Components for IT-Based Ideas Competition. In: Journal of Management Information Systems (2009), Volume: 26, Issue: 1, Publisher: M.E. Sharpe Inc., Pages: 197–224, ISSN 0742-1222, doi:10.2753/MIS0742-1222260108 . Winfried Ebner; Jan Marco Leimeister; Helmut Krcmar (2009): Community Engineering for Innovations – The Ideas Competition as a method to nurture a Virtual Community for Innovations. In: R&D Management, 39 (4), pp 342–356 doi:10.1111/j.1467-9310.2009.00564.x

[152] Ebner, Winfried; Leimeister, Jan Marco; Krcmar, Helmut (2009-09-01). "Community engineering for innovations: the ideas competition as a method to nurture a virtual community for innovations". *R&D Management*. **39** (4): 342–356. doi:10.1111/j.1467-9310.2009.00564.x.

[153] Abdul-Karim, Kashif (2015-11-23). *In Search of The Black Seed*. Lulu.com. ISBN 9781329694897.

[154] "Capitalism and Social Progress by Phillip Brown and Hugh Lauder (cont...)". *The Guardian*. 2001-06-18. Retrieved 2016-12-12.

[155] "John Zerzan: anti-civilization theorist, writer and speaker". *www.johnzerzan.net*. Retrieved 2016-12-12.

[156] Moore, Jason W. (2001-01-01). Arrighi, Giovanni; Silver, Beverly J., eds. "Globalization in Historical Perspective". *Science & Society*. **65** (3): 386–397. JSTOR 40403938. doi:10.1521/siso.65.3.386.17767.

[157] "metamorphoptics". *metamorphoptics.blogspot.com*. Retrieved 2016-12-12.

[158] Joy, Bill. "Why the Future Doesn't Need Us". *WIRED*. Retrieved 2016-12-12.

7.9 Bibliography

- Brown, Philip; Lauder, Hugh (2000). "Collective intelligence". In S. Baron, J. Field & T Schuller. *Social Capital: Critical Perspectives*. New York: Oxford University Press.

- Brown, Philip; Lauder, Hugh (2001). "Collective intelligence (chapter 13)". In Brown & Lauder. *Capitalism and social progress: the future of society in a global economy*. Palgrave.

- Fadul, Jose A. (2009). "Collective Learning: Applying Distributed Cognition for Collective Intelligence". *The International Journal of Learning*. Common Ground. **16** (4): 211–220. doi:10.18848/1447-9494/cgp/v16i04/46223.

- CIA. (2008). *The World Factbook*. (accessed 3 September 2008)

- Flew, Terry (2008). *New Media: an introduction*. Melbourne: Oxford University Press.

- Hofstadter, Douglas (1979). *Gödel, Escher, Bach: an Eternal Golden Braid*. New York: Basic Books.

- Leiner, Barry, Cerf, Vinton, Clark, David, Kahn, Robert, Kleinrock, Leonard, Lynch, Daniel, Postel, Jon, Roberts, Larry and Wolff, Stephen. 2003. *A Brief History of the Internet*. Version 3.32 (accessed 3 September 2008)

- Noubel, Jean-François; (2004, rev. 2007), "Collective Intelligence: the Invisible Revolution"

- Por, George (1995). "The Quest for Collective intelligence". In K. Gozdz. *Community Building: Renewing Spirit and Learning in Business*. San Francisco: New Leaders Press.

- Rheingold, Howard (2002). *Smart Mobs: The Next Social Revolution*. Basic Books.

- Ron, Sun (1979). *Cognition and Multi-Agent Interaction*. Cambridge University Press.

- Rosenberg, L. (2015). Human Swarms, a real time method for Collective Intelligence. Proceedings of the European Conference on Artificial Life (ECAL 2015), pp. 658–659.

- Riedl, Christoph; Blohm, Ivo; Leimeister, Jan Marco; Krcmar, Helmut (2010). "Rating Scales for Collective Intelligence in Innovation Communities: Why Quick and Easy Decision Making Does Not Get It Right" (PDF).

- Leimeister, Jan Marco (2010). *Collective "Intelligence"*.

- Roy Chowdhury, Soudip; Rodriguez, Carlos; Daniel, Florian; Casati, Fabio (2010). "Wisdom-aware computing: on the interactive recommendation of composition knowledge".

- Stephen R. Diasio, Nuria Agell. "The evolution of expertise in decision support technologies: A challenge for organizations," cscwd, pp. 692–697, 2009 13th International Conference on Computer Supported Cooperative Work in Design, 2009. http://www.computer.org/portal/web/csdl/doi/10.1109/CSCWD.2009.4968139

- Raven, Jean, ed. (2008). *Uses and Abuses of Intelligence*. Unionville (NY): Royal Fireworks Press. ISBN 978-0-89824-356-7. Lay summary (6 July 2010).

- Kaiser, C., Kröckel, J., Bodendorf, F. (2010). Swarm Intelligence for Analyzing Opinions in Online Communities. Proceedings of the 43rd Hawaii International Conference on System Sciences, pp. 1–9.

- Hideyasu Sasaki (2010). International Journal of Organizational and Collective Intelligence (IJOCI), vol 1 No. 1.

- Olivier Zara, *Managing Collective Intelligence, Toward a New Corporate Governance*, Axiopole editions, 2004

- The collective intelligence framework, open-source framework for leveraging collective intelligence

- Raimund Minichbauer (2012). Fragmented Collectives. On the Politics of "Collective Intelligence" in Electronic Networks, transversal 01 12, 'unsettling knowledges'

7.10 External links

- Blog of Collective Intelligence

- GFIS – Global Futures Intelligence System

- CIRI – the Collective Intelligence Research Institute - a R&D non-profit organization on collective intelligence

- An application of Collective Intelligence for the Global Climate Change Situation Room designed and implemented by The Millennium Project in Gimcheon, South Korea in 2009.

- MIT Handbook of Collective Intelligence

- Cultivating Society's Civic Intelligence Doug Schuler *Journal of Society, Information and Communication*, vol 4 No. 2.

- Jennifer H. Watkins (2007). Prediction Markets as an Aggregation Mechanism for Collective Intelligence Los Alamos National Laboratory article on Collective Intelligence

Chapter 8

Collective action

Collective action refers to action taken together by a group of people whose goal is to enhance their status and achieve a common objective.[1] It is enacted by a representative of the group.[2] It is a term that has formulations and theories in many areas of the social sciences including psychology, sociology, anthropology, political science and economics.

8.1 The social identity model

Researchers Martijn van Zomeren, Tom Postmes, and Russell Spears conducted a meta-analysis of over 180 studies of collective action, in an attempt to integrate three dominant socio-psychological perspectives explaining antecedent conditions to this phenomenon – injustice, efficacy, and identity.[3] In their resultant 2008 review article, an integrative Social Identity Model of Collective Action (SIMCA) was proposed which accounts for interrelationships among the three predictors as well as their predictive capacities for collective action.[3] An important assumption of this approach is that people tend to respond to subjective states of disadvantage, which may or may not flow from objective physical and social reality.

8.1.1 Perceived injustice

Examining collective action through perceived injustice was initially guided by relative deprivation theory (RDT). RDT focuses on a subjective state of unjust disadvantage, proposing that engaging in fraternal (group-based) social comparisons with others may result in feelings of relative deprivation that foster collective action. Group-based emotions resulting from perceived injustice, such as anger, are thought to motivate collective action in an attempt to rectify the state of unfair deprivation.[3] The extent to which individuals respond to this deprivation involves several different factors and varies from extremely high to extremely low across different settings.[4] Meta-analysis results confirm that effects of injustice causally predict collective action, highlighting the theoretical importance of this variable.[3]

8.1.2 Perceived efficacy

Moving beyond RDT, scholars suggested that in addition to a sense of injustice, people must also have the objective, structural resources necessary to mobilize change through social protest. An important psychological development saw this research instead directed towards subjective expectations and beliefs that unified effort (collective action) is a viable option for achieving group-based goals – this is referred to as perceived collective efficacy. Empirically, collective efficacy is shown to causally affect collective action among a number of populations across varied contexts.[3]

8.1.3 Social identity

Social identity theory (SIT) suggests that people strive to achieve and maintain positive social identities associated with their group memberships.[5] Where a group membership is disadvantaged (for example, low status), SIT implicates three variables in the evocation of collective action to improve conditions for the group – permeability of group boundaries,[2] legitimacy of the intergroup structures, and the stability of these relationships. For example, when disadvantaged groups perceive intergroup status relationships as illegitimate and unstable, collective action is predicted to occur, in an attempt to change status structures for the betterment of the disadvantaged group.

Meta-analysis results also confirm that social identity causally predicts collective action across a number of diverse contexts. Additionally, the integrated SIMCA affords another important role to social identity – that of a psychological bridge forming the collective base from which both collective efficacy and group injustice may be conceived.

8.1.4 Model refinement

While there is sound empirical support for the causal importance of SIMCA's key theoretical variables on collective action,[3] more recent literature has addressed the issue of reverse causation, finding support for a related, yet distinct, encapsulation model of social identity in collective action (EMSICA).[6] This model suggests that perceived group efficacy and perceived injustice provide the basis from which social identity emerges, highlighting an alternative causal pathway to collective action. Recent research has sought to integrate SIMCA with intergroup contact theory (see Cakal, Hewstone, Schwär, & Heath[7]) and others have extended SIMCA through bridging morality research with the collective action literature (see van Zomeren, Postmes, & Spears[8] for a review).

8.2 Public good

The economic theory of collective action is concerned with the provision of public goods (and other collective consumption) through the collaboration of two or more individuals, and the impact of externalities on group behavior. It is more commonly referred to as Public Choice. Mancur Olson's 1965 book *The Logic of Collective Action: Public Goods and the Theory of Groups*, is an important early analysis of the problems of public good cost.

Besides economics, the theory has found many applications in political science, sociology, communication, anthropology and environmentalism.

8.2.1 Collective action problem

The term "collective action problem" describes the situation in which multiple individuals would all benefit from a certain action, but has an associated cost making it implausible that any individual can or will undertake and solve it alone. The ideal solution is then to undertake this as a collective action, the cost of which is shared. Situations like this include the prisoner's dilemma, a collective action problem in which no communication is allowed, the free rider problem, and the tragedy of the commons, also known as the problem with open access.[9] An allegorical metaphor often used to describe the problem is "belling the cat".[10]

Solutions to collective action problems include mutually binding agreements, government regulation, privatisation, and assurance contracts, also known as crowdacting.[11]

8.2.2 Exploitation of the great by the small

Mancur Olson made the claim that individual rational choice leads to situations where individuals with more resources will carry a higher burden in the provision of the public good than poorer ones.[12] Poorer individuals will usually have little choice but to opt for the free rider strategy, i.e., they will attempt to benefit from the public good without contributing to its provision. This may also encourage the under-production (inefficient production) of the public good.

8.2.3 Institutional design

While public goods are often provided by governments, this is not always the case. Various institutional designs have been studied with the aim of reducing the collaborative failure. The best design for a given situation depends on the production costs, the utility function, and the collaborative effects, amongst other things. Here are only some examples:

Joint products

A joint-product model analyzes the collaborative effect of joining a private good to a public good. For example, a tax deduction (private good) can be tied to a donation to a charity (public good).

It can be shown that the provision of the public good increases when tied to the private good, as long as the private good is provided by a monopoly (otherwise the private good would be provided by competitors without the link to the public good).

Clubs

Some institutional design, e.g., intellectual property rights, can introduce an exclusion mechanism and turn a pure public good into an impure public good artificially.

If the costs of the exclusion mechanism are not higher than the gain from the collaboration, clubs can emerge. James M. Buchanan showed in his seminal paper that clubs can be an efficient alternative to government interventions.[13]

A nation can be seen as a club whose members are its citizens. Government would then be the manager of this club.

Federated structure

In some cases, theory shows that collaboration emerges spontaneously in smaller groups rather than in large ones (see e.g. Dunbar's number). This explains why labor unions or charities often have a federated structure.

8.3 In philosophy

Over the past twenty years or so analytic philosophers have been exploring the nature of collective action in the sense of acting together, as when people paint a house together, go for a walk together, or together execute a pass play. These particular examples have been central for three of the philosophers who have made well known contributions to this literature: Michael Bratman, Margaret Gilbert, and John Searle, respectively.

In (Gilbert 1989) and subsequent articles and book chapters including Gilbert (2006, chapter 7) Gilbert argues for an account of collective action according to which this rests on a special kind of interpersonal commitment, what Gilbert calls a "joint commitment". A joint commitment in Gilbert's sense is not a matter of a set of personal commitments independently created by each of the participants, as when each makes a personal decision to do something. Rather, it is a single commitment to whose creation each participant makes a contribution. Thus suppose that one person says "Shall we go for a walk?" and the other says "Yes, let's". Gilbert proposes that as a result of this exchange the parties are jointly committed to go for a walk, and thereby obligated to one another to act as if they were parts of a single person taking a walk. Joint commitments can be created less explicitly and through processes that are more extended in time. One merit of a joint commitment account of collective action, in Gilbert's view, is that it explains the fact that those who are out on a walk together, for instance, understand that each of them is in a position to demand corrective action of the other if he or she acts in ways that affect negatively the completion of their walk. In (Gilbert 2006a) she discusses the pertinence of joint commitment to collective actions in the sense of the theory of rational choice.

In Searle (1990) Searle argues that what lies at the heart of a collective action is the presence in the mind of each participant of a "we-intention". Searle does not give an account of we-intentions or, as he also puts it, "collective intentionality", but insists that they are distinct from the "I-intentions" that animate the actions of persons acting alone.

In Bratman (1993) Bratman proposed that, roughly, two people "share an intention" to paint a house together when each intends that the house is painted by virtue of the activity of each, and also intends that it is so painted by virtue of the intention of each that it is so painted. That these conditions obtain must also be "common knowledge" between the participants.

Discussion in this area continues to expand, and has influenced discussions in other disciplines including anthropology, developmental psychology, and economics. One general question is whether it is necessary to think in terms that go beyond the personal intentions of individual human beings properly to characterize what it is to act together. Bratman's account does not go beyond such personal intentions. Gilbert's account, with its invocation of joint commitment, does go beyond them. Searle's account does also, with its invocation of collective intentionality. The question of whether and how one must account for the existence of mutual obligations when there is a collective intention is another of the issues in this area of inquiry.

8.4 Spontaneous consensus

In addition to the psychological mechanisms of collective action as explained by the social identity model, researchers have developed sociological models of why collective action exists and have studied under what conditions collective action emerges.[14] Along this social dimension, a special case of the general collective action problem is one of collective agreement: how does a group of agents (humans, animals, robots, etc.) reach consensus about a decision or belief in the absence of central organization? Common examples can be found from domains as diverse as biology (flocking, shoaling and schooling, and general collective animal behavior), economics (stock market bubbles), and sociology (social conventions and norms) among others.

Consensus is distinct from the collective action problem in that there often is not an explicit goal, benefit, or cost of action but rather it concerns itself with a social equilibrium of the individuals involved (and their beliefs). And it can be considered *spontaneous* when it emerges without the presence of a centralized institution among self-interested individuals.[15]

8.4.1 Dimensions

Spontaneous consensus can be considered along 4 dimensions involving the social structure of the individuals participating (local versus global) in the consensus as well as the processes (competitive vs cooperative) involved in reaching consensus:[14]

- Competitive
- Cooperative
- Local
- Global

Competitive versus cooperative

The underlying processes of spontaneous consensus can be viewed either as *cooperation* among individuals trying to coordinate themselves through their interactions or as *competition* between the alternatives or choices to be decided upon.[14] Depending on the dynamics of the individuals involved as well as the context of the alternatives considered for consensus, the process can be wholly cooperative, wholly competitive, or a mix of the two.

Local versus global

The distinction between local and global consensus can be viewed in terms of the social structure underlying the network of individuals participating in the consensus making process. *Local consensus* occurs when there is agreement between groups of neighboring nodes while *global consensus* refers to the state in which most of the population has reached an agreement.[14] How and why consensus is reached is dependent on both the structure of the social network of individuals as well as the presence (or lack) of centralized institutions.

8.4.2 Equilibrium mechanisms

There are many mechanisms (social and psychological) that have been identified to underlie the consensus making process.[14] They have been used to both explain the emergence of spontaneous consensus and understand how to facilitate an equilibrium between individuals and can be grouped according to their role in the process.

- Facilitation of Equilibrium
    - Communication[16]
    - Punishment of Deviants[17]
    - Positive Payoffs[18]
    - Conformity Bias[19]
- Selection of Alternatives
    - Logical Reflection[20]
    - Psychological and shared biases[18]
    - Chance (when all alternatives are equivalent)[21]

8.4.3 Methods and techniques

Due to the interdisciplinary nature of both the mechanisms as well as the applications of spontaneous consensus, a variety of techniques have been developed to study the emergence and evolution of spontaneous cooperation. Two of the most widely used are game theory and social network analysis.

Game theory

Traditionally game theory has been used to study zero-sum games but has been extended to many different types of games. Relevant to the study of spontaneous consensus are cooperative and non-cooperative games. Since a consensus must be reached without the presence of any external authoritative institution for it to be considered *spontaneous*, non-cooperative games and nash equilibrium have been the dominant paradigm for which to study its emergence.

In the context of non-cooperative games, a consensus is a formal nash equilibrium that all players tend towards through self-enforcing alliances or agreements.

Social network analysis

An alternative approach to studying the emergence of spontaneous consensus—that avoids many of the unnatural or overly constrained assumptions of game theoretic models—is the use of network based methods and social network analysis (SNA). These SNA models are theoretically grounded in the communication mechanism[16] of facilitating consensus and describe its emergence through the information propagation processes of the network (behavioral contagion). Through the spread of influence (and ideas) between agents participating in the consensus, local and global consensus can emerge if the agents in the network achieve a shared equilibrium state. Leveraging this model of consensus, researchers have shown that local peer influence can be used to reach a global consensus and cooperation across the entire network.[22] While this model of consensus and cooperation has been shown to be successful in certain contexts, research suggest that communication and social influence cannot be fully captured by simple contagion models[23] and as such a pure contagion based model of consensus may have limits.

8.5 See also

- Collaborative innovation network
- Collective intelligence
- Collective intentionality
- Common property resource
- Constitutional economics
- Coordination good

- Free rider problem
- Group action (sociology)
- Mass collaboration
- Nash equilibrium
- Pareto efficiency
- Polytely
- Prisoner's dilemma
- Private-collective model of innovation
- Public good
- Social fact
- Tragedy of the commons
- Tragedy of the anticommons

8.6 Footnotes

[1] "collective action problem - collective action". *Encyclopædia Britannica*.

[2] Stephen C. Wright; Donald M. Taylor; Fathali M. Moghaddam (June 1990). "Responding to Membership in a Disadvantaged Group: From Acceptance to Collective Protest". *Journal of Personality and Social Psychology*. **58** (6): 994–1003. doi:10.1037/0022-3514.58.6.994.

[3] van Zomeren, M.; Postmes, T.; Spears, R. (2008). "Toward an integrative social identity model of collective action: A quantitative research synthesis of three socio-psychological perspectives". *Psychological Bulletin*. **134** (4): 504–535. PMID 18605818. doi:10.1037/0033-2909.134.4.504.

[4] Ostrom, Elinor. "Collective Action and the Evolution of Social Norms". *The Journal of Economics Perspective*. **14** (3): 137–158. doi:10.1257/jep.14.3.137.

[5] Tajfel, H.; Turner, J.C. (1979). "An integrative theory of inter-group conflict. In W.G. Austin & S. Worchel (Eds.)". *The social psychology of inter-group relations*. Monterey, CA: Brooks/Cole: 33–47.

[6] Thomas, E.F.; Mavor, K.I.; McGarty, C. (2011). "Social identities facilitate and encapsulate action-relevant constructs: A test of the social identity model of collective action". *Group Processes and Intergroup Relations*. **15** (1): 75–88. doi:10.1177/1368430211413619.

[7] Cakal, H.; Hewstone, M.; Schwär, G.; Heath, A. (2011). "An investigation of the social identity model of collective action and the 'sedative' effect of intergroup contact among Black and White students in South Africa". *British Journal of Social Psychology*. **50** (4): 606–627. PMID 22122025. doi:10.1111/j.2044-8309.2011.02075.x.

[8] van Zomeren, M.; Postmes, T.; Spears, R. (2012). "On conviction's collective consequences: Integrating moral conviction with the social identity model of collective action". *British Journal of Social Psychology*. **51** (1): 52–71. PMID 22435846. doi:10.1111/j.2044-8309.2010.02000.x.

[9] Agar, Jesse. "Tragedy of the Commons | The Problem with Open Access". *Youtube*. This Place (youtube channel). Retrieved Jun 9, 2015.

[10] Dowding, Keith (1996). *Power*. University of Minnesota Press. pp. 31 *ff*. ISBN 0-8166-2941-2

[11] van den Akker, Ron. "Crowdacting". *YouTube*. Bord&Stift. Retrieved Sep 7, 2016.

[12] Olson, Mancur (1965). "logic of collective action: Public goods and the theory of groups". Cambridge, MA: Harvard University Press

[13] Buchanan, James M. (1965). "An Economic Theory of Clubs". *Economica*. **32** (125): 1–14. JSTOR 2552442. doi:10.2307/2552442.

[14] Baronchelli, Andrea (2017). "The Emergence of Consensus". arXiv:1704.07767 [physics.soc-ph].

[15] Sugden, Robert (1989). "Spontaneous order". *The Journal of Economic Perspectives*. **3** (4): 85–97. doi:10.1257/jep.3.4.85.

[16] Garrod, Simon; Doherty, Gwyneth (1994). "Conversation, co-ordination and convention: An empirical investigation of how groups establish linguistic conventions". *Cognition*. **53** (3): 181–215. PMID 7842633. doi:10.1016/0010-0277(94)90048-5.

[17] Boyd, Robert; Richerson, Peter (1992). "Punishment allows the evolution of cooperation (or anything else) in sizable groups". *Ethology and sociobiology*. **13** (3): 171–195. doi:10.1016/0162-3095(92)90032-Y.

[18] Schelling, Thomas (1960). *The strategy of conflict*. Harvard University Press. ISBN 9780674840317.

[19] Asch, Solomon (1955). "Opinions and social pressure". In Aronson, Elliot. *Readings about the social animal*. Macmillan. pp. 17–26. ISBN 9780716759669.

[20] Harsanyi, John; Selten, Reinhard (1988). *A general theory of equilibrium selection in games*. MIT Press Books. p. 1. ISBN 9780262582384.

[21] Young, H Peyton (1996). "The economics of convention". *The Journal of Economic Perspectives*. **10** (2): 105–122. JSTOR 2138484. doi:10.1257/jep.10.2.105.

[22] Mani, Ankur; Rahwan, Iyad; Pentland, Alex (2013). "Inducing Peer Pressure to Promote Cooperation". *Scientific Reports*. **3** (1735). doi:10.1038/srep01735.

[23] Alshamsi, Aamena; Pianesi, Fabio; Lepri, Bruno; Pentland, Alex; Rahwan, Iyad (2015). "Beyond Contagion: Reality Mining Reveals Complex Patterns of Social Influence". *PLOS ONE.* **10** (8): e0135740. doi:10.1371/journal.pone.0135740.

8.7 Bibliography

- Bratman, Michael (October 1993). "Shared intention". *Ethics.* Oxford Journals. **104** (1): 97–113. JSTOR 2381695. doi:10.1086/293577.

- Dolata, Ulrich; Schrape, Jan-Felix (2015). "Masses, Crowds, Communities, Movements: Collective Action in the Internet Age". *Social Movement Studies.* **15**: 1–18. doi:10.1080/14742837.2015.1055722.

- Gilbert, Margaret (1989). *On social facts.* London New York: Routledge. ISBN 9780415024440.

- Gilbert, Margaret (2006a). "Rationality in Collective Action". *Philosophy of the Social Sciences.* **36** (1): 3–17. doi:10.1177/0048393105284167

- Gilbert, Margaret (2006). *A theory of political obligation : membership, commitment, and the bonds of society.* Oxford Oxford New York: Clarendon Press Oxford University Press. ISBN 9780199274956.

- Hardin, Russell (1982). *Collective action.* Baltimore: Johns Hopkins University Press. ISBN 9780801828195.

- Meinzen-Dick, Ruth Suseela; di Gregorio, Monica, eds. (2004). *Collective action and property rights for sustainable development.* Washington, DC: International Food Policy Research Institute. 2020 Focus No. 11. Pdf.

- Olson, Mancur (1965). *The logic of collective action: public goods and the theory of groups.* Cambridge, Massachusetts: Harvard University Press. ISBN 9780674537514.

- Ostrom, Elinor (1990). *Governing the commons: the evolution of institutions for collective action.* Cambridge New York: Cambridge University Press. ISBN 9780521405997.

- Searle, John R. (2002), "Collective intentions and actions", in Searle, John R., *Consciousness and language*, New York: Cambridge University Press, pp. 90–105, ISBN 9780521597449.

- van Winden, Frans (December 2015). "Political economy with affect: on the role of emotions and relationships in political economics". *European Journal of Political Economy.* Elsevier. **40** (B): 298–311. doi:10.1016/j.ejpoleco.2015.05.005.

Chapter 9

Self-organized criticality

In physics, **self-organized criticality** (SOC) is a property of dynamical systems that have a critical point as an attractor. Their macroscopic behaviour thus displays the spatial and/or temporal scale-invariance characteristic of the critical point of a phase transition, but without the need to tune control parameters to a precise value, because the system, effectively, tunes itself as it evolves towards criticality.

The concept was put forward by Per Bak, Chao Tang and Kurt Wiesenfeld ("BTW") in a paper[1] published in 1987 in *Physical Review Letters*, and is considered to be one of the mechanisms by which complexity[2] arises in nature. Its concepts have been enthusiastically applied across fields as diverse as geophysics,[3] physical cosmology, evolutionary biology and ecology, bio-inspired computing and optimization (mathematics), economics, quantum gravity, sociology, solar physics, plasma physics, neurobiology[4][5][6] and others.

SOC is typically observed in slowly driven non-equilibrium systems with extended degrees of freedom and a high level of nonlinearity. Many individual examples have been identified since BTW's original paper, but to date there is no known set of general characteristics that *guarantee* a system will display SOC.

9.1 Overview

Self-organized criticality is one of a number of important discoveries made in statistical physics and related fields over the latter half of the 20th century, discoveries which relate particularly to the study of complexity in nature. For example, the study of cellular automata, from the early discoveries of Stanislaw Ulam and John von Neumann through to John Conway's Game of Life and the extensive work of Stephen Wolfram, made it clear that complexity could be generated as an emergent feature of extended systems with simple local interactions. Over a similar period of time, Benoît Mandelbrot's large body of work on fractals showed that much complexity in nature could be described by certain ubiquitous mathematical laws, while the extensive study of phase transitions carried out in the 1960s and 1970s showed how scale invariant phenomena such as fractals and power laws emerged at the critical point between phases. However, the term Self-Organized Criticality was firstly introduced by Bak, Tang and Wiesenfeld's 1987 paper which clearly linked together these factors: a simple cellular automaton was shown to produce several characteristic features observed in natural complexity (fractal geometry, pink (1/f) noise and power laws) in a way that could be linked to critical-point phenomena. Crucially, however, the paper emphasized that the complexity observed emerged in a robust manner that did not depend on finely tuned details of the system: variable parameters in the model could be changed widely without affecting the emergence of critical behaviour (hence, *self-organized* criticality). Thus, the key result of BTW's paper was its discovery of a mechanism by which the emergence of complexity from simple local interactions could be *spontaneous* — and therefore plausible as a source of natural complexity — rather than something that was only possible in the lab (or lab computer) where it was possible to tune control parameters to precise values. The publication of this research sparked considerable interest from both theoreticians and experimentalists, and important papers on the subject are among the most cited papers in the scientific literature.

Due to BTW's metaphorical visualization of their model as a "sandpile" on which new sand grains were being slowly sprinkled to cause "avalanches", much of the initial experimental work tended to focus on examining real avalanches in granular matter, the most famous and extensive such study probably being the Oslo ricepile experiment. Other experiments include those carried out on magnetic-domain patterns, the Barkhausen effect and vortices in superconductors. Early theoretical work included the development of a variety of alternative SOC-generating dynamics distinct from the BTW model, attempts to prove model properties analytically (including calculating the critical exponents[7][8]), and examination of the necessary conditions for SOC to emerge. One of the important issues for the latter investigation was whether conservation of energy was required in the local dynamical exchanges of models: the answer in general is no, but with (minor) reservations, as some exchange dynamics (such as those of BTW) do require local conservation at least on average. In the long term, key theoretical issues yet to be resolved include the calculation of the possible universality classes of SOC behaviour and the question of whether it is possible to derive a general rule for determining if an arbitrary algorithm displays SOC.

Alongside these largely lab-based approaches, many other investigations have centered around large-scale natural or social systems that are known (or suspected) to display scale-invariant behavior. Although these approaches were not always welcomed (at least initially) by specialists in the subjects examined, SOC has nevertheless become established as a strong candidate for explaining a number of natural phenomena, including: earthquakes (which, long before SOC was discovered, were known as a source of scale-invariant behavior such as the Gutenberg–Richter law describing the statistical distribution of earthquake sizes and the Omori law describing the frequency of aftershocks, and where models that displayed SOC were proposed and analyzed prior to the BTW 87 paper;[9][3]); solar flares; fluctuations in economic systems such as financial markets (references to SOC are common in econophysics); landscape formation; forest fires; landslides; epidemics; neuronal avalanches in cortex;[5][10] 1/f noise in the amplitude envelope of electrophysiological signals;[4] and biological evolution (where SOC has been invoked, for example, as the dynamical mechanism behind the theory of "punctuated equilibria" put forward by Niles Eldredge and Stephen Jay Gould). These "applied" investigations of SOC have included both attempts at modelling (either developing new models or adapting existing ones to the specifics of a given natural system), and extensive data analysis to determine the existence and/or characteristics of natural scaling laws.

The recent excitement generated by scale-free networks has raised some interesting new questions for SOC-related research: a number of different SOC models have been shown to generate such networks as an emergent phenomenon, as opposed to the simpler models proposed by network researchers where the network tends to be assumed to exist independently of any physical space or dynamics. While many single phenomena have been shown to exhibit scale free properties over narrow ranges, by far the most extensive data concern the solvent-accessible surface areas of globular proteins, plotted for modular segments centered on a specific amino acid, against lengths L, with $9 <= L <= 35$.[11] Because this fractal linearity holds for all 20 amino acids over the same range, it exists independently of any model. It can be regarded as the basic mechanism of life. These studies quantify the differential geometry of proteins, and resolve many evolutionary puzzles regarding the biological emergence of complexity.[12]

Despite the considerable interest and research output generated from the SOC hypothesis there remains no general agreement with regards to its mathematical mechanisms. Bak Tang and Wiesenfeld based their hypothesis on the behavior of their sandpile model.[1] However, this model was subsequently shown to actually generate $1/f^2$ noise rather than $1/f$ noise.[13] Other simulation models were proposed later that could produce true $1/f$ noise,[14], and experimental sandpile models were observed to yield $1/f$ noise.[15] In addition to the nonconservative theoretical model mentioned above, other theoretical models for SOC have been based upon information theory[16] and mean field theory.[17]

9.2 Examples of self-organized critical dynamics

In chronological order of development:

- Bak–Tang–Wiesenfeld sandpile
- Forest-fire model
- Olami–Feder–Christensen model
- Bak–Sneppen model

9.3 See also

- 1/f noise
- Complex systems
- Detrended fluctuation analysis, a method to detect power-law scaling in time series.
- Dual-phase evolution, another process that contributes to self-organization in complex systems.
- Fractals
- Power laws
- Scale invariance
- Self-organization
- Critical exponents
- Ilya Prigogine, a systems scientist who helped formalize dissipative system behavior in general terms.
- Red Queen hypothesis

- Self-organized criticality control
- Critical brain hypothesis

9.4 References

[1] Bak, P., Tang, C. and Wiesenfeld, K. (1987). "Self-organized criticality: an explanation of 1/f noise". *Physical Review Letters*. **59** (4): 381–384. Bibcode:1987PhRvL..59..381B. doi:10.1103/PhysRevLett.59.381. Papercore summary: http://papercore.org/Bak1987.

[2] Bak, P., and Paczuski, M. (1995). "Complexity, contingency, and criticality". *Proc Natl Acad Sci U S A*. **92** (15): 6689–6696. Bibcode:1995PNAS...92.6689B. PMC 41396 ⊚. PMID 11607561. doi:10.1073/pnas.92.15.6689.

[3] Smalley, R. F., Jr.; Turcotte, D. L.; Solla, S. A. (1985). "A renormalization group approach to the stick-slip behavior of faults". *Journal of Geophysical Research*. **90** (B2): 1894. Bibcode:1985JGR....90.1894S. doi:10.1029/JB090iB02p01894.

[4] K. Linkenkaer-Hansen; V. V. Nikouline; J. M. Palva & R. J. Ilmoniemi. (2001). "Long-Range Temporal Correlations and Scaling Behavior in Human Brain Oscillations". *J. Neurosci*. **21** (4): 1370–1377. PMID 11160408.

[5] J. M. Beggs & D. Plenz (2006). "Neuronal Avalanches in Neocortical Circuits". *J. Neurosci*. 23.

[6] Chialvo, D. R. (2004). "Critical brain networks". *Physica A*. **340** (4): 756–765. Bibcode:2004PhyA..340..756C. arXiv:cond-mat/0402538 ⊚. doi:10.1016/j.physa.2004.05.064.

[7] Tang, C. and Bak, P. (1988). "Critical exponents and scaling relations for self-organized critical phenomena". *Physical Review Letters*. **60** (23): 2347–2350. Bibcode:1988PhRvL..60.2347T. PMID 10038328. doi:10.1103/PhysRevLett.60.2347.

[8] Tang, C. and Bak, P. (1988). "Mean field theory of self-organized critical phenomena". *Journal of Statistical Physics*. **51** (5–6): 797–802. Bibcode:1988JSP....51..797T. doi:10.1007/BF01014884.

[9] Turcotte, D. L.; Smalley, R. F., Jr.; Solla, S. A. (1985). "Collapse of loaded fractal trees". *Nature*. **313** (6004): 671–672. Bibcode:1985Natur.313..671T. doi:10.1038/313671a0.

[10] Poil, SS; Hardstone, R; Mansvelder, HD; Linkenkaer-Hansen, K (Jul 2012). "Critical-state dynamics of avalanches and oscillations jointly emerge from balanced excitation/inhibition in neuronal networks". *Journal of Neuroscience*. **32** (29): 9817–23. PMC 3553543 ⊚. PMID 22815496. doi:10.1523/JNEUROSCI.5990-11.2012.

[11] Moret, M. A. and Zebende, G. (2007). "Amino acid hydrophobicity and accessible surface area". *Phys. Rev. E*. **75**: 011920. Bibcode:2007PhRvE..75a1920M. doi:10.1103/PhysRevE.75.011920.

[12] Phillips, J. C. (2014). "Fractals and self-organized criticality in proteins". *Physica A*. **415**: 440–448. Bibcode:2014PhyA..415..440P. doi:10.1016/j.physa.2014.08.034.

[13] Jensen, H. J., Christensen, K. and Fogedby, H. C. (1989). "1/f noise, distribution of lifetimes, and a pile of sand". *Phys. Rev. B*. **40**: 7425–7427. Bibcode:1989PhRvB..40.7425J. doi:10.1103/physrevb.40.7425.

[14] Maslov, S., Tang, C. and Zhang, Y. - C. (1999). "1/f noise in Bak-Tang-Wiesenfeld models on narrow stripes". *Phys. Rev. Lett*. **83**: 2449–2452. Bibcode:1999PhRvL..83.2449M. arXiv:cond-mat/9902074 ⊚. doi:10.1103/physrevlett.83.2449.

[15] Frette, V., Christinasen, K., Malthe-Sørenssen, A., Feder, J. Jossang, T and Meaken, P (1996). "Avalanche dynamics in a pile of rice". *Nature*. **379**: 49–52. Bibcode:1996Natur.379...49F. doi:10.1038/379049a0.

[16] Dewar, R. (2003). "Information theory explanation of the fluctuation theorem, maximum entropy production and self-organized criticality in non-equilibrium stationary states". *J. Phys. A: Math. Gen*. **36**: 631–641. Bibcode:2003JPhA...36..631D. arXiv:cond-mat/0005382 ⊚. doi:10.1088/0305-4470/36/3/303.

[17] Vespignani, A., and Zapperi, S. (1998). "How self-organized criticality works: a unified mean-field picture". *Phys. Rev. E*. **57**: 6345–6362. Bibcode:1998PhRvE..57.6345V. arXiv:cond-mat/9709192 ⊚. doi:10.1103/physreve.57.6345.

9.5 Further reading

- Adami, C. (1995). "Self-organized criticality in living systems". *Physics Letters A*. **203** (1): 29–32. Bibcode:1995PhLA..203...29A. doi:10.1016/0375-9601(95)00372-A.

- Bak, P. (1996). *How Nature Works: The Science of Self-Organized Criticality*. New York: Copernicus. ISBN 0-387-94791-4.

- Bak, P. and Paczuski, M. (1995). "Complexity, contingency, and criticality". *Proceedings of the National Academy of Sciences of the USA*. **92** (15): 6689–6696. Bibcode:1995PNAS...92.6689B. PMC 41396 ⊚. PMID 11607561. doi:10.1073/pnas.92.15.6689.

9.5. FURTHER READING

- Bak, P. and Sneppen, K. (1993). "Punctuated equilibrium and criticality in a simple model of evolution". *Physical Review Letters*. **71** (24): 4083–4086. Bibcode:1993PhRvL..71.4083B. PMID 10055149. doi:10.1103/PhysRevLett.71.4083.

- Bak, P., Tang, C. and Wiesenfeld, K. (1987). "Self-organized criticality: an explanation of $1/f$ noise". *Physical Review Letters*. **59** (4): 381–384. Bibcode:1987PhRvL..59..381B. doi:10.1103/PhysRevLett.59.381.

- Bak, P., Tang, C. and Wiesenfeld, K. (1988). "Self-organized criticality". *Physical Review A*. **38** (1): 364–374. Bibcode:1988PhRvA..38..364B. doi:10.1103/PhysRevA.38.364. Papercore summary.

- Buchanan, M. (2000). *Ubiquity*. London: Weidenfeld & Nicolson. ISBN 0-7538-1297-5.

- Jensen, H. J. (1998). *Self-Organized Criticality*. Cambridge: Cambridge University Press. ISBN 0-521-48371-9.

- Katz, J. I. (1986). "A model of propagating brittle failure in heterogeneous media". *Journal of Geophysical Research*. **91** (B10): 10412. Bibcode:1986JGR....9110412K. doi:10.1029/JB091iB10p10412.

- Kron, T./Grund, T. (2009). "Society as a Selforganized Critical System". *Cybernetics and Human Knowing*. **16**: 65–82.

- Paczuski, M. (2005). "Networks as renormalized models for emergent behavior in physical systems". The Science and Culture Series – Physics. Bibcode:2005cmn..conf..363P. ISBN 978-981-256-525-9. arXiv:physics/0502028. doi:10.1142/9789812701558_0042.

- Turcotte, D. L. (1997). *Fractals and Chaos in Geology and Geophysics*. Cambridge: Cambridge University Press. ISBN 0-521-56733-5.

- Turcotte, D. L. (1999). "Self-organized criticality". *Reports on Progress in Physics*. **62** (10): 1377–1429. Bibcode:1999RPPh...62.1377T. doi:10.1088/0034-4885/62/10/201.

- Md. Nurujjaman/A. N. Sekar Iyengar (2007). "Realization of {SOC} behavior in a dc glow discharge plasma". *Physics Letters A*. **360**: 717–721. Bibcode:2007PhLA..360..717N. arXiv:physics/0611069. doi:10.1016/j.physleta.2006.09.005.

- Self-organized criticality on arxiv.org

Chapter 10

Herd mentality

"Mob Mentality" and "Mob mentality" redirect here. For album, see Mob Mentality (album).

Herd mentality, or **mob mentality**, describes how people are influenced by their peers to adopt certain behaviors. Examples of the herd mentality include nationalism, globalism, stock market trends, superstition, and home décor. Social psychologists study the related topics of group intelligence, crowd wisdom, and decentralized decision making.

10.1 History

Herd mentality and herd behavior have been prevalent descriptors for human behavior since people began to form tribes, migrate in groups, and perform cooperative marketing and agricultural functions. The idea of a "group mind" or "mob behavior" was first put forward by 19th-century French social psychologists Gabriel Tarde and Gustave Le Bon. Herd behavior in human societies has also been studied by Sigmund Freud and Wilfred Trotter, whose book *Instincts of the Herd in Peace and War* is a classic in the field of social psychology. Sociologist and Economist Thorstein Veblen's *The Theory of the Leisure Class* illustrates how individuals imitate other group members of higher social status in their consumer behavior. More recently, Malcolm Gladwell in *The Tipping Point*, examines how cultural, social, and economic factors converge to create trends in consumer behavior. In 2004, the *New Yorker*'s financial columnist James Suroweicki published *The Wisdom of Crowds*.

21st-century academic fields such as marketing and behavioral finance attempt to identify and predict the rational and irrational behavior of investors. (See the work of Daniel Kahneman, Robert Shiller, Vernon L. Smith, and Amos Tversky.) Driven by emotional reactions such as greed and fear, investors can be seen to join in frantic purchasing and sales of stocks, creating bubbles and crashes. As a result, herd behavior is closely studied by behavioral finance experts in order to help predict future economic crises.[1]

10.2 Evidence based examples

Researchers at Leeds University performed a group experiments where volunteers were told to randomly walk around a large hall without talking to each other. A select few were then given more detailed instructions on where to walk. The scientists discovered that people end up blindly following one or two instructed people who appear to know where they're going. The results of this experiments showed that it only takes 5% of confident looking and instructed people to influence the direction of the 95% of people in the crowd and the 200 volunteers did this without even realizing it.[2]

10.3 See also

- Anonymity
- Argumentum ad populum
- Asch conformity experiments
- Bandwagon effect
- Collective intelligence
- Conformity
- Critical mass (sociodynamics)
- Crowd abuse
- Crowd psychology
- Decentralized decision making
- Delphi method
- Early adopter
- Freethought
- Group intelligence
- Groupthink

- Herd behavior
- Information cascade
- Monkey see, monkey do
- Opinion leadership
- Peer pressure
- Predictive market
- Religious paranoia
- Sheeple
- Social network
- *The Wisdom of Crowds*
- Trial by media

Philosophers

- Søren Kierkegaard
- Friedrich Nietzsche
- José Ortega y Gasset
- Everett Dean Martin

10.4 References

[1] Fromlet, Hubert. "Predictability of Financial Crises: Lessons from Sweden for Other Countries." Business Economics 47.4

[2] "Sheep in human clothing - scientists reveal our flock mentality". University of Leeds Press Office. 14 February 2008.

10.4.1 Further reading

- Bloom, Howard, *The Global Brain: The Evolution of Mass Mind from the Big Bang to the 21st Century*. (2000) John Wiley & Sons, New York.
- Freud, Sigmund's *Massenpsychologie und Ich-Analyse* (1921; English translation *Group Psychology and the Analysis of the Ego*, *1922). Reprinted 1959 Liveright, New York.
- Gladwell, Malcolm, *The Tipping Point: How Little Things Can Make a Big Difference*. (2002) Little, Brown & Co., Boston.
- Le Bon, Gustav, *Les Lois psychologiques de l'évolution des peuples*. (1894) National Library of France, Paris.
- Le Bon, Gustave. *The Crowd: A Study of the Popular Mind*. (1895) Project Gutenberg.
- Martin, Everett Dean, *The Behavior of Crowds* (1920).
- McPhail, Clark. The Myth of the Madding Crowd (1991) Aldine-DeGruyter.
- Trotter, Wilfred, *Instincts of the Herd in Peace and War*. (1915) Macmillan, New York.
- Suroweicki, James: *The Wisdom of Crowds: Why the Many Are Smarter Than the Few and How Collective Wisdom Shapes Business, Economies, *Societies and Nations*. (2004) Little, Brown, Boston.
- Sunstein, Cass, *Infotopia: How Many Minds Produce Knowledge*. (2006) Oxford University Press, Oxford, United Kingdom.

10.5 External links

- The Wisdom of Crowds and Iowa Electronic Market Statistics

Chapter 11

Phase transition

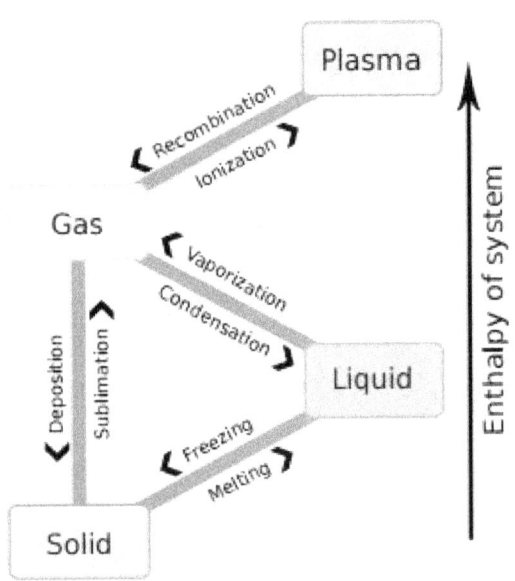

This diagram shows the nomenclature for the different phase transitions.

11.1 Types of phase transition

Examples of phase transitions include:

- The transitions between the solid, liquid, and gaseous phases of a single component, due to the effects of temperature and/or pressure:

  - (see also vapor pressure and phase diagram)

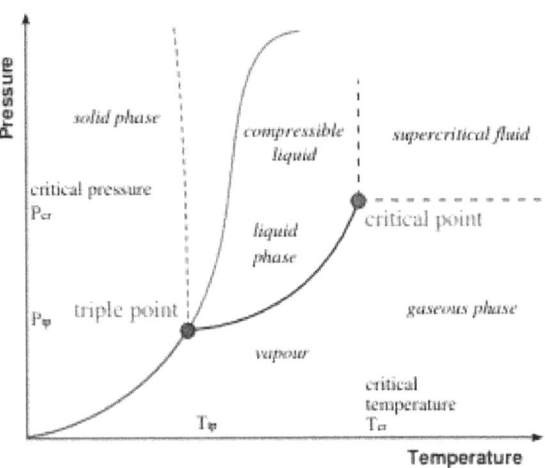

A typical phase diagram. The dotted line gives the anomalous behavior of water.

- A eutectic transformation, in which a two component single phase liquid is cooled and transforms into two solid phases. The same process, but beginning with a solid instead of a liquid is called a eutectoid transformation.

- A peritectic transformation, in which a two component single phase solid is heated and transforms into a solid phase and a liquid phase.

The term **phase transition** (or **phase change**) is most commonly used to describe transitions between solid, liquid and gaseous states of matter, and, in rare cases, plasma (physics). A phase of a thermodynamic system and the states of matter have uniform physical properties. During a phase transition of a given medium certain properties of the medium change, often discontinuously, as a result of the change of some external condition, such as temperature, pressure, or others. For example, a liquid may become gas upon heating to the boiling point, resulting in an abrupt change in volume. The measurement of the external conditions at which the transformation occurs is termed the phase transition. Phase transitions are common in nature and used today in many technologies.

11.1. TYPES OF PHASE TRANSITION

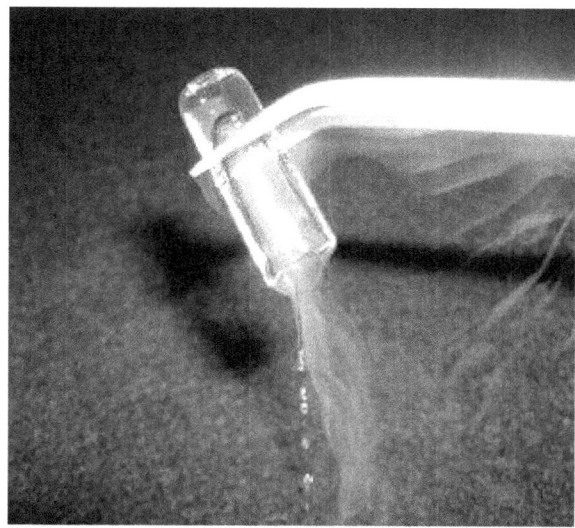

A small piece of rapidly melting solid argon simultaneously shows the transitions from solid to liquid and liquid to gas.

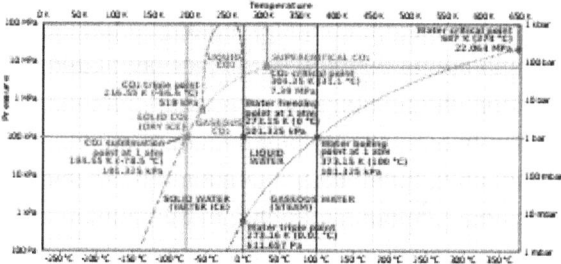

Comparison of phase diagrams of carbon dioxide (red) and water (blue) explaining their different phase transitions at 1 atmosphere

- A spinodal decomposition, in which a single phase is cooled and separates into two different compositions of that same phase.
- Transition to a mesophase between solid and liquid, such as one of the "liquid crystal" phases.
- The transition between the ferromagnetic and paramagnetic phases of magnetic materials at the Curie point.
- The transition between differently ordered, commensurate or incommensurate, magnetic structures, such as in cerium antimonide.
- The martensitic transformation which occurs as one of the many phase transformations in carbon steel and stands as a model for displacive phase transformations.
- Changes in the crystallographic structure such as between ferrite and austenite of iron.
- Order-disorder transitions such as in alpha-titanium aluminides.

- The dependence of the adsorption geometry on coverage and temperature, such as for hydrogen on iron (110).
- The emergence of superconductivity in certain metals and ceramics when cooled below a critical temperature.
- The transition between different molecular structures (polymorphs, allotropes or polyamorphs), especially of solids, such as between an amorphous structure and a crystal structure, between two different crystal structures, or between two amorphous structures.
- Quantum condensation of bosonic fluids (Bose–Einstein condensation). The superfluid transition in liquid helium is an example of this.
- The breaking of symmetries in the laws of physics during the early history of the universe as its temperature cooled.
- Isotope fractionation occurs during a phase transition, the ratio of light to heavy isotopes in the involved molecules changes. When water vapor condenses (an equilibrium fractionation), the heavier water isotopes (18O and 2H) become enriched in the liquid phase while the lighter isotopes (16O and 1H) tend toward the vapor phase.[1]

Phase transitions occur when the thermodynamic free energy of a system is non-analytic for some choice of thermodynamic variables (cf. phases). This condition generally stems from the interactions of a large number of particles in a system, and does not appear in systems that are too small. It is important to note that phase transitions can occur and are defined for non-thermodynamic systems, where temperature is not a parameter. Examples include: quantum phase transitions, dynamic phase transitions, and topological (structural) phase transitions. In these types of systems other parameters take the place of temperature. For instance, connection probability replaces temperature for percolating networks.

At the phase transition point (for instance, boiling point) the two phases of a substance, liquid and vapor, have identical free energies and therefore are equally likely to exist. Below the boiling point, the liquid is the more stable state of the two, whereas above the gaseous form is preferred.

It is sometimes possible to change the state of a system diabatically (as opposed to adiabatically) in such a way that it can be brought past a phase transition point without undergoing a phase transition. The resulting state is metastable, i.e., less stable than the phase to which the transition would have occurred, but not unstable either. This occurs in superheating, supercooling, and supersaturation, for example.

11.2 Classifications

11.2.1 Ehrenfest classification

Paul Ehrenfest classified phase transitions based on the behavior of the thermodynamic free energy as a function of other thermodynamic variables.[2] Under this scheme, phase transitions were labeled by the lowest derivative of the free energy that is discontinuous at the transition. *First-order phase transitions* exhibit a discontinuity in the first derivative of the free energy with respect to some thermodynamic variable.[3] The various solid/liquid/gas transitions are classified as first-order transitions because they involve a discontinuous change in density, which is the (inverse of the) first derivative of the free energy with respect to pressure. *Second-order phase transitions* are continuous in the first derivative (the order parameter, which is the first derivative of the free energy with respect to the external field, is continuous across the transition) but exhibit discontinuity in a second derivative of the free energy.[3] These include the ferromagnetic phase transition in materials such as iron, where the magnetization, which is the first derivative of the free energy with respect to the applied magnetic field strength, increases continuously from zero as the temperature is lowered below the Curie temperature. The magnetic susceptibility, the second derivative of the free energy with the field, changes discontinuously. Under the Ehrenfest classification scheme, there could in principle be third, fourth, and higher-order phase transitions.

Though useful, Ehrenfest's classification has been found to be an incomplete method of classifying phase transitions, for it does not take into account the case where a derivative of free energy diverges (which is only possible in the thermodynamic limit). For instance, in the ferromagnetic transition, the heat capacity diverges to infinity. The same phenomenon is also seen in superconducting phase transition.

11.2.2 Modern classifications

In the modern classification scheme, phase transitions are divided into two broad categories, named similarly to the Ehrenfest classes:[2]

First-order phase transitions are those that involve a latent heat. During such a transition, a system either absorbs or releases a fixed (and typically large) amount of energy per volume. During this process, the temperature of the system will stay constant as heat is added: the system is in a "mixed-phase regime" in which some parts of the system have completed the transition and others have not. Familiar examples are the melting of ice or the boiling of water (the water does not instantly turn into vapor, but forms a turbulent mixture of liquid water and vapor bubbles). Imry and Wortis showed that quenched disorder can broaden a first-order transition in that the transformation is completed over a finite range of temperatures, but phenomena like supercooling and superheating survive and hysteresis is observed on thermal cycling.[4][5][6]

Second-order phase transitions are also called *continuous phase transitions*. They are characterized by a divergent susceptibility, an infinite correlation length, and a power-law decay of correlations near criticality. Examples of second-order phase transitions are the ferromagnetic transition, superconducting transition (for a Type-I superconductor the phase transition is second-order at zero external field and for a Type-II superconductor the phase transition is second-order for both normal-state—mixed-state and mixed-state—superconducting-state transitions) and the superfluid transition. In contrast to viscosity, thermal expansion and heat capacity of amorphous materials show a relatively sudden change at the glass transition temperature[7] which enables accurate detection using differential scanning calorimetry measurements. Lev Landau gave a phenomenological theory of second-order phase transitions.

Apart from isolated, simple phase transitions, there exist transition lines as well as multicritical points, when varying external parameters like the magnetic field or composition.

Several transitions are known as the *infinite-order phase transitions*. They are continuous but break no symmetries. The most famous example is the Kosterlitz–Thouless transition in the two-dimensional XY model. Many quantum phase transitions, e.g., in two-dimensional electron gases, belong to this class.

The liquid–glass transition is observed in many polymers and other liquids that can be supercooled far below the melting point of the crystalline phase. This is atypical in several respects. It is not a transition between thermodynamic ground states: it is widely believed that the true ground state is always crystalline. Glass is a *quenched disorder* state, and its entropy, density, and so on, depend on the thermal history. Therefore, the glass transition is primarily a dynamic phenomenon: on cooling a liquid, internal degrees of freedom successively fall out of equilibrium. Some theoretical methods predict an underlying phase transition in the hypothetical limit of infinitely long relaxation times.[8][9] No direct experimental evidence supports the existence of these transitions.

11.3 Characteristic properties

11.3.1 Phase coexistence

A disorder-broadened first-order transition occurs over a finite range of temperatures where the fraction of the low-temperature equilibrium phase grows from zero to one (100%) as the temperature is lowered. This continuous variation of the coexisting fractions with temperature raised interesting possibilities. On cooling, some liquids vitrify into a glass rather than transform to the equilibrium crystal phase. This happens if the cooling rate is faster than a critical cooling rate, and is attributed to the molecular motions becoming so slow that the molecules cannot rearrange into the crystal positions.[10] This slowing down happens below a glass-formation temperature T_g, which may depend on the applied pressure.[7][11] If the first-order freezing transition occurs over a range of temperatures, and T_g falls within this range, then there is an interesting possibility that the transition is arrested when it is partial and incomplete. Extending these ideas to first-order magnetic transitions being arrested at low temperatures, resulted in the observation of incomplete magnetic transitions, with two magnetic phases coexisting, down to the lowest temperature. First reported in the case of a ferromagnetic to anti-ferromagnetic transition,[12] such persistent phase coexistence has now been reported across a variety of first-order magnetic transitions. These include colossal-magnetoresistance manganite materials,[13][14] magnetocaloric materials,[15] magnetic shape memory materials,[16] and other materials.[17] The interesting feature of these observations of T_g falling within the temperature range over which the transition occurs is that the first-order magnetic transition is influenced by magnetic field, just like the structural transition is influenced by pressure. The relative ease with which magnetic field can be controlled, in contrast to pressure, raises the possibility that one can study the interplay between T_g and T_c in an exhaustive way. Phase coexistence across first-order magnetic transitions will then enable the resolution of outstanding issues in understanding glasses.

11.3.2 Critical points

In any system containing liquid and gaseous phases, there exists a special combination of pressure and temperature, known as the critical point, at which the transition between liquid and gas becomes a second-order transition. Near the critical point, the fluid is sufficiently hot and compressed that the distinction between the liquid and gaseous phases is almost non-existent. This is associated with the phenomenon of critical opalescence, a milky appearance of the liquid due to density fluctuations at all possible wavelengths (including those of visible light).

11.3.3 Symmetry

Phase transitions often involve a symmetry breaking process. For instance, the cooling of a fluid into a crystalline solid breaks continuous translation symmetry: each point in the fluid has the same properties, but each point in a crystal does not have the same properties (unless the points are chosen from the lattice points of the crystal lattice). Typically, the high-temperature phase contains more symmetries than the low-temperature phase due to spontaneous symmetry breaking, with the exception of certain accidental symmetries (e.g. the formation of heavy virtual particles, which only occurs at low temperatures).[18]

11.3.4 Order parameters

An **order parameter** is a measure of the degree of order across the boundaries in a phase transition system; it normally ranges between zero in one phase (usually above the critical point) and nonzero in the other.[19] At the critical point, the order parameter susceptibility will usually diverge.

An example of an order parameter is the net magnetization in a ferromagnetic system undergoing a phase transition. For liquid/gas transitions, the order parameter is the difference of the densities.

From a theoretical perspective, order parameters arise from symmetry breaking. When this happens, one needs to introduce one or more extra variables to describe the state of the system. For example, in the ferromagnetic phase, one must provide the net magnetization, whose direction was spontaneously chosen when the system cooled below the Curie point. However, note that order parameters can also be defined for non-symmetry-breaking transitions. Some phase transitions, such as superconducting and ferromagnetic, can have order parameters for more than one degree of freedom. In such phases, the order parameter may take the form of a complex number, a vector, or even a tensor, the magnitude of which goes to zero at the phase transition.

There also exist dual descriptions of phase transitions in terms of disorder parameters. These indicate the presence of line-like excitations such as vortex- or defect lines.

11.3.5 Relevance in cosmology

Symmetry-breaking phase transitions play an important role in cosmology. It has been speculated by Lee Smolin and Benjamin and Jeremy Bernstein that, in the hot early universe, the vacuum (i.e. the various quantum fields that fill space) possessed a large number of symmetries. As the universe expanded and cooled, the vacuum underwent a series of symmetry-breaking phase transitions. For example, the electroweak transition broke the SU(2)×U(1) symmetry of the electroweak field into the U(1) symmetry of the present-day electromagnetic field. This transition is important to understanding the asymmetry between the amount of matter and antimatter in the present-day universe (see electroweak baryogenesis.)

Progressive phase transitions in an expanding universe are implicated in the development of order in the universe, as is illustrated by the work of Eric Chaisson[20] and David Layzer.[21] See also Relational order theories.

See also: Order–disorder

11.3.6 Critical exponents and universality classes

Main article: critical exponent

Continuous phase transitions are easier to study than first-order transitions due to the absence of latent heat, and they have been discovered to have many interesting properties. The phenomena associated with continuous phase transitions are called critical phenomena, due to their association with critical points.

It turns out that continuous phase transitions can be characterized by parameters known as critical exponents. The most important one is perhaps the exponent describing the divergence of the thermal correlation length by approaching the transition. For instance, let us examine the behavior of the heat capacity near such a transition. We vary the temperature T of the system while keeping all the other thermodynamic variables fixed, and find that the transition occurs at some critical temperature T_c. When T is near T_c, the heat capacity C typically has a power law behavior,

$$C \propto |T_c - T|^{-\alpha}.$$

The heat capacity of amorphous materials has such a behaviour near the glass transition temperature where the universal critical exponent $\alpha = 0.59$[22] A similar behavior, but with the exponent ν instead of α, applies for the correlation length.

The exponent ν is positive. This is different with α. Its actual value depends on the type of phase transition we are considering.

It is widely believed that the critical exponents are the same above and below the critical temperature. It has now been shown that this is not necessarily true: When a continuous symmetry is explicitly broken down to a discrete symmetry by irrelevant (in the renormalization group sense) anisotropies, then some exponents (such as γ, the exponent of the susceptibility) are not identical.[23]

For $-1 < \alpha < 0$, the heat capacity has a "kink" at the transition temperature. This is the behavior of liquid helium at the lambda transition from a normal state to the superfluid state, for which experiments have found $\alpha = -0.013 \pm 0.003$. At least one experiment was performed in the zero-gravity conditions of an orbiting satellite to minimize pressure differences in the sample.[24] This experimental value of α agrees with theoretical predictions based on variational perturbation theory.[25]

For $0 < \alpha < 1$, the heat capacity diverges at the transition temperature (though, since $\alpha < 1$, the enthalpy stays finite). An example of such behavior is the 3D ferromagnetic phase transition. In the three-dimensional Ising model for uniaxial magnets, detailed theoretical studies have yielded the exponent $\alpha \sim +0.110$.

Some model systems do not obey a power-law behavior. For example, mean field theory predicts a finite discontinuity of the heat capacity at the transition temperature, and the two-dimensional Ising model has a logarithmic divergence. However, these systems are limiting cases and an exception to the rule. Real phase transitions exhibit power-law behavior.

Several other critical exponents, β, γ, δ, ν, and η, are defined, examining the power law behavior of a measurable physical quantity near the phase transition. Exponents are related by scaling relations, such as

$$\beta = \gamma/(\delta - 1), \qquad \nu = \gamma/(2 - \eta)$$

It can be shown that there are only two independent exponents, e.g. ν and η.

It is a remarkable fact that phase transitions arising in different systems often possess the same set of critical exponents. This phenomenon is known as *universality*. For example, the critical exponents at the liquid–gas critical point have been found to be independent of the chemical composition of the fluid.

More impressively, but understandably from above, they are an exact match for the critical exponents of the ferromagnetic phase transition in uniaxial magnets. Such systems are said to be in the same universality class. Universality is a prediction of the renormalization group theory of phase transitions, which states that the thermodynamic properties of a system near a phase transition depend only on a small number of features, such as dimensionality and symmetry, and are insensitive to the underlying microscopic properties of the system. Again, the divergence of the correlation length is the essential point.

11.3.7 Critical slowing down and other phenomena

There are also other critical phenomena; e.g., besides *static functions* there is also *critical dynamics*. As a consequence, at a phase transition one may observe critical slowing down or *speeding up*. The large *static universality classes* of a continuous phase transition split into smaller *dynamic universality* classes. In addition to the critical exponents, there are also universal relations for certain static or dynamic functions of the magnetic fields and temperature differences from the critical value.

11.3.8 Percolation theory

Another phenomenon which shows phase transitions and critical exponents is percolation. The simplest example is perhaps percolation in a two dimensional square lattice. Sites are randomly occupied with probability p. For small values of p the occupied sites form only small clusters. At a certain threshold p_c a giant cluster is formed and we have a second-order phase transition.[26] The behavior of P∞ near p_c is, P∞~$(p-p_c)^\beta$, where β is a critical exponent.

11.3.9 Phase transitions in biological systems

Phase transitions play many important roles in biological systems. Examples include the lipid bilayer formation, the coil-globule transition in the process of protein folding and DNA melting, liquid crystal-like transitions in the process of DNA condensation, and cooperative ligand binding to DNA and proteins with the character of phase transition.[27]

In *biological membranes*, gel to liquid crystalline phase transitions play a critical role in physiological functioning of biomembranes. In gel phase, due to low fluidity of membrane lipid fatty-acyl chains, membrane proteins have restricted movement and thus are restrained in exercise of their physiological role. Plants depend critically on photosynthesis by chloroplast thylakoid membranes which are exposed cold environmental temperatures. Thylakoid membranes retain innate fluidity even at relatively low temperatures because of high degree of fatty-acyl disorder allowed by their high content of linolenic acid, 18-carbon chain with 3-double bonds.[28] Gel-to-liquid crystalline phase transition temperature of biological membranes can be determined by many techniques including calorimetry, flourescence, spin label electron paramagnetic resonance and NMR by recording measurements of the concerned parameter by at series of sample temperatures. A simple method for its determination from 13-C NMR line intensities has also been proposed.[29]

It has been proposed that some biological systems might lie near critical points. Examples include neural networks in the salamander retina,[30] bird flocks[31] gene expression networks in Drosophila,[32] and protein folding.[33] However, it is not clear whether or not alternative reasons could explain some of the phenomena supporting arguments for criticality.[34] It has also been suggested that biological organisms share two key properties of phase transitions: the change of macroscopic behavior and the coherence of a system at a critical point.[35]

11.4 See also

- Allotropy
- Autocatalytic reactions and order creation
- Crystal growth
    - Abnormal grain growth
- Differential scanning calorimetry
- Diffusionless transformations
- Ehrenfest equations
- Jamming (physics)
- Kelvin probe force microscope
- Landau theory of second order phase transitions
- Laser-heated pedestal growth
- List of states of matter
- Micro-pulling-down
- Percolation theory
    - Continuum percolation theory
- Superfluid film
- Superradiant phase transition
- Topological quantum field theory

11.5 References

[1] Carol Kendall (2004). "Fundamentals of Stable Isotope Geochemistry". USGS. Retrieved 10 April 2014.

[2] Jaeger, Gregg (1 May 1998). "The Ehrenfest Classification of Phase Transitions: Introduction and Evolution". *Archive for History of Exact Sciences*. **53** (1): 51–81. doi:10.1007/s004070050021.

[3] Blundell, Stephen J.; Katherine M. Blundell (2008). *Concepts in Thermal Physics*. Oxford University Press. ISBN 978-0-19-856770-7.

[4] Imry, Y.; Wortis, M. (1979). "Influence of quenched impurities on first-order phase transitions". *Phys. Rev. B*. **19** (7): 3580–3585. Bibcode:1979PhRvB..19.3580I. doi:10.1103/physrevb.19.3580.

[5] Kumar, Kranti; Pramanik, A. K.; Banerjee, A.; Chaddah, P.; Roy, S. B.; Park, S.; Zhang, C. L.; Cheong, S.-W. (2006). "Relating supercooling and glass-like arrest of kinetics for phase separated systems: DopedCeFe2and(La,Pr,Ca)MnO3". *Physical Review B*. **73** (18). Bibcode:2006PhRvB..73r4435K. ISSN 1098-0121. arXiv:cond-mat/0602627. doi:10.1103/PhysRevB.73.184435.

[6] Pasquini, G.; Daroca, D. Pérez; Chiliotte, C.; Lozano, G. S.; Bekeris, V. (2008). "Ordered, Disordered, and Coexistent Stable Vortex Lattices inNbSe2Single Crystals". *Physical Review Letters*. **100** (24). Bibcode:2008PhRvL..100x7003P. ISSN 0031-9007. arXiv:0803.0307. doi:10.1103/PhysRevLett.100.247003.

[7] Ojovan, M.I. (2013). "Ordering and structural changes at the glass-liquid transition". *J. Non-Cryst. Solids*. **382**: 79–86. Bibcode:2013JNCS..382...79O. doi:10.1016/j.jnoncrysol.2013.10.016.

[8] Gotze, Wolfgang. "Complex Dynamics of Glass-Forming Liquids: A Mode-Coupling Theory."

[9] Lubchenko, V. Wolynes; Wolynes, Peter G. (2007). "Theory of Structural Glasses and Supercooled Liquids". *Annual Review of Physical Chemistry*. **58**: 235–266. Bibcode:2007ARPC...58..235L. PMID 17067282. arXiv:cond-mat/0607349. doi:10.1146/annurev.physchem.58.032806.104653.

[10] Greer, A. L. (1995). "Metallic Glasses". *Science*. **267** (5206): 1947–1953. Bibcode:1995Sci...267.1947G. PMID 17770105. doi:10.1126/science.267.5206.1947.

[11] Tarjus, G. (2007). "Materials science: Metal turned to glass". *Nature*. **448** (7155): 758–759. Bibcode:2007Natur.448..758T. PMID 17700684. doi:10.1038/448758a.

[12] Manekar, M. A.; Chaudhary, S.; Chattopadhyay, M. K.; Singh, K. J.; Roy, S. B.; Chaddah, P. (2001). "First-order transition from antiferromagnetism to ferromagnetism inCe(Fe0.96Al0.04)2". *Physical Review B*. **64** (10). Bibcode:2001PhRvB..64j4416M. ISSN 0163-1829. arXiv:cond-mat/0012472. doi:10.1103/PhysRevB.64.104416.

[13] Banerjee, A; Pramanik, A K; Kumar, Kranti; Chaddah, P (2006). "Coexisting tunable fractions of glassy and equilibrium long-range-order phases in manganites". *Journal of Physics: Condensed Matter*. **18** (49): L605. Bibcode:2006JPCM...18L.605B. arXiv:cond-mat/0611152. doi:10.1088/0953-8984/18/49/L02.

[14] Wu W, Israel C, Hur N, Park S, Cheong SW, de Lozanne A (2006). "Magnetic imaging of a supercooling glass transition in a weakly disordered ferromagnet". *Nature Materials*. **5** (11): 881–886. Bibcode:2006NatMa...5..881W. PMID 17028576. doi:10.1038/nmat1743.

[15] Roy, S. B.; Chattopadhyay, M. K.; Chaddah, P.; Moore, J. D.; Perkins, G. K.; Cohen, L. F.; Gschneidner, K. A.; Pecharsky, V. K. (2006). "Evidence of a magnetic glass state in the magnetocaloric materialGd5Ge4". *Physical Review B*. **74** (1). Bibcode:2006PhRvB..74a2403R. ISSN 1098-0121. doi:10.1103/PhysRevB.74.012403.

[16] Lakhani, Archana; Banerjee, A; Chaddah, P; Chen, X; Ramanujan, R V (2012). "Magnetic glass in shape memory alloy: Ni45Co5Mn38Sn12". *Journal of Physics: Condensed Matter*. **24** (38): 386004. Bibcode:2012JPCM...24L6004L. ISSN 0953-8984. arXiv:1206.2024. doi:10.1088/0953-8984/24/38/386004.

[17] Kushwaha, Pallavi; Lakhani, Archana; Rawat, R.; Chaddah, P. (2009). "Low-temperature study of field-induced antiferromagnetic-ferromagnetic transition in Pd-doped Fe-Rh". *Physical Review B*. **80** (17). Bibcode:2009PhRvB..80q4413K. ISSN 1098-0121. arXiv:0911.4552. doi:10.1103/PhysRevB.80.174413.

[18] Ivancevic, Vladimir G.; Ivancevic, Tijiana, T. (2008). *Complex Nonlinearity*. Berlin: Springer. pp. 176–177. ISBN 978-3-540-79357-1. Retrieved 12 October 2014.

[19] A. D. McNaught and A. Wilkinson (ed.). "Compendium of Chemical Terminology". IUPAC. ISBN 0-86542-684-8. Retrieved 2007-10-23.

[20] Chaisson, *Cosmic Evolution*, Harvard, 2001

[21] David Layzer, *Cosmogenesis, The Development of Order in the Universe*, Oxford Univ. Press, 1991

[22] Ojovan, Michael I; Lee, William E (2006). "Topologically disordered systems at the glass transition" (PDF). *Journal of Physics: Condensed Matter*. **18** (50): 11507–11520. Bibcode:2006JPCM...1811507O. doi:10.1088/0953-8984/18/50/007.

[23] Leonard, F.; Delamotte, B. (2015). "Critical exponents can be different on the two sides of a transition". *Phys. Rev. Lett*. **115** (20): 200601. Bibcode:2015PhRvL.115t0601L. arXiv:1508.07852. doi:10.1103/PhysRevLett.115.200601.

[24] Lipa, J.; Nissen, J.; Stricker, D.; Swanson, D.; Chui, T. (2003). "Specific heat of liquid helium in zero gravity very near the lambda point". *Physical Review B*. **68** (17): 174518. Bibcode:2003PhRvB..68q4518L. arXiv:cond-mat/0310163. doi:10.1103/PhysRevB.68.174518.

[25] Kleinert, Hagen (1999). "Critical exponents from seven-loop strong-coupling $\varphi 4$ theory in three dimensions". *Physical Review D.* **60** (8): 085001. Bibcode:1999PhRvD..60h5001K. arXiv:hep-th/9812197. doi:10.1103/PhysRevD.60.085001.

[26] Armin Bunde and Shlomo Havlin (1996). *Fractals and Disordered Systems.* Springer.

[27] D.Y. Lando and V.B. Teif (2000). "Long-range interactions between ligands bound to a DNA molecule give rise to adsorption with the character of phase transition of the first kind". *J. Biomol. Struct. Dynam.* **17** (5): 903–911. doi:10.1080/07391102.2000.10506578.

[28] YashRoy, R.C. (1987). "13-C NMR studies of lipid fatty acyl chains of chloroplast membranes". *Indian Journal of Biochemistry and Biophysics.* **24** (6): 177–178.

[29] YashRoy, R C (1990). "Determination of membrane lipid phase transition temperature from 13-C NMR intensities". *Journal of Biochemical and Biophysical Methods.* **20** (4): 353–356. PMID 2365951. doi:10.1016/0165-022X(90)90097-V.

[30] Tkacik, Gasper; Mora, Thierry; Marre, Olivier; Amodei, Dario; Berry II, Michael J.; Bialek, William (2014). "Thermodynamics for a network of neurons: Signatures of criticality". arXiv:1407.5946 [q-bio.NC].

[31] Bialek, W; Cavagna, A; Giardina, I (2014). "Social interactions dominate speed control in poising natural flocks near criticality". *PNAS.* **111** (20): 7212–7217. Bibcode:2014PNAS..111.7212B. PMC 4034227. PMID 24785504. arXiv:1307.5563. doi:10.1073/pnas.1324045111.

[32] Krotov, D; Dubuis, J O; Gregor, T; Bialek, W (2014). "Morphogenesis at criticality". *PNAS.* **111**: 3683–3688. Bibcode:2014PNAS..111.3683K. arXiv:1309.2614. doi:10.1073/pnas.1324186111.

[33] Mora, Thierry; Bialek, William (2011). "Are biological systems poised at criticality?". *Journal of Statistical Physics.* **144** (2): 268–302. Bibcode:2011JSP...144..268M. arXiv:1012.2242. doi:10.1007/s10955-011-0229-4.

[34] Schwab, David J; Nemenman, Ilya; Mehta, Pankaj (2013). "Zipf's law and criticality in multivariate data without fine-tuning". *Physical Review Letters.* **113** (6): 068102. Bibcode:2014PhRvL.113f8102S. PMID 25148352. arXiv:1310.0448. doi:10.1103/PhysRevLett.113.068102.

[35] Longo, G.; Montévil, M. (2011-08-01). "From physics to biology by extending criticality and symmetry breakings". *Progress in Biophysics and Molecular Biology. Systems Biology and Cancer.* **106** (2): 340–347. doi:10.1016/j.pbiomolbio.2011.03.005.

11.6 Further reading

- Anderson, P.W., *Basic Notions of Condensed Matter Physics*, Perseus Publishing (1997).

- Fisher, M.E. (1974). "The renormalization group in the theory of critical behavior". *Rev. Mod. Phys.* **46** (4): 597–616. Bibcode:1974RvMP...46..597F. doi:10.1103/revmodphys.46.597.

- Goldenfeld, N., *Lectures on Phase Transitions and the Renormalization Group*, Perseus Publishing (1992).

- Ivancevic, Vladimir G; Ivancevic, Tijana T (2008), *Chaos, Phase Transitions, Topology Change and Path Integrals*, Berlin: Springer, ISBN 978-3-540-79356-4, retrieved 14 March 2013 e-ISBN 978-3-540-79357-1

- Kogut, J.; Wilson, K (1974). "The Renormalization Group and the epsilon-Expansion". *Phys. Rep.* **12** (2): 75–199. Bibcode:1974PhR....12...75W. doi:10.1016/0370-1573(74)90023-4.

- Krieger, Martin H., *Constitutions of matter : mathematically modelling the most everyday of physical phenomena*, University of Chicago Press, 1996. Contains a detailed pedagogical discussion of Onsager's solution of the 2-D Ising Model.

- Landau, L.D. and Lifshitz, E.M., *Statistical Physics Part 1*, vol. 5 of *Course of Theoretical Physics*, Pergamon Press, 3rd Ed. (1994).

- Kleinert, H., *Gauge Fields in Condensed Matter*, Vol. I, "Superfluid and Vortex lines; Disorder Fields, Phase Transitions,", pp. 1–742, World Scientific (Singapore, 1989); Paperback ISBN 9971-5-0210-0 (readable online physik.fu-berlin.de)

- Kleinert, H. and Verena Schulte-Frohlinde, *Critical Properties of φ^4-Theories*, World Scientific (Singapore, 2001); Paperback ISBN 981-02-4659-5 *(readable online here)*.

- Mussardo G., "Statistical Field Theory. An Introduction to Exactly Solved Models of Statistical Physics", Oxford University Press, 2010.

- Schroeder, Manfred R., *Fractals, chaos, power laws : minutes from an infinite paradise*, New York: W. H. Freeman, 1991. Very well-written book in "semi-popular" style—not a textbook—aimed at an audience with some training in mathematics and the physical sciences. Explains what scaling in phase transitions is all about, among other things.

- Yeomans J. M., *Statistical Mechanics of Phase Transitions*, Oxford University Press, 1992.

- H. E. Stanley, *Introduction to Phase Transitions and Critical Phenomena* (Oxford University Press, Oxford and New York 1971).

- M.R.Khoshbin-e-Khoshnazar, *Ice Phase Transition as a sample of finite system phase transition*, (Physics Education(India)Volume 32. No. 2, Apr - Jun 2016)

11.7 External links

- Interactive Phase Transitions on lattices with Java applets
- Universality classes from Sklogwiki

Chapter 12

Agent-based model

An **agent-based model** (ABM) is one of a class of computational models for simulating the actions and interactions of autonomous agents (both individual or collective entities such as organizations or groups) with a view to assessing their effects on the system as a whole. It combines elements of game theory, complex systems, emergence, computational sociology, multi-agent systems, and evolutionary programming. Monte Carlo methods are used to introduce randomness. Particularly within ecology, ABMs are also called **individual-based models (IBMs)**,[1] and individuals within IBMs may be simpler than fully autonomous agents within ABMs. A review of recent literature on individual-based models, agent-based models, and multiagent systems shows that ABMs are used on non-computing related scientific domains including biology, ecology and social science.[2] Agent-based modeling is related to, but distinct from, the concept of **multi-agent systems** or **multi-agent simulation** in that the goal of ABM is to search for explanatory insight into the collective behavior of agents obeying simple rules, typically in natural systems, rather than in designing agents or solving specific practical or engineering problems.[2]

Agent-based models are a kind of microscale model[3] that simulate the simultaneous operations and interactions of multiple agents in an attempt to re-create and predict the appearance of complex phenomena. The process is one of emergence from the lower (micro) level of systems to a higher (macro) level. As such, a key notion is that simple behavioral rules generate complex behavior. This principle, known as K.I.S.S. ("Keep it simple, stupid"), is extensively adopted in the modeling community. Another central tenet is that the whole is greater than the sum of the parts. Individual agents are typically characterized as boundedly rational, presumed to be acting in what they perceive as their own interests, such as reproduction, economic benefit, or social status,[4] using heuristics or simple decision-making rules. ABM agents may experience "learning", adaptation, and reproduction.[5]

Most agent-based models are composed of: (1) numerous agents specified at various scales (typically referred to as agent-granularity); (2) decision-making heuristics; (3) learning rules or adaptive processes; (4) an interaction topology; and (5) an environment. ABMs are typically implemented as computer simulations, either as custom software, or via ABM toolkits, and this software can be then used to test how changes in individual behaviors will affect the system's emerging overall behavior.

12.1 History

The idea of agent-based modeling was developed as a relatively simple concept in the late 1940s. Since it requires computation-intensive procedures, it did not become widespread until the 1990s.

12.1.1 Early developments

The history of the agent-based model can be traced back to the Von Neumann machine, a theoretical machine capable of reproduction. The device von Neumann proposed would follow precisely detailed instructions to fashion a copy of itself. The concept was then built upon by von Neumann's friend Stanislaw Ulam, also a mathematician; Ulam suggested that the machine be built on paper, as a collection of cells on a grid. The idea intrigued von Neumann, who drew it up—creating the first of the devices later termed cellular automata. Another advance was introduced by the mathematician John Conway. He constructed the well-known Game of Life. Unlike von Neumann's machine, Conway's Game of Life operated by tremendously simple rules in a virtual world in the form of a 2-dimensional checkerboard.

12.1.2 1970s and 1980s: the first models

One of the earliest agent-based models in concept was Thomas Schelling's segregation model,[6] which was discussed in his paper "Dynamic Models of Segregation" in

1971. Though Schelling originally used coins and graph paper rather than computers, his models embodied the basic concept of agent-based models as autonomous agents interacting in a shared environment with an observed aggregate, emergent outcome.

In the early 1980s, Robert Axelrod hosted a tournament of Prisoner's Dilemma strategies and had them interact in an agent-based manner to determine a winner. Axelrod would go on to develop many other agent-based models in the field of political science that examine phenomena from ethnocentrism to the dissemination of culture.[7] By the late 1980s, Craig Reynolds' work on flocking models contributed to the development of some of the first biological agent-based models that contained social characteristics. He tried to model the reality of lively biological agents, known as artificial life, a term coined by Christopher Langton.

The first use of the word "agent" and a definition as it is currently used today is hard to track down. One candidate appears to be John Holland and John H. Miller's 1991 paper "Artificial Adaptive Agents in Economic Theory",[8] based on an earlier conference presentation of theirs.

At the same time, during the 1980s, social scientists, mathematicians, operations researchers, and a scattering of people from other disciplines developed Computational and Mathematical Organization Theory (CMOT). This field grew as a special interest group of The Institute of Management Sciences (TIMS) and its sister society, the Operations Research Society of America (ORSA).

12.1.3 1990s: expansion

With the appearance of StarLogo in 1990, Swarm and NetLogo in the mid-1990s and RePast and AnyLogic in 2000, or GAMA[9] in 2007 as well as some custom-designed code, modelling software became widely available and the range of domains that ABM was applied to, grew. Bonabeau (2002) is a good survey of the potential of agent-based modeling as of the time[10]

The 1990s were especially notable for the expansion of ABM within the social sciences, one notable effort was the large-scale ABM, Sugarscape, developed by Joshua M. Epstein and Robert Axtell to simulate and explore the role of social phenomena such as seasonal migrations, pollution, sexual reproduction, combat, and transmission of disease and even culture.[11] Other notable 1990s developments included Carnegie Mellon University's Kathleen Carley ABM,[12] to explore the co-evolution of social networks and culture. During this 1990s timeframe Nigel Gilbert published the first textbook on Social Simulation: Simulation for the social scientist (1999) and established a journal from the perspective of social sciences: the *Journal of Artificial Societies and Social Simulation* (JASSS). Other than JASSS, agent-based models of any discipline are within scope of SpringerOpen journal *Complex Adaptive Systems Modeling* (CASM).[13]

Through the mid-1990s, the social sciences thread of ABM began to focus on such issues as designing effective teams, understanding the communication required for organizational effectiveness, and the behavior of social networks. CMOT—later renamed Computational Analysis of Social and Organizational Systems (CASOS)—incorporated more and more agent-based modeling. Samuelson (2000) is a good brief overview of the early history,[14] and Samuelson (2005) and Samuelson and Macal (2006) trace the more recent developments.[15][16]

In the late 1990s, the merger of TIMS and ORSA to form INFORMS, and the move by INFORMS from two meetings each year to one, helped to spur the CMOT group to form a separate society, the North American Association for Computational Social and Organizational Sciences (NAACSOS). Kathleen Carley was a major contributor, especially to models of social networks, obtaining National Science Foundation funding for the annual conference and serving as the first President of NAACSOS. She was succeeded by David Sallach of the University of Chicago and Argonne National Laboratory, and then by Michael Prietula[17] of Emory University. At about the same time NAACSOS began, the European Social Simulation Association (ESSA) and the Pacific Asian Association for Agent-Based Approach in Social Systems Science (PAAA), counterparts of NAACSOS, were organized. As of 2013, these three organizations collaborate internationally. The First World Congress on Social Simulation was held under their joint sponsorship in Kyoto, Japan, in August 2006. The Second World Congress was held in the northern Virginia suburbs of Washington, D.C., in July 2008, with George Mason University taking the lead role in local arrangements.

12.1.4 2000s and later

More recently, Ron Sun developed methods for basing agent-based simulation on models of human cognition, known as cognitive social simulation.[18] Bill McKelvey, Suzanne Lohmann, Dario Nardi, Dwight Read and others at UCLA have also made significant contributions in organizational behavior and decision-making. Since 2001, UCLA has arranged a conference at Lake Arrowhead, California, that has become another major gathering point for practitioners in this field. In 2014, Sadegh Asgari from Columbia University and his colleagues developed an agent-based model of the construction competitive bidding.[19] While his model was used to analyze the low-bid lump-sum construction bids, it could be applied to other bidding methods with little modifications to the model.

12.2 Theory

Most computational modeling research describes systems in equilibrium or as moving between equilibria. Agent-based modeling, however, using simple rules, can result in different sorts of complex and interesting behavior. The three ideas central to agent-based models are agents as objects, emergence, and complexity.

Agent-based models consist of dynamically interacting rule-based agents. The systems within which they interact can create real-world-like complexity. Typically agents are situated in space and time and reside in networks or in lattice-like neighborhoods. The location of the agents and their responsive behavior are encoded in algorithmic form in computer programs. In some cases, though not always, the agents may be considered as intelligent and purposeful. In ecological ABM (often referred to as "individual-based models" in ecology), agents may, for example, be trees in forest, and would not be considered intelligent, although they may be "purposeful" in the sense of optimizing access to a resource (such as water). The modeling process is best described as inductive. The modeler makes those assumptions thought most relevant to the situation at hand and then watches phenomena emerge from the agents' interactions. Sometimes that result is an equilibrium. Sometimes it is an emergent pattern. Sometimes, however, it is an unintelligible mangle.

In some ways, agent-based models complement traditional analytic methods. Where analytic methods enable humans to characterize the equilibria of a system, agent-based models allow the possibility of generating those equilibria. This generative contribution may be the most mainstream of the potential benefits of agent-based modeling. Agent-based models can explain the emergence of higher-order patterns—network structures of terrorist organizations and the Internet, power-law distributions in the sizes of traffic jams, wars, and stock-market crashes, and social segregation that persists despite populations of tolerant people. Agent-based models also can be used to identify lever points, defined as moments in time in which interventions have extreme consequences, and to distinguish among types of path dependency.

Rather than focusing on stable states, many models consider a system's robustness—the ways that complex systems adapt to internal and external pressures so as to maintain their functionalities. The task of harnessing that complexity requires consideration of the agents themselves—their diversity, connectedness, and level of interactions.

12.2.1 Framework

Recent work on the Modeling and simulation of Complex Adaptive Systems has demonstrated the need for combining agent-based and complex network based models.[20][21][22] describe a framework consisting of four levels of developing models of complex adaptive systems described using several example multidisciplinary case studies:

1. Complex Network Modeling Level for developing models using interaction data of various system components.

2. Exploratory Agent-based Modeling Level for developing agent-based models for assessing the feasibility of further research. This can e.g. be useful for developing proof-of-concept models such as for funding applications without requiring an extensive learning curve for the researchers.

3. Descriptive Agent-based Modeling (DREAM) for developing descriptions of agent-based models by means of using templates and complex network-based models. Building DREAM models allows model comparison across scientific disciplines.

4. Validated agent-based modeling using Virtual Overlay Multiagent system (VOMAS) for the development of verified and validated models in a formal manner.

Other methods of describing agent-based models include code templates[23] and text-based methods such as the ODD (Overview, Design concepts, and Design Details) protocol.[24]

The role of the environment where agents live, both macro and micro,[25] is also becoming an important factor in agent-based modelling and simulation work. Simple environment affords simple agents, but complex environments generates diversity of behaviour.[26]

12.3 Applications

12.3.1 In biology

Main article: Agent-based model in biology

Agent-based modeling has been used extensively in biology, including the analysis of the spread of epidemics,[27] and the threat of biowarfare, biological applications including population dynamics,[28] vegetation ecology,[29] landscape diversity,[30] the growth and decline of ancient civilizations, evolution of ethnocentric behavior,[31] forced displacement/migration,[32] language choice dynamics,[33] cognitive modeling, and biomedical applications including modeling 3D breast tissue formation/morphogenesis,[34] the effects of ionizing radiation on mammary stem cell subpopulation dynamics,[35] inflammation,[36][37] and the human immune system. Agent-based models have also been used for developing decision support systems such as for breast cancer.[38] Agent-based models are increasingly being used to model pharmacological systems in early stage and pre-clinical research to aid in drug development and gain insights into biological systems that would not be possible *a priori*.[39] Military applications have also been evaluated.[40] Moreover, agent-based models have been recently employed to study molecular-level biological systems.[41][42][43]

12.3.2 In business, technology and network theory

Agent-based models have been used since the mid-1990s to solve a variety of business and technology problems. Examples of applications include the modeling of organizational behaviour and cognition,[44] team working,[45] supply chain optimization and logistics, modeling of consumer behavior, including word of mouth, social network effects, distributed computing, workforce management, and portfolio management. They have also been used to analyze traffic congestion.[46]

Recently, agent based modelling and simulation has been applied to various domains such as studying the impact of publication venues by researchers in the computer science domain (journals versus conferences).[47] In addition, ABMs have been used to simulate information delivery in ambient assisted environments.[48] A November 2016 article in arXiv analyzed an agent based simulation of posts spread in the Facebook online social network.[49] In the domain of peer-to-peer, ad-hoc and other self-organizing and complex networks, the usefulness of agent based modeling and simulation has been shown.[50] The use of a computer science-based formal specification framework coupled with wireless sensor networks and an agent-based simulation has recently been demonstrated.[51]

Agent based evolutionary search or algorithm is a new research topic for solving complex optimization problems.[52]

12.3.3 In economics and social sciences

Main articles: Agent-based computational economics and Agent-based social simulation

Prior to, and in the wake of the financial crisis, interest has grown in ABMs as possible tools for economic

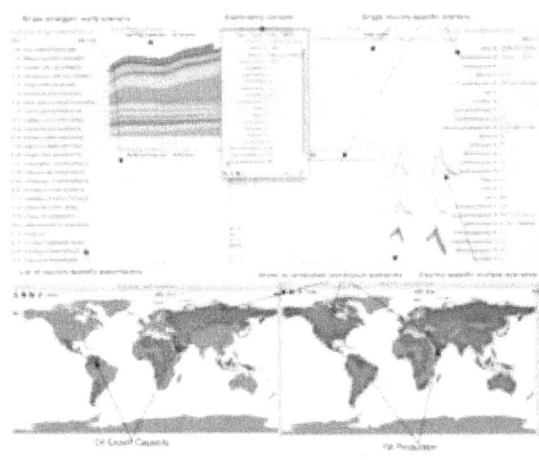

Graphic user interface for an agent-based modeling tool.

analysis.[53][54] ABMs do not assume the economy can achieve equilibrium and "representative agents" are replaced by agents with diverse, dynamic, and interdependent behavior including herding. ABMs take a "bottom-up" approach and can generate extremely complex and volatile simulated economies. ABMs can represent unstable systems with crashes and booms that develop out of non-linear (disproportionate) responses to proportionally small changes.[55] A July 2010 article in *The Economist* looked at ABMs as alternatives to DSGE models.[55] The journal *Nature* also encouraged agent-based modeling with an editorial that suggested ABMs can do a better job of representing financial markets and other economic complexities than standard models[56] along with an essay by J. Doyne Farmer and Duncan Foley that argued ABMs could fulfill both the desires of Keynes to represent a complex economy and of Robert Lucas to construct models based on microfoundations.[57] Farmer and Foley pointed to progress that has been made using ABMs to model parts of an economy, but argued for the creation of a very large model that incorporates low level models.[58] By modeling a complex system of analysts based on three distinct behavioral profiles – imitating, anti-imitating, and indifferent – financial markets were simulated to high accuracy. Results showed a correlation between network morphology and the stock market index.[59]

Since the beginning of the 21st century ABMs have been deployed in architecture and urban planning to evaluate design and to simulate pedestrian flow in the urban environment.[60] There is also a growing field of socio-economic analysis of infrastructure investment impact using ABM's ability to discern systemic impacts upon a socio-economic network.[61]

12.3.4 Organizational ABM: agent-directed simulation

The agent-directed simulation (ADS) metaphor distinguishes between two categories, namely "Systems for Agents" and "Agents for Systems."[62] Systems for Agents (sometimes referred to as agents systems) are systems implementing agents for the use in engineering, human and social dynamics, military applications, and others. Agents for Systems are divided in two subcategories. Agent-supported systems deal with the use of agents as a support facility to enable computer assistance in problem solving or enhancing cognitive capabilities. Agent-based systems focus on the use of agents for the generation of model behavior in a system evaluation (system studies and analyses).

12.4 Implementation

Many agent-based modeling software are designed for serial von-Neumann computer architectures. This limits the speed and scalability of these systems. A recent development is the use of data-parallel algorithms on Graphics Processing Units GPUs for ABM simulation.[63][64][65] The extreme memory bandwidth combined with the sheer number crunching power of multi-processor GPUs has enabled simulation of millions of agents at tens of frames per second.

12.5 Verification and validation

Verification and validation (V&V) of simulation models is extremely important.[66][67] Verification involves the model being debugged to ensure it works correctly, whereas validation ensures that the right model has been built. Face validation, sensitivity analysis, calibration and statistical validation have also been demonstrated.[68] A discrete-event simulation framework approach for the validation of agent-based systems has been proposed.[69] A comprehensive resource on empirical validation of agent-based models can be found here.[70]

As an example of V&V technique, consider VOMAS (virtual overlay multi-agent system),[71] a software engineering based approach, where a virtual overlay multi-agent system is developed alongside the agent-based model. The agents in the multi-agent system are able to gather data by generation of logs as well as provide run-time validation and verification support by watch agents and also agents to check any violation of invariants at run-time. These are set by the Simulation Specialist with help from the SME (subject-matter expert). Muazi et al. also provide an example of using VOMAS for verification and validation of a forest fire simulation model.[72]

VOMAS provides a formal way of validation and verification. To develop a VOMAS, one must design VOMAS agents along with the agents in the actual simulation, preferably from the start. In essence, by the time the simulation model is complete, one can essentially consider it to be one model containing two models:

1. An agent-based model of the intended system
2. An agent-based model of the VOMAS

Unlike all previous work on verification and validation, VOMAS agents ensure that the simulations are validated in-simulation i.e. even during execution. In case of any exceptional situations, which are programmed on the directive of the Simulation Specialist (SS), the VOMAS agents can report them. In addition, the VOMAS agents can be used to log key events for the sake of debugging and subsequent analysis of simulations. In other words, VOMAS allows for a flexible use of any given technique for the sake of verification and validation of an agent-based model in any domain.

Details of validated agent-based modeling using VOMAS along with several case studies are given in.[73] This thesis also gives details of "exploratory agent-based modeling", "descriptive agent-based modeling" and "validated agent-based modeling", using several worked case study examples.

12.5.1 Complex systems modelling

Mathematical models of complex systems are of three types: black-box (phenomenological), white-box (mechanistic, based on the first principles) and grey-box (mixtures of phenomenological and mechanistic models) [74] [75]. In black-box models, the individual-based (mechanistic) mechanisms of a complex dynamic system remain hidden.

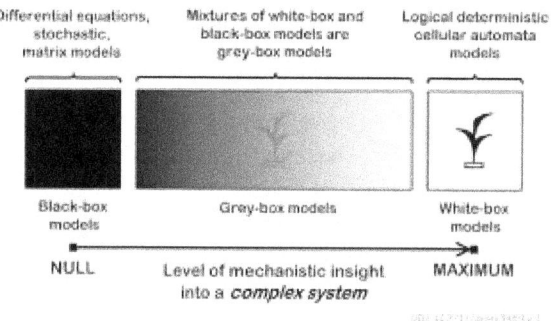

Mathematical models for complex systems

Black-box models are completely nonmechanistic. They are phenomenological and ignore a composition and internal structure of a complex system. We cannot investigate interactions of subsystems of such a non-transparent model. A white-box model of complex dynamic system has 'transparent walls' and directly shows underlying mechanisms. All events at micro-, meso- and macro-levels of a dynamic system are directly visible at all stages of its white-box model evolution. In most cases mathematical modelers use the heavy black-box mathematical methods, which cannot produce mechanistic models of complex dynamic systems. Grey-box models are intermediate and combine black-box and white-box approaches.

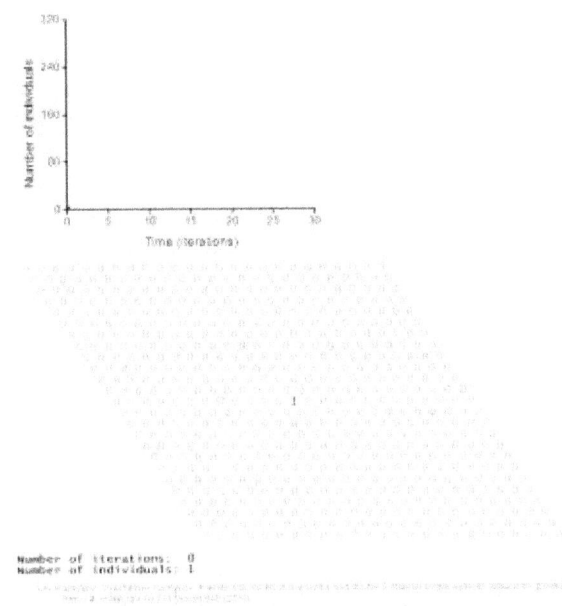

Logical deterministic individual-based cellular automata model of single species population growth

Creation of a white-box model of complex system is associated with the problem of the necessity of an a priori basic knowledge of the modeling subject. The deterministic logical cellular automata are necessary but not sufficient condition of a white-box model. The second necessary prerequisite of a white-box model is the presence of the physical ontology of the object under study. The white-box modeling represents an automatic hyper-logical inference from the first principles because it is completely based on the deterministic logic and axiomatic theory of the subject. The purpose of the white-box modeling is to derive from the basic axioms a more detailed, more concrete mechanistic knowledge about the dynamics of the object under study. The necessity to formulate an intrinsic axiomatic system of the subject before creating its white-box model distinguishes the cellular automata models of white-box type from cellular automata models based on arbitrary logical rules. If cellular automata rules have not been formulated from the first principles of the subject, then such a model may have a weak relevance to the real problem [75].

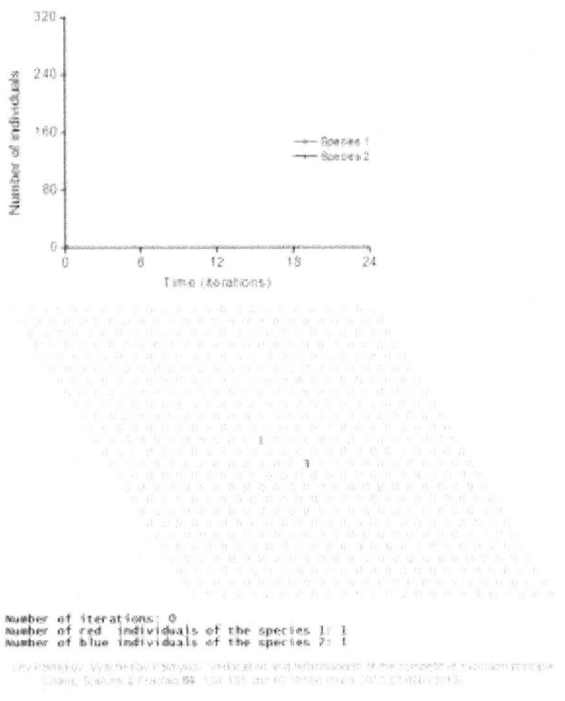

Logical deterministic individual-based cellular automata model of interspecific competition for a single limited resource

12.6 See also

- Agent-based computational economics
- Agent-based model in biology
- Agent-based social simulation (ABSS)
- Artificial society
- Boids
- Comparison of agent-based modeling software
- Complex system
- Complex adaptive system
- Computational sociology
- Conway's Game of Life
- Dynamic network analysis
- Emergence
- Evolutionary algorithm

12.7. REFERENCES

- Flocking
- Kinetic exchange models of markets
- Multi-agent system
- Simulated reality
- Social complexity
- Social simulation
- Sociophysics
- Software agent
- Internet bot
- Swarming behaviour
- Web-based simulation

12.7 References

12.7.1 Inline

[1] Grimm, Volker; Railsback, Steven F. (2005). *Individual-based Modeling and Ecology*. Princeton University Press. p. 485. ISBN 978-0-691-09666-7.

[2] Niazi, Muaz; Hussain, Amir (2011). "Agent-based Computing from Multi-agent Systems to Agent-Based Models: A Visual Survey" (PDF). *Scientometrics*. Springer. **89** (2): 479–499. doi:10.1007/s11192-011-0468-9. Archived from the original (PDF) on October 12, 2013.

[3] Gustafsson, Leif; Sternad, Mikael (2010). "Consistent micro, macro, and state-based population modelling". *Mathematical Biosciences*. **225** (2): 94–107. PMID 20171974. doi:10.1016/j.mbs.2010.02.003.

[4] "Agent-Based Models of Industrial Ecosystems". Rutgers University. October 6, 2003. Archived from the original on July 20, 2011.

[5] Bonabeau, E. (May 14, 2002). "Agent-based modeling: Methods and techniques for simulating human systems". *Proceedings of the National Academy of Sciences of the United States of America*. National Academy of Sciences. **99**: 7280–7. Bibcode:2002PNAS...99.7280B. PMC 128598. PMID 12011407. doi:10.1073/pnas.082080899.

[6] Schelling, Thomas C. (1971). "Dynamic Models of Segregation" (PDF). *Journal of Mathematical Sociology*. **1** (2): 143–186. doi:10.1080/0022250x.1971.9989794.

[7] Axelrod, Robert (1997). *The Complexity of Cooperation: Agent-Based Models of Competition and Collaboration*. Princeton: Princeton University Press. ISBN 978-0-691-01567-5.

[8] Holland, J.H.; Miller, J.H. (1991). "Artificial Adaptive Agents in Economic Theory" (PDF). *American Economic Review*. **81** (2): 365–71.

[9] GAMA

[10] Bonabeau, Eric (2002). "Agent-based modeling: methods and techniques for simulating human systems". *Proceedings of the National Academy of Sciences of the United States of America*. **99** (3): 7280–7287. Bibcode:2002PNAS...99.7280B. PMC 128598. PMID 12011407. doi:10.1073/pnas.082080899.

[11] Epstein, Joshua M.; Axtell, Robert (October 11, 1996). *Growing artificial societies: social science from the bottom up*. Brookings Institution Press. p. 224. ISBN 978-0-262-55025-3.

[12] "Construct". Computational Analysis of Social Organizational Systems.

[13] Springer Complex Adaptive Systems Modeling Journal (CASM)

[14] Samuelson, Douglas A. (December 2000). "Designing Organizations". *OR/MS Today*. Institute for Operations Research and the Management Sciences.

[15] Samuelson, Douglas A. (February 2005). "Agents of Change". *OR/MS Today*. Institute for Operations Research and the Management Sciences.

[16] Samuelson, Douglas A.; Macal, Charles M. (August 2006). "Agent-Based Modeling Comes of Age". *OR/MS Today*. Institute for Operations Research and the Management Sciences.

[17] Michael Prietula

[18] Sun, Ron (2006). *Cognition and Multi-Agent Interaction: From Cognitive Modeling to Social Simulation*. Cambridge University Press. ISBN 0-521-83964-5.

[19] Awwad, R.; Asgari, S.; Kandil, A. (2014-10-06). "Developing a Virtual Laboratory for Construction Bidding Environment Using Agent-Based Modeling". *Journal of Computing in Civil Engineering*. **29** (6): 04014105. doi:10.1061/(ASCE)CP.1943-5487.0000440.

[20] Aditya Kurve; Khashayar Kotobi; George Kesidis (2013). "An agent-based framework for performance modeling of an optimistic parallel discrete event simulator". *Complex Adaptive Systems Modeling*. **1**: 12. doi:10.1186/2194-3206-1-12.

[21] Niazi, Muaz A. K. "Towards A Novel Unified Framework for Developing Formal, Network and Validated Agent-Based Simulation Models of Complex Adaptive Systems". (PhD Thesis)

[22] Niazi, M.A. and Hussain, A (2012), Cognitive Agent-based Computing-I: A Unified Framework for Modeling Complex Adaptive Systems using Agent-based & Complex Network-based Methods Cognitive Agent-based Computing

[23] "Swarm code templates for model comparison". Swarm Development Group. Archived from the original on August 3, 2008.

[24] Volker Grimm, Uta Berger, Finn Bastiansen, Sigrunn Eliassen, Vincent Ginot, Jarl Giske, John Goss-Custard, Tamara Grand, Simone K. Heinz, Geir Huse, Andreas Huth, Jane U. Jepsen, Christian Jørgensen, Wolf M. Mooij, Birgit Müller, Guy Pe'er, Cyril Piou, Steven F. Railsback, Andrew M. Robbins, Martha M. Robbins, Eva Rossmanith, Nadja Rüger, Espen Strand, Sami Souissi, Richard A. Stillman, Rune Vabø, Ute Visser, Donald L. DeAngelis (September 15, 2006). "A standard protocol for describing individual-based and agent-based models". *Ecological Modelling*. **198** (1–2): 115–126. doi:10.1016/j.ecolmodel.2006.04.023. (ODD Paper)

[25] Ch'ng, E. (2012) Macro and Micro Environment for Diversity of Behaviour in Artificial Life Simulation. Artificial Life Session, The 6th International Conference on Soft Computing and Intelligent Systems, The 13th International Symposium on Advanced Intelligent Systems, November 20–24, 2012, Kobe, Japan. Macro and Micro Environment

[26] Simon, Herbert A. The sciences of the artificial. MIT press, 1996.

[27] Situngkir, Hokky (2004). "Epidemiology Through Cellular Automata: Case of Study Avian Influenza in Indonesia". arXiv:nlin/0403035.

[28] Caplat, Paul; Anand, Madhur; Bauch, Chris (March 10, 2008). "Symmetric competition causes population oscillations in an individual-based model of forest dynamics". *Ecological Modelling*. **211** (3–4): 491–500. doi:10.1016/j.ecolmodel.2007.10.002.

[29] Ch'ng, E. (2009) An Artificial Life-Based Vegetation Modelling Approach for Biodiversity Research, in Nature-Inspired informatics for Intelligent Applications and Knowledge Discovery: Implications in Business, Science and Engineering, R. Chiong, Editor. 2009, IGI Global: Hershey, PA. http://complexity.io/Publications/NII-alifeVeg-eCHNG.pdf

[30] Wirth, E.; Szabó, Gy.; Czinkóczky, A. (2016-06-07). "MEASURE OF LANDSCAPE HETEROGENEITY BY AGENT-BASED METHODOLOGY". *ISPRS Annals of Photogrammetry, Remote Sensing and Spatial Information Sciences*. III-8: 145–151. Bibcode:2016ISPAnIII8..145W. doi:10.5194/isprs-annals-iii-8-145-2016.

[31] Lima, Francisco W. S.; Hadzibeganovic, Tarik; Stauffer, Dietrich (2009). "Evolution of ethnocentrism on undirected and directed Barabási–Albert networks". *Physica A*. **388** (24): 4999–5004. Bibcode:2009PhyA..388.4999L. arXiv:0905.2672. doi:10.1016/j.physa.2009.08.029.

[32] Edwards, Scott (June 9, 2009). *The Chaos of Forced Migration: A Modeling Means to an Humanitarian End*. VDM Verlag. p. 168. ISBN 978-3-639-16516-6.

[33] Hadzibeganovic, Tarik; Stauffer, Dietrich; Schulze, Christian (2009). "Agent-based computer simulations of language choice dynamics". *Annals of the New York Academy of Sciences*. **1167**: 221–229. Bibcode:2009NYASA1167..221H. PMID 19580569. doi:10.1111/j.1749-6632.2009.04507.x.

[34] Tang, Jonathan; Enderling, Heiko; Becker-Weimann, Sabine; Pham, Christopher; Polyzos, Aris; Chen, Charlie; Costes, Sylvain (2011). "Phenotypic transition maps of 3D breast acini obtained by imaging-guided agent-based modeling". *Integrative Biology*. **3** (4): 408–21. PMC 4009383. PMID 21373705. doi:10.1039/c0ib00092b.

[35] Tang, Jonathan; Fernando-Garcia, Ignacio; Vijayakumar, Sangeetha; Martinez-Ruis, Haydeliz; Illa-Bochaca, Irineu; Nguyen, David; Mao, Jian-Hua; Costes, Sylvain; Barcellos-Hoff, Mary Helen (2014). "Irradiation of juvenile, but not adult, mammary gland increases stem cell self-renewal and estrogen receptor negative tumors". *Stem Cells*. **32** (3): 649–61. PMID 24038768. doi:10.1002/stem.1533.

[36] Tang, Jonathan; Ley, Klaus; Hunt, C. Anthony (2007). "Dynamics of in silico leukocyte rolling, activation, and adhesion". *BMC Systems Biology*. **1** (14): 14. PMC 1839892. PMID 17408504. doi:10.1186/1752-0509-1-14.

[37] Tang, Jonathan; Hunt, C. Anthony (2010). "Identifying the rules of engagement enabling leukocyte rolling, activation, and adhesion". *PLoS Computational Biology*. **6** (2): e1000681. Bibcode:2010PLSCB...6E0681T. PMC 2824748. PMID 20174606. doi:10.1371/journal.pcbi.1000681.

[38] Amnah Siddiqah; Muaz Niazi; Farah Mustafa; Habib Bokhari; Amir Hussain; Noreen Akram; Shabnum Shaheen; Fouzia Ahmed; Sarah Iqbal (August 15–16, 2009). "A new hybrid agent-based modeling decision support system for breast cancer research" (PDF). *Ieee Icict*. Karachi: IBA. (Breast Cancer DSS)

[39] Template:Cite journa\

[40] Barathy, Gnana; Yilmaz, Levent; Tolk, Andreas (March 2012). "Agent Directed Simulation for Combat Modeling and Distributed Simulation". *Engineering Principles of Combat Modeling and Distributed Simulation*. Hoboken, NJ: Wiley. pp. 669–714. ISBN 9781118180310. doi:10.1002/9781118180310.ch27.

[41] Azimi, Mohammad; Jamali, Yousef; Mofrad, Mohammad R. K. (2011). "Accounting for Diffusion in Agent Based Models of Reaction-Diffusion Systems with Application to Cytoskeletal Diffusion". *PLoS ONE*. **6** (9): e25306. Bibcode:2011PLoSO...625306A. PMC 3179499. PMID 21966493. doi:10.1371/journal.pone.0025306.

[42] Azimi, Mohammad; Mofrad, Mohammad R. K. (2013). "Higher Nucleoporin-Importinβ Affinity at the Nuclear Basket Increases Nucleocytoplasmic Import". *PLoS ONE*. **8** (11): e81741.

12.7. REFERENCES

Bibcode:2013PLoSO...881741A. PMC 3840022. PMID 24282617. doi:10.1371/journal.pone.0081741.

[43] Azimi, Mohammad; Bulat, Evgeny; Weis, Karsten; Mofrad, Mohammad R. K. (2014-11-05). "An agent-based model for mRNA export through the nuclear pore complex". *Molecular Biology of the Cell.* **25** (22): 3643–3653. PMC 4230623. PMID 25253717. doi:10.1091/mbc.E14-06-1065.

[44] Hughes, H. P. N.; Clegg, C. W.; Robinson, M. A.; Crowder, R. M. (2012). "Agent-based modelling and simulation: The potential contribution to organizational psychology". *Journal of Occupational and Organizational Psychology.* **85** (3): 487–502. doi:10.1111/j.2044-8325.2012.02053.x.

[45] Crowder, R. M.; Robinson, M. A.; Hughes, H. P. N.; Sim, Y. W. (2012). "The development of an agent-based modeling framework for simulating engineering team work". *IEEE Transactions on Systems, Man, and Cybernetics – Part A: Systems and Humans.* **42** (6): 1425–1439. doi:10.1109/TSMCA.2012.2199304.

[46] "Application of Agent Technology to Traffic Simulation". United States Department of Transportation. May 15, 2007.

[47] Niazi, M.; Baig, A. R.; Hussain, A.; Bhatti, S. (2008). Mason, S.; Hill, R.; Mönch, L.; Rose, O.; Jefferson, T.; Fowler, J. W., eds. "Simulation of the Research Process" (PDF). *Proceedings of the 40th Conference on Winter Simulation (Miami, Florida, December 7 – 10, 2008)*: 1326–1334.

[48] Niazi, Muaz A. (2008). "Self-Organized Customized Content Delivery Architecture for Ambient Assisted Environments" (PDF). *UPGRADE '08: Proceedings of the third international workshop on Use of P2P, grid and agents for the development of content networks*: 45–54.

[49] Nasrinpour, Hamid Reza; Friesen, Marcia R.; McLeod, Robert D. (2016-11-22). "An Agent-Based Model of Message Propagation in the Facebook Electronic Social Network". arXiv:1611.07454 [cs.SI].

[50] Niazi, Muaz; Hussain, Amir (March 2009). "Agent based Tools for Modeling and Simulation of Self-Organization in Peer-to-Peer, Ad-Hoc and other Complex Networks" (PDF). *IEEE Communications Magazine.* **47** (3): 163–173. doi:10.1109/MCOM.2009.4804403.

[51] Niazi, Muaz; Hussain, Amir (2011). "A Novel Agent-Based Simulation Framework for Sensing in Complex Adaptive Environments" (PDF). *IEEE Sensors Journal.* **11** (2): 404–412. doi:10.1109/JSEN.2010.2068044.

[52] Sarker, R. A.; Ray, T. (2010). "Agent Based Evolutionary Approach: An Introduction". *Agent-Based Evolutionary Search.* Adaptation, Learning, and Optimization. **5**. p. 1. ISBN 978-3-642-13424-1. doi:10.1007/978-3-642-13425-8_1.

[53] Page, Scott E. (2008). *Agent-Based Models. The New Palgrave Dictionary of Economics* (2 ed.).

[54] Testfatsion, Leigh; Judd, Kenneth, eds. (May 2006). *Handbook of Computational Economics.* **2**. Elsevier. p. 904. ISBN 978-0-444-51253-6. (Chapter preview)

[55] "Agents of change". *The Economist.* July 22, 2010. Retrieved February 16, 2011.

[56] "A model approach" (Editorial). *Nature.* **460** (7256): 667. August 6, 2009. Bibcode:2009Natur.460Q.667.. doi:10.1038/460667a.

[57] Farmer & Foley 2009, p. 685.

[58] Farmer & Foley 2009, p. 686.

[59] Stefan, F., & Atman, A. (2015). Is there any connection between the network morphology and the fluctuations of the stock market index? Physica A: Statistical Mechanics and Its Applications, (419), 630-641.

[60] Aschwanden, G.D.P.A; Wullschleger, Tobias; Müller, Hanspeter; Schmitt, Gerhard (2009). "Evaluation of 3D city models using automatic placed urban agents". *Automation in Construction.* **22**: 81–89. doi:10.1016/j.autcon.2011.07.001.

[61] Smetanin, Paul; Stiff, David (2015). Investing in Ontario's Public Infrastructure: A Prosperity at Risk Perspective, with an analysis of the Greater Toronto and Hamilton Area (PDF). *The Canadian Centre for Economic Analysis* (Report).

[62] "Agent-Directed Simulation".

[63] Isaac Rudomin; et al. (2006). "Large Crowds in the GPU". Monterrey Institute of Technology and Higher Education. Archived from the original on January 11, 2014.

[64] D'Souza, Roshan M. "Mega-Scale Interactive Agent-Based Model Simulations on the GPU". Michigan Technological University.

[65] Richmond, Paul; Romano, Daniela M. (2008). "Agent Based GPU, a Real-time 3D Simulation and Interactive Visualisation Framework for Massive Agent Based Modelling on the GPU" (PDF). *Proceedings International Workshop on Super Visualisation (IWSV08)*. Retrieved April 27, 2012.

[66] Sargent, R. G. (2000). "Verification, validation and accreditation of simulation models". *2000 Winter Simulation Conference Proceedings (Cat. No.00CH37165).* **1**. pp. 50–59. ISBN 0-7803-6579-8. doi:10.1109/WSC.2000.899697.

[67] Galán, José Manuel; Izquierdo, Luis; Izquierdo, Segismundo S.; Santos, José Ignacio; del Olmo, Ricardo; López-Paredes, Adolfo; Edmonds, Bruce (2009). "Errors and Artefacts in Agent-Based Modelling". *Journal of Artificial Societies and Social Simulation.* **12** (1): 1.

[68] Klügl, F. (2008). "A validation methodology for agent-based simulations". *Proceedings of the 2008 ACM symposium on Applied computing - SAC '08*. p. 39. ISBN 9781595937537. doi:10.1145/1363686.1363696.

[69] Fortino, G.; Garro, A.; Russo, W. (2005). "A Discrete-Event Simulation Framework for the Validation of Agent-Based and Multi-Agent Systems" (PDF).

[70] Tesfatsion, Leigh. "Empirical Validation: Agent-Based Computational Economics". Iowa State University.

[71] Niazi, Muaz; Hussain, Amir; Kolberg, Mario. "Verification and Validation of Agent-Based Simulations using the VOMAS approach" (PDF). *Proceedings of the Third Workshop on Multi-Agent Systems and Simulation '09 (MASS '09), as part of MALLOW 09, Sep 7–11, 2009, Torino, Italy.* Archived from the original (PDF) on June 14, 2011.

[72] Niazi, Muaz; Siddique, Qasim; Hussain, Amir; Kolberg, Mario (April 11–15, 2010). "Verification & Validation of an Agent-Based Forest Fire Simulation Model" (PDF). *Proceedings of the Agent Directed Simulation Symposium 2010, as part of the ACM SCS Spring Simulation Multiconference.* Orlando, FL.: 142–149. Archived from the original (PDF) on July 25, 2011.

[73] Niazi, Muaz A. K. (June 11, 2011). "Towards A Novel Unified Framework for Developing Formal, Network and Validated Agent-Based Simulation Models of Complex Adaptive Systems". University of Stirling. PhD Thesis

[74] Kalmykov, Lev V.; Kalmykov, Vyacheslav L. (2015). "A Solution to the Biodiversity Paradox by Logical Deterministic Cellular Automata", *Acta Biotheoretica*: 1–19, doi:10.1007/s10441-015-9257-9

[75] Kalmykov, Lev V.; Kalmykov, Vyacheslav L. (2015). "A white-box model of S-shaped and double S-shaped single-species population growth", *PeerJ*, 3:e948, doi:10.7717/peerj.948

12.7.2 General

- Barnes, D.J.; Chu, D. (2010). *Introduction to Modelling for Biosciences (chapter 2 & 3)*. Springer Verlag. ISBN 978-1-84996-325-1.

- Carley, Kathleen M. "Smart Agents and Organizations of the Future". In Lievrouw, Leah; Livingstone, Sonia. *Handbook of New Media*. Thousand Oaks, CA.: Sage. pp. 206–220.

- Farmer, J. Doyne; Foley, Duncan (August 6, 2009). "Nature". *Nature*. **460** (7256): 685–686. Bibcode:2009Natur.460..685F. PMID 19661896. doi:10.1038/460685a.

- Gilbert, Nigel; Troitzsch, Klaus (2005). *Simulation for the Social Scientist* (2 ed.). Open University Press. ISBN 978-0-335-21600-0. first edition, 1999.

- Gilbert, Nigel (2008). *Agent-based Models*. SAGE. ISBN 9781412949644.

- Helbing, Dirk; Balietti, Stefano. Helbing, Dirk, ed. "Agent-Based Modeling" (PDF). *Social Self-Organization*. Berlin: Springer: 25–70.

- Holland, John H. (1992). "Genetic Algorithms". *Scientific American*. **267** (1): 66–72. doi:10.1038/scientificamerican0792-66.

- Holland, John H. (September 1, 1996). *Hidden Order: How Adaptation Builds Complexity* (1 ed.). Reading, Mass.: Addison-Wesley. ISBN 978-0-201-44230-4.

- Miller, John H.; Page, Scott E. (March 5, 2007). *Complex Adaptive Systems: An Introduction to Computational Models of Social Life*. Princeton, NJ: Princeton University Press. ISBN 978-0-691-12702-6.

- Murthy, V. K.; Krishnamurthy, E. V. (2009). "Multiset of Agents in a Network for Simulation of Complex Systems". *Recent Advances in Nonlinear Dynamics and Synchronization*. Studies in Computational Intelligence. **254**. p. 153. ISBN 978-3-642-04226-3. doi:10.1007/978-3-642-04227-0_6.

- O'Sullivan, D.; Haklay, M. (2000). "Agent-based models and individualism: Is the world agent-based?". *Environment and Planning A*. **32** (8): 1409–1425. doi:10.1068/a32140.

- Naldi, G.; Pareschi, L.; Toscani, G. (2010). *Mathematical modeling of collective behavior in socio-economic and life sciences*. Birkhauser. ISBN 978-0-8176-4945-6.

- Preis, T.; Golke, S.; Paul, W.; Schneider, J. J. (2006). "Multi-agent-based Order Book Model of financial markets". *Europhysics Letters (EPL)*. **75** (3): 510–516. Bibcode:2006EL......75..510P. doi:10.1209/epl/i2006-10139-0.

- Rudomín, I.; Millán, E.; Hernández, B. N. (November 2005). "Fragment shaders for agent animation using finite state machines". *Simulation Modelling Practice and Theory*. Elsevier. **13** (8): 741–751. doi:10.1016/j.simpat.2005.08.008.

- Salamon, Tomas (2011). *Design of Agent-Based Models: Developing Computer Simulations for a Better Understanding of Social Processes*. Bruckner Publishing. ISBN 978-80-904661-1-1.

- Sallach, David; Macal, Charles (2001). "The simulation of social agents: an introduction". *Social Science Computer Review*. **19** (33): 245–248. doi:10.1177/089443930101900301.

- Shoham, Yoav; Leyton-Brown, Kevin (2009). *Multiagent Systems: Algorithmic, Game-Theoretic, and*

Logical Foundations. Cambridge University Press. p. 504. ISBN 978-0-521-89943-7.

12.8 External links

12.8.1 Articles/general Information

- Agent-based models of social networks, java applets.
- On-Line Guide for Newcomers to Agent-Based Modeling in the Social Sciences
- Introduction to Agent-based Modeling and Simulation. Argonne National Laboratory, November 29, 2006.
- Agent-based models in Ecology – Using computer models as theoretical tools to analyze complex ecological systems
- Open Agent-Based Modeling Consortium's Agent Based Modeling FAQ
- Multiagent Information Systems – Article on the convergence of SOA, BPM and Multi-Agent Technology in the domain of the Enterprise Information Systems. Jose Manuel Gomez Alvarez, Artificial Intelligence, Technical University of Madrid – 2006
- Artificial Life Framework
- Article providing methodology for moving real world human behaviors into a simulation model where agent behaviors are represented
- Agent-based Modeling Resources, an information hub for modelers, methods, and philosophy for agent-based modeling
- An Agent-Based Model of the Flash Crash of May 6, 2010, with Policy Implications, Tommi A. Vuorenmaa (Valo Research and Trading), Liang Wang (University of Helsinki - Department of Computer Science), October, 2013

12.8.2 Simulation models

- Collection of Agent-Based Models at RunTheModel.com
- Multi-agent Meeting Scheduling System Model by Qasim Siddique
- Multi-firm market simulation by Valentino Piana

Chapter 13

Synchronization

"Synchronised" redirects here. For the racehorse, see Synchronised (horse).
For other uses, see Synchronization (disambiguation).
Synchronization is the coordination of events to operate

Firefighters marching in a parade

a system in unison. The conductor of an orchestra serves to keep the orchestra *in time*. Systems operating with all their parts in synchrony are said to be *synchronous* or *in sync*; those which are not are *asynchronous*.

Today, synchronization can occur on a global basis through satellite navigation signals.

13.1 Transport

Time-keeping and synchronization of clocks was a critical problem in long-distance ocean navigation; accurate time is required in conjunction with astronomical observations to determine how far East or West a vessel has traveled. The invention of an accurate marine chronometer revolutionized marine navigation. By the end of the 19th century, time signals in the form of a signal gun, flag, or dropping time ball, were provided at important ports so that mariners could check their chronometers for error.

Synchronization was important in the operation of 19th century railways, these being the first major means of transport fast enough for the differences in local time between adjacent towns to be noticeable. Each line handled the problem by synchronizing all its stations to headquarters as a standard railroad time. In some territories, sharing of single railroad tracks was controlled by the timetable. The need for strict timekeeping led the companies to settle on one standard, and civil authorities eventually abandoned local mean solar time in favor of that standard.

13.2 Communication

Further information: Synchronization in telecommunications

In electrical engineering terms, for digital logic and data transfer, a synchronous circuit requires a clock signal. However, the use of the word "clock" in this sense is different from the typical sense of a clock as a device that keeps track of time-of-day; the clock signal simply signals the start and/or end of some time period, often very minute (measured in microseconds or nanoseconds), that has an arbitrary relationship to sidereal, solar, or lunar time, or to any other system of measurement of the passage of minutes, hours, and days.

In a different sense, electronic systems are sometimes synchronized to make events at points far apart appear simultaneous or near-simultaneous from a certain perspective. (Albert Einstein proved in 1905 in his first relativity paper that there actually are no such things as absolutely simultaneous events.) Timekeeping technologies such as the GPS satellites and Network Time Protocol (NTP) provide real-time access to a close approximation to the UTC timescale and are used for many terrestrial synchronization applications of this kind.

Synchronization is an important concept in the following fields:

- Computer science (In computer science, especially parallel computing, synchronization refers to the coordination of simultaneous threads or processes to complete a task with correct runtime order and no unexpected race conditions.)
- Cryptography
- Multimedia
- Music (rhythm)
- Neuroscience
- Photography
- Physics (The idea of simultaneity has many difficulties, both in practice and theory.)
- Synthesizers
- Telecommunication

13.3 Dynamical systems

Synchronization of multiple interacting dynamical systems can occur when the systems are autonomous oscillators. For instance, integrate-and-fire oscillators with either two-way (symmetric) or one-way coupling can synchronize when the strength of the coupling (in frequency units) is greater than the differences among the free-running natural oscillator frequencies. Poincare phase oscillators are model systems that can interact and partially synchronize within random or regular networks.[1] In the case of global synchronization of phase oscillators, an abrupt transition from unsynchronized to full synchronization takes place when the coupling strength exceeds a critical threshold. This is known as the Kuramoto model phase transition. Synchronization is an emergent property that occurs in a broad range of dynamical systems, including neural signaling, the beating of the heart and the synchronization of fire-fly light waves.

13.4 Human movement

Synchronization of movement is defined as similar movements between two or more people which are temporally aligned.[2] This is different to mimicry, as these movements occur after a short delay.[3] Muscular bonding is the idea that moving in time evokes particular emotions.[4] This sparked some of the first research into movement synchronization and its effects on human emotion.

In groups, synchronization of movement has been shown to increase conformity,[5] cooperation and trust[6] however more research on group synchronization is needed to determine its effects on the group as a whole and on individuals within a group. In dyads, groups of two people, synchronization has been demonstrated to increase affiliation,[7] self-esteem,[8] compassion and altruistic behaviour[9] and increase rapport.[10] During arguments, synchrony between the arguing pair has been noted to decrease, however it is not clear whether this is due to the change in emotion or other factors.[11] There is evidence to show that movement synchronization requires other people to cause its beneficial effects, as the effect on affiliation does not occur when one of the dyad is synchronizing their movements to something outside the dyad.[7] This is known as interpersonal synchrony.

There has been dispute regarding the true effect of synchrony in these studies. Research in this area detailing the positive effects of synchrony, have attributed this to synchrony alone; however, many of the experiments incorporate a shared intention to achieve synchrony. Indeed, the Reinforcement of Cooperation Model suggests that perception of synchrony leads to reinforcement that cooperation is occurring, which leads to the pro-social effects of synchrony.[12] More research is required to separate the effect of intentionality from the beneficial effect of synchrony.[13]

13.5 Uses

- Film synchronization of image and sound in sound film.
- Synchronization is important in fields such as digital telephony, video and digital audio where streams of sampled data are manipulated.
- In electric power systems, alternator synchronization is required when multiple generators are connected to an electrical grid.
- Arbiters are needed in digital electronic systems such as microprocessors to deal with asynchronous inputs.

There are also electronic digital circuits called *synchronizers* that attempt to perform arbitration in one clock cycle. Synchronizers, unlike arbiters, are prone to failure. (See metastability in electronics).

- Encryption systems usually require some synchronization mechanism to ensure that the receiving cipher is decoding the right bits at the right time.

- Automotive transmissions contain synchronizers that bring the toothed rotating parts (gears and splined shaft) to the same rotational velocity before engaging the teeth.

- Film, video, and audio applications use time code to synchronize audio and video.

- Flash photography, see Flash synchronization

Some systems may be only approximately synchronized, or plesiochronous. Some applications require that relative offsets between events be determined. For others, only the order of the event is important.

13.6 See also

- Alaska yo-yo
- Asynchrony
- Atomic clock
- Clock synchronization
- Data synchronization
- Double-ended synchronization
- Einstein synchronization
- Entrainment
- File synchronization
- Flywheel
- Homochronous
- Kuramoto model
- Mutual exclusion
- Neural synchronization
- Phase-locked loops
- Phase synchronization
- Reciprocal socialization
- Synchronism
- Synchronization (alternating current)
- Synchronization in telecommunications
- Synchronization of chaos
- Synchronization rights
- Synchronizer (disambiguation)
- Synchronous conferencing
- Time
- Timing synchronization function (TSF)
- Time transfer
- Timecode
- Tuning fork

Order synchronization and related topics

- Rendezvous problem
- Interlocking
- Race condition
- Concurrency control
- Room synchronization
- Comparison of synchronous and asynchronous signalling

Video and audio engineering

- Genlock
- Jam sync
- Word sync

Aircraft gun engineering

- Synchronization gear

Compare with

- Synchronicity, an alternative organizing principle to causality conceived by Carl Jung.

13.7 References

[1] Nolte, David (2015). *Introduction to Modern Dynamics: Chaos, Networks, Space and Time*. Oxford University Press.

[2] Condon, W. S.; Ogston, W. D. (1 October 1966). "Sound film analysis of normal and pathological behavior patterns". *The Journal of Nervous and Mental Disease*. **143** (4): 338–347. ISSN 0022-3018. PMID 5958766. doi:10.1097/00005053-196610000-00005.

[3] Richardson, Michael J.; Marsh, Kerry L.; Schmidt, R. C. (1 February 2005). "Effects of visual and verbal interaction on unintentional interpersonal coordination". *Journal of Experimental Psychology. Human Perception and Performance*. **31** (1): 62–79. ISSN 0096-1523. PMID 15709863. doi:10.1037/0096-1523.31.1.62.

[4] McNeill, William Hardy (30 September 1997). *Keeping Together in Time*. ISBN 978-0-674-50230-7. hdl:2027/heb.04002.0001.001.

[5] Dong, Ping; Dai, Xianchi; Wyer, Robert S. (1 January 2015). "Actors conform, observers react: the effects of behavioral synchrony on conformity". *Journal of Personality and Social Psychology*. **108** (1): 60–75. ISSN 1939-1315. PMID 25437130. doi:10.1037/pspi0000001.

[6] "Synchrony and Cooperation – PubMed – Search Results". Retrieved 2017-02-02.

[7] Hove, Michael J.; Risen, Jane L. (2009). "It's All in the Timing: Interpersonal Synchrony Increases Affiliation". *Social Cognition*. **27** (6): 949. doi:10.1521/soco.2009.27.6.949.

[8] Lumsden, Joanne; Miles, Lynden K.; Macrae, C. Neil (1 January 2014). "Sync or sink? Interpersonal synchrony impacts self-esteem". *Frontiers in Psychology*. **5**: 1064. PMC 4168669. PMID 25285090. doi:10.3389/fpsyg.2014.01064.

[9] Valdesolo, Piercarlo; Desteno, David (1 April 2011). "Synchrony and the social tuning of compassion". *Emotion (Washington, D.C.)*. **11** (2): 262–266. ISSN 1931-1516. PMID 21500895. doi:10.1037/a0021302.

[10] Vacharkulksemsuk, Tanya; Fredrickson, Barbara L. (1 January 2012). "Strangers in sync: Achieving embodied rapport through shared movements". *Journal of Experimental Social Psychology*. **48** (1): 399–402. ISSN 0022-1031. PMC 3290409. PMID 22389521. doi:10.1016/j.jesp.2011.07.015.

[11] Paxton, Alexandra; Dale, Rick (1 January 2013). "Argument disrupts interpersonal synchrony". *Quarterly Journal of Experimental Psychology (2006)*. **66** (11): 2092–2102. ISSN 1747-0226. PMID 24303888. doi:10.1080/17470218.2013.853089.

[12] Reddish, Paul; Fischer, Ronald; Bulbulia, Joseph (1 January 2013). "Let's dance together: synchrony, shared intentionality and cooperation". *PloS One*. **8** (8): e71182. ISSN 1932-6203. PMC 3737148. PMID 23951106. doi:10.1371/journal.pone.0071182.

[13] Ellamil, Melissa; Berson, Josh; Margulies, Daniel S. (1 January 2016). "Influences on and Measures of Unintentional Group Synchrony". *Frontiers in Psychology*. **7**: 1744. PMC 5101201. PMID 27881968. doi:10.3389/fpsyg.2016.01744.

13.8 External links

- J. Domański "Mathematical synchronization of image and sound in an animated film"

Chapter 14

Ant colony optimization algorithms

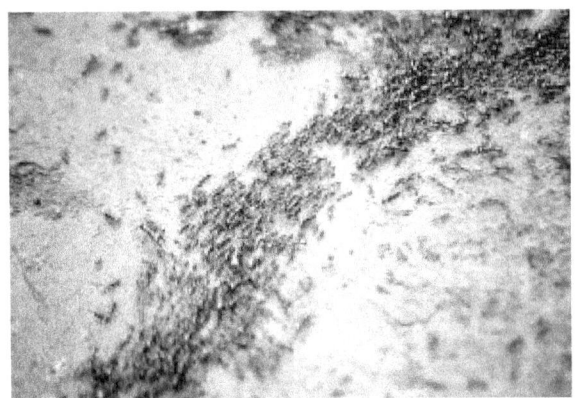

Ant behavior was the inspiration for the metaheuristic optimization technique

In computer science and operations research, the **ant colony optimization** algorithm (**ACO**) is a probabilistic technique for solving computational problems which can be reduced to finding good paths through graphs.

This algorithm is a member of the **ant colony algorithms** family, in swarm intelligence methods, and it constitutes some metaheuristic optimizations. Initially proposed by Marco Dorigo in 1992 in his PhD thesis,[1][2] the first algorithm was aiming to search for an optimal path in a graph, based on the behavior of ants seeking a path between their colony and a source of food. The original idea has since diversified to solve a wider class of numerical problems, and as a result, several problems have emerged, drawing on various aspects of the behavior of ants. From a broader perspective, ACO performs a model-based search [3] and share some similarities with Estimation of Distribution Algorithms.

14.1 Overview

In the natural world, ants of some species (initially) wander randomly, and upon finding food return to their colony while laying down pheromone trails. If other ants find such a path, they are likely not to keep travelling at random, but instead to follow the trail, returning and reinforcing it if they eventually find food (see Ant communication).

Over time, however, the pheromone trail starts to evaporate, thus reducing its attractive strength. The more time it takes for an ant to travel down the path and back again, the more time the pheromones have to evaporate. A short path, by comparison, gets marched over more frequently, and thus the pheromone density becomes higher on shorter paths than longer ones. Pheromone evaporation also has the advantage of avoiding the convergence to a locally optimal solution. If there were no evaporation at all, the paths chosen by the first ants would tend to be excessively attractive to the following ones. In that case, the exploration of the solution space would be constrained. The influence of pheromone evaporation in real ant systems is unclear, but it is very important in artificial systems.[4]

The overall result is that when one ant finds a good (i.e., short) path from the colony to a food source, other ants are more likely to follow that path, and positive feedback eventually leads to all the ants following a single path. The idea of the ant colony algorithm is to mimic this behavior with "simulated ants" walking around the graph representing the problem to solve.

14.2 Common extensions

Here are some of the most popular variations of ACO algorithms.

14.2.1 Elitist ant system

The global best solution deposits pheromone on every iteration along with all the other ants.

14.2.2 Max-min ant system (MMAS)

Added maximum and minimum pheromone amounts [τ_{max}, τ_{min}]. Only global best or iteration best tour deposited pheromone <MAZ>. All edges are initialized to τ_{min} and reinitialized to τ_{max} when nearing stagnation.[5]

14.2.3 Ant colony system

It has been presented above.[6]

14.2.4 Rank-based ant system (ASrank)

All solutions are ranked according to their length. The amount of pheromone deposited is then weighted for each solution, such that solutions with shorter paths deposit more pheromone than the solutions with longer paths.

14.2.5 Continuous orthogonal ant colony (COAC)

The pheromone deposit mechanism of COAC is to enable ants to search for solutions collaboratively and effectively. By using an orthogonal design method, ants in the feasible domain can explore their chosen regions rapidly and efficiently, with enhanced global search capability and accuracy.

The orthogonal design method and the adaptive radius adjustment method can also be extended to other optimization algorithms for delivering wider advantages in solving practical problems.[7]

14.2.6 Recursive ant colony optimization

It is a recursive form of ant system which divides the whole search domain into several sub-domains and solves the objective on these subdomains.[8] The results from all the subdomains are compared and the best few of them are promoted for the next level. The subdomains corresponding to the selected results are further subdivided and the process is repeated until an output of desired precision is obtained. This method has been tested on ill-posed geophysical inversion problems and works well.[9]

14.3 Convergence

For some versions of the algorithm, it is possible to prove that it is convergent (i.e., it is able to find the global optimum in finite time). The first evidence of a convergence ant colony algorithm was made in 2000, the graph-based ant system algorithm, and then algorithms for ACS and MMAS. Like most metaheuristics, it is very difficult to estimate the theoretical speed of convergence. In 2004, Zlochin and his colleagues[10] showed that COA-type algorithms could be assimilated methods of stochastic gradient descent, on the cross-entropy and estimation of distribution algorithm. They proposed these metaheuristics as a "research-based model". A performance analysis of continuous ant colony algorithm based on its various parameter suggest its sensitivity of convergence on parameter tuning.[11]

14.4 Example pseudo-code and formula

procedure ACO_MetaHeuristic while(not_termination) generateSolutions() daemonActions() pheromoneUpdate() end while end procedure

14.4.1 Edge selection

An ant is a simple computational agent in the ant colony optimization algorithm. It iteratively constructs a solution for the problem at hand. The intermediate solutions are referred to as solution states. At each iteration of the algorithm, each ant moves from a state x to state y, corresponding to a more complete intermediate solution. Thus, each ant k computes a set $A_k(x)$ of feasible expansions to its current state in each iteration, and moves to one of these in probability. For ant k, the probability p_{xy}^k of moving from state x to state y depends on the combination of two values, viz., the *attractiveness* η_{xy} of the move, as computed by some heuristic indicating the *a priori* desirability of that move and the *trail level* τ_{xy} of the move, indicating how proficient it has been in the past to make that particular move.

The *trail level* represents a posteriori indication of the desirability of that move. Trails are updated usually when all ants have completed their solution, increasing or decreasing the level of trails corresponding to moves that were part of "good" or "bad" solutions, respectively.

In general, the k th ant moves from state x to state y with probability

$$p_{xy}^k = \frac{(\tau_{xy}^\alpha)(\eta_{xy}^\beta)}{\sum_{z \in allowed_y} (\tau_{xz}^\alpha)(\eta_{xz}^\beta)}$$

where

τ_{xy} is the amount of pheromone deposited for transition from state x to y, $0 \le \alpha$ is a parameter to control the influence of τ_{xy}. η_{xy} is the desirability of state transition xy (*a priori* knowledge, typically $1/d_{xy}$, where d is the distance) and $\beta \geq 1$ is a parameter to control the influence of η_{xy}. τ_{xz} and η_{xz} represent the attractiveness and trail level for the other possible state transitions.

14.4.2 Pheromone update

When all the ants have completed a solution, the trails are updated by $\tau_{xy} \leftarrow (1-\rho)\tau_{xy} + \sum_k \Delta\tau^k_{xy}$

where τ_{xy} is the amount of pheromone deposited for a state transition xy, ρ is the *pheromone evaporation coefficient* and $\Delta\tau^k_{xy}$ is the amount of pheromone deposited by k th ant, typically given for a TSP problem (with moves corresponding to arcs of the graph) by

$$\Delta\tau^k_{xy} = \begin{cases} Q/L_k & \text{if ant } k \text{ uses curve } xy \text{ in its tour} \\ 0 & \text{otherwise} \end{cases}$$

where L_k is the cost of the k th ant's tour (typically length) and Q is a constant.

14.5 Applications

Knapsack problem: The ants prefer the smaller drop of honey over the more abundant, but less nutritious, sugar

Ant colony optimization algorithms have been applied to many combinatorial optimization problems, ranging from quadratic assignment to protein folding or routing vehicles and a lot of derived methods have been adapted to dynamic problems in real variables, stochastic problems, multi-targets and parallel implementations. It has also been used to produce near-optimal solutions to the travelling salesman problem. They have an advantage over simulated annealing and genetic algorithm approaches of similar problems when the graph may change dynamically; the ant colony algorithm can be run continuously and adapt to changes in real time. This is of interest in network routing and urban transportation systems.

The first ACO algorithm was called the ant system[12] and it was aimed to solve the travelling salesman problem, in which the goal is to find the shortest round-trip to link a series of cities. The general algorithm is relatively simple and based on a set of ants, each making one of the possible round-trips along the cities. At each stage, the ant chooses to move from one city to another according to some rules:

1. It must visit each city exactly once;
2. A distant city has less chance of being chosen (the visibility);
3. The more intense the pheromone trail laid out on an edge between two cities, the greater the probability that that edge will be chosen;
4. Having completed its journey, the ant deposits more pheromones on all edges it traversed, if the journey is short;
5. After each iteration, trails of pheromones evaporate.

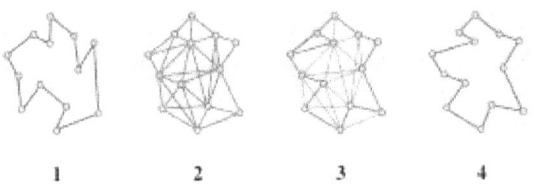

14.5.1 Scheduling problem

- Job-shop scheduling problem (JSP)[13]
- Open-shop scheduling problem (OSP)[14][15]
- Permutation flow shop problem (PFSP)[16]
- Single machine total tardiness problem (SMTTP)[17]
- Single machine total weighted tardiness problem (SMTWTP)[18][19][20]
- Resource-constrained project scheduling problem (RCPSP)[21]
- Group-shop scheduling problem (GSP)[22]

14.5. APPLICATIONS

- Single-machine total tardiness problem with sequence dependent setup times (SMTTPDST)[23]
- Multistage flowshop scheduling problem (MFSP) with sequence dependent setup/changeover times[24]

14.5.2 Vehicle routing problem

- Capacitated vehicle routing problem (CVRP)[25][26][27]
- Multi-depot vehicle routing problem (MDVRP)[28]
- Period vehicle routing problem (PVRP)[29]
- Split delivery vehicle routing problem (SDVRP)[30]
- Stochastic vehicle routing problem (SVRP)[31]
- Vehicle routing problem with pick-up and delivery (VRPPD)[32][33]
- Vehicle routing problem with time windows (VRPTW)[34][35][36]
- Time dependent vehicle routing problem with time windows (TDVRPTW)[37]
- Vehicle routing problem with time windows and multiple service workers (VRPTWMS)

14.5.3 Assignment problem

- Quadratic assignment problem (QAP)[38]
- Generalized assignment problem (GAP)[39][40]
- Frequency assignment problem (FAP)[41]
- Redundancy allocation problem (RAP)[42]

14.5.4 Set problem

- Set cover problem (SCP)[43][44]
- Partition problem (SPP)[45]
- Weight constrained graph tree partition problem (WCGTPP)[46]
- Arc-weighted l-cardinality tree problem (AWlCTP)[47]
- Multiple knapsack problem (MKP)[48]
- Maximum independent set problem (MIS)[49]

14.5.5 Device sizing problem in nanoelectronics physical design

- Ant colony optimization (ACO) based optimization of 45 nm CMOS-based sense amplifier circuit could converge to optimal solutions in very minimal time.[50]
- Ant colony optimization (ACO) based reversible circuit synthesis could improve efficiency significantly.[51]

14.5.6 Antennas Optimization and Synthesis

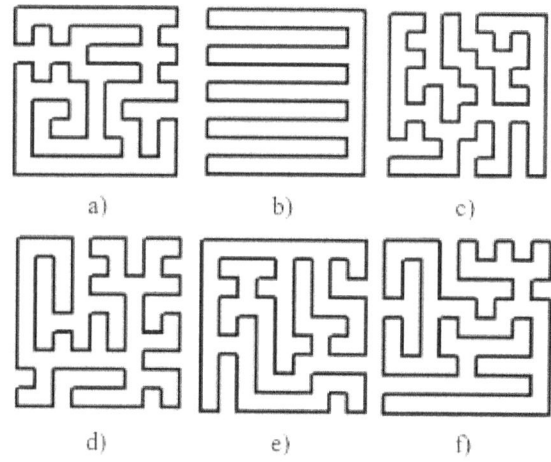

Loopback vibrators 10×10, synthesized by means of ACO algorithm[52]

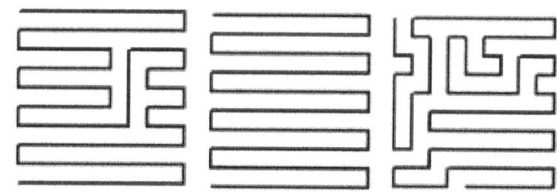

Unloopback vibrators 10×10, synthesized by means of ACO algorithm[52]

To optimize the form of an antennas ant colony algorithm can be used. As example can be considare antennas RFID-tags based on ant colony algorithms (ACO) [53], loopback and unloopback vibrators 10×10[52].

14.5.7 Image processing

The ACO algorithm is used in image processing for image edge detection and edge linking.[54][55]

- **Edge detection:**

The graph here is the 2-D image and the ants traverse from one pixel depositing pheromone. The movement of ants from one pixel to another is directed by the local variation of the image's intensity values. This movement causes the highest density of the pheromone to be deposited at the edges.

The following are the steps involved in edge detection using ACO:[56][57][58]

Step1: Initialization:
Randomly place K ants on the image $I_{M_1 M_2}$ where $K = (M_1 * M_2)^{\frac{1}{2}}$. Pheromone matrix $\tau_{(i,j)}$ are initialized with a random value. The major challenge in the initialization process is determining the heuristic matrix.

There are various methods to determine the heuristic matrix. For the below example the heuristic matrix was calculated based on the local statistics: the local statistics at the pixel position (i,j).

$$\eta_{(i,j)} = \tfrac{1}{Z} * Vc * I_{(i,j)}$$

Where I is the image of size $M_1 * M_2$
$Z = \sum_{i=1:M_1} \sum_{j=1:M_2} Vc(I_{i,j})$,which is a normalization factor

$$Vc(I_{i,j}) = f(|I_{(i-2,j-1)} - I_{(i+2,j+1)}| + |I_{(i-2,j+1)} - I_{(i+2,j-1)}|$$
$$+ |I_{(i-1,j-2)} - I_{(i+1,j+2)}| + |I_{(i-1,j-1)} - I_{(i+1,j+1)}|$$
$$+ |I_{(i-1,j)} - I_{(i+1,j)}| + |I_{(i-1,j+1)} - I_{(i-1,j-1)}|$$
$$+ |I_{(i-1,j+2)} - I_{(i-1,j-2)}| + |I_{(i,j-1)} - I_{(i,j+1)}|)$$

$f(\cdot)$ can be calculated using the following functions:
$f(x) = \lambda x.$ (1) $0; \geq x$ for $f(x) = \lambda x^2.$ (2) $0; \geq x$ for

$$f(x) = \begin{cases} \sin(\tfrac{\pi x}{2\lambda}), & \leq x \leq 0 \text{ for } \lambda (3) \\ 0, & \text{else} \end{cases} \quad f(x) =$$

$$\begin{cases} \pi x \sin(\tfrac{\pi x}{2\lambda}), & \leq x \leq 0 \text{ for } \lambda (4) \\ 0, & \text{else} \end{cases}$$
The parameter λ in each of above functions adjusts the functions' respective shapes.

Step 2 Construction process:
The ant's movement is based on 4-connected pixels or 8-connected pixels. The probability with which the ant moves is given by the probability equation $P_{x,y}$

Step 3 and Step 5 Update process:
The pheromone matrix is updated twice. in step 3 the trail of the ant (given by $\tau_{(x,y)}$) is updated where as in step 5 the evaporation rate of the trail is updated which is given by the below equation.
$\tau_{new} \leftarrow (1 - \psi)\tau_{old} + \psi\tau_0$, where ψ is the pheromone decay coefficient $0 < \tau < 1$

Step 7 Decision Process:
Once the K ants have moved a fixed distance L for N iteration, the decision whether it is an edge or not is based on the threshold T on the pheromone matrixτ. Threshold for the below example is calculated based on Otsu's method.

Image Edge detected using ACO:
The images below are generated using different functions given by the equation (1) to (4).[59]

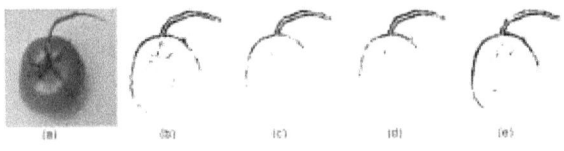

- **Edge linking:**[60]

ACO has also been proven effective in edge linking algorithms too.

14.5.8 Others

- Classification[61]
- Connection-oriented network routing[62]
- Connectionless network routing[63][64]
- Data mining[61][65][66][67]
- Discounted cash flows in project scheduling[68]
- Distributed information retrieval[69][70]
- Grid workflow scheduling problem[71]
- Intelligent testing system[72]
- System identification[73][74]
- Protein folding[75][76][77]
- Power electronic circuit design[78]
- bankruptcy prediction[79]
- Inhibitory peptide design for protein - protein interactions[80]

14.6 Definition difficulty

With an ACO algorithm, the shortest path in a graph, between two points A and B, is built from a combination of several paths.[81] It is not easy to give a precise definition of what algorithm is or is not an ant colony, because the definition may vary according to the authors and uses. Broadly speaking, ant colony algorithms are regarded as

14.7 Stigmergy algorithms

There is in practice a large number of algorithms claiming to be "ant colonies", without always sharing the general framework of optimization by canonical ant colonies (COA).[86] In practice, the use of an exchange of information between ants via the environment (a principle called "stigmergy") is deemed enough for an algorithm to belong to the class of ant colony algorithms. This principle has led some authors to create the term "value" to organize methods and behavior based on search of food, sorting larvae, division of labour and cooperative transportation.[87]

14.8 Related methods

- Genetic algorithms (GA) maintain a pool of solutions rather than just one. The process of finding superior solutions mimics that of evolution, with solutions being combined or mutated to alter the pool of solutions, with solutions of inferior quality being discarded.

- Estimation of Distribution Algorithm (EDA) is an Evolutionary Algorithm that substitutes traditional reproduction operators by model-guided operators. Such models are learned from the population by employing machine learning techniques and represented as Probabilistic Graphical Models, from which new solutions can be sampled[88][89] or generated from guided-crossover.[90][91]

- Simulated annealing (SA) is a related global optimization technique which traverses the search space by generating neighboring solutions of the current solution. A superior neighbor is always accepted. An inferior neighbor is accepted probabilistically based on the difference in quality and a temperature parameter. The temperature parameter is modified as the algorithm progresses to alter the nature of the search.

- Reactive search optimization focuses on combining machine learning with optimization, by adding an internal feedback loop to self-tune the free parameters of an algorithm to the characteristics of the problem, of the instance, and of the local situation around the current solution.

- Tabu search (TS) is similar to simulated annealing in that both traverse the solution space by testing mutations of an individual solution. While simulated annealing generates only one mutated solution, tabu search generates many mutated solutions and moves to the solution with the lowest fitness of those generated. To prevent cycling and encourage greater movement through the solution space, a tabu list is maintained of

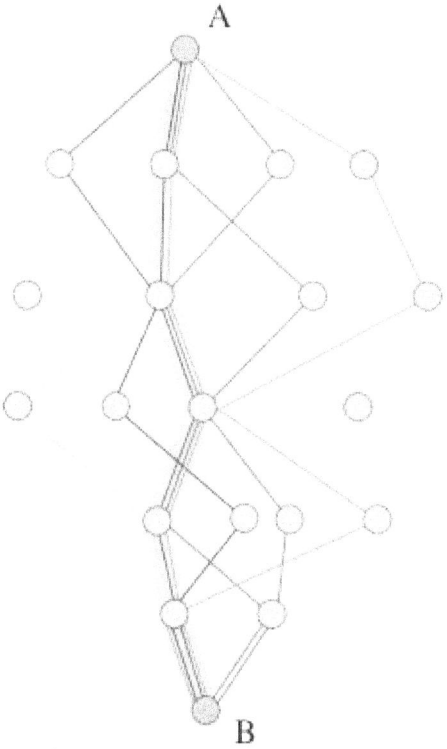

populated metaheuristics with each solution represented by an ant moving in the search space.[82] Ants mark the best solutions and take account of previous markings to optimize their search. They can be seen as probabilistic multi-agent algorithms using a probability distribution to make the transition between each iteration.[83] In their versions for combinatorial problems, they use an iterative construction of solutions.[84] According to some authors, the thing which distinguishes ACO algorithms from other relatives (such as algorithms to estimate the distribution or particle swarm optimization) is precisely their constructive aspect. In combinatorial problems, it is possible that the best solution eventually be found, even though no ant would prove effective. Thus, in the example of the Travelling salesman problem, it is not necessary that an ant actually travels the shortest route: the shortest route can be built from the strongest segments of the best solutions. However, this definition can be problematic in the case of problems in real variables, where no structure of 'neighbours' exists. The collective behaviour of social insects remains a source of inspiration for researchers. The wide variety of algorithms (for optimization or not) seeking self-organization in biological systems has led to the concept of "swarm intelligence",[85] which is a very general framework in which ant colony algorithms fit.

partial or complete solutions. It is forbidden to move to a solution that contains elements of the tabu list, which is updated as the solution traverses the solution space.

- Artificial immune system (AIS) algorithms are modeled on vertebrate immune systems.

- Particle swarm optimization (PSO), a swarm intelligence method

- Intelligent water drops (IWD), a swarm-based optimization algorithm based on natural water drops flowing in rivers

- Gravitational search algorithm (GSA), a swarm intelligence method

- Ant colony clustering method (ACCM), a method that make use of clustering approach, extending the ACO.

- Stochastic diffusion search (SDS), an agent-based probabilistic global search and optimization technique best suited to problems where the objective function can be decomposed into multiple independent partial-functions

14.9 History

Chronology of COA algorithms

Chronology of ant colony optimization algorithms.

- 1959, Pierre-Paul Grassé invented the theory of stigmergy to explain the behavior of nest building in termites;[92]

- 1983, Deneubourg and his colleagues studied the collective behavior of ants;[93]

- 1988, and Moyson Manderick have an article on **self-organization** among ants;[94]

- 1989, the work of Goss, Aron, Deneubourg and Pasteels on the **collective behavior of Argentine ants**, which will give the idea of ant colony optimization algorithms;[95]

- 1989, implementation of a model of behavior for food by Ebling and his colleagues;[96]

- 1991, M. Dorigo proposed the **ant system** in his doctoral thesis (which was published in 1992[2]). A technical report extracted from the thesis and co-authored by V. Maniezzo and A. Colorni[97] was published five years later;[12]

- 1994, Appleby and Steward of British Telecommunications Plc published the first application to telecommunications networks[98]

- 1996, publication of the article on ant system;[12]

- 1996, Hoos and Stützle invent the **max-min ant system**;[5]

- 1997, Dorigo and Gambardella publish the **ant colony system**;[6]

- 1997, Schoonderwoerd and his colleagues published an improved application to telecommunication networks;[99]

- 1998, Dorigo launches first conference dedicated to the ACO algorithms;[100]

- 1998, Stützle proposes initial **parallel implementations**;[101]

- 1999, Bonabeau, Dorigo and Theraulaz publish a book dealing mainly with artificial ants[102]

- 2000, special issue of the Future Generation Computer Systems journal on ant algorithms[103]

- 2000, first applications to the scheduling, scheduling sequence and the satisfaction of constraints;

- 2000, Gutjahr provides the first evidence of convergence for an algorithm of ant colonies[104]

- 2001, the first use of COA algorithms by companies (Eurobios and AntOptima);

- 2001, IREDA and his colleagues published the first **multi-objective** algorithm[105]

- 2002, first applications in the design of schedule, Bayesian networks;

- 2002, Bianchi and her colleagues suggested the first algorithm for stochastic problem;[106]

- 2004, Zlochin and Dorigo show that some algorithms are equivalent to the stochastic gradient descent, the cross-entropy method and algorithms to estimate distribution[10]

- 2005, first applications to protein folding problems.

- 2012, Prabhakar and colleagues publish research relating to the operation of individual ants communicating in tandem without pheromones, mirroring the principles of computer network organization. The communication model has been compared to the Transmission Control Protocol.[107]

- 2016, first application to peptide sequence design.[80]

14.10 References

[1] A. Colorni, M. Dorigo et V. Maniezzo, *Distributed Optimization by Ant Colonies*, actes de la première conférence européenne sur la vie artificielle, Paris, France, Elsevier Publishing, 134-142, 1991.

[2] M. Dorigo, *Optimization, Learning and Natural Algorithms*, PhD thesis, Politecnico di Milano, Italy, 1992.

[3] Zlochin, Mark; Birattari, Mauro; Meuleau, Nicolas; Dorigo, Marco (1 October 2004). "Model-Based Search for Combinatorial Optimization: A Critical Survey". *Annals of Operations Research*. 131 (1-4): 373–395. ISSN 0254-5330. doi:10.1023/B:ANOR.0000039526.52305.af.

[4] Marco Dorigo and Thomas Stützle, Ant Colony Optimization, p.12, 2004.

[5] T. Stützle et H.H. Hoos, *MAX MIN Ant System*, Future Generation Computer Systems, volume 16, pages 889-914, 2000.

[6] M. Dorigo et L.M. Gambardella, *Ant Colony System : A Cooperative Learning Approach to the Traveling Salesman Problem*, IEEE Transactions on Evolutionary Computation, volume 1, numéro 1, pages 53-66, 1997.

[7] X Hu, J Zhang, and Y Li (2008). Orthogonal methods based ant colony search for solving continuous optimization problems. *Journal of Computer Science and Technology*, 23(1), pp.2-18.

[8] Gupta, D.K.; Arora, Y.; Singh, U.K.; Gupta, J.P., "Recursive Ant Colony Optimization for estimation of parameters of a function," Recent Advances in Information Technology (RAIT), 2012 1st International Conference on , vol., no., pp.448-454, 15–17 March 2012

[9] Gupta, D.K.; Gupta, J.P.; Arora, Y.; Shankar, U., "Recursive ant colony optimization: a new technique for the estimation of function parameters from geophysical field data," Near Surface Geophysics , vol. 11, no. 3, pp.325-339

[10] M. Zlochin, M. Birattari, N. Meuleau, et M. Dorigo, *Model-based search for combinatorial optimization: A critical survey*, Annals of Operations Research, vol. 131, pp. 373-395, 2004.

[11] V.K.Ojha, A. Abraham and V. Snasel, ACO for Continuous Function Optimization: A Performance Analysis, 14th International Conference on Intelligent Systems Design and Applications (ISDA), Japan, Page 145 - 150 978-1-4799-7938-7/14 2014 IEEE

[12] M. Dorigo, V. Maniezzo, et A. Colorni, *Ant system: optimization by a colony of cooperating agents*, IEEE Transactions on Systems, Man, and Cybernetics--Part B , volume 26, numéro 1, pages 29-41, 1996.

[13] D. Martens, M. De Backer, R. Haesen, J. Vanthienen, M. Snoeck, B. Baesens, *Classification with Ant Colony Optimization*, IEEE Transactions on Evolutionary Computation, volume 11, number 5, pages 651—665, 2007.

[14] B. Pfahring, "Multi-agent search for open scheduling: adapting the Ant-Q formalism," Technical report TR-96-09, 1996.

[15] C. Blem, "Beam-ACO, Hybridizing ant colony optimization with beam search. An application to open shop scheduling," Technical report TR/IRIDIA/2003-17, 2003.

[16] T. Stützle, "An ant approach to the flow shop problem," Technical report AIDA-97-07, 1997.

[17] A. Bauer, B. Bullnheimer, R. F. Hartl and C. Strauss, "Minimizing total tardiness on a single machine using ant colony optimization," Central European Journal for Operations Research and Economics, vol.8, no.2, pp.125-141, 2000.

[18] M. den Besten, "Ants for the single machine total weighted tardiness problem," Master's thesis, University of Amsterdam, 2000.

[19] M. den Bseten, T. Stützle and M. Dorigo, "Ant colony optimization for the total weighted tardiness problem," Proceedings of PPSN-VI, Sixth International Conference on Parallel Problem Solving from Nature, vol. 1917 of Lecture Notes in Computer Science, pp.611-620, 2000.

[20] D. Merkle and M. Middendorf, "An ant algorithm with a new pheromone evaluation rule for total tardiness problems," Real World Applications of Evolutionary Computing, vol. 1803 of Lecture Notes in Computer Science, pp.287-296, 2000.

[21] D. Merkle, M. Middendorf and H. Schmeck, "Ant colony optimization for resource-constrained project scheduling," Proceedings of the Genetic and Evolutionary Computation Conference (GECCO 2000), pp.893-900, 2000.

[22] C. Blum, "ACO applied to group shop scheduling: a case study on intensification and diversification," Proceedings of ANTS 2002, vol. 2463 of Lecture Notes in Computer Science, pp.14-27, 2002.

[23] C. Gagné, W. L. Price and M. Gravel, "Comparing an ACO algorithm with other heuristics for the single machine scheduling problem with sequence-dependent setup times," Journal of the Operational Research Society, vol.53, pp.895-906, 2002.

[24] A. V. Donati, V. Darley, B. Ramachandran, "An Ant-Bidding Algorithm for Multistage Flowshop Scheduling Problem: Optimization and Phase Transitions", book chapter in Advances in Metaheuristics for Hard Optimization, Springer, ISBN 978-3-540-72959-4, pp.111-138, 2008.

[25] P. Toth, D. Vigo, "Models, relaxations and exact approaches for the capacitated vehicle routing problem," Discrete Applied Mathematics, vol.123, pp.487-512, 2002.

[26] J. M. Belenguer, and E. Benavent, "A cutting plane algorithm for capacitated arc routing problem," Computers & Operations Research, vol.30, no.5, pp.705-728, 2003.

[27] T. K. Ralphs, "Parallel branch and cut for capacitated vehicle routing," Parallel Computing, vol.29, pp.607-629, 2003.

[28] S. Salhi and M. Sari, "A multi-level composite heuristic for the multi-depot vehicle fleet mix problem," European Journal for Operations Research, vol.103, no.1, pp.95-112, 1997.

[29] E. Angelelli and M. G. Speranza, "The periodic vehicle routing problem with intermediate facilities," European Journal for Operations Research, vol.137, no.2, pp.233-247, 2002.

[30] S. C. Ho and D. Haugland, "A tabu search heuristic for the vehicle routing problem with time windows and split deliveries," Computers & Operations Research, vol.31, no.12, pp.1947-1964, 2004.

[31] N. Secomandi, "Comparing neuro-dynamic programming algorithms for the vehicle routing problem with stochastic demands," Computers & Operations Research, vol.27, no.11, pp.1201-1225, 2000.

[32] W. P. Nanry and J. W. Barnes, "Solving the pickup and delivery problem with time windows using reactive tabu search," Transportation Research Part B, vol.34, no. 2, pp.107-121, 2000.

[33] R. Bent and P.V. Hentenryck, "A two-stage hybrid algorithm for pickup and delivery vehicle routing problems with time windows," Computers & Operations Research, vol.33, no.4, pp.875-893, 2003.

[34] A. Bachem, W. Hochstattler and M. Malich, "The simulated trading heuristic for solving vehicle routing problems," Discrete Applied Mathematics, vol. 65, pp.47-72, 1996..

[35] [57] S. C. Hong and Y. B. Park, "A heuristic for bi-objective vehicle routing with time window constraints," International Journal of Production Economics, vol.62, no.3, pp.249-258, 1999.

[36] R. A. Rusell and W. C. Chiang, "Scatter search for the vehicle routing problem with time windows," European Journal for Operations Research, vol.169, no.2, pp.606-622, 2006.

[37] A. V. Donati, R. Montemanni, N. Casagrande, A. E. Rizzoli, L. M. Gambardella, "Time Dependent Vehicle Routing Problem with a Multi Ant Colony System", European Journal of Operational Research, vol.185, no.3, pp.1174–1191, 2008.

[38] T. Stützle, "MAX-MIN Ant System for quadratic assignment problems," Technical Report AIDA-97-4, FB Informatik, TU Darmstadt, Germany, 1997.

[39] R. Lourenço and D. Serra "Adaptive search heuristics for the generalized assignment problem," Mathware & soft computing, vol.9, no.2-3, 2002.

[40] M. Yagiura, T. Ibaraki and F. Glover, "An ejection chain approach for the generalized assignment problem," INFORMS Journal on Computing, vol. 16, no. 2, pp. 133–151, 2004.

[41] K. I. Aardal, S. P. M. van Hoesel, A. M. C. A. Koster, C. Mannino and Antonio. Sassano, "Models and solution techniques for the frequency assignment problem," A Quarterly Journal of Operations Research, vol.1, no.4, pp.261-317, 2001.

[42] Y. C. Liang and A. E. Smith, "An ant colony optimization algorithm for the redundancy allocation problem (RAP)," IEEE Transactions on Reliability, vol.53, no.3, pp.417-423, 2004.

[43] G. Leguizamon and Z. Michalewicz, "A new version of ant system for subset problems," Proceedings of the 1999 Congress on Evolutionary Computation(CEC 99), vol.2, pp.1458-1464, 1999.

[44] R. Hadji, M. Rahoual, E. Talbi and V. Bachelet "Ant colonies for the set covering problem," Abstract proceedings of ANTS2000, pp.63-66, 2000.

[45] V Maniezzo and M Milandri, "An ant-based framework for very strongly constrained problems," Proceedings of ANTS2000, pp.222-227, 2002.

[46] R. Cordone and F. Maffioli,"Colored Ant System and local search to design local telecommunication networks," Applications of Evolutionary Computing: Proceedings of Evo Workshops, vol.2037, pp.60-69, 2001.

[47] C. Blum and M.J. Blesa, "Metaheuristics for the edge-weighted k-cardinality tree problem," Technical Report TR/IRIDIA/2003-02, IRIDIA, 2003.

[48] S. Fidanova, "ACO algorithm for MKP using various heuristic information", Numerical Methods and Applications, vol.2542, pp.438-444, 2003.

[49] G. Leguizamon, Z. Michalewicz and Martin Schutz, "An ant system for the maximum independent set problem," Proceedings of the 2001 Argentinian Congress on Computer Science, vol.2, pp.1027-1040, 2001.

[50] O. Okobiah, S. P. Mohanty, and E. Kougianos, "Ordinary Kriging Metamodel-Assisted Ant Colony Algorithm for Fast Analog Design Optimization Archived March 4, 2016, at the Wayback Machine.", in Proceedings of the 13th IEEE International Symposium on Quality Electronic Design (ISQED), pp. 458-–463, 2012.

[51] M. Sarkar, P. Ghosal, and S. P. Mohanty, "Reversible Circuit Synthesis Using ACO and SA based Quinne-McCluskey Method Archived July 29, 2014, at the Wayback Machine.", in Proceedings of the 56th IEEE International Midwest Symposium on Circuits & Systems (MWSCAS), 2013, pp. 416-–419.

[52] Ermolaev S.Y., Slyusar V.I. Antenna synthesis based on the ant colony optimization algorithm.// Proc. ICATT'2009, Lviv, Ukraine 6 - 9 Octobre, 2009. - Pages 298 - 300

[53] Marcus Randall, Andrew Lewis, Amir Galehdar, David Thiel. Using Ant Colony Optimisation to Improve the Efficiency of Small Meander Line RFID Antennas.// In 3rd

14.10. REFERENCES

IEEE International e-Science and Grid Computing Conference, 2007

[54] S. Meshoul and M Batouche, "Ant colony system with extremal dynamics for point matching and pose estimation," Proceeding of the 16th International Conference on Pattern Recognition, vol.3, pp.823-826, 2002.

[55] H. Nezamabadi-pour, S. Saryazdi, and E. Rashedi, " Edge detection using ant algorithms", Soft Computing, vol. 10, no.7, pp. 623-628, 2006.

[56] Tian, Jing; Yu, Weiyu; Xie, Shengli. "An Ant Colony Optimization Algorithm For Image Edge Detection".

[57] Gupta, Charu; Gupta, Sunanda. "Edge Detection of an Image based on Ant ColonyOptimization Technique".

[58] Jevtić, A.; Quintanilla-Dominguez, J.; Cortina-Januchs, M.G.; Andina, D. (2009). "Edge detection using ant colony search algorithm and multiscale contrast enhancement". *IEEE International Conference on Systems, Man and Cybernetics, 2009. SMC 2009*: 2193–2198. doi:10.1109/ICSMC.2009.5345922.

[59] "File Exchange – Ant Colony Optimization (ACO)". *MATLAB Central*.

[60] Jevtić, A.; Melgar, I.; Andina, D. (2009). "Ant based edge linking algorithm". 35th Annual Conference of IEEE Industrial Electronics, 2009. IECON '09. pp. 3353–3358.

[61] D. Martens, M. De Backer, R. Haesen, J. Vanthienen, M. Snoeck, B. Baesens, "Classification with Ant Colony Optimization", IEEE Transactions on Evolutionary Computation, volume 11, number 5, pages 651—665, 2007.

[62] G. D. Caro and M. Dorigo, "Extending AntNet for best-effort quality-of-service routing," Proceedings of the First International Workshop on Ant Colony Optimization (ANTS'98), 1998.

[63] G.D. Caro and M. Dorigo "AntNet: a mobile agents approach to adaptive routing." Proceedings of the Thirty-First Hawaii International Conference on System Science, vol.7, pp.74-83, 1998.

[64] G. D. Caro and M. Dorigo, "Two ant colony algorithms for best-effort routing in datagram networks." Proceedings of the Tenth IASTED International Conference on Parallel and Distributed Computing and Systems (PDCS'98), pp.541-546, 1998.

[65] D. Martens, B. Baesens, T. Fawcett "Editorial Survey: Swarm Intelligence for Data Mining," Machine Learning, volume 82, number 1, pp. 1-42, 2011

[66] R. S. Parpinelli, H. S. Lopes and A. A Freitas, "An ant colony algorithm for classification rule discovery," Data Mining: A heuristic Approach, pp.191-209, 2002.

[67] R. S. Parpinelli, H. S. Lopes and A. A Freitas, "Data mining with an ant colony optimization algorithm," IEEE Transaction on Evolutionary Computation, vol.6, no.4, pp.321-332, 2002.

[68] W. N. Chen, J. ZHANG and H. Chung, "Optimizing Discounted Cash Flows in Project Scheduling--An Ant Colony Optimization Approach", IEEE Transactions on Systems, Man, and Cybernetics--Part C: Applications and Reviews Vol.40 No.5 pp.64-77, Jan. 2010.

[69] D. Picard, A. Revel, M. Cord, "An Application of Swarm Intelligence to Distributed Image Retrieval", Information Sciences, 2010

[70] D. Picard, M. Cord, A. Revel, "Image Retrieval over Networks : Active Learning using Ant Algorithm", IEEE Transactions on Multimedia, vol. 10, no. 7, pp. 1356-–1365 - nov 2008

[71] W. N. Chen and J. ZHANG "Ant Colony Optimization Approach to Grid Workflow Scheduling Problem with Various QoS Requirements", IEEE Transactions on Systems, Man, and Cybernetics--Part C: Applications and Reviews, Vol. 31, No. 1,pp.29-43.Jan 2009.

[72] Xiao. M.Hu, J. ZHANG, and H. Chung. "An Intelligent Testing System Embedded with an Ant Colony Optimization Based Test Composition Method", IEEE Transactions on Systems, Man, and Cybernetics--Part C: Applications and Reviews, Vol. 39, No. 6, pp. 659-669, Dec 2009.

[73] L. Wang and Q. D. Wu, "Linear system parameters identification based on ant system algorithm," Proceedings of the IEEE Conference on Control Applications, pp. 401-406, 2001.

[74] K. C. Abbaspour, R. Schulin, M. T. Van Genuchten, "Estimating unsaturated soil hydraulic parameters using ant colony optimization," Advances In Water Resources, vol. 24, no. 8, pp. 827-841, 2001.

[75] X. M. Hu, J. ZHANG,J. Xiao and Y. Li, "Protein Folding in Hydrophobic-Polar Lattice Model: A Flexible Ant- Colony Optimization Approach ", Protein and Peptide Letters, Volume 15, Number 5, 2008, Pp. 469-477.

[76] A. Shmygelska, R. A. Hernández and H. H. Hoos, "An ant colony optimization algorithm for the 2D HP protein folding problem," Proceedings of the 3rd International Workshop on Ant Algorithms/ANTS 2002, Lecture Notes in Computer Science, vol.2463, pp.40-52, 2002.

[77] M. Nardelli; L. Tedesco; A. Bechini (2013). "Cross-lattice behavior of general ACO folding for proteins in the HP model". *Proc. of ACM SAC 2013*: 1320–1327. doi:10.1145/2480362.2480611.

[78] J. ZHANG, H. Chung, W. L. Lo, and T. Huang, "Extended Ant Colony Optimization Algorithm for Power Electronic Circuit Design", IEEE Transactions on Power Electronic. Vol.24,No.1, pp.147-162, Jan 2009.

[79] Zhang, Y. (2013). "A Rule-Based Model for Bankruptcy Prediction Based on an Improved Genetic Ant Colony Algorithm". *Mathematical Problems in Engineering*. **2013**: 753251.

[80] Zaidman, Daniel; Wolfson, Haim J. (2016-08-01). "PinaColada: peptide–inhibitor ant colony ad-hoc design algorithm". *Bioinformatics*. **32** (15): 2289–2296. ISSN 1367-4803. doi:10.1093/bioinformatics/btw133.

[81] http://bmcbioinformatics.biomedcentral.com/articles/10.1186/1471-2105-6-30

[82] Fred W. Glover,Gary A. Kochenberger, *Handbook of Metaheuristics*, , Springer (2003)

[83] http://www.multiagent.fr/extensions/ICAPManager/pdf/LauriCharpillet2006.pdf

[84] WJ Gutjahr , *ACO algorithms with guaranteed convergence to the optimal solution*, , (2002)

[85] Waldner, Jean-Baptiste (2008). *Nanocomputers and Swarm Intelligence*. London: ISTE John Wiley & Sons. p. 214. ISBN 1-84704-002-0.

[86] Santpal Singh Dhillon , *Ant Routing, Searching and Topology Estimation Algorithms for Ad Hoc Networks*, , IOS Press, (2008)

[87] A. Ajith; G. Crina; R. Vitorino (éditeurs), *Stigmergic Optimization*, Studies in Computational Intelligence , volume 31, 299 pages, 2006. ISBN 978-3-540-34689-0

[88] Pelikan, Martin; Goldberg, David E.; Cantú-Paz, Erick (1 January 1999). "BOA: The Bayesian Optimization Algorithm". *Proceedings of the 1st Annual Conference on Genetic and Evolutionary Computation - Volume 1*. Morgan Kaufmann Publishers Inc.: 525–532.

[89] Pelikan, Martin (2005). *Hierarchical Bayesian optimization algorithm : toward a new generation of evolutionary algorithms* (1st ed.). Berlin [u.a.]: Springer. ISBN 978-3-540-23774-7.

[90] Thierens, Dirk (11 September 2010). "The Linkage Tree Genetic Algorithm". *Parallel Problem Solving from Nature, PPSN XI*. Springer Berlin Heidelberg: 264–273. doi:10.1007/978-3-642-15844-5_27.

[91] Martins, Jean P.; Fonseca, Carlos M.; Delbem, Alexandre C. B. (25 December 2014). "On the performance of linkage-tree genetic algorithms for the multidimensional knapsack problem". *Neurocomputing*. **146**: 17–29. doi:10.1016/j.neucom.2014.04.069.

[92] P.-P. Grassé, *La reconstruction du nid et les coordinations inter-individuelles chez Belicositermes natalensis et Cubitermes sp. La théorie de la Stigmergie : Essai d'interprétation du comportement des termites constructeurs*, Insectes Sociaux, numéro 6, p. 41-80, 1959.

[93] J.L. Denebourg, J.M. Pasteels et J.C. Verhaeghe, *Probabilistic Behaviour in Ants : a Strategy of Errors?*, Journal of Theoretical Biology, numéro 105, 1983.

[94] F. Moyson, B. Manderick, *The collective behaviour of Ants : an Example of Self-Organization in Massive Parallelism*, Actes de AAAI Spring Symposium on Parallel Models of Intelligence, Stanford, Californie, 1988.

[95] S. Goss, S. Aron, J.-L. Deneubourg et J.-M. Pasteels, *Self-organized shortcuts in the Argentine ant*, Naturwissenschaften, volume 76, pages 579-581, 1989

[96] M. Ebling, M. Di Loreto, M. Presley, F. Wieland, et D. Jefferson,*An Ant Foraging Model Implemented on the Time Warp Operating System*, Proceedings of the SCS Multiconference on Distributed Simulation, 1989

[97] Dorigo M., V. Maniezzo et A. Colorni, *Positive feedback as a search strategy*, rapport technique numéro 91-016, Dip. Elettronica, Politecnico di Milano, Italy, 1991

[98] Appleby, S. & Steward, S. Mobile software agents for control in telecommunications networks, BT Technol. J., 12(2):104–113, April 1994

[99] R. Schoonderwoerd, O. Holland, J. Bruten et L. Rothkrantz, *Ant-based load balancing in telecommunication networks*, Adaptive Behaviour, volume 5, numéro 2, pages 169-207, 1997

[100] M. Dorigo, ANTS' 98, *From Ant Colonies to Artificial Ants : First International Workshop on Ant Colony Optimization*, ANTS 98, Bruxelles, Belgique, octobre 1998.

[101] T. Stützle, *Parallelization Strategies for Ant Colony Optimization*, Proceedings of PPSN-V, Fifth International Conference on Parallel Problem Solving from Nature, Springer-Verlag, volume 1498, pages 722-731, 1998.

[102] É. Bonabeau, M. Dorigo et G. Theraulaz, *Swarm intelligence*, Oxford University Press, 1999.

[103] M. Dorigo , G. Di Caro et T. Stützle, *Special issue on "Ant Algorithms"*, Future Generation Computer Systems, volume 16, numéro 8, 2000

[104] W.J. Gutjahr, *A graph-based Ant System and its convergence*, Future Generation Computer Systems, volume 16, pages 873-888, 2000.

[105] S. Iredi, D. Merkle et M. Middendorf, *Bi-Criterion Optimization with Multi Colony Ant Algorithms*, Evolutionary Multi-Criterion Optimization, First International Conference (EMO'01), Zurich, Springer Verlag, pages 359-372, 2001.

[106] L. Bianchi, L.M. Gambardella et M.Dorigo, *An ant colony optimization approach to the probabilistic traveling salesman problem*, PPSN-VII, Seventh International Conference on Parallel Problem Solving from Nature, Lecture Notes in Computer Science, Springer Verlag, Berlin, Allemagne, 2002.

[107] B. Prabhakar, K. N. Dektar, D. M. Gordon, "The regulation of ant colony foraging activity without spatial information", PLOS Computational Biology, 2012. URL: http://www.ploscompbiol.org/article/info%3Adoi%2F10.1371%2Fjournal.pcbi.1002670

14.11 Publications (selected)

- M. Dorigo, 1992. *Optimization, Learning and Natural Algorithms*, PhD thesis, Politecnico di Milano, Italy.

- M. Dorigo, V. Maniezzo & A. Colorni, 1996. "Ant System: Optimization by a Colony of Cooperating Agents", IEEE Transactions on Systems, Man, and Cybernetics–Part B, 26 (1): 29–41.

- M. Dorigo & L. M. Gambardella, 1997. "Ant Colony System: A Cooperative Learning Approach to the Traveling Salesman Problem". IEEE Transactions on Evolutionary Computation, 1 (1): 53–66.

- M. Dorigo, G. Di Caro & L. M. Gambardella, 1999. "Ant Algorithms for Discrete Optimization". Artificial Life, 5 (2): 137–172.

- E. Bonabeau, M. Dorigo et G. Theraulaz, 1999. *Swarm Intelligence: From Natural to Artificial Systems*. Oxford University Press. ISBN 0-19-513159-2

- M. Dorigo & T. Stützle, 2004. *Ant Colony Optimization*. MIT Press. ISBN 0-262-04219-3

- M. Dorigo, 2007. "Ant Colony Optimization". Scholarpedia.

- C. Blum, 2005 "Ant colony optimization: Introduction and recent trends". Physics of Life Reviews, 2: 353-373

- M. Dorigo, M. Birattari & T. Stützle, 2006 *Ant Colony Optimization: Artificial Ants as a Computational Intelligence Technique*. TR/IRIDIA/2006-023

- Mohd Murtadha Mohamad,"Articulated Robots Motion Planning Using Foraging Ant Strategy",Journal of Information Technology - Special Issues in Artificial Intelligence, Vol.20, No. 4 pp. 163–181, December 2008. ISSN 0128-3790.

- N. Monmarché, F. Guinand & P. Siarry (eds), "Artificial Ants", August 2010 Hardback 576 pp. ISBN 978-1-84821-194-0.

- A. Kazharov, V. Kureichik, 2010. "Ant colony optimization algorithms for solving transportation problems". Journal of Computer and Systems Sciences International, Vol. 49. No. 1. pp. 30–43.

- C-M. Pintea, 2014. Advances in Bio-inspired Computing for Combinatorial Optimization Problem, Springer ISBN 978-3-642-40178-7

- K. Saleem, N. Fisal, M. A. Baharudin, A. A. Ahmed, S. Hafizah and S. Kamilah, "Ant colony inspired self-optimized routing protocol based on cross layer architecture for wireless sensor networks", WSEAS Trans. Commun., vol. 9, no. 10, pp. 669–678, 2010. ISBN 978-960-474-200-4

- K. Saleem and N. Fisal, "Enhanced Ant Colony algorithm for self-optimized data assured routing in wireless sensor networks", Networks (ICON) 2012 18th IEEE International Conference on, pp. 422–427. ISBN 978-1-4673-4523-1

14.12 External links

- Ant Colony Optimization Home Page

- "Ant Colony Optimization" - Russian scientific and research community

- AntSim - Simulation of Ant Colony Algorithms

- MIDACO-Solver General purpose optimization software based on ant colony optimization (Matlab, Excel, VBA, C/C++, R, C#, Java, Fortran and Python)

- University of Kaiserslautern, Germany, AG Wehn: Ant Colony Optimization Applet Visualization of Traveling Salesman solved by ant system with numerous options and parameters (Java Applet)

- Ant Farm Simulator

- Ant algorithm simulation (Java Applet)

- Java Ant Colony System Framework

Chapter 15

Particle swarm optimization

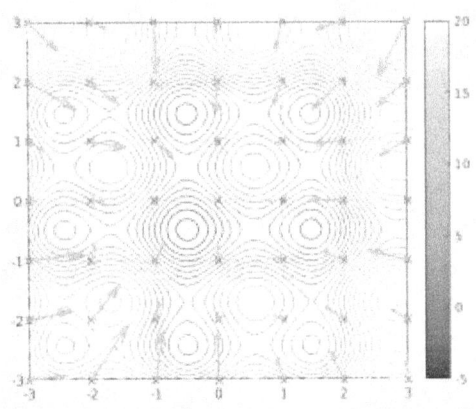

A particle swarm searching for the global minimum of a function

In computer science, **particle swarm optimization** (**PSO**) is a computational method that optimizes a problem by iteratively trying to improve a candidate solution with regard to a given measure of quality. It solves a problem by having a population of candidate solutions, here dubbed particles, and moving these particles around in the search-space according to simple mathematical formulae over the particle's position and velocity. Each particle's movement is influenced by its local best known position, but is also guided toward the best known positions in the search-space, which are updated as better positions are found by other particles. This is expected to move the swarm toward the best solutions.

PSO is originally attributed to Kennedy, Eberhart and Shi[1][2] and was first intended for simulating social behaviour,[3] as a stylized representation of the movement of organisms in a bird flock or fish school. The algorithm was simplified and it was observed to be performing optimization. The book by Kennedy and Eberhart[4] describes many philosophical aspects of PSO and swarm intelligence. An extensive survey of PSO applications is made by Poli.[5][6] Recently, a comprehensive review on theoretical and experimental works on PSO has been published by Bonyadi and Michalewicz.[7]

PSO is a metaheuristic as it makes few or no assumptions about the problem being optimized and can search very large spaces of candidate solutions. However, metaheuristics such as PSO do not guarantee an optimal solution is ever found. Also, PSO does not use the gradient of the problem being optimized, which means PSO does not require that the optimization problem be differentiable as is required by classic optimization methods such as gradient descent and quasi-newton methods.

15.1 Algorithm

A basic variant of the PSO algorithm works by having a population (called a swarm) of candidate solutions (called particles). These particles are moved around in the search-space according to a few simple formulae.[8] The movements of the particles are guided by their own best known position in the search-space as well as the entire swarm's best known position. When improved positions are being discovered these will then come to guide the movements of the swarm. The process is repeated and by doing so it is hoped, but not guaranteed, that a satisfactory solution will eventually be discovered.

Formally, let $f: \mathbb{R}^n \to \mathbb{R}$ be the cost function which must be minimized. The function takes a candidate solution as argument in the form of a vector of real numbers and produces a real number as output which indicates the objective function value of the given candidate solution. The gradient of f is not known. The goal is to find a solution $\mathbf{a}$ for which $f(\mathbf{a}) \leq f(\mathbf{b})$ for all $\mathbf{b}$ in the search-space, which would mean $\mathbf{a}$ is the global minimum. Maximization can be performed by considering the function $h = -f$ instead.

Let S be the number of particles in the swarm, each having a position $\mathbf{x}_i \in \mathbb{R}^n$ in the search-space and a velocity $\mathbf{v}_i \in \mathbb{R}^n$. Let $\mathbf{p}_i$ be the best known position of particle i and let $\mathbf{g}$ be the best known position of the entire swarm. A basic PSO algorithm is then:[9]

for each particle $i = 1, ..., S$ **do** Initialize the particle's position with a uniformly distributed random vector: $x_i \sim U(b_{lo}, b_{up})$ Initialize the particle's best known position to its initial position: $p_i \leftarrow x_i$ if $f(p_i) < f(g)$ then update the swarm's best known position: $g \leftarrow p_i$ Initialize the particle's velocity: $v_i \sim U(-|b_{up}-b_{lo}|, |b_{up}-b_{lo}|)$ while a termination criterion is not met do: for each particle $i = 1, ..., S$ do for each dimension $d = 1, ..., n$ do Pick random numbers: $r_p, r_g \sim U(0,1)$ Update the particle's velocity: $v_{i,d} \leftarrow \omega\, v_{i,d} + \varphi_p\, r_p\, (p_{i,d}-x_{i,d}) + \varphi_g\, r_g\, (g_d-x_{i,d})$ Update the particle's position: $x_i \leftarrow x_i + v_i$ if $f(x_i) < f(p_i)$ then Update the particle's best known position: $p_i \leftarrow x_i$ if $f(p_i) < f(g)$ then Update the swarm's best known position: $g \leftarrow p_i$

The values b_{lo} and b_{up} are respectively the lower and upper boundaries of the search-space. The termination criterion can be number of iterations performed, or a solution with adequate objective function value is found.[10] The parameters ω, φ_p, and φ_g are selected by the practitioner and control the behaviour and efficacy of the PSO method, see below.

15.2 Parameter selection

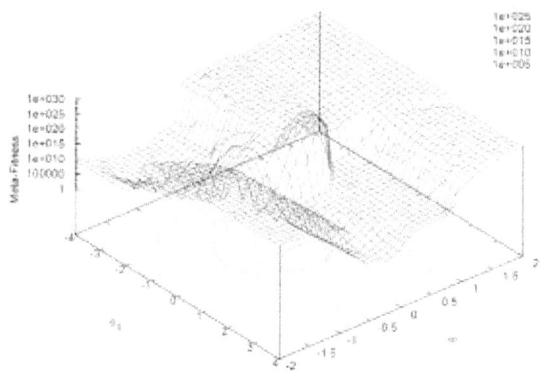

Performance landscape showing how a simple PSO variant performs in aggregate on several benchmark problems when varying two PSO parameters.

The choice of PSO parameters can have a large impact on optimization performance. Selecting PSO parameters that yield good performance has therefore been the subject of much research.[11][12][13][14][15][16][17][18][19]

The PSO parameters can also be tuned by using another overlaying optimizer, a concept known as meta-optimization.[20][21][22] or even fine-tuned during the optimization, e.g., by means of fuzzy logic.[23]

Parameters have also been tuned for various optimization scenarios.[24][25]

15.3 Neighbourhoods and topologies

The topology of the swarm defines the subset of particles with which each particle can exchange information.[26] The basic version of the algorithm uses the global topology as the swarm communication structure.[10] This topology allows all particles to communicate with all the other particles, thus the whole swarm share the same best position g from a single particle. However, this approach might lead the swarm to be trapped into a local minimum,[27] thus different topologies have been used to control the flow of information among particles. For instance, in local topologies, particles only share information with a subset of particles.[10] This subset can be a geometrical one[28] – for example "the m nearest particles" – or, more often, a social one, i.e. a set of particles that is not depending on any distance. In such cases, the PSO variant is said to be local best (vs global best for the basic PSO).

A commonly used swarm topology is the ring, in which each particle has just two neighbours, but there are many others.[10] The topology is not necessarily static. In fact, since the topology is related to the diversity of communication of the particles,[29] some efforts have been done to create adaptive topologies (SPSO,[30] stochastic star,[31] TRIBES,[32] Cyber Swarm,[33] and C-PSO[34]).

15.4 Inner workings

There are several schools of thought as to why and how the PSO algorithm can perform optimization.

A common belief amongst researchers is that the swarm behaviour varies between exploratory behaviour, that is, searching a broader region of the search-space, and exploitative behaviour, that is, a locally oriented search so as to get closer to a (possibly local) optimum. This school of thought has been prevalent since the inception of PSO.[2][3][12][16] This school of thought contends that the PSO algorithm and its parameters must be chosen so as to properly balance between exploration and exploitation to avoid premature convergence to a local optimum yet still ensure a good rate of convergence to the optimum. This belief is the precursor of many PSO variants, see below.

Another school of thought is that the behaviour of a PSO swarm is not well understood in terms of how it affects actual optimization performance, especially for higher-dimensional search-spaces and optimization problems that may be discontinuous, noisy, and time-varying. This school of thought merely tries to find PSO algorithms and parameters that cause good performance regardless of how the swarm behaviour can be interpreted in relation to e.g. exploration and exploitation. Such studies have led to the simplification of the PSO algorithm, see below.

15.4.1 Convergence

In relation to PSO the word *convergence* typically refers to two different definitions:

- Convergence of the sequence of solutions (aka, stability analysis, converging) in which all particles have converged to a point in the search-space, which may or may not be the optimum.

- Convergence to a local optimum where all personal bests **p** or, alternatively, the swarm's best known position **g**, approaches a local optimum of the problem, regardless of how the swarm behaves.

Convergence of the sequence of solutions has been investigated for PSO.[15][16][17] These analyses have resulted in guidelines for selecting PSO parameters that are believed to cause convergence to a point and prevent divergence of the swarm's particles (particles do not move unboundedly and will converge to somewhere). However, the analyses were criticized by Pedersen[22] for being oversimplified as they assume the swarm has only one particle, that it does not use stochastic variables and that the points of attraction, that is, the particle's best known position **p** and the swarm's best known position **g**, remain constant throughout the optimization process. However, it was shown[35] that these simplifications do not affect the boundaries found by these studies for parameter where the swarm is convergent.

Convergence to a local optimum has been analyzed for PSO in[36] and.[37] It has been proven that PSO need some modification to guarantee to find a local optimum.

This means that determining convergence capabilities of different PSO algorithms and parameters therefore still depends on empirical results. One attempt at addressing this issue is the development of an "orthogonal learning" strategy for an improved use of the information already existing in the relationship between **p** and **g**, so as to form a leading converging exemplar and to be effective with any PSO topology. The aims are to improve the performance of PSO overall, including faster global convergence, higher solution quality, and stronger robustness.[38] However, such studies do not provide theoretical evidence to actually prove their claims.

15.4.2 Biases

As the basic PSO works dimension by dimension, the solution point is easier found when it lies on an axis of the search space, on a diagonal, and even easier if it is right on the centre.[39][40]

One approach is to modify the algorithm so that it is not any more sensitive to the system of coordinates.[41][42][43][44] Note that some of these methods have a higher computational complexity (are in O(n^2) where n is the number of dimensions) that make the algorithm very slow for large scale optimization.[37]

The only currently existing PSO variant that is not sensitive to the rotation of the coordinates while is locally convergent has been proposed at 2014.[37] The method has shown a very good performance on many benchmark problems while its rotation invariance and local convergence have been mathematically proven.

15.5 Variants

Numerous variants of even a basic PSO algorithm are possible. For example, there are different ways to initialize the particles and velocities (e.g. start with zero velocities instead), how to dampen the velocity, only update p_i and **g** after the entire swarm has been updated, etc. Some of these choices and their possible performance impact have been discussed in the literature.[14]

A series of standard implementations have been created by leading researchers, "intended for use both as a baseline for performance testing of improvements to the technique, as well as to represent PSO to the wider optimization community. Having a well-known, strictly-defined standard algorithm provides a valuable point of comparison which can be used throughout the field of research to better test new advances."[10] The latest is Standard PSO 2011 (SPSO-2011).[45]

15.5.1 Hybridization

New and more sophisticated PSO variants are also continually being introduced in an attempt to improve optimization performance. There are certain trends in that research; one is to make a hybrid optimization method using PSO combined with other optimizers,[46][47][48] e.g., combined PSO with biogeography-based optimization,[49] and the incorporation of an effective learning method.[38]

15.5.2 Alleviate premature

Another research trend is to try and alleviate premature convergence (that is, optimization stagnation), e.g. by reversing or perturbing the movement of the PSO particles.[19][50][51][52] another approach to deal with premature convergence is the use of multiple swarms[53] (multi-swarm optimization). The multi-swarm approach can also be used to implement multi-objective optimization.[54] Finally, there are developments in adapting the behavioural parameters of PSO during optimization.[55]

15.5.3 Simplifications

Another school of thought is that PSO should be simplified as much as possible without impairing its performance; a general concept often referred to as Occam's razor. Simplifying PSO was originally suggested by Kennedy[3] and has been studied more extensively,[18][21][22][56] where it appeared that optimization performance was improved, and the parameters were easier to tune and they performed more consistently across different optimization problems.

Another argument in favour of simplifying PSO is that metaheuristics can only have their efficacy demonstrated empirically by doing computational experiments on a finite number of optimization problems. This means a metaheuristic such as PSO cannot be proven correct and this increases the risk of making errors in its description and implementation. A good example of this[57] presented a promising variant of a genetic algorithm (another popular metaheuristic) but it was later found to be defective as it was strongly biased in its optimization search towards similar values for different dimensions in the search space, which happened to be the optimum of the benchmark problems considered. This bias was because of a programming error, and has now been fixed.[58]

Initialization of velocities may require extra inputs. The Bare Bones PSO variant[59] has been proposed in 2003 by James Kennedy, and does not need to use velocity at all.

Another simpler variant is the accelerated particle swarm optimization (APSO),[60] which also does not need to use velocity and can speed up the convergence in many applications. A simple demo code of APSO is available.[61]

15.5.4 Multi-objective optimization

PSO has also been applied to multi-objective problems,[62][63] in which the objective function comparison takes pareto dominance into account when moving the PSO particles and non-dominated solutions are stored so as to approximate the pareto front.

15.5.5 Binary, discrete, and combinatorial

As the PSO equations given above work on real numbers, a commonly used method to solve discrete problems is to map the discrete search space to a continuous domain, to apply a classical PSO, and then to demap the result. Such a mapping can be very simple (for example by just using rounded values) or more sophisticated.[64]

However, it can be noted that the equations of movement make use of operators that perform four actions:

- computing the difference of two positions. The result is a velocity (more precisely a displacement)
- multiplying a velocity by a numerical coefficient
- adding two velocities
- applying a velocity to a position

Usually a position and a velocity are represented by n real numbers, and these operators are simply -, *, +, and again +. But all these mathematical objects can be defined in a completely different way, in order to cope with binary problems (or more generally discrete ones), or even combinatorial ones.[65][66][67][68] One approach is to redefine the operators based on sets.[69]

15.6 See also

- Bees algorithm / Artificial bee colony algorithm
- Derivative-free optimization
- Multi-swarm optimization
- Particle filter
- Swarm intelligence
- Fish School Search

15.7 References

[1] Kennedy, J.; Eberhart, R. (1995). "Particle Swarm Optimization". *Proceedings of IEEE International Conference on Neural Networks.* **IV.** pp. 1942–1948. doi:10.1109/ICNN.1995.488968.

[2] Shi, Y.; Eberhart, R.C. (1998). "A modified particle swarm optimizer". *Proceedings of IEEE International Conference on Evolutionary Computation.* pp. 69–73.

[3] Kennedy, J. (1997). "The particle swarm: social adaptation of knowledge". *Proceedings of IEEE International Conference on Evolutionary Computation.* pp. 303–308.

[4] Kennedy, J.; Eberhart, R.C. (2001). *Swarm Intelligence.* Morgan Kaufmann. ISBN 1-55860-595-9.

[5] Poli, R. (2007). "An analysis of publications on particle swarm optimisation applications" (PDF). *Technical Report CSM-469*. Department of Computer Science, University of Essex, UK.

[6] Poli, R. (2008). "Analysis of the publications on the applications of particle swarm optimisation" (PDF). *Journal of Artificial Evolution and Applications*. **2008**: 1–10. doi:10.1155/2008/685175.

[7] Bonyadi, M. R.; Michalewicz, Z. (2017). "Particle swarm optimization for single objective continuous space problems: a review". *Evolutionary Computation*. **25** (1): 1–54. doi:10.1162/EVCO_r_00180.

[8] Zhang, Y. (2015). "A Comprehensive Survey on Particle Swarm Optimization Algorithm and Its Applications". *Mathematical Problems in Engineering*. **2015**: 931256.

[9] Clerc, M. (2012). "Standard Particle Swarm Optimisation" (PDF). *HAL open access archive*.

[10] Bratton, Daniel; Kennedy, James (2007). "Defining a Standard for Particle Swarm Optimization" (PDF). *Proceedings of the 2007 IEEE Swarm Intelligence Symposium (SIS 2007)*.

[11] Taherkhani, M.; Safabakhsh, R. (2016). "A novel stability-based adaptive inertia weight for particle swarm optimization". *Applied Soft Computing*. **38**: 281–295. doi:10.1016/j.asoc.2015.10.004.

[12] Shi, Y.; Eberhart, R.C. (1998). "Parameter selection in particle swarm optimization". *Proceedings of Evolutionary Programming VII (EP98)*. pp. 591–600.

[13] Eberhart, R.C.; Shi, Y. (2000). "Comparing inertia weights and constriction factors in particle swarm optimization". *Proceedings of the Congress on Evolutionary Computation*. **1**. pp. 84–88.

[14] Carlisle, A.; Dozier, G. (2001). "An Off-The-Shelf PSO" (PDF). *Proceedings of the Particle Swarm Optimization Workshop*. pp. 1–6.

[15] van den Bergh, F. (2001). *An Analysis of Particle Swarm Optimizers* (PhD thesis). University of Pretoria, Faculty of Natural and Agricultural Science.

[16] Clerc, M.; Kennedy, J. (2002). "The particle swarm - explosion, stability, and convergence in a multidimensional complex space". *IEEE Transactions on Evolutionary Computation*. **6** (1): 58–73. doi:10.1109/4235.985692.

[17] Trelea, I.C. (2003). "The Particle Swarm Optimization Algorithm: convergence analysis and parameter selection". *Information Processing Letters*. **85** (6): 317–325. doi:10.1016/S0020-0190(02)00447-7.

[18] Bratton, D.; Blackwell, T. (2008). "A Simplified Recombinant PSO". *Journal of Artificial Evolution and Applications*.

[19] Evers, G. (2009). *An Automatic Regrouping Mechanism to Deal with Stagnation in Particle Swarm Optimization* (Master's thesis). The University of Texas - Pan American, Department of Electrical Engineering.

[20] Meissner, M.; Schmuker, M.; Schneider, G. (2006). "Optimized Particle Swarm Optimization (OPSO) and its application to artificial neural network training". *BMC Bioinformatics*. **7** (1): 125. PMC 1464136. PMID 16529661. doi:10.1186/1471-2105-7-125.

[21] Pedersen, M.E.H. (2010). *Tuning & Simplifying Heuristical Optimization* (PhD thesis). University of Southampton, School of Engineering Sciences, Computational Engineering and Design Group.

[22] Pedersen, M.E.H.; Chipperfield, A.J. (2010). "Simplifying particle swarm optimization" (PDF). *Applied Soft Computing*. **10** (2): 618–628. doi:10.1016/j.asoc.2009.08.029.

[23] Nobile, M.S.; Pasi, G.; Cazzaniga, P.; Besozzi, D.; Colombo, R.; Mauri, G. (2015). "Proactive particles in swarm optimization: a self-tuning algorithm based on fuzzy logic". *Proceedings of the 2015 IEEE International Conference on Fuzzy Systems (FUZZ-IEEE 2015), Istanbul (Turkey)*. pp. 1–8.

[24] Cazzaniga, P.; Nobile, M.S.; Besozzi, D. (2015). "The impact of particles initialization in PSO: parameter estimation as a case in point, (Canada)". *Proceedings of IEEE Conference on Computational Intelligence in Bioinformatics and Computational Biology*.

[25] Pedersen, M.E.H. (2010). "Good parameters for particle swarm optimization" (PDF). *Technical Report HL1001*. Hvass Laboratories.

[26] Kennedy, J.; Mendes, R. (2002). "Population structure and particle swarm performance". *Evolutionary Computation, 2002. CEC'02. Proceedings of the 2002 Congress on*. doi:10.1109/CEC.2002.1004493.

[27] Mendes, R. (2004). Population Topologies and Their Influence in Particle Swarm Performance (PhD thesis). Universidade do Minho.

[28] Suganthan, Ponnuthurai N. "Particle swarm optimiser with neighbourhood operator." Evolutionary Computation, 1999. CEC 99. Proceedings of the 1999 Congress on. Vol. 3. IEEE, 1999.

[29] Oliveira, M.; Pinheiro, D.; Andrade, B.; Bastos-Filho, C.; Menezes, R. (2016). "Communication Diversity in Particle Swarm Optimizers". *International Conference on Swarm Intelligence*. doi:10.1007/978-3-319-44427-7_7.

[30] SPSO Particle Swarm Central

[31] Miranda, V., Keko, H. and Duque, Á. J. (2008). Stochastic Star Communication Topology in Evolutionary Particle Swarms (EPSO). International Journal of Computational Intelligence Research (IJCIR), Volume 4, Number 2, pp. 105-116

15.7. REFERENCES

[32] Clerc, M. (2006). Particle Swarm Optimization. ISTE (International Scientific and Technical Encyclopedia), 2006

[33] Yin, P., Glover, F., Laguna, M., & Zhu, J. (2011). A Complementary Cyber Swarm Algorithm. International Journal of Swarm Intelligence Research (IJSIR), 2(2), 22-41

[34] Elshamy, W.; Rashad, H.; Bahgat, A. (2007). "Clubs-based Particle Swarm Optimization" (PDF). IEEE Swarm Intelligence Symposium 2007 (SIS2007). Honolulu, HI. pp. 289–296.

[35] Cleghorn, Christopher W (2014). "Particle Swarm Convergence: Standardized Analysis and Topological Influence". Swarm Intelligence Conference.

[36] Van den Bergh, F. "A convergence proof for the particle swarm optimiser". Fundamenta Informaticae.

[37] Bonyadi, Mohammad reza.; Michalewicz, Z. (2014). "A locally convergent rotationally invariant particle swarm optimization algorithm". Swarm intelligence. 8 (3): 159–198. doi:10.1007/s11721-014-0095-1.

[38] Zhan, Z.-H.; Zhang, J.; Li, Y; Shi, Y.-H. (2011). "Orthogonal Learning Particle Swarm Optimization" (PDF). IEEE Transactions on Evolutionary Computation. 15 (6): 832–847. doi:10.1109/TEVC.2010.2052054.

[39] Monson, C. K. & Seppi, K. D. (2005). Exposing Origin-Seeking Bias in PSO GECCO'05, pp. 241-248

[40] Spears, W. M., Green, D. T. & Spears, D. F. (2010). Biases in Particle Swarm Optimization. International Journal of Swarm Intelligence Research, Vol. 1(2), pp. 34-57

[41] Wilke, D. N., Kok, S. & Groenwold, A. A. (2007). Comparison of linear and classical velocity update rules in particle swarm optimization: notes on scale and frame invariance. International Journal for Numerical Methods in Engineering. John Wiley & Sons, Ltd., 70, pp. 985-1008

[42] SPSO 2011, Particle Swarm Central

[43] Bonyadi, Mohammad reza; Michalewicz, Z. (2014). "SPSO 2011 analysis of stability; local convergence; and rotation sensitivity.". GECCO2014 (the best paper award in the track ACSI): 9–16.

[44] Bonyadi, Mohammad reza.; Michalewicz, Z. (2014). "An analysis of the velocity updating rule of the particle swarm optimization algorithm". Journal of Heuristics. 20 (4): 417–452. doi:10.1007/s10732-014-9245-2.

[45] Zambrano-Bigiarini, M.; Clerc, M.; Rojas, R. (2013). "Standard Particle Swarm Optimisation 2011 at CEC-2013: A baseline for future PSO improvements". Evolutionary Computation (CEC), 2013 IEEE Congress on.

[46] Lovbjerg, M.; Krink, T. (2002). "The LifeCycle Model: combining particle swarm optimisation, genetic algorithms and hillclimbers". Proceedings of Parallel Problem Solving from Nature VII (PPSN). pp. 621–630.

[47] Niknam, T.; Amiri, B. (2010). "An efficient hybrid approach based on PSO, ACO and k-means for cluster analysis". Applied Soft Computing. 10 (1): 183–197. doi:10.1016/j.asoc.2009.07.001.

[48] Zhang, Wen-Jun; Xie, Xiao-Feng (2003). DEPSO: hybrid particle swarm with differential evolution operator. IEEE International Conference on Systems, Man, and Cybernetics (SMCC), Washington, DC, USA: 3816-3821.

[49] Zhang, Y.; Wang, S. (2015). "Pathological Brain Detection in Magnetic Resonance Imaging Scanning by Wavelet Entropy and Hybridization of Biogeography-based Optimization and Particle Swarm Optimization". Progress in Electromagnetics Research – Pier. 152: 41–58.

[50] Lovbjerg, M.; Krink, T. (2002). "Extending Particle Swarm Optimisers with Self-Organized Criticality". Proceedings of the Fourth Congress on Evolutionary Computation (CEC). 2. pp. 1588–1593.

[51] Xinchao, Z. (2010). "A perturbed particle swarm algorithm for numerical optimization". Applied Soft Computing. 10 (1): 119–124. doi:10.1016/j.asoc.2009.06.010.

[52] Xie, Xiao-Feng; Zhang, Wen-Jun; Yang, Zhi-Lian (2002). A dissipative particle swarm optimization. Congress on Evolutionary Computation (CEC), Honolulu, HI, USA: 1456-1461.

[53] Cheung, N. J., Ding, X.-M., & Shen, H.-B. (2013). OptiFel: A Convergent Heterogeneous Particle Sarm Optimization Algorithm for Takagi-Sugeno Fuzzy Modeling. IEEE Transactions on Fuzzy Systems. doi:10.1109/TFUZZ.2013.2278972

[54] Nobile, M.; Besozzi, D.; Cazzaniga, P.; Mauri, G.; Pescini, D. (2012). "A GPU-Based Multi-Swarm PSO Method for Parameter Estimation in Stochastic Biological Systems Exploiting Discrete-Time Target Series". Evolutionary Computation, Machine Learning and Data Mining in Bioinformatics. Lecture Notes in Computer Science. 7264. pp. 74–85.

[55] Zhan, Z.-H.; Zhang, J.; Li, Y; Chung, H.S-H. (2009). "Adaptive Particle Swarm Optimization" (PDF). IEEE Transactions on Systems, Man, and Cybernetics. 39 (6): 1362–1381. doi:10.1109/TSMCB.2009.2015956.

[56] Yang, X.S. (2008). Nature-Inspired Metaheuristic Algorithms. Luniver Press. ISBN 978-1-905986-10-1.

[57] Tu, Z.; Lu, Y. (2004). "A robust stochastic genetic algorithm (StGA) for global numerical optimization". IEEE Transactions on Evolutionary Computation. 8 (5): 456–470. doi:10.1109/TEVC.2004.831258.

[58] Tu, Z.; Lu, Y. (2008). "Corrections to "A Robust Stochastic Genetic Algorithm (StGA) for Global Numerical Optimization". IEEE Transactions on Evolutionary Computation. 12 (6): 781–781. doi:10.1109/TEVC.2008.926734.

[59] Kennedy, James (2003). "Bare Bones Particle Swarms". *Proceedings of the 2003 IEEE Swarm Intelligence Symposium*.

[60] X. S. Yang, S. Deb and S. Fong, Accelerated particle swarm optimization and support vector machine for business optimization and applications, NDT 2011, Springer CCIS 136, pp. 53-66 (2011).

[61] http://www.mathworks.com/matlabcentral/fileexchange/?term=APSO

[62] Parsopoulos, K.; Vrahatis, M. (2002). "Particle swarm optimization method in multiobjective problems". *Proceedings of the ACM Symposium on Applied Computing (SAC)*. pp. 603–607.

[63] Coello Coello, C.; Salazar Lechuga, M. (2002). "MOPSO: A Proposal for Multiple Objective Particle Swarm Optimization". *Congress on Evolutionary Computation (CEC'2002)*. pp. 1051–1056.

[64] Roy, R., Dehuri, S., & Cho, S. B. (2012). A Novel Particle Swarm Optimization Algorithm for Multi-Objective Combinatorial Optimization Problem. 'International Journal of Applied Metaheuristic Computing (IJAMC)', 2(4), 41-57

[65] Kennedy, J. & Eberhart, R. C. (1997). A discrete binary version of the particle swarm algorithm, Conference on Systems, Man, and Cybernetics, Piscataway, NJ: IEEE Service Center, pp. 4104-4109

[66] Clerc, M. (2004). Discrete Particle Swarm Optimization, illustrated by the Traveling Salesman Problem, New Optimization Techniques in Engineering, Springer, pp. 219-239

[67] Clerc, M. (2005). Binary Particle Swarm Optimisers: toolbox, derivations, and mathematical insights, Open Archive HAL

[68] Jarboui, B., Damak, N., Siarry, P., and Rebai, A.R. (2008). A combinatorial particle swarm optimization for solving multi-mode resource-constrained project scheduling problems. In Proceedings of Applied Mathematics and Computation. pp. 299-308.

[69] Chen, Wei-neng; Zhang, Jun (2010). "A novel set-based particle swarm optimization method for discrete optimization problem". *IEEE Transactions on Evolutionary Computation*. **14** (2): 278–300. doi:10.1109/tevc.2009.2030331.

- Applications of PSO.
- Automatic Calibration of a Rainfall-Runoff Model Using a Fast and Elitist Multi-objective Particle Swarm Algorithm
- Particle Swarm Optimization (see and listen to Lecture 27)
- Links to PSO source code

15.8 External links

- Particle Swarm Central is a repository for information on PSO. Several source codes are freely available.
- A brief video of particle swarms optimizing three benchmark functions.
- Simulation of PSO convergence in a two-dimensional space (Matlab).

Chapter 16

Swarm behaviour

"Swarm" redirects here. For other uses, see Swarm (disambiguation).

Swarm behaviour, or **swarming**, is a collective be-

A flock of auklets exhibit swarm behaviour

haviour exhibited by entities, particularly animals, of similar size which aggregate together, perhaps milling about the same spot or perhaps moving *en masse* or migrating in some direction. It is a highly interdisciplinary topic.[1] As a term, *swarming* is applied particularly to insects, but can also be applied to any other entity or animal that exhibits swarm behaviour. The term *flocking* or 'murmuration' are usually used to refer specifically to swarm behaviour in birds, *herding* to refer to swarm behaviour in quadrupeds, *shoaling* or *schooling* to refer to swarm behaviour in fish. Phytoplankton also gather in huge swarms called *blooms*, although these organisms are algae and are not self-propelled the way animals are. By extension, the term swarm is applied also to inanimate entities which exhibit parallel behaviours, as in a robot swarm, an earthquake swarm, or a swarm of stars.

From a more abstract point of view, swarm behaviour is the collective motion of a large number of self-propelled entities.[2] From the perspective of the mathematical modeller, it is an emergent behaviour arising from simple rules that are followed by individuals and does not involve any central coordination. Swarm behaviour is also studied by active matter physicists as a phenomenon which is not in thermodynamic equilibrium, and as such requires the development of tools beyond those available from the statistical physics of systems in thermodynamic equilibrium.

Swarm behaviour was first simulated on a computer in 1986 with the simulation program boids.[3] This program simulates simple agents (boids) that are allowed to move according to a set of basic rules. The model was originally designed to mimic the flocking behaviour of birds, but it can be applied also to schooling fish and other swarming entities.

16.1 Models

See also: Collective animal behaviour

In recent decades, scientists have turned to modeling swarm behaviour to gain a deeper understanding of the behaviour.

16.1.1 Mathematical models

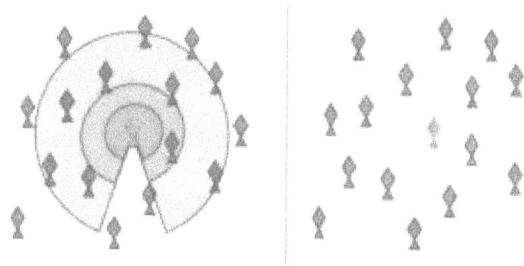

In the metric distance model of a fish school (left), the focal fish (yellow) pays attention to all fish within the small zone of repulsion (red), the zone of alignment (lighter red) and the larger zone of attraction (lightest red). In the topological distance model (right), the focal fish only pays attention to the six or seven closest fish (green), regardless of their distance.

Early studies of swarm behaviour employed mathematical

models to simulate and understand the behaviour. The simplest mathematical models of animal swarms generally represent individual animals as following three rules:

1. Move in the same direction as your neighbours
2. Remain close to your neighbours
3. Avoid collisions with your neighbours

The boids computer program, created by Craig Reynolds in 1986, simulates swarm behaviour following the above rules.[3] Many subsequent and current models use variations on these rules, often implementing them by means of concentric "zones" around each animal. In the *zone of repulsion*, very close to the animal, the focal animal will seek to distance itself from its neighbours to avoid collision. Slightly further away, in the *zone of alignment*, the focal animal will seek to align its direction of motion with its neighbours. In the outermost *zone of attraction*, which extends as far away from the focal animal as it is able to sense, the focal animal will seek to move towards a neighbour.

The shape of these zones will necessarily be affected by the sensory capabilities of the given animal. For example, the visual field of a bird does not extend behind its body. Fish rely on both vision and on hydrodynamic perceptions relayed through their lateral line, while Antarctic krill rely both on vision and hydrodynamic signals relayed through antennae.

However recent studies of starling flocks have shown that each bird modifies its position, relative to the six or seven animals directly surrounding it, no matter how close or how far away those animals are.[4] Interactions between flocking starlings are thus based on a topological rule rather than a metric rule. It remains to be seen whether this applies to other animals. Another recent study, based on an analysis of high speed camera footage of flocks above Rome and assuming minimal behavioural rules, has convincingly simulated a number of aspects of flock behaviour.[5][6][7][8]

16.1.2 Evolutionary models

In order to gain insight into why animals evolve swarming behaviour, scientists have turned to evolutionary models that simulate populations of evolving animals. Typically these studies use a genetic algorithm to simulate evolution over many generations in the model. These studies have investigated a number of hypotheses explaining why animals evolve swarming behaviour, such as the selfish herd theory[9][10][11][12] the predator confusion effect,[13][14] the dilution effect,[15][16] and the many eyes theory.[17]

16.1.3 Agents

Main article: Agent-based model in biology
See also: Agent-based models, Intelligent agent, Autonomous agent, and Quorum sensing

- Mach, Robert; Schweitzer, Frank (2003). "Multi-Agent Model of Biological Swarming". *Advances In Artificial Life*. Lecture Notes in Computer Science. **2801**. pp. 810–820. CiteSeerX 10.1.1.87.8022. ISBN 978-3-540-20057-4. doi:10.1007/978-3-540-39432-7_87.

16.1.4 Self-organization

Flocking birds are an example of self-organization in biology

See also: Self-organization and Biological organisation

16.1.5 Emergence

Main article: Emergence

The concept of emergence—that the properties and functions found at a hierarchical level are not present and are irrelevant at the lower levels–is often a basic principle behind self-organizing systems.[18] An example of self-organization in biology leading to emergence in the natural world occurs in ant colonies. The queen does not give direct orders and does not tell the ants what to do. Instead, each ant reacts to stimuli in the form of chemical scent from larvae, other ants, intruders, food and buildup of waste, and leaves behind a chemical trail, which, in turn, provides a stimulus to other ants. Here each ant is an autonomous unit that reacts depending only on its local environment and the genetically encoded rules for its variety of ant. Despite the lack of centralized decision making, ant colonies exhibit complex behaviour and have even been able to demonstrate the ability to solve geometric problems. For example, colonies routinely find the maximum distance from all colony entrances to dispose of dead bodies.

16.1.6 Stigmergy

Main article: Stigmergy

A further key concept in the field of swarm intelligence is stigmergy.[19][20] Stigmergy is a mechanism of indirect coordination between agents or actions. The principle is that the trace left in the environment by an action stimulates the performance of a next action, by the same or a different agent. In that way, subsequent actions tend to reinforce and build on each other, leading to the spontaneous emergence of coherent, apparently systematic activity. Stigmergy is a form of self-organization. It produces complex, seemingly intelligent structures, without need for any planning, control, or even direct communication between the agents. As such it supports efficient collaboration between extremely simple agents, who lack any memory, intelligence or even awareness of each other.[20]

16.1.7 Swarm intelligence

Main article: Swarm intelligence

Swarm intelligence is the collective behaviour of decentralized, self-organized systems, natural or artificial. The concept is employed in work on artificial intelligence. The expression was introduced by Gerardo Beni and Jing Wang in 1989, in the context of cellular robotic systems.[21]

Swarm intelligence systems are typically made up of a population of simple agents such as boids interacting locally with one another and with their environment. The agents follow very simple rules, and although there is no centralized control structure dictating how individual agents should behave, local, and to a certain degree random, interactions between such agents lead to the emergence of intelligent global behaviour, unknown to the individual agents.

Swarm intelligence research is multidisciplinary. It can be divided into natural swarm research studying biological systems and artificial swarm research studying human artefacts. There is also a scientific stream attempting to model the swarm systems themselves and understand their underlying mechanisms, and an engineering stream focused on applying the insights developed by the scientific stream to solve practical problems in other areas.[22]

16.1.8 Algorithms

Swarm algorithms follow a Lagrangian approach or an Eulerian approach.[23] The Eulerian approach views the swarm as a field, working with the density of the swarm and deriving mean field properties. It is a hydrodynamic approach, and can be useful for modelling the overall dynamics of large swarms.[24][25][26] However, most models work with the Lagrangian approach, which is an agent-based model following the individual agents (points or particles) that make up the swarm. Individual particle models can follow information on heading and spacing that is lost in the Eulerian approach.[23][27]

Ant colony optimization

Main article: Ant colony optimization

Ant colony optimization is a widely used algorithm which was inspired by the behaviours of ants, and has been effective solving discrete optimization problems related to swarming.[29] The algorithm was initially proposed by Marco Dorigo in 1992,[30][31] and has since been diversified to solve a wider class of numerical problems. Species that have multiple queens may have a queen leaving the nest along with some workers to found a colony at a new site, a process akin to swarming in honeybees.[32][33]

- Ants are behaviourally unsophisticated; collectively they perform complex tasks. Ants have highly developed sophisticated sign-based communication.

- Ants communicate using pheromones; trails are laid that can be followed by other ants.

- Routing problem ants drop different pheromones used to compute the "shortest" path from source to destination(s).

- Rauch, EM; Millonas, MM; Chialvo, DR (1995). "Pattern formation and functionality in swarm models". *Physics Letters A*. **207**: 185. Bibcode:1995PhLA..207..185R. arXiv:adap-org/9507003. doi:10.1016/0375-9601(95)00624-c.

Self-propelled particles

Main article: Self-propelled particles

Self-propelled particles (SPP) is a concept introduced in 1995 by Vicsek et al.[35] as a special case of the boids model introduced in 1986 by Reynolds.[3] A swarm is modelled in SPP by a collection of particles that move with a constant speed but respond to a random perturbation by adopting at each time increment the average direction of motion of the other particles in their local neighbourhood.[36]

Simulations demonstrate that a suitable "nearest neighbour rule" eventually results in all the particles swarming together, or moving in the same direction. This emerges, even though there is no centralized coordination, and even though the neighbours for each particle constantly change over time (see the interactive simulation in the box on the right).[35] SPP models predict that swarming animals share certain properties at the group level, regardless of the type of animals in the swarm.[37] Swarming systems give rise to emergent behaviours which occur at many different scales, some of which are turning out to be both universal and robust. It has become a challenge in theoretical physics to find minimal statistical models that capture these behaviours.[38][39]

Particle swarm optimization

Main article: Particle swarm optimization

Particle swarm optimization is another algorithm widely used to solve problems related to swarms. It was developed in 1995 by Kennedy and Eberhart and was first aimed at simulating the social behaviour and choreography of bird flocks and fish schools.[40][41] The algorithm was simplified and it was observed to be performing optimization. The system initially seeds a population with random solutions. It then searches in the problem space through successive generations using stochastic optimization to find the best solutions. The solutions it finds are called particles. Each particle stores its position as well as the best solution it has achieved so far. The particle swarm optimizer tracks the best local value obtained so far by any particle in the local neighbourhood. The remaining particles then move through the problem space following the lead of the optimum particles. At each time iteration, the particle swarm optimiser accelerates each particle toward its optimum locations according to simple mathematical rules. Particle swarm optimization has been applied in many areas. It has few parameters to adjust, and a version that works well for a specific applications can also work well with minor modifications across a range of related applications.[42] A book by Kennedy and Eberhart describes some philosophical aspects of particle swarm optimization applications and swarm intelligence.[43] An extensive survey of applications is made by Poli.[44][45]

Altruism

Researchers in Switzerland have developed an algorithm based on Hamilton's rule of kin selection. The algorithm shows how altruism in a swarm of entities can, over time, evolve and result in more effective swarm behaviour.[46][47]

16.2 Biological swarming

Bats swarming out of a cave in Thailand

Examples of biological swarming are found in bird flocks,[48] fish schools,[49][50] insect swarms,[51] bacteria swarms,[52][53] molds,[54] molecular motors,[55] quadruped herds[56] and people.[57][58][59]

- Swarm Theory *National Geographic*. Feature article, July 2007.

- Beekman M, Sword GA and Simpson SK (2008) Biological Foundations of Swarm Intelligence. In *Swarm intelligence: introduction and applications*, Eds Blum C and Merkle D. ▓▓▓▓▓▓▓▓▓▓▓▓▓▓▓▓. Page 3–43. ISBN 978-3-540-74088-9

- Parrish JK, Edelstein-Keshet L (1999). "Complexity, pattern and evolutionary trade-offs in animal aggregation" (PDF). *Science*. **284** (5411): 99–101. Bibcode:1999Sci...284...99P. PMID 10102827. doi:10.1126/science.284.5411.99.

- Spawning

- Synchronous spawning

- External fertilization

16.2.1 Insects

The behaviour of insects that live in colonies, such as ants, bees, wasps and termites, has always been a source of fascination for children, naturalists and artists. Individual insects seem to do their own thing without any central control, yet the colony as a whole behaves in a highly coordinated manner.[60] Researchers have found that cooperation at the colony level is largely self-organized. The group coordination that emerges is often just a consequence of the way individuals in the colony interact. These interactions can be remarkably simple, such as one ant merely following the trail left by another ant. Yet put together, the cumulative effect of such behaviours can solve highly complex problems, such as locating the shortest route in a network of possible paths to a food source. The organised behaviour that emerges in this way is sometimes called swarm intelligence.[60]

are found simultaneously, the pheromone trail to the better one will be stronger. Ants in the nest follow another simple rule, to favor stronger trails, on average. More ants then follow the stronger trail, so more ants arrive at the high quality food source, and a positive feedback cycle ensures, resulting in a collective decision for the best food source. If there are two paths from the ant nest to a food source, then the colony usually selects the shorter path. This is because the ants that first return to the nest from the food source are more likely to be those that took the shorter path. More ants then retrace the shorter path, reinforcing the pheromone trail.[63]

The successful techniques used by ant colonies have been studied in computer science and robotics to produce distributed and fault-tolerant systems for solving problems. This area of biomimetics has led to studies of ant locomotion, search engines that make use of "foraging trails", fault-tolerant storage and networking algorithms.[64]

Ants

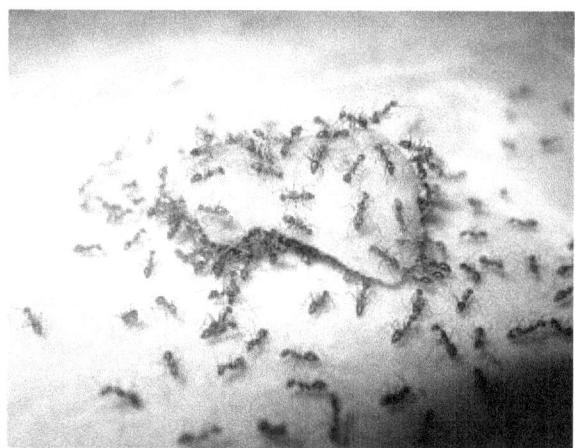

A swarm of ants which have discovered a food source

See also: Ant colony, Ant colony optimization, Ant mill, Ant robotics, and Artificial ants

Individual ants do not exhibit complex behaviours, yet a colony of ants collectively achieves complex tasks such as constructing nests, taking care of their young, building bridges and foraging for food. A colony of ants can collectively select (i.e. send most workers towards) the best, or closest, food source from several in the vicinity.[61] Such collective decisions are achieved using positive feedback mechanisms. Selection of the best food source is achieved by ants following two simple rules. First, ants which find food return to the nest depositing a pheromone chemical. More pheromone is laid for higher quality food sources.[62] Thus, if two equidistant food sources of different qualities

Bees

Bees swarming on a shrub

Main articles: Bees algorithm and Swarming (honey bee)

When a honey bee swarm emerges from a hive they do not fly far at first. They may gather in a tree or on a branch only a few meters from the hive. In this new location, the bees cluster about the queen and send 20 –50 scout bees out to find a suitable new nest locations. The scout bees are the most experienced foragers in the cluster. An individual scout returning to the cluster promotes a location she has found. She uses a dance similar to the waggle dance to indicate direction and distance to others in the cluster. The more excited she is about her findings the more excitedly she dances. If she can convince other scouts to check out the location she found, they may take off, check out the proposed site and promote the site further upon their return. Several different sites may be promoted by different scouts at first. After several hours and sometimes days, slowly a favourite location emerges from this decision making process. When all scouts agree on a final location the whole cluster takes off and flies to it. Sometimes, if no decision is reached, the swarm will separate, some bees going in one direction; others, going in another. This usually results in failure, with both groups dying. A swarm may fly for a kilometre or more to the scouted out location, though some species may establish new colonies within as little as 500 meters from the natal nest, such as *Apis dorsata*.[65] This collective decision making process is remarkably successful in identifying the most suitable new nest site and keeping the swarm intact. A good nest site has to be large enough to accommodate the swarm (about 15 litres in volume), has to be well protected from the elements, receive a certain amount of warmth from the sun, be some height above the ground, have a small entrance and resist the infestation of ants - hence why trees are often selected.[66][67][68][69][70]

Cockroaches

Similar to ants, cockroaches leave chemical trails in their faeces as well as emitting airborne pheromones for swarming and mating. Other cockroaches will follow these trails to discover sources of food and water, and also discover where other cockroaches are hiding. Thus, cockroaches can exhibit emergent behaviour,[71] in which group or swarm behaviour emerges from a simple set of individual interactions.

Cockroaches are mainly nocturnal and will run away when exposed to light. A study tested the hypothesis that cockroaches use just two pieces of information to decide where to go under those conditions: how dark it is and how many other cockroaches there are. The study conducted by José Halloy and colleagues at the Free University of Brussels and other European institutions created a set of tiny robots that appear to the roaches as other roaches and can thus alter the roaches' perception of critical mass. The robots were also specially scented so that they would be accepted by the real roaches.[72]

Locusts

A 19th century depiction of a swarm of desert locusts

See also: Marching locusts

Locusts are the swarming phase of the short-horned grasshoppers of the family Acrididae. Some species can breed rapidly under suitable conditions and subsequently become gregarious and migratory. They form bands as nymphs and swarms as adults—both of which can travel great distances, rapidly stripping fields and greatly damaging crops. The largest swarms can cover hundreds of square miles and contain billions of locusts. A locust can eat its own weight (about 2 grams) in plants every day. That means one million locusts can eat about one ton of food each day, and the largest swarms can consume over 100,000 tonnes each day.[73]

Swarming in locusts has been found to be associated with increased levels of serotonin which causes the locust to change colour, eat much more, become mutually attracted, and breed much more easily. Researchers propose that swarming behaviour is a response to overcrowding and studies have shown that increased tactile stimulation of the hind legs or, in some species, simply encountering other individuals causes an increase in levels of serotonin. The transformation of the locust to the swarming variety can be induced by several contacts per minute over a four-hour period.[74][75][76][77] Notably, an innate predisposition to aggregate has been found in hatchlings of the desert locust, *Schistocerca gregaria*, independent of their parental phase.[78]

An individual locust's response to a loss of alignment in the group appears to increase the randomness of its motion, until an aligned state is again achieved. This noise-induced alignment appears to be an intrinsic characteristic of collective coherent motion.[79]

Insect migration

Main article: Insect migration
See also: Lepidoptera migration

Insect migration is the seasonal movement of insects, particularly those by species of dragonflies, beetles, butterflies and moths. The distance can vary from species to species, but in most cases these movements involve large numbers of individuals. In some cases the individuals that migrate in one direction may not return and the next generation may instead migrate in the opposite direction. This is a significant difference from bird migration.

Monarch butterflies are especially noted for their lengthy annual migration. In North America they make massive southward migrations starting in August until the first frost. A northward migration takes place in the spring. The monarch is the only butterfly that migrates both north and south as the birds do on a regular basis. But no single individual makes the entire round trip. Female monarchs deposit eggs for the next generation during these migrations.[80] The length of these journeys exceeds the normal lifespan of most monarchs, which is less than two months for butterflies born in early summer. The last generation of the summer enters into a non-reproductive phase known as diapause and may live seven months or more.[81] During diapause, butterflies fly to one of many overwintering sites. The generation that overwinters generally does not reproduce until it leaves the overwintering site sometime in February and March. It is the second, third and fourth generations that return to their northern locations in the United States and Canada in the spring. How the species manages to return to the same overwintering spots over a gap of several generations is still a subject of research; the flight patterns appear to be inherited, based on a combination of the position of the sun in the sky[82] and a time-compensated Sun compass that depends upon a circadian clock that is based in their antennae.[83][84]

16.2.2 Birds

Recent studies of starling flocks have shown that each bird modifies its position, relative to the six or seven animals directly surrounding it, no matter how close or how far away those animals are.[4]
---- Murmurations of starlings

Main article: Flocking (behaviour)
See also: Flock (birds), Bird landings, Bird strike, Mixed-species foraging flock, and Mobbing behaviour

- Nagy, M; Akos Zs, Biro D; Vicsek, T (2010). "Hierarchical group dynamics in pigeon flocks" (PDF). *Nature*. **464**: 890–893. doi:10.1038/nature08891. Supplementary pdf

Bird migration

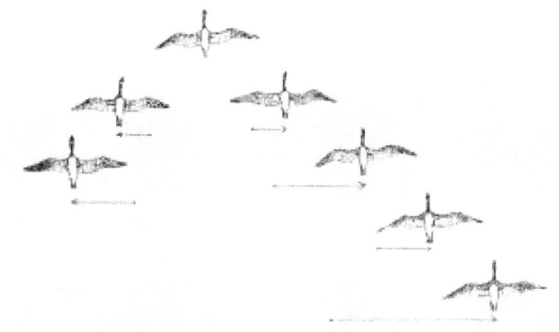

Large bird typically migrate in V echelon formations. There are significant aerodynamic gains. All birds can see ahead, and towards one side, making a good arrangement for protection.

Main article: Bird migration
See also: Reverse migration (birds)

Approximately 1800 of the world's 10,000 bird species are long-distance migrants.[85] The primary motivation for migration appears to be food; for example, some hummingbirds choose not to migrate if fed through the winter. Also, the longer days of the northern summer provide extended time for breeding birds to feed their young. This helps diurnal birds to produce larger clutches than related non-migratory species that remain in the tropics. As the days shorten in autumn, the birds return to warmer regions where the available food supply varies little with the season. These advantages offset the high stress, physical exertion costs, and other risks of the migration such as predation.

Many birds migrate in flocks. For larger birds, it is assumed that flying in flocks reduces energy costs. The V formation is often supposed to boost the efficiency and range of flying birds, particularly over long migratory routes. All the birds except the first fly in the upwash from one of the wingtip vortices of the bird ahead. The upwash assists each bird in supporting its own weight in flight, in the same way a glider can climb or maintain height indefinitely in rising air. Geese flying in a V formation save energy by flying in the updraft of the wingtip vortex generated by the previous animal in the formation. Thus, the birds flying behind do not need to work as hard to achieve lift. Studies show that birds in a V formation place themselves roughly at the optimum distance predicted by simple aerodynamic theory.[86] Geese in a V-formation may conserve 12–20% of the energy they would need to fly alone.[87][88] Red knots and dunlins were found in radar studies to fly 5 km per hour faster in flocks than when they were flying alone.[89] The birds flying at the tips and at the front are rotated in a timely cyclical fashion to spread flight fatigue equally among the flock members. The formation also makes communication easier and allows the birds to maintain visual contact with each other.

Other animals may use similar drafting techniques when migrating. Lobsters, for example, migrate in close single-file formation "lobster trains", sometimes for hundreds of miles.

The Mediterranean and other seas present a major obstacle to soaring birds, which must cross at the narrowest points. Massive numbers of large raptors and storks pass through areas such as Gibraltar, Falsterbo, and the Bosphorus at migration times. More common species, such as the European honey buzzard, can be counted in hundreds of thousands in autumn. Other barriers, such as mountain ranges, can also cause funnelling, particularly of large diurnal migrants. This is a notable factor in the Central American migratory bottleneck. This concentration of birds during migration can put species at risk. Some spectacular migrants have already gone extinct, the most notable being the passenger pigeon. During migration the flocks were a mile (1.6 km) wide and 300 miles (500 km) long, taking several days to pass and containing up to a billion birds.

16.2.3 Marine life

Fish

Main article: Shoaling and schooling
The term "shoal" can be used to describe any group of fish,

Schooling predator fish size up schooling anchovies

including mixed-species groups, while "school" is used for more closely knit groups of the same species swimming in a highly synchronised and polarised manner.

Fish derive many benefits from shoaling behaviour including defence against predators (through better predator detection and by diluting the chance of capture), enhanced foraging success, and higher success in finding a mate.[91] It is also likely that fish benefit from shoal membership through increased hydrodynamic efficiency.[92]

Fish use many traits to choose shoalmates. Generally they prefer larger shoals, shoalmates of their own species, shoalmates similar in size and appearance to themselves, healthy fish, and kin (when recognised). The "oddity effect" posits that any shoal member that stands out in appearance will be preferentially targeted by predators. This may explain why fish prefer to shoal with individuals that resemble them. The oddity effect would thus tend to homogenise shoals.[93]

One puzzling aspect of shoal selection is how a fish can choose to join a shoal of animals similar to themselves, given that it cannot know its own appearance. Experiments with zebrafish have shown that shoal preference is a learned ability, not innate. A zebrafish tends to associate with shoals that resemble shoals in which it was reared, a form of imprinting.[94]

Other open questions of shoaling behaviour include identifying which individuals are responsible for the direction of shoal movement. In the case of migratory movement, most members of a shoal seem to know where they are going. In the case of foraging behaviour, captive shoals of golden shiner (a kind of minnow) are led by a small number of experienced individuals who knew when and where food was available.[95]

Radakov estimated herring schools in the North Atlantic can occupy up to 4.8 cubic kilometres with fish densities between 0.5 and 1.0 fish/cubic metre. That's several billion fish in one school.[96]

See also: Eel life history

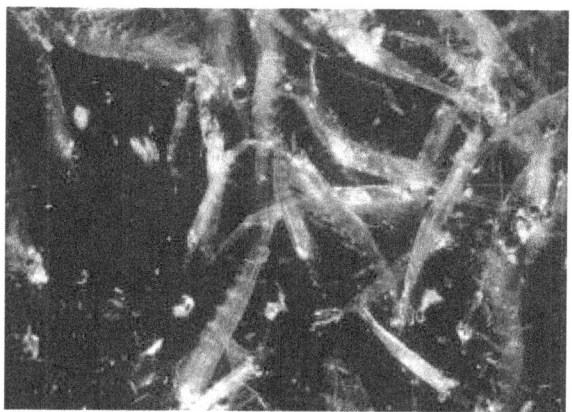

Swarming krill

- Partridge BL (1982) "The structure and function of fish schools" *Scientific American*, June:114–123.
- Parrish JK, Viscido SV, Grunbaum D (2002). "Self-Organized Fish Schools: An Examination of Emergent Properties" (PDF). *Biol. Bull.* **202** (3): 296–305. PMID 12087003. doi:10.2307/1543482.

Fish migration

Main article: Fish migration
See also: Sardine run and Salmon run

Between May and July huge numbers of sardines spawn in the cool waters of the Agulhas Bank and then follow a current of cold water northward along the east coast of South Africa. This great migration, called the sardine run, creates spectacular feeding frenzies along the coastline as marine predators, such as dolphins, sharks and gannets attack the schools.

Krill

Most krill, small shrimp-like crustaceans, form large swarms, sometimes reaching densities of 10,000–60,000 individual animals per cubic metre.[98][99][100] Swarming is a defensive mechanism, confusing smaller predators that would like to pick out single individuals. The largest swarms are visible from space and can be tracked by satellite.[101] One swarm was observed to cover an area of 450 square kilometers (175 square miles) of ocean, to a depth of 200 meters (650 feet) and was estimated to contain over 2 million tons of krill.[102] Recent research suggests that krill do not simply drift passively in these currents but actually modify them.[102] Krill typically follow a diurnal vertical migration. By moving vertically through the ocean on a 12-hour cycle, the swarms play a major part in mixing deeper, nutrient-rich water with nutrient-poor water at the surface.[102] Until recently it has been assumed that they spend the day at greater depths and rise during the night toward the surface. It has been found that the deeper they go, the more they reduce their activity,[103] apparently to reduce encounters with predators and to conserve energy. Later work suggested that swimming activity in krill varied with stomach fullness. Satiated animals that had been feeding at the surface swim less actively and therefor sink below the mixed layer.[104] As they sink they produce faeces which may mean that they have an important role to play in the Antarctic carbon cycle. Krill with empty stomachs were found to swim more actively and thus head towards the surface. This implies that vertical migration may be a bi- or tri-daily occurrence. Some species form surface swarms during the day for feeding and reproductive purposes even though such behaviour is dangerous because it makes them extremely vulnerable to predators.[105] Dense swarms may elicit a feeding frenzy among fish, birds and mammal predators, especially near the surface. When disturbed, a swarm scatters, and some individuals have even been observed to moult instantaneously, leaving the exuvia behind as a decoy.[106] In 2012, Gandomi and Alavi presented what appears to be a successful stochastic algorithm for modelling the behaviour of krill swarms. The algorithm is based on three main factors: " (i) movement induced by the presence of other individuals (ii) foraging activity, and (iii) random diffusion."[107]

Copepods

See also: Hunting copepods

Copepods are a group of tiny crustaceans found in the sea and lakes. Many species are planktonic (drifting in sea waters), and others are benthic (living on the ocean floor). Copepods are typically 1 to 2 millimetres (0.04 to 0.08

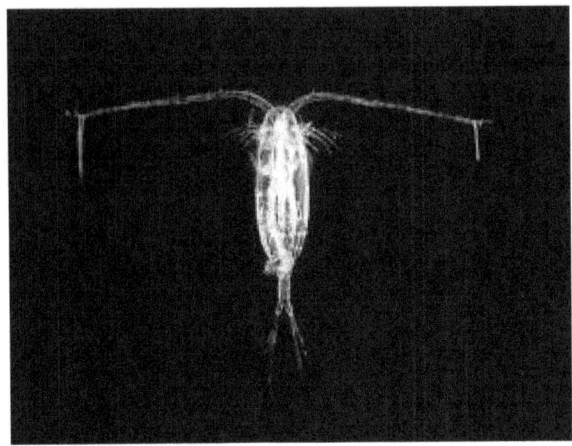

This copepod has its antenna spread (click to enlarge). The antenna detects the pressure wave of an approaching fish.

in) long, with a teardrop shaped body and large antennae. Although like other crustaceans they have an armoured exoskeleton, they are so small that in most species this thin armour, and the entire body, is almost totally transparent. Copepods have a compound, median single eye, usually bright red, in the centre of the transparent head.

Copepods also swarm. For example, monospecific swarms have been observed regularly around coral reefs and sea grass, and in lakes. Swarms densities were about one million copepods per cubic metre. Typical swarms were one or two metres in diameter, but some exceeded 30 cubic metres. Copepods need visual contact to keep together, and they disperse at night.[108]

Spring produces blooms of swarming phytoplankton which provide food for copepods. Planktonic copepods are usually the dominant members of the zooplankton, and are in turn major food organisms for many other marine animals. In particular, copepods are prey to forage fish and jellyfish, both of which can assemble in vast, million-strong swarms. Some copepods have extremely fast escape responses when a predator is sensed and can jump with high speed over a few millimetres (see animated image below).

- 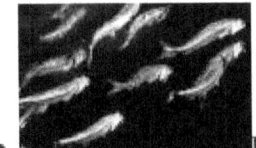Photo: School of herrings ram feeding on a swarm of copepods.

- Animation showing how herrings hunting in a synchronised way can capture the very alert and evasive copepod (click to view).

- 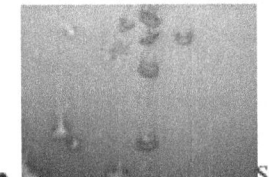Swarms of jellyfish also prey on copepods

Planktonic copepods are important to the carbon cycle. Some scientists say they form the largest animal biomass on earth.[109] They compete for this title with Antarctic krill. Because of their smaller size and relatively faster growth rates, however, and because they are more evenly distributed throughout more of the world's oceans, copepods almost certainly contribute far more to the secondary productivity of the world's oceans, and to the global ocean carbon sink than krill, and perhaps more than all other groups of organisms together. The surface layers of the oceans are currently believed to be the world's largest carbon sink, absorbing about 2 billion tons of carbon a year, the equivalent to perhaps a third of human carbon emissions, thus reducing their impact. Many planktonic copepods feed near the surface at night, then sink into deeper water during the day to avoid visual predators. Their moulted exoskeletons, faecal pellets and respiration at depth all bring carbon to the deep sea.

Algal blooms

Many single-celled organisms called phytoplankton live in oceans and lakes. When certain conditions are present, such as high nutrient or light levels, these organisms reproduce explosively. The resulting dense swarm of phytoplankton is called an algal bloom. Blooms can cover hundreds of square kilometres and are easily seen in satellite images. Individual phytoplankton rarely live more than a few days, but blooms can last weeks.[110][111]

16.2.4 Plants

Scientists have attributed swarm behavior to plants for hundreds of years. In his 1800 book, *Phytologia: or, The philosophy of agriculture and gardening*, Erasmus Darwin wrote that plant growth resembled swarms observed elsewhere in nature.[112] While he was referring to more broad observations of plant morphology, and was focused on both root and shoot behavior, recent research has supported this claim.

Roots, in particular, display observable swarm behavior, growing in patterns that exceed the statistical threshold for random probability, and indicate the presence of communication between individual root apexes. The primary function of plant roots is the uptake of soil nutrients, and it is this purpose which drives swarm behavior. Plants growing in close proximity have adapted their growth to assure optimal nutrient availability. This is accomplished by growing in a direction that optimizes the distance between nearby roots, thereby increasing their chance of exploiting untapped nutrient reserves. The action of this behavior takes two forms: maximization of distance from, and repulsion by, neighboring root apexes.[113] The transition zone of a root tip is largely responsible for monitoring for the presence of soil-borne hormones, signaling responsive growth patterns as appropriate. Plant responses are often complex, integrating multiple inputs to inform an autonomous response. Additional inputs that inform swarm growth includes light and gravity, both of which are also monitored in the transition zone of a root's apex.[114] These forces act to inform any number of growing "main" roots, which exhibit their own independent releases of inhibitory chemicals to establish appropriate spacing, thereby contributing to a swarm behavior pattern. Horizontal growth of roots, whether in response to high mineral content in soil or due to stolon growth, produces branched growth that establish to also form their own, independent root swarms.[115]

16.2.5 Other organisms

Bacteria

See also: Swarming motility and Microbial intelligence

Swarming is also used to describe groupings of some kinds of bacteria such as myxobacteria. Myxobacteria swarm together in "wolf packs", actively moving using a process known as bacterial gliding and keeping together with the help of intercellular molecular signals.[52][116]

Quadrupeds

See also: Herd, Herd behaviour, and Animal migration

- Parrish JK, Edelstein-Keshet L (1999). "Complexity, pattern and evolutionary trade-offs in animal aggregation" (PDF). *Science*. **284** (5411): 99–101. Bibcode:1999Sci...284...99P. PMID 10102827. doi:10.1126/science.284.5411.99.

Sheep dogs (here a Border Collie) control the flocking behaviour of sheep

16.3 People

Police protect Nick Altrock from an adoring crowd during baseball's 1906 World Series

See also: Crowd and Crowd simulation

A collection of people can also exhibit swarm behaviour, such as pedestrians[119] or soldiers swarming the parapets. In Cologne, Germany, two biologists from the University of Leeds demonstrated flock like behaviour in humans. The group of people exhibited similar behavioural pattern to a flock, where if five percent of the flock changed direction the others would follow. If one person was designated as a predator and everyone else was to avoid him, the flock behaved very much like a school of fish.[120][121] Understanding how humans interact in crowds is important if crowd management is to effectively avoid casualties at football grounds, music concerts and subway stations.[122]

The mathematical modelling of flocking behaviour is a common technology, and has found uses in animation. Flocking simulations have been used in many films[123] to generate crowds which move realistically. Tim Burton's *Batman Returns* was the first movie to make use of swarm technology for rendering, realistically depicting the movements of a group of bats using the boids system. *The Lord of the Rings* film trilogy made use of similar technology, known as massive, during battle scenes. Swarm technology is particularly attractive because it is cheap, robust, and simple.

An ant-based computer simulation using only six interaction rules has also been used to evaluate aircraft boarding behaviour.[124] Airlines have also used ant-based routing in assigning aircraft arrivals to airport gates. An airline system developed by Douglas A. Lawson uses swarm theory, or swarm intelligence—the idea that a colony of ants works better than one alone. Each pilot acts like an ant searching for the best airport gate. "The pilot learns from his experience what's the best for him, and it turns out that that's the best solution for the airline," Lawson explains. As a result, the "colony" of pilots always go to gates they can arrive and depart quickly. The program can even alert a pilot of plane back-ups before they happen. "We can anticipate that it's going to happen, so we'll have a gate available," says Lawson.[125]

Swarm behaviour occurs also in traffic flow dynamics, such as the traffic wave. Bidirectional traffic can be observed in ant trails.[126][127] In recent years this behaviour has been researched for insight into pedestrian and traffic models.[128][129] Simulations based on pedestrian models have also been applied to crowds which stampede because of panic.[130]

Herd behaviour in marketing has been used to explain the dependencies of customers' mutual behaviour. *The Economist* reported a recent conference in Rome on the subject of the simulation of adaptive human behaviour.[131] It shared mechanisms to increase impulse buying and get people "to buy more by playing on the herd instinct." The basic idea is that people will buy more of products that are seen to be popular, and several feedback mechanisms to get product popularity information to consumers are mentioned, including smart card technology and the use of Radio Frequency Identification Tag technology. A "swarm-moves" model was introduced by a Florida Institute of Technology researcher, which is appealing to supermarkets because it can "increase sales without the need to give people discounts."

- Helbing D, Keltsch J, Molnar P (1997). "Modelling the evolution of human trail systems". *Nature*. **388** (6637): 47–50. Bibcode:1997Natur.388...47H. PMID 9214501. arXiv:cond-mat/9805158. doi:10.1038/40353.

- Helbing D, Farkas I, Vicsek T (2000). "Simulating dynamical features of escape panic". *Nature*. **407** (6803): 487–490. Bibcode:2000Natur.407..487H. PMID 11028994. arXiv:cond-mat/0009448. doi:10.1038/35035023.

- Helbing D, Farkas IJ, Vicsek T (2000). "Freezing by heating in a driven mesoscopic system". *Physical Review Letters*. **84** (6): 1240–1243. Bibcode:2000PhRvL..84.1240H. PMID 11017488. arXiv:cond-mat/9904326. doi:10.1103/PhysRevLett.84.1240.

16.4 Robotics

Main article: Swarm robotics
See also: Ant robotics and Robotic materials

The application of swarm principles to robots is called

Kilobot thousand robot swarm developed by Radhika Nagpal and Michael Rubenstein at Harvard University.

swarm robotics, while swarm intelligence refers to the more general set of algorithms.

Partially inspired by colonies of insects such as ants and bees, researchers are modelling the behaviour of swarms of thousands of tiny robots which together perform a useful task, such as finding something hidden, cleaning, or spying. Each robot is quite simple, but the emergent behaviour of the swarm is more complex.[1] The whole set of robots can be considered as one single distributed system, in the same way an ant colony can be considered a superorganism, exhibiting swarm intelligence. The largest swarms so far created is the 1024 robot Kilobot swarm.[133] Other large swarms include the iRobot swarm, the SRI International/ActivMedia Robotics Centibots project,[134] and the Open-source Micro-robotic Project swarm, which are being used to research collective behaviours.[135][136] Swarms are also more resistant to failure. Whereas one large robot may fail and ruin a mission, a swarm can continue even if several robots fail. This could make them attractive for space exploration missions, where failure is normally extremely costly.[137] In addition to ground vehicles, swarm robotics includes also research of swarms of aerial robots[132][138] and heterogeneous teams of ground and aerial vehicles.[139][140]

16.5 Military

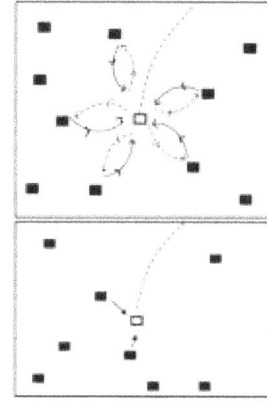

Contrast between guerrilla ambush and true swarming (Edwards-2003)

Main article: Swarming (military)

Military swarming is a behaviour where autonomous or partially autonomous units of action attack an enemy from several different directions and then regroup. *Pulsing*, where the units shift the point of attack, is also a part of military swarming. Military swarming involves the use of a decentralized force against an opponent, in a manner that emphasizes mobility, communication, unit autonomy and coordination or synchronization.[141] Historically military forces used principles of swarming without really examining them explicitly, but now active research consciously examines military doctrines that draw ideas from swarming.

Merely because multiple units converge on a target, they are not necessarily swarming. Siege operations do not involve swarming, because there is no manoeuvre; there is convergence but on the besieged fortification. Nor do guerrilla ambushes constitute swarms, because they are "hit-and-run". Even though the ambush may have several points of attack on the enemy, the guerillas withdraw when they either have inflicted adequate damage, or when they are endangered.

In 2014 the U. S. Office of Naval Research released a video showing tests of a swarm of small autonomous drone attack boats that can steer and take coordinated offensive action as a group.[142]

16.6 Gallery

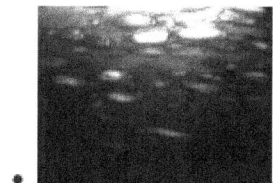
A swarm of migrating herrings

A swarm of bees

Salps arranged in chains form huge swarms.[143]

People swarming through an exit do not always behave like a fluid.[144][145]

A swarm of ladybirds

A swarm of robots

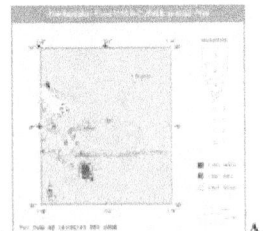
- A swarm of earthquakes

- A swarm of ancient stars

16.7 Myths

- There is a popular myth that lemmings commit mass suicide by swarming off cliffs when they migrate. Driven by strong biological urges, some species of lemmings may migrate in large groups when population density becomes too great. Lemmings can swim and may choose to cross a body of water in search of a new habitat. In such cases, many may drown if the body of water is so wide as to stretch their physical capability to the limit. This fact combined with some unexplained fluctuations in the population of Norwegian lemmings gave rise to the myth.[146]

- Piranha have a reputation as fearless fish that swarm in ferocious and predatory packs. However, recent research, which started "with the premise that they school as a means of cooperative hunting", discovered that they were in fact rather fearful fish, like other fish, who schooled for protection from their predators, such as cormorants, caimans and dolphins. A researcher described them as "basically like regular fish with large teeth".[147]

16.8 See also

- Dyson swarm
- List of collective nouns
- Mobile Bay jubilee
- Swarm (simulation)
- Traffic wave

16.9 References

[1] Bouffanais, Roland. *Design and Control of Swarm Dynamics* (First ed.). Springer. ISBN 978-981-287-750-5.

[2] O'Loan; Evans (1998). "Alternating steady state in one-dimensional flocking". *Journal of Physics A: Mathematical and General*. **32** (8): L99–L105. Bibcode:1999JPhA...32L..99O. arXiv:cond-mat/9811336. doi:10.1088/0305-4470/32/8/002.

[3] Reynolds CW (1987). "Flocks, herds and schools: A distributed behavioral model". *Computer Graphics*. **21** (4): 25–34. CiteSeerX 10.1.1.103.7187. ISBN 0-89791-227-6. doi:10.1145/37401.37406.

[4] Ballerini M, Cabibbo N, Candelier R, Cavagna A, Cisbani E, Giardina I, Lecomte V, Orlandi A, Parisi G, Procaccini A, Viale M, Zdravkovic V (2008). "Interaction ruling animal collective behavior depends on topological rather than metric distance: Evidence from a field study". *Proc. Natl. Acad. Sci. U.S.A.* **105** (4): 1232–7. Bibcode:2008PNAS..105.1232B. PMC 2234121. PMID 18227508. arXiv:0709.1916. doi:10.1073/pnas.0711437105.

[5] Hildenbrandt H, Carere C, Hemelrijk CK (2010). "Self-organized aerial displays of thousands of starlings: a model". *Behavioral Ecology*. **21** (6): 1349–1359. doi:10.1093/beheco/arq149.

[6] Hemelrijk CK, Hildenbrandt H (2011). "Some causes of the variable shape of flocks of birds". *PLoS ONE*. **6** (8): e22479. PMC 3150374. PMID 21829627. doi:10.1371/journal.pone.0022479.

[7] "Zwermen en scholen - Swarming - Permanente expo - Bezoek onze expo's & workshops! - Science LinX - Rijksuniversiteit Groningen".

[8] "Onderzoek aan de Faculteit Wiskunde en Natuurwetenschappen - Faculteit Wiskunde en Natuurwetenschappen - Over ons - Rijksuniversiteit Groningen".

[9] Olson RS, Knoester DB, Adami C (2013). "Critical Interplay Between Density-dependent Predation and Evolution of the Selfish Herd". *Proceedings of GECCO 2013*: 247–254. doi:10.1145/2463372.2463394.

[10] Ward CR, Gobet F, Kendall G (2001). "Evolving collective behavior in an artificial ecology". *Artificial Life*. **7** (2): 191–209. PMID 11580880. doi:10.1162/106454601753139005.

[11] Reluga TC, Viscido S (2005). "Simulated evolution of selfish herd behavior". *Journal of Theoretical Biology*. **234** (2): 213–225. PMID 15757680. doi:10.1016/j.jtbi.2004.11.035.

16.9. REFERENCES

[12] Wood AJ, Ackland GJ (2007). "Evolving the selfish herd: emergence of distinct aggregating strategies in an individual-based model". *Proc Biol Sci.* **274** (1618): 1637–1642. PMC 2169279. PMID 17472913. doi:10.1098/rspb.2007.0306.

[13] Olson RS, Hintze A, Dyer FC, Knoester DB, Adami C (2013). "Predator confusion is sufficient to evolve swarming behaviour". *J. R. Soc. Interface.* **10** (85): 20130305. PMC 4043163. PMID 23740485. doi:10.1098/rsif.2013.0305.

[14] Demsar J, Hemelrijk CK, Hildenbrandt H, Bajec IL (2015). "Simulating predator attacks on schools: Evolving composite tactics". *Ecological Modelling.* **304**: 22–33. doi:10.1016/j.ecolmodel.2015.02.018.

[15] Tosh CR (2011). "Which conditions promote negative density dependent selection on prey aggregations?". *Journal of Theoretical Biology.* **281** (1): 24–30. PMID 21540037. doi:10.1016/j.jtbi.2011.04.014.

[16] Ioannou CC, Guttal V, Couzin ID (2012). "Predatory Fish Select for Coordinated Collective Motion in Virtual Prey". *Science.* **337** (6099): 1212–1215. Bibcode:2012Sci...337.1212I. PMID 22903520. doi:10.1126/science.1218919.

[17] Olson RS, Haley PB, Dyer FC, Adami C (2015). "Exploring the evolution of a trade-off between vigilance and foraging in group-living organisms". *Royal Society Open Science.* **2**: 150135. PMC 4593673. PMID 26473039. doi:10.1098/rsos.150135.

[18] "Hierarchy of Life". 14 September 2008. Retrieved 6 October 2009

[19] Parunak, H. v D. (2003). "Making swarming happen" In: Proceedings of Conference on Swarming and Network Enabled Command, Control, Communications, Computers, Intelligence, Surveillance and Reconnaissance (C4ISR), McLean, Virginia, USA, 3 January 2003.

[20] Marsh L.; Onof C. (2008). "Stigmergic epistemology, stigmergic cognition" (PDF). *Cognitive Systems Research.* **9** (1): 136–149. doi:10.1016/j.cogsys.2007.06.009.

[21] Beni, G., Wang, J. Swarm Intelligence in Cellular Robotic Systems, Proceed. NATO Advanced Workshop on Robots and Biological Systems, Tuscany, Italy, June 26–30 (1989)

[22] Dorigo, M; Birattari, M (2007). "Swarm intelligence". *Scholarpedia.* **2** (9): 1462. doi:10.4249/scholarpedia.1462.

[23] Li YX, Lukeman R, Edelstein-Keshet L, et al. (2007). "Minimal mechanisms for school formation in self-propelled particles" (PDF). *Physica D: Nonlinear Phenomena.* **237** (5): 699–720. Bibcode:2008PhyD..237..699L. doi:10.1016/j.physd.2007.10.009.

[24] Toner J and Tu Y (1995) "Long-range order in a two-dimensional xy model: how birds fly together" *Physical Revue Letters,* **75 (23)**(1995), **4326–4329.**

[25] Topaz C, Bertozzi A (2004). "Swarming patterns in a two-dimensional kinematic model for biological groups". *SIAM J Appl Math.* **65** (1): 152–174. doi:10.1137/S0036139903437424.

[26] Topaz C, Bertozzi A, Lewis M (2006). "A nonlocal continuum model for biological aggregation". *Bull Math Bio.* **68** (7): 1601–1623. doi:10.1007/s11538-006-9088-6.

[27] Carrillo, J; Fornasier, M; Toscani, G (2010). "Particle, kinetic, and hydrodynamic models of swarming" (PDF). *Modeling and Simulation in Science, Engineering and Technology.* **3**: 297–336. doi:10.1007/978-0-8176-4946-3_12.

[28] "Swarmanoid project".

[29] Ant colony optimization Retrieved 15 December 2010.

[30] A. Colorni, M. Dorigo et V. Maniezzo, *Distributed Optimization by Ant Colonies*, actes de la première conférence européenne sur la vie artificielle, Paris, France, Elsevier Publishing, 134-142, 1991.

[31] M. Dorigo, *Optimization, Learning and Natural Algorithms*, PhD thesis, Politecnico di Milano, Italie, 1992.

[32] Hölldobler & Wilson (1990), pp. 143–179

[33] DORIGO, M.; DI CARO, G.; GAMBERELLA, L. M. (1999). *Ant Algorithms for Discrete Optimization, Artificial Life.* MIT Press.

[34] Self driven particle model Interactive simulations, 2005, University of Colorado. Retrieved 10 April 2011.

[35] Vicsek T, Czirok A, Ben-Jacob E, Cohen I, Shochet O (1995). "Novel type of phase transition in a system of self-driven particles". *Physical Review Letters.* **75** (6): 1226–1229. Bibcode:1995PhRvL..75.1226V. PMID 10060237. arXiv:cond-mat/0611743. doi:10.1103/PhysRevLett.75.1226.

[36] Czirók A, Vicsek T (2006). "Collective behavior of interacting self-propelled particles". *Physica A.* **281** (1–4): 17–29. Bibcode:2000PhyA..281...17C. arXiv:cond-mat/0611742. doi:10.1016/S0378-4371(00)00013-3.

[37] Buhl J, Sumpter DJT, Couzin D, Hale JJ, Despland E, Miller ER, Simpson SJ, et al. (2006). "From disorder to order in marching locusts" (PDF). *Science.* **312** (5778): 1402–1406. Bibcode:2006Sci...312.1402B. PMID 16741126. doi:10.1126/science.1125142.

[38] Toner J, Tu Y, Ramaswamy S (2005). "Hydrodynamics and phases of flocks" (PDF). *Annals of Physics.* **318** (1): 170–244. Bibcode:2005AnPhy.318..170T. doi:10.1016/j.aop.2005.04.011.

[39] Bertin, E; Droz; Grégoire, G (2009). "Hydrodynamic equations for self-propelled particles: microscopic derivation and stability analysis". *J. Phys. A.* **42** (44): 445001. arXiv:0907.4688. doi:10.1088/1751-8113/42/44/445001.

[40] Kennedy, J.; Eberhart, R. (1995). "Particle Swarm Optimization". *Proceedings of IEEE International Conference on Neural Networks*. **IV**. pp. 1942–1948.

[41] Kennedy, J. (1997). "The particle swarm: social adaptation of knowledge". *Proceedings of IEEE International Conference on Evolutionary Computation*. pp. 303–308.

[42] Hu X Particle swarm optimization: Tutorial. Retrieved 15 December 2010.

[43] Kennedy, J.; Eberhart, R.C. (2001). *Swarm Intelligence*. Morgan Kaufmann. ISBN 1-55860-595-9.

[44] Poli, R. (2007). "An analysis of publications on particle swarm optimisation applications" (PDF). *Technical Report CSM-469*. Department of Computer Science, University of Essex, UK.

[45] Poli, R. (2008). "Analysis of the publications on the applications of particle swarm optimisation" (PDF). *Journal of Artificial Evolution and Applications*. **2008**: 1–10. doi:10.1155/2008/685175.

[46] Altruism helps swarming robots fly better *genevalunch.com*, 4 May 2011.

[47] Waibel, M; Floreano, D; Keller, L (2011). "A quantitative test of Hamilton's rule for the evolution of altruism". *PLoS Biology*. **9** (5): 1000615. PMC 3086867. PMID 21559320. doi:10.1371/journal.pbio.1000615.

[48] Feare C (1984) *The Starling*. Oxford University Press. ISBN 978-0-19-217705-6.

[49] Partridge BL (1982). "The structure and function of fish schools" (PDF). *Scientific American*. Vol. 246 no. 6. pp. 114–123. PMID 7201674. doi:10.1038/scientificamerican0682-114.

[50] Hubbard S, Babak P, Sigurdsson S, Magnusson K (2004). "A model of the formation of fish schools and migrations of fish". *Ecol. Model.* **174** (4): 359–374. doi:10.1016/j.ecolmodel.2003.06.006.

[51] Rauch E, Millonas M, Chialvo D (1995). "Pattern formation and functionality in swarm models". *Physics Letters A*. **207** (3–4): 185–193. Bibcode:1995PhLA..207..185R. arXiv:adap-org/9507003. doi:10.1016/0375-9601(95)00624-C.

[52] Allison C, Hughes C (1991). "Bacterial swarming: an example of prokaryotic differentiation and multicellular behaviour". *Science Progress*. **75** (298 Pt 3-4): 403–422. PMID 1842857.

[53] Ben-Jacob E, Cohen I, Shochet O, Czirok A, Vicsek T (1995). "Cooperative Formation of Chiral Patterns during Growth of Bacterial Colonies". *Physical Review Letters*. **75** (15): 2899–2902. Bibcode:1995PhRvL..75.2899B. PMID 10059433. doi:10.1103/PhysRevLett.75.2899.

[54] Rappel WJ, Nicol A, Sarkissian A, Levine H, Loomis WF (1999). "Self-organized vortex state in two-dimensional Dictyostelium dynamics". *Physical Review Letters*. **83** (6): 1247–1250. Bibcode:1999PhRvL..83.1247R. arXiv:patt-sol/9811001. doi:10.1103/PhysRevLett.83.1247.

[55] Chowdhury, D (2006). "Collective effects in intracellular molecular motor transport: coordination, cooperation and competetion". *Physica A*. **372** (1): 84–95. doi:10.1016/j.physa.2006.05.005.

[56] Parrish JK and Hamner WM (eds) (1997) *Animal Groups in Three Dimensions* Cambridge University Press. ISBN 978-0-521-46024-8.

[57] Helbing D, Keltsch J, Molnar P (1997). "Modelling the evolution of human trail systems". *Nature*. **388** (6637): 47–50. Bibcode:1997Natur.388...47H. PMID 9214501. arXiv:cond-mat/9805158. doi:10.1038/40353.

[58] Helbing D, Farkas I, Vicsek T (2000). "Simulating dynamical features of escape panic". *Nature*. **407** (6803): 487–490. Bibcode:2000Natur.407..487H. PMID 11028994. arXiv:cond-mat/0009448. doi:10.1038/35035023.

[59] Helbing D, Farkas IJ, Vicsek T (2000). "Freezing by heating in a driven mesoscopic system". *Physical Review Letters*. **84** (6): 1240–1243. Bibcode:2000PhRvL..84.1240H. PMID 11017488. arXiv:cond-mat/9904326. doi:10.1103/PhysRevLett.84.1240.

[60] Bonabeau E and Theraulaz G (2008) "Swarm Smarts". In *Your Future with Robots* Scientific American Special Editions.

[61] Czaczkes, T.J.; Grüter, C.; Ratnieks, F. L. W. (2015). "Trail pheromones: an integrative view of their role in colony organisation". *Annual Review of Entomology*. **60**: 581–599. PMID 25386724. doi:10.1146/annurev-ento-010814-020627.

[62] Beckers, R.; Deneubourg, J. L.; Goss, S (1993). "Modulation of trail laying in the ant Lasius niger (Hymenoptera: Formicidae) and its role in the collective selection of a food source". *Journal of Insect Behavior*. **6** (6): 751–759. doi:10.1007/BF01201674.

[63] Goss, S.; Aron, S.; Deneubourg, J. L.; Pasteels, J. M. (1989). "Self-organized shortcuts in the Argentine ant". *Naturwissenschaften*. **76**: 579–581. Bibcode:1989NW.....76..579G. doi:10.1007/BF00462870.

[64] Dicke E, Byde A, Cliff D, Layzell P (2004). A. J. Ispeert, M. Murata, N. Wakamiya, eds. "Proceedings of Biologically Inspired Approaches to Advanced Information Technology: First International Workshop, BioADIT 2004 LNCS 3141": 364–379. |chapter= ignored (help)

[65] Oldroyd, BP (1998). "Colony relatedness in aggregations of Apis dorsata Fabricius (Hymenoptera, Apidae)". *Insectes Sociaux*. **47**: 94–95. doi:10.1007/s000400050015.

[66] Milius, Susan Swarm Savvy. How bees, ants and other animals avoid dumb collective decisions; Science News, May 9th, 2009; Vol.175 #10 (p. 16)

[67] Bee Swarms Follow High-speed 'Streaker' Bees To Find A New Nest; ScienceDaily (Nov. 24, 2008)

[68] Seeley, Thomas D.; Visscher, P. Kirk (2003). "Choosing a home: how the scouts in a honey bee swarm perceive the completion of their group decision making" (PDF). Behavioral Ecology and Sociobiology. 54 (5): 511–520. doi:10.1007/s00265-003-0664-6.

[69] Morse, R.A. (1963). "Swarm orientation in honeybees". Science. 141 (3578): 357–358. PMID 17815993. doi:10.1126/science.141.3578.357.

[70] Seeley, Thomas (2003). "Consensus building during nest-site selection in honey bee swarms: The expiration of dissent". Behavioral Ecology and Sociobiology. 53 (6): 417–424. doi:10.1007/s00265-003-0598-z.

[71] Jennifer Viegas. "Cockroaches Make Group Decisions". Discovery Channel. Retrieved 10 June 2006.

[72] Lemonick, Michael D. (2007-11-15). "Robotic Roaches Do the Trick". Time Magazine.

[73] Locust Locustidae National Geographic. Retrieved 12 December 2010.

[74] "Locust swarms 'high' on serotonin". 29 January 2009 – via bbc.co.uk.

[75] Rogers SM, Matheson T, Despland E, Dodgson T, Burrows M, Simpson SJ (November 2003). "Mechanosensory-induced behavioural gregarization in the desert locust Schistocerca gregaria". J. Exp. Biol. 206 (Pt 22): 3991–4002. PMID 14555739. doi:10.1242/jeb.00648.

[76] Stevenson, PA (2009). "The Key to Pandora's Box". Science. 323 (5914): 594–5. PMID 19179520. doi:10.1126/science.1169280.

[77] Blocking 'happiness' chemical may prevent locust plagues, New scientist, 2009-01-29, accessed 2009-01-31

[78] Moshe Guershon; Amir Ayali (May 2012). "Innate phase behavior in the desert locust, Schistocerca gregaria". Insect Science. 19 (6): 649–656. doi:10.1111/j.1744-7917.2012.01518.x.

[79] Yates, CA; Erban, R; Escudero, C; Couzin, ID; Buhl, J; Kevrekidis, IG; Maini, PK; Sumpter, DJ (2009). "Inherent noise can facilitate coherence in collective swarm motion". Proc. Natl. Acad. Sci. U.S.A. 106 (14): 5464–9. Bibcode:2009PNAS..106.5464Y. PMC 2667078. PMID 19336580. doi:10.1073/pnas.0811195106.

[80] Pyle, Robert Michael. "National Audubon Society Field Guide to North American Butterflies", p712-713, Alfred A. Knopf, New York, ISBN 0-394-51914-0

[81] "Monarch, Danaus plexippus". Retrieved 2008-08-27.

[82] Gugliotta, Guy (2003): Butterflies Guided By Body Clocks, Sun Scientists Shine Light on Monarchs' Pilgrimage. Washington Post, May 23, 2003, page A03. Retrieved 2006-JAN-07.

[83] Merlin C, Gegear RJ, Reppert SM (2009). "Antennal Circadian Clocks Coordinate Sun Compass Orientation in Migratory Monarch Butterflies". Science. 325 (5948): 1700–1704. Bibcode:2009Sci...325.1700M. PMC 2754321. PMID 19779201. doi:10.1126/science.1176221.

[84] Kyriacou CP (2009). "Unraveling Traveling". Science. 325 (5948): 1629–1630. PMID 19779177. doi:10.1126/science.1178935.

[85] Sekercioglu, C.H. (2007). "Conservation ecology: area trumps mobility in fragment bird extinctions". Current Biology. 17 (8): R283–R286. PMID 17437705. doi:10.1016/j.cub.2007.02.019.

[86] Drag Reduction from Formation Flight. Flying Aircraft in Bird-Like Formations Could Significantly Increase Range; Defense Technical Information Center; April 2002; Retrieved February 27, 2008

[87] Hummel D.; Beukenberg M. (1989). "Aerodynamische Interferenzeffekte beim Formationsflug von Vögeln". J. Ornithol. 130 (1): 15–24. doi:10.1007/BF01647158.

[88] Cutts, C. J. & J R Speakman (1994). "Energy savings in formation flight of Pink-footed Geese" (PDF). J. Exp. Biol. 189 (1): 251–261. PMID 9317742.

[89] Newton, I. (2008). The Migration Ecology of Birds. Elsevier. ISBN 978-0-12-517367-4.

[90] Pitcher et al. 1982.

[91] Pitcher TJ and Parish JK (1993) "Functions of shoaling behaviour in teleosts" In: Pitcher TJ (ed) Behaviour of teleost fishes. Chapman and Hall, New York, pp 363–440

[92] Hoare DJ, Krause J, Peuhkuri N and Godin JGJ (2000) Body size and shoaling in fish Journal of Fish Biology, 57(6) 1351-1366.

[93] Snekser JL, Ruhl N, Bauer K, McRobert SP (2010). "The Influence of Sex and Phenotype on Shoaling Decisions in Zebrafish" (PDF). International Journal of Comparative Psychology. 23: 70–81.

[94] Engeszer RE, Ryan MJ, Parichy DM (2004). "Learned Social Preference in Zebrafish" (PDF). Current Biology. 14 (10): 881–884. PMID 15186744. doi:10.1016/j.cub.2004.04.042.

[95] Reebs, S.G. (2000). "Can a minority of informed leaders determine the foraging movements of a fish shoal?". Animal Behaviour. 59 (2): 403–9. PMID 10675263. doi:10.1006/anbe.1999.1314.

[96] Radakov DV (1973) *Schooling in the ecology of fish*. Israel Program for Scientific Translation, translated by Mill H. Halsted Press, New York. ISBN 978-0-7065-1351-6

[97] Photographer: Mark van Coller

[98] Hamner, WM; Hamner, PP; Strand, SW; Gilmer, RW (1983). "Behavior of Antarctic Krill, *Euphausia superba*: Chemoreception, Feeding, Schooling and Molting'". *Science*. **220** (4595): 433–5. Bibcode:1983Sci...220..433H. PMID 17831417. doi:10.1126/science.220.4595.433.

[99] U. Kils; P. Marshall (1995). "Der Krill, wie er schwimmt und frisst – neue Einsichten mit neuen Methoden ("*The Antarctic krill – how it swims and feeds – new insights with new methods*")". In I. Hempel; G. Hempel. *Biologie der Polarmeere – Erlebnisse und Ergebnisse* (Biology of the Polar Oceans Experiences and Results). Fischer Verlag. pp. 201–210. ISBN 3-334-60950-2.

[100] R. Piper (2007). *Extraordinary Animals: An Encyclopedia of Curious and Unusual Animals*. Greenwood Press. ISBN 0-313-33922-8.

[101] Hoare, Ben (2009). Animal Migration. London: Natural History Museum. p. 107. ISBN 978-0-565-09243-6.

[102] Hoare, Ben (2009). Animal Migration. London: Natural History Museum. p. 107. ISBN 978-0-565-09243-6

[103] J.S. Jaffe; M.D. Ohmann; A. de Robertis (1999). "Sonar estimates of daytime activity levels of *Euphausia pacifica* in Saanich Inlet" (PDF). *Canadian Journal of Fisheries and Aquatic Sciences*. **56** (11): 2000–10. doi:10.1139/cjfas-56-11-2000.

[104] Geraint A. Tarling & Magnus L. Johnson (2006). "Satiation gives krill that sinking feeling". *Current Biology*. **16** (3): 83–4. PMID 16461267. doi:10.1016/j.cub.2006.01.044.

[105] Howard, D.: "Krill", pp. 133–140 in Karl, H.A. et al. (eds): *Beyond the Golden Gate – Oceanography, Geology, Biology, and Environmental Issues in the Gulf of the Farallones*, USGS Circular 1198, 2001. URLs last accessed 2010-06-04.

[106] D. Howard. "Krill in Cordell Bank National Marine Sanctuary". NOAA. Retrieved 15 June 2005.

[107] Gandomi, A.H.; Alavi, A.H. (2012). "Krill Herd Algorithm: A New Bio-Inspired Optimization Algorithm". *Communications in Nonlinear Science and Numerical Simulation*. **17** (12): 4831–4845. Bibcode:2012CNSNS..17.4831G. doi:10.1016/j.cnsns.2012.05.010.

[108] Hamner, WM; Carleton, JH (1979). "Copepod swarms: Attributes and role in coral reef ecosystems" (PDF). *Limnol. Oceanogr*. **24** (1): 1–14. doi:10.4319/lo.1979.24.1.0001.

[109] Johannes Dürbaum & Thorsten Künnemann (November 5, 1997). "Biology of Copepods: An Introduction". Carl von Ossietzky University of Oldenburg. Retrieved December 8, 2009.

[110] Lindsey R and Scott M (2010) What are phytoplankton NASA Earth Observatory.

[111] Harmful algal blooms in the Great Lakes 2009, NOAA, Center of Excellence for Great Lakes and Human Health.

[112] Darwin, Erasmus (1800-01-01). *Phytologia: Or, The Philosophy of Agriculture and Gardening. With the Theory of Draining Morasses and with an Improved Construction of the Drill Plough*. P. Byrne.

[113] Ciszak, Marzena; Comparini, Diego; Mazzolai, Barbara; Baluska, Frantisek; Arecchi, F. Tito; Vicsek, Tamás; Mancuso, Stefano (2012-01-17). "Swarming Behavior in Plant Roots". *PLoS ONE*. **7** (1): e29759. ISSN 1932-6203. PMC 3260168. PMID 22272246. doi:10.1371/journal.pone.0029759.

[114] Baluška, František; Mancuso, Stefano; Volkmann, Dieter; Barlow, Peter W. (2010-07-01). "Root apex transition zone: a signalling–response nexus in the root". *Trends in Plant Science*. **15** (7): 402–408. doi:10.1016/j.tplants.2010.04.007.

[115] J., Trewavas, A. (2014). *Plant behaviour and intelligence*. Oxford university press. ISBN 9780199539543. OCLC 961862730.

[116] Reichenbach H (2001). "Myxobacteria, producers of novel bioactive substances". *J Ind Microbiol Biotechnol*. **27** (3): 149–56. PMID 11780785. doi:10.1038/sj.jim.7000025.

[117] Farkas I, Helbing D, Vicsek T (2002). "Mexican waves in an excitable medium" (PDF). *Nature*. **419** (6903): 131–132. Bibcode:2002Natur.419..131F. PMID 12226653. arXiv:cond-mat/0210073. doi:10.1038/419131a.

[118] Neda Z, Ravasz E, Brechet Y, Vicsek T, Barabasi AL (2002). "Physics of Rhythmic Applause" (PDF). *Physical Review E*. **61** (6): 6987–6992. doi:10.1103/physreve.61.6987.

[119] Helbing, D; Keltsch, J; Molnar, P (1997). "Modelling the evolution of human trail systems". *Nature*. **388**: 47–50. PMID 9214501. arXiv:cond-mat/9805158. doi:10.1038/40353.

[120] "http://psychcentral.com/news/2008/02/15/herd-mentality-explained/1922.html". Retrieved on October 31st 2008.

[121] "Danger in numbers during Haj". The National.

[122] Couzin ID, Krause J (2003). "Self-organization and collective behavior in vertebrates" (PDF). *Advances in the Study of Behavior*. Advances in the Study of Behavior. **32**: 1–75. ISBN 978-0-12-004532-7. doi:10.1016/S0065-3454(03)01001-5.

[123] Gabbai, J.M.E. (2005). "Complexity and the Aerospace Industry: Understanding Emergence by Relating Structure to Performance using Multi-Agent Systems". Manchester: University of Manchester Doctoral Thesis.

[124] Livermore R (2008) "A multi-agent system approach to a simulation study comparing the performance of aircraft boarding using pre-assigned seating and free-for-all strategies" *Open University*, Technical report No 2008/25.

[125] "Planes, Trains and Ant Hills: Computer scientists simulate activity of ants to reduce airline delays" *Science Daily*, 1 April 2008.

[126] Burd, Martin; N. Aranwela (February 2003). "Head-on encounter rates and walking speed of foragers in leaf-cutting ant traffic". *Insectes Sociaux*. **50** (1): 3–8. doi:10.1007/s000400030000l.

[127] Ribeiro, Pedro; André Frazão Helene; Gilberto Xavier; Carlos Navas; Fernando Leite Ribeiro (2009-04-01). Dornhaus, Anna, ed. "Ants can learn to forage on one-way trails". *PLoS ONE*. **4** (4): e5024. Bibcode:2009PLoSO...4.5024R. PMC 2659768. PMID 19337369. doi:10.1371/journal.pone.0005024.

[128] John, Alexander; Andreas Schadschneider; Debashish Chowdhury; Katsuhiro Nishinari (March 2008). "Characteristics of ant-inspired traffic flow". *Swarm Intelligence*. **2** (1): 25–41. doi:10.1007/s11721-008-0010-8.

[129] Are we nearly there yet? Motorists could learn a thing or two from ants *The Economist*, 10 July 2009.

[130] Helbing, Dirk; Farkas, Illés; Vicsek, Tamás (2000). "Simulating dynamical features of escape panic". *Nature*. **407** (6803): 487–490. Bibcode:2000Natur.407..487H. PMID 11028994. arXiv:cond-mat/0009448. doi:10.1038/35035023.

[131] "Swarming the shelves: How shops can exploit people's herd mentality to increase sales?". The Economist. 2006-11-11. p. 90.

[132] Kushleyev, Alex; Mellinger, Daniel; Powers, Caitlin; Kumar, Vijay (2013). "Towards a swarm of agile micro quadrotors". *Autonomous Robots*. **35** (4): 287–300. doi:10.1007/s10514-013-9349-9.

[133] "Self-organizing Systems Research Group".

[134] "Centibots 100-Robot Collaborative Reconnaissance Project". ActivMedia Robotics. Archived from the original on 2011-07-14.

[135] "Open-source micro-robotic project". Retrieved 2007-10-28.

[136] "Swarm", iRobot Corporation. Archived from the original on 2007-09-27. Retrieved 2007-10-28.

[137] Knapp, Louise (2000-12-21). "Look, Up in the Sky: Robofly". Wired Magazine. Retrieved 2008-09-25.

[138] Saska, Martin; Jan, Vakula; Libor, Preucil (2014). *Swarms of micro aerial vehicles stabilized under a visual relative localization*. IEEE International Conference on Robotics and Automation (ICRA).

[139] Saska, Martin; Vonasek, Vojtech; Krajnik, Tomas; Preucil, Libor (2014). "Coordination and navigation of heterogeneous MAV–UGV formations localized by a hawk-eye-like approach under a model predictive control scheme.". *International Journal of Robotics Research*. **33** (10): 1393–1412. doi:10.1177/0278364914530482.

[140] Saska, Martin; Vonasek, Vojtech; Krajnik, Tomas; Preucil, Libor (2012). *Coordination and Navigation of Heterogeneous UAVs-UGVs Teams Localized by a Hawk-Eye Approach*. IEEE/RSJ International Conference on Intelligent Robots and Systems (IROS).

[141] Edwards, Sean J.A. (2000). *Swarming on the Battlefield: Past, Present, and Future*. Rand Monograph MR-1100. Rand Corporation. ISBN 0-8330-2779-4.

[142] U.S. Navy could 'swarm' foes with robot boats, *CNN*, 13 October 2014.

[143] "Dive and Discover: Scientific Expedition 10: Antarctica". Retrieved 2008-09-03.

[144] Crowd modelling: Simulating the behaviour of crowds of people, or swarms of animals, has both frivolous and important uses *The Economist*, 5 March 2009.

[145] Fisher, Len (2009) *The perfect swarm: the science of complexity in everyday life* Page 57. Basic Books. ISBN 978-0-465-01884-0

[146] Woodford, Riley. "Lemming Suicide Myth Disney Film Faked Bogus Behavior". Archived from the original on January 3, 2010.

[147] Red-Bellied Piranha Is Really Yellow *New York Times*, 24 May 2005.

16.10 Sources

- Blum C and Merkle D (2008) *Swarm intelligence: introduction and applications* Springer. ISBN 978-3-540-74088-9.

- Camazine S. Deneubourg JL, Franks NR, Sneyd J, Theraulaz G and Bonabeau E (2003) *Self-Organization in Biological Systems* Princeton University Press. ISBN 978-0-691-11624-2.

- Fisher L (2009) *The perfect swarm: the science of complexity in everyday life* Basic Books. ISBN 978-0-465-01884-0.

- Kennedy JF, Kennedy J, Eberhart RC and Shi Y (2001) *Swarm intelligence* Morgan Kaufmann. ISBN 978-1-55860-595-4.

- Krause, J (2005) *Living in Groups* Oxford University Press. ISBN 978-0-19-850818-2

- Lim CP, Jain LC and Dehuri S (2009) *Innovations in Swarm Intelligence* Springer. ISBN 978-3-642-04224-9.

- Miller, Peter (2010) *The Smart Swarm: How understanding flocks, schools, and colonies can make us better at communicating, decision making, and getting things done* Penguin. ISBN 978-1-58333-390-7

- Nedjah N and Mourelle LdM (2006) *Swarm intelligent systems* Springer. ISBN 978-3-540-33868-0.

- Sumpter, David JT (2010) *Collective Animal Behavior* Princeton University Press. ISBN 978-0-691-14843-4.

- Vicsek A, Zafeiris A (2012). "Collective motion". *Physics Reports*. **517** (3–4): 71–140. Bibcode:2012PhR...517...71V. arXiv:1010.5017. doi:10.1016/j.physrep.2012.03.004.

16.11 External links

- New York times article on investigations into swarming

- From the *Wolfram Demonstrations Project* — requires CDF player (free):

    - Model of a Firefly Swarm.
    - Garbage Collection by Ants
    - Beverton and Merging Schools of Fish
    - Propp Circles

Chapter 17

Network science

For other uses, see Network (disambiguation).

Network science is an academic field which studies complex networks such as telecommunication networks, computer networks, biological networks, cognitive and semantic networks, and social networks, considering distinct elements or actors represented by *nodes* (or *vertices*) and the connections between the elements or actors as *links* (or *edges*). The field draws on theories and methods including graph theory from mathematics, statistical mechanics from physics, data mining and information visualization from computer science, inferential modeling from statistics, and social structure from sociology. The United States National Research Council defines network science as "the study of network representations of physical, biological, and social phenomena leading to predictive models of these phenomena."[1]

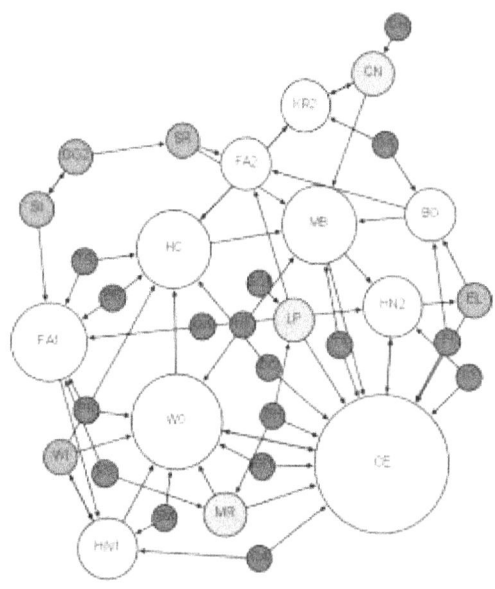

Moreno's sociogram of a 1st grade class.

17.1 Background and history

The study of networks has emerged in diverse disciplines as a means of analyzing complex relational data. The earliest known paper in this field is the famous Seven Bridges of Königsberg written by Leonhard Euler in 1736. Euler's mathematical description of vertices and edges was the foundation of graph theory, a branch of mathematics that studies the properties of pairwise relations in a network structure. The field of graph theory continued to develop and found applications in chemistry (Sylvester, 1878).

In the 1930s Jacob Moreno, a psychologist in the Gestalt tradition, arrived in the United States. He developed the sociogram and presented it to the public in April 1933 at a convention of medical scholars. Moreno claimed that "before the advent of sociometry no one knew what the interpersonal structure of a group 'precisely' looked like (Moreno, 1953). The sociogram was a representation of the social structure of a group of elementary school students. The boys were friends of boys and the girls were friends of girls with the exception of one boy who said he liked a single girl. The feeling was not reciprocated. This network representation of social structure was found so intriguing that it was printed in The New York Times (April 3, 1933, page 17). The sociogram has found many applications and has grown into the field of social network analysis.

Probabilistic theory in network science developed as an offshoot of graph theory with Paul Erdős and Alfréd Rényi's eight famous papers on random graphs. For social networks the exponential random graph model or p* is a notational framework used to represent the probability space of a tie occurring in a social network. An alternate approach to network probability structures is the network probability matrix, which models the probability of edges occurring in a network, based on the historic presence or absence of the edge in a sample of networks.

In 1998, David Krackhardt and Kathleen Carley introduced the idea of a meta-network with the PCANS Model. They suggest that "all organizations are structured along these three domains, Individuals, Tasks, and Resources". Their paper introduced the concept that networks occur across multiple domains and that they are interrelated. This field has grown into another sub-discipline of network science called dynamic network analysis.

More recently other network science efforts have focused on mathematically describing different network topologies. Duncan Watts reconciled empirical data on networks with mathematical representation, describing the small-world network. Albert-László Barabási and Reka Albert developed the scale-free network which is a loosely defined network topology that contains hub vertices with many connections, that grow in a way to maintain a constant ratio in the number of the connections versus all other nodes. Although many networks, such as the internet, appear to maintain this aspect, other networks have long tailed distributions of nodes that only approximate scale free ratios.

17.1.1 Department of Defense initiatives

The U.S. military first became interested in network-centric warfare as an operational concept based on network science in 1996. John A. Parmentola, the U.S. Army Director for Research and Laboratory Management, proposed to the Army's Board on Science and Technology (BAST) on December 1, 2003 that Network Science become a new Army research area. The BAST, the Division on Engineering and Physical Sciences for the National Research Council (NRC) of the National Academies, serves as a convening authority for the discussion of science and technology issues of importance to the Army and oversees independent Army-related studies conducted by the National Academies. The BAST conducted a study to find out whether identifying and funding a new field of investigation in basic research, Network Science, could help close the gap between what is needed to realize Network-Centric Operations and the current primitive state of fundamental knowledge of networks.

As a result, the BAST issued the NRC study in 2005 titled Network Science (referenced above) that defined a new field of basic research in Network Science for the Army. Based on the findings and recommendations of that study and the subsequent 2007 NRC report titled Strategy for an Army Center for Network Science, Technology, and Experimentation, Army basic research resources were redirected to initiate a new basic research program in Network Science. To build a new theoretical foundation for complex networks, some of the key Network Science research efforts now ongoing in Army laboratories address:

- Mathematical models of network behavior to predict performance with network size, complexity, and environment

- Optimized human performance required for network-enabled warfare

- Networking within ecosystems and at the molecular level in cells.

As initiated in 2004 by Frederick I. Moxley with support he solicited from David S. Alberts, the Department of Defense helped to establish the first Network Science Center in conjunction with the U.S. Army at the United States Military Academy (USMA). Under the tutelage of Dr. Moxley and the faculty of the USMA, the first interdisciplinary undergraduate courses in Network Science were taught to cadets at West Point. In order to better instill the tenets of network science among its cadre of future leaders, the USMA has also instituted a five-course undergraduate minor in Network Science.

In 2006, the U.S. Army and the United Kingdom (UK) formed the Network and Information Science International Technology Alliance, a collaborative partnership among the Army Research Laboratory, UK Ministry of Defense and a consortium of industries and universities in the U.S. and UK. The goal of the alliance is to perform basic research in support of Network-Centric Operations across the needs of both nations.

In 2009, the U.S. Army formed the Network Science CTA, a collaborative research alliance among the Army Research Laboratory, CERDEC, and a consortium of about 30 industrial R&D labs and universities in the U.S. The goal of the alliance is to develop a deep understanding of the underlying commonalities among intertwined social/cognitive, information, and communications networks, and as a result improve our ability to analyze, predict, design, and influence complex systems interweaving many kinds of networks.

Subsequently, as a result of these efforts, the U.S. Department of Defense has sponsored numerous research projects that support Network Science.

17.2 Network properties

Often, networks have certain attributes that can be calculated to analyze the properties & characteristics of the network. These network properties often define network models and can be used to analyze how certain models contrast to each other. Many of the definitions for other terms used in network science can be found in Glossary of graph theory.

17.2.1 Size

The size of a network can refer to the number of nodes N or, less commonly, the number of edges E which can range from $N-1$ (a tree) to E_{max} (a complete graph).

17.2.2 Density

The density D of a network is defined as a ratio of the number of edges E to the number of possible edges in a network with N nodes, given by the binomial coefficient $\binom{N}{2}$, giving $D = \frac{2(E-N+1)}{N(N-3)+2}$. Another possible equation is $D = \frac{T-2N+2}{N(N-3)+2}$, whereas the ties T are unidirectional (Wasserman & Faust 1994).[2] This gives a better overview over the network density, because unidirectional relationships can be measured.

17.2.3 Planar Network Density

The density D of a network, where there is no intersection between edges, is defined as a ratio of the number of edges E to the number of possible edges in a network with N nodes, given by a graph with no intersecting edges ($E_{max} = 3N - 6$), giving $D = \frac{E-N+1}{2N-5}$.

17.2.4 Average degree

The degree k of a node is the number of edges connected to it. Closely related to the density of a network is the average degree, $\langle k \rangle = \frac{2E}{N}$. In the ER random graph model, we can compute $\langle k \rangle = p(N-1)$ where p is the probability of two nodes being connected.

17.2.5 Average path length (or characteristic path length)

Average path length is calculated by finding the shortest path between all pairs of nodes, adding them up, and then dividing by the total number of pairs. This shows us, on average, the number of steps it takes to get from one member of the network to another.

17.2.6 Diameter of a network

As another means of measuring network graphs, we can define the diameter of a network as the longest of all the calculated shortest paths in a network. It is the shortest distance between the two most distant nodes in the network. In other words, once the shortest path length from every node to all other nodes is calculated, the diameter is the longest of all the calculated path lengths. The diameter is representative of the linear size of a network.

17.2.7 Clustering coefficient

The clustering coefficient is a measure of an "all-my-friends-know-each-other" property. This is sometimes described as the friends of my friends are my friends. More precisely, the clustering coefficient of a node is the ratio of existing links connecting a node's neighbors to each other to the maximum possible number of such links. The clustering coefficient for the entire network is the average of the clustering coefficients of all the nodes. A high clustering coefficient for a network is another indication of a small world.

The clustering coefficient of the i'th node is

$$C_i = \frac{2e_i}{k_i(k_i-1)}.$$

where k_i is the number of neighbours of the i'th node, and e_i is the number of connections between these neighbours. The maximum possible number of connections between neighbors is, then,

$$\binom{k}{2} = \frac{k(k-1)}{2}.$$

17.2.8 Connectedness

The way in which a network is connected plays a large part into how networks are analyzed and interpreted. Networks are classified in four different categories:

- *Clique/Complete Graph*: a completely connected network, where all nodes are connected to every other node. These networks are symmetric in that all nodes have in-links and out-links from all others.

- *Giant Component*: A single connected component which contains most of the nodes in the network.

- *Weakly Connected Component*: A collection of nodes in which there exists a path from any node to any other, ignoring directionality of the edges.

- *Strongly Connected Component*: A collection of nodes in which there exists a *directed* path from any node to any other.

17.2.9 Node centrality

Main article: Centrality

Centrality indices produce rankings which seek to identify the most important nodes in a network model. Different centrality indices encode different contexts for the word "importance." The betweenness centrality, for example, considers a node highly important if it form bridges between many other nodes. The eigenvalue centrality, in contrast, considers a node highly important if many other highly important nodes link to it. Hundreds of such measures have been proposed in the literature.

Centrality indices are only accurate for identifying the most central nodes. The measures are seldom, if ever, meaningful for the remainder of network nodes. [3] [4] Also, their indications are only accurate within their assumed context for importance, and tend to "get it wrong" for other contexts.[5] For example, imagine two separate communities whose only link is an edge between the most junior member of each community. Since any transfer from one community to the other must go over this link, the two junior members will have high betweenness centrality. But, since they are junior, (presumably) they have few connections to the "important" nodes in their community, meaning their eigenvalue centrality would be quite low.

The concept of centrality in the context of static networks was extended, based on empirical and theoretical research, to dynamic centrality[6] in the context of time-dependent and temporal networks.[7][8][9]

17.2.10 Node influence

Main article: Node influence metric

Limitations to centrality measures have led to the development of more general measures. Two examples are the **accessibility**, which uses the diversity of random walks to measure how accessible the rest of the network is from a given start node,[10] and the **expected force**, derived from the expected value of the force of infection generated by a node.[3] Both of these measures can be meaningfully computed from the structure of the network alone.

17.3 Network models

Network models serve as a foundation to understanding interactions within empirical complex networks. Various random graph generation models produce network structures that may be used in comparison to real-world complex networks.

17.3.1 Erdős–Rényi random graph model

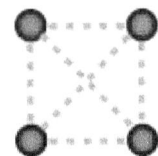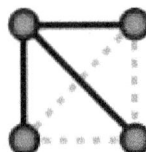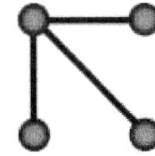

This Erdős–Rényi model is generated with N = 4 nodes. For each edge in the complete graph formed by all N nodes, a random number is generated and compared to a given probability. If the random number is less than p, an edge is formed on the model.

The **Erdős–Rényi model**, named for Paul Erdős and Alfréd Rényi, is used for generating random graphs in which edges are set between nodes with equal probabilities. It can be used in the probabilistic method to prove the existence of graphs satisfying various properties, or to provide a rigorous definition of what it means for a property to hold for almost all graphs.

To generate an Erdős–Rényi model two parameters must be specified: the number of nodes in the graph generated as N and the probability that a link should be formed between any two nodes as p. A constant $\langle k \rangle$ may derived from these two components with the formula $\langle k \rangle = 2 \cdot E / N = p \cdot (N - 1)$, where E is the expected number of edges.

The Erdős–Rényi model has several interesting characteristics in comparison to other graphs. Because the model is generated without bias to particular nodes, the degree distribution is binomial in nature with regards to the formula:

$$P(\deg(v) = k) = \binom{n-1}{k} p^k (1-p)^{n-1-k}.$$

Also as a result of this characteristic, the clustering coefficient tends to 0. The model tends to form a giant component in situations where $\langle k \rangle > 1$ in a process called percolation. The average path length is relatively short in this model and tends to log N.

17.3.2 Watts–Strogatz small world model

The Watts and Strogatz model is a random graph generation model that produces graphs with small-world properties.

An initial lattice structure is used to generate a Watts-Strogatz model. Each node in the network is initially linked to its $\langle k \rangle$ closest neighbors. Another parameter is specified as the rewiring probability. Each edge has a probability p that it will be rewired to the graph as a random

17.3. NETWORK MODELS

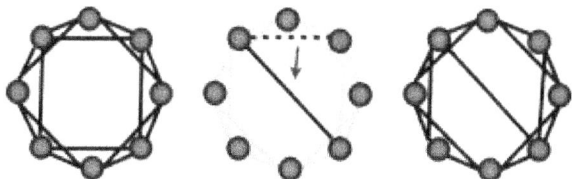

The Watts and Strogatz model uses the concept of rewiring to achieve its structure. The model generator will iterate through each edge in the original lattice structure. An edge may changed its connected vertices according to a given rewiring probability. $\langle k \rangle = 4$ in this example.

edge. The expected number of rewired links in the model is $pE = pN\langle k \rangle / 2$.

As the Watts–Strogatz model begins as non-random lattice structure, it has a very high clustering coefficient along with high average path length. Each rewire is likely to create a shortcut between highly connected clusters. As the rewiring probability increases, the clustering coefficient decreases slower than the average path length. In effect, this allows the average path length of the network to decrease significantly with only slightly decreases in clustering coefficient. Higher values of p force more rewired edges, which in effect makes the Watts–Strogatz model a random network.

17.3.3 Barabási–Albert (BA) preferential attachment model

The Barabási–Albert model is a random network model used to demonstrate a preferential attachment or a "rich-get-richer" effect. In this model, an edge is most likely to attach to nodes with higher degrees. The network begins with an initial network of m_0 nodes. $m_0 \geq 2$ and the degree of each node in the initial network should be at least 1, otherwise it will always remain disconnected from the rest of the network.

In the BA model, new nodes are added to the network one at a time. Each new node is connected to m existing nodes with a probability that is proportional to the number of links that the existing nodes already have. Formally, the probability p_i that the new node is connected to node i is[11]

$$p_i = \frac{k_i}{\sum_j k_j}.$$

where k_i is the degree of node i. Heavily linked nodes ("hubs") tend to quickly accumulate even more links, while nodes with only a few links are unlikely to be chosen as the destination for a new link. The new nodes have a "preference" to attach themselves to the already heavily linked nodes.

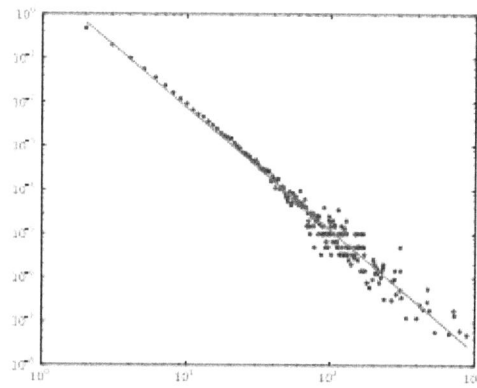

The degree distribution of the BA Model, which follows a power law. In loglog scale the power law function is a straight line.[12]

The degree distribution resulting from the BA model is scale free, in particular, it is a power law of the form:

$$P(k) \sim k^{-3}$$

Hubs exhibit high betweenness centrality which allows short paths to exist between nodes. As a result, the BA model tends to have very short average path lengths. The clustering coefficient of this model also tends to 0. While the diameter, D, of many models including the Erdős Rényi random graph model and several small world networks is proportional to log N, the BA model exhibits D~loglogN (ultra-small world).[13] Note that the average path length scales with N as the diameter.

Mediation-driven attachment (MDA) model

In the mediation-driven attachment (MDA) model in which a new node coming with m edges picks an existing connected node at random and then connects itself not with that one but with m of its neighbors chosen also at random. The probability $\Pi(i)$ that the node i of the existing node picked is

$$\Pi(i) = \frac{k_i}{N} \frac{\sum_{j=1}^{k_i} \frac{1}{k_j}}{k_i}.$$

The factor $\frac{\sum_{j=1}^{k_i} \frac{1}{k_j}}{k_i}$ is the inverse of the harmonic mean (IHM) of degrees of the k_i neighbors of a node i. Extensive numerical investigation suggest that for an approximately $m > 14$ the mean IHM value in the large N limit becomes a constant which means $\Pi(i) \propto k_i$. It implies that the higher the links (degree) a node has, the higher its chance of gaining more links since they can be reached in a larger number of ways through mediators which essentially embodies the intuitive idea of rich get richer mechanism (or the preferential attachment rule of the Barabasi–Albert model). Therefore, the MDA network can be seen to follow the PA rule but in disguise.[14]

However, for $m = 1$ it describes the winner takes it all mechanism as we find that almost 99% of the total nodes has degree one and one is super-rich in degree. As m value increases the disparity between the super rich and poor decreases and as $m > 14$ we find a transition from rich get super richer to rich get richer mechanism.

17.3.4 Fitness model

Another model where the key ingredient is the nature of the vertex has been introduced by Caldarelli et al.[15] Here a link is created between two vertices i, j with a probability given by a linking function $f(\eta_i, \eta_j)$ of the fitnesses of the vertices involved. The degree of a vertex i is given by [16]

$$k(\eta_i) = N \int_0^\infty f(\eta_i, \eta_j) \rho(\eta_j) \, d\eta_j$$

If $k(\eta_i)$ is an invertible and increasing function of η_i, then the probability distribution $P(k)$ is given by

$$P(k) = \rho(\eta(k)) \cdot \eta'(k)$$

As a result, if the fitnesses η are distributed as a power law, then also the node degree does.

Less intuitively with a fast decaying probability distribution as $\rho(\eta) = e^{-\eta}$ together with a linking function of the kind

$$f(\eta_i, \eta_j) = \Theta(\eta_i + \eta_j - Z)$$

with Z a constant and Θ the Heavyside function, we also obtain scale-free networks.

Such model has been successfully applied to describe trade between nations by using GDP as fitness for the various nodes i, j and a linking function of the kind [17] [18]

$$\frac{\delta \eta_i \eta_j}{1 + \delta \eta_i \eta_j}.$$

17.4 Network analysis

17.4.1 Social network analysis

Social network analysis examines the structure of relationships between social entities.[19] These entities are often persons, but may also be groups, organizations, nation states, web sites, scholarly publications.

Since the 1970s, the empirical study of networks has played a central role in social science, and many of the mathematical and statistical tools used for studying networks have been first developed in sociology.[20] Amongst many other applications, social network analysis has been used to understand the diffusion of innovations, news and rumors. Similarly, it has been used to examine the spread of both diseases and health-related behaviors. It has also been applied to the study of markets, where it has been used to examine the role of trust in exchange relationships and of social mechanisms in setting prices. Similarly, it has been used to study recruitment into political movements and social organizations. It has also been used to conceptualize scientific disagreements as well as academic prestige. More recently, network analysis (and its close cousin traffic analysis) has gained a significant use in military intelligence, for uncovering insurgent networks of both hierarchical and leaderless nature.[21][22]

17.4.2 Dynamic network analysis

Dynamic network analysis examines the shifting structure of relationships among different classes of entities in complex socio-technical systems effects, and reflects social stability and changes such as the emergence of new groups, topics, and leaders.[6][7][8][9][23] Dynamic Network Analysis focuses on meta-networks composed of multiple types of nodes (entities) and multiple types of links. These entities can be highly varied.[6] Examples include people, organizations, topics, resources, tasks, events, locations, and beliefs.

Dynamic network techniques are particularly useful for assessing trends and changes in networks over time, identification of emergent leaders, and examining the co-evolution of people and ideas.

17.4.3 Biological network analysis

With the recent explosion of publicly available high throughput biological data, the analysis of molecular networks has gained significant interest. The type of analysis in this content are closely related to social network analysis, but often focusing on local patterns in the network. For example, network motifs are small subgraphs that are over-represented in the network. Activity motifs are similar over-represented patterns in the attributes of nodes and

17.4.4 Link analysis

Link analysis is a subset of network analysis, exploring associations between objects. An example may be examining the addresses of suspects and victims, the telephone numbers they have dialed and financial transactions that they have partaken in during a given timeframe, and the familial relationships between these subjects as a part of police investigation. Link analysis here provides the crucial relationships and associations between very many objects of different types that are not apparent from isolated pieces of information. Computer-assisted or fully automatic computer-based link analysis is increasingly employed by banks and insurance agencies in fraud detection, by telecommunication operators in telecommunication network analysis, by medical sector in epidemiology and pharmacology, in law enforcement investigations, by search engines for relevance rating (and conversely by the spammers for spamdexing and by business owners for search engine optimization), and everywhere else where relationships between many objects have to be analyzed.

Network robustness

The structural robustness of networks[25] is studied using percolation theory. When a critical fraction of nodes is removed the network becomes fragmented into small clusters. This phenomenon is called percolation[26] and it represents an order-disorder type of phase transition with critical exponents.

Pandemic analysis

The SIR model is one of the most well known algorithms on predicting the spread of global pandemics within an infectious population.

Susceptible to infected

$$S = \beta \left(\frac{1}{N} \right)$$

The formula above describes the "force" of infection for each susceptible unit in an infectious population, where β is equivalent to the transmission rate of said disease.

To track the change of those susceptible in an infectious population:

$$\Delta S = \beta \times S \frac{1}{N} \Delta t$$

Infected to recovered

$$\Delta I = \mu I \, \Delta t$$

Over time, the number of those infected fluctuates by: the specified rate of recovery, represented by μ but deducted to one over the average infectious period $\frac{1}{\tau}$, the numbered of infecious individuals, I, and the change in time, Δt.

Infectious period Whether a population will be overcome by a pandemic, with regards to the SIR model, is dependent on the value of R_0 or the "average people infected by an infected individual."

$$R_0 = \beta \tau = \frac{\beta}{\mu}$$

Web link analysis

Several Web search ranking algorithms use link-based centrality metrics, including (in order of appearance) Marchiori's Hyper Search, Google's PageRank, Kleinberg's HITS algorithm, the CheiRank and TrustRank algorithms. Link analysis is also conducted in information science and communication science in order to understand and extract information from the structure of collections of web pages. For example, the analysis might be of the interlinking between politicians' web sites or blogs.

PageRank PageRank works by randomly picking "nodes" or websites and then with a certain probability, "randomly jumping" to other nodes. By randomly jumping to these other nodes, it helps PageRank completely traverse the network as some webpages exist on the periphery and would not as readily be assessed.

Each node, x_i, has a PageRank as defined by the sum of pages j that link to i times one over the outlinks or "outdegree" of j times the "importance" or PageRank of j.

$$x_i = \sum_{j \to i} \frac{1}{N_j} x_j^{(k)}$$

Random jumping As explained above, PageRank enlists random jumps in attempts to assign PageRank to every website on the internet. These random jumps find websites that might not be found during the normal search methodologies such as Breadth-First Search and Depth-First Search.

In an improvement over the aforementioned formula for determining PageRank includes adding these random jump components. Without the random jumps, some pages would receive a PageRank of 0 which would not be good.

The first is α, or the probability that a random jump will occur. Contrasting is the "damping factor", or $1 - \alpha$.

$$R(p) = \frac{\alpha}{N} + (1-\alpha) \sum_{j \to i} \frac{1}{N_j} x_j^{(k)}$$

Another way of looking at it:

$$R(A) = \sum \frac{R_B}{B_{(outlinks)}} + \cdots + \frac{R_n}{n_{(outlinks)}}$$

17.4.5 Centrality measures

Information about the relative importance of nodes and edges in a graph can be obtained through centrality measures, widely used in disciplines like sociology. Centrality measures are essential when a network analysis has to answer questions such as: "Which nodes in the network should be targeted to ensure that a message or information spreads to all or most nodes in the network?" or conversely, "Which nodes should be targeted to curtail the spread of a disease?". Formally established measures of centrality are degree centrality, closeness centrality, betweenness centrality, eigenvector centrality, and katz centrality. The objective of network analysis generally determines the type of centrality measure(s) to be used.[19]

- **Degree centrality** of a node in a network is the number of links (vertices) incident on the node.

- **Closeness centrality** determines how "close" a node is to other nodes in a network by measuring the sum of the shortest distances (geodesic paths) between that node and all other nodes in the network.

- **Betweenness centrality** determines the relative importance of a node by measuring the amount of traffic flowing through that node to other nodes in the network. This is done by measuring the fraction of paths connecting all pairs of nodes and containing the node of interest. Group Betweenness centrality measures the amount of traffic flowing through a group of nodes.[27]

- **Eigenvector centrality** is a more sophisticated version of degree centrality where the centrality of a node not only depends on the number of links incident on the node but also the quality of those links. This quality factor is determined by the eigenvectors of the adjacency matrix of the network.

- **Katz centrality** of a node is measured by summing the geodesic paths between that node and all (reachable) nodes in the network. These paths are weighted, paths connecting the node with its immediate neighbors carry higher weights than those which connect with nodes farther away from the immediate neighbors.

17.5 Spread of content in networks

Content in a complex network can spread via two major methods: conserved spread and non-conserved spread.[28] In conserved spread, the total amount of content that enters a complex network remains constant as it passes through. The model of conserved spread can best be represented by a pitcher containing a fixed amount of water being poured into a series of funnels connected by tubes . Here, the pitcher represents the original source and the water is the content being spread. The funnels and connecting tubing represent the nodes and the connections between nodes, respectively. As the water passes from one funnel into another, the water disappears instantly from the funnel that was previously exposed to the water. In non-conserved spread, the amount of content changes as it enters and passes through a complex network. The model of non-conserved spread can best be represented by a continuously running faucet running through a series of funnels connected by tubes . Here, the amount of water from the original source is infinite Also, any funnels that have been exposed to the water continue to experience the water even as it passes into successive funnels. The non-conserved model is the most suitable for explaining the transmission of most infectious diseases.

17.5.1 The SIR model

In 1927, W. O. Kermack and A. G. McKendrick created a model in which they considered a fixed population with only three compartments, susceptible: $S(t)$, infected, $I(t)$, and recovered, $R(t)$. The compartments used for this model consist of three classes:

- $S(t)$ is used to represent the number of individuals not yet infected with the disease at time t, or those susceptible to the disease

- $I(t)$ denotes the number of individuals who have been infected with the disease and are capable of spreading the disease to those in the susceptible category

- $R(t)$ is the compartment used for those individuals who have been infected and then recovered from the disease. Those in this category are not able to be infected again or to transmit the infection to others.

The flow of this model may be considered as follows:

$$S \to I \to R$$

Using a fixed population, $N = S(t) + I(t) + R(t)$, Kermack and McKendrick derived the following equations:

$$\frac{dS}{dt} = -\beta SI$$

$$\frac{dI}{dt} = \beta SI - \gamma I$$

$$\frac{dR}{dt} = \gamma I$$

Several assumptions were made in the formulation of these equations: First, an individual in the population must be considered as having an equal probability as every other individual of contracting the disease with a rate of β, which is considered the contact or infection rate of the disease. Therefore, an infected individual makes contact and is able to transmit the disease with βN others per unit time and the fraction of contacts by an infected with a susceptible is S/N. The number of new infections in unit time per infective then is $\beta N(S/N)$, giving the rate of new infections (or those leaving the susceptible category) as $\beta N(S/N)I = \beta SI$ (Brauer & Castillo-Chavez, 2001). For the second and third equations, consider the population leaving the susceptible class as equal to the number entering the infected class. However, a number equal to the fraction (γ which represents the mean recovery rate, or $1/\gamma$ the mean infective period) of infectives are leaving this class per unit time to enter the removed class. These processes which occur simultaneously are referred to as the Law of Mass Action, a widely accepted idea that the rate of contact between two groups in a population is proportional to the size of each of the groups concerned (Daley & Gani, 2005). Finally, it is assumed that the rate of infection and recovery is much faster than the time scale of births and deaths and therefore, these factors are ignored in this model.

More can be read on this model on the Epidemic model page.

17.5.2 The master equation approach

A master equation can express the behaviour of an undirected growing network where, at each time step, a new node is added to the network, linked to an old node (randomly chosen and without preference). The initial network is formed by two nodes and two links between them at time $t = 2$, this configuration is necessary only to simplify further calculations, so at time $t = n$ the network have n nodes and n links.

The master equation for this network is:

$$p(k, s, t+1) = \frac{1}{t} p(k-1, s, t) + \left(1 - \frac{1}{t}\right) p(k, s, t),$$

where $p(k, s, t)$ is the probability to have the node s with degree k at time $t + 1$, and s is the time step when this node was added to the network. Note that there are only two ways for an old node s to have k links at time $t + 1$:

- The node s have degree $k - 1$ at time t and will be linked by the new node with probability $1/t$

- Already has degree k at time t and will not be linked by the new node.

After simplifying this model, the degree distribution is $P(k) = 2^{-k}$. [29]

Based on this growing network, an epidemic model is developed following a simple rule: Each time the new node is added and after choosing the old node to link, a decision is made: whether or not this new node will be infected. The master equation for this epidemic model is:

$$p_r(k, s, t) = r_t \frac{1}{t} p_r(k-1, s, t) + \left(1 - \frac{1}{t}\right) p_r(k, s, t).$$

where r_t represents the decision to infect ($r_t = 1$) or not ($r_t = 0$). Solving this master equation, the following solution is obtained: $\tilde{P}_r(k) = \left(\frac{r}{2}\right)^k$. [30]

17.6 Interdependent networks

Main article: Interdependent networks

An interdependent network is a system of coupled networks where nodes of one or more networks depend on nodes in other networks. Such dependencies are enhanced by the developments in modern technology. Dependencies may lead to cascading failures between the networks and a relatively small failure can lead to a catastrophic breakdown of the system. Blackouts are a fascinating demonstration of the important role played by the dependencies between networks. A recent study developed a framework to study the cascading failures in an interdependent networks system.[31][32]

17.7 Multilayer networks

Main article: Multidimensional network

Multilayer networks are networks with multiple kinds of relations.[33][34][35][36][37][38] Increasingly sophisticated attempts to model real-world systems as multidimensional networks have yielded valuable insight in the fields of social network analysis,[34][35][39][40][41][42] economics, history,[43] urban and international transport,[44][45][46][47] ecology,[48][49] psychology,[50] medicine, biology,[51] commerce, climatology, physics,[52][53] computational neuroscience[54],[55][56] operations management, and finance.

17.8 Network optimization

Network problems that involve finding an optimal way of doing something are studied under the name of combinatorial optimization. Examples include network flow, shortest path problem, transport problem, transshipment problem, location problem, matching problem, assignment problem, packing problem, routing problem, Critical Path Analysis and PERT (Program Evaluation & Review Technique).

17.9 See also

- Climate as complex networks
- Collaborative innovation network
- Communicative ecology
- Complex network
- Constructal law[57]
- Core-periphery structures in networks
- Dual-phase evolution
- Erdős–Rényi model
- Glossary of graph theory
- Higher category theory
- Immune network theory
- Irregular warfare
- Network analyzer
- Network dynamics
- Network formation
- Network theory in risk assessment
- Network topology
- Networks in labor economics
- Non-linear preferential attachment
- Percolation
- Policy network analysis
- Polytely
- Quantum complex network
- Random networks
- Rumor spread in social network
- Scale-free networks
- Sequential dynamical system
- Service network
- Small-world networks
- Structural cut-off
- Systems theory

17.10 Further reading

- "A Very Short Introduction to Networks." G. Caldarelli, M. Catanzaro (Oxford University Press, Oxford 2007)
- "Network Science Center," http://www.dodccrp.org/files/Network_Science_Center.asf
- "Connected: The Power of Six Degrees," http://ivl.slis.indiana.edu/km/movies/2008-talas-connected.mov
- Cohen, R.; Erez, K.; Havlin, S. (2000). "Resilience of the Internet to random breakdown". *Phys. Rev. Lett.* **85** (21): 4626–4628. Bibcode:2000PhRvL..85.4626C. PMID 11082612. doi:10.1103/physrevlett.85.4626.

- Pu, Cun-Lai; Wen-; Pei, Jiang; Michaelson, Andrew (2012). "Robustness analysis of network controllability" (PDF). *Physica A*. **391** (18): 4420–4425. Bibcode:2012PhyA..391.4420P. doi:10.1016/j.physa.2012.04.019.

- "Leader Profile: The Burgeoning Field of Network Science. The Military Engineer recently had the opportunity to speak with Frederick I. Moxley, Ph.D." http://themilitaryengineer.com/index.php/item/160-leader-profile-the-burgeoning-field-of-network-science

- S.N. Dorogovtsev and J.F.F. Mendes, *Evolution of Networks: From biological networks to the Internet and WWW*, Oxford University Press, 2003, ISBN 0-19-851590-1

- *Linked: The New Science of Networks*, A.-L. Barabási (Perseus Publishing, Cambridge)

- 'Scale-Free Networks, G. Caldarelli (Oxford University Press, Oxford)

- *Network Science*, Committee on Network Science for Future Army Applications, National Research Council. 2005. The National Academies Press (2005)ISBN 0-309-10026-7

- *Network Science Bulletin*, USMA (2007) ISBN 978-1-934808-00-9

- *The Structure and Dynamics of Networks* Mark Newman, Albert-László Barabási, & Duncan J. Watts (The Princeton Press, 2006) ISBN 0-691-11357-2

- *Dynamical processes on complex networks*, Alain Barrat, Marc Barthelemy, Alessandro Vespignani (Cambridge University Press, 2008) ISBN 978-0-521-87950-7

- *Network Science: Theory and Applications*, Ted G. Lewis (Wiley, March 11, 2009) ISBN 0-470-33188-7

- *Nexus: Small Worlds and the Groundbreaking Theory of Networks*, Mark Buchanan (W. W. Norton & Company, June 2003) ISBN 0-393-32442-7

- *Six Degrees: The Science of a Connected Age*, Duncan J. Watts (W. W. Norton & Company, February 17, 2004) ISBN 0-393-32542-3

- Kitsak, M.; Gallos, L. K.; Havlin, S.; Liljeros, F.; Muchnik, L.; Stanley, H. E.; Makse, H.A. (2010). "Influential Spreaders in Networks". *Nature Physics*. **6**: 888–893. doi:10.1038/nphys1746.

17.11 Notes

[1] Committee on Network Science for Future Army Applications (2006). *Network Science*. National Research Council. ISBN 0309653886.

[2] http://psycnet.apa.org/journals/prs/9/4/172/

[3] Lawyer, Glenn (March 2015). "Understanding the spreading power of all nodes in a network". *Scientific Reports*. **5** (O8665): 8665. Bibcode:2015NatSR...5E8665L. PMC 4345333. PMID 25727453. doi:10.1038/srep08665.

[4] Sikic, Mile; Lancic, Alen; Antulov-Fantulin, Nino; Stefancic, Hrvoje (October 2013). "Epidemic centrality -- is there an underestimated epidemic impact of network peripheral nodes?". *European Physical Journal B*. **86** (10): 1–13. Bibcode:2013EPJB...86..440S. arXiv:1110.2558. doi:10.1140/epjb/e2013-31025-5.

[5] Borgatti, Stephen P. (2005). "Centrality and Network Flow". *Social Networks*. Elsevier. **27**: 55–71. doi:10.1016/j.socnet.2004.11.008.

[6] Braha, D.; Bar-Yam, Y. (2006). "From Centrality to Temporary Fame: Dynamic Centrality in Complex Networks". *Complexity*. **12**: 59–63. doi:10.1002/cplx.20156.

[7] Hill, S.A.; Braha, D. (2010). "Dynamic Model of Time-Dependent Complex Networks". *Physical Review E*. **82**: 046105. doi:10.1103/physreve.82.046105.

[8] Gross, T. and Sayama, H. (Eds.). 2009. *Adaptive Networks: Theory, Models and Applications*. Springer.

[9] Holme, P. and Saramäki, J. 2013. *Temporal Networks*. Springer.

[10] Travençolo, B. A. N.; da F. Costa, L. (2008). "Accessibility in complex networks". *Physics Letters A*. **373** (1): 89–95. Bibcode:2008PhLA..373...89T. doi:10.1016/j.physleta.2008.10.069.

[11] R. Albert; A.-L. Barabási (2002). "Statistical mechanics of complex networks" (PDF). *Reviews of Modern Physics*. **74**: 47–97. Bibcode:2002RvMP...74...47A. arXiv:cond-mat/0106096. doi:10.1103/RevModPhys.74.47.

[12] Albert-László Barabási & Réka Albert (October 1999). "Emergence of scaling in random networks" (PDF). *Science*. **286** (5439): 509–512. Bibcode:1999Sci...286..509B. PMID 10521342. arXiv:cond-mat/9910332. doi:10.1126/science.286.5439.509.

[13] Cohen, R.; Havlin, S. (2003). "Scale-free networks are ultrasmall". *Phys. Rev. Lett.* **90** (5): 058701. Bibcode:2003PhRvL..90e8701C. PMID 12633404. doi:10.1103/PhysRevLett.90.058701.

[14] Hassan, M. K.; Islam, Liana; Arefinul Haque, Syed (2017;). "Degree distribution, rank-size distribution, and leadership persistence in mediation-driven attachment networks". *Physica A*. **469**: 23–30. doi:10.1016/j.physa.2016.11.001. Check date values in: |date= (help)

[15] Caldarelli G., A. Capocci, P. De Los Rios, M.A. Muñoz. Physical Review Letters 89, 258702 (2002)

[16] Servedio V.D.P., G. Caldarelli, P. Buttà. Physical Review E 70, 056126 (2004)

[17] Garlaschelli D., M I Loffredo Physical Review Letters 93, 188701 (2004)

[18] Cimini G., T. Squartini, D. Garlaschelli and A. Gabrielli, Scientific Reports 5, 15758 (2015)

[19] Wasserman, Stanley and Katherine Faust. 1994. *Social Network Analysis: Methods and Applications.* Cambridge: Cambridge University Press.

[20] Newman, M.E.J. *Networks: An Introduction*. Oxford University Press. 2010. ISBN 978-0199206650

[21] "Toward a Complex Adaptive Intelligence Community The Wiki and the Blog". *D. Calvin Andrus*. cia.gov. Retrieved 25 August 2012.

[22] Network analysis of terrorist networks

[23] Xanthos, Aris, Pante, Isaac, Rochat, Yannick, Grandjean, Martin (2016). Visualising the Dynamics of Character Networks. In Digital Humanities 2016: Jagiellonian University & Pedagogical University, Kraków, pp. 417–419.

[24] Barabási, A. L.; Gulbahce, N.; Loscalzo, J. (2011). "Network medicine: a network-based approach to human disease". *Nature Reviews Genetics*. **12** (1): 56–68. PMC 3140052. PMID 21164525. doi:10.1038/nrg2918.

[25] R. Cohen; S. Havlin (2010). *Complex Networks: Structure, Robustness and Function*. Cambridge University Press.

[26] A. Bunde; S. Havlin (1996). *Fractals and Disordered Systems*. Springer.

[27] Puzis, R.; Yagil, D.; Elovici, Y.; Braha, D. (2009). "Collaborative attack on Internet users' anonymity" (PDF). *Internet Research*. **19**: 1.

[28] Newman, M., Barabási, A.-L., Watts, D.J. [eds.] (2006) The Structure and Dynamics of Networks. Princeton, N.J.: Princeton University Press.

[29] Dorogovtsev, S N; Mendes, J F F (2003). *Evolution of Networks: From Biological Nets to the Internet and WWW*. New York, NY, USA: Oxford University Press, Inc. ISBN 0198515901.

[30] Cotacallapa, M; Hase, M O (2016). "Epidemics in networks: a master equation approach". *Journal of Physics A*. **49** (6): 065001. Bibcode:2016JPhA...49f5001C. doi:10.1088/1751-8113/49/6/065001. Retrieved 2 August 2016.

[31] S. V. Buldyrev; R. Parshani; G. Paul; H. E. Stanley; S. Havlin (2010). "Catastrophic cascade of failures in interdependent networks". *Nature*. **464** (7291): 1025–28. Bibcode:2010Natur.464.1025B. PMID 20393559. arXiv:0907.1182. doi:10.1038/nature08932.

[32] Gao, Jianxi; Buldyrev, Sergey V.; Havlin, Shlomo; Stanley, H. Eugene (2011). "Robustness of a Network of Networks". *Phys. Rev. Lett.* **107** (19): 195701. Bibcode:2011PhRvL.107s5701G. PMID 22181627. arXiv:1010.5829. doi:10.1103/PhysRevLett.107.195701.

[33] Coscia, Michele; Rossetti, Giulio; Pennacchioli, Diego; Ceccarelli, Damiano; Giannotti, Fosca (2013). ""You Know Because I Know": A Multidimensional Network Approach to Human Resources Problem". *Advances in Social Network Analysis and Mining (ASONAM)*. **2013**: 434. ISBN 9781450322409. arXiv:1305.7146. doi:10.1145/2492517.2492537.

[34] De Domenico, M.; Solé-Ribalta, A.; Cozzo, E.; Kivelä, M.; Moreno, Y.; Porter, M.; Gómez, S.; Arenas, A. (2013). "Mathematical Formulation of Multilayer Networks" (PDF). *Physical Review X*. **3** (4): 041022. Bibcode:2013PhRvX...3d1022D. doi:10.1103/PhysRevX.3.041022.

[35] Battiston, F.; Nicosia, V.; Latora, V. (2014). "Structural measures for multiplex networks". *Physical Review E*. **89** (3): 032804. Bibcode:2014PhRvE..89c2804B. doi:10.1103/PhysRevE.89.032804.

[36] Kivela, M.; Arenas, A.; Barthelemy, M.; Gleeson, J. P.; Moreno, Y.; Porter, M. A. (2014). "Multilayer networks". *Journal of Complex Networks*. **2** (3): 203–271. doi:10.1093/comnet/cnu016.

[37] Boccaletti, S.; Bianconi, G.; Criado, R.; del Genio, C. I.; Gómez-Gardeñes, J.; Romance, M.; Sendiña-Nadal, I.; Wang, Z.; Zanin, M. (2014). "The structure and dynamics of multilayer networks". *Physics Reports*. **544** (1): 1–122. Bibcode:2014PhR...544....1B. arXiv:1407.0742. doi:10.1016/j.physrep.2014.07.001.

[38] Battiston, Federico; Nicosia, Vincenzo; Latora, Vito (2017-02-01). "The new challenges of multiplex networks: Measures and models". *The European Physical Journal Special Topics*. **226** (3): 401–416. ISSN 1951-6355. doi:10.1140/epjst/e2016-60274-8.

[39] Mucha, P.; et al. (2010). "Community structure in time-dependent, multiscale, and multiplex networks" (PDF). *Science*. **328** (5980): 876–878. Bibcode:2010Sci...328..876M. PMID 20466926. arXiv:0911.1824. doi:10.1126/science.1184819.

[40] De Domenico, M.; Lancichinetti, A.; Arenas, A.; Rosvall, M. (2015). "Identifying Modular Flows on

Multilayer Networks Reveals Highly Overlapping Organization in Interconnected Systems". *Physical Review X*. **5**: 011027. Bibcode:2015PhRvX...5a1027D. arXiv:1408.2925. doi:10.1103/PhysRevX.5.011027.

[41] De Domenico, M.; Sole-Ribalta, A.; Omodei, E.; Gomez, S.; Arenas, A. (2015). "Ranking in interconnected multilayer networks reveals versatile nodes". *Nature Communications*. **6**: 6868. Bibcode:2015NatCo...6E6868D. PMID 25904405. doi:10.1038/ncomms7868.

[42] Battiston, Federico; Iacovacci, Jacopo; Nicosia, Vincenzo; Bianconi, Ginestra; Latora, Vito (2016-01-27). "Emergence of Multiplex Communities in Collaboration Networks". *PLOS ONE*. **11** (1): e0147451. ISSN 1932-6203. PMC 4731389. PMID 26815700. doi:10.1371/journal.pone.0147451.

[43] Grandjean, Martin (2016). Archives Distant Reading: Mapping the Activity of the League of Nations' Intellectual Cooperation. In Digital Humanities 2016: Jagiellonian University & Pedagogical University. Kraków. pp. 531–534.

[44] Cardillo, A.; et al. (2013). "Emergence of network features from multiplexity". *Scientific Reports*. **3**: 1344. Bibcode:2013NatSR...3E1344C. PMC 3583169. PMID 23446838. doi:10.1038/srep01344.

[45] Boeing, G. (2017). "OSMnx: New Methods for Acquiring, Constructing, Analyzing, and Visualizing Complex Street Networks". *Computers, Environment and Urban Systems*. **65**: 126–139. SSRN 2865501. arXiv:1611.01890. doi:10.1016/j.compenvurbsys.2017.05.004. Retrieved 2017-08-26.

[46] Gallotti, R.; Barthelemy, M. (2014). "Anatomy and efficiency of urban multimodal mobility". *Scientific Reports*. **4**: 6911. Bibcode:2014NatSR...4E6911G. PMC 4220282. PMID 25371238. arXiv:1411.1274. doi:10.1038/srep06911.

[47] De Domenico, M.; Sole-Ribalta, A.; Gomez, S.; Arenas, A. (2014). "Navigability of interconnected networks under random failures". *PNAS*. **111** (23): 8351–8356. Bibcode:2014PNAS..111.8351D. PMC 4060702. PMID 24912174. doi:10.1073/pnas.1318469111.

[48] Pilosof, S.; Porter, M.A.; Pascual, M.; Kefi, S. (2015). "The Multilayer Nature of Ecological Networks". *Nature Ecology & Evolution*. **1**: 0101. arXiv:1511.04453 [q-bio.QM]. doi:10.1038/s41559-017-0101.

[49] Kouvaris, N.E.; Hata, S.; Diaz-Guilera, A. (2015). "Pattern Formation in Multiplex Networks". *Scientific Reports*. **5** (1): 10840. doi:10.1038/srep10840.

[50] Fiori, K. L.; Smith, J; Antonucci, T. C. (2007). "Social network types among older adults: A multidimensional approach". *The Journals of Gerontology Series B*. **62** (6): P322–30. PMID 18079416. doi:10.1093/geronb/62.6.p322.

[51] De Domenico, M.; Nicosia, V.; Arenas, A.; Latora, V. (2015). "Structural reducibility of multilayer networks". *Nature Communications*. **6**: 6864. Bibcode:2015NatCo...6E6864D. PMID 25904309. doi:10.1038/ncomms7864.

[52] Gao; Buldyrev; Stanley; Havlin (22 December 2011). "Networks formed from interdependent networks". *Nature Physics*. **8**: 40–48. Bibcode:2012NatPh...8...40G. doi:10.1038/nphys2180.

[53] De Domenico, M.; Granell, C.; Porter, Mason A.; Arenas, A. (7 April 2016). "The physics of multilayer networks". *Nature Physics*. **12**: 901–906. arXiv:1604.02021 [physics.soc-ph]. doi:10.1038/nphys3865.

[54] Timme, N.; Ito, S.; Myroshnychenko, M.; Yeh, F.C.; Hiolski, E.; Hottowy, P.; Beggs, J.M. (2014). "Multiplex Networks of Cortical and Hippocampal Neurons Revealed at Different Timescales". *PLoS ONE*. **9** (12): e115764. Bibcode:2014PLoSO...9k5764T. PMC 4275261. PMID 25536059. doi:10.1371/journal.pone.0115764.

[55] De Domenico, M.; Sasai, S.; Arenas, A. (2016). "Mapping multiplex hubs in human functional brain networks". *Frontiers in Neuroscience*. **10**. doi:10.3389/fnins.2016.00326.

[56] Battiston, F.; Nicosia, V.; Chavez, M.; Latora, V. (2016). "Multilayer motif analysis of brain networks". *Chaos: An Interdisciplinary Journal of Nonlinear Science*. **27**: 047404. arXiv:1606.09115 [physics.soc-ph]. doi:10.1063/1.4979282.

[57] Bejan A., Lorente S., The Constructal Law of Design and Evolution in Nature. *Philosophical Transactions of the Royal Society B*, Biological Science, Vol. 365, 2010, pp. 1335–1347.

Chapter 18

Scale-free network

A **scale-free network** is a network whose degree distribution follows a power law, at least asymptotically. That is, the fraction $P(k)$ of nodes in the network having k connections to other nodes goes for large values of k as

$$P(k) \sim k^{-\gamma}$$

where γ is a parameter whose value is typically in the range $2 < \gamma < 3$, although occasionally it may lie outside these bounds.[1][2]

Many networks have been reported to be scale-free, although statistical analysis has refuted many of these claims and seriously questioned others.[3] Preferential attachment and the fitness model have been proposed as mechanisms to explain conjectured power law degree distributions in real networks.

18.1 History

In studies of the networks of citations between scientific papers, Derek de Solla Price showed in 1965 that the number of links to papers—i.e., the number of citations they receive—had a heavy-tailed distribution following a Pareto distribution or power law, and thus that the citation network is scale-free. He did not however use the term "scale-free network", which was not coined until some decades later. In a later paper in 1976, Price also proposed a mechanism to explain the occurrence of power laws in citation networks, which he called "cumulative advantage" but which is today more commonly known under the name preferential attachment.

Recent interest in scale-free networks started in 1999 with work by Albert-László Barabási and colleagues at the University of Notre Dame who mapped the topology of a portion of the World Wide Web,[4] finding that some nodes, which they called "hubs", had many more connections than others and that the network as a whole had a power-law distribution of the number of links connecting to a node. After finding that a few other networks, including some social and biological networks, also had heavy-tailed degree distributions, Barabási and collaborators coined the term "scale-free network" to describe the class of networks that exhibit a power-law degree distribution. Amaral et al. showed that most of the real-world networks can be classified into two large categories according to the decay of degree distribution $P(k)$ for large k.

Barabási and Albert proposed a generative mechanism to explain the appearance of power-law distributions, which they called "preferential attachment" and which is essentially the same as that proposed by Price. Analytic solutions for this mechanism (also similar to the solution of Price) were presented in 2000 by Dorogovtsev, Mendes and Samukhin [5] and independently by Krapivsky, Redner, and Leyvraz, and later rigorously proved by mathematician Béla Bollobás.[6] Notably, however, this mechanism only produces a specific subset of networks in the scale-free class, and many alternative mechanisms have been discovered since.[7]

The history of scale-free networks also includes some disagreement. On an empirical level, the scale-free nature of several networks has been called into question. For instance, the three brothers Faloutsos believed that the Internet had a power law degree distribution on the basis of traceroute data; however, it has been suggested that this is a layer 3 illusion created by routers, which appear as high-degree nodes while concealing the internal layer 2 structure of the ASes they interconnect. [8]

On a theoretical level, refinements to the abstract definition of scale-free have been proposed. For example, Li et al. (2005) recently offered a potentially more precise "scale-free metric". Briefly, let G be a graph with edge set E, and denote the degree of a vertex v (that is, the number of edges incident to v) by $\deg(v)$. Define

$$s(G) = \sum_{(u,v) \in E} \deg(u) \cdot \deg(v).$$

This is maximized when high-degree nodes are connected

to other high-degree nodes. Now define

$$S(G) = \frac{s(G)}{s_{max}}.$$

where s_{max} is the maximum value of $s(H)$ for H in the set of all graphs with degree distribution identical to that of G. This gives a metric between 0 and 1, where a graph G with small $S(G)$ is "scale-rich", and a graph G with $S(G)$ close to 1 is "scale-free". This definition captures the notion of self-similarity implied in the name "scale-free".

18.2 Characteristics

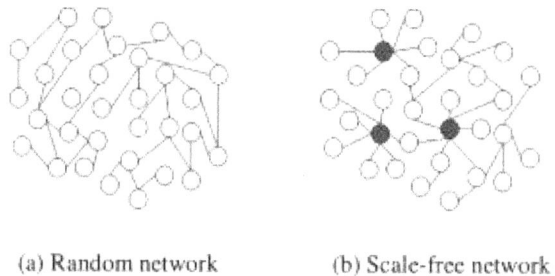

(a) Random network   (b) Scale-free network

Random network (a) and scale-free network (b). In the scale-free network, the larger hubs are highlighted.

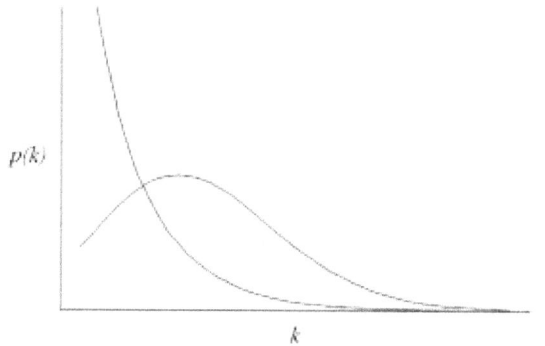

Complex network degree distribution of random and scale-free

The most notable characteristic in a scale-free network is the relative commonness of vertices with a degree that greatly exceeds the average. The highest-degree nodes are often called "hubs", and are thought to serve specific purposes in their networks, although this depends greatly on the domain.

The scale-free property strongly correlates with the network's robustness to failure. It turns out that the major hubs are closely followed by smaller ones. These smaller hubs, in turn, are followed by other nodes with an even smaller degree and so on. This hierarchy allows for a fault tolerant behavior. If failures occur at random and the vast majority of nodes are those with small degree, the likelihood that a hub would be affected is almost negligible. Even if a hub-failure occurs, the network will generally not lose its connectedness, due to the remaining hubs. On the other hand, if we choose a few major hubs and take them out of the network, the network is turned into a set of rather isolated graphs. Thus, hubs are both a strength and a weakness of scale-free networks. These properties have been studied analytically using percolation theory by Cohen et al.[9][10] and by Callaway et al.[11] It was proven by Cohen [12] that for a broad range of scale free networks the critical percolation threshold, p_c=0. This means that removing randomly any fraction of nodes from the network will not destroy the network. This is in contrast to Erdös–Rényi graph where p_c =1/<k>, where <k> is the average degree.

Another important characteristic of scale-free networks is the clustering coefficient distribution, which decreases as the node degree increases. This distribution also follows a power law. This implies that the low-degree nodes belong to very dense sub-graphs and those sub-graphs are connected to each other through hubs. Consider a social network in which nodes are people and links are acquaintance relationships between people. It is easy to see that people tend to form communities, i.e., small groups in which everyone knows everyone (one can think of such community as a complete graph). In addition, the members of a community also have a few acquaintance relationships to people outside that community. Some people, however, are connected to a large number of communities (e.g., celebrities, politicians). Those people may be considered the hubs responsible for the small-world phenomenon.

At present, the more specific characteristics of scale-free networks vary with the generative mechanism used to create them. For instance, networks generated by preferential attachment typically place the high-degree vertices in the middle of the network, connecting them together to form a core, with progressively lower-degree nodes making up the regions between the core and the periphery. The random removal of even a large fraction of vertices impacts the overall connectedness of the network very little, suggesting that such topologies could be useful for security, while targeted attacks destroys the connectedness very quickly. Other scale-free networks, which place the high-degree vertices at the periphery, do not exhibit these properties. Similarly, the clustering coefficient of scale-free networks can vary significantly depending on other topological details.

A final characteristic concerns the average distance between two vertices in a network. As with most disordered networks, such as the small world network model, this distance is very small relative to a highly ordered network such as a lattice graph. Notably, an uncorrelated power-law graph having $2 < \gamma < 3$ will have ultrasmall diameter $d \sim \ln \ln N$ where N is the number of nodes in the network, as proved by Cohen and Havlin.[13] The diameter of a growing scale-free network might be considered almost constant in practice.

Properties of random graph may change or remain invariant under graph transformations. Mashaghi A. et al., for example, demonstrated that a transformation which converts random graphs to their edge-dual graphs (or line graphs) produces an ensemble of graphs with nearly the same degree distribution, but with degree correlations and a significantly higher clustering coefficient. Scale free graphs, as such, remain scale free under such transformations.[14]

18.3 Examples

Although many real-world networks are thought to be scale-free, the evidence often remains inconclusive, primarily due to the developing awareness of more rigorous data analysis techniques.[3] As such, the scale-free nature of many networks is still being debated by the scientific community. A few examples of networks claimed to be scale-free include:

- Social networks, including collaboration networks. Two examples that have been studied extensively are the collaboration of movie actors in films and the co-authorship by mathematicians of papers.
- Many kinds of computer networks, including the internet and the webgraph of the World Wide Web.
- Some financial networks such as interbank payment networks [15][16]
- Protein-protein interaction networks.
- Semantic networks.[17]
- Airline networks.

Scale free topology has been also found in high temperature superconductors.[18] The qualities of a high-temperature superconductor — a compound in which electrons obey the laws of quantum physics, and flow in perfect synchrony, without friction — appear linked to the fractal arrangements of seemingly random oxygen atoms and lattice distortion.[19]

A space-filling cellular structure, weighted planar stochastic lattice (WPSL) has recently been proposed whose co-ordination number distribution follow a power-law. It implies that the lattice has a few blocks which have astonishingly large number neighbors with whom they share common borders. Its construction starts with an initiator, say a

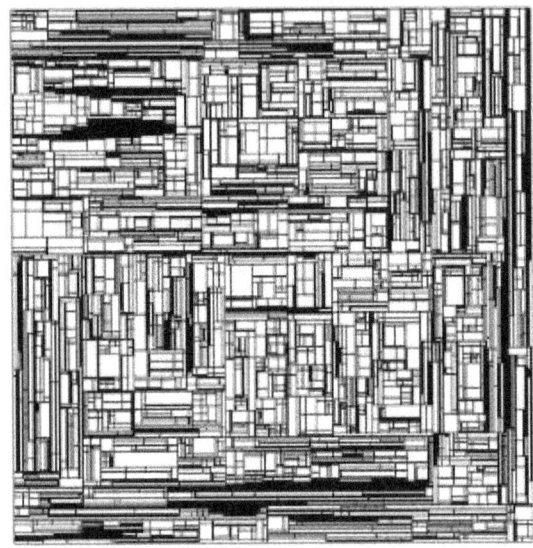

A snapshot of the weighted planar stochastic lattice (WPSL).

square of unit area, and a generator that divides it randomly into four blocks. The generator thereafter is sequentially applied over and over again to only one of the available blocks picked preferentially with respect to their areas. It results in the partitioning of the square into ever smaller mutually exclusive rectangular blocks. the dual of the WPSL (DW-PSL) obtained by replacing each block with a node at its center and common border between blocks with an edge joining the two corresponding vertices emerges as a network whose degree distribution follows a power-law.[20][21] The reason for it is that it grows following mediation-driven attachment model rule which also embodies preferential attachment rule but in disguise.

18.4 Generative models

Scale-free networks do not arise by chance alone. Erdös and Rényi (1960) studied a model of growth for graphs in which, at each step, two nodes are chosen uniformly at random and a link is inserted between them. The properties of these random graphs are different from the properties found in scale-free networks, and therefore a model for this growth process is needed.

The most widely known generative model for a subset of scale-free networks is Barabási and Albert's (1999) rich get richer generative model in which each new Web page creates links to existing Web pages with a probability distribution which is not uniform, but proportional to the current in-degree of Web pages. This model was originally invented by Derek J. de Solla Price in 1965 under the term **cumulative** advantage, but did not reach popularity until Barabási rediscovered the results under its current name (BA Model). According to this process, a page with many in-links will attract more in-links than a regular page. This generates a power-law but the resulting graph differs from the actual Web graph in other properties such as the presence of small tightly connected communities. More general models and network characteristics have been proposed and studied (for a review see the book by Dorogovtsev and Mendes).

A somewhat different generative model for Web links has been suggested by Pennock et al. (2002). They examined communities with interests in a specific topic such as the home pages of universities, public companies, newspapers or scientists, and discarded the major hubs of the Web. In this case, the distribution of links was no longer a power law but resembled a normal distribution. Based on these observations, the authors proposed a generative model that mixes preferential attachment with a baseline probability of gaining a link.

Another generative model is the **copy** model studied by Kumar et al.[22] (2000), in which new nodes choose an existent node at random and copy a fraction of the links of the existent node. This also generates a power law.

Interestingly, the *growth* of the networks (adding new nodes) is not a necessary condition for creating a scale-free network. Dangalchev[23] (2004) gives examples of generating static scale-free networks. Another possibility (Caldarelli et al. 2002) is to consider the structure as static and draw a link between vertices according to a particular property of the two vertices involved. Once specified the statistical distribution for these vertex properties (fitnesses), it turns out that in some circumstances also static networks develop scale-free properties.

18.5 Generalized scale-free model

There has been a burst of activity in the modeling of scale-free complex networks. The recipe of Barabási and Albert[24] has been followed by several variations and generalizations[25][26][27][28] and the revamping of previous mathematical works.[29] As long as there is a power law distribution in a model, it is a scale-free network, and a model of that network is a scale-free model.

18.5.1 Features

Many real networks are (approximately) scale-free and hence require scale-free models to describe them. In Price's scheme, there are two ingredients needed to build up a scale-free model:

1. Adding or removing nodes. Usually we concentrate on growing the network, i.e. adding nodes.

2. Preferential attachment: The probability Π that new nodes will be connected to the "old" node.

Note that Fitness models (see below) could work also statically, without changing the number of nodes. It should also be kept in mind that the fact that "preferential attachment" models give rise to scale-free networks does not prove that this is the mechanism underlying the evolution of real-world scale-free networks, as there might exist different mechanisms at work in real-world systems that nevertheless give rise to scaling.

18.5.2 Examples

There have been several attempts to generate scale-free network properties. Here are some examples:

The Barabási–Albert model

For example, the first scale-free model, the Barabási–Albert model, has a linear preferential attachment $\Pi(k_i) = \frac{k_i}{\sum_j k_j}$ and adds one new node at every time step.

(Note, another general feature of $\Pi(k)$ in real networks is that $\Pi(0) \neq 0$, i.e. there is a nonzero probability that a new node attaches to an isolated node. Thus in general $\Pi(k)$ has the form $\Pi(k) = A + k^\alpha$, where A is the initial attractiveness of the node.)

Two-level network model

Dangalchev[23] builds a 2-L model by adding a second-order preferential attachment. The attractiveness of a node in the 2-L model depends not only on the number of nodes linked to it but also on the number of links in each of these nodes.

$$\Pi(k_i) = \frac{k_i + C \sum_{(i,j)} k_j}{\sum_j k_j + C \sum_j k_j^2}.$$

where C is a coefficient between 0 and 1.

Mediation-driven attachment (MDA) model

In the mediation-driven attachment (MDA) model in which a new node coming with m edges picks an existing connected node at random and then connects itself not with that one but with m of its neighbors chosen also at random. The probability $\Pi(i)$ that the node i of the existing node picked is

$$\Pi(i) = \frac{k_i}{N} \frac{\sum_{j=1}^{k_i} \frac{1}{k_j}}{k_i}.$$

The factor $\frac{\sum_{j=1}^{k_i} \frac{1}{k_j}}{k_i}$ is the inverse of the harmonic mean (IHM) of degrees of the k_i neighbors of a node i. Extensive numerical investigation suggest that for a approximately $m > 14$ the mean IHM value in the large N limit becomes a constant which means $\Pi(i) \propto k_i$. It implies that the higher the links (degree) a node has, the higher its chance of gaining more links since they can be reached in a larger number of ways through mediators which essentially embodies the intuitive idea of rich get richer mechanism (or the preferential attachment rule of the Barabasi–Albert model). Therefore, the MDA network can be seen to follow the PA rule but in disguise.[30]

However, for $m = 1$ it describes the winner takes it all mechanism as we find that almost 99% of the total nodes has degree one and one is super-rich in degree. As m value increases the disparity between the super rich and poor decreases and as $m > 14$ we find a transition from rich get super richer to rich get richer mechanism.

Non-linear preferential attachment

See also: Non-linear preferential attachment

The Barabási–Albert model assumes that the probability $\Pi(k)$ that a node attaches to node i is proportional to the degree k of node i. This assumption involves two hypotheses: first, that $\Pi(k)$ depends on k, in contrast to random graphs in which $\Pi(k) = p$, and second, that the functional form of $\Pi(k)$ is linear in k. The precise form of $\Pi(k)$ is not necessarily linear, and recent studies have demonstrated that the degree distribution depends strongly on $\Pi(k)$

Krapivsky, Redner, and Leyvraz[27] demonstrate that the scale-free nature of the network is destroyed for nonlinear preferential attachment. The only case in which the topology of the network is scale free is that in which the preferential attachment is asymptotically linear, i.e. $\Pi(k_i) \sim a_\infty k_i$ as $k_i \to \infty$. In this case the rate equation leads to

$$P(k) \sim k^{-\gamma} \text{ with } \gamma = 1 + \frac{\mu}{a_\infty}.$$

This way the exponent of the degree distribution can be tuned to any value between 2 and ∞.

Hierarchical network model

There is another kind of scale-free model, which grows according to some patterns, such as the hierarchical network model.[31]

The iterative construction leading to a hierarchical network. Starting from a fully connected cluster of five nodes, we create four identical replicas connecting the peripheral nodes of each cluster to the central node of the original cluster. From this, we get a network of 25 nodes ($N = 25$). Repeating the same process, we can create four more replicas of the original cluster – the four peripheral nodes of each one connect to the central node of the nodes created in the first step. This gives $N = 125$, and the process can continue indefinitely.

Fitness model

The idea is that the link between two vertices is assigned not randomly with a probability p equal for all the couple of vertices. Rather, for every vertex j there is an intrinsic *fitness xj* and a link between vertex i and j is created with a probability $p(x_i, x_j)$.[32] In the case of World Trade Web it is possible to reconstruct all the properties by using as fitnesses of the country their GDP, and taking

$$p(x_i, x_j) = \frac{\delta x_i x_j}{1+\delta x_i x_j}.\text{[33]}$$

Hyperbolic geometric graphs

Main article: Hyperbolic geometric graph

Assuming that a network has an underlying hyperbolic geometry, one can use the framework of spatial networks to generate scale-free degree distributions. This heterogeneous degree distribution then simply reflects the negative curvature and metric properties of the underlying hyperbolic geometry.[34]

Edge dual transformation to generate scale free graphs with desired properties

Starting with scale free graphs with low degree correlation and clustering coefficient, one can generate new graphs with much higher degree correlations and clustering coefficients by applying edge-dual transformation.[14]

18.6 Scale-free ideal network

In the context of network theory a **scale-free ideal network** is a random network with a degree distribution following the scale-free ideal gas density distribution. These networks are able to reproduce city-size distributions and electoral results by unraveling the size distribution of social groups with information theory on complex networks when a competitive cluster growth process is applied to the network.[35][36] In models of scale-free ideal networks it is possible to demonstrate that Dunbar's number is the cause of the phenomenon known as the 'six degrees of separation'.

18.7 See also

- Random graph
- Erdős–Rényi model
- Non-linear preferential attachment
- Bose–Einstein condensation: a network theory approach
- Scale invariance
- Complex network
- Webgraph
- Barabási–Albert model
- Bianconi–Barabási model

18.8 References

[1] Onnela, J. -P.; Saramaki, J.; Hyvonen, J.; Szabo, G.; Lazer, D.; Kaski, K.; Kertesz, J.; Barabasi, A. -L. (2007). "Structure and tie strengths in mobile communication networks". *Proceedings of the National Academy of Sciences*. **104** (18): 7332–7336. Bibcode:2007PNAS..104.7332O. PMC 1863470. PMID 17456605. arXiv:physics/0610104. doi:10.1073/pnas.0610245104.

[2] Choromański, K.; Matuszak, M.; MięKisz, J. (2013). "Scale-Free Graph with Preferential Attachment and Evolving Internal Vertex Structure". *Journal of Statistical Physics*. **151** (6): 1175–1183. Bibcode:2013JSP...151.1175C. doi:10.1007/s10955-013-0749-1.

[3] Clauset, Aaron; Cosma Rohilla Shalizi; M. E. J Newman (2007-06-07). "Power-law distributions in empirical data". *SIAM Review*. **51**: 661–703. Bibcode:2009SIAMR..51..661C. arXiv:0706.1062. doi:10.1137/070710111.

[4] Barabási, Albert-László; Albert, Réka. (October 15, 1999). "Emergence of scaling in random networks". *Science*. **286** (5439): 509–512. Bibcode:1999Sci...286..509B. MR 2091634. PMID 10521342. arXiv:cond-mat/9910332. doi:10.1126/science.286.5439.509.

[5] Dorogovtsev, S.; Mendes, J.; Samukhin, A. (2000). "Structure of Growing Networks with Preferential Linking". *Physical Review Letters*. **85** (21): 4633–4636. Bibcode:2000PhRvL..85.4633D. PMID 11082614. arXiv:cond-mat/0004434. doi:10.1103/PhysRevLett.85.4633.

[6] Bollobás, B.; Riordan, O.; Spencer, J.; Tusnády, G. (2001). "The degree sequence of a scale-free random graph process". *Random Structures and Algorithms*. **18** (3): 279–290. MR 1824277. doi:10.1002/rsa.1009.

[7] Dorogovtsev, S. N.; Mendes, J. F. F. (2002). "Evolution of networks". *Advances in Physics*. **51** (4): 1079–1187. Bibcode:2002AdPhy..51.1079D. doi:10.1080/00018730110112519.

[8] Willinger, Walter; David Alderson; John C. Doyle (May 2009). "Mathematics and the Internet: A Source of Enormous Confusion and Great Potential" (PDF). *Notices of the AMS*. American Mathematical Society. **56** (5): 586–599. Retrieved 2011-02-03.

[9] Cohen, Reoven; Erez, K.; ben-Avraham, D.; Havlin, S. (2000). "Resilience of the Internet to Random Breakdowns". *Physical Review Letters*. **85**: 4626–8. Bibcode:2000PhRvL..85.4626C. PMID 11082612. arXiv:cond-mat/0007048. doi:10.1103/PhysRevLett.85.4626.

[10] Cohen, Reoven; Erez, K.; ben-Avraham, D.; Havlin, S. (2001). "Breakdown of the Internet under Intentional Attack". *Physical Review Letters*. **86**: 3682–5. Bibcode:2001PhRvL..86.3682C. PMID 11328053. arXiv:cond-mat/0010251. doi:10.1103/PhysRevLett.86.3682.

[11] Callaway, Duncan S.; Newman, M. E. J.; Strogatz, S. H.; Watts, D. J. (2000). "Network Robustness and Fragility: Percolation on Random Graphs". *Physical Review Letters*. **85**: 5468–71. Bibcode:2000PhRvL..85.5468C. PMID 11136023. arXiv:cond-mat/0007300. doi:10.1103/PhysRevLett.85.5468.

[12] Cohen, Reuven; Erez, Keren; ben-Avraham, Daniel; Havlin, Shlomo (2000). "Resilience of the Internet to Random Breakdowns". *Physical Review Letters*. **85** (21): 4626–4628. Bibcode:2000PhRvL..85.4626C. PMID 11082612. doi:10.1103/PhysRevLett.85.4626.

[13] Cohen, Reuven; Havlin, Shlomo (2003). "Scale-Free Networks Are Ultrasmall". *Physical Review Letters*. **90** (5): 058701. PMID 12633404. doi:10.1103/PhysRevLett.90.058701.

[14] Ramezanpour, A.; Karimipour, V.; Mashaghi, A. (2003). "Generating correlated networks from uncorrelated ones". *Phys. Rev. E.* **67**: 046107. doi:10.1103/PhysRevE.67.046107.

[15] De Masi, Giulia; et. al (2006). "Fitness model for the Italian interbank money market". *Physical Review E.* **74**: 066112. doi:10.1103/PhysRevE.74.066112.

[16] Soramäki, Kimmo; et. al (2007). "The topology of interbank payment flows". *Physica A: Statistical Mechanics and its Applications.* **379** (1): 317–333. Bibcode:2007PhyA..379..317S. doi:10.1016/j.physa.2006.11.093.

[17] Steyvers, Mark; Joshua B. Tenenbaum (2005). "The Large-Scale Structure of Semantic Networks: Statistical Analyses and a Model of Semantic Growth". *Cognitive Science.* **29** (1): 41–78. doi:10.1207/s15516709cog2901_3.

[18] Fratini, Michela, Poccia, Nicola, Ricci, Alessandro, Campi, Gaetano, Burghammer, Manfred, Aeppli, Gabriel Bianconi, Antonio (2010). "Scale-free structural organization of oxygen interstitials in La2CuO4+y". *Nature.* **466** (7308): 841–4. Bibcode:2010Natur.466..841F. PMID 20703301. arXiv:1008.2015. doi:10.1038/nature09260.

[19] Poccia, Nicola, Ricci, Alessandro, Campi, Gaetano, Fratini, Michela, Puri, Alessandro, Di Gioacchino, Daniele, Marcelli, Augusto, Reynolds, Michael, Burghammer, Manfred, Saini, Naurang L., Aeppli, Gabriel Bianconi, Antonio, (2012). "Optimum inhomogeneity of local lattice distortions in La2CuO4+y". *PNAS.* **109** (39): 15685–15690. arXiv:1208.0101. doi:10.1073/pnas.1208492109.

[20] M. K. Hassan, M. Z. Hassan and N. I. Pavel, "Scale-free network topology and multifractality in a weighted planar stochastic lattice" New Journal of Physics **12** 093045 (2010) doi:10.1088/1367-263/12/9/093045.

[21] M. K. Hassan, M. Z. Hassan and N. I. Pavel, Scale-free coordination number disorder and multifractal size disorder in weighted planar stochastic lattice, J. Phys: Conf. Ser. **297** 012010 (2011).

[22] Kumar, Ravi; Raghavan, Prabhakar (2000). *Stochastic Models for the Web Graph* (PDF). Foundations of Computer Science, 41st Annual Symposium on. pp. 57–65. doi:10.1109/SFCS.2000.892065.

[23] Dangalchev Ch., Generation models for scale-free networks, Physica A **338**, 659 (2004).

[24] Barabási, A.-L. and R. Albert, Science **286**, 509 (1999).

[25] R. Albert, and A.L. Barabási, Phys. Rev. Lett. **85**, 5234(2000).

[26] S. N. Dorogovtsev, J. F. F. Mendes, and A. N. Samukhin, cond-mat/0011115.

[27] P.L. Krapivsky, S. Redner, and F. Leyvraz, Phys. Rev. Lett. **85**, 4629 (2000).

[28] B. Tadic, Physica A **293**, 273(2001).

[29] S. Bornholdt and H. Ebel, cond-mat/0008465; H.A. Simon, Bimetrika **42**, 425(1955).

[30] Hassan, M. K.; Islam, Liana; Arefinul Haque, Syed (2017). "Degree distribution, rank-size distribution, and leadership persistence in mediation-driven attachment networks". *Physica A.* **469**: 23–30. doi:10.1016/j.physa.2016.11.001.

[31] Ravasz, E.; Barabási (2003). "Hierarchical organization in complex networks". *Phys. Rev. E.* **67**: 026112. doi:10.1103/physreve.67.026112.

[32] Caldarelli, G.; et al. (2002). "Scale-free networks from varying vertex intrinsic fitness". *Phys. Rev. Lett.* **89**: 258702. PMID 12484927. doi:10.1103/physrevlett.89.258702.

[33] Garlaschelli, D.; et al. (2004). "Fitness-Dependent Topological Properties of the World Trade Web". *Phys. Rev. Lett.* **93**: 188701. doi:10.1103/physrevlett.93.188701.

[34] Krioukov, Dmitri; Papadopoulos, Fragkiskos; Kitsak, Maksim; Vahdat, Amin; Boguñá, Marián. "Hyperbolic geometry of complex networks". *Physical Review E.* **82** (3). doi:10.1103/PhysRevE.82.036106.

[35] A. Hernando; D. Villuendas; C. Vesperinas; M. Abad; A. Plastino (2009). "Unravelling the size distribution of social groups with information theory on complex networks". arXiv:0905.3704 [physics.soc-ph]., submitted to *European Physics Journal B*

[36] André A. Moreira; Demétrius R. Paula; Raimundo N. Costa Filho; José S. Andrade, Jr. (2006). "Competitive cluster growth in complex networks". *Physical Review E.* **73**. arXiv:cond-mat/0603272 [cond-mat.dis-nn]. doi:10.1103/PhysRevE.73.065101.

18.9 Further reading

- Albert R.; Barabási A.-L. (2002). "Statistical mechanics of complex networks". *Rev. Mod. Phys.* **74**: 47–97. Bibcode:2002RvMP...74...47A. arXiv:cond-mat/0106096. doi:10.1103/RevModPhys.74.47.

- Amaral, LAN, Scala, A., Barthelemy, M., Stanley, HE. (2000). "Classes of behavior of small-world networks". *PNAS.* **97** (21): 11149–52. Bibcode:2000PNAS...9711149A. PMC 17168. PMID 11005838. arXiv:cond-mat/0001458. doi:10.1073/pnas.200327197.

- Barabási, Albert-László (2004). *Linked: How Everything is Connected to Everything Else.* ISBN 0-452-28439-2.

18.9. FURTHER READING

- Barabási, Albert-László; Bonabeau, Eric (May 2003). "Scale-Free Networks" (PDF). *Scientific American*. **288** (5): 50–9. doi:10.1038/scientificamerican0503-60.

- Dan Braha; Yaneer Bar-Yam (2004). "Topology of Large-Scale Engineering Problem-Solving Networks" (PDF). *Phys. Rev. E.* **69**: 016113. Bibcode:2004PhRvE..69a6113B. doi:10.1103/PhysRevE.69.016113.

- Caldarelli G. "Scale-Free Networks" Oxford University Press, Oxford (2007).

- Caldarelli G.; Capocci A.; De Los Rios P.; Muñoz M.A. (2002). "Scale-free networks from varying vertex intrinsic fitness". *Physical Review Letters*. **89** (25): 258702. Bibcode:2002PhRvL..89y8702C. PMID 12484927. arXiv:cond-mat/0207366. doi:10.1103/PhysRevLett.89.258702.

- R. Cohen, K. Erez, D. ben-Avraham and S. Havlin (2000). "Resilience of the Internet to Random Breakdowns". *Phys. Rev. Lett.* **85**: 4626–8. Bibcode:2000PhRvL..85.4626C. PMID 11082612. arXiv:cond-mat/0007048. doi:10.1103/PhysRevLett.85.4626.

- R. Cohen, K. Erez, D. ben-Avraham and S. Havlin (2001). "Breakdown of the Internet under Intentional Attack". *Phys. Rev. Lett.* **86**: 3682–5. Bibcode:2001PhRvL..86.3682C. PMID 11328053. arXiv:cond-mat/0010251. doi:10.1103/PhysRevLett.86.3682.

- A.F. Rozenfeld, R. Cohen, D. ben-Avraham, S. Havlin (2002). "Scale-free networks on lattices". *Phys. Rev. Lett.* **89**. doi:10.1103/physrevlett.89.218701.

- Dangalchev, Ch. (2004). "Generation models for scale-free networks". *Physica A*. **338**: 659–671. doi:10.1016/j.physa.2004.01.056.

- Dorogovtsev, Mendes, J.F.F., Samukhin, A.N. (2000). "Structure of Growing Networks: Exact Solution of the Barabási—Albert's Model". *Phys. Rev. Lett.* **85** (21): 4633–6. Bibcode:2000PhRvL..85.4633D. PMID 11082614. arXiv:cond-mat/0004434. doi:10.1103/PhysRevLett.85.4633.

- Dorogovtsev, S.N., Mendes, J.F.F. (2003). *Evolution of Networks: from biological networks to the Internet and WWW*. Oxford University Press. ISBN 0-19-851590-1.

- Dorogovtsev, S.N., Goltsev A. V., Mendes, J.F.F. (2008). "Critical phenomena in complex networks". *Rev. Mod. Phys.* **80**: 1275–1335. Bibcode:2008RvMP...80.1275D. arXiv:0705.0010. doi:10.1103/RevModPhys.80.1275.

- Dorogovtsev, S.N., Mendes, J.F.F. (2002). "Evolution of networks". *Advances in Physics*. **51**: 1079–1187. Bibcode:2002AdPhy..51.1079D. arXiv:cond-mat/0106144. doi:10.1080/00018730110112519.

- Erdős, P.; Rényi, A. (1960). *On the Evolution of Random Graphs* (PDF). **5**. Publication of the Mathematical Institute of the Hungarian Academy of Science. pp. 17–61.

- Faloutsos, M., Faloutsos, P., Faloutsos, C. (1999). "On power-law relationships of the internet topology". *Comp. Comm. Rev.* **29**: 251–262. doi:10.1145/316194.316229.

- Li, L., Alderson, D., Tanaka, R., Doyle, J.C., Willinger, W. (2005). "Towards a Theory of Scale-Free Graphs: Definition, Properties, and Implications (Extended Version)". arXiv:cond-mat/0501169 [cond-mat.dis-nn].

- Kumar, R., Raghavan, P., Rajagopalan, S., Sivakumar, D., Tomkins, A., Upfal, E. (2000). "Stochastic models for the web graph" (PDF). *Proceedings of the 41st Annual Symposium on Foundations of Computer Science (FOCS)*. Redondo Beach, CA: IEEE CS Press. pp. 57–65.

- Manev R.; Manev H. (2005). "The meaning of mammalian adult neurogenesis and the function of newly added neurons: the "small-world" network". *Med. Hypotheses*. **64** (1): 114–7. PMID 15533625. doi:10.1016/j.mehy.2004.05.013.

- Matlis, Jan (November 4, 2002). "Scale-Free Networks".

- Newman, Mark E.J. (2003). "The structure and function of complex networks". *SIAM Review*. **45**: 167–256. Bibcode:2003SIAMR..45..167N. arXiv:cond-mat/0303516 [cond-mat.stat-mech]. doi:10.1137/S003614450342480.

- Pastor-Satorras, R., Vespignani, A. (2004). *Evolution and Structure of the Internet: A Statistical Physics Approach*. Cambridge University Press. ISBN 0-521-82698-5.

- Pennock, D.M., Flake, G.W., Lawrence, S., Glover, E.J., Giles, C.L. (2002). "Winners don't take all: Characterizing the competition for links on the web". *PNAS*. **99** (8): 5207–11. Bibcode:2002PNAS...99.5207P. PMC 122747. PMID 16578867. doi:10.1073/pnas.032085699.

- Robb, John. Scale-Free Networks and Terrorism, 2004.

- Keller, E.F. (2005). "Revisiting "scale-free" networks". *BioEssays*. **27** (10): 1060–8. PMID 16163729. doi:10.1002/bies.20294.

- Onody, R.N., de Castro, P.A. (2004). "Complex Network Study of Brazilian Soccer Player". *Phys. Rev. E*. **70**: 037103. Bibcode:2004PhRvE..70c7103O. arXiv:cond-mat/0409609. doi:10.1103/PhysRevE.70.037103.

- Reuven Cohen; Shlomo Havlin (2003). "Scale-Free Networks are Ultrasmall". *Phys. Rev. Lett*. **90** (5): 058701. Bibcode:2003PhRvL..90e8701C. PMID 12633404. arXiv:cond-mat/0205476. doi:10.1103/PhysRevLett.90.058701.

- Kasthurirathna, D., Piraveenan, M. (2015). "Complex Network Study of Brazilian Soccer Player". *Sci. Rep*. In Press.

Chapter 19

Social network analysis

This article is about the theoretical concept. For social networking sites, see social networking service. For other uses, see Social network (disambiguation).

Social network analysis (SNA) is the process of in-

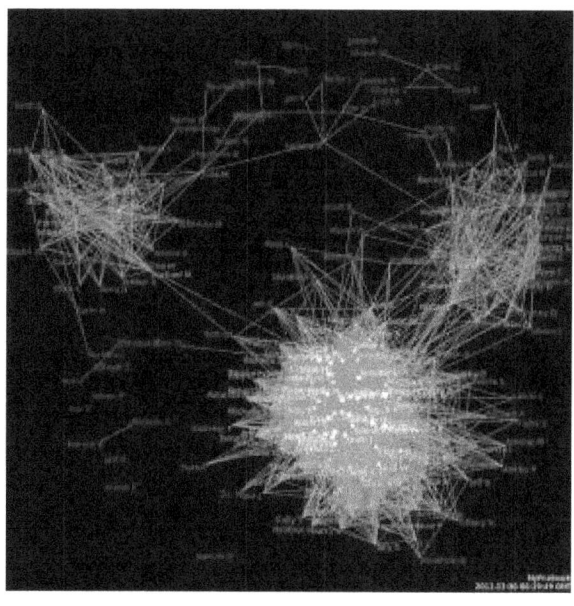

A social network diagram displaying friendship ties among a set of Facebook users.

vestigating social structures through the use of networks and graph theory.[1] It characterizes networked structures in terms of *nodes* (individual actors, people, or things within the network) and the *ties*, *edges*, or *links* (relationships or interactions) that connect them. Examples of social structures commonly visualized through social network analysis include social media networks,[2] memes spread,[3] friendship and acquaintance networks, collaboration graphs, kinship, disease transmission, and sexual relationships.[4][5] These networks are often visualized through *sociograms* in which nodes are represented as points and ties are represented as lines.

Social network analysis has emerged as a key technique in modern sociology. It has also gained a significant follow-

ing in anthropology, biology, demography, communication studies, economics, geography, history, information science, organizational studies, political science, social psychology, development studies, sociolinguistics, and computer science and is now commonly available as a consumer tool.[6][7][8][9]

19.1 History

Social network analysis has its theoretical roots in the work of early sociologists such as Georg Simmel and Émile Durkheim, who wrote about the importance of studying patterns of relationships that connect social actors. Social scientists have used the concept of "social networks" since early in the 20th century to connote complex sets of relationships between members of social systems at all scales, from interpersonal to international. In the 1930s Jacob Moreno and Helen Jennings introduced basic analytical methods.[10] In 1954, John Arundel Barnes started using the term systematically to denote patterns of ties, encompassing concepts traditionally used by the public and those used by social scientists: bounded groups (e.g., tribes, families) and social categories (e.g., gender, ethnicity). Scholars such as Ronald Burt, Kathleen Carley, Mark Granovetter, David Krackhardt, Edward Laumann, Anatol Rapoport, Barry Wellman, Douglas R. White, and Harrison White expanded the use of systematic social network analysis.[11] Even in the study of literature, network analysis has been applied by Anheier, Gerhards and Romo,[12] Wouter De Nooy,[13] and Burgert Senekal.[14] Indeed, social network analysis has found applications in various academic disciplines, as well as practical applications such as countering money laundering and terrorism.

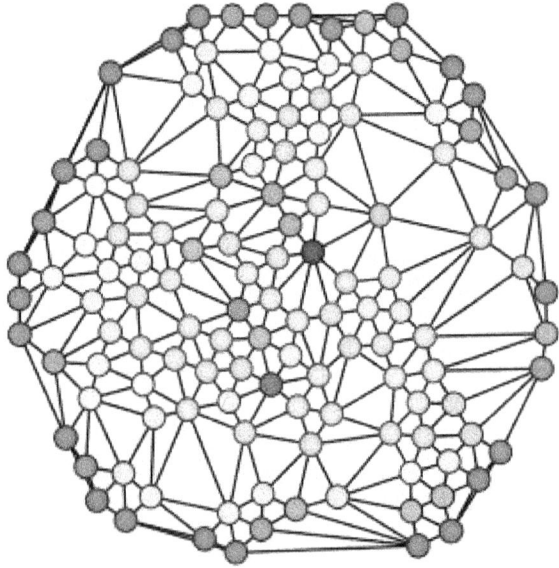

Hue (from red=0 to blue=max) indicates each node's betweenness centrality.

19.2 Metrics

19.2.1 Connections

Homophily: The extent to which actors form ties with similar versus dissimilar others. Similarity can be defined by gender, race, age, occupation, educational achievement, status, values or any other salient characteristic.[15] Homophily is also referred to as assortativity.

Multiplexity: The number of content-forms contained in a tie.[16] For example, two people who are friends and also work together would have a multiplexity of 2.[17] Multiplexity has been associated with relationship strength.

Mutuality/Reciprocity: The extent to which two actors reciprocate each other's friendship or other interaction.[18]

Network Closure: A measure of the completeness of relational triads. An individual's assumption of network closure (i.e. that their friends are also friends) is called transitivity. Transitivity is an outcome of the individual or situational trait of Need for Cognitive Closure.[19]

Propinquity: The tendency for actors to have more ties with geographically close others.[18]

19.2.2 Distributions

Bridge: An individual whose weak ties fill a structural hole, providing the only link between two individuals or clusters. It also includes the shortest route when a longer one is unfeasible due to a high risk of message distortion or delivery failure.[20]

Centrality: Centrality refers to a group of metrics that aim to quantify the "importance" or "influence" (in a variety of senses) of a particular node (or group) within a network.[21][22][23][24] Examples of common methods of measuring "centrality" include betweenness centrality,[25] closeness centrality, eigenvector centrality, alpha centrality, and degree centrality.[26]

Density: The proportion of direct ties in a network relative to the total number possible.[27][28]

Distance: The minimum number of ties required to connect two particular actors, as popularized by Stanley Milgram's small world experiment and the idea of 'six degrees of separation'.

Structural holes: The absence of ties between two parts of a network. Finding and exploiting a structural hole can give an entrepreneur a competitive advantage. This concept was developed by sociologist Ronald Burt, and is sometimes referred to as an alternate conception of social capital.

Tie Strength: Defined by the linear combination of time, emotional intensity, intimacy and reciprocity (i.e. mutuality).[20] Strong ties are associated with homophily, propinquity and transitivity, while weak ties are associated with bridges.

19.2.3 Segmentation

Groups are identified as 'cliques' if every individual is directly tied to every other individual, 'social circles' if there is less stringency of direct contact, which is imprecise, or as structurally cohesive blocks if precision is wanted.[29]

Clustering coefficient: A measure of the likelihood that two associates of a node are associates. A higher clustering coefficient indicates a greater 'cliquishness'.[30]

Cohesion: The degree to which actors are connected directly to each other by cohesive bonds. Structural cohesion refers to the minimum number of members who, if removed from a group, would disconnect the group.[31][32]

19.3 Modelling and visualization of networks

Visual representation of social networks is important to understand the network data and convey the result of the analysis.[33] Numerous methods of visualization for data produced by social network analysis have been presented.[34][35][36] Many of the analytic software have modules for network visualization. Exploration of the data is done through displaying nodes and ties in various layouts, and attributing colors, size and other advanced properties to nodes. Visual representations of networks may be a powerful method for conveying complex information, but care should be taken in interpreting node and graph properties from visual displays alone, as they may misrepresent structural properties better captured through quantitative analyses.[37]

Signed graphs can be used to illustrate good and bad relationships between humans. A positive edge between two nodes denotes a positive relationship (friendship, alliance, dating) and a negative edge between two nodes denotes a negative relationship (hatred, anger). Signed social network graphs can be used to predict the future evolution of the graph. In signed social networks, there is the concept of "balanced" and "unbalanced" cycles. A balanced cycle is defined as a cycle where the product of all the signs are positive. According to balance theory, balanced graphs represent a group of people who are unlikely to change their opinions of the other people in the group. Unbalanced graphs represent a group of people who are very likely to change their opinions of the people in their group. For example, a group of 3 people (A, B, and C) where A and B have a positive relationship, B and C have a positive relationship, but C and A have a negative relationship is an unbalanced cycle. This group is very likely to morph into a balanced cycle, such as one where B only has a good relationship with A, and both A and B have a negative relationship with C. By using the concept of balanced and unbalanced cycles, the evolution of signed social network graphs can be predicted.[38]

Especially when using social network analysis as a tool for facilitating change, different approaches of participatory network mapping have proven useful. Here participants / interviewers provide network data by actually mapping out the network (with pen and paper or digitally) during the data collection session. An example of a pen-and-paper network mapping approach, which also includes the collection of some actor attributes (perceived influence and goals of actors) is the * Net-map toolbox. One benefit of this approach is that it allows researchers to collect qualitative data and ask clarifying questions while the network data is collected.[39]

19.3.1 Social networking potential

Social networking potential (SNP) is a numeric coefficient, derived through algorithms[40][41] to represent both the size of an individual's social network and their ability to influence that network. A close synonym is the Alpha User, a person with a high SNP.

SNP coefficients have two primary functions:

1. the classification of individuals based on their social networking potential, and
2. the weighting of respondents in quantitative marketing research studies.

By calculating the SNP of respondents and by targeting High SNP respondents, the strength and relevance of quantitative marketing research used to drive viral marketing strategies is enhanced.

Variables used to calculate an individual's SNP include but are not limited to: participation in Social Networking activities, group memberships, leadership roles, recognition, publication/editing/contributing to non-electronic media, publication/editing/contributing to electronic media (websites, blogs), and frequency of past distribution of information within their network. The acronym "SNP" and some of the first algorithms developed to quantify an individual's social networking potential were described in the white paper "Advertising Research is Changing" (Gerstley, 2003) See Viral Marketing.[42]

The first book[43] to discuss the commercial use of Alpha Users among mobile telecoms audiences was 3G Marketing by Ahonen, Kasper and Melkko in 2004. The first book to discuss Alpha Users more generally in the context of social marketing intelligence was Communities Dominate Brands by Ahonen & Moore in 2005. In 2012, Nicola Greco (UCL) presents at TEDx the Social Networking Potential as a parallelism to the potential energy that users generate and companies should use, stating that "SNP is the new asset that every company should aim to have".[44]

19.4 Practical applications

See also: Social network analysis (criminology)

Social network analysis is used extensively in a wide range of applications and disciplines. Some common network analysis applications include data aggregation and mining, network propagation modeling, network modeling and sampling, user attribute and behavior analysis, community-maintained resource support, location-based interaction analysis, social sharing and filtering, recommender systems development, and link prediction and entity resolution.[45] In the private sector, businesses use social network analysis to support activities such as customer interaction and analysis, information system development analysis,[46] marketing, and business intelligence needs. Some public sector uses include development of leader engagement strategies, analysis of individual and group engagement and media use, and community-based problem solving.

Social network analysis is also used in intelligence, counter-intelligence and law enforcement activities. This technique allows the analysts to map a clandestine or covert organization such as a espionage ring, an organized crime family or a street gang. The National Security Agency (NSA) uses its clandestine mass electronic surveillance programs to generate the data needed to perform this type of analysis on terrorist cells and other networks deemed relevant to national security. The NSA looks up to three nodes deep during this network analysis.[47] After the initial mapping of the social network is complete, analysis is performed to determine the structure of the network and determine, for example, the leaders within the network.[48] This allows military or law enforcement assets to launch capture-or-kill decapitation attacks on the high-value targets in leadership positions to disrupt the functioning of the network. The NSA has been performing social network analysis on call detail records (CDRs), also known as metadata, since shortly after the September 11 attacks.[49][50]

Large textual corpora can be turned into networks and then analysed with the method of social network analysis. In these networks, the nodes are Social Actors, and the links are Actions. The extraction of these networks can be automated, by using parsers.

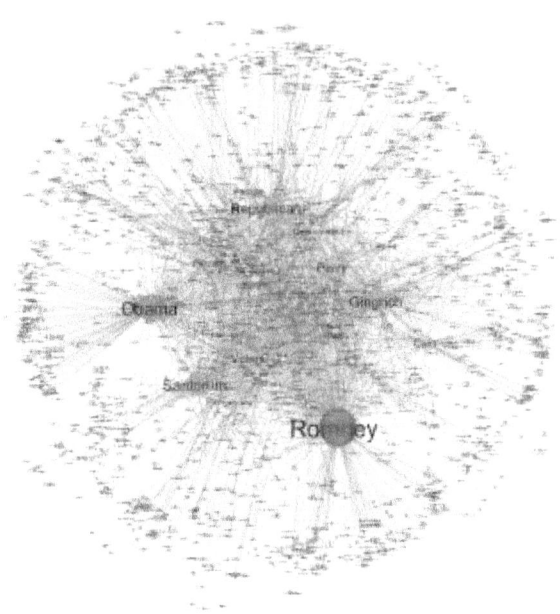

Narrative network of US Elections 2012[51]

The resulting networks, which can contain thousands of nodes, are then analysed by using tools from network theory to identify the key actors, the key communities or parties, and general properties such as robustness or structural stability of the overall network, or centrality of certain nodes.[52] This automates the approach introduced by Quantitative Narrative Analysis,[53] whereby subject-verb-object triplets are identified with pairs of actors linked by an action, or pairs formed by actor-object.[51]

19.4.1 In computer-supported collaborative learning

One of the most current methods of the application of SNA is to the study of computer-supported collaborative learning (CSCL). When applied to CSCL, SNA is used to help understand how learners collaborate in terms of amount, frequency, and length, as well as the quality, topic, and strategies of communication.[54] Additionally, SNA can focus on specific aspects of the network connection, or the entire network as a whole. It uses graphical representations, written representations, and data representations to help examine the connections within a CSCL network.[54] When applying SNA to a CSCL environment the interactions of the participants are treated as a social network. The focus of the analysis is on the "connections" made among the participants – how they interact and communicate – as opposed to how each participant behaved on his or her own.

Key terms

There are several key terms associated with social network analysis research in computer-supported collaborative learning such as: **density**, **centrality**, **indegree**, **outdegree**, and **sociogram**.

- **Density** refers to the "connections" between participants. Density is defined as the number of connections a participant has, divided by the total possible connections a participant could have. For example, if there are 20 people participating, each person could potentially connect to 19 other people. A density of 100% (19/19) is the greatest density in the system. A density of 5% indicates there is only 1 of 19 possible connections.[54]

- **Centrality** focuses on the behavior of individual participants within a network. It measures the extent to which an individual interacts with other individuals in the network. The more an individual connects to others in a network, the greater their centrality in the network.[54]

In-degree and out-degree variables are related to centrality.

- **In-degree** centrality concentrates on a specific individual as the point of focus; centrality of all other individuals is based on their relation to the focal point of the "in-degree" individual.[54]

- **Out-degree** is a measure of centrality that still focuses on a single individual, but the analytic is concerned with the out-going interactions of the individual; the measure of out-degree centrality is how many times the focus point individual interacts with others.[54]

- A **sociogram** is a visualization with defined boundaries of connections in the network. For example, a sociogram which shows out-degree centrality points for Participant A would illustrate all outgoing connections Participant A made in the studied network.[54]

Unique capabilities

Researchers employ social network analysis in the study of computer-supported collaborative learning in part due to the unique capabilities it offers. This particular method allows the study of interaction patterns within a networked learning community and can help illustrate the extent of the participants' interactions with the other members of the group.[54] The graphics created using SNA tools provide visualizations of the connections among participants and the strategies used to communicate within the group. Some authors also suggest that SNA provides a method of easily analyzing changes in participatory patterns of members over time.[55]

A number of research studies have applied SNA to CSCL across a variety of contexts. The findings include the correlation between a network's density and the teacher's presence,[54] a greater regard for the recommendations of "central" participants,[56] infrequency of cross-gender interaction in a network,[57] and the relatively small role played by an instructor in an asynchronous learning network.[58]

Other methods used alongside SNA

Although many studies have demonstrated the value of social network analysis within the computer-supported collaborative learning field,[54] researchers have suggested that SNA by itself is not enough for achieving a full understanding of CSCL. The complexity of the interaction processes and the myriad sources of data make it difficult for SNA to provide an in-depth analysis of CSCL.[59] Researchers indicate that SNA needs to be complemented with other methods of analysis to form a more accurate picture of collaborative learning experiences.[60]

A number of research studies have combined other types of analysis with SNA in the study of CSCL. This can be referred to as a multi-method approach or data triangulation, which will lead to an increase of evaluation reliability in CSCL studies.

- Qualitative method – The principles of qualitative case study research constitute a solid framework for the integration of SNA methods in the study of CSCL experiences.[61]

    - *Ethnographic data* such as student questionnaires and interviews and classroom non-participant observations[60]
    - *Case studies*: comprehensively study particular CSCL situations and relate findings to general schemes[60]
    - *Content analysis:* offers information about the content of the communication among members[60]

- Quantitative method – This includes simple descriptive statistical analyses on occurrences to identify particular attitudes of group members who have not been able to be tracked via SNA in order to detect general tendencies.

    - *Computer log files*: provide automatic data on how collaborative tools are used by learners[60]
    - *Multidimensional scaling (MDS)*: charts similarities among actors, so that more similar input data is closer together[60]
    - *Software tools*: QUEST, SAMSA (System for Adjacency Matrix and Sociogram-based Analysis), and Nud*IST[60]

19.5 See also

- Actor-network theory
- Community structure
- Complex network
- Digital humanities
- Dynamic network analysis
- Friendship paradox
- Individual mobility
- Mathematical sociology
- Metcalfe's law
- Network science
- Organizational patterns

- Small world phenomenon
- Social media mining
- Social network
- Social network analysis software
- Social networking service
- Social software
- Social web

19.6 References

[1] Otte, Evelien; Rousseau, Ronald (2002). "Social network analysis: a powerful strategy, also for the information sciences". *Journal of Information Science*. **28** (6): 441–453. doi:10.1177/016555150202800601. Retrieved 2015-03-23.

[2] Grandjean, Martin (2016). "A social network analysis of Twitter: Mapping the digital humanities community". *Cogent Arts & Humanities*. **3** (1): 1171458. doi:10.1080/23311983.2016.1171458.

[3] Nasrinpour, Hamid Reza; Friesen, Marcia R.; McLeod, Robert D. (2016-11-22). "An Agent-Based Model of Message Propagation in the Facebook Electronic Social Network". arXiv:1611.07454 [cs.SI].

[4] Pinheiro, Carlos A.R. (2011). *Social Network Analysis in Telecommunications*. John Wiley & Sons. p. 4. ISBN 978-1-118-01094-5.

[5] D'Andrea, Alessia; et al. (2009). "An Overview of Methods for Virtual Social Network Analysis". In Abraham, Ajith. *Computational Social Network Analysis: Trends, Tools and Research Advances*. Springer. p. 8. ISBN 978-1-84882-228-3.

[6] "Facebook friends mapped by Wolfram Alpha app". *BBC News*. September 24, 2012. Retrieved July 25, 2016.

[7] Frederic Lardinois (August 30, 2012). "Wolfram Alpha Launches Personal Analytics Reports For Facebook". *Tech Crunch*. Retrieved July 25, 2016.

[8] Institute of Reproductive Health

[9] Ivaldi M.; Ferreri L.; Daolio F.; Giacobini M.; Tomassini M.; Rainoldi A. "We-Sport: from academy spin-off to database for complex network analysis; an innovative approach to a new technology". *J Sports Med and Phys Fitnes*. **51** (suppl. 1 to issue 3). The social network analysis was used to analyze properties of the network We-Sport.com allowing a deep interpretation and analysis of the level of aggregation phenomena in the specific context of sport and physical exercise.

[10] Freeman, L. C. (2004). *The development of social network analysis: a study in the sociology of science*. Vancouver, B. C.: Empirical Press.

[11] Linton Freeman (2006). *The Development of Social Network Analysis*. Vancouver: Empirical Press.

[12] Anheier, H.K.; Gerhards, J.; Romo, F.P. (1995). "Forms of capital and social structure of fields: examining Bourdieu's social topography". *American Journal of Sociology*. **100** (4): 859–903. doi:10.1086/230603.

[13] De Nooy, W (2003). "Fields and networks: Correspondence analysis and social network analysis in the framework of Field Theory". *Poetics*. **31** (5–6): 305–27. doi:10.1016/s0304-422x(03)00035-4.

[14] Senekal, B. A. 2012. Die Afrikaanse literêre sisteem: ń Eksperimentele benadering met behulp van Sosiale-netwerk-analise (SNA). LitNet Akademies 9(3)

[15] McPherson, N.; Smith-Lovin, L.; Cook, J.M. (2001). *Birds of a feather: Homophily in social networks*. Annual Review of Sociology. **27**. pp. 415–444. doi:10.1146/annurev.soc.27.1.415.

[16] Podolny, J.M. & Baron, J.N. (1997). "Resources and relationships: Social networks and mobility in the workplace". *American Sociological Review*. **62** (5): 673–693. JSTOR 2657354.

[17] Kilduff, M.; Tsai, W. (2003). *Social networks and organisations*. Sage Publications.

[18] Kadushin, C. (2012). *Understanding social networks: Theories, concepts, and findings*. Oxford: Oxford University Press.

[19] Flynn, F.J.; Reagans, R.E.; Guillory, L. (2010). "Do you two know each other? Transitivity, homophily, and the need for (network) closure". *Journal of Personality and Social Psychology*. **99** (5): 855–869. PMID 20954787. doi:10.1037/a0020961.

[20] Granovetter, M. (1973). *The strength of weak ties*. American Journal of Sociology. **78**. pp. 1360–1380. doi:10.1086/225469.

[21] Hansen, Derek; et al. (2010). *Analyzing Social Media Networks with NodeXL*. Morgan Kaufmann. p. 32. ISBN 978-0-12-382229-1.

[22] Liu, Bing (2011). *Web Data Mining: Exploring Hyperlinks, Contents, and Usage Data*. Springer. p. 271. ISBN 978-3-642-19459-7.

[23] Hanneman, Robert A. & Riddle, Mark (2011). "Concepts and Measures for Basic Network Analysis". *The Sage Handbook of Social Network Analysis*. SAGE. pp. 364–367. ISBN 978-1-84787-395-8.

[24] Tsvetovat, Maksim & Kouznetsov, Alexander (2011). *Social Network Analysis for Startups: Finding Connections on the Social Web*. O'Reilly. p. 45. ISBN 978-1-4493-1762-1.

19.6. REFERENCES

[25] The most comprehensive reference is: Wasserman, Stanley & Faust, Katherine (1994). *Social Networks Analysis: Methods and Applications*. Cambridge: Cambridge University Press. A short, clear basic summary is in Krebs, Valdis (2000). "The Social Life of Routers". *Internet Protocol Journal*. 3 (December): 14–25.

[26] Opsahl, Tore; Agneessens, Filip; Skvoretz, John (2010). "Node centrality in weighted networks: Generalizing degree and shortest paths". *Social Networks*. **32** (3): 245–251. doi:10.1016/j.socnet.2010.03.006.

[27] "Social Network Analysis". *Field Manual 3-24: Counterinsurgency* (PDF). Headquarters, Department of the Army. pp. B-11 – B-12.

[28] Xu, Guandong; et al. (2010). *Web Mining and Social Networking: Techniques and Applications*. Springer. p. 25. ISBN 978-1-4419-7734-2.

[29] Cohesive.blocking is the R program for computing structural cohesion according to the Moody-White (2003) algorithm. This wiki site provides numerous examples and a tutorial for use with R.

[30] Hanneman, Robert A. & Riddle, Mark (2011). "Concepts and Measures for Basic Network Analysis". *The Sage Handbook of Social Network Analysis*. SAGE. pp. 346–347. ISBN 978-1-84787-395-8.

[31] Moody, James & Douglas R. White (2003). "Structural Cohesion and Embeddedness: A Hierarchical Concept of Social Groups" (PDF). *American Sociological Review*. **68** (1): 103–127. CiteSeerX 10.1.1.18.5695. doi:10.2307/3088904.

[32] Pattillo, Jeffrey; et al. (2011). "Clique relaxation models in social network analysis". In Thai, My T. & Pardalos, Panos M. *Handbook of Optimization in Complex Networks: Communication and Social Networks*. Springer. p. 149. ISBN 978-1-4614-0856-7.

[33] Linton C. Freeman, "Visualizing Social Networks". *Journal of Social Structure*. **1**.

[34] Hamdaqa, Mohammad; Tahvildari, Ladan; LaChapelle, Neil; Campbell, Brian (2014). "Cultural Scene Detection Using Reverse Louvain Optimization". *Science of Computer Programming*. **95**: 44–72. doi:10.1016/j.scico.2014.01.006.

[35] Bacher, R. (1995). *Graphical Interaction and Visualization for the Analysis and Interpretation of Contingency Analysis Result*. Proceedings of the 1995 Power Industry Computer Applications. Salt Lake City, USA: IEEE Power Engineering Society. pp. 128–134. doi:10.1109/PICA.1995.515175.

[36] Caschera, M. C.; Ferri, F.; Grifoni, P. (2008). "SIM: A dynamic multidimensional visualization method for social networks". *PsychNology Journal*. **6** (3): 291–320.

[37] McGrath; Blythe & Krackhardt (1997). "The effect of spatial arrangement on judgements and errors in interpreting graphs" (PDF). *Social Networks*. **19** (3): 223–242. doi:10.1016/S0378-8733(96)00299-7.

[38] Cartwright, D.; Frank Harary (1956). "Structural balance: a generalization of Heider's theory" (PDF). *Psychological Review*. **63** (5): 277–293. doi:10.1037/h0046049. Link from Stanford University.

[39] Bernie Hogan; Juan-Antonio Carrasco & Barry Wellman (May 2007). "Visualizing Personal Networks: Working with Participant-Aided Sociograms" (PDF). *Field Methods*. **19** (2): 116–144. doi:10.1177/1525822X06298589.

[40] e.g., Anger, I., & Kittl, C. (2011, September). Measuring influence on Twitter. In Proceedings of the 11th International Conference on Knowledge Management and Knowledge Technologies (p. 31). ACM.

[41] Riquelme, F., & González-Cantergiani, P. (2016). Measuring user influence on Twitter: A survey. Information Processing & Management. 52, p. 949-975.

[42] (Hrsg.), Sara Rosengren (2013). *The Changing Roles of Advertising*. Wiesbaden: Springer Fachmedien Wiesbaden GmbH. ISBN 9783658023645. Retrieved 22 October 2015.

[43] Ahonen, T. T., Kasper, T., & Melkko, S. (2005). 3G marketing: communities and strategic partnerships. John Wiley & Sons.

[44] "technology" "Watch "TEDxMilano - Nicola Greco - on math and social network" Video at TEDxTalks". *TEDxTalks*.

[45] Golbeck, J. (2013). *Analyzing the Social Web*. Morgan Kaufmann. ISBN 0-12-405856-6.

[46] Aram, Michael; Neumann, Gustaf (2015-07-01). "Multilayered analysis of co-development of business information systems" (PDF). *Journal of Internet Services and Applications*. **6** (1). doi:10.1186/s13174-015-0030-8.

[47] "NSA warned to rein in surveillance as agency reveals even greater scope". 17 July 2013. Retrieved 19 July 2013.

[48] "How The NSA Uses Social Network Analysis To Map Terrorist Networks". 12 June 2013. Retrieved 19 Jul 2013.

[49] "NSA Using Social Network Analysis". 12 May 2006. Retrieved 19 July 2013.

[50] "NSA has massive database of Americans' phone calls". 11 May 2006. Retrieved 19 July 2013.

[51] Sudhahar S, Veltri GA, Cristianini N (2015). "Automated analysis of the US presidential elections using Big Data and network analysis". *Big Data & Society*. **2** (1): 1–28. doi:10.1177/2053951715572916.

[52] Sudhahar S, De Fazio G, Franzosi R, Cristianini N (2013). "Network analysis of narrative content in large corpora". *Natural Language Engineering*. **21** (1): 1–32. doi:10.1017/S1351324913000247.

[53] Quantitative Narrative Analysis; Roberto Franzosi; Emory University © 2010

[54] Laat, Maarten de; Lally, Vic; Lipponen, Lasse; Simons, Robert-Jan (2007-03-08). "Investigating patterns of interaction in networked learning and computer-supported collaborative learning: A role for Social Network Analysis". *International Journal of Computer-Supported Collaborative Learning*. **2** (1): 87–103. doi:10.1007/s11412-007-9006-4.

[55] Palonen, T. & Hakkarainen, K. B. Fishman & S. O'Connor-Divelbiss, eds. *Patterns of Interaction in Computer-Supported Learning: A Social Network Analysis* (PDF). Fourth International Conference of the Learning Sciences. Mahwah, NJ: Erlbaum. pp. 334–339.

[56] Martínez, A.; Dimitriadis, Y.; Rubia, B.; Gómez, E.; de la Fuente, P. (2003-12-01). "Combining qualitative evaluation and social network analysis for the study of classroom social interactions". *Computers & Education*. Documenting Collaborative Interactions: Issues and Approaches. **41** (4): 353–368. doi:10.1016/j.compedu.2003.06.001.

[57] Cho, H.; Stefanone, M. & Gay, G (2002). *Social information sharing in a CSCL community*. Computer support for collaborative learning: Foundations for a CSCL community. Hillsdale, NJ: Lawrence Erlbaum. pp. 43–50.

[58] Aviv, R.; Erlich, Z.; Ravid, G. & Geva, A. (2003). "Network analysis of knowledge construction in asynchronous learning networks". *Journal of Asynchronous Learning Networks*. **7** (3): 1–23. CiteSeerX 10.1.1.2.9044.

[59] Daradoumis, Thanasis; Martínez-Monés, Alejandra; Xhafa, Fatos (2004-09-05). Vreede, Gert-Jan de; Guerrero, Luis A.; Raventós, Gabriela Marín, eds. *Groupware: Design, Implementation, and Use*. Lecture Notes in Computer Science. Springer Berlin Heidelberg. pp. 289–304. ISBN 9783540230168. doi:10.1007/978-3-540-30112-7_25.

[60] Martínez, A.; Dimitriadis, Y.; Rubia, B.; Gómez, E.; de la Fuente, P. (2003-12-01). "Combining qualitative evaluation and social network analysis for the study of classroom social interactions". *Computers & Education*. Documenting Collaborative Interactions: Issues and Approaches. **41** (4): 353–368. doi:10.1016/j.compedu.2003.06.001.

[61] Johnson, Karen E. (1996-01-01). "Review of The Art of Case Study Research". *The Modern Language Journal*. **80** (4): 556–557. JSTOR 329758. doi:10.2307/329758.

19.7 External links

19.7.1 Further reading

- Awesome Network Analysis (200+ links to books, conferences, courses, journals, research groups, software, tutorials and more)

- Introduction to Stochastic Actor-Based Models for Network Dynamics - Snijders et al.

- Center for Computational Analysis of Social and Organizational Systems (CASOS) at Carnegie Mellon

- NetLab at the University of Toronto, studies the intersection of social, communication, information and computing networks

- Netwiki (wiki page devoted to social networks; maintained at University of North Carolina at Chapel Hill)

- Program on Networked Governance – Program on Networked Governance, Harvard University

- The International Workshop on Social Network Analysis and Mining (SNA-KDD) - An annual workshop on social network analysis and mining, with participants from computer science, social science, and related disciplines.

- Historical Dynamics in a time of Crisis: Late Byzantium, 1204–1453 (a discussion of social network analysis from the point of view of historical studies)

- Social Network Analysis: A Systematic Approach for Investigating

19.7.2 Organizations

- International Network for Social Network Analysis

19.7.3 Peer-reviewed journals

- *Social Networks*
- *Network Science*
- *Journal of Social Structure*
- *Journal of Complex Networks*
- *Journal of Mathematical Sociology*
- *Social Network Analysis and Mining (SNAM)*
- "REDES". Spain: Universidad Autónoma de Barcelona y Universidad de Sevilla.
- "Connections". International Network for Social Network Analysis.

19.7.4 Textbooks and educational resources

- *Networks, Crowds, and Markets* (2010) by D. Easley & J. Kleinberg
- *Introduction to Social Networks Methods* (2005) by R. Hanneman & M. Riddle
- *Social Network Analysis with Applications* (2013) by I. McCulloh, H. Armstrong & A. Johnson

Chapter 20

Small-world network

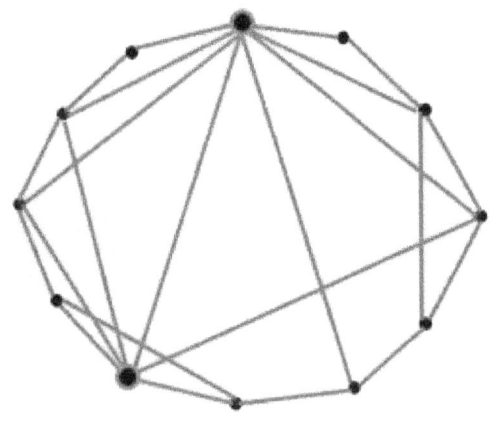

Small-world network example
Hubs *are bigger than other nodes Average degree*= *3.833*
Average shortest path length = *1.803.*
Clustering coefficient = *0.522*

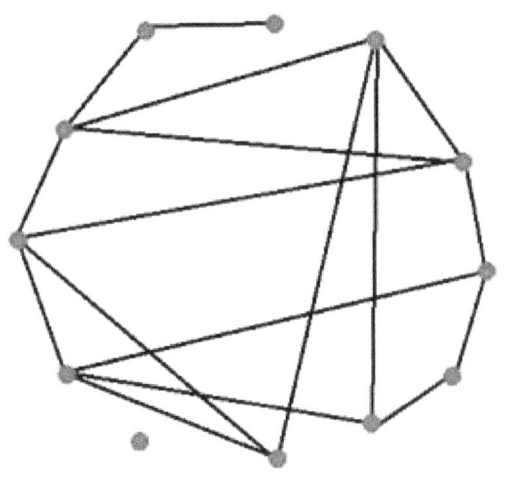

Random graph
Average degree = *2.833*
Average shortest path length = *2.109.*
Clustering coefficient = *0.167*

A **small-world network** is a type of mathematical graph in which most nodes are not neighbors of one another, but the neighbors of any given node are likely to be neighbors of each other and most nodes can be reached from every other node by a small number of hops or steps. Specifically, a small-world network is defined to be a network where the typical distance L between two randomly chosen nodes (the number of steps required) grows proportionally to the logarithm of the number of nodes N in the network, that is:[1]

$$L \propto \log N$$

while the clustering coefficient is not small. In the context of a social network, this results in the small world phenomenon of strangers being linked by a short chain of acquaintances. Many empirical graphs show the small-world effect, e.g., social networks, the underlying architecture of the Internet, wikis such as Wikipedia, and gene networks.

A certain category of small-world networks were identified as a class of random graphs by Duncan Watts and Steven Strogatz in 1998.[2] They noted that graphs could be classified according to two independent structural features, namely the clustering coefficient, and average node-to-node distance (also known as average shortest path length). Purely random graphs, built according to the Erdős–Rényi (ER) model, exhibit a small average shortest path length (varying typically as the logarithm of the number of nodes) along with a small clustering coefficient. Watts and Strogatz measured that in fact many real-world networks have a small average shortest path length, but also a clustering coefficient significantly higher than expected by random chance. Watts and Strogatz then proposed a novel graph model, currently named the Watts and Strogatz model, with (i) a small average shortest path length, and (ii) a large clustering coefficient. The crossover in the Watts–Strogatz model between a "large world" (such as a lattice) and a small world was first described by Barthelemy and Amaral in 1999.[3] This work was followed by a large number of studies, including exact results (Barrat and Weigt, 1999; Dorogovtsev and Mendes; Barmpoutis and Murray, 2010). Braunstein [4] found that for weighted ER networks where the weights have a very broad distribution, the optimal path scales becomes significantly longer and scales as $N^{1/3}$.

20.1 Properties of small-world networks

Small-world networks tend to contain cliques, and near-cliques, meaning sub-networks which have connections between almost any two nodes within them. This follows from the defining property of a high clustering coefficient. Secondly, most pairs of nodes will be connected by at least one short path. This follows from the defining property that the mean-shortest path length be small. Several other properties are often associated with small-world networks. Typically there is an over-abundance of *hubs* – nodes in the network with a high number of connections (known as high degree nodes). These hubs serve as the common connections mediating the short path lengths between other edges. By analogy, the small-world network of airline flights has a small mean-path length (i.e. between any two cities you are likely to have to take three or fewer flights) because many flights are routed through hub cities.

This property is often analyzed by considering the fraction of nodes in the network that have a particular number of connections going into them (the degree distribution of the network). Networks with a greater than expected number of hubs will have a greater fraction of nodes with high degree, and consequently the degree distribution will be enriched at high degree values. This is known colloquially as a fat-tailed distribution. Graphs of very different topology qualify as small-world networks as long as they satisfy the two definitional requirements above.

Network small-worldness has been quantified by a small-coefficient, σ, calculated by comparing clustering and path length of a given network to an equivalent random network with same degree on average.[5][6]

$$\sigma = \frac{\frac{C}{C_r}}{\frac{L}{L_r}}$$

if $\sigma > 1$ ($C \gg C_r$ and $L \approx L_r$), network is small-world.

Another method for quantifying network small-worldness utilizes the original definition of the small-world network comparing the clustering of a given network to an equivalent lattice network and its path length to an equivalent random network.[7] The small-world measure (ω) is defined as[8]

$$\omega = \frac{L_r}{L} - \frac{C}{C_\ell}$$

Where the characteristic path length L and clustering coefficient C are calculated from the network you are testing. $C\ell$ is the clustering coefficient for an equivalent lattice network and Lr is the characteristic path length for an equivalent random network.

R. Cohen and Havlin[9][10] showed analytically that scale-free networks are ultra-small worlds. In this case, due to hubs, the shortest paths become significantly smaller and scale as

$$L \propto \log \log N$$

20.2 Examples of small-world networks

Small-world properties are found in many real-world phenomena, including websites with navigation menus, food chains, electric power grids, metabolite processing networks, networks of brain neurons, voter networks, telephone call graphs, and social influence networks. Cultural networks[11] and word co-occurrence networks[12] have also been shown to be small-world networks.

Networks of connected proteins have small world properties such as power-law obeying degree distributions.[13] Similarly transcriptional networks, in which the nodes are genes, and they are linked if one gene has an up or down-regulatory genetic influence on the other, have small world network properties.[14]

20.3 Examples of non-small-world networks

In another example, the famous theory of "six degrees of separation" between people tacitly presumes that the domain of discourse is the set of people alive at any one time. The number of degrees of separation between Albert Einstein and Alexander the Great is almost certainly greater than 30[15] and this network does not have small-world properties. A similarly constrained network would be the "went to school with" network: if two people went to the same college ten years apart from one another, it is unlikely that they have acquaintances in common amongst the student body.

Similarly, the number of relay stations through which a message must pass was not always small. In the days when the post was carried by hand or on horseback, the number of times a letter changed hands between its source and destination would have been much greater than it is today. The number of times a message changed hands in the days of the visual telegraph (circa 1800–1850) was determined by the requirement that two stations be connected by line-of-sight.

Tacit assumptions, if not examined, can cause a bias in the literature on graphs in favor of finding small-world networks (an example of the file drawer effect resulting from the publication bias).

20.4 Network robustness

It is hypothesized by some researchers such as Barabási that the prevalence of small world networks in biological systems may reflect an evolutionary advantage of such an architecture. One possibility is that small-world networks are more robust to perturbations than other network architectures. If this were the case, it would provide an advantage to biological systems that are subject to damage by mutation or viral infection.

In a small world network with a degree distribution following a power-law, deletion of a random node rarely causes a dramatic increase in mean-shortest path length (or a dramatic decrease in the clustering coefficient). This follows from the fact that most shortest paths between nodes flow through hubs, and if a peripheral node is deleted it is unlikely to interfere with passage between other peripheral nodes. As the fraction of peripheral nodes in a small world network is much higher than the fraction of hubs, the probability of deleting an important node is very low. For example, if the small airport in Sun Valley, Idaho was shut down, it would not increase the average number of flights that other passengers traveling in the United States would have to take to arrive at their respective destinations. However, if random deletion of a node hits a hub by chance, the average path length can increase dramatically. This can be observed annually when northern hub airports, such as Chicago's O'Hare airport, are shut down because of snow; many people have to take additional flights.

By contrast, in a random network, in which all nodes have roughly the same number of connections, deleting a random node is likely to increase the mean-shortest path length slightly but significantly for almost any node deleted. In this sense, random networks are vulnerable to random perturbations, whereas small-world networks are robust. However, small-world networks are vulnerable to targeted attack of hubs, whereas random networks cannot be targeted for catastrophic failure.

Appropriately, viruses have evolved to interfere with the activity of hub proteins such as p53, thereby bringing about the massive changes in cellular behavior which are conducive to viral replication. A useful method to analyze network robustness is the percolation theory.

20.5 Construction of small-world networks

The main mechanism to construct small-world networks is the Watts–Strogatz mechanism.

Small-world networks can also be introduced with time-delay,[16] which will not only produces fractals but also chaos[17] under the right conditions, or transition to chaos in dynamics networks.[18]

Degree–diameter graphs are constructed such that the number of neighbors each vertex in the network has is bounded, while the distance from any given vertex in the network to any other vertex (the diameter of the network) is minimized. Constructing such small-world networks is done as part of the effort to find graphs of order close to the Moore bound.

Another way to construct a small world network from scratch is given in Barmpoutis *et al.*,[19] where a network with very small average distance and very large average clustering is constructed. A fast algorithm of constant complexity is given, along with measurements of the robustness of the resulting graphs. Depending on the application of each network, one can start with one such "ultra small-world" network, and then rewire some edges, or use several small such networks as subgraphs to a larger graph.

Small-world properties can arise naturally in social networks and other real-world systems via the process of dual-phase evolution. This is particularly common where time or spatial constraints limit the addition of connections between vertices The mechanism generally involves periodic shifts between phases, with connections being added during a "global" phase and being reinforced or removed during a "local" phase.

See also: Diffusion-limited aggregation, Pattern formation

20.6 Applications

20.6.1 Applications to sociology

The advantages to small world networking for social movement groups are their resistance to change due to the filtering apparatus of using highly connected nodes, and its better effectiveness in relaying information while keeping the number of links required to connect a network to a minimum.[20]

The small world network model is directly applicable to affinity group theory represented in sociological arguments by William Finnegan. Affinity groups are social movement groups that are small and semi-independent pledged to a larger goal or function. Though largely unaffiliated at the node level, a few members of high connectivity function as connectivity nodes, linking the different groups through networking. This small world model has proven an extremely effective protest organization tactic against police action.[21] Clay Shirky argues that the larger the social network created through small world networking, the more valuable the nodes of high connectivity within the network.[20] The same can be said for the affinity group model, where the few people within each group connected to outside groups allowed for a large amount of mobilization and adaptation. A practical example of this is small world networking through affinity groups that William Finnegan outlines in reference to the 1999 Seattle WTO protests.

20.6.2 Applications to earth sciences

Many networks studied in geology and geophysics have been shown to have characteristics of small-world networks. Networks defined in fracture systems and porous substances have demonstrated these characteristics.[22] The seismic network in the Southern California region may be a small-world network.[23] The examples above occur on very different spatial scales, demonstrating the scale invariance of the phenomenon in the earth sciences. Climate networks may be regarded as small world networks where the links are of different length scales.[24]

20.6.3 Applications to computing

Small-world networks have been used to estimate the usability of information stored in large databases. The measure is termed the Small World Data Transformation Measure.[25][26] The greater the database links align to a small-world network the more likely a user is going to be able to extract information in the future. This usability typically comes at the cost of the amount of information that can be stored in the same repository.

The Freenet peer-to-peer network has been shown to form a small-world network in simulation,[27] allowing information to be stored and retrieved in a manner that scales efficiency as the network grows.

20.6.4 Small-world neural networks in the brain

Both anatomical connections in the brain[28] and the synchronization networks of cortical neurons[29] exhibit small-world topology.

A small-world network of neurons can exhibit short-term memory. A computer model developed by Solla et al.[30][31] had two stable states, a property (called bistability) thought to be important in memory storage. An activating pulse generated self-sustaining loops of communication activity among the neurons. A second pulse ended this activity. The pulses switched the system between stable states: flow (recording a "memory"), and stasis (holding it). Small world neuronal networks have also been used as models to understand seizures.[32]

On a more general level, many large-scale neural networks in the brain, such as the visual system and brain stem, exhibit small-world properties.[5]

20.7 Small world with a distribution of link length

The WS model includes a uniform distribution of long-range links. When the distribution of link lengths follows a power law distribution, the mean distance between two sites changes depending on the power of the distribution.[33]

20.8 See also

- Barabási–Albert model
- Climate network
- Dual-phase evolution
- Dunbar's number
- Erdős number
- Erdős–Rényi (ER) model
- Percolation theory
- Scale-free network
- Six degrees of Kevin Bacon
- Small world experiment
- Social network
- Watts and Strogatz model

20.9 References

[1] http://www.nature.com/nature/journal/v393/n6684/full/393440a0.html

[2] Watts, Duncan J.; Strogatz, Steven H. (June 1998). "Collective dynamics of 'small-world' networks". *Nature*. **393** (6684): 440–442. Bibcode:1998Natur.393..440W. PMID 9623998. doi:10.1038/30918. Papercore Summary http://www.papercore.org/Watts1998

[3] Barthelemy, M.; Amaral, LAN (1999). "Small-world networks: Evidence for a crossover picture". *Phys. Rev. Lett.* **82** (15): 3180–3183. Bibcode:1999PhRvL..82.3180B. arXiv:cond-mat/9903108 ⊙. doi:10.1103/PhysRevLett.82.3180.

[4] Braunstein, Lidia A.; Buldyrev, Sergey V.; Cohen, Reuven; Havlin, Shlomo; Stanley, H. Eugene (2003). "Optimal Paths in Disordered Complex Networks". *Physical Review Letters*. **91** (16). Bibcode:2003PhRvL..91p8701B. ISSN 0031-9007. doi:10.1103/PhysRevLett.91.168701.

[5] The brainstem reticular formation is a small-world, not scale-free, network M. D. Humphries, K. Gurney and T. J. Prescott, Proc. Roy. Soc. B 2006 273, 503–511, doi:10.1098/rspb.2005.3354

[6] Humphries and Gurney (2008). "Network 'Small-World-Ness': A Quantitative Method for Determining Canonical Network Equivalence". *PLOS ONE*. **3** (4): e0002051. PMC 2323569 ⊙. PMID 18446219. doi:10.1371/journal.pone.0002051.

[7] The ubiquity of small-world networks Q.K. Telesford, K.E. Joyce, S. Hayasaka, J.H. Burdette, P.J. Laurienti, Brain Connect. 2011;1(5):367–75, doi:10.1089/brain.2011.0038

[8] Telesford, Joyce, Hayasaka, Burdette, and Laurienti (2011). "The Ubiquity of Small-World Networks". *Brain Connectivity*. **1** (0038): 367–75. PMC 3604768 ⊙. PMID 22432451. doi:10.1089/brain.2011.0038.

[9] R. Cohen, S. Havlin, and D. ben-Avraham (2002). "Structural properties of scale free networks". *Handbook of graphs and networks*. Wiley-VCH. 2002 (Chap. 4).

[10] R. Cohen, S. Havlin (2003). "Scale-free networks are ultrasmall". *Phys. Rev. Lett.* **90** (5): 058701. Bibcode:2003PhRvL..90e8701C. PMID 12633404. arXiv:cond-mat/0205476 ⊙. doi:10.1103/PhysRevLett.90.058701.

[11] "'n Kwantifisering van kleinwêreldsheid in Afrikaanse kultuurnetwerke in vergelyking met ander komplekse netwerke | LitNet". *LitNet*. 2015-11-05. Retrieved 2017-02-27.

[12] "Die statistiese eienskappe van geskrewe Afrikaans as 'n komplekse netwerk | LitNet". *LitNet*. 2017-02-09. Retrieved 2017-02-27.

[13] Bork, P.; Jensen, LJ; von Mering, C.; Ramani, A.; Lee, I.; Marcotte, EM. (2004). "Protein interaction networks from yeast to human" (PDF). *Current Opinion in Structural Biology*. **14** (3): 292–299. PMID 15193308. doi:10.1016/j.sbi.2004.05.003.

[14] Van Noort, V; Snel, B; Huynen, MA. (Mar 2004). "The yeast coexpression network has a small-world, scale-free architecture and can be explained by a simple model". *EMBO Rep*. **5** (3): 280–4. PMC 1299002 ⊙. PMID 14968131. doi:10.1038/sj.embor.7400090.

[15] Einstein and Alexander the Great lived 2202 years apart. Assuming an age difference of 70 years between any two connected people in the chain that connects the two, this would necessitate at least 32 connections between Einstein and Alexander the Great.

[16] X. S. Yang, Fractals in small-world networks with time-delay, Chaos, Solitons & Fractals, vol. 13, 215–219 (2002)

[17] X. S. Yang, Chaos in small-world networks, Phys. Rev. E 63, 046206 (2001)

[18] W. Yuan, X. S. Luo, P. Jiang, B. Wang, J. Fang, Transition to chaos in small-world dynamical network

[19] D.Barmpoutis and R.M. Murray (2010). "Networks with the Smallest Average Distance and the Largest Average Clustering". arXiv:1007.4031 ⊙ [q-bio.MN].

[20] Shirky, Clay. 2008. *Here Comes Everybody*

[21] Finnegan, William "Affinity Groups and the Movement Against Corporate Globalization"

[22] X. S. Yang, Small-world networks in geophysics, Geophys. Res. Lett., 28(13), 2549–2552 (2001)

[23] A. Jimenez, K. F. Tiampo, and A. M. Posadas, Small-world in a seismic network: the California case, Nonlin. Processes Geophys., 15, 389–395 (2008)

[24] Gozolchiani, A.; Havlin, S.; Yamasaki, K. (2011). "Emergence of El Niño as an Autonomous Component in the Climate Network". *Physical Review Letters*. **107** (14): 148501. Bibcode:2011PhRvL.107n8501G. ISSN 0031-9007. PMID 22107243. doi:10.1103/PhysRevLett.107.148501.

[25] http://mike2.openmethodology.org/wiki/Small_Worlds_Data_Transformation_Measure

[26] Hillard, Robert (2010). *Information-Driven Business*. Wiley. ISBN 978-0-470-62577-4.

[27] Sandberg, Oskar. "Searching in a Small World" (PDF).

[28] Sporns, Olaf; Chialvo DR; Kaiser M; Hilgetag CC (2004). "Organization, development and function of complex brain networks". *Trends Cogn Sci*. **8** (9): 418–425. PMID 15350243. doi:10.1016/j.tics.2004.07.008.

[29] Yu, Shan; D. Huang; W. Singer; D. Nikolić (2008). "A Small World of Neuronal Synchrony". *Cerebral Cortex.* **18** (12): 2891–2901. PMC 2583154. PMID 18400792. doi:10.1093/cercor/bhn047.

[30] Cohen, Philip. Small world networks key to memory. *New Scientist.* 26 May 2004.

[31] Sara Solla's Lecture & Slides: Self-Sustained Activity in a Small-World Network of Excitable Neurons

[32] Ponten, S.C.; Bartolomei, F.; Stam, C.J. (April 2007). "Small-world networks and epilepsy: Graph theoretical analysis of intracerebrally recorded mesial temporal lobe seizures". *Clinical Neurophysiology.* **118** (4): 918–927. doi:10.1016/j.clinph.2006.12.002.

[33] D. Li, K. Kosmidis, A. Bunde, S. Havlin (2011). "Dimension of spatially embedded networks". *Nature Physics.* **7**: 481–484. Bibcode:2011NatPh...7..481D. doi:10.1038/nphys1932.

20.9.1 Books

- Buchanan, Mark (2003). *Nexus: Small Worlds and the Groundbreaking Theory of Networks.* Norton, W. W. & Company, Inc. ISBN 0-393-32442-7.

- Dorogovtsev, S.N. & Mendes, J.F.F. (2003). *Evolution of Networks: from biological networks to the Internet and WWW.* Oxford University Press. ISBN 0-19-851590-1.

- Watts, D. J. (1999). *Small Worlds: The Dynamics of Networks Between Order and Randomness.* Princeton University Press. ISBN 0-691-00541-9.

- Fowler, JH. (2005) "Turnout in a Small World," in Alan Zuckerman, ed., *Social Logic of Politics*, Temple University Press, 269–287

- Reuven Cohen and Shlomo Havlin (2010). *Complex Networks: Structure, Robustness and Function.* Cambridge University Press.

20.9.2 Journal articles

- Albert, R.; Barabási A.L. (2002). "Statistical mechanics of complex networks". *Rev. Mod. Phys.* **74**: 47–97. Bibcode:2002RvMP...74...47A. arXiv:cond-mat/0106096. doi:10.1103/RevModPhys.74.47.

- Albert, R.; Barabási A.L. (1999). "Emergence of scaling in random networks". *Science.* **286** (5439): 509–12. Bibcode:1999Sci...286..509B. PMID 10521342. arXiv:cond-mat/9910332. doi:10.1126/science.286.5439.509.

- Barthelemy, M.; Amaral, LAN. (1999). "Small-world networks: Evidence for a crossover picture". *Phys. Rev. Lett.* **82** (15): 3180–3183. Bibcode:1999PhRvL..82.3180B. arXiv:cond-mat/9903108. doi:10.1103/PhysRevLett.82.3180.

- Dorogovtsev, S.N.; Mendes, J.F.F. (2000). "Exactly solvable analogy of small-world networks". *Europhys. Lett.* **50**: 1–7. Bibcode:2000EL......50....1D. arXiv:cond-mat/9907445. doi:10.1209/epl/i2000-00227-1.

- Milgram, Stanley (1967). "The Small World Problem". *Psychology Today.* **1** (1): 60–67.

- Newman, Mark (2003). "The Structure and Function of Complex Networks". *SIAM Review.* **45** (2): 167–256. Bibcode:2003SIAMR..45..167N. arXiv:cond-mat/0303516. doi:10.1137/S003614450342480.pdf

- Ravid, D.; Rafaeli, S. (2004). "Asynchronous discussion groups as Small World and Scale Free Networks". *First Monday.* **9** (9). doi:10.5210/fm.v9i9.1170.

- R. Parshani, S.V. Buldyrev, S. Havlin (2011). "Critical effect of dependency groups on the function of networks". *PNAS.* **108**: 1007–1010. Bibcode:2011PNAS..108.1007P. PMC 3024657. PMID 21191103. arXiv:1010.4498. doi:10.1073/pnas.1008404108.

- S. V. Buldyrev, R. Parshani, G. Paul, H. E. Stanley, S. Havlin (2010). "Catastrophic cascade of failures in interdependent networks". *Nature.* **464** (7291): 1025–8. Bibcode:2010Natur.464.1025B. PMID 20393559. arXiv:0907.1182. doi:10.1038/nature08932.

20.10 External links

- Dynamic Proximity Networks by Seth J. Chandler, The Wolfram Demonstrations Project.

- Small-World Networks entry on Scholarpedia (by Mason A. Porter)

Chapter 21

Centrality

For the statistical concept, see Central tendency.

In graph theory and network analysis, indicators of **cen-trality** identify the most important vertices within a graph. Applications include identifying the most influential person(s) in a social network, key infrastructure nodes in the Internet or urban networks, and super-spreaders of disease. Centrality concepts were first developed in social network analysis, and many of the terms used to measure centrality reflect their sociological origin.[1] They should not be confused with node influence metrics, which seek to quantify the influence of every node in the network.

Examples of A) Betweenness centrality, B) Closeness centrality, C) Eigenvector centrality, D) Degree centrality, E) Harmonic Centrality and F) Katz centrality of the same graph.

21.1 Definition and characterization of centrality indices

Centrality indices are answers to the question "What characterizes an important vertex?" The answer is given in terms of a real-valued function on the vertices of a graph, where the values produced are expected to provide a ranking which identifies the most important nodes.[2][3]

The word "importance" has a wide number of meanings, leading to many different definitions of centrality. Two categorization schemes have been proposed. "Importance" can be conceived in relation to a type of flow or transfer across the network. This allows centralities to be classified by the type of flow they consider important.[3] "Importance" can alternately be conceived as involvement in the cohesiveness of the network. This allows centralities to be classified based on how they measure cohesiveness.[4] Both of these approaches divide centralities in distinct categories. A further conclusion is that a centrality which is appropriate for one category will often "get it wrong" when applied to a different category.[3]

When centralities are categorized by their approach to cohesiveness, it becomes apparent that the majority of centralities inhabit one category. The count of the number of walks starting from a given vertex differs only in how walks are defined and counted. Restricting consideration to this group allows for a soft characterization which places centralities on a spectrum from walks of length one (degree centrality) to infinite walks (eigenvalue centrality).[2][5] The observation that many centralities share this familial relationships perhaps explains the high rank correlations between these indices.

21.1. DEFINITION AND CHARACTERIZATION OF CENTRALITY INDICES

21.1.1 Characterization by network flows

A network can be considered a description of the paths along which something flows. This allows a characterization based on the type of flow and the type of path encoded by the centrality. A flow can be based on transfers, where each undivisible item goes from one node to another, like a package delivery which goes from the delivery site to the client's house. A second case is the serial duplication, where this is a replication of the item which goes to the next node, so both the source and the target have it. An example is the propagation of information through gossip, with the information being propagated in a private way and with both the source and the target nodes being informed at the end of the process. The last case is the parallel duplication, with the item being duplicated to several links at the same time, like a radio broadcast which provides the same information to many listeners at once.[3]

Likewise, the type of path can be constrained to: Geodesics (shortest paths), paths (no vertex is visited more than once), trails (vertices can be visited multiple times, no edge is traversed more than once), or walks (vertices and edges can be visited/traversed multiple times).[3]

21.1.2 Characterization by walk structure

An alternate classification can be derived from how the centrality is constructed. This again splits into two classes. Centralities are either *Radial* or *Medial*. Radial centralities count walks which start/end from the given vertex. The degree and eigenvalue centralities are examples of radial centralities, counting the number of walks of length one or length infinity. Medial centralities count walks which pass through the given vertex. The canonical example is Freeman's betweenness centrality, the number of shortest paths which pass through the given vertex.[4]

Likewise, the counting can capture either the *volume* or the *length* of walks. Volume is the total number of walks of the given type. The three examples from the previous paragraph fall into this category. Length captures the distance from the given vertex to the remaining vertices in the graph. Freeman's closeness centrality, the total geodesic distance from a given vertex to all other vertices, is the best known example.[4] Note that this classification is independent of the type of walk counted (i.e. walk, trail, path, geodesic).

Borgatti and Everett propose that this typology provides insight into how best to compare centrality measures. Centralities placed in the same box in this 2×2 classification are similar enough to make plausible alternatives; one can reasonably compare which is better for a given application. Measures from different boxes, however, are categorically distinct. Any evaluation of relative fitness can only occur within the context of predetermining which category is more applicable, rendering the comparison moot.[4]

21.1.3 Radial-volume centralities exist on a spectrum

The characterization by walk structure shows that almost all centralities in wide use are radial-volume measures. These encode the belief that a vertex's centrality is a function of the centrality of the vertices it is associated with. Centralities distinguish themselves on how association is defined.

Bonacich showed that if association is defined in terms of walks, then a family of centralities can be defined based on the length of walk considered.[2] The degree counts walks of length one, the eigenvalue centrality counts walks of length infinity. Alternate definitions of association are also reasonable. The alpha centrality allows vertices to have an external source of influence. Estrada's subgraph centrality proposes only counting closed paths (triangles, squares, ...).

The heart of such measures is the observation that powers of the graph's adjacency matrix gives the number of walks of length given by that power. Similarly, the matrix exponential is also closely related to the number of walks of a given length. An initial transformation of the adjacency matrix allows differing definition of the type of walk counted. Under either approach, the centrality of a vertex can be expressed as an infinite sum, either

$$\sum_{k=0}^{\infty} A_R^k \beta^k$$

for matrix powers or

$$\sum_{k=0}^{\infty} \frac{(A_R \beta)^k}{k!}$$

for matrix exponentials, where

- k is walk length,
- A_R is the transformed adjacency matrix, and
- β is a discount parameter which ensures convergence of the sum.

Bonacich's family of measures does not transform the adjacency matrix. The alpha centrality replaces the adjacency matrix with its resolvent. The subgraph centrality replaces the adjacency matrix with its trace. A startling conclusion is that regardless of the initial transformation of the adjacency matrix, all such approaches have common limiting behavior. As β approaches zero, the indices converge to the degree centrality. As β approaches its maximal value, the indices converge to the eigenvalue centrality.[5]

21.2 Important limitations

Centrality indices have two important limitations, one obvious and the other subtle. The obvious limitation is that a centrality which is optimal for one application is often suboptimal for a different application. Indeed, if this were not so, we would not need so many different centralities.

The more subtle limitation is the commonly held fallacy that vertex centrality indicates the relative importance of vertices. Centrality indices are explicitly designed to produce a ranking which allows indication of the most important vertices.[2][3] This they do well, under the limitation just noted. They are not designed to measure the influence of nodes in general. Recently, network physicists have begun developing node influence metrics to address this problem.

The error is two-fold. Firstly, a ranking only orders vertices by importance, it does not quantify the difference in importance between different levels of the ranking. This may be mitigated by applying Freeman centralization to the centrality measure in question, which provide some insight to the importance of nodes depending on the differences of their centralization scores. Furthermore, Freeman centralization enables one to compare several networks by comparing their highest centralization scores.[6] This approach, however, is seldom seen in practice.

Secondly, the features which (correctly) identify the most important vertices in a given network/application do not necessarily generalize to the remaining vertices. For the majority of other network nodes the rankings may be meaningless.[7][8][9][10] This explains why, for example, only the first few results of a Google image search appear in a reasonable order. The pagerank is a highly unstable measure, showing frequent rank reversals after small adjustments of the jump parameter.[11]

While the failure of centrality indices to generalize to the rest of the network may at first seem counter-intuitive, it follows directly from the above definitions. Complex networks have heterogeneous topology. To the extent that the optimal measure depends on the network structure of the most important vertices, a measure which is optimal for such vertices is sub-optimal for the remainder of the network.[7]

21.3 Degree centrality

Main article: Degree (graph theory)

Historically first and conceptually simplest is **degree centrality**, which is defined as the number of links incident upon a node (i.e., the number of ties that a node has). The degree can be interpreted in terms of the immediate risk of a node for catching whatever is flowing through the network (such as a virus, or some information). In the case of a directed network (where ties have direction), we usually define two separate measures of degree centrality, namely indegree and outdegree. Accordingly, indegree is a count of the number of ties directed to the node and outdegree is the number of ties that the node directs to others. When ties are associated to some positive aspects such as friendship or collaboration, indegree is often interpreted as a form of popularity, and outdegree as gregariousness.

The degree centrality of a vertex v, for a given graph $G := (V, E)$ with $|V|$ vertices and $|E|$ edges, is defined as

$$C_D(v) = \deg(v)$$

Calculating degree centrality for all the nodes in a graph takes $\Theta(V^2)$ in a dense adjacency matrix representation of the graph, and for edges takes $\Theta(E)$ in a sparse matrix representation.

The definition of centrality on the node level can be extended to the whole graph, in which case we are speaking of *graph centralization*.[12] Let $v*$ be the node with highest degree centrality in G. Let $X := (Y, Z)$ be the $|Y|$ node connected graph that maximizes the following quantity (with $y*$ being the node with highest degree centrality in X):

$$H = \sum_{j=1}^{|Y|}[C_D(y*) - C_D(y_j)]$$

Correspondingly, the degree centralization of the graph G is as follows:

$$C_D(G) = \frac{\sum_{i=1}^{|V|}[C_D(v*) - C_D(v_i)]}{H}$$

The value of H is maximized when the graph X contains one central node to which all other nodes are connected (a star graph), and in this case $H = (n-1) \cdot ((n-1)-1) = n^2 - 3n + 2$.

21.4 Closeness centrality

Main article: Closeness centrality

In a connected graph, the normalized **closeness centrality** (or **closeness**) of a node is the average length of the shortest path between the node and all other nodes in the graph. Thus the more central a node is, the closer it is to all other nodes.

Closeness was defined by Bavelas (1950) as the reciprocal of the **farness**,[13][14] that is:

$$C(x) = \frac{1}{\sum_y d(y,x)}$$

where $d(y,x)$ is the distance between vertices x and y. However, when speaking of closeness centrality, people usually refer to its normalized form, generally given by the previous formula multiplied by $N-1$, where N is the number of nodes in the graph. This adjustment allows comparisons between nodes of graphs of different sizes.

Taking distances *from* or *to* all other nodes is irrelevant in undirected graphs, whereas it can produce totally different results in directed graphs (e.g. a website can have a high closeness centrality from outgoing link, but low closeness centrality from incoming links).

21.4.1 Harmonic centrality

In a (not necessarily connected) graph, the **harmonic centrality** reverses the sum and reciprocal operations in the definition of closeness centrality:

$$H(x) = \sum_{y \neq x} \frac{1}{d(y,x)}$$

where $1/d(y,x) = 0$ if there is no path from y to x. Harmonic centrality can be normalized by dividing by $N-1$, where N is the number of nodes in the graph.

Harmonic centrality was proposed independently by Dekker (2005), using the name "valued centrality,"[15] and by Rochat (2009).[16]

21.5 Betweenness centrality

Main article: Betweenness centrality

Betweenness is a centrality measure of a vertex within a graph (there is also edge betweenness, which is not discussed here). Betweenness centrality quantifies the number of times a node acts as a bridge along the shortest path between two other nodes. It was introduced as a measure for quantifying the control of a human on the communication between other humans in a social network by Linton Freeman[17] In his conception, vertices that have a high probability to occur on a randomly chosen shortest path between two randomly chosen vertices have a high betweenness.

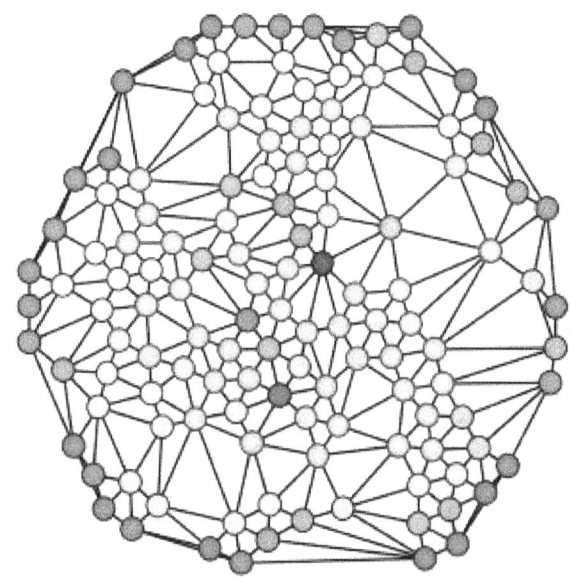

Hue (from red = 0 to blue = max) shows the node betweenness.

The betweenness of a vertex v in a graph $G := (V, E)$ with V vertices is computed as follows:

1. For each pair of vertices (s,t), compute the shortest paths between them.

2. For each pair of vertices (s,t), determine the fraction of shortest paths that pass through the vertex in question (here, vertex v).

3. Sum this fraction over all pairs of vertices (s,t).

More compactly the betweenness can be represented as:[18]

$$C_B(v) = \sum_{s \neq v \neq t \in V} \frac{\sigma_{st}(v)}{\sigma_{st}}$$

where σ_{st} is total number of shortest paths from node s to node t and $\sigma_{st}(v)$ is the number of those paths that pass through v. The betweenness may be normalised by dividing through the number of pairs of vertices not including v, which for directed graphs is $(n-1)(n-2)$ and for undirected graphs is $(n-1)(n-2)/2$. For example, in an undirected star graph, the center vertex (which is contained in every possible shortest path) would have a betweenness of $(n-1)(n-2)/2$ (1, if normalised) while the leaves (which are contained in no shortest paths) would have a betweenness of 0.

From a calculation aspect, both betweenness and closeness centralities of all vertices in a graph involve calculating the shortest paths between all pairs of vertices on a graph, which requires $\Theta(V^3)$ time with the Floyd–Warshall algorithm. However, on sparse graphs, Johnson's algorithm may be more efficient, taking $O(V^2 \log V + VE)$ time. In the case of unweighted graphs the calculations can be done with Brandes' algorithm[18] which takes $O(VE)$ time. Normally, these algorithms assume that graphs are undirected and connected with the allowance of loops and multiple edges. When specifically dealing with network graphs, often graphs are without loops or multiple edges to maintain simple relationships (where edges represent connections between two people or vertices). In this case, using Brandes' algorithm will divide final centrality scores by 2 to account for each shortest path being counted twice.[18]

21.6 Eigenvector centrality

Main article: Eigenvector centrality

Eigenvector centrality (also called **eigencentrality**) is a measure of the influence of a node in a network. It assigns relative scores to all nodes in the network based on the concept that connections to high-scoring nodes contribute more to the score of the node in question than equal connections to low-scoring nodes. Google's PageRank and the Katz centrality are variants of the eigenvector centrality.[19]

21.6.1 Using the adjacency matrix to find eigenvector centrality

For a given graph $G := (V, E)$ with $|V|$ number of vertices let $A = (a_{v,t})$ be the adjacency matrix, i.e. $a_{v,t} = 1$ if vertex v is linked to vertex t, and $a_{v,t} = 0$ otherwise. The relative centrality score of vertex v can be defined as:

$$x_v = \frac{1}{\lambda} \sum_{t \in M(v)} x_t = \frac{1}{\lambda} \sum_{t \in G} a_{v,t} x_t$$

where $M(v)$ is a set of the neighbors of v and λ is a constant. With a small rearrangement this can be rewritten in vector notation as the eigenvector equation

$$\mathbf{Ax} = \lambda \mathbf{x}$$

In general, there will be many different eigenvalues λ for which a non-zero eigenvector solution exists. Since the entries in the adjacency matrix are non-negative, there is a unique largest eigenvalue, which is real and positive, by the Perron–Frobenius theorem. This greatest eigenvalue results in the desired centrality measure.[20] The vth component of the related eigenvector then gives the relative centrality score of the vertex v in the network. The eigenvector is only defined up to a common factor, so only the ratios of the centralities of the vertices are well defined. To define an absolute score one must normalise the eigen vector e.g. such that the sum over all vertices is 1 or the total number of vertices n. Power iteration is one of many eigenvalue algorithms that may be used to find this dominant eigenvector.[19] Furthermore, this can be generalized so that the entries in A can be real numbers representing connection strengths, as in a stochastic matrix.

21.7 Katz centrality

Main article: Katz centrality

Katz centrality[21] is a generalization of degree centrality. Degree centrality measures the number of direct neighbors, and Katz centrality measures the number of all nodes that can be connected through a path, while the contributions of distant nodes are penalized. Mathematically, it is defined as $x_i = \sum_{k=1}^{\infty} \sum_{j=1}^{N} \alpha^k (A^k)_{ji}$ where α is an attenuation factor in $(0, 1)$.

Katz centrality can be viewed as a variant of eigenvector centrality. Another form of Katz centrality is $x_i = \alpha \sum_{j=1}^{N} a_{ij}(x_j + 1)$. Compared to the expression of eigenvector centrality, x_j is replaced by $x_j + 1$.

It is shown that[22] the principal eigenvector (associated with the largest eigenvalue of A, the adjacency matrix) is the limit of Katz centrality as α approaches $1/\lambda$ from below.

21.8 PageRank centrality

Main article: PageRank

PageRank satisfies the following equation

$x_i = \alpha \sum_j a_{ji} \frac{x_j}{L(j)} + \frac{1-\alpha}{N}$, where $L(j) = \sum_i a_{ji}$ is the number of neighbors of node j (or number of outbound links in a directed graph). Compared to eigenvector centrality and Katz centrality, one major difference is the scaling factor $L(j)$. Another difference between PageRank and eigenvector centrality is that the PageRank vector is a left hand eigenvector (note the factor a_{ji} has indices

21.9 Percolation centrality

A slew of centrality measures exist to determine the 'importance' of a single node in a complex network. However, these measures quantify the importance of a node in purely topological terms, and the value of the node does not depend on the 'state' of the node in any way. It remains constant regardless of network dynamics. This is true even for the weighted betweenness measures. However, a node may very well be centrally located in terms of betweenness centrality or another centrality measure, but may not be 'centrally' located in the context of a network in which there is percolation. Percolation of a 'contagion' occurs in complex networks in a number of scenarios. For example, viral or bacterial infection can spread over social networks of people, known as contact networks. The spread of disease can also be considered at a higher level of abstraction, by contemplating a network of towns or population centres, connected by road, rail or air links. Computer viruses can spread over computer networks. Rumours or news about business offers and deals can also spread via social networks of people. In all of these scenarios, a 'contagion' spreads over the links of a complex network, altering the 'states' of the nodes as it spreads, either recoverably or otherwise. For example, in an epidemiological scenario, individuals go from 'susceptible' to 'infected' state as the infection spreads. The states the individual nodes can take in the above examples could be binary (such as received/not received a piece of news), discrete (susceptible/infected/recovered), or even continuous (such as the proportion of infected people in a town), as the contagion spreads. The common feature in all these scenarios is that the spread of contagion results in the change of node states in networks. Percolation centrality (PC) was proposed with this in mind, which specifically measures the importance of nodes in terms of aiding the percolation through the network. This measure was proposed by Piraveenan et al.[24]

The Percolation Centrality is defined for a given node, at a given time, as the proportion of 'percolated paths' that go through that node. A 'percolated path' is a shortest path between a pair of nodes, where the source node is percolated (e.g., infected). The target node can be percolated or non-percolated, or in a partially percolated state.

$$PC^t(v) = \frac{1}{N-2} \sum_{s \neq v \neq r} \frac{\sigma_{sr}(v)}{\sigma_{sr}} \frac{x^t_s}{\sum [x^t_i] - x^t_v}$$

where σ_{sr} is total number of shortest paths from node s to node r and $\sigma_{sr}(v)$ is the number of those paths that pass through v. The percolation state of the node i at time t is denoted by x^t_i and two special cases are when $x^t_i = 0$ which indicates a non-percolated state at time t whereas when $x^t_i = 1$ which indicates a fully percolated state at time t. The values in between indicate partially percolated states (e.g., in a network of townships, this would be the percentage of people infected in that town).

The attached weights to the percolation paths depend on the percolation levels assigned to the source nodes, based on the premise that the higher the percolation level of a source node is, the more important are the paths that originate from that node. Nodes which lie on shortest paths originating from highly percolated nodes are therefore potentially more important to the percolation. The definition of PC may also be extended to include target node weights as well. Percolation centrality calculations run in $O(NM)$ time with an efficient implementation adopted from Brandes' fast algorithm and if the calculation needs to consider target nodes weights, the worst case time is $O(N^3)$.

21.10 Cross-clique centrality

Cross-clique centrality of a single node, in a complex graph determines the connectivity of a node to different cliques. A node with high cross-clique connectivity facilitates the propagation of information or disease in a graph. Cliques are subgraphs in which every node is connected to every other node in the clique. The cross-clique connectivity of a node v for a given graph $G := (V, E)$ with $|V|$ vertices and $|E|$ edges, is defined as $X(v)$ where $X(v)$ is the number of cliques to which vertex v belongs. This measure was used in [25] but was first proposed by Everett and Borgatti in 1998 where they called it clique-overlap centrality.

21.11 Freeman Centralization

The *centralization* of any network is a measure of how central its most central node is in relation to how central all the other nodes are.[6] Centralization measures then (a) calculate the sum in differences in centrality between the most central node in a network and all other nodes; and (b) divide this quantity by the theoretically largest such sum of differences in any network of the same size.[6] Thus, every centrality measure can have its own centralization measure. Defined formally, if $C_x(p_i)$ is any centrality measure of point i, if $C_x(p_*)$ is the largest such measure in the network, and if $\max \sum_{i=1}^{N} C_x(p_*) - C_x(p_i)$ is the largest sum of differences in point centrality C_x for any graph with the same number of nodes, then the centralization of the network is:[6] $C_x = \frac{\sum_{i=1}^{N} C_x(p_*) - C_x(p_i)}{\max \sum_{i=1}^{N} C_x(p_*) - C_x(p_i)}$

21.12 Dissimilarity based centrality measures

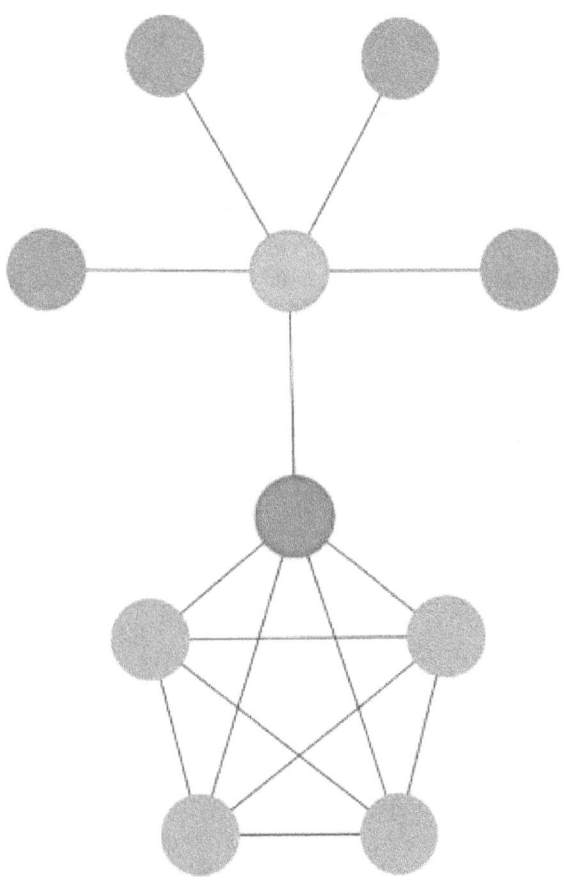

In the illustrated network, green and red nodes are the most dissimilar because they do not share neighbors between them. So, the green one contributes more to the centrality of the red one than the gray ones, because the red one can access to the blue ones only through the green, and the gray nodes are redundant for the red one, because it can access directly to each gray node without any intermediary.

In order to obtain better results in the ranking of the nodes of a given network, in [26] are used dissimilarity measures (specific to the theory of classification and data mining) to enrich the centrality measures in complex networks. This is illustrated with the Eigenvector centrality, calculating the centrality of each node through the solution of the eigenvalue problem

$$W\mathbf{c} = \lambda \mathbf{c}$$

where $W_{ij} = A_{ij} D_{ij}$ (coordinate-to-coordinate product) and D_{ij} is an arbitrary dissimilarity matrix, defined through a dissimilarity measure, e.g., Jaccard dissimilarity given by

$$D_{ij} = 1 - \frac{|V^+(i) \cap V^+(j)|}{|V^+(i) \cup V^+(j)|}$$

Where this measure permits us to quantify the topological contribution (which is why is called contribution centrality) of each node to the centrality of a given node, having more weight/relevance those nodes with greater dissimilarity, since these allow to the given node access to nodes that which themselves can not access directly.

Is noteworthy that W is non-negative because A and D are non-negative matrices, so we can use the Perron–Frobenius theorem to ensure that the above problem has a unique solution for $\lambda = \lambda max$ with $\mathbf{c}$ non-negative, allowing us to infer the centrality of each node in the network. Therefore, the centrality of the i-th node is

$$c_i = \frac{1}{n} \sum_{j=1}^{n} W_{ij} c_j, \quad j = 1, \cdots, n$$

where n is the number of the nodes in the network. Several dissimilarity measures and networks where tested in [27] obtaining improved results in the studied cases.

21.13 Extensions

Empirical and theoretical research have extended the concept of centrality in the context of static networks to dynamic centrality[28] in the context of time-dependent and temporal networks.[29][30][31]

For generalizations to weighted networks, see Opsahl et al. (2010).[32]

The concept of centrality was extended to a group level as well. For example, **group betweenness** centrality shows the proportion of geodesics connecting pairs of non-group members that pass through the group.[33][34]

21.14 See also

- Alpha centrality
- Core-Periphery Structures in Networks
- Distance in graphs

21.15 Notes and references

[1] Newman, M.E.J. 2010. *Networks: An Introduction*. Oxford, UK: Oxford University Press.

[2] Bonacich, Phillip (1987). "Power and Centrality: A Family of Measures". *American Journal of Sociology*. University of Chicago Press. **92**: 1170–1182. doi:10.1086/228631.

[3] Borgatti, Stephen P. (2005). "Centrality and Network Flow". *Social Networks*. Elsevier. **27**: 55–71. doi:10.1016/j.socnet.2004.11.008.

[4] Borgatti, Stephen P.; Everett, Martin G. (2006). "A Graph-Theoretic Perspective on Centrality". *Social Networks*. Elsevier. **28**: 466–484. doi:10.1016/j.socnet.2005.11.005.

[5] Benzi, Michele; Klymko, Christine (2013). "A matrix analysis of different centrality measures". *SIAM Journal on Matrix Analysis and Applications*. **36**: 686–706. arXiv:1312.6722. doi:10.1137/130950550.

[6] Freeman, Linton C. (1979), "centrality in social networks: Conceptual clarification" (PDF). *Social Networks*, **1** (3): 215–239, doi:10.1016/0378-8733(78)90021-7

[7] Lawyer, Glenn (2015). "Understanding the spreading power of all nodes in a network: a continuous-time perspective". *Sci Rep*. **5**: 8665. PMC 4345333. PMID 25727453. doi:10.1038/srep08665.

[8] da Silva, Renato; Viana, Matheus; da F. Costa, Luciano (2012). "Predicting epidemic outbreak from individual features of the spreaders". *J. Stat Mech Theor Exp*. **2012** (07): P07005. doi:10.1088/1742-5468/2012/07/p07005.

[9] Bauer, Frank; Lizier, Joseph (2012). "Identifying influential spreaders and efficiently estimating infection numbers in epidemic models: A walk counting approach". *Europhys Lett*. **99** (6): 68007. doi:10.1209/0295-5075/99/68007.

[10] Sikic, Mile; Lancic, Alen; Antulov-Fantulin, Nino; Stefanic, Hrvoje (2013). "Epidemic centrality -- is there an underestimated epidemic impact of network peripheral nodes?". *The European Physical Journal B*. **86** (10): 1–13. doi:10.1140/epjb/e2013-31025-5.

[11] Ghoshal, G.; Barabsi, A L (2011). "Ranking stability and super-stable nodes in complex networks.". *Nat Commun*. **2** (394).

[12] Freeman, Linton C. "Centrality in social networks conceptual clarification." Social networks 1.3 (1979): 215–239.

[13] Alex Bavelas. Communication patterns in task-oriented groups. *J. Acoust. Soc. Am*, **22**(6):725–730, 1950.

[14] Sabidussi, G (1966). "The centrality index of a graph". *Psychometrika*. **31**: 581–603. doi:10.1007/bf02289527.

[15] Dekker, Anthony (2005). "Conceptual Distance in Social Network Analysis". *Journal of Social Structure*. **6** (3).

[16] Yannick Rochat. *Closeness centrality extended to unconnected graphs: The harmonic centrality index* (PDF). Applications of Social Network Analysis, ASNA 2009.

[17] Freeman, Linton (1977). "A set of measures of centrality based upon betweenness". *Sociometry*. **40**: 35–41. doi:10.2307/3033543.

[18] Brandes, Ulrik (2001). "A faster algorithm for betweenness centrality" (PDF). *Journal of Mathematical Sociology*. **25**: 163–177. doi:10.1080/0022250x.2001.9990249. Retrieved October 11, 2011.

[19] http://www.ams.org/samplings/feature-column/fcarc-pagerank

[20] M. E. J. Newman. "The mathematics of networks" (PDF). Retrieved 2006-11-09.

[21] Katz, L. 1953. A New Status Index Derived from Sociometric Index. Psychometrika, 39–43.

[22] Bonacich, P (1991). "Simultaneous group and individual centralities". *Social Networks*. **13**: 155–168. doi:10.1016/0378-8733(91)90018-o.

[23] How does Google rank webpages? Archived January 31, 2012, at the Wayback Machine. 20Q: About Networked Life

[24] Piraveenan, Mahendra (2013). "Percolation Centrality: Quantifying Graph-Theoretic Impact of Nodes during Percolation in Networks". *PLOS ONE*. **8** (1): e53095. PMC 3551907. PMID 23349699. doi:10.1371/journal.pone.0053095.

[25] Faghani, Mohamamd Reza (2013). "A Study of XSS Worm Propagation and Detection Mechanisms in Online Social Networks". *IEEE Trans. Inf. Forensics and Security*.

[26] Alvarez-Socorro, A. J.; Herrera-Almarza, G. C.; González-Díaz, L. A. (2015-11-25). "Eigencentrality based on dissimilarity measures reveals central nodes in complex networks". *Scientific Reports*. **5**: 17095. PMC 4658528. PMID 26603652. doi:10.1038/srep17095.

[27] Alvarez-Socorro, A.J.; Herrera-Almarza; González-Díaz, L. A. "Supplementary Information for Eigencentrality based on dissimilarity measures reveals central nodes in complex networks" (PDF). Nature Publishing Group.

[28] Braha, D.; Bar-Yam, Y. (2006). "From Centrality to Temporary Fame: Dynamic Centrality in Complex Networks". *Complexity*. **12**: 59–63. doi:10.1002/cplx.20156.

[29] Hill, S.A.; Braha, D. (2010). "Dynamic Model of Time-Dependent Complex Networks". *Physical Review E*. **82**: 046105. doi:10.1103/physreve.82.046105.

[30] Gross, T. and Sayama, H. (Eds.). 2009. *Adaptive Networks: Theory, Models and Applications*. Springer.

[31] Holme, P. and Saramäki, J. 2013. *Temporal Networks*. Springer.

[32] Opsahl, Tore; Agneessens, Filip; Skvoretz, John (2010). "Node centrality in weighted networks: Generalizing degree and shortest paths". *Social Networks*. **32** (3): 245–251. doi:10.1016/j.socnet.2010.03.006.

[33] Everett, M. G. and Borgatti, S. P. (2005). Extending centrality. In P. J. Carrington, J. Scott and S. Wasserman (Eds.), *Models and methods in social network analysis* (pp. 57–76). New York: Cambridge University Press.

[34] Puzis, R., Yagil, D., Elovici, Y., Braha, D. (2009).Collaborative attack on Internet users' anonymity. *Internet Research* **19**(1)

21.16 Further reading

- Koschützki, D.; Lehmann, K. A.; Peeters, L.; Richter, S.; Tenfelde-Podehl, D. and Zlotowski, O. (2005) Centrality Indices. In Brandes, U. and Erlebach, T. (Eds.) *Network Analysis: Methodological Foundations*, pp. 16–61, LNCS 3418, Springer-Verlag.

Chapter 22

Network motif

All networks, including biological networks, social networks, technological networks (e.g., computer networks and electrical circuits) and more, can be represented as graphs, which include a wide variety of subgraphs. One important local property of networks are so-called **network motifs**, which are defined as recurrent and statistically significant sub-graphs or patterns.

Network motifs are sub-graphs that repeat themselves in a specific network or even among various networks. Each of these sub-graphs, defined by a particular pattern of interactions between vertices, may reflect a framework in which particular functions are achieved efficiently. Indeed, motifs are of notable importance largely because they may reflect functional properties. They have recently gathered much attention as a useful concept to uncover structural design principles of complex networks.[1] Although network motifs may provide a deep insight into the network's functional abilities, their detection is computationally challenging.

22.1 Definition

Let $G = (V, E)$ and $G' = (V', E')$ be two graphs. Graph G' is a *sub-graph* of graph G (written as $G' \subseteq G$) if $V' \subseteq V$ and $E' \subseteq E \cap (V' \times V')$. If $G' \subseteq G$ and G' contains all of the edges $\langle u, v \rangle \in E$ with $u, v \in V'$, then G' is an *induced sub-graph* of G. We call G' and G isomorphic (written as $G' \leftrightarrow G$), if there exists a bijection (one-to-one) $f: V' \rightarrow V$ with $\langle u, v \rangle \in E' \Leftrightarrow \langle f(u), f(v) \rangle \in E$ for all $u, v \in V'$. The mapping f is called an isomorphism between G and G'.[2]

When $G'' \subset G$ and there exists an isomorphism between the sub-graph G'' and a graph G', this mapping represents an *appearance* of G' in G. The number of appearances of graph G' in G is called the frequency F_G of G' in G. A graph is called *recurrent* (or *frequent*) in G, when its *frequency* $F_G(G')$ is above a predefined threshold or cut-off value. We use terms *pattern* and *frequent sub-graph* in this review interchangeably. There is an ensemble $\Omega(G)$ of random graphs corresponding to the null-model associated to G. We should choose N random graphs uniformly from $\Omega(G)$ and calculate the frequency for a particular frequent sub-graph G' in G. If the frequency of G' in G is higher than its arithmetic mean frequency in N random graphs R_i, where $1 \leq i \leq N$, we call this recurrent pattern *significant* and hence treat G' as a *network motif* for G. For a small graph G', the network G and a set of randomized networks $R(G) \subseteq \Omega(R)$, where , the *Z-Score* that has been defined by the following formula:

$$Z(G') = \frac{F_G(G') - \mu_R(G')}{\sigma_R(G')}$$

where $\mu R(G')$ and $\sigma R(G')$ stand for mean and standard deviation frequency in set $R(G)$, respectively.[3][4][5][6][7][8] The larger the $Z(G')$, the more significant is the sub-graph G' as a motif. Alternatively, another measurement in statistical hypothesis testing that can be considered in motif detection is the P-Value, given as the probability of $F_R(G') \geq F_G(G')$ (as its null-hypothesis), where $F_R(G')$ indicates the frequency of G' in a randomized network.[6] A sub-graph with P-value less than a threshold (commonly 0.01 or 0.05) will be treated as a significant pattern. The P-value is defined as

$$P(G') = \frac{1}{N} \sum_{i=1}^{N} \delta(c(i)); c(i) : F_R^i(G') \geq F_G(G')$$

Where N indicates number of randomized networks, i is defined over an ensemble of randomized networks and the Kronecker delta function $\delta(c(i))$ is one if the condition $c(i)$ holds. The concentration [9][10] of a particular n-size sub-graph G' in network G refers to the ratio of the sub-graph appearance in the network to the total *n*-size non-isomorphic sub-graphs' frequencies, which is formulated by

$$C_G(G') = \frac{F_G(G')}{\sum_i F_G(G_i)}$$

where index i is defined over the set of all non-isomorphic n-size graphs. Another statistical measurement is defined for evaluating network motifs, but it is rarely used in known algorithms. This measurement is introduced by Picard *et al.* in 2008 and used the Poisson distribution, rather than the Gaussian normal distribution that is implicitly being used above.[11]

In addition, three specific concepts of sub-graph frequency

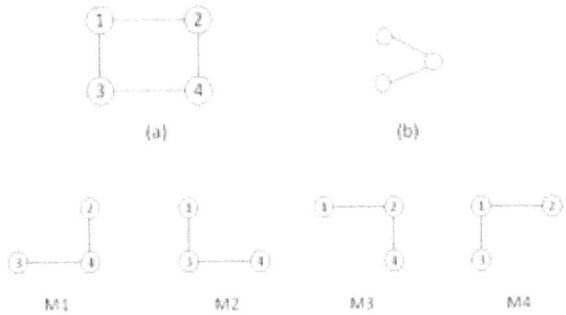

Different occurrences of a sub-graph in a graph. (M1 – M4) are different occurrences of sub-graph (b) in graph (a). For frequency concept F_1, the set M1, M2, M3, M4 represent all matches, so $F_1 = 4$. For F_2, one of the two set M1, M4 or M2, M3 are possible matches, $F_2 = 2$. Finally, for frequency concept F_3, merely one of the matches (M1 to M4) is allowed, therefore $F_3 = 1$. The frequency of these three frequency concepts decrease as the usage of network elements are restricted.

have been proposed.[12] As figure illustrates, the first frequency concept F_1 considers all matches of a graph in original network. This definition is similar to what we have introduced above. The second concept F_2 is defined as the maximum number of edge-disjoint instances of a given graph in original network. And finally, the frequency concept F_3 entails matches with disjoint edges and nodes. Therefore, the two concepts F_2 and F_3 restrict the usage of elements of the graph, and as can be inferred, the frequency of a sub-graph declines by imposing restrictions on network element usage. As a result, a network motif detection algorithm would pass over more candidate sub-graphs if we insist on frequency concepts F_2 and F_3.

22.2 History

This idea was first presented in 2002 by Uri Alon and his group [13] when network motifs were discovered in the gene regulation (transcription) network of the bacteria E. coli and then in a large set of natural networks. Since then, a considerable number of studies have been conducted on the subject. Some of these studies focus on the biological applications, while others focus on the computational theory of network motifs.

The biological studies endeavor to interpret the motifs detected for biological networks. For example, in work following,[13] the network motifs found in E. coli were discovered in the transcription networks of other bacteria[14] as well as yeast[15][16] and higher organisms.[17][18][19] A distinct set of network motifs were identified in other types of biological networks such as neuronal networks and protein interaction networks.[5][20][21]

The computational research has focused on improving existing motif detection tools to assist the biological investigations and allow larger networks to be analyzed. Several different algorithms have been provided so far, which are elaborated in the next section in chronological order.

Most recently, the acc-MOTIF tool to detect network motifs was released.[22]

22.3 Motif Discovery Algorithms

Various solutions have been proposed for the challenging problem of motif discovery. These algorithms can be classified under various paradigms such as exact counting methods, sampling methods, pattern growth methods and so on. However, motif discovery problem comprises two main steps: first, calculating the number of occurrences of a sub-graph and then, evaluating the sub-graph significance. The recurrence is significant if it is detectably far more than expected. Roughly speaking, the expected number of appearances of a sub-graph can be determined by a Null-model, which is defined by an ensemble of random networks with some of the same properties as the original network.

Here, a review on computational aspects of major algorithms is given and their related benefits and drawbacks from an algorithmic perspective are discussed.

22.3.1 mfinder

mfinder, the first motif-mining tool, implements two kinds of motif finding algorithms: a full enumeration and a sampling method. Until 2004, the only exact counting method for NM (network motif) detection was the brute-force one proposed by Milo *et al.*.[3] This algorithm was successful for discovering small motifs, but using this method for finding even size 5 or 6 motifs was not computationally feasible. Hence, a new approach to this problem was needed.

Kashtan *et al.* [9] presented the first sampling NM discovery algorithm, which was based on *edge sampling* throughout the network. This algorithm estimates concentrations of induced sub-graphs and can be utilized for motif discovery in directed or undirected networks. The sampling procedure of the algorithm starts from an arbitrary edge of the network that leads to a sub-graph of size two, and then expands the sub-graph by choosing a random edge that is incident to the current sub-graph. After that, it continues choosing random neighboring edges until a sub-graph of size n is obtained. Finally, the sampled sub-graph is expanded to include all of the edges that exist in the network between these n nodes. When an algorithm uses a sampling approach, taking unbiased samples is the most important issue that the algorithm might address. The sampling procedure, however, does not take samples uniformly and therefore Kashtan et al. proposed a weighting scheme that assigns different weights to the different sub-graphs within network.[9] The underlying principle of weight allocation is exploiting the information of the sampling probability for each sub-graph, i.e. the probable sub-graphs will obtain comparatively less weights in comparison to the improbable sub-graphs; hence, the algorithm must calculate the sampling probability of each sub-graph that has been sampled. This weighting technique assists *mfinder* to determine sub-graph concentrations impartially.

In expanded to include sharp contrast to exhaustive search, the computational time of the algorithm surprisingly is asymptotically independent of the network size. An analysis of the computational time of the algorithm has shown that it takes $O(n^n)$ for each sample of a sub-graph of size n from the network. On the other hand, there is no analysis in [9] on the classification time of sampled sub-graphs that requires solving the *graph isomorphism* problem for each sub-graph sample. Additionally, an extra computational effort is imposed on the algorithm by the sub-graph weight calculation. But it is unavoidable to say that the algorithm may sample the same sub-graph multiple times – spending time without gathering any information.[10] In conclusion, by taking the advantages of sampling, the algorithm performs more efficiently than an exhaustive search algorithm; however, it only determines sub-graphs concentrations approximately. This algorithm can find motifs up to size 6 because of its main implementation, and as result it gives the most significant motif, not all the others too. Also, it is necessary to mention that this tool has no option of visual presentation. The sampling algorithm is shown briefly:

22.3.2 FPF (Mavisto)

Schreiber and Schwöbbermeyer [12] proposed an algorithm named *flexible pattern finder (FPF)* for extracting frequent sub-graphs of an input network and implemented it in a system named *Mavisto*.[23] Their algorithm exploits the *downward closure* property which is applicable for frequency concepts F_2 and F_3. The downward closure property asserts that the frequency for sub-graphs decrease monotonically by increasing the size of sub-graphs; however, this property does not hold necessarily for frequency concept F_1. FPF is based on a *pattern tree* (see figure) consisting of nodes that represents different graphs (or patterns), where the parent of each node is a sub-graph of its children nodes; in other words, the corresponding graph of each pattern tree's node is expanded by adding a new edge to the graph of its parent node.

At first, the FPF algorithm enumerates and maintains the in-

Illustration of the pattern tree in FPF algorithm.[12]

formation of all matches of a sub-graph located at the root of the pattern tree. Then, one-by-one it builds child nodes of the previous node in the pattern tree by adding one edge supported by a matching edge in the target graph, and tries to expand all of the previous information about matches to the new sub-graph (child node). In next step, it decides whether the frequency of the current pattern is lower than a predefined threshold or not. If it is lower and if downward closure holds, FPF can abandon that path and not traverse further in this part of the tree; as a result, unnecessary computation is avoided. This procedure is continued until there is no remaining path to traverse.

The advantage of the algorithm is that it does not consider infrequent sub-graphs and tries to finish the enumeration process as soon as possible; therefore, it only spends time for promising nodes in the pattern tree and discards all other nodes. As an added bonus, the pattern tree notion permits FPF to be implemented and executed in a parallel manner since it is possible to traverse each path of the pattern tree independently. However, FPF is most useful for frequency concepts F_2 and F_3, because downward closure is not applicable to F_1. Nevertheless, the pattern tree is still practical for F_1 if the algorithm runs in parallel. Another advantage of the algorithm is that the implementation of this algorithm has no limitation on motif size, which makes it more amenable to improvements. The pseudocode of FPF (Mavisto) is shown below:

22.3.3 ESU (FANMOD)

The sampling bias of Kashtan et al. [9] provided great impetus for designing better algorithms for the NM discovery problem. Although Kashtan et al. tried to settle this drawback by means of a weighting scheme, this method imposed an undesired overhead on the running time as well a more complicated implementation. This tool is one of the most useful ones, as it supports visual options and also is an efficient algorithm with respect to time. But, it has a limitation on motif size as it does not allow searching for motifs of size 9 or higher because of the way the tool is implemented.

Wernicke [10] introduced an algorithm named *RAND-ESU* that provides a significant improvement over *mfinder*.[9] This algorithm, which is based on the exact enumeration algorithm *ESU*, has been implemented as an application called *FANMOD*.[10] *RAND-ESU* is a NM discovery algorithm applicable for both directed and undirected networks, effectively exploits an unbiased node sampling throughout the network, and prevents overcounting sub-graphs more than once. Furthermore, *RAND-ESU* uses a novel analytical approach called *DIRECT* for determining sub-graph significance instead of using an ensemble of random networks as a Null-model. The *DIRECT* method estimates the sub-graph concentration without explicitly generating random networks.[10] Empirically, the DIRECT method is more efficient in comparison with the random network ensemble in case of sub-graphs with a very low concentration; however, the classical Null-model is faster than the *DIRECT* method for highly concentrated sub-graphs.[3][10] In the following, we detail the *ESU* algorithm and then we show how this exact algorithm can be modified efficiently to *RAND-ESU* that estimates sub-graphs concentrations.

The algorithms *ESU* and *RAND-ESU* are fairly simple, and hence easy to implement. *ESU* first finds the set of all induced sub-graphs of size k, let S_k be this set. *ESU* can be implemented as a recursive function; the running of this function can be displayed as a tree-like structure of depth k, called the ESU-Tree (see figure). Each of the ESU-Tree nodes indicate the status of the recursive function that entails two consecutive sets SUB and EXT. SUB refers to nodes in the target network that are adjacent and establish a partial sub-graph of size |SUB| ≤ k. If |SUB| = k, the algorithm has found an induced complete sub-graph, so S_k = SUB ∪ S_k. However, if |SUB| < k, the algorithm must expand SUB to achieve cardinality k. This is done by the EXT set that contains all the nodes that satisfy two conditions: First, each of the nodes in EXT must be adjacent to at least one of the nodes in SUB; second, their numerical labels must be larger than the label of first element in SUB. The first condition makes sure that the expansion of SUB nodes yields a connected graph and the second condition causes ESU-Tree leaves (see figure) to be distinct; as a result, it prevents overcounting. Note that, the EXT set is not a static set, so in each step it may expand by some new nodes that do not breach the two conditions. The next step of ESU involves classification of sub-graphs placed in the ESU-Tree leaves into non-isomorphic size-k graph classes; consequently, ESU determines sub-graphs frequencies and concentrations. This stage has been implemented simply by employing McKay's *nauty* algorithm,[24][25] which classifies each sub-graph by performing a graph isomorphism test. Therefore, ESU finds the set of all induced k-size sub-graphs in a target graph by a recursive algorithm and then determines their frequency using an efficient tool.

How has the exact *ESU* been algorithm modified to *RAND-ESU* that estimates sub-graph concentrations? The procedure of implementing *RAND-ESU* is quite straightforward and is one of the main advantages of *FANMOD*. One can change the *ESU* algorithm to explore just a portion of the ESU-Tree leaves by applying a probability value $0 \leq p_d \leq 1$ for each level of the ESU-Tree and oblige *ESU* to traverse each child node of a node in level d-1 with probability p_d. This new algorithm is called *RAND-ESU*. Evidently, when $p_d = 1$ for all levels, *RAND-ESU* acts like *ESU*. For $p_d = 0$ the algorithm finds nothing. Note that, this procedure ensures that the chances of visiting each leaf of the ESU-Tree are the same, resulting in *unbiased* sampling of sub-graphs through the network. The probability of visiting each leaf is $\prod_d p_d$ and this is identical for all of the ESU-Tree leaves; therefore, this method guarantees unbiased sampling of sub-graphs from the network. Nonetheless, determining the value of p_d for $1 \leq d \leq k$ is another issue that must be determined manually by an expert to get precise results of sub-graph concentrations.[8] While there is no lucid prescript for this matter, the Wernicke provides some general observations that may help in determining p_d values. In summary, *RAND-ESU* is a very fast algorithm for NM discovery in the case of induced sub-graphs supporting unbiased sampling method. Although, the main *ESU* algorithm and so the *FANMOD* tool is known for discovering induced sub-graphs, there is trivial modification to *ESU* which makes it possible for finding non-induced sub-graphs, too. The pseudo code of *ESU (FANMOD)* is shown below:

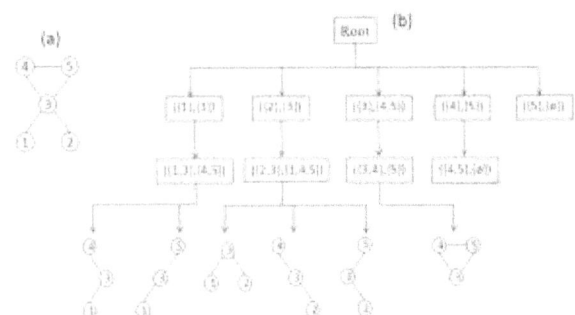

(a) A target graph of size 5, (b) the ESU-tree of depth k that is associated to the extraction of sub-graphs of size 3 in the target graph. *Leaves correspond to set S3 or all of the size-3 induced sub-graphs of the target graph (a). Nodes in the ESU-tree include two adjoining sets, the first set contains adjacent nodes called SUB and the second set named EXT holds all nodes that are adjacent to at least one of the SUB nodes and where their numerical labels are larger than the SUB nodes labels. The EXT set is utilized by the algorithm to expand a SUB set until it reaches a desired sub-graph size that are placed at the lowest level of ESU-Tree (or its leaves).*

22.3.4 NeMoFinder

Chen et al. [26] introduced a new NM discovery algorithm called *NeMoFinder*, which adapts the idea in *SPIN* [27] to extract frequent trees and after that expands them into non-isomorphic graphs.[8] *NeMoFinder* utilizes frequent size-n trees to partition the input network into a collection of size-n graphs, afterward finding frequent size-n sub-graphs by expansion of frequent trees edge-by-edge until getting a complete size-n graph K_n. The algorithm finds NMs in undirected networks and is not limited to extracting only induced sub-graphs. Furthermore, *NeMoFinder* is an exact enumeration algorithm and is not based on a sampling method. As Chen et al. claim, *NeMoFinder* is applicable for detecting relatively large NMs, for instance, finding NMs up to size-12 from the whole *S. cerevisiae* (yeast) PPI network as the authors claimed.[28]

NeMoFinder consists of three main steps. First, finding frequent size-n trees, then utilizing repeated size-n trees to divide the entire network into a collection of size-n graphs, finally, performing sub-graph join operations to find frequent size-n sub-graphs.[26] In the first step, the algorithm detects all non-isomorphic size-n trees and mappings from a tree to the network. In the second step, the ranges of these mappings are employed to partition the network into size-n graphs. Up to this step, there is no distinction between *NeMoFinder* and an exact enumeration method. However, a large portion of non-isomorphic size-n graphs still remain. *NeMoFinder* exploits a heuristic to enumerate non-tree size-n graphs by the obtained information from the preceding steps. The main advantage of the algorithm is in the third step, which generates candidate sub-graphs from previously enumerated sub-graphs. This generation of new size-n sub-graphs is done by joining each previous sub-graph with derivative sub-graphs from itself called *cousin sub-graphs*. These new sub-graphs contain one additional edge in comparison to the previous sub-graphs. However, there exist some problems in generating new sub-graphs: There is no clear method to derive cousins from a graph, joining a sub-graph with its cousins leads to redundancy in generating particular sub-graph more than once, and cousin determination is done by a canonical representation of the adjacency matrix which is not closed under join operation. *NeMoFinder* is an efficient network motif finding algorithm for motifs up to size 12 only for protein-protein interaction networks, which are presented as undirected graphs. And it is not able to work on directed networks which are so important in the field of complex and biological networks. The pseudocode of *NeMoFinder* is shown below:

22.3.5 Grochow-Kellis

Grochow and Kellis [29] proposed an *exact* algorithm for enumerating sub-graph appearances. The algorithm is based on a *motif-centric* approach, which means that the frequency of a given sub-graph, called the *query graph*, is exhaustively determined by searching for all possible mappings from the query graph into the larger network. It is claimed [29] that a *motif-centric* method in comparison to *network-centric* methods has some beneficial features. First of all it avoids the increased complexity of sub-graph enumeration. Also, by using mapping instead of enumerating, it enables an improvement in the isomorphism test. To improve the performance of the algorithm, since it is an inefficient exact enumeration algorithm, the authors introduced a fast method which is called *symmetry-breaking conditions*. During straightforward sub-graph isomorphism tests, a sub-graph may be mapped to the same sub-graph of the query graph multiple times. In the Grochow-Kellis (GK) algorithm symmetry-breaking is used to avoid such multiple mappings. Here we introduce the GK algorithm and the symmetry-breaking condition which eliminates redundant isomorphism tests.

(a) graph G. (b) illustration of all automorphisms of G that is showed in (a). From set AutG we can obtain a set of symmetry-breaking conditions of G given by SymG in (c). Only the first mapping in AutG satisfies the SynG conditions; as a result, by applying SymG in the Isomorphism Extension module the algorithm only enumerate each match-able sub-graph in the network to G once. Note that SynG is not necessarily a unique set for an arbitrary graph G.

The GK algorithm discovers the whole set of mappings of a given query graph to the network in two major steps. It starts with the computation of symmetry-breaking conditions of the query graph. Next, by means of a branch-and-bound method, the algorithm tries to find every possible mapping from the query graph to the network that meets the associated symmetry-breaking conditions. An example of the usage of symmetry-breaking conditions in GK algorithm is demonstrated in figure.

As it is mentioned above, the symmetry-breaking technique is a simple mechanism that precludes spending time finding a sub-graph more than once due to its symmetries.[29][30] Note that, computing symmetry-breaking conditions requires finding all automorphisms of a given query graph. Even though, there is no efficient (or polynomial time) algorithm for the graph automorphism problem, this problem can be tackled efficiently in practice by McKay's tools.[24][25] As it is claimed, using symmetry-breaking conditions in NM detection lead to save a great deal of running time. Moreover, it can be inferred from the results in [29][30] that using the symmetry-breaking conditions results in high efficiency particularly for directed networks in comparison to undirected networks. The symmetry-breaking conditions used in the GK algorithm are similar to the restriction which *ESU* algorithm applies to the labels in EXT and SUB sets. In conclusion, the GK algorithm computes the exact number of appearance of a given query graph in a large complex network and exploiting symmetry-breaking conditions improves the algorithm performance. Also, GK algorithm is one of the known algorithms having no limitation for motif size in implementation and potentially it can find motifs of any size.

22.3.6 Color-Coding Approach

Most algorithms in the field of NM discovery are used to find induced sub-graphs of a network. In 2008, Noga Alon *et al.* [31] introduced an approach for finding non-induced sub-graphs too. Their technique works on undirected networks such as PPI ones. Also, it counts non-induced trees and bounded treewidth sub-graphs. This method is applied for sub-graphs of size up to 10.

This algorithm counts the number of non-induced occurrences of a tree T with k = O(logn) vertices in a network G with n vertices as follows:

1. **Color coding.** Color each vertex of input network G independently and uniformly at random with one of the k colors.

2. **Counting.** Apply a dynamic programming routine to count the number of non-induced occurrences of T in which each vertex has a unique color. For more details on this step, see.[31]

3. Repeat the above two steps $O(e^k)$ times and add up the number of occurrences of T to get an estimate on the number of its occurrences in G.

As available PPI networks are far from complete and error free, this approach is suitable for NM discovery for such networks. As Grochow-Kellis Algorithm and this one are the ones popular for non-induced sub-graphs, it is worth to mention that the algorithm introduced by Alon *et al.* is less time consuming than the Grochow-Kellis Algorithm.[31]

22.3.7 MODA

Omidi *et al.* [32] introduced a new algorithm for motif detection named *MODA* which is applicable for induced and non-induced NM discovery in undirected networks. It is based on the motif-centric approach discussed in the Grochow-Kellis algorithm section. It is very important to distinguish motif-centric algorithms such as MODA and GK algorithm because of their ability to work as query-finding algorithms. This feature allows such algorithms to be able to find a single motif query or a small number of motif queries (not all possible sub-graphs of a given size) with larger sizes. As the number of possible non-isomorphic sub-graphs increases exponentially with sub-graph size, for large size motifs (even larger than 10), the network-centric algorithms, those looking for all possible sub-graphs, face a problem. Although motif-centric algorithms also have problems in discovering all possible large size sub-graphs, but their ability to find small numbers of them is sometimes a significant property.

Using a hierarchical structure called an *expansion tree*, the *MODA* algorithm is able to extract NMs of a given size systematically and similar to *FPF* that avoids enumerating unpromising sub-graphs; *MODA* takes into consideration potential queries (or candidate sub-graphs) that would result in frequent sub-graphs. Despite the fact that *MODA* resembles *FPF* in using a tree like structure, the expansion tree is applicable merely for computing frequency concept F_1. As we will discuss next, the advantage of this algorithm is that it does not carry out the sub-graph isomorphism test for *non-tree* query graphs. Additionally, it utilizes a sampling method in order to speed up the running time of the algorithm.

Here is the main idea: by a simple criterion one can generalize a mapping of a k-size graph into the network to its same size supergraphs. For example, suppose there is mapping f(G) of graph G with k nodes into the network and we have a same size graph G' with one more edge ⟨u, v⟩; fG will map G' into the network, if there is an edge ⟨fG(u), fG(v)⟩ in the network. As a result, we can exploit the mapping set of a graph to determine the frequencies of its same order supergraphs simply in O(1) time without carrying out sub-graph isomorphism testing. The algorithm starts ingeniously with minimally connected query graphs of size k and finds their mappings in the network via sub-graph isomorphism. After that, with conservation of the graph size, it expands previously considered query graphs edge-by-edge and computes the frequency of these expanded graphs as mentioned above. The expansion process continues until reaching a complete graph K_k (fully connected with $k(k-1)/2$ edge).

As discussed above, the algorithm starts by computing sub-tree frequencies in the network and then expands sub-trees edge by edge. One way to implement this idea is called the

22.3. MOTIF DISCOVERY ALGORITHMS

expansion tree T_k for each k. Figure shows the expansion tree for size-4 sub-graphs. T_k organizes the running process and provides query graphs in a hierarchical manner. Strictly speaking, the expansion tree T_k is simply a directed acyclic graph or DAG, with its root number k indicating the graph size existing in the expansion tree and each of its other nodes containing the adjacency matrix of a distinct k-size query graph. Nodes in the first level of T_k are all distinct k-size trees and by traversing T_k in depth query graphs expand with one edge at each level. A query graph in a node is a sub-graph of the query graph in a node's child with one edge difference. The longest path in T_k consists of $(k^2-3k+4)/2$ edges and is the path from the root to the leaf node holding the complete graph. Generating expansion trees can be done by a simple routine which is explained in.[32]

MODA traverses T_k and when it extracts query trees from the first level of T_k it computes their mapping sets and saves these mappings for the next step. For non-tree queries from T_k, the algorithm extracts the mappings associated with the parent node in T_k and determines which of these mappings can support the current query graphs. The process will continue until the algorithm gets the complete query graph. The query tree mappings are extracted using the Grochow-Kellis algorithm. For computing the frequency of non-tree query graphs, the algorithm employs a simple routine that takes $O(1)$ steps. In addition, MODA exploits a sampling method where the sampling of each node in the network is linearly proportional to the node degree, the probability distribution is exactly similar to the well-known Barabási-Albert preferential attachment model in the field of complex networks.[33] This approach generates approximations; however, the results are almost stable in different executions since sub-graphs aggregate around highly connected nodes.[34] The pseudocode of MODA is shown below:

22.3.8 Kavosh

A recently introduced algorithm named Kavosh[35] aims at improved main memory usage. Kavosh is usable to detect NM in both directed and undirected networks. The main idea of the enumeration is similar to the GK and MODA algorithms, which first find all k-size sub-graphs that a particular node participated in, then remove the node, and subsequently repeat this process for the remaining nodes.[35]

For counting the sub-graphs of size k that include a particular node, trees with maximum depth of k, rooted at this node and based on neighborhood relationship are implicitly built. Children of each node include both incoming and outgoing adjacent nodes. To descend the tree, a child is chosen at each level with the restriction that a particular child can be included only if it has not been included at any upper level. After having descended to the lowest level possible,

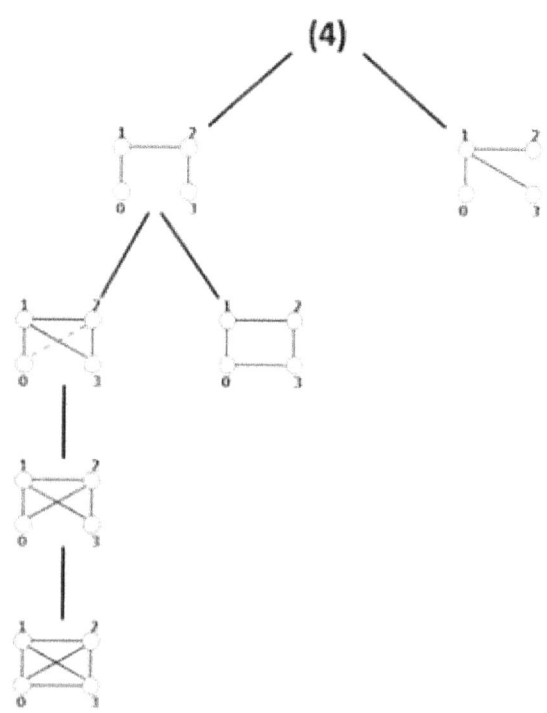

Illustration of the expansion tree T4 for 4-node query graphs. At the first level, there are non-isomorphic k-size trees and at each level, an edge is added to the parent graph to form a child graph. In the second level, there is a graph with two alternative edges that is shown by a dashed red edge. In fact, this node represents two expanded graphs that are isomorphic.[32]

the tree is again ascended and the process is repeated with the stipulation that nodes visited in earlier paths of a descendent are now considered unvisited nodes. A final restriction in building trees is that all children in a particular tree must have numerical labels larger than the label of the root of the tree. The restrictions on the labels of the children are similar to the conditions which GK and ESU algorithm use to avoid overcounting sub-graphs.

The protocol for extracting sub-graphs makes use of the compositions of an integer. For the extraction of sub-graphs of size k, all possible compositions of the integer k-1 must be considered. The compositions of k-1 consist of all possible manners of expressing k-1 as a sum of positive integers. Summations in which the order of the summands differs are considered distinct. A composition can be expressed as $k_2, k_3, \ldots, k_m$ where $k_2 + k_3 + \ldots + k_m = k-1$. To count sub-graphs based on the composition, k_i nodes are selected from the i-th level of the tree to be nodes of the sub-graphs ($i = 2, 3, \ldots, m$). The k-1 selected nodes along with the node at the root define a sub-graph within the network. After discovering a sub-graph involved as a match in the target network, in order to be able to evaluate the size of each class according to the target network, *Kavosh* employs the *nauty* algorithm [24][25] in the same way as *FANMOD*. The enumeration part of Kavosh algorithm is shown below:

Recently a *Cytoscape* plugin called *CytoKavosh* [36] is developed for this software. It is available via *Cytoscape* web page .

22.3.9 G-Tries

In 2010, Pedro Ribeiro and Fernando Silva proposed a novel data structure for storing a collection of sub-graphs, called a *g-trie*.[37] This data structure, which is conceptually akin to a prefix tree, stores sub-graphs according to their structures and finds occurrences of each of these sub-graphs in a larger graph. One of the noticeable aspects of this data structure is that coming to the network motif discovery, the sub-graphs in the main network are needed to be evaluated. So, there is no need to find the ones in random network which are not in the main network. This can be one of the time-consuming parts in the algorithms in which all sub-graphs in random networks are derived.

A *g-trie* is a multiway tree that can store a collection of graphs. Each tree node contains information about a single graph vertex and its corresponding edges to ancestor nodes. A path from the root to a leaf corresponds to one single graph. Descendants of a g-trie node share a common sub-graph. Constructing a *g-trie* is well described in.[37] After constructing a *g-trie*, the counting part takes place. The main idea in counting process is to backtrack by all possible sub-graphs, but at the same time do the isomorphism tests. This backtracking technique is essentially the same technique employed by other motif-centric approaches like *MODA* and *GK* algorithms. Taking advantage of common substructures in the sense that at a given time there is a partial isomorphic match for several different candidate subgraphs.

Among the mentioned algorithms, *G-Tries* is the fastest. But, the excessive use of memory is the drawback of this algorithm, which might limit the size of discoverable motifs by a personal computer with average memory.

22.3.10 Comparison

Tables and figure below show the results of running the mentioned algorithms on different standard networks. These results are taken from the corresponding sources.[32][35][37] thus they should be treated individually.

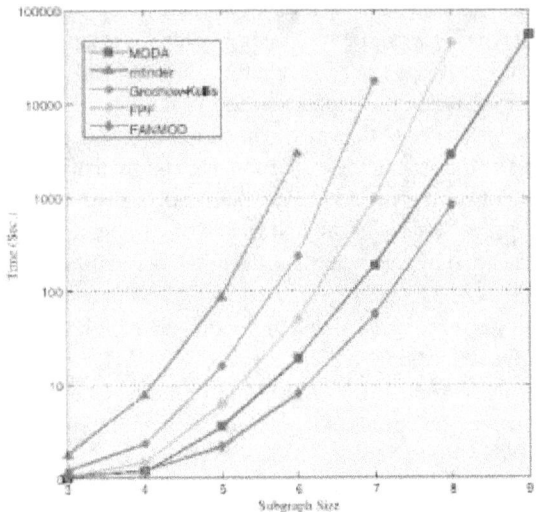

Runtimes of Grochow-Kellis, mfinder, FANMOD, FPF and MODA for subgraphs from three nodes up to nine nodes.[32]

22.3.11 Classification of Algorithms

As seen in the table, motif discovery algorithms can be divided into two general categories: those based on exact counting and those using statistical sampling and estimations instead. Because the second group does not count all the occurrences of a subgraph in the main network, the algorithms belonging to this group are faster, but they might yield in biased and unrealistic results.

In the next level, the exact counting algorithms can be classified to network-centric and subgraph-centric methods. The algorithms of the first class search the given network for all subgraphs of a given size, while the algorithms falling into the second class first generate different possible non-isomorphic graphs of the given size, and then explore the network for each generated subgraph separately. Each approach has its advantages and disadvantages which are discussed above.

On the other hand, estimation methods might utilize color-coding approach as described before. Other approaches used in this category usually skip some subgraphs during enumeration (e.g., as in FANMOD) and base their estimation on the enumerated subgraphs.

Furthermore, table indicates whether an algorithm can be used for directed or undirected networks as well as induced or non-induced subgraphs. For more information refer to the provided web links or lab addresses.

22.4 Well-Established Motifs and Their Functions

Much experimental work has been devoted to understanding network motifs in gene regulatory networks. These networks control which genes are expressed in the cell in response to biological signals. The network is defined such that genes are nodes, and directed edges represent the control of one gene by a transcription factor (regulatory protein that binds DNA) encoded by another gene. Thus, network motifs are patterns of genes regulating each other's transcription rate. When analyzing transcription networks, it is seen that the same network motifs appear again and again in diverse organisms from bacteria to human. The transcription network of *E. coli* and yeast, for example, is made of three main motif families, that make up almost the entire network. The leading hypothesis is that the network motif were independently selected by evolutionary processes in a converging manner,[38][39] since the creation or elimination of regulatory interactions is fast on evolutionary time scale, relative to the rate at which genes change.[38][39][40] Furthermore, experiments on the dynamics generated by network motifs in living cells indicate that they have characteristic dynamical functions. This suggests that the network motif serve as building blocks in gene regulatory networks that are beneficial to the organism.

The functions associated with common network motifs in transcription networks were explored and demonstrated by several research projects both theoretically and experimentally. Below are some of the most common network motifs and their associated function.

Schematic representation of an auto-regulation motif

NAR motif this motif slows the response time compared to simple regulation.[47] In the case of a strong PAR the motif may lead to a bimodal distribution of protein levels in cell populations.[48]

22.4.1 Negative auto-regulation (NAR)

One of simplest and most abundant network motifs in *E. coli* is negative auto-regulation in which a transcription factor (TF) represses its own transcription. This motif was shown to perform two important functions. The first function is response acceleration. NAR was shown to speed-up the response to signals both theoretically [41] and experimentally. This was first shown in a synthetic transcription network[42] and later on in the natural context in the SOS DNA repair system of E .coli.[43] The second function is increased stability of the auto-regulated gene product concentration against stochastic noise, thus reducing variations in protein levels between different cells.[44][45][46]

22.4.2 Positive auto-regulation (PAR)

Positive auto-regulation (PAR) occurs when a transcription factor enhances its own rate of production. Opposite to the

22.4.3 Feed-forward loops (FFL)

This motif is commonly found in many gene systems and organisms. The motif consists of three genes and three regulatory interactions. The target gene C is regulated by 2 TFs A and B and in addition TF B is also regulated by TF A . Since each of the regulatory interactions may either be positive or negative there are possibly eight types of FFL motifs.[49] Two of those eight types: the coherent type 1 FFL (C1-FFL) (where all interactions are positive) and the incoherent type 1 FFL (I1-FFL) (A activates C and also activates B which represses C) are found much more frequently in the transcription network of *E. coli* and yeast than the other six types.[49][50] In addition to the structure of the circuitry the way in which the signals from A and B are integrated by the C promoter should also be considered. In most of the cases the FFL is either an AND gate (A and B are required for C activation) or OR gate (either A or B are sufficient for C activation) but other input function are also possible.

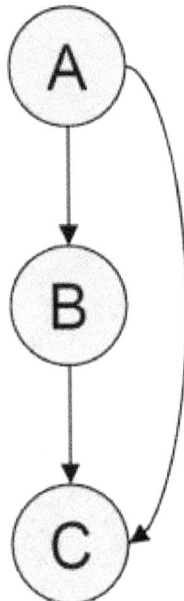

Schematic representation of a Feed-forward motif

22.4.4 Coherent type 1 FFL (C1-FFL)

The C1-FFL with an AND gate was shown to have a function of a 'sign-sensitive delay' element and a persistence detector both theoretically [49] and experimentally [51] with the arabinose system of *E. coli*. This means that this motif can provide pulse filtration in which short pulses of signal will not generate a response but persistent signals will generate a response after short delay. The shut off of the output when a persistent pulse is ended will be fast. The opposite behavior emerges in the case of a sum gate with fast response and delayed shut off as was demonstrated in the flagella system of *E. coli*.[52]

22.4.5 Incoherent type 1 FFL (I1-FFL)

The I1-FFL is a pulse generator and response accelerator. The two signal pathways of the I1-FFL act in opposite directions where one pathway activates Z and the other represses it. When the repression is complete this leads to a pulse-like dynamics. It was also demonstrated experimentally that the I1-FFL can serve as response accelerator in a way which is similar to the NAR motif. The difference is that the I1-FFL can speed-up the response of any gene and not necessarily a transcription factor gene.[53] An additional function was assigned to the I1-FFL network motif: it was shown both theoretically and experimentally that the I1-FFL can generate non-monotonic input function in both a synthetic [54] and native systems.[55] Finally, expression units that incorporate incoherent feedforward control of the gene product provide adaptation to the amount of DNA template and can be superior to simple combinations of constitutive promoters.[56] Feedforward regulation displayed better adaptation than negative feedback, and circuits based on RNA interference were the most robust to variation in DNA template amounts.[56]

22.4.6 Multi-output FFLs

In some cases the same regulators X and Y regulate several Z genes of the same system. By adjusting the strength of the interactions this motif was shown to determine the temporal order of gene activation. This was demonstrated experimentally in the flagella system of *E. coli*.[57]

22.4.7 Single-input modules (SIM)

This motif occurs when a single regulator regulates a set of genes with no additional regulation. This is useful when the genes are cooperatively carrying out a specific function and therefore always need to be activated in a synchronized manner. By adjusting the strength of the interactions it can create temporal expression program of the genes it regulates.[58]

In the literature, Multiple-input modules (MIM) arose as a generalization of SIM. However, the precise definitions of SIM and MIM have been a source of inconsistency. There are attempts to provide orthogonal definitions for canonical motifs in biological networks and algorithms to enumerate them, especially SIM, MIM and Bi-Fan (2x2 MIM).[59]

22.4.8 Dense overlapping regulons (DOR)

This motif occurs in the case that several regulators combinatorially control a set of genes with diverse regulatory combinations. This motif was found in *E. coli* in various systems such as carbon utilization, anaerobic growth, stress response and others.[13][18] In order to better understand the function of this motif one has to obtain more information about the way the multiple inputs are integrated by the genes. Kaplan et al.[60] has mapped the input functions of the sugar utilization genes in *E. coli*, showing diverse shapes.

22.5 Activity motifs

An interesting generalization of the network-motifs, **activity motifs** are over occurring patterns that can be found when nodes and edges in the network are annotated with quantitative features. For instance, when edges in a metabolic pathways are annotated with the magnitude or timing of the corresponding gene expression, some patterns are over occurring **given** the underlying network structure.[61]

22.6 Criticism

An assumption (sometimes more sometimes less implicit) behind the preservation of a topological sub-structure is that it is of a particular functional importance. This assumption has recently been questioned. Some authors have argued that motifs, like *bi-fan motifs*, might show a variety depending on the network context, and therefore,[62] structure of the motif does not necessarily determine function. Network structure certainly does not always indicate function; this is an idea that has been around for some time, for an example see the Sin operon.[63]

Most analyses of motif function are carried out looking at the motif operating in isolation. Recent research[64] provides good evidence that network context, i.e. the connections of the motif to the rest of the network, is too important to draw inferences on function from local structure only — the cited paper also reviews the criticisms and alternative explanations for the observed data. An analysis of the impact of a single motif module on the global dynamics of a network is studied in.[65] Yet another recent work suggests that certain topological features of biological networks naturally give rise to the common appearance of canonical motifs, thereby questioning whether frequencies of occurrences are reasonable evidence that the structures of motifs are selected for their functional contribution to the operation of networks.[66][67]

22.7 See also

Clique (graph theory)

Graphical model

22.8 References

[1] Masoudi-Nejad A, Schreiber F, Razaghi MK Z (2012). "Building Blocks of Biological Networks: A Review on Major Network Motif Discovery Algorithms". *IET Systems Biology, in press.*

[2] Diestel R (2005). "Graph Theory (Graduate Texts in Mathematics)". **173**. New York: Springer-Verlag Heidelberg.

[3] Milo R, Shen-Orr SS, Itzkovitz S, Kashtan N, Chklovskii D, Alon U (2002). "Network motifs: simple building blocks of complex networks". *Science.* **298** (5594): 824–827. Bibcode:2002Sci...298..824M. PMID 12399590. doi:10.1126/science.298.5594.824.

[4] Albert R, Barabási AL (2002). "Statistical mechanics of complex networks". *Reviews of Modern Physics.* **74**: 47–49. Bibcode:2002RvMP...74...47A. doi:10.1103/RevModPhys.74.47.

[5] Milo R, Itzkovitz S, Kashtan N, Levitt R, Shen-Orr S, Ayzenshtat I, Sheffer M, Alon U (2004). "Superfamilies of designed and evolved networks". *Science.* **303** (5663): 1538–1542. Bibcode:2004Sci...303.1538M. PMID 15001784. doi:10.1126/science.1089167.

[6] Schwöbbermeyer, H (2008). "Network Motifs". In Junker BH, Schreiber F. *Analysis of Biological Networks.* Hoboken, New Jersey: John Wiley & Sons. pp. 85–108.

[7] Bornholdt, S, Schuster, HG (2003). "Handbook of graphs and networks : from the genome to the Internet". *Handbook of Graphs and Networks: from the Genome to the Internet.* Weinheim: Wiley-VCH. p. 417. Bibcode:2003hgnf.book.....B.

[8] Ciriello G, Guerra C (2008). "A review on models and algorithms for motif discovery in protein-protein interaction networks". *Briefings in Functional Genomics and Proteomics.* **7** (2): 147–156. PMID 18443014. doi:10.1093/bfgp/eln015.

[9] Kashtan N, Itzkovitz S, Milo R, Alon U (2004). "Efficient sampling algorithm for estimating sub-graph concentrations and detecting network motifs". *Bioinformatics.* **20** (11): 1746–1758. PMID 15001476. doi:10.1093/bioinformatics/bth163.

[10] Wernicke S (2006). "Efficient detection of network motifs". *IEEE/ACM Transactions on Computational Biology and Bioinformatics.* **3** (4): 347–359. doi:10.1109/tcbb.2006.51.

[11] Picard F, Daudin JJ, Schbath S, Robin S (2005). "Assessing the Exceptionality of Network Motifs". *J. Comp. Bio.* **15** (1): 1–20.

[12] Schreiber F, Schwöbbermeyer H (2005). "Frequency concepts and pattern detection for the analysis of motifs in networks". *Transactions on Computational Systems Biology III*: 89–104.

[13] Shen-Orr SS, Milo R, Mangan S, Alon U (May 2002). "Network motifs in the transcriptional regulation network of *Escherichia coli*". *Nat. Genet.* **31** (1): 64–8. PMID 11967538. doi:10.1038/ng881.

[14] Eichenberger P, Fujita M, Jensen ST, et al. (October 2004). "The program of gene transcription for a single differentiating cell type during sporulation in *Bacillus subtilis*". *PLOS Biology.* **2** (10): e328. PMC 517825. PMID 15383836. doi:10.1371/journal.pbio.0020328.

[15] Milo R, Shen-Orr S, Itzkovitz S, Kashtan N, Chklovskii D, Alon U (October 2002). "Network motifs: simple building blocks of complex networks". *Science.* **298** (5594): 824–7. Bibcode:2002Sci...298..824M. PMID 12399590. doi:10.1126/science.298.5594.824.

[16] Lee TI, Rinaldi NJ, Robert F, et al. (October 2002). "Transcriptional regulatory networks in Saccharomyces cerevisiae". *Science.* **298** (5594): 799–804. Bibcode:2002Sci...298..799L. PMID 12399584. doi:10.1126/science.1075090.

[17] Odom DT, Zizlsperger N, Gordon DB, et al. (February 2004). "Control of pancreas and liver gene expression by HNF transcription factors". *Science.* **303** (5662): 1378–81. Bibcode:2004Sci...303.1378O. PMC 3012624. PMID 14988562. doi:10.1126/science.1089769.

[18] Boyer LA, Lee TI, Cole MF, et al. (September 2005). "Core transcriptional regulatory circuitry in human embryonic stem cells". *Cell.* **122** (6): 947–56. PMC 3006442. PMID 16153702. doi:10.1016/j.cell.2005.08.020.

[19] Iranfar N, Fuller D, Loomis WF (February 2006). "Transcriptional regulation of post-aggregation genes in Dictyostelium by a feed-forward loop involving GBF and LagC". *Dev. Biol.* **290** (2): 460–9. PMID 16386729. doi:10.1016/j.ydbio.2005.11.035.

[20] Ma'ayan A, Jenkins SL, Neves S, et al. (August 2005). "Formation of regulatory patterns during signal propagation in a Mammalian cellular network". *Science.* **309** (5737): 1078–83. Bibcode:2005Sci...309.1078M. PMC 3032439. PMID 16099987. doi:10.1126/science.1108876.

[21] Ptacek J, Devgan G, Michaud G, et al. (December 2005). "Global analysis of protein phosphorylation in yeast". *Nature.* **438** (7068): 679–84. Bibcode:2005Natur.438..679P. PMID 16319894. doi:10.1038/nature04187.

[22] http://www.ft.unicamp.br/docentes/meira/accmotifs/

[23] Schreiber F, Schwobbermeyer H (2005). "MAVisto: a tool for the exploration of network motifs". *Bioinformatics.* **21** (17): 3572–3574. PMID 16020473. doi:10.1093/bioinformatics/bti556.

[24] McKay BD (1981). "Practical graph isomorphism". *Congressus Numerantium.* **30**: 45–87. Bibcode:2013arXiv1301.1493M. arXiv:1301.1493 [cs.DM].

[25] McKay BD (1998). "Isomorph-free exhaustive generation". *Journal of Algorithms.* **26** (2): 306–324. doi:10.1006/jagm.1997.0898.

[26] Chen J, Hsu W, Li Lee M, et al. (2006). *NeMoFinder: dissecting genome-wide protein-protein interactions with mesoscale network motifs.* the 12th ACM SIGKDD international conference on Knowledge discovery and data mining. Philadelphia, Pennsylvania, USA. pp. 106–115.

[27] Huan J, Wang W, Prins J, et al. (2004). *SPIN: mining maximal frequent sub-graphs from graph databases.* the 10th ACM SIGKDD international conference on Knowledge discovery and data mining. pp. 581–586.

[28] Uetz P, Giot L, Cagney G, et al. (2000). "A comprehensive analysis of protein-protein interactions in Saccharomyces cerevisiae". *Nature.* **403** (6770): 623–627. Bibcode:2000Natur.403..623U. PMID 10688190. doi:10.1038/35001009.

[29] Grochow JA, Kellis M (2007). *Network Motif Discovery Using Sub-graph Enumeration and Symmetry-Breaking* (PDF). RECOMB. pp. 92–106. doi:10.1007/978-3-540-71681-5_7.

[30] Grochow JA (2006). *On the structure and evolution of protein interaction networks* (PDF). Thesis M. Eng., Massachusetts Institute of Technology, Dept. of Electrical Engineering and Computer Science.

[31] Alon N; Dao P; Hajirasouliha I; Hormozdiari F; Sahinalp S.C (2008). "Biomolecular network motif counting and discovery by color coding". *Bioinformatics.* **24** (13): i241–i249. PMC 2718641. PMID 18586721. doi:10.1093/bioinformatics/btn163.

[32] Omidi S, Schreiber F, Masoudi-Nejad A (2009). "MODA: an efficient algorithm for network motif discovery in biological networks". *Genes Genet Syst.* **84** (5): 385–395. PMID 20154426. doi:10.1266/ggs.84.385.

[33] Barabasi AL, Albert R (1999). "Emergence of scaling in random networks". *Science.* **286** (5439): 509–512. Bibcode:1999Sci...286..509B. PMID 10521342. doi:10.1126/science.286.5439.509.

[34] Vázquez A, Dobrin R, Sergi D, et al. (2004). "The topological relationship between the large-scale attributes and local interaction patterns of complex networks". *PNAS.* **101** (52): 17940–17945. Bibcode:2004PNAS..10117940V. PMC 539752. PMID 15598746. doi:10.1073/pnas.0406024101.

[35] Kashani ZR, Ahrabian H, Elahi E, Nowzari-Dalini A, Ansari ES, Asadi S, Mohammadi S, Schreiber F, Masoudi-Nejad A (2009). "Kavosh: a new algorithm for finding network motifs". *BMC Bioinformatics.* **10** (318): 318. PMC 2765973. PMID 19799800. doi:10.1186/1471-2105-10-318.

[36] Ali Masoudi-Nejad; Mitra Anasariola; Ali Salehzadeh-Yazdi; Sahand Khakabimamaghani (2012). "CytoKavosh: a Cytoscape Plug-in for Finding Network Motifs in Large Biological Networks". *PLoS ONE.* **7** (8): e43287. Bibcode:2012PLoSO...743287M. PMC 3430699. PMID 22952659. doi:10.1371/journal.pone.0043287.

[37] Ribeiro P, Silva F (2010). *G-Tries: an efficient data structure for discovering network motifs.* ACM 25th Symposium On Applied Computing - Bioinformatics Track. Sierre, Switzerland. pp. 1559–1566.

[38] Babu MM, Luscombe NM, Aravind L, Gerstein M, Teichmann SA (June 2004). "Structure and evolution of transcriptional regulatory networks". *Current Opinion in Structural Biology*. **14** (3): 283–91. PMID 15193307. doi:10.1016/j.sbi.2004.05.004.

[39] Conant GC, Wagner A (July 2003). "Convergent evolution of gene circuits". *Nat. Genet.* **34** (3): 264–6. PMID 12819781. doi:10.1038/ng1181.

[40] Dekel E, Alon U (July 2005). "Optimality and evolutionary tuning of the expression level of a protein". *Nature*. **436** (7050): 588–92. Bibcode:2005Natur.436..588D. PMID 16049495. doi:10.1038/nature03842.

[41] Zabet NR (September 2011). "Negative feedback and physical limits of genes". *Journal of Theoretical Biology*. **284** (1): 82–91. PMID 21723295. doi:10.1016/j.jtbi.2011.06.021.

[42] Rosenfeld N, Elowitz MB, Alon U (November 2002). "Negative autoregulation speeds the response times of transcription networks". *J. Mol. Biol.* **323** (5): 785–93. PMID 12417193. doi:10.1016/S0022-2836(02)00994-4.

[43] Camas FM, Blázquez J, Poyatos JF (August 2006). "Autogenous and nonautogenous control of response in a genetic network". *Proc. Natl. Acad. Sci. U.S.A.* **103** (34): 12718–23. Bibcode:2006PNAS..10312718C. PMC 1568915. PMID 16908855. doi:10.1073/pnas.0602119103.

[44] Becskei A, Serrano L (June 2000). "Engineering stability in gene networks by autoregulation". *Nature*. **405** (6786): 590–3. PMID 10850721. doi:10.1038/35014651.

[45] Dublanche Y, Michalodimitrakis K, Kümmerer N, Foglierini M, Serrano L (2006). "Noise in transcription negative feedback loops: simulation and experimental analysis". *Mol. Syst. Biol.* **2** (1): 41. PMC 1681513. PMID 16883354. doi:10.1038/msb4100081.

[46] Shimoga V, White J, Li Y, Sontag E, Bleris L (2013). "Synthetic mammalian transgene negative autoregulation". *Mol. Syst. Biol.* **9**: 670. PMC 3964311. PMID 23736683. doi:10.1038/msb.2013.27.

[47] Maeda YT, Sano M (June 2006). "Regulatory dynamics of synthetic gene networks with positive feedback". *J. Mol. Biol.* **359** (4): 1107–24. PMID 16701695. doi:10.1016/j.jmb.2006.03.064.

[48] Becskei A, Séraphin B, Serrano L (May 2001). "Positive feedback in eukaryotic gene networks: cell differentiation by graded to binary response conversion". *EMBO J.* **20** (10): 2528–35. PMC 125456. PMID 11350942. doi:10.1093/emboj/20.10.2528.

[49] Mangan S, Alon U (October 2003). "Structure and function of the feed-forward loop network motif". *Proc. Natl. Acad. Sci. U.S.A.* **100** (21): 11980–5. Bibcode:2003PNAS..10011980M. PMC 218699. PMID 14530388. doi:10.1073/pnas.2133841100.

[50] Ma HW, Kumar B, Ditges U, Gunzer F, Buer J, Zeng AP (2004). "An extended transcriptional regulatory network of *Escherichia coli* and analysis of its hierarchical structure and network motifs". *Nucleic Acids Res.* **32** (22): 6643–9. PMC 545451. PMID 15604458. doi:10.1093/nar/gkh1009.

[51] Mangan S, Zaslaver A, Alon U (November 2003). "The coherent feedforward loop serves as a sign-sensitive delay element in transcription networks". *J. Mol. Biol.* **334** (2): 197–204. PMID 14607112. doi:10.1016/j.jmb.2003.09.049.

[52] Kalir S, Mangan S, Alon U (2005). "A coherent feed-forward loop with a SUM input function prolongs flagella expression in *Escherichia coli*". *Mol. Syst. Biol.* **1** (1): 2005.0006. PMC 1681456. PMID 16729041. doi:10.1038/msb4100010.

[53] Mangan S, Itzkovitz S, Zaslaver A, Alon U (March 2006). "The incoherent feed-forward loop accelerates the response-time of the gal system of *Escherichia coli*". *J. Mol. Biol.* **356** (5): 1073–81. PMID 16406067. doi:10.1016/j.jmb.2005.12.003.

[54] Entus R, Aufderheide B, Sauro HM (August 2007). "Design and implementation of three incoherent feed-forward motif based biological concentration sensors". *Syst Synth Biol*. **1** (3): 119–28. PMC 2398716. PMID 19003446. doi:10.1007/s11693-007-9008-6.

[55] Kaplan S, Bren A, Dekel E, Alon U (2008). "The incoherent feed-forward loop can generate non-monotonic input functions for genes". *Mol. Syst. Biol.* **4** (1): 203. PMC 2516365. PMID 18628744. doi:10.1038/msb.2008.43.

[56] Bleris L, Xie Z, Glass D, Adadey A, Sontag E, Benenson Y (2011). "Synthetic incoherent feedforward circuits show adaptation to the amount of their genetic template". *Mol. Syst. Biol.* **7** (1): 519. PMC 3202791. PMID 21811230. doi:10.1038/msb.2011.49.

[57] Kalir S, McClure J, Pabbaraju K, et al. (June 2001). "Ordering genes in a flagella pathway by analysis of expression kinetics from living bacteria". *Science*. **292** (5524): 2080–3. PMID 11408658. doi:10.1126/science.1058758.

[58] Zaslaver A, Mayo AE, Rosenberg R, et al. (May 2004). "Just-in-time transcription program in metabolic pathways". *Nat. Genet.* **36** (5): 486–91. PMID 15107854. doi:10.1038/ng1348.

[59] Konagurthu AS, Lesk AM (2008). "Single and Multiple Input Modules in regulatory networks". *Proteins*. **73** (2): 320–324. PMID 18433061. doi:10.1002/prot.22053.

[60] Kaplan S, Bren A, Zaslaver A, Dekel E, Alon U (March 2008). "Diverse two-dimensional input functions control bacterial sugar genes". *Mol. Cell*. **29** (6): 786–92. PMC 2366073. PMID 18374652. doi:10.1016/j.molcel.2008.01.021.

[61] Chechik G, Oh E, Rando O, Weissman J, Regev A, Koller D (November 2008). "Activity motifs reveal principles of timing in transcriptional control of the yeast metabolic network". *Nat. Biotechnol.* **26** (11): 1251–9. PMC 2651818. PMID 18953355. doi:10.1038/nbt.1499.

[62] Ingram PJ, Stumpf MP, Stark J (2006). "Network motifs: structure does not determine function". *BMC Genomics.* **7**: 108. PMC 1488845. PMID 16677373. doi:10.1186/1471-2164-7-108.

[63] Voigt CA, Wolf DM, Arkin AP (March 2005). "The *Bacillus subtilis* sin operon: an evolvable network motif". *Genetics.* **169** (3): 1187–202. PMC 1449569. PMID 15466432. doi:10.1534/genetics.104.031955.

[64] Knabe JF, Nehaniv CL, Schilstra MJ (2008). "Do motifs reflect evolved function?—No convergent evolution of genetic regulatory network subgraph topologies". *BioSystems.* **94** (1–2): 68–74. PMID 18611431. doi:10.1016/j.biosystems.2008.05.012.

[65] Taylor D, Restrepo JG (2011). "Network connectivity during mergers and growth: Optimizing the addition of a module". *Physical Review E.* **83** (6): 66112. Bibcode:2011PhRvE..83f6112T. doi:10.1103/PhysRevE.83.066112.

[66] Konagurthu, Arun S.; Lesk, Arthur M. (23 April 2008). "Single and multiple input modules in regulatory networks". *Proteins: Structure, Function, and Bioinformatics.* **73** (2): 320–324. doi:10.1002/prot.22053.

[67] Konagurthu AS, Lesk AM (2008). "On the origin of distribution patterns of motifs in biological networks". *BMC Syst Biol.* **2**: 73. PMC 2538512. PMID 18700017. doi:10.1186/1752-0509-2-73.

22.9 External links

- Uri Alon's web page
- A software tool that can detect network motifs
- bio-physics-wiki NETWORK MOTIFS
- FANMOD: a tool for fast network motif detection
- MAVisto: network motif analysis and visualisation tool
- NeMoFinder
- Grochow-Kellis
- Noga Alon's web page
- MODA
- Kavosh
- CytoKavosh
- G-Tries
- acc-MOTIF detection tool

Chapter 23

Graph theory

This article is about sets of vertices connected by edges. For graphs of mathematical functions, see Graph of a function. For other uses, see Graph (disambiguation).

In mathematics, **graph theory** is the study of *graphs*,

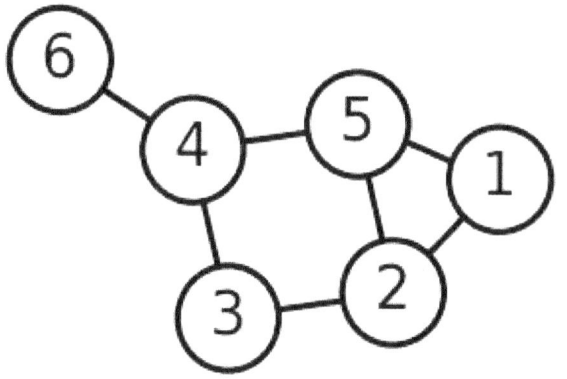

A drawing of a graph

which are mathematical structures used to model pairwise relations between objects. A graph in this context is made up of *vertices*, *nodes*, or *points* which are connected by *edges*, *arcs*, or *lines*. A graph may be *undirected*, meaning that there is no distinction between the two vertices associated with each edge, or its edges may be *directed* from one vertex to another; see Graph (discrete mathematics) for more detailed definitions and for other variations in the types of graph that are commonly considered. Graphs are one of the prime objects of study in discrete mathematics.

Refer to the glossary of graph theory for basic definitions in graph theory.

23.1 Definitions

Definitions in graph theory vary. The following are some of the more basic ways of defining graphs and related mathematical structures.

23.1.1 Graph

In the most common sense of the term,[1] a **graph** is an ordered pair $G = (V, E)$ comprising a set V of *vertices* or *nodes* or *points* together with a set E of *edges* or *arcs* or *lines*, which are 2-element subsets of V (i.e. an edge is associated with two vertices, and that association takes the form of the unordered pair comprising those two vertices). To avoid ambiguity, this type of graph may be described precisely as undirected and simple.

Other senses of *graph* stem from different conceptions of the edge set. In one more generalized notion,[2] V is a set together with a relation of *incidence* that associates with each edge two vertices. In another generalized notion, E is a multiset of unordered pairs of (not necessarily distinct) vertices. Many authors call this type of object a multigraph or pseudograph.

All of these variants and others are described more fully below.

The vertices belonging to an edge are called the *ends* or *end vertices* of the edge. A vertex may exist in a graph and not belong to an edge.

V and E are usually taken to be finite, and many of the well-known results are not true (or are rather different) for infinite graphs because many of the arguments fail in the infinite case. The *order* of a graph is $|V|$, its number of vertices. The *size* of a graph is $|E|$, its number of edges. The *degree* or *valency* of a vertex is the number of edges that connect to it, where an edge that connects a vertex to itself (a loop) is counted twice.

For an edge $\{x, y\}$, graph theorists usually use the somewhat shorter notation xy.

23.2 Applications

Graphs can be used to model many types of relations and processes in physical, biological,[4] social and information

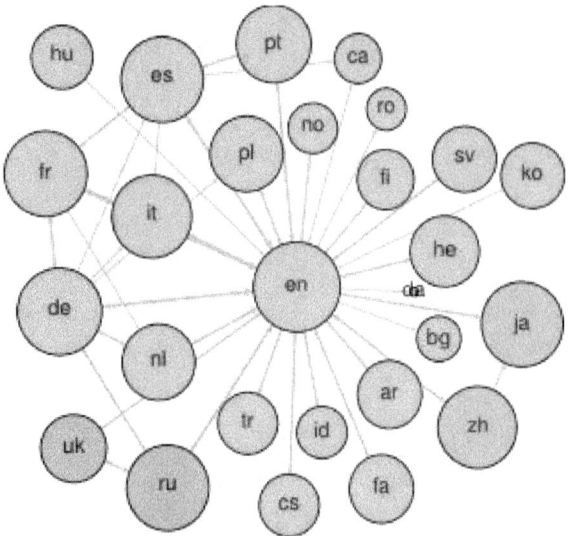

The network graph formed by Wikipedia editors (edges) contributing to different Wikipedia language versions (vertices) during one month in summer 2013[3]

systems. Many practical problems can be represented by graphs. Emphasizing their application to real-world systems, the term *network* is sometimes defined to mean a graph in which attributes (e.g. names) are associated with the nodes and/or edges.

In computer science, graphs are used to represent networks of communication, data organization, computational devices, the flow of computation, etc. For instance, the link structure of a website can be represented by a directed graph, in which the vertices represent web pages and directed edges represent links from one page to another. A similar approach can be taken to problems in social media,[5] travel, biology, computer chip design, and many other fields. The development of algorithms to handle graphs is therefore of major interest in computer science. The transformation of graphs is often formalized and represented by graph rewrite systems. Complementary to graph transformation systems focusing on rule-based in-memory manipulation of graphs are graph databases geared towards transaction-safe, persistent storing and querying of graph-structured data.

Graph-theoretic methods, in various forms, have proven particularly useful in linguistics, since natural language often lends itself well to discrete structure. Traditionally, syntax and compositional semantics follow tree-based structures, whose expressive power lies in the principle of compositionality, modeled in a hierarchical graph. More contemporary approaches such as head-driven phrase structure grammar model the syntax of natural language using typed feature structures, which are directed acyclic graphs. Within lexical semantics, especially as applied to comput-

ers, modeling word meaning is easier when a given word is understood in terms of related words; semantic networks are therefore important in computational linguistics. Still other methods in phonology (e.g. optimality theory, which uses lattice graphs) and morphology (e.g. finite-state morphology, using finite-state transducers) are common in the analysis of language as a graph. Indeed, the usefulness of this area of mathematics to linguistics has borne organizations such as TextGraphs, as well as various 'Net' projects, such as WordNet, VerbNet, and others.

Graph theory is also used to study molecules in chemistry and physics. In condensed matter physics, the three-dimensional structure of complicated simulated atomic structures can be studied quantitatively by gathering statistics on graph-theoretic properties related to the topology of the atoms. In chemistry a graph makes a natural model for a molecule, where vertices represent atoms and edges bonds. This approach is especially used in computer processing of molecular structures, ranging from chemical editors to database searching. In statistical physics, graphs can represent local connections between interacting parts of a system, as well as the dynamics of a physical process on such systems. Similarly, in computational neuroscience graphs can be used to represent functional connections between brain areas that interact to give rise to various cognitive processes, where the vertices represent different areas of the brain and the edges represent the connections between those areas. Graph theory plays an important role in electrical modelling of electrical networks, here, weights are associated with resistance of the wire segments to obtain electrical properties of network structures.[6] Graphs are also used to represent the micro-scale channels of porous media, in which the vertices represent the pores and the edges represent the smaller channels connecting the pores.

Graph theory is also widely used in sociology as a way, for example, to measure actors' prestige or to explore rumor spreading, notably through the use of social network analysis software. Under the umbrella of social networks are many different types of graphs.[8] Acquaintanceship and friendship graphs describe whether people know each other. Influence graphs model whether certain people can influence the behavior of others. Finally, collaboration graphs model whether two people work together in a particular way, such as acting in a movie together.

Likewise, graph theory is useful in biology and conservation efforts where a vertex can represent regions where certain species exist (or inhabit) and the edges represent migration paths, or movement between the regions. This information is important when looking at breeding patterns or tracking the spread of disease, parasites or how changes to the movement can affect other species.

In mathematics, graphs are useful in geometry and certain

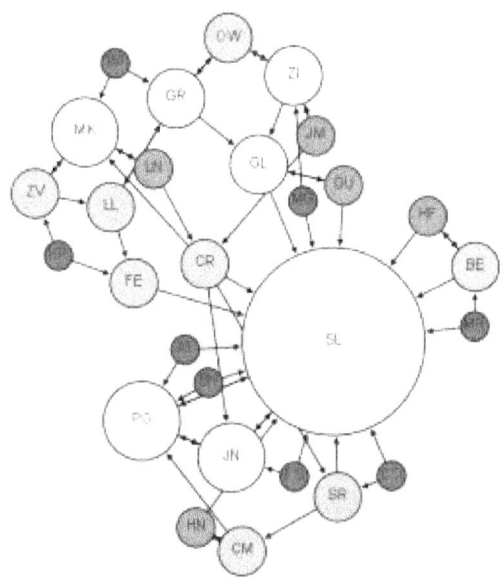

Graph theory in sociology: Moreno Sociogram (1953)[7]

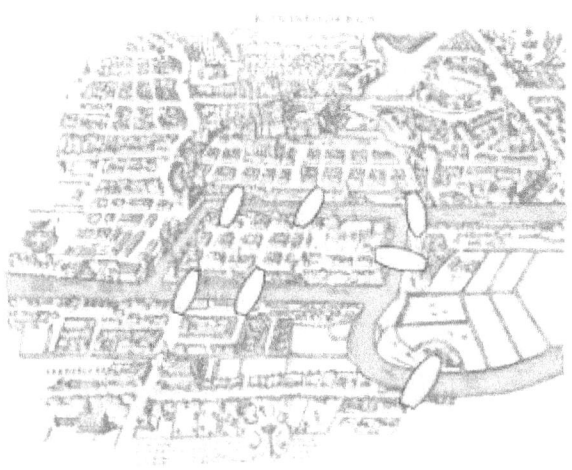

The Königsberg Bridge problem

parts of topology such as knot theory. Algebraic graph theory has close links with group theory.

A graph structure can be extended by assigning a weight to each edge of the graph. Graphs with weights, or weighted graphs, are used to represent structures in which pairwise connections have some numerical values. For example, if a graph represents a road network, the weights could represent the length of each road. There may be several weights associated with each edge, including distance (as in the previous example), travel time, or monetary cost. Such weighted graphs are commonly used to program GPS's, and travel-planning search engines that compare flight times and costs.

23.3 History

The paper written by Leonhard Euler on the *Seven Bridges of Königsberg* and published in 1736 is regarded as the first paper in the history of graph theory.[9] This paper, as well as the one written by Vandermonde on the *knight problem*, carried on with the *analysis situs* initiated by Leibniz. Euler's formula relating the number of edges, vertices, and faces of a convex polyhedron was studied and generalized by Cauchy[10] and L'Huillier,[11] and represents the beginning of the branch of mathematics known as topology.

More than one century after Euler's paper on the bridges of Königsberg and while Listing was introducing the concept of topology, Cayley was led by an interest in particular analytical forms arising from differential calculus to study a particular class of graphs, the *trees*.[12] This study had many implications for theoretical chemistry. The techniques he used mainly concern the enumeration of graphs with particular properties. Enumerative graph theory then arose from the results of Cayley and the fundamental results published by Pólya between 1935 and 1937. These were generalized by De Bruijn in 1959. Cayley linked his results on trees with contemporary studies of chemical composition.[13] The fusion of ideas from mathematics with those from chemistry began what has become part of the standard terminology of graph theory.

In particular, the term "graph" was introduced by Sylvester in a paper published in 1878 in *Nature*, where he draws an analogy between "quantic invariants" and "co-variants" of algebra and molecular diagrams:[14]

> "[...] Every invariant and co-variant thus becomes expressible by a *graph* precisely identical with a Kekuléan diagram or chemicograph. [...] I give a rule for the geometrical multiplication of graphs, *i.e.* for constructing a *graph* to the product of in- or co-variants whose separate graphs are given. [...]" (italics as in the original).

The first textbook on graph theory was written by Dénes Kőnig, and published in 1936.[15] Another book by Frank Harary, published in 1969, was "considered the world over to be the definitive textbook on the subject",[16] and enabled mathematicians, chemists, electrical engineers and social scientists to talk to each other. Harary donated all of the royalties to fund the Pólya Prize.[17]

One of the most famous and stimulating problems in graph theory is the four color problem: "Is it true that any map drawn in the plane may have its regions colored with four colors, in such a way that any two regions having a common border have different colors?" This problem was first posed by Francis Guthrie in 1852 and its first written record is in a letter of De Morgan addressed to Hamilton the same year. Many incorrect proofs have been proposed, including those by Cayley, Kempe, and others. The study and the generalization of this problem by Tait, Heawood, Ramsey and Hadwiger led to the study of the colorings of the graphs embedded on surfaces with arbitrary genus. Tait's reformulation generated a new class of problems, the *factorization problems*, particularly studied by Petersen and Kőnig. The works of Ramsey on colorations and more specially the results obtained by Turán in 1941 was at the origin of another branch of graph theory, *extremal graph theory*.

The four color problem remained unsolved for more than a century. In 1969 Heinrich Heesch published a method for solving the problem using computers.[18] A computer-aided proof produced in 1976 by Kenneth Appel and Wolfgang Haken makes fundamental use of the notion of "discharging" developed by Heesch.[19][20] The proof involved checking the properties of 1,936 configurations by computer, and was not fully accepted at the time due to its complexity. A simpler proof considering only 633 configurations was given twenty years later by Robertson, Seymour, Sanders and Thomas.[21]

The autonomous development of topology from 1860 and 1930 fertilized graph theory back through the works of Jordan, Kuratowski and Whitney. Another important factor of common development of graph theory and topology came from the use of the techniques of modern algebra. The first example of such a use comes from the work of the physicist Gustav Kirchhoff, who published in 1845 his Kirchhoff's circuit laws for calculating the voltage and current in electric circuits.

The introduction of probabilistic methods in graph theory, especially in the study of Erdős and Rényi of the asymptotic probability of graph connectivity, gave rise to yet another branch, known as *random graph theory*, which has been a fruitful source of graph-theoretic results.

23.4 Graph drawing

Main article: Graph drawing

Graphs are represented visually by drawing a dot or circle for every vertex, and drawing an arc between two vertices if they are connected by an edge. If the graph is directed, the direction is indicated by drawing an arrow.

A graph drawing should not be confused with the graph itself (the abstract, non-visual structure) as there are several ways to structure the graph drawing. All that matters is which vertices are connected to which others by how many edges and not the exact layout. In practice it is often difficult to decide if two drawings represent the same graph. Depending on the problem domain some layouts may be better suited and easier to understand than others.

The pioneering work of W. T. Tutte was very influential in the subject of graph drawing. Among other achievements, he introduced the use of linear algebraic methods to obtain graph drawings.

Graph drawing also can be said to encompass problems that deal with the crossing number and its various generalizations. The crossing number of a graph is the minimum number of intersections between edges that a drawing of the graph in the plane must contain. For a planar graph, the crossing number is zero by definition.

Drawings on surfaces other than the plane are also studied.

23.5 Graph-theoretic data structures

Main article: Graph (abstract data type)

There are different ways to store graphs in a computer system. The data structure used depends on both the graph structure and the algorithm used for manipulating the graph. Theoretically one can distinguish between list and matrix structures but in concrete applications the best structure is often a combination of both. List structures are often preferred for sparse graphs as they have smaller memory requirements. Matrix structures on the other hand provide faster access for some applications but can consume huge amounts of memory.

List structures include the incidence list, an array of pairs of vertices, and the adjacency list, which separately lists the neighbors of each vertex: Much like the incidence list, each vertex has a list of which vertices it is adjacent to.

Matrix structures include the incidence matrix, a matrix of 0's and 1's whose rows represent vertices and whose columns represent edges, and the adjacency matrix, in which both the rows and columns are indexed by vertices. In both cases a 1 indicates two adjacent objects and a 0 indicates two non-adjacent objects. The Laplacian matrix is a modified form of the adjacency matrix that incorporates information about the degrees of the vertices, and is useful in some calculations such as Kirchhoff's theorem on the number of spanning trees of a graph. The distance matrix, like the adjacency matrix, has both its rows and columns indexed by vertices, but rather than containing a 0 or a 1 in each cell it contains the length of a shortest path between two vertices.

23.6 Problems

23.6.1 Enumeration

There is a large literature on graphical enumeration: the problem of counting graphs meeting specified conditions. Some of this work is found in Harary and Palmer (1973).

23.6.2 Subgraphs, induced subgraphs, and minors

A common problem, called the subgraph isomorphism problem, is finding a fixed graph as a subgraph in a given graph. One reason to be interested in such a question is that many graph properties are *hereditary* for subgraphs, which means that a graph has the property if and only if all subgraphs have it too. Unfortunately, finding maximal subgraphs of a certain kind is often an NP-complete problem. For example:

- Finding the largest complete subgraph is called the clique problem (NP-complete).

A similar problem is finding induced subgraphs in a given graph. Again, some important graph properties are hereditary with respect to induced subgraphs, which means that a graph has a property if and only if all induced subgraphs also have it. Finding maximal induced subgraphs of a certain kind is also often NP-complete. For example:

- Finding the largest edgeless induced subgraph or independent set is called the independent set problem (NP-complete).

Still another such problem, the minor containment problem, is to find a fixed graph as a minor of a given graph. A minor or subcontraction of a graph is any graph obtained by taking a subgraph and contracting some (or no) edges. Many graph properties are hereditary for minors, which means that a graph has a property if and only if all minors have it too. For example, Wagner's Theorem states:

- A graph is planar if it contains as a minor neither the complete bipartite graph $K_{3,3}$ (see the Three-cottage problem) nor the complete graph K_5.

A similar problem, the subdivision containment problem, is to find a fixed graph as a subdivision of a given graph. A subdivision or homeomorphism of a graph is any graph obtained by subdividing some (or no) edges. Subdivision containment is related to graph properties such as planarity. For example, Kuratowski's Theorem states:

- A graph is planar if it contains as a subdivision neither the complete bipartite graph $K_{3,3}$ nor the complete graph K_5.

Another problem in subdivision containment is Kelmans-Seymour conjecture:

- Every 5-vertex-connected graph that is not planar contains a subdivision of the 5-vertex complete graph K_5.

Another class of problems has to do with the extent to which various species and generalizations of graphs are determined by their *point-deleted subgraphs*. For example:

- The reconstruction conjecture

23.6.3 Graph coloring

Main article: Graph coloring

Many problems and theorems in graph theory have to do with various ways of coloring graphs. Typically, one is interested in coloring a graph so that no two adjacent vertices have the same color, or with other similar restrictions. One may also consider coloring edges (possibly so that no two coincident edges are the same color), or other variations. Among the famous results and conjectures concerning graph coloring are the following:

- Four-color theorem
- Strong perfect graph theorem
- Erdős–Faber–Lovász conjecture (unsolved)
- Total coloring conjecture, also called Behzad's conjecture (unsolved)
- List coloring conjecture (unsolved)
- Hadwiger conjecture (graph theory) (unsolved)

23.6.4 Subsumption and unification

Constraint modeling theories concern families of directed graphs related by a partial order. In these applications, graphs are ordered by specificity, meaning that more constrained graphs—which are more specific and thus contain a greater amount of information—are subsumed by those that are more general. Operations between graphs include evaluating the direction of a subsumption relationship between two graphs, if any, and computing graph unification. The unification of two argument graphs is defined as the most general graph (or the computation thereof) that is consistent with (i.e. contains all of the information in) the inputs, if such a graph exists; efficient unification algorithms are known.

For constraint frameworks which are strictly compositional, graph unification is the sufficient satisfiability and combination function. Well-known applications include automatic theorem proving and modeling the elaboration of linguistic structure.

23.6.5 Route problems

- Hamiltonian path problem
- Minimum spanning tree
- Route inspection problem (also called the "Chinese postman problem")
- Seven bridges of Königsberg
- Shortest path problem
- Steiner tree
- Three-cottage problem
- Traveling salesman problem (NP-hard)

23.6.6 Network flow

There are numerous problems arising especially from applications that have to do with various notions of flows in networks, for example:

- Max flow min cut theorem

23.6.7 Visibility problems

- Museum guard problem

23.6.8 Covering problems

Covering problems in graphs are specific instances of subgraph-finding problems, and they tend to be closely related to the clique problem or the independent set problem.

- Set cover problem
- Vertex cover problem

23.6.9 Decomposition problems

Decomposition, defined as partitioning the edge set of a graph (with as many vertices as necessary accompanying the edges of each part of the partition), has a wide variety of question. Often, it is required to decompose a graph into subgraphs isomorphic to a fixed graph; for instance, decomposing a complete graph into Hamiltonian cycles. Other problems specify a family of graphs into which a given graph should be decomposed, for instance, a family of cycles, or decomposing a complete graph Kn into $n - 1$ specified trees having, respectively, 1, 2, 3, ..., $n - 1$ edges.

Some specific decomposition problems that have been studied include:

- Arboricity, a decomposition into as few forests as possible
- Cycle double cover, a decomposition into a collection of cycles covering each edge exactly twice
- Edge coloring, a decomposition into as few matchings as possible
- Graph factorization, a decomposition of a regular graph into regular subgraphs of given degrees

23.6.10 Graph classes

Many problems involve characterizing the members of various classes of graphs. Some examples of such questions are below:

- Enumerating the members of a class
- Characterizing a class in terms of forbidden substructures
- Ascertaining relationships among classes (e.g. does one property of graphs imply another)
- Finding efficient algorithms to decide membership in a class
- Finding representations for members of a class

23.7 See also

- Gallery of named graphs
- Glossary of graph theory
- List of graph theory topics
- List of unsolved problems in graph theory
- Publications in graph theory

23.7.1 Related topics

- Algebraic graph theory
- Citation graph
- Conceptual graph
- Data structure
- Disjoint-set data structure
- Dual-phase evolution
- Entitative graph
- Existential graph
- Graph algebra
- Graph automorphism
- Graph coloring
- Graph database
- Graph data structure
- Graph drawing
- Graph equation
- Graph rewriting
- Graph sandwich problem
- Graph property
- Intersection graph
- Logical graph
- Loop
- Network theory
- Null graph
- Pebble motion problems
- Percolation
- Perfect graph
- Quantum graph
- Random regular graphs
- Semantic networks
- Spectral graph theory
- Strongly regular graphs
- Symmetric graphs
- Transitive reduction
- Tree data structure

23.7.2 Algorithms

- Bellman–Ford algorithm
- Dijkstra's algorithm
- Ford–Fulkerson algorithm
- Kruskal's algorithm
- Nearest neighbour algorithm
- Prim's algorithm
- Depth-first search
- Breadth-first search

23.7.3 Subareas

- Algebraic graph theory
- Geometric graph theory
- Extremal graph theory
- Probabilistic graph theory
- Topological graph theory

23.7.4 Related areas of mathematics

- Combinatorics
- Group theory
- Knot theory
- Ramsey theory

23.7.5 Generalizations

- Hypergraph
- Abstract simplicial complex

23.7.6 Prominent graph theorists

- Alon, Noga
- Berge, Claude
- Bollobás, Béla
- Bondy, Adrian John
- Brightwell, Graham
- Chudnovsky, Maria

- Chung, Fan
- Dirac, Gabriel Andrew
- Erdős, Paul
- Euler, Leonhard
- Faudree, Ralph
- Golumbic, Martin
- Graham, Ronald
- Harary, Frank
- Heawood, Percy John
- Kotzig, Anton
- Kőnig, Dénes
- Lovász, László
- Murty, U. S. R.
- Nešetřil, Jaroslav
- Rényi, Alfréd
- Ringel, Gerhard
- Robertson, Neil
- Seymour, Paul
- Sudakov, Benny
- Szemerédi, Endre
- Thomas, Robin
- Thomassen, Carsten
- Turán, Pál
- Tutte, W. T.
- Whitney, Hassler

23.8 Notes

[1] See, for instance, Iyanaga and Kawada, **69 J**, p. 234 or Biggs, p. 4.

[2] See, for instance, Graham et al., p. 5.

[3] Hale, Scott A. (2013). "Multilinguals and Wikipedia Editing". *Proceedings of the 2014 ACM conference on Web science - WebSci '14*. arXiv:1312.0976 [cs.CY]. doi:10.1145/2615569.2615684.

[4] Mashaghi, A.; et al. (2004). "Investigation of a protein complex network". *European Physical Journal B*. **41** (1): 113–121. doi:10.1140/epjb/e2004-00301-0.

[5] Grandjean, Martin (2016). "A social network analysis of Twitter: Mapping the digital humanities community". *Cogent Arts & Humanities*. **3** (1): 1171458. doi:10.1080/23311983.2016.1171458.

[6] Kumar, Ankush; Kulkarni, G. U. (2016-01-04). "Evaluating conducting network based transparent electrodes from geometrical considerations". *Journal of Applied Physics*. **119** (1): 015102. ISSN 0021-8979. doi:10.1063/1.4939280.

[7] Grandjean, Martin (2015). "Social network analysis and visualization: Moreno's Sociograms revisited". Redesigned network strictly based on Moreno (1934), *Who Shall Survive*.

[8] Rosen, Kenneth H. *Discrete mathematics and its applications* (7th ed.). New York: McGraw-Hill. ISBN 978-0-07-338309-5.

[9] Biggs, N.; Lloyd, E.; Wilson, R. (1986), *Graph Theory, 1736-1936*, Oxford University Press

[10] Cauchy, A.L. (1813). "Recherche sur les polyèdres - premier mémoire", *[[:fr:Journal de l'École polytechnique]]*, 9 (Cahier 16): 66–86.

[11] L'Huillier, S.-A.-J. (1861), "Mémoire sur la polyèdrométrie", *Annales de Mathématiques*, **3**: 169–189.

[12] Cayley, A. (1857). "On the theory of the analytical forms called trees", *Philosophical Magazine*, Series IV, **13** (85): 172–176, doi:10.1017/CBO9780511703690.046

[13] Cayley, A. (1875). "Ueber die Analytischen Figuren, welche in der Mathematik Bäume genannt werden und ihre Anwendung auf die Theorie chemischer Verbindungen", *Berichte der Deutschen Chemischen Gesellschaft*, **8** (2): 1056–1059, doi:10.1002/cber.18750080252.

[14] Sylvester, James Joseph (1878). "Chemistry and Algebra". *Nature*. **17**: 284. doi:10.1038/017284a0.

[15] Tutte, W.T. (2001), *Graph Theory*, Cambridge University Press, p. 30, ISBN 978-0-521-79489-3, retrieved 2016-03-14

[16] Gardner, Martin (1992), *Fractal Music, Hypercards, and more...Mathematical Recreations from Scientific American*, W. H. Freeman and Company, p. 203

[17] Society for Industrial and Applied Mathematics (2002), "The George Polya Prize", *Looking Back, Looking Ahead: A SIAM History* (PDF), p. 26, retrieved 2016-03-14

[18] Heinrich Heesch: Untersuchungen zum Vierfarbenproblem. Mannheim: Bibliographisches Institut 1969.

[19] Appel, K.; Haken, W. (1977), "Every planar map is four colorable. Part I. Discharging", *Illinois J. Math.*, **21**: 429–490.

[20] Appel, K.; Haken, W. (1977), "Every planar map is four colorable. Part II. Reducibility", *Illinois J. Math.*, **21**: 491–567.

[21] Robertson, N.; Sanders, D.; Seymour, P.; Thomas, R. (1997), "The four color theorem", *Journal of Combinatorial Theory Series B*, **70**: 2–44, doi:10.1006/jctb.1997.1750.

23.9 References

- Berge, Claude (1958), *Théorie des graphes et ses applications*, Collection Universitaire de Mathématiques, II, Paris: Dunod. English edition, Wiley 1961; Methuen & Co, New York 1962; Russian, Moscow 1961; Spanish, Mexico 1962; Roumanian, Bucharest 1969; Chinese, Shanghai 1963; Second printing of the 1962 first English edition, Dover, New York 2001.

- Biggs, N.; Lloyd, E.; Wilson, R. (1986), *Graph Theory, 1736–1936*, Oxford University Press.

- Bondy, J.A.; Murty, U.S.R. (2008), *Graph Theory*, Springer, ISBN 978-1-84628-969-9.

- Bollobás, Béla; Riordan, O.M (2003), *Mathematical results on scale-free random graphs in "Handbook of Graphs and Networks" (S. Bornholdt and H.G. Schuster (eds)), Wiley VCH, Weinheim, 1st ed.*.

- Chartrand, Gary (1985), *Introductory Graph Theory*, Dover, ISBN 0-486-24775-9.

- Gibbons, Alan (1985), *Algorithmic Graph Theory*, Cambridge University Press.

- Reuven Cohen, Shlomo Havlin (2010), *Complex Networks: Structure, Robustness and Function*, Cambridge University Press.

- Golumbic, Martin (1980), *Algorithmic Graph Theory and Perfect Graphs*, Academic Press.

- Harary, Frank (1969), *Graph Theory*, Reading, MA: Addison-Wesley.

- Harary, Frank; Palmer, Edgar M. (1973), *Graphical Enumeration*, New York, NY: Academic Press.

- Mahadev, N.V.R.; Peled, Uri N. (1995), *Threshold Graphs and Related Topics*, North-Holland.

- Mark Newman (2010), *Networks: An Introduction*, Oxford University Press.

23.10 External links

- Hazewinkel, Michiel, ed. (2001) [1994], "Graph theory", *Encyclopedia of Mathematics*, Springer Science+Business Media B.V. / Kluwer Academic Publishers, ISBN 978-1-55608-010-4
- Graph theory tutorial
- A searchable database of small connected graphs
- Image gallery: graphs at the Wayback Machine (archived February 6, 2006)
- Concise, annotated list of graph theory resources for researchers
- rocs — a graph theory IDE
- The Social Life of Routers — non-technical paper discussing graphs of people and computers
- Graph Theory Software — tools to teach and learn graph theory
- Online books, and library resources in your library and in other libraries about graph theory
- A list of graph algorithms with references and links to graph library implementations

23.10.1 Online textbooks

- Phase Transitions in Combinatorial Optimization Problems, Section 3: Introduction to Graphs (2006) by Hartmann and Weigt
- Digraphs: Theory Algorithms and Applications 2007 by Jorgen Bang-Jensen and Gregory Gutin
- Graph Theory, by Reinhard Diestel

Chapter 24

Scalability

Scalability is the capability of a system, network, or process to handle a growing amount of work, or its potential to be enlarged to accommodate that growth.[1] For example, a system is considered scalable if it is capable of increasing its total output under an increased load when resources (typically hardware) are added. An analogous meaning is implied when the word is used in an economic context, where a company's scalability implies that the underlying business model offers the potential for economic growth within the company.

Scalability, as a property of systems, is generally difficult to define[2] and in any particular case it is necessary to define the specific requirements for scalability on those dimensions that are deemed important. It is a highly significant issue in electronics systems, databases, routers, and networking. A system whose performance improves after adding hardware, proportionally to the capacity added, is said to be a **scalable system**.

An algorithm, design, networking protocol, program, or other system is said to *scale* if it is suitably efficient and practical when applied to large situations (e.g. a large input data set, a large number of outputs or users, or a large number of participating nodes in the case of a distributed system). If the design or system fails when a quantity increases, it *does not scale*. In practice, if there are a large number of things (n) that affect scaling, then resource requirements (for example, algorithmic time-complexity) must grow less than n^2 as n increases. An example is a search engine, which scales not only for the number of users, but also for the number of objects it indexes. Scalability refers to the ability of a site to increase in size as demand warrants.[3]

The concept of scalability is desirable in technology as well as business settings. The base concept is consistent – the ability for a business or technology to accept increased volume without impacting the contribution margin (= revenue – variable costs). For example, a given piece of equipment may have a capacity for 1–1000 users, while beyond 1000 users additional equipment is needed or performance will decline (variable costs will increase and reduce contribution margin).

24.1 Measures

Scalability can be measured in various dimensions, such as:

- *Administrative scalability*: The ability for an increasing number of organizations or users to easily share a single distributed system.

- *Functional scalability*: The ability to enhance the system by adding new functionality at minimal effort.

- *Geographic scalability*: The ability to maintain performance, usefulness, or usability regardless of expansion from concentration in a local area to a more distributed geographic pattern.

- *Load scalability*: The ability for a distributed system to easily expand and contract its resource pool to accommodate heavier or lighter loads or number of inputs. Alternatively, the ease with which a system or component can be modified, added, or removed, to accommodate changing load.

- *Generation scalability*: The ability of a system to scale up by using new generations of components. Thereby, *heterogeneous scalability* is the ability to use the components from different vendors.[4]

24.2 Examples

- A routing protocol is considered scalable with respect to network size, if the size of the necessary routing table on each node grows as $O(\log N)$, where N is the number of nodes in the network.

- A scalable online transaction processing system or database management system is one that can be upgraded to process more transactions by adding new processors, devices and storage, and which can be upgraded easily and transparently without shutting it down.

- Some early peer-to-peer (P2P) implementations of Gnutella had scaling issues. Each node query flooded its requests to all peers. The demand on each peer would increase in proportion to the total number of peers, quickly overrunning the peers' limited capacity. Other P2P systems like BitTorrent scale well because the demand on each peer is independent of the total number of peers. There is no centralized bottleneck, so the system may expand indefinitely without the addition of supporting resources (other than the peers themselves).

- The distributed nature of the Domain Name System allows it to work efficiently even when all hosts on the worldwide Internet are served, so it is said to "scale well".

24.3 Horizontal and vertical scaling

Methods of adding more resources for a particular application fall into two broad categories: horizontal and vertical scaling.[5]

- To **scale horizontally** (or **scale out/in**) means to add more nodes to (or remove nodes from) a system, such as adding a new computer to a distributed software application. An example might involve scaling out from one Web server system to three. As computer prices have dropped and performance continues to increase, high-performance computing applications such as seismic analysis and biotechnology workloads have adopted low-cost "commodity" systems for tasks that once would have required supercomputers. System architects may configure hundreds of small computers in a cluster to obtain aggregate computing power that often exceeds that of computers based on a single traditional processor. The development of high-performance interconnects such as Gigabit Ethernet, InfiniBand and Myrinet further fueled this model. Such growth has led to demand for software that allows efficient management and maintenance of multiple nodes, as well as hardware such as shared data storage with much higher I/O performance. *Size scalability* is the maximum number of processors that a system can accommodate.[4]

- To **scale vertically** (or **scale up/down**) means to add resources to (or remove resources from) a single node in a system, typically involving the addition of CPUs or memory to a single computer. Such vertical scaling of existing systems also enables them to use virtualization technology more effectively, as it provides more resources for the hosted set of operating system and application modules to share. Taking advantage of such resources can also be called "scaling up", such as expanding the number of Apache daemon processes currently running. *Application scalability* is the improved performance of running applications on a scaled-up version of the system.[4]

There are tradeoffs between the two models. Larger numbers of computers means increased management complexity, as well as a more complex programming model and issues such as throughput and latency between nodes; also, some applications do not lend themselves to a distributed computing model. In the past, the price difference between the two models has favored "scale up" computing for those applications that fit its paradigm, but recent advances in virtualization technology have blurred that advantage, since deploying a new virtual system over a hypervisor (where possible) is often less expensive than actually buying and installing a real one. Configuring an existing idle system has always been less expensive than buying, installing, and configuring a new one, regardless of the model.

Note that NFV defines these terms differently: scaling out/in is the ability to scale by add/remove resource instances (e.g. virtual machine), whereas scaling up/down is the ability to scale by changing allocated resources (e.g. memory/CPU/storage capacity).[6]

24.4 Database scalability

A number of different approaches enable databases to grow to very large size while supporting an ever-increasing rate of transactions per second. Not to be discounted, of course, is the rapid pace of hardware advances in both the speed and capacity of mass storage devices, as well as similar advances in CPU and networking speed.

One technique supported by most of the major database management system (DBMS) products is the partitioning of large tables, based on ranges of values in a key field. In this manner, the database can be *scaled out* across a cluster of separate database servers. Also, with the advent of 64-bit microprocessors, multi-core CPUs, and large SMP multiprocessors, DBMS vendors have been at the forefront of supporting multi-threaded implementations that substantially *scale up* transaction processing capacity.

Network-attached storage (NAS) and Storage area networks (SANs) coupled with fast local area networks and Fibre Channel technology enable still larger, more loosely coupled configurations of databases and distributed computing power. The widely supported X/Open XA standard employs a global transaction monitor to coordinate distributed transactions among semi-autonomous XA-compliant database resources. Oracle RAC uses a different model to achieve scalability, based on a "shared-everything" architecture that relies upon high-speed connections between servers.

While DBMS vendors debate the relative merits of their favored designs, some companies and researchers question the inherent limitations of relational database management systems. GigaSpaces, for example, contends that an entirely different model of distributed data access and transaction processing, space-based architecture, is required to achieve the highest performance and scalability. On the other hand, Base One makes the case for extreme scalability without departing from mainstream relational database technology.[7] For specialized applications, NoSQL architectures such as Google's BigTable can further enhance scalability. Google's massively distributed Spanner technology, positioned as a successor to BigTable, supports general-purpose database transactions and provides a more conventional SQL-based query language.[8]

24.5 Strong versus eventual consistency (storage)

In the context of scale-out data storage, scalability is defined as the maximum storage cluster size which guarantees full data consistency, meaning there is only ever one valid version of stored data in the whole cluster, independently from the number of redundant physical data copies. Clusters which provide "lazy" redundancy by updating copies in an asynchronous fashion are called 'eventually consistent'. This type of scale-out design is suitable when availability and responsiveness are rated higher than consistency, which is true for many web file hosting services or web caches (*if you want the latest version, wait some seconds for it to propagate*). For all classical transaction-oriented applications, this design should be avoided.[9]

Many open source and even commercial scale-out storage clusters, especially those built on top of standard PC hardware and networks, provide eventual consistency only. Idem some NoSQL databases like CouchDB and others mentioned above. Write operations invalidate other copies, but often don't wait for their acknowledgements. Read operations typically don't check every redundant copy prior to answering, potentially missing the preceding write operation. The large amount of metadata signal traffic would require specialized hardware and short distances to be handled with acceptable performance (i.e. act like a non-clustered storage device or database).

Whenever strong data consistency is expected, look for these indicators:

- the use of InfiniBand, Fibrechannel or similar low-latency networks to avoid performance degradation with increasing cluster size and number of redundant copies.
- short cable lengths and limited physical extent, avoiding signal runtime performance degradation.
- majority / quorum mechanisms to guarantee data consistency whenever parts of the cluster become inaccessible.

Indicators for eventually consistent designs (not suitable for transactional applications!) are:

- write performance increases linearly with the number of connected devices in the cluster.
- while the storage cluster is partitioned, all parts remain responsive. There is a risk of conflicting updates.

24.6 Performance tuning versus hardware scalability

It is often advised to focus system design on hardware scalability rather than on capacity. It is typically cheaper to add a new node to a system in order to achieve improved performance than to partake in performance tuning to improve the capacity that each node can handle. But this approach can have diminishing returns (as discussed in performance engineering). For example: suppose 70% of a program can be sped up if parallelized and run on multiple CPUs instead of one. If α is the fraction of a calculation that is sequential, and $1 - \alpha$ is the fraction that can be parallelized, the maximum speedup that can be achieved by using P processors is given according to Amdahl's Law:

$$\frac{1}{\alpha + \frac{1-\alpha}{P}}.$$

Substituting the value for this example, using 4 processors we get

$$\frac{1}{0.3 + \frac{1-0.3}{4}} = 2.105.$$

If we double the compute power to 8 processors we get

$$\frac{1}{0.3 + \frac{1-0.3}{8}} = 2.581.$$

Doubling the processing power has only improved the speedup by roughly one-fifth. If the whole problem was parallelizable, we would, of course, expect the speed up to double also. Therefore, throwing in more hardware is not necessarily the optimal approach.

24.7 Weak versus strong scaling

In the context of high performance computing there are two common notions of scalability:

- The first is *strong scaling*, which is defined as how the solution time varies with the number of processors for a fixed *total* problem size.

- The second is *weak scaling*, which is defined as how the solution time varies with the number of processors for a fixed problem size *per processor*.[10]

24.8 See also

- Computational complexity theory
- Extensibility
- Gustafson's law
- List of system quality attributes
- Load balancing (computing)
- Lock (computer science)
- NoSQL
- Scalable Video Coding (SVC)
- Similitude (model)

24.9 References

[1] Bondi, André B. (2000). *Characteristics of scalability and their impact on performance*. Proceedings of the second international workshop on Software and performance – WOSP '00. p. 195. ISBN 158113195X. doi:10.1145/350391.350432.

[2] See for instance, Hill, Mark D. (1990). "What is scalability?". *ACM SIGARCH Computer Architecture News*. **18** (4): 18. doi:10.1145/121973.121975. and Duboc, Leticia; Rosenblum, David S.; Wicks, Tony (2006). *A framework for modelling and analysis of software systems scalability*. Proceeding of the 28th international conference on Software engineering – ICSE '06. p. 949. ISBN 1595933751. doi:10.1145/1134285.1134460.

[3] Laudon, Kenneth Craig; Traver, Carol Guercio (2008). *E-commerce: Business, Technology, Society*. Pearson Prentice Hall/Pearson Education. ISBN 9780136006459.

[4] Hesham El-Rewini and Mostafa Abd-El-Barr (April 2005). *Advanced Computer Architecture and Parallel Processing*. John Wiley & Sons. p. 66. ISBN 978-0-471-47839-3.

[5] Michael, Maged; Moreira, Jose E.; Shiloach, Doron; Wisniewski, Robert W. (March 26, 2007). *Scale-up x Scale-out: A Case Study using Nutch/Lucene*. 2007 IEEE International Parallel and Distributed Processing Symposium. p. 1. ISBN 1-4244-0909-8. doi:10.1109/IPDPS.2007.370631.

[6] "Network Functions Virtualisation (NFV); Terminology for Main Concepts in NFV" (PDF).

[7] Base One (2007). "Database Scalability - Dispelling myths about the limits of database-centric architecture". Retrieved May 23, 2007.

[8] "Spanner: Google's Globally-Distributed Database" (PDF). OSDI'12 Proceedings of the 10th USENIX conference on Operating Systems Design and Implementation. 2012: 251–264. ISBN 978-1-931971-96-6. Retrieved September 30, 2012.

[9] Sadek Drobi (January 11, 2008). "Eventual consistency by Werner Vogels". InfoQ. Retrieved April 8, 2017.

[10] "The Weak Scaling of DL_POLY 3". STFC Computational Science and Engineering Department. Archived from the original on March 7, 2014. Retrieved March 8, 2014.

24.10 External links

- Architecture of a Highly Scalable NIO-Based Server – an article about writing scalable server in Java (java.net).

- Links to diverse learning resources – page curated by the memcached project.

- Scalable Definition – by The Linux Information Project (LINFO)

- Scale in Distributed Systems B. Clifford Neumann, In: *Readings in Distributed Computing Systems*, IEEE Computer Society Press, 1994

Chapter 25

Robustness (computer science)

See also: Fault-tolerant computer system

In computer science, **robustness** is the ability of a computer system to cope with errors during execution[1][2] and cope with erroneous input.[2] Robustness can encompass many areas of computer science, such as robust programming, robust machine learning, and Robust Security Network. Formal techniques, such as fuzz testing, are essential to showing robustness since this type of testing involves invalid or unexpected inputs. Alternatively, fault injection can be used to test robustness. Various commercial products perform robustness testing of software analysis.[3]

25.1 Introduction

In general, building robust systems that encompass every point of possible failure is difficult because of the vast quantity of possible inputs and input combinations.[4] Since all inputs and input combinations would require too much time to test, developers cannot run through all cases exhaustively. Instead, the developer will try to generalize such cases.[5] For example, imagine inputting some integer values. Some selected inputs might consist of a negative number, zero, and a positive number. When using these numbers to test software in this way, the developer generalizes the set of all reals into three numbers. This is a more efficient and manageable method, but more prone to failure. Generalizing test cases is an example of just one technique to deal with failure—specifically, failure due to invalid user input. Systems generally may also fail due to other reasons as well, such as disconnecting from a network.

Regardless, complex systems should still handle any errors encountered gracefully. There are many examples of such successful systems. Some of the most robust systems are evolvable and can be easily adapted to new situations.[4]

25.2 Challenges

Programs and software are tools focused on a very specific task, and thus aren't generalized and flexible.[4] However, observations in systems such as the internet or biological systems demonstrate adaptation to their environments. One of the ways biological systems adapt to environments is through the use of redundancy.[4] Many organs are redundant in humans. The kidney is one such example. Humans generally only need one kidney, but having a second kidney allows room for failure. This same principle may be taken to apply to software, but there are some challenges. When applying the principle of redundancy to computer science, blindly adding code is not suggested. Blindly adding code introduces more errors, makes the system more complex, and renders it harder to understand.[6] Code that doesn't provide any reinforcement to the already existing code is unwanted. The new code must instead possess equivalent functionality, so that if a function is broken, another providing the same function can replace it. To do so, the new code must know how and when to accommodate the failure point.[4] This means more logic needs to be added to the system. But as a system adds more logic, components, and increases in size, it becomes more complex. Thus, when making a more redundant system, the system also becomes more complex and developers must consider balancing redundancy with complexity.

Currently, computer science practices do not focus on building robust systems.[4] Rather, they tend to focus on scalability and efficiency. One of the main reasons why there is no focus on robustness today is because it is hard to do in a general way.[4]

25.3 Areas

25.3.1 Robust programming

Robust programming is a style of programming that focuses on handling unexpected termination and unexpected actions.[7] It requires code to handle these terminations and actions gracefully by displaying accurate and unambiguous error messages. These error messages allow the user to more easily debug the program.

Principles

Paranoia - When building software, the programmer assumes users are out to break their code.[7] The programmer also assumes that his or her own written code may fail or work incorrectly.[7]

Stupidity - The programmer assumes users will try incorrect, bogus and malformed inputs.[7] As a consequence, the programmer returns to the user an unambiguous, intuitive error message that does not require looking up error codes. The error message should try to be as accurate as possible without being misleading to the user, so that the problem can be fixed with ease.

Dangerous implements - Users should not gain access to libraries, data structures, or pointers to data structures.[7] This information should be hidden from the user so that the user doesn't accidentally modify them and introduce a bug in the code. When such interfaces are correctly built, users use them without finding loopholes to modify the interface. The interface should already be correctly implemented, so the user does not need to make modifications. The user therefore focuses solely on his or her own code.

Can't happen - Very often, code is modified and may introduce a possibility that an "impossible" case occurs. Impossible cases are therefore assumed to be highly unlikely instead.[7] The developer thinks about how to handle the case that is highly unlikely, and implements the handling accordingly.

25.3.2 Robust machine learning

Robust machine learning typically refers to the robustness of machine learning algorithms. For a machine learning algorithm to be considered robust, either the testing error has to be consistent with the training error, or the performance is stable after adding some noise to the dataset.[8]

25.3.3 Robust network design

Robust network design is the study of network design in the face of variable or uncertain demands.[9] In a sense, robustness in network design is broad just like robustness in software design because of the vast possibilities of changes or inputs.

25.4 Examples

- Examples Robustness Requirements

25.5 See also

- Defensive programming
- Non-functional requirement

25.6 References

[1] "A Model-Based Approach for Robustness Testing" (PDF). *Dl.ifip.org*. Retrieved 2016-11-13.

[2] 1990. IEEE Standard Glossary of Software Engineering Terminology, IEEE Std 610.12-1990 defines robustness as "The degree to which a system or component can function correctly in the presence of invalid inputs or stressful environmental conditions"

[3] "doi:10.1016/j.strusafe.2006.11.004" (PDF). *Stanford.edu*. Retrieved 2016-11-13.

[4] Gerald Jay Sussman (January 13, 2007). "Building Robust Systems an essay" (PDF). *Groups.csail.mit.edu*. Retrieved 2016-11-13.

[5] Joseph, Joby (2009-09-21). "Importance of Making Generalized Testcases - Software Testing Club - An Online Software Testing Community". Software Testing Club. Retrieved 2016-11-13.

[6] Agents on the wEb : Robust Software. "Building Robust Systems an essay" (PDF). *Cse.sc.edu*. Retrieved 2016-11-13.

[7] "Robust Programming". *Nob.cs.ucdavis.edu*. Retrieved 2016-11-13.

[8] El Sayed Mahmoud. "What is the definition of the robustness of a machine learning algorithm?". ResearchGate. Retrieved 2016-11-13.

[9] "Robust Network Design" (PDF). *Math.mit.edu*. Retrieved 2016-11-13.

Chapter 26

Systems biology

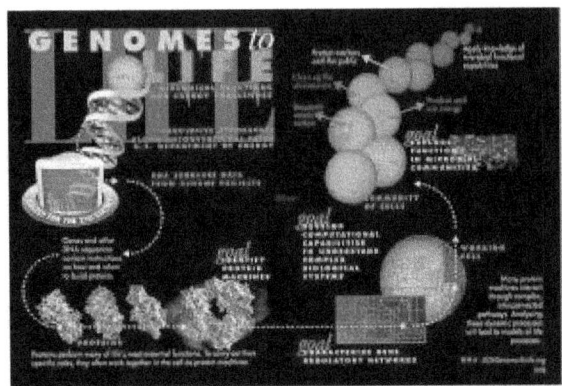

An illustration of the systems approach to biology

Systems biology is the computational and mathematical modeling of complex biological systems. It is a biology-based interdisciplinary field of study that focuses on complex interactions within biological systems, using a holistic approach (holism instead of the more traditional reductionism) to biological research.

Particularly from year 2000 onwards, the concept has been used widely in biology in a variety of contexts. The Human Genome Project is an example of applied systems thinking in biology which has led to new, collaborative ways of working on problems in the biological field of genetics.[1] One of the aims of systems biology is to model and discover emergent properties, properties of cells, tissues and organisms functioning as a system whose theoretical description is only possible using techniques of systems biology.[2] These typically involve metabolic networks or cell signaling networks.[3]

26.1 Overview

Systems biology can be considered from a number of different aspects:

- As a field of study, particularly, the study of the interactions between the components of biological systems, and how these interactions give rise to the function and behavior of that system (for example, the enzymes and metabolites in a metabolic pathway or the heart beats).[4][5][6]

- As a paradigm, usually defined in antithesis to the so-called reductionist paradigm (biological organisation), although fully consistent with the scientific method. The distinction between the two paradigms is referred to in these quotations:

"The reductionist approach has successfully identified most of the components and many of the interactions but, unfortunately, offers no convincing concepts or methods to understand how system properties emerge...the pluralism of causes and effects in biological networks is better addressed by observing, through quantitative measures, multiple components simultaneously and by rigorous data integration with mathematical models" (Sauer et al.).[7]

"Systems biology...is about putting together rather than taking apart, integration rather than reduction. It requires that we develop ways of thinking about integration that are as rigorous as our reductionist programmes, but different....It means changing our philosophy, in the full sense of the term" (Denis Noble).[6]

- As a series of operational protocols used for performing research, namely a cycle composed of theory, analytic or computational modelling to propose specific testable hypotheses about a biological system, experimental validation, and then using the newly acquired quantitative description of cells or cell processes to refine the computational model or theory.[8] Since the objective is a model of the interactions in a system, the experimental techniques that most suit systems biology are those that are system-wide and attempt to be as complete as possible. Therefore, transcriptomics, metabolomics, proteomics and high-throughput techniques are used to collect quantitative data for the construction and validation of models.[9]

- As the application of dynamical systems theory to molecular biology. Indeed, the focus on the dynamics of the studied systems is the main conceptual difference between systems biology and bioinformatics.[10]

- As a socioscientific phenomenon defined by the strategy of pursuing integration of complex data about the interactions in biological systems from diverse experimental sources using interdisciplinary tools and personnel.[11]

This variety of viewpoints is illustrative of the fact that systems biology refers to a cluster of peripherally overlapping concepts rather than a single well-delineated field. However, the term has widespread currency and popularity as of 2007, with chairs and institutes of systems biology proliferating worldwide.

26.2 History

Systems biology finds its roots in:

- the quantitative modeling of enzyme kinetics, a discipline that flourished between 1900 and 1970,
- the mathematical modeling of population dynamics,
- the simulations developed to study neurophysiology,
- control theory and cybernetics,
- synergetics.

One of the theorists who can be seen as one of the precursors of systems biology is Ludwig von Bertalanffy with his general systems theory.[12] One of the first numerical simulations in cell biology was published in 1952 by the British neurophysiologists and Nobel prize winners Alan Lloyd Hodgkin and Andrew Fielding Huxley, who constructed a mathematical model that explained the action potential propagating along the axon of a neuronal cell.[13] Their model described a cellular function emerging from the interaction between two different molecular components, a potassium and a sodium channel, and can therefore be seen as the beginning of computational systems biology.[14] Also in 1952, Alan Turing published The Chemical Basis of Morphogenesis, describing how non-uniformity could arise in an initially homogeneous biological system.[15]

In 1960, Denis Noble developed the first computer model of the heart pacemaker.[16]

The formal study of systems biology, as a distinct discipline, was launched by systems theorist Mihajlo Mesarovic in 1966 with an international symposium at the Case Institute of Technology in Cleveland, Ohio, entitled "Systems Theory and Biology".[17][18]

The 1960s and 1970s saw the development of several approaches to study complex molecular systems, such as the metabolic control analysis and the biochemical systems theory. The successes of molecular biology throughout the 1980s, coupled with a skepticism toward theoretical biology, that then promised more than it achieved, caused the quantitative modeling of biological processes to become a somewhat minor field.[19]

However, the birth of functional genomics in the 1990s meant that large quantities of high-quality data became available, while the computing power exploded, making more realistic models possible. In 1992, then 1994, serial articles [20][21][22][23][24] on systems medicine, systems genetics, and systems biological engineering by B. J. Zeng was published in China and was giving a lecture on biosystems theory and systems-approach research at the First International Conference on Transgenic Animals, Beijing, 1996. In 1997, the group of Masaru Tomita published the first quantitative model of the metabolism of a whole (hypothetical) cell.[25]

Around the year 2000, after Institutes of Systems Biology were established in Seattle and Tokyo, systems biology emerged as a movement in its own right, spurred on by the completion of various genome projects, the large increase in data from the omics (e.g., genomics and proteomics) and the accompanying advances in high-throughput experiments and bioinformatics.

In 2002, the National Science Foundation (NSF) put forward a grand challenge for systems biology in the 21st century to build a mathematical model of the whole cell.[26] In 2003, work at the Massachusetts Institute of Technology was begun on CytoSolve, a method to model the whole cell by dynamically integrating multiple molecular pathway models.[27][28] Since then, various research institutes dedicated to systems biology have been developed. For example, the NIGMS of NIH established a project grant that is currently supporting over ten systems biology centers in the United States.[29] As of summer 2006, due to a shortage of people in systems biology[30] several doctoral training programs in systems biology have been established in many parts of the world. In that same year, the National Science Foundation (NSF) put forward a grand challenge for systems biology in the 21st century to build a mathematical model of the whole cell.[31] In 2012 the first whole-cell model of Mycoplasma Genitalium was achieved by the Karr

Laboratory at the Mount Sinai School of Medicine in New York. The whole-cell model is able to predict viability of M. Genitalium cells in response to genetic mutations.[32]

An important milestone in the development of systems biology has become the international project Physiome.

26.3 Associated disciplines

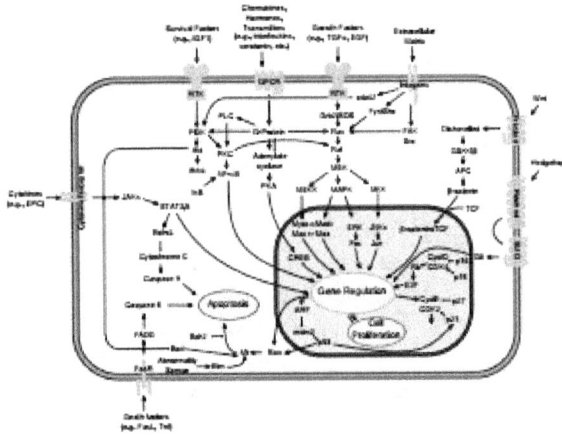

Overview of signal transduction pathways

According to the interpretation of Systems Biology as the ability to obtain, integrate and analyze complex data sets from multiple experimental sources using interdisciplinary tools, some typical technology platforms are:

- Phenomics

    Organismal variation in phenotype as it changes during its life span.

- Genomics

    Organismal deoxyribonucleic acid (DNA) sequence, including intra-organisamal cell specific variation. (i.e., telomere length variation)

- Epigenomics / Epigenetics

    Organismal and corresponding cell specific transcriptomic regulating factors not empirically coded in the genomic sequence. (i.e., DNA methylation, Histone acetylation and deacetylation, etc.).

- Transcriptomics

    Organismal, tissue or whole cell gene expression measurements by DNA microarrays or serial analysis of gene expression

- Interferomics

    Organismal, tissue, or cell-level transcript correcting factors (i.e., RNA interference)

- Proteomics

    Organismal, tissue, or cell level measurements of proteins and peptides via two-dimensional gel electrophoresis, mass spectrometry or multi-dimensional protein identification techniques (advanced HPLC systems coupled with mass spectrometry). Sub disciplines include phosphoproteomics, glycoproteomics and other methods to detect chemically modified proteins.

- Metabolomics

    Organismal, tissue, or cell-level measurements of small molecules known as metabolites

- Glycomics

    Organismal, tissue, or cell-level measurements of carbohydrates

- Lipidomics

    Organismal, tissue, or cell level measurements of lipids.

In addition to the identification and quantification of the above given molecules further techniques analyze the dynamics and interactions within a cell. This includes:

- Interactomics

    Organismal, tissue, or cell level study of interactions between molecules. Currently, the authoritative molecular discipline in this field of study is protein-protein interactions (PPI), although the working definition does not preclude inclusion of other molecular disciplines such as those defined here.

- NeuroElectroDynamics

Organismal, brain computing function as a dynamic system, underlying biophysical mechanisms and emerging computation by electrical interactions.

- Fluxomics

  Organismal, tissue, or cell level measurements of molecular dynamic changes over time.

- Biomics

  Systems analysis of the biome.

- Molecular Biokinematics

  The study of "biology in motion" focused on how cells transit between steady states. Various technologies utilized to capture dynamic changes in mRNA, proteins, and post-translational modifications.

- Semiomics

  Analysis of the system of sign relations of an organism or other biosystems.

- Physiomics

  A systematic study of physiome in biology.

Cancer Systems Biology is an example of the systems biology approach, which can be distinguished by the specific object of study (tumorigenesis and treatment of cancer). It works with the specific data (patient samples, high-throughput data with particular attention to characterizing cancer genome in patient tumour samples) and tools (immortalized cancer cell lines, mouse models of tumorigenesis, xenograft models, Next Generation Sequencing methods, siRNA-based gene knocking down screenings, computational modeling of the consequences of somatic mutations and genome instability).[33] The long-term objective of the systems biology of cancer is ability to better diagnose cancer, classify it and better predict the outcome of a suggested treatment, which is a basis for personalized cancer medicine and virtual cancer patient in more distant prospective. Significant efforts in Computational systems Biology of Cancer have been made in creating realistic multi-scale *in silico* models of various tumours.[34]

The investigations are frequently combined with large-scale perturbation methods, including gene-based (RNAi, misexpression of wild type and mutant genes) and chemical approaches using small molecule libraries. Robots and automated sensors enable such large-scale experimentation and data acquisition. These technologies are still emerging and many face problems that the larger the quantity of data produced, the lower the quality. A wide variety of quantitative scientists (computational biologists, statisticians, mathematicians, computer scientists and physicists) are working to improve the quality of these approaches and to create, refine, and retest the models to accurately reflect observations.

The systems biology approach often involves the development of mechanistic models, such as the reconstruction of dynamic systems from the quantitative properties of their elementary building blocks.[35][36][37][38] For instance, a cellular network can be modelled mathematically using methods coming from chemical kinetics and control theory. Due to the large number of parameters, variables and constraints in cellular networks, numerical and computational techniques are often used (e.g., flux balance analysis).[37]

26.4 Bioinformatics and data analysis

Other aspects of computer science, informatics, and statistics are also used in systems biology. These include:

- New forms of computational models, such as the use of process calculi to model biological processes (notable approaches include stochastic π-calculus, BioAmbients, Beta Binders, BioPEPA, and Brane calculus) and constraint-based modeling.

- Integration of information from the literature, using techniques of information extraction and text mining.[39]

- Development of online databases and repositories for sharing data and models, approaches to database integration and software interoperability via loose coupling of software, websites and databases, or commercial suits.

- Development of syntactically and semantically sound ways of representing biological models.

- Network-based approaches for analyzing high dimensional genomic data sets. For example, weighted correlation network analysis is often used for identifying clusters (referred to as modules), modeling the relationship between clusters, calculating fuzzy measures of cluster (module) membership, identifying intramodular hubs, and for studying cluster preservation in other data sets.

- Pathway-based methods for omics data analysis, e.g. approaches to identify and score pathways with differential activity of their gene, protein, or metabolite members.[40]

26.5 See also

- Biological computation
- Computational biology
- Interactome
- Exposome
- Network Biology
- Weighted correlation network analysis
- Synthetic biology
- List of omics topics in biology
- Systems biologists
- Systems biomedicine
- Flux balance analysis
- Metabolic network modelling
- Molecular pathological epidemiology
- Systems pharmacology
- Cancer systems biology
- Network medicine

26.6 References

[1] Zewail, Ahmed (2008). *Physical Biology: From Atoms to Medicine*. Imperial College Press. p. 339.

[2] Longo, Giuseppe; Montévil, Maël. *Perspectives on Organisms - Springer*. doi:10.1007/978-3-642-35938-5.

[3] Bu Z, Callaway DJ (2011). "Proteins MOVE! Protein dynamics and long-range allostery in cell signaling". *Advances in Protein Chemistry and Structural Biology*. Advances in Protein Chemistry and Structural Biology. **83**: 163–221. ISBN 978-0-123-81262-9. PMID 21570668. doi:10.1016/B978-0-12-381262-9.00005-7.

[4] Snoep, Jacky L; Westerhoff, Hans V (2005). Alberghina, Lilia; Westerhoff, Hans V, eds. "Systems Biology: Definitions and Perspectives". *Topics in Current Genetics*. Topics in Current Genetics. Berlin: Springer-Verlag. **13**: 13–30. ISBN 978-3-540-22968-1. doi:10.1007/b106456. |chapter= ignored (help)

[5] "Systems Biology: the 21st Century Science". Institute for Systems Biology. Retrieved 15 June 2011.

[6] Noble, Denis (2006). *The music of life: Biology beyond the genome*. Oxford: Oxford University Press. p. 176. ISBN 978-0-19-929573-9.

[7] Sauer, Uwe; Heinemann, Matthias; Zamboni, Nicola (27 April 2007). "Genetics: Getting Closer to the Whole Picture". *Science*. **316** (5824): 550–551. PMID 17463274. doi:10.1126/science.1142502.

[8] Kholodenko, Boris N; Sauro, Herbert M (2005). Alberghina, Lilia; Westerhoff, Hans V, eds. "Systems Biology: Definitions and Perspectives". *Topics in Current Genetics*. Topics in Current Genetics. Berlin: Springer-Verlag. **13**: 357–451. ISBN 978-3-540-22968-1. doi:10.1007/b136809. |chapter= ignored (help)

[9] Chiara Romualdi; Gerolamo Lanfranchi (2009). "Statistical Tools for Gene Expression Analysis and Systems Biology and Related Web Resources". In Stephen Krawetz. *Bioinformatics for Systems Biology* (2nd ed.). Humana Press. pp. 181–205. ISBN 978-1-59745-440-7.

[10] Voit, Eberhard (2012). *A First Course in Systems Biology*. Garland Science. ISBN 9780815344674.

[11] Baitaluk, M. (2009). "System Biology of Gene Regulation". *Biomedical Informatics*. Methods in Molecular Biology. **569**. pp. 55–87. ISBN 978-1-934115-63-3. PMID 19623486. doi:10.1007/978-1-59745-524-4_4.

[12] von Bertalanffy, Ludwig (28 March 1976) [1968]. *General System theory: Foundations, Development, Applications*. George Braziller. p. 295. ISBN 978-0-8076-0453-3.

[13] Hodgkin, Alan L; Huxley, Andrew F (28 August 1952). "A quantitative description of membrane current and its application to conduction and excitation in nerve". *Journal of Physiology*. **117** (4): 500–544. PMC 1392413. PMID 12991237. doi:10.1113/jphysiol.1952.sp004764.

[14] Le Novère, Nicolas (13 June 2007). "The long journey to a Systems Biology of neuronal function". *BMC Systems Biology*. **1**: 28. PMC 1904462. PMID 17567903. doi:10.1186/1752-0509-1-28.

[15] Turing, A. M. (1952). "The Chemical Basis of Morphogenesis" (PDF). *Philosophical Transactions of the Royal Society B: Biological Sciences*. **237** (641): 37–72. JSTOR 92463. doi:10.1098/rstb.1952.0012.

[16] Noble, Denis (5 November 1960). "Cardiac action and pacemaker potentials based on the Hodgkin-Huxley equations". *Nature*. **188** (4749): 495–497. Bibcode:1960Natur.188..495N. PMID 13729365. doi:10.1038/188495b0.

[17] Mesarovic, Mihajlo D. (1968). *Systems Theory and Biology*. Berlin: Springer-Verlag.

[18] Rosen, Robert (5 July 1968). "A Means Toward a New Holism". *Science*. **161** (3836): 34–35. Bibcode:1968Sci...161...34M. JSTOR 1724368. doi:10.1126/science.161.3836.34.

[19] Hunter, Philip (May 2012). "Back down to Earth: Even if it has not yet lived up to its promises, systems biology has now matured and is about to deliver its first results.". *EMBO Reports*. **13** (5): 408–411. PMC 3343359. PMID 22491028. doi:10.1038/embor.2012.49.

[20] B. J. Zeng, "On the holographic model of human body", 1st National Conference of Comparative Studies Traditional Chinese Medicine and West Medicine, Medicine and Philosophy, April 1992 ("systems medicine and pharmacology" termed).

[21] Zeng (B.) J., On the concept of system biological engineering, Communication on Transgenic Animals, No. 6, June, 1994.

[22] B. J. Zeng, "Transgenic animal expression system – transgenic egg plan (goldegg plan)", *Communication on Transgenic Animal*, Vol.1, No.11, 1994 (on the concept of system genetics and term coined).

[23] B. J. Zeng, "From positive to synthetic science", *Communication on Transgenic Animals*, No. 11, 1995 (on systems medicine).

[24] B. J. Zeng, "The structure theory of self-organization systems", *Communication on Transgenic Animals*, No.8-10, 1996. Etc.

[25] Tomita, Masaru; Hashimoto, Kenta; Takahashi, Kouichi; Shimizu, Thomas S; Matsuzaki, Yuri; Miyoshi, Fumihiko; Saito, Kanako; Tanida, Sakura; et al. (1997). "E-CELL: Software Environment for Whole Cell Simulation". *Genome Inform Ser Workshop Genome Inform*. **8**: 147–155. PMID 11072314. Retrieved 15 June 2011.

[26] American Association for the Advancement of Science, , *Science*

[27] Ayyadurai, VA; Dewey, CF (March 2011). "CytoSolve: A Scalable Computational Method for Dynamic Integration of Multiple Molecular Pathway Models". *Cell Mol Bioeng*. **4** (1): 28–45. PMC 3032229. PMID 21423324. doi:10.1007/s12195-010-0143-x.

[28] Massachusetts Institute of Technology

[29] "Systems Biology - National Institute of General Medical Sciences". Retrieved 12 December 2012.

[30] Kling, Jim (3 March 2006). "Working the Systems". Science. Retrieved 15 June 2011.

[31] Omenn, Gilbert S. (December 2006). "Grand Challenges and Great Opportunities in Science, Technology, and Public Policy". *Science*. **314** (5806): 1696–1704. doi:10.1126/science.1135003.

[32] Karr, Jonathan R.; Sanghvi, Jayodita C.; Macklin, Derek N.; Gutschow, Miriam V.; Jacobs, Jared M.; Bolival, Benjamin; Assad-Garcia, Nacyra; Glass, John I.; Covert, Markus W. (July 2012). "A Whole-Cell Computational Model Predicts Phenotype from Genotype". *Cell*. **150** (2): 389–401. PMC 3413483. PMID 22817898. doi:10.1016/j.cell.2012.05.044.

[33] Barillot, Emmanuel; Calzone, Laurence; Hupe, Philippe; Vert, Jean-Philippe; Zinovyev, Andrei (2012). *Computational Systems Biology of Cancer*. Chapman & Hall/CRCMathematical & Computational Biology. p. 461. ISBN 978-1439831441.

[34] Byrne, Helen M. (2010). "Dissecting cancer through mathematics: from the cell to the animal model". *Nature Reviews Cancer*. **10** (3): 221–230. PMID 20179714. doi:10.1038/nrc2808.

[35] Gardner, Timothy S; di Bernardo, Diego; Lorenz, David; Collins, James J. (4 July 2003). "Inferring Genetic Networks and Identifying Compound Mode of Action via Expression Profiling". *Science*. **301** (5629): 102–105. Bibcode:2003Sci...301..102G. PMID 12843395. doi:10.1126/science.1081900.

[36] di Bernardo, Diego; Thompson, Michael J.; Gardner, Timothy S.; Chobot, Sarah E.; Eastwood, Erin L.; Wojtovich, Andrew P.; Elliott, Sean J.; Schaus, Scott E.; Collins, James J. (March 2005). "Chemogenomic profiling on a genome-wide scale using reverse-engineered gene networks". *Nature Biotechnology*. **23** (3): 377–383. PMID 15765094. doi:10.1038/nbt1075.

[37] Tavassoly, Iman (2015). *Dynamics of Cell Fate Decision Mediated by the Interplay of Autophagy and Apoptosis in Cancer Cells*. Springer International Publishing. ISBN 978-3-319-14961-5.

[38] Korkut, A; Wang, W; Demir, E; Aksoy, BA; Jing, X; Molinelli, EJ; Babur, Ö; Bemis, DL; Onur Sumer, S; Solit, DB; Pratilas, CA; Sander, C (18 August 2015). "Perturbation biology nominates upstream-downstream drug combinations in RAF inhibitor resistant melanoma cells.". *eLife*. **4**. PMC 4539601. PMID 26284497. doi:10.7554/eLife.04640.

[39] Ananadou, Sophia; Kell, Douglas; Tsujii, Jun-ichi (December 2006). "Text mining and its potential applications in systems biology". *Trends in Biotechnology*. Elsevier. **24** (12): 571–579. doi:10.1016/j.tibtech.2006.10.002. Retrieved 23 July 2017.

[40] Glaab, Enrico; Schneider, Reinhard (2012). "PathVar: analysis of gene and protein expression variance in cellular pathways using microarray data". *Bioinformatics*. **28** (3): 446–447. PMC 3268235. PMID 22123829. doi:10.1093/bioinformatics/btr656.

26.7 Further reading

- Klipp, Edda; Liebermeister, Wolfram; Wierling, Christoph; Kowald, Axel (2016). *Systems Biology - A Textbook, 2nd edition*. Wiley. ISBN 978-3-527-33636-4.

- Asfar S. Azmi, ed. (2012). "Systems Biology in Cancer Research and Drug Discovery". ISBN 978-94-007-4819-4.

- Kitano, Hiroaki (15 October 2001). *Foundations of Systems Biology*. MIT Press. ISBN 978-0-262-11266-6.

- Werner, Eric (29 March 2007). "All systems go". *Nature*. **446** (7135): 493–494. Bibcode:2007Natur.446..493W. doi:10.1038/446493a. provides a comparative review of three books:

- Alon, Uri (7 July 2006). *An Introduction to Systems Biology: Design Principles of Biological Circuits*. Chapman & Hall. ISBN 978-1-58488-642-6.

- Kaneko, Kunihiko (15 September 2006). *Life: An Introduction to Complex Systems Biology*. Springer-Verlag. ISBN 978-3-540-32666-3.

- Palsson, Bernhard O. (16 January 2006). *Systems Biology: Properties of Reconstructed Networks*. Cambridge University Press. ISBN 978-0-521-85903-5.

- Werner Dubitzky; Olaf Wolkenhauer; Hiroki Yokota; Kwan-Hyun Cho, eds. (13 August 2013). *Encyclopedia of Systems Biology*. Springer-Verlag. ISBN 978-1-4419-9864-4.

26.8 External links

- Biological Systems in bio-physics-wiki

Chapter 27

Dynamic network analysis

Dynamic network analysis (DNA) is an emergent scientific field that brings together traditional social network analysis (SNA), link analysis (LA), social simulation and multi-agent systems (MAS) within network science and network theory. There are two aspects of this field. The first is the statistical analysis of DNA data. The second is the utilization of simulation to address issues of network dynamics. DNA networks vary from traditional social networks in that they are larger, dynamic, multi-mode, multi-plex networks, and may contain varying levels of uncertainty. The main difference of DNA to SNA is that DNA takes interactions of social features conditioning structure and behavior of networks into account. DNA is tied to temporal analysis but temporal analysis is not necessarily tied to DNA, as changes in networks sometimes result from external factors which are independent of social features found in networks. One of the most notable and earliest of cases in the use of DNA is in Sampson's monastery study, where he took snapshots of the same network from different intervals and observed and analyzed the evolution of the network.[1]

DNA statistical tools are generally optimized for large-scale networks and admit the analysis of multiple networks simultaneously in which, there are multiple types of nodes (multi-node) and multiple types of links (multi-plex). Multi-node multi-plex networks are generally referred to as meta-networks or high-dimensional networks. In contrast, SNA statistical tools focus on single or at most two mode data and facilitate the analysis of only one type of link at a time.

DNA statistical tools tend to provide more measures to the user, because they have measures that use data drawn from multiple networks simultaneously. Latent space models (Sarkar and Moore, 2005)[2] and agent-based simulation are often used to examine dynamic social networks (Carley et al., 2009).[3] From a computer simulation perspective, nodes in DNA are like atoms in quantum theory, nodes can be, though need not be, treated as probabilistic. Whereas nodes in a traditional SNA model are static, nodes in a DNA model have the ability to learn. Properties change over time; nodes can adapt: A company's employees can learn new skills and increase their value to the network; or,

capture one terrorist and three more are forced to improvise. Change propagates from one node to the next and so on. DNA adds the element of a network's evolution and considers the circumstances under which change is likely to occur.

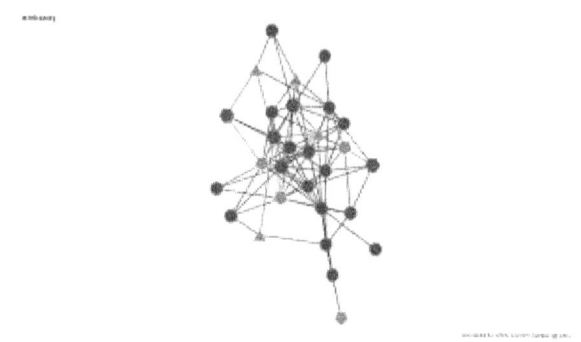

An example of a multi-entity, multi-network, dynamic network diagram

There are three main features to dynamic network analysis that distinguish it from standard social network analysis. First, rather than just using social networks, DNA looks at meta-networks. Second, agent-based modeling and other forms of simulations are often used to explore how networks evolve and adapt as well as the impact of interventions on those networks. Third, the links in the network are not binary; in fact, in many cases they represent the probability that there is a link.

27.1 Meta-network

A meta-network is a multi-mode, multi-link, multi-level network. Multi-mode means that there are many types of nodes; e.g., nodes people and locations. Multi-link means that there are many types of links; e.g., friendship and advice. Multi-level means that some nodes may be members of other nodes, such as a network composed of people and organizations and one of the links is who is a member of which organization.

While different researchers use different modes, common modes reflect who, what, when, where, why and how. A simple example of a meta-network is the PCANS formulation with people, tasks, and resources.[4] A more detailed formulation considers people, tasks, resources, knowledge, and organizations.[5] The ORA tool was developed to support meta-network analysis.[6]

27.2 Illustrative problems that people in the DNA area work on

- Developing metrics and statistics to assess and identify change within and across networks.
- Developing and validating simulations to study network change, evolution, adaptation, decay. See Computer simulation and organizational studies
- Developing and testing theory of network change, evolution, adaptation, decay[7]
- Developing and validating formal models of network generation and evolution
- Developing techniques to visualize network change overall or at the node or group level
- Developing statistical techniques to see whether differences observed over time in networks are due to simply different samples from a distribution of links and nodes or changes over time in the underlying distribution of links and nodes
- Developing control processes for networks over time
- Developing algorithms to change distributions of links in networks over time
- Developing algorithms to track groups in networks over time
- Developing tools to extract or locate networks from various data sources such as texts
- Developing statistically valid measurements on networks over time
- Examining the robustness of network metrics under various types of missing data
- Empirical studies of multi-mode multi-link multi-time period networks
- Examining networks as probabilistic time-variant phenomena
- Forecasting change in existing networks
- Identifying trails through time given a sequence of networks
- Identifying changes in node criticality given a sequence of networks anything else related to multi-mode multi-link multi-time period networks
- Studying random walks on temporal networks[8]
- Quantifying structural properties of contact sequences in dynamic networks, which influence dynamical processes[9]
- Assessment of covert activity[10] and dark networks[11]
- Citational analysis[12]
- Social media analysis[13]
- Assessment of public health systems[14]
- Analysis of hospital safety outcomes[15]
- Assessment of the structure of ethnic violence from news data[16]
- Assessment of terror groups[17]
- Online social decay of social interactions[18]
- Visualization of large financial networks over time[19]

27.3 See also

- Graph dynamical system
- International Network for Social Network Analysis
- Kathleen M. Carley
- Network dynamics
- Network science
- Sequential dynamical system

27.4 References

[1] Identity and control: a structural theory of social action By Harrison C. White

[2] Purnamrita Sarkar and Andrew W. Moore. 2005. Dynamic social network analysis using latent space models. SIGKDD Explor. Newsl. 7, 2 (December 2005), 31-40.

[3] Kathleen M. Carley, Michael K. Martin and Brian Hirshman, 2009, "The Etiology of Social Change," Topics in Cognitive Science, 1.4:621-650

[4] David Krackhardt and Kathleen M. Carley. 1998. "A PCANS Model of Structure in Organization." In proceedings of the 1998 International Symposium on Command and Control Research and Technology, Monterey, CA, June 1998, Evidence Based Research, Vienna, VA. Pp. 113-119.

[5] Kathleen M. Carley, 2002, "Smart Agents and Organizations of the Future," The Handbook of New Media. Edited by Leah Lievrouw and Sonia Livingstone (Eds.), Thousand Oaks, CA, Sage, Ch. 12: 206-220.

[6] Kathleen M. Carley. 2014. "ORA: A Toolkit for Dynamic Network Analysis and Visualization," In Reda Alhajj and Jon Rokne (Eds.) Encyclopedia of Social Network Analysis and Mining. Springer.

[7] Majdandzic, A.; et al. (2013). "Spontaneous recovery in dynamical networks". *Nature Physics*. doi:10.1038/nphys2819.

[8] Michele Starnini, Andrea Baronchelli, Alain Barrat, 2012, Random walks on temporal networks. Phys. Rev. E 85, 056115, http://link.aps.org/doi/10.1103/PhysRevE.85.056115

[9] René Pfitzner, Ingo Scholtes, Antonios Garas, Claudio Juan Tessone, Frank Schweitzer, 2012, "Betweenness Preference: Quantifying Correlations in the Topological Dynamics of Temporal Networks". Physical Review Letters, Vol. 110, May 10, 2013.

[10] Carley, Kathleen M., Michael K., Martin and John P. Hancock, 2009, "Dynamic Network Analysis Applied to Experiments from the Decision Architectures Research Environment." Advanced Decision Architectures for the Warfighter: Foundation and Technology, Ch. 4.

[11] Everton, Sean, 2012, Disrupting Dark Networks, Cambridge University Press, New York, NY

[12] Kas, Miray, Kathleen M. Carley and L. Richard Carley, 2012, "Who was Where, When? Spatiotemporal Analysis of Researcher Mobility in Nuclear Science," In proceedings of the International Workshop on Spatio Temporal data Integration and Retrieval (STIR 2012), held in conjunction with ICDE 2012, April 1, 2012, Washington D.C.

[13] Carley, Kathleen. M., Jürgen Pfeffer, Huan Liu, Fred Morstatter, Rebecca Goolsby, 2013, Near Real Time Assessment of Social Media Using Geo-Temporal Network Analytics, In Proceedings of 2013 IEEE/ACM International Conference on Advances in Social Networks Analysis and Mining (ASONAM), August 25–28, 2013, Niagara Falls, Canada.

[14] Merrill, Jacqueline, Mark G. Orr, Christie Y. Jeon, Rosalind V. Wilson, Jonathan Storrick and Kathleen M. Carley, 2012, "Topology of Local Health Officials' Advice Networks: Mind the Gaps," Journal of Public Health Management Practice, 18(6): 602–608

[15] Effken, Judith A.,Sheila Gephart and Kathleen M. Carley. 2013. "Using ORA to Assess the Relationship of Handoffs to Quality and Safety Outcomes," CIN: Computers, Informatics, Nursing. 31(1): 36-44.

[16] Van Holt, Tracy, Jeffrey C. Johnson, Jamie Brinkley, Kathleen M. Carley and Janna Caspersen, 2012, "Structure of ethnic violence in Sudan: an automated content, meta-network and geospatial analytical approach," Computational and Mathematical Organization Theory, 18:340-355.

[17] Kenney, Michael J., John Horgan, Cale Horne, Peter Vining, Kathleen M. Carley, Michael Bigrigg, Mia Bloom, Kurt Braddock, 2012, Organizational adaptation in an activist network: Social networks, leadership, and change in al-Muhajiroun, Applied Ergonomics, 44(5):739-747.

[18] M. Abufouda, K. A. Zweig ."A Theoretical Model for Understanding the Dynamics of Online Social Networks Decay". arXiv preprint arXiv:1610.01538.

[19] Heijmans, Ronald; Heuver, Richard; Levallois, Clement; van Lelyveld, Iman (2016). "Dynamic visualization of large financial networks". *The Journal of Network Theory in Finance*. **2** (2): 57–79. ISSN 2055-7795. doi:10.21314/JNTF.2016.017.

27.5 Further reading

- Kathleen M. Carley, 2003, "Dynamic Network Analysis" in Dynamic Social Network Modeling and Analysis: Workshop Summary and Papers, Ronald Breiger, Kathleen Carley, and Philippa Pattison, (Eds.) Committee on Human Factors, National Research Council. National Research Council. Pp. 133–145, Washington, DC.

- Kathleen M. Carley, 2002, "Smart Agents and Organizations of the Future" The Handbook of New Media. Edited by Leah Lievrouw and Sonia Livingstone, Ch. 12, pp. 206–220, Thousand Oaks, CA, Sage.

- Kathleen M. Carley, Jana Diesner, Jeffrey Reminga, Maksim Tsvetovat, 2008, Toward an Interoperable Dynamic Network Analysis Toolkit, DSS Special Issue on Cyberinfrastructure for Homeland Security: Advances in Information Sharing, Data Mining, and Collaboration Systems. Decision Support Systems 43(4):1324-1347 (article 20)

- Terrill L. Frantz, Kathleen M. Carley, 2009, Toward A Confidence Estimate For The Most-Central-Actor Finding. Academy of Management Annual Conference, Chicago, IL, USA, 7–11 August. (Awarded the Sage Publications/RM Division Best Student Paper Award)

- Petter Holme, Jari Saramäki, 2011, "Temporal networks". http://arxiv.org/abs/1108.1780

- C. Aggarwal, K. Subbian, 2014, "Evolutionary Network Analysis: A Survey". ACM Computing Surveys, 47(1). (pdf)

27.6 External links

- Radcliffe Exploratory Seminar on Dynamic Networks
- Center for Computational Analysis of Social and Organizational Systems (CASOS)

Chapter 28

Complex adaptive system

A **complex adaptive system** is a system in which a perfect understanding of the individual parts does not automatically convey a perfect understanding of the whole system's behavior.[1] The study of complex adaptive systems is highly interdisciplinary and blends insights from the natural and social sciences to develop system-level models and insights that allow for heterogeneous agents, phase transition, and emergent behavior.[2]

They are *complex* in that they are dynamic networks of interactions, and their relationships are not aggregations of the individual static entities, i.e., the behavior of the ensemble is not predicted by the behavior of the components. They are *adaptive* in that the individual and collective behavior mutate and self-organize corresponding to the change-initiating micro-event or collection of events.[3][4][1] They are a "complex macroscopic collection" of relatively "similar and partially connected micro-structures" formed in order to adapt to the changing environment and increase their survivability as a macro-structure.[3][4][5]

28.1 Overview

The term *complex adaptive systems*, or *complexity science*, is often used to describe the loosely organized academic field that has grown up around the study of such systems. Complexity science is not a single theory—it encompasses more than one theoretical framework and is highly interdisciplinary, seeking the answers to some fundamental questions about living, adaptable, changeable systems. The study of CAS focuses on complex, emergent and macroscopic properties of the system.[5][6][7] John H. Holland said that CAS "are systems that have a large numbers of components, often called agents, that interact and adapt or learn."[8]

Typical examples of complex adaptive systems include: climate; cities; firms; markets; governments; industries; ecosystems; social networks; power grids; animal swarms; traffic flows; social insect (e.g. ant) colonies;[9] the brain and the immune system; and the cell and the developing embryo. Human social group-based endeavors, such as political parties, communities, geopolitical organizations, war, and terrorist networks are also considered CAS.[9][10][11] The internet and cyberspace—composed, collaborated, and managed by a complex mix of human–computer interactions, is also regarded as a complex adaptive system.[12][13][14] CAS can be hierarchical, but more often exhibit aspects of "self-organization."[15]

28.1.1 General properties

What distinguishes a CAS from a pure multi-agent system (MAS) is the focus on top-level properties and features like self-similarity, complexity, emergence and self-organization. A MAS is defined as a system composed of multiple interacting agents; whereas in CAS, the agents as well as the system are adaptive and the system is self-similar. A CAS is a complex, self-similar collectivity of interacting, adaptive agents. Complex Adaptive Systems are characterized by a high degree of adaptive capacity, giving them resilience in the face of perturbation.

Other important properties are adaptation (or homeostasis), communication, cooperation, specialization, spatial and temporal organization, and reproduction. They can be found on all levels: cells specialize, adapt and reproduce themselves just like larger organisms do. Communication and cooperation take place on all levels, from the agent to the system level. The forces driving co-operation between agents in such a system, in some cases, can be analyzed with game theory.

28.1.2 Characteristics

Some of the most important characteristics of complex systems are:[16]

- The number of elements is sufficiently large that conventional descriptions (e.g. a system of differential equations) are not only impractical, but cease to assist in understanding the system. Moreover, the elements interact dynamically, and the interactions can be physical or involve the exchange of information

- Such interactions are rich, i.e. any element or subsystem in the system is affected by and affects several other elements or sub-systems

- The interactions are non-linear: small changes in inputs, physical interactions or stimuli can cause large effects or very significant changes in outputs

- Interactions are primarily but not exclusively with immediate neighbours and the nature of the influence is modulated

- Any interaction can feed back onto itself directly or after a number of intervening stages. Such feedback can vary in quality. This is known as *recurrency*

- The overall behavior of the system of elements is not predicted by the behavior of the individual elements

- Such systems may be open and it may be difficult or impossible to define system boundaries

- Complex systems operate under far from equilibrium conditions. There has to be a constant flow of energy to maintain the organization of the system

- Complex systems have a history. They evolve and their past is co-responsible for their present behaviour

- Elements in the system may be ignorant of the behaviour of the system as a whole, responding only to the information or physical stimuli available to them locally

Robert Axelrod & Michael D. Cohen[17] identify a series of key terms from a modeling perspective:

- **Strategy**, a conditional action pattern that indicates what to do in which circumstances

- **Artifact**, a material resource that has definite location and can respond to the action of agents

- **Agent**, a collection of properties, strategies & capabilities for interacting with artifacts & other agents

- **Population**, a collection of agents, or, in some situations, collections of strategies

- **System**, a larger collection, including one or more populations of agents and possibly also artifacts

- **Type**, all the agents (or strategies) in a population that have some characteristic in common

- **Variety**, the diversity of types within a population or system

- **Interaction pattern**, the recurring regularities of contact among types within a system

- **Space (physical)**, location in geographical space & time of agents and artifacts

- **Space (conceptual)**, "location" in a set of categories structured so that "nearby" agents will tend to interact

- **Selection**, processes that lead to an increase or decrease in the frequency of various types of agent or strategies

- **Success criteria** or **performance measures**, a "score" used by an agent or designer in attributing credit in the selection of relatively successful (or unsuccessful) strategies or agents

28.2 Modeling and simulation

CAS are occasionally modeled by means of agent-based models and complex network-based models.[18] Agent-based models are developed by means of various methods and tools primarily by means of first identifying the different agents inside the model.[19] Another method of developing models for CAS involves developing complex network models by means of using interaction data of various CAS components.[20]

Recently, SpringerOpen/BioMed Central has launched an online open-access journal on the topic of *complex adaptive systems modeling* (CASM).[21]

28.3 Evolution of complexity

Main article: Evolution of biological complexity

Living organisms are complex adaptive systems. Although complexity is hard to quantify in biology, evolution has produced some remarkably complex organisms.[22] This observation has led to the common misconception of evolution being progressive and leading towards what are viewed as "higher organisms".[23]

If this were generally true, evolution would possess an active trend towards complexity. As shown below, in this type of process the value of the most common amount of complexity would increase over time.[24] Indeed, some artificial life simulations have suggested that the generation of CAS is an inescapable feature of evolution.[25][26]

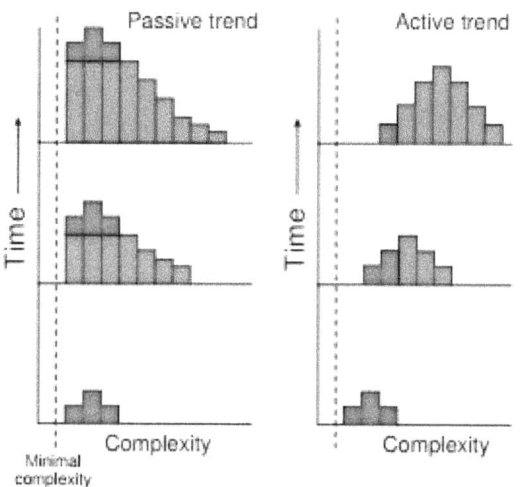

Passive versus active trends in the evolution of complexity. CAS at the beginning of the processes are colored red. Changes in the number of systems are shown by the height of the bars, with each set of graphs moving up in a time series.

However, the idea of a general trend towards complexity in evolution can also be explained through a passive process.[24] This involves an increase in variance but the most common value, the mode, does not change. Thus, the maximum level of complexity increases over time, but only as an indirect product of there being more organisms in total. This type of random process is also called a bounded random walk.

In this hypothesis, the apparent trend towards more complex organisms is an illusion resulting from concentrating on the small number of large, very complex organisms that inhabit the right-hand tail of the complexity distribution and ignoring simpler and much more common organisms. This passive model emphasizes that the overwhelming majority of species are microscopic prokaryotes,[27] which comprise about half the world's biomass[28] and constitute the vast majority of Earth's biodiversity.[29] Therefore, simple life remains dominant on Earth, and complex life appears more diverse only because of sampling bias.

This lack of an overall trend towards complexity in biology does not preclude the existence of forces driving systems towards complexity in a subset of cases. These minor trends are balanced by other evolutionary pressures that drive systems towards less complex states.

28.4 See also

- Artificial life
- Chaos theory
- Cognitive science
- Command and Control Research Program
- Complex system
- Computational economics
- Computational sociology
- Dual-phase evolution
- Enterprise systems engineering
- Generative sciences
- Open system (systems theory)
- Santa Fe Institute
- Simulated reality
- Sociology and complexity science
- Super wicked problem
- Swarm Development Group
- Universal Darwinism

28.5 References

[1] Miller, John H., and Scott E. Page (2007-01-01). *Complex adaptive systems : an introduction to computational models of social life*. Princeton University Press. ISBN 9781400835522. OCLC 760073369.

[2] Auerbach, David (2016-01-19). "The Theory of Everything and Then Some". *Slate*. ISSN 1091-2339. Retrieved 2017-03-07.

[3] "Insights from Complexity Theory: Understanding Organisations better". by Assoc. Prof. Amit Gupta, Student contributor - S. Anish, IIM Bangalore. Retrieved 1 June 2012.

[4] "Ten Principles of Complexity & Enabling Infrastructures". by Professor Eve Mitleton-Kelly, Director Complexity Research Programme, London School of Economics. Retrieved 1 June 2012.

[5] "Evolutionary Psychology, Complex Systems, and Social Theory" (PDF). *Bruce MacLennan, Department of Electrical Engineering & Computer Science, University of Tennessee, Knoxville*. eecs.utk.edu. Retrieved 25 August 2012.

[6] "A Complex Adaptive Organization Under the Lens of the LIFE Model:The Case of Wikipedia". Retrieved 25 August 2012.

[7] "Complex Adaptive Systems as a Model for Evaluating Organisational : Change Caused by the Introduction of Health Information Systems" (PDF). *Kieren Diment, Ping Yu, Karin Garrety, Health Informatics Research Lab, Faculty of Informatics, University of Wollongong, School of Management, University of Wollongong, NSW.* uow.edu.au. Retrieved 25 August 2012.

[8] Holland John H (2006). "Studying Complex Adaptive Systems". *Journal of Systems Science and Complexity.* **19** (1): 1–8. doi:10.1007/s11424-006-0001-z.

[9] Steven Strogatz, Duncan J. Watts and Albert-Laszlo Barabasi "explaining synchronicity *(at 6:08)*, network theory, self-adaptation mechanism of complex systems, Six Degrees of separation, Small world phenomenon, events are never isolated as they depend upon each other *(at 27:07)* in the BBC / Discovery Documentary". *BBC / Discovery.* Retrieved 11 June 2012. "Unfolding the science behind the idea of six degrees of separation"

[10] "Toward a Complex Adaptive Intelligence Community The Wiki and the Blog". *D. Calvin Andrus.* cia.gov. Retrieved 25 August 2012.

[11] Solvit, Samuel (2012). "Dimensions of War: Understanding War as a Complex Adaptive System". L'Harmattan. Retrieved 25 August 2013.

[12] "The Internet Analyzed as a Complex Adaptive System". Retrieved 25 August 2012.

[13] "Cyberspace: The Ultimate Complex Adaptive System" (PDF). The International C2 Journal. Retrieved 25 August 2012. by Paul W. Phister Jr

[14] "Complex Adaptive Systems" (PDF). mit.edu. 2001. Retrieved 25 August 2012. by Serena Chan, Research Seminar in Engineering Systems

[15] Holland, John H. (John Henry). (1996). *Hidden order : how adaptation builds complexity.* Addison-Wesley. ISBN 0201442302. OCLC 970420200.

[16] Paul Cilliers (1998) *Complexity and Postmodernism: Understanding Complex Systems*

[17] Robert Axelrod & Michael D. Cohen, *Harnessing Complexity.* Basic Books, 2001

[18] Muaz A. K. Niazi, Towards A Novel Unified Framework for Developing Formal, Network and Validated Agent-Based Simulation Models of Complex Adaptive Systems PhD Thesis

[19] John H. Miller & Scott E. Page, Complex Adaptive Systems: An Introduction to Computational Models of Social Life, Princeton University Press Book page

[20] Melanie Mitchell, Complexity A Guided Tour, Oxford University Press, Book page

[21] Springer *Complex Adaptive Systems Modeling Journal* (CASM)

[22] Adami C (2002). "What is complexity?". *BioEssays.* **24** (12): 1085–94. PMID 12447974. doi:10.1002/bies.10192.

[23] McShea D (1991). "Complexity and evolution: What everybody knows". *Biology and Philosophy.* **6** (3): 303–24. doi:10.1007/BF00132234.

[24] Carroll SB (2001). "Chance and necessity: the evolution of morphological complexity and diversity". *Nature.* **409** (6823): 1102–9. Bibcode:2001Natur.409.1102C. PMID 11234024. doi:10.1038/35059227.

[25] Furusawa C, Kaneko K (2000). "Origin of complexity in multicellular organisms". *Phys. Rev. Lett.* **84** (26 Pt 1): 6130–3. Bibcode:2000PhRvL..84.6130F. PMID 10991141. arXiv:nlin/0009008. doi:10.1103/PhysRevLett.84.6130.

[26] Adami C, Ofria C, Collier TC (2000). "Evolution of biological complexity". *Proc. Natl. Acad. Sci. U.S.A.* **97** (9): 4463–8. Bibcode:2000PNAS...97.4463A. PMC 18257. PMID 10781045. arXiv:physics/0005074. doi:10.1073/pnas.97.9.4463.

[27] Oren A (2004). "Prokaryote diversity and taxonomy: current status and future challenges". *Philos. Trans. R. Soc. Lond., B, Biol. Sci.* **359** (1444): 623–38. PMC 1693353. PMID 15253349. doi:10.1098/rstb.2003.1458.

[28] Whitman W, Coleman D, Wiebe W (1998). "Prokaryotes: the unseen majority". *Proc Natl Acad Sci USA.* **95** (12): 6578–83. Bibcode:1998PNAS...95.6578W. PMC 33863. PMID 9618454. doi:10.1073/pnas.95.12.6578.

[29] Schloss P, Handelsman J (2004). "Status of the microbial census". *Microbiol Mol Biol Rev.* **68** (4): 686–91. PMC 539005. PMID 15590780. doi:10.1128/MMBR.68.4.686-691.2004.

28.6 Literature

- Ahmed E, Elgazzar AS, Hegazi AS (28 June 2005). "An overview of complex adaptive systems". *Mansoura J. Math.* **32**: 6059. Bibcode:2005nlin......6059A. arXiv:nlin/0506059. arXiv:nlin/0506059v1 [nlin.AO].

- Bullock S, Cliff D (2004). "Complexity and Emergent Behaviour in ICT Systems". Hewlett-Packard Labs. HP-2004-187.; commissioned as a report by the UK government's Foresight Programme.

- Dooley, K., *Complexity in Social Science* glossary a research training project of the European Commission.

- Edwin E. Olson; Glenda H. Eoyang (2001). *Facilitating Organization Change.* San Francisco: Jossey-Bass. ISBN 0-7879-5330-X.

- Gell-Mann, Murray (1994). *The quark and the jaguar: adventures in the simple and the complex*. San Francisco: W.H. Freeman. ISBN 0-7167-2581-9.

- Holland, John H. (1992). *Adaptation in natural and artificial systems: an introductory analysis with applications to biology, control, and artificial intelligence*. Cambridge, Mass: MIT Press. ISBN 0-262-58111-6.

- Holland, John H. (1999). *Emergence: from chaos to order*. Reading, Mass: Perseus Books. ISBN 0-7382-0142-1.

- Solvit, Samuel (2012). *Dimensions of War: Understanding War as a Complex Adaptive System*. Paris, France: L'Harmattan. ISBN 978-2-296-99721-9.

- Kelly, Kevin (1994). *Out of control: the new biology of machines, social systems and the economic world* (Full text available online). Boston: Addison-Wesley. ISBN 0-201-48340-8.

- Pharaoh, M.C. (online). Looking to systems theory for a reductive explanation of phenomenal experience and evolutionary foundations for higher order thought Retrieved 15 January 2008.

- Hobbs, George & Scheepers, Rens (2010)."Agility in Information Systems: Enabling Capabilities for the IT Function," *Pacific Asia Journal of the Association for Information Systems*: Vol. 2: Iss. 4, Article 2. Link

- Sidney Dekker (2011). *Drift into Failure: From Hunting Broken Components to Understanding Complex Systems*. CRC Press.

28.7 External links

- Complexity Digest comprehensive digest of latest CAS related news and research.

- Complex Adaptive Systems Group loosely coupled group of scientists and software engineers interested in complex adaptive systems

- DNA Wales Research Group Current Research in Organisational change CAS/CES related news and free research data. Also linked to the Business Doctor & BBC documentary series

- A description of complex adaptive systems on the Principia Cybernetica Web.

- Quick reference single-page description of the 'world' of complexity and related ideas hosted by the Center for the Study of Complex Systems at the University of Michigan.

- Complex systems research network

- The Open Agent-Based Modeling Consortium

- A group of multidisciplinary researchers interested in the Modeling and Simulation of Complex Adaptive Systems

- TEDxRotterdam - Igor Nikolic - Complex adaptive systems, and The emergence of universal consciousness: Brendan Hughes at TEDxPretoria . Talks discussing various practical examples of complex adaptive systems, including Wikipedia, star galaxies, genetic mutation, and other examples

Chapter 29

Evolution

This article is about evolution in biology. For related articles, see Outline of evolution. For other uses, see Evolution (disambiguation).

For a more accessible and less technical introduction to this topic, see Introduction to evolution.

Evolution is change in the heritable characteristics of biological populations over successive generations.[1][2] Evolutionary processes give rise to biodiversity at every level of biological organisation, including the levels of species, individual organisms, and molecules.[3]

Repeated formation of new species (speciation), change within species (anagenesis), and loss of species (extinction) throughout the evolutionary history of life on Earth are demonstrated by shared sets of morphological and biochemical traits, including shared DNA sequences.[4] These shared traits are more similar among species that share a more recent common ancestor, and can be used to reconstruct a biological "tree of life" based on evolutionary relationships (phylogenetics), using both existing species and fossils. The fossil record includes a progression from early biogenic graphite,[5] to microbial mat fossils,[6][7][8] to fossilised multicellular organisms. Existing patterns of biodiversity have been shaped both by speciation and by extinction.[9]

In the mid-19th century, Charles Darwin formulated the scientific theory of evolution by natural selection, published in his book *On the Origin of Species* (1859). Evolution by natural selection is a process demonstrated by the observation that more offspring are produced than can possibly survive, along with three facts about populations: 1) traits vary among individuals with respect to morphology, physiology, and behaviour (phenotypic variation), 2) different traits confer different rates of survival and reproduction (differential fitness), and 3) traits can be passed from generation to generation (heritability of fitness).[10] Thus, in successive generations members of a population are replaced by progeny of parents better adapted to survive and reproduce in the biophysical environment in which natural selection takes place.

This teleonomy is the quality whereby the process of natural selection creates and preserves traits that are seemingly fitted for the functional roles they perform.[11] The processes by which the changes occur, from one generation to another, are called evolutionary processes or mechanisms.[12] The four most widely recognised evolutionary processes are natural selection (including sexual selection), genetic drift, mutation and gene migration due to genetic admixture.[12] Natural selection and genetic drift sort variation; mutation and gene migration create variation.[12]

Consequences of selection can include meiotic drive[13] (unequal transmission of certain alleles), nonrandom mating[14] and genetic hitchhiking. In the early 20th century the modern evolutionary synthesis integrated classical genetics with Darwin's theory of evolution by natural selection through the discipline of population genetics. The importance of natural selection as a cause of evolution was accepted into other branches of biology. Moreover, previously held notions about evolution, such as orthogenesis, evolutionism, and other beliefs about innate "progress" within the largest-scale trends in evolution, became obsolete.[15] Scientists continue to study various aspects of evolutionary biology by forming and testing hypotheses, constructing mathematical models of theoretical biology and biological theories, using observational data, and performing experiments in both the field and the laboratory.

All life on Earth shares a common ancestor known as the last universal common ancestor (LUCA),[16][17][18] which lived approximately 3.5–3.8 billion years ago.[19] This should not be assumed to be the first living organism on Earth; a study in 2015 found "remains of biotic life" from 4.1 billion years ago in ancient rocks in Western Australia.[20][21] In July 2016, scientists reported identifying a set of 355 genes from the LUCA of all organisms living on Earth.[22] More than 99 percent of all species that ever lived on Earth are estimated to be extinct.[23][24] Estimates of Earth's current species range from 10 to 14 million,[25][26] of which about 1.9 million are estimated to have been named[27] and 1.6 million documented in a central database to date.[28] More recently, in May 2016, scientists reported that 1 trillion species are estimated to be on Earth currently with only one-thousandth of one percent described.[29]

In terms of practical application, an understanding of evolution has been instrumental to developments in numerous scientific and industrial fields, including agriculture, human and veterinary medicine, and the life sciences in general.[30][31][32] Discoveries in evolutionary biology have made a significant impact not just in the traditional branches of biology but also in other academic disciplines, including biological anthropology, and evolutionary psychology.[33][34] Evolutionary computation, a sub-field of artificial intelligence, involves the application of Darwinian principles to problems in computer science.

29.1 History of evolutionary thought

Alfred Russel Wallace

29.1.1 Classical times

The proposal that one type of organism could descend from another type goes back to some of the first pre-Socratic Greek philosophers, such as Anaximander and Empedocles.[36] Such proposals survived into Roman times. The poet and philosopher Lucretius followed Empedocles in his masterwork *De rerum natura* (*On the Nature of Things*).[37][38]

29.1.2 Medieval

In contrast to these materialistic views, Aristotelianism considered all natural things as actualisations of fixed natural possibilities, known as forms.[39][40] This was part of a medieval teleological understanding of nature in which all things have an intended role to play in a divine cosmic order. Variations of this idea became the standard understanding of the Middle Ages and were integrated into Christian learning, but Aristotle did not demand that real types of organisms always correspond one-for-one with exact metaphysical forms and specifically gave examples of how new types of living things could come to be.[41]

Lucretius

Main article: History of evolutionary thought

Thomas Robert Malthus

In 1842, Charles Darwin penned his first sketch of On the Origin of Species.[35]

29.1.3 Pre-Darwinian

In the 17th century, the new method of modern science rejected the Aristotelian approach. It sought explanations of natural phenomena in terms of physical laws that were the same for all visible things and that did not require the existence of any fixed natural categories or divine cosmic order. However, this new approach was slow to take root in the biological sciences, the last bastion of the concept of fixed natural types. John Ray applied one of the previously more general terms for fixed natural types, "species," to plant and animal types, but he strictly identified each type of living thing as a species and proposed that each species could be defined by the features that perpetuated themselves generation after generation.[42] The biological classification introduced by Carl Linnaeus in 1735 explicitly recognised the hierarchical nature of species relationships, but still viewed species as fixed according to a divine plan.[43]

Other naturalists of this time speculated on the evolutionary change of species over time according to natural laws. In 1751, Pierre Louis Maupertuis wrote of natural modifications occurring during reproduction and accumulating over many generations to produce new species.[44] Georges-Louis Leclerc, Comte de Buffon suggested that species could degenerate into different organisms, and Erasmus Darwin proposed that all warm-blooded animals could have descended from a single microorganism (or "filament").[45] The first full-fledged evolutionary scheme was Jean-Baptiste Lamarck's "transmutation" theory of 1809,[46] which envisaged spontaneous generation continually producing simple forms of life that developed greater complexity in parallel lineages with an inherent progressive tendency, and postulated that on a local level these lineages adapted to the environment by inheriting changes caused by their use or disuse in parents.[47][48] (The latter process was later called Lamarckism.)[47][49][50][51] These ideas were condemned by established naturalists as speculation lacking empirical support. In particular, Georges Cuvier insisted that species were unrelated and fixed, their similarities reflecting divine design for functional needs. In the meantime, Ray's ideas of benevolent design had been developed by William Paley into the *Natural Theology or Evidences of the Existence and Attributes of the Deity* (1802), which proposed complex adaptations as evidence of divine design and which was admired by Charles Darwin.[52][53][54]

29.1.4 Darwinian revolution

The crucial break from the concept of constant typological classes or types in biology came with the theory of evolution through natural selection, which was formulated by Charles Darwin in terms of variable populations. Partly influenced by *An Essay on the Principle of Population* (1798) by Thomas Robert Malthus, Darwin noted that population growth would lead to a "struggle for existence" in which favorable variations prevailed as others perished. In each generation, many offspring fail to survive to an age of reproduction because of limited resources. This could explain the diversity of plants and animals from a common ancestry through the working of natural laws in the same way for all types of organism.[55][56][57][58] Darwin developed his theory of "natural selection" from 1838 onwards and was writing up his "big book" on the subject when Alfred Russel Wallace sent him a version of virtually the same theory in 1858. Their separate papers were presented together at a 1858 meeting of the Linnean Society of London.[59] At the end of 1859, Darwin's publication of his "abstract" as On the Origin of Species explained natural selection in detail and in a way that led to an increasingly wide acceptance of Darwin's concepts of evolution at the expense of alternative theories. Thomas Henry Huxley applied Darwin's ideas to humans, using paleontology and comparative anatomy to provide strong evidence that humans and apes shared a common ancestry. Some were disturbed by this since it implied that humans did not have a special place in the universe.[60]

29.1.5 Pangenesis and heredity

The mechanisms of reproductive heritability and the origin of new traits remained a mystery. Towards this end, Darwin developed his provisional theory of pangenesis.[61] In 1865, Gregor Mendel reported that traits were inherited in a predictable manner through the independent assortment and segregation of elements (later known as genes). Mendel's laws of inheritance eventually supplanted most of Darwin's pangenesis theory.[62] August Weismann made the important distinction between germ cells that give rise to gametes (such as sperm and egg cells) and the somatic cells of the body, demonstrating that heredity passes through the germ line only. Hugo de Vries connected Darwin's pangenesis theory to Weismann's germ/soma cell distinction and proposed that Darwin's pangenes were concentrated in the cell nucleus and when expressed they could move into the cytoplasm to change the cells structure. De Vries was also one of the researchers who made Mendel's work well-known, believing that Mendelian traits corresponded to the transfer of heritable variations along the germline.[63] To explain how new variants originate, de Vries developed a mutation theory that led to a temporary rift between those who accepted Darwinian evolution and biometricians who allied with de Vries.[48][64][65] In the 1930s, pioneers in the field of population genetics, such as Ronald Fisher, Sewall Wright and J. B. S. Haldane set the foundations of evolution onto a robust statistical philosophy. The false contradiction between Darwin's theory, genetic mutations, and Mendelian inheritance was thus reconciled.[66]

29.1.6 The 'modern synthesis'

Main article: Modern synthesis

In the 1920s and 1930s the so-called modern synthesis connected natural selection and population genetics, based on Mendelian inheritance, into a unified theory that applied generally to any branch of biology. The modern synthesis explained patterns observed across species in populations, through fossil transitions in palaeontology, and complex cellular mechanisms in developmental biology.[48][67] The publication of the structure of DNA by James Watson and Francis Crick in 1953 demonstrated a physical mechanism for inheritance.[68] Molecular biology improved our understanding of the relationship between genotype and phenotype. Advancements were also made in phylogenetic systematics, mapping the transition of traits into a comparative and testable framework through the publication and use of evolutionary trees.[69][70] In 1973, evolutionary biologist Theodosius Dobzhansky penned that "nothing in biology makes sense except in the light of evolution," because it has brought to light the relations of what first seemed disjointed facts in natural history into a coherent explanatory body of knowledge that describes and predicts many observable facts about life on this planet.[71]

29.1.7 Further syntheses

Since then, the modern synthesis has been further extended to explain biological phenomena across the full and integrative scale of the biological hierarchy, from genes to species. This extension, known as evolutionary developmental biology and informally called "evo-devo," emphasises how changes between generations (evolution) acts on patterns of change within individual organisms (development).[72][73][74] Since the beginning of the 21st century and in light of discoveries made in recent decades, some biologists have argued for an extended evolutionary synthesis, which would account for the effects of non-genetic inheritance modes, such as epigenetics, parental effects, ecological and cultural inheritance, and evolvability.[75][76]

29.2 Heredity

Further information: Introduction to genetics, Genetics, Heredity, and Reaction norm

Evolution in organisms occurs through changes in heritable traits—the inherited characteristics of an organism. In

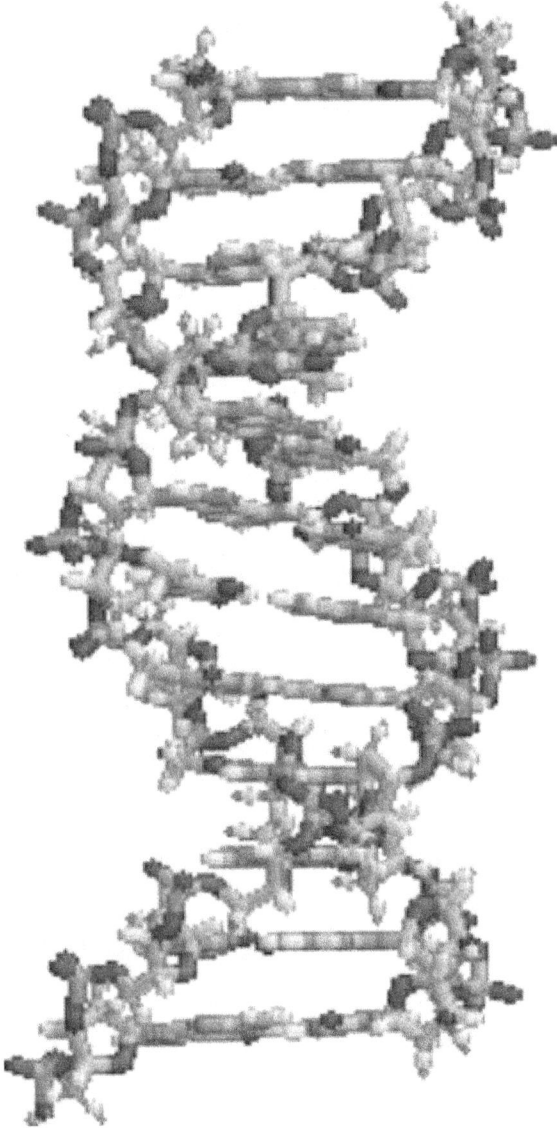

DNA structure. Bases are in the centre, surrounded by phosphate–sugar chains in a double helix.

humans, for example, eye colour is an inherited characteristic and an individual might inherit the "brown-eye trait" from one of their parents.[77] Inherited traits are controlled by genes and the complete set of genes within an organism's genome (genetic material) is called its genotype.[78]

The complete set of observable traits that make up the structure and behaviour of an organism is called its phenotype. These traits come from the interaction of its genotype with the environment.[79] As a result, many aspects of an organism's phenotype are not inherited. For example, suntanned skin comes from the interaction between a person's genotype and sunlight; thus, suntans are not passed on to people's children. However, some people tan more easily than others, due to differences in genotypic variation; a striking example are people with the inherited trait of albinism, who do not tan at all and are very sensitive to sunburn.[80]

Heritable traits are passed from one generation to the next via DNA, a molecule that encodes genetic information.[78] DNA is a long biopolymer composed of four types of bases. The sequence of bases along a particular DNA molecule specify the genetic information, in a manner similar to a sequence of letters spelling out a sentence. Before a cell divides, the DNA is copied, so that each of the resulting two cells will inherit the DNA sequence. Portions of a DNA molecule that specify a single functional unit are called genes; different genes have different sequences of bases. Within cells, the long strands of DNA form condensed structures called chromosomes. The specific location of a DNA sequence within a chromosome is known as a locus. If the DNA sequence at a locus varies between individuals, the different forms of this sequence are called alleles. DNA sequences can change through mutations, producing new alleles. If a mutation occurs within a gene, the new allele may affect the trait that the gene controls, altering the phenotype of the organism.[81] However, while this simple correspondence between an allele and a trait works in some cases, most traits are more complex and are controlled by quantitative trait loci (multiple interacting genes).[82][83]

Recent findings have confirmed important examples of heritable changes that cannot be explained by changes to the sequence of nucleotides in the DNA. These phenomena are classed as epigenetic inheritance systems.[84] DNA methylation marking chromatin, self-sustaining metabolic loops, gene silencing by RNA interference and the three-dimensional conformation of proteins (such as prions) are areas where epigenetic inheritance systems have been discovered at the organismic level.[85][86] Developmental biologists suggest that complex interactions in genetic networks and communication among cells can lead to heritable variations that may underlay some of the mechanics in developmental plasticity and canalisation.[87] Heritability may also occur at even larger scales. For example, ecological inheritance through the process of niche construction is defined by the regular and repeated activities of organisms in their environment. This generates a legacy of effects that modify and feed back into the selection regime of subsequent generations. Descendants inherit genes plus environmental characteristics generated by the ecological actions of ancestors.[88] Other examples of heritability in evolution that are not under the direct control of genes include the inheritance of cultural traits and symbiogenesis.[89][90]

29.3 Variation

White peppered moth

Black morph in peppered moth evolution

Main article: Genetic variation

Further information: Genetic diversity and Population genetics

An individual organism's phenotype results from both its genotype and the influence from the environment it has lived in. A substantial part of the phenotypic variation in a population is caused by genotypic variation.[83] The modern evolutionary synthesis defines evolution as the change over time in this genetic variation. The frequency of one particular allele will become more or less prevalent relative to other forms of that gene. Variation disappears when a new allele reaches the point of fixation—when it either disappears from the population or replaces the ancestral allele entirely.[91]

Natural selection will only cause evolution if there is enough genetic variation in a population. Before the discovery of Mendelian genetics, one common hypothesis was blending inheritance. But with blending inheritance, genetic variance would be rapidly lost, making evolution by natural selection implausible. The Hardy–Weinberg principle provides the solution to how variation is maintained in a population with Mendelian inheritance. The frequencies of alleles (variations in a gene) will remain constant in the absence of selection, mutation, migration and genetic drift.[92]

Variation comes from mutations in the genome, reshuffling of genes through sexual reproduction and migration between populations (gene flow). Despite the constant introduction of new variation through mutation and gene flow, most of the genome of a species is identical in all individuals of that species.[93] However, even relatively small differences in genotype can lead to dramatic differences in phenotype: for example, chimpanzees and humans differ in only about 5% of their genomes.[94]

29.3.1 Mutation

Main article: Mutation

Mutations are changes in the DNA sequence of a cell's genome. When mutations occur, they may alter the product of a gene, or prevent the gene from functioning, or have no effect. Based on studies in the fly *Drosophila melanogaster*, it has been suggested that if a mutation changes a protein produced by a gene, this will probably be harmful, with about 70% of these mutations having damaging effects, and the remainder being either neutral or weakly beneficial.[95]

Mutations can involve large sections of a chromosome becoming duplicated (usually by genetic recombination), which can introduce extra copies of a gene into a genome.[96] Extra copies of genes are a major source of the raw material needed for new genes to evolve.[97] This is important because most new genes evolve within gene families from pre-existing genes that share common ancestors.[98] For example, the human eye uses four genes to make structures that sense light: three for colour vision and one for night vision; all four are descended from a single ancestral gene.[99]

New genes can be generated from an ancestral gene when a duplicate copy mutates and acquires a new function. This process is easier once a gene has been duplicated because it increases the redundancy of the system; one gene in the pair can acquire a new function while the other copy continues to perform its original function.[100][101] Other types of mutations can even generate entirely new genes from previously noncoding DNA.[102][103]

The generation of new genes can also involve small parts of several genes being duplicated, with these fragments then recombining to form new combinations with new functions.[104][105] When new genes are assembled from shuffling pre-existing parts, domains act as modules with simple independent functions, which can be mixed together to produce new combinations with new and complex functions.[106] For example, polyketide synthases are large enzymes that make antibiotics; they contain up to one hundred independent domains that each catalyse one step in the overall process, like a step in an assembly line.[107]

29.3.2 Sex and recombination

Further information: Sexual reproduction, Genetic recombination, and Evolution of sexual reproduction

In asexual organisms, genes are inherited together, or *linked*, as they cannot mix with genes of other organisms during reproduction. In contrast, the offspring of sexual organisms contain random mixtures of their parents' chromosomes that are produced through independent assortment. In a related process called homologous recombination, sexual organisms exchange DNA between two matching chromosomes.[108] Recombination and reassortment do not alter allele frequencies, but instead change which alleles are associated with each other, producing offspring with new combinations of alleles.[109] Sex usually increases genetic variation and may increase the rate of evolution.[110][111]

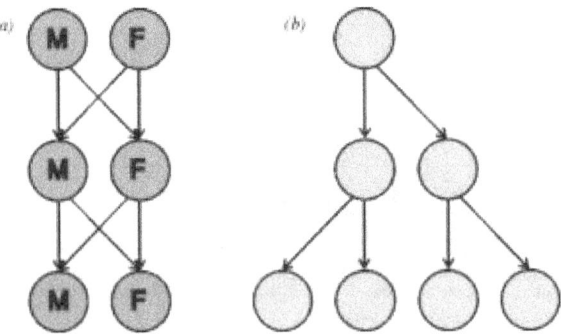

This diagram illustrates the twofold cost of sex. *If each individual were to contribute to the same number of offspring (two), (a) the sexual population remains the same size each generation, where the (b) Asexual reproduction population doubles in size each generation.*

The two-fold cost of sex was first described by John Maynard Smith.[112] The first cost is that in sexually dimorphic species only one of the two sexes can bear young. (This cost does not apply to hermaphroditic species, like most plants and many invertebrates.) The second cost is that any individual who reproduces sexually can only pass on 50% of its genes to any individual offspring, with even less passed on as each new generation passes.[113] Yet sexual reproduction is the more common means of reproduction among eukaryotes and multicellular organisms. The Red Queen hypothesis has been used to explain the significance of sexual reproduction as a means to enable continual evolution and adaptation in response to coevolution with other species in an ever-changing environment.[113][114][115][116]

29.3.3 Gene flow

Further information: Gene flow

Gene flow is the exchange of genes between populations and between species.[117] It can therefore be a source of variation that is new to a population or to a species. Gene flow can be caused by the movement of individuals between separate populations of organisms, as might be caused by the movement of mice between inland and coastal populations, or the movement of pollen between heavy metal tolerant and heavy metal sensitive populations of grasses.

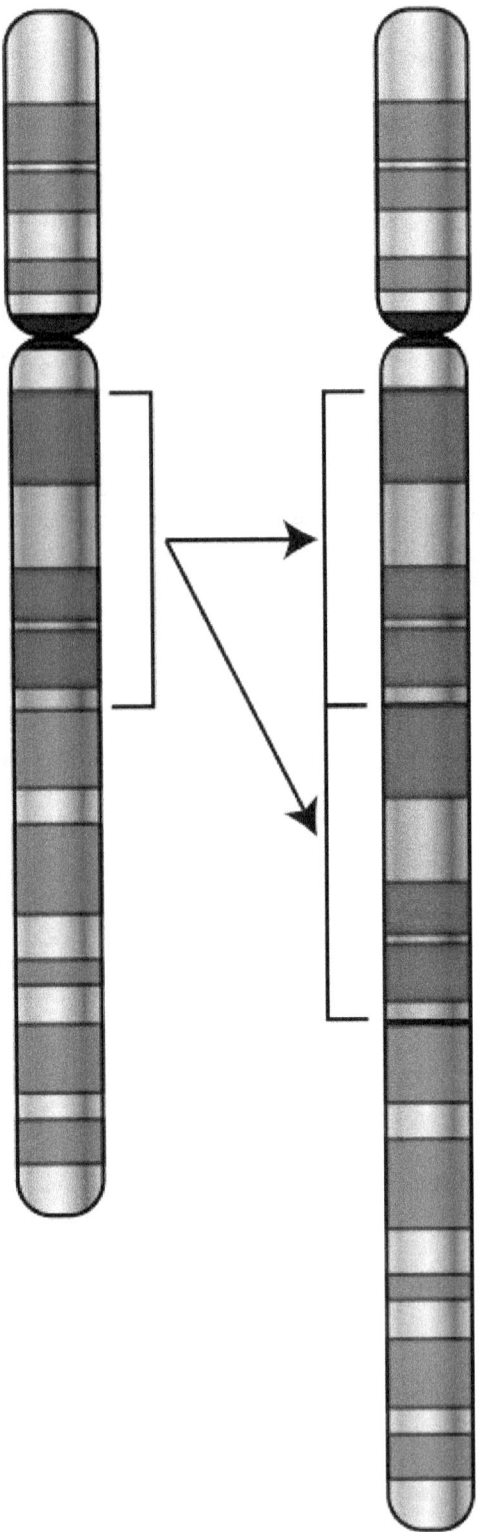

Duplication of part of a chromosome

29.4. MECHANISMS

Gene transfer between species includes the formation of hybrid organisms and horizontal gene transfer. Horizontal gene transfer is the transfer of genetic material from one organism to another organism that is not its offspring; this is most common among bacteria.[118] In medicine, this contributes to the spread of antibiotic resistance, as when one bacteria acquires resistance genes it can rapidly transfer them to other species.[119] Horizontal transfer of genes from bacteria to eukaryotes such as the yeast *Saccharomyces cerevisiae* and the adzuki bean weevil *Callosobruchus chinensis* has occurred.[120][121] An example of larger-scale transfers are the eukaryotic bdelloid rotifers, which have received a range of genes from bacteria, fungi and plants.[122] Viruses can also carry DNA between organisms, allowing transfer of genes even across biological domains.[123]

Large-scale gene transfer has also occurred between the ancestors of eukaryotic cells and bacteria, during the acquisition of chloroplasts and mitochondria. It is possible that eukaryotes themselves originated from horizontal gene transfers between bacteria and archaea.[124]

29.4 Mechanisms

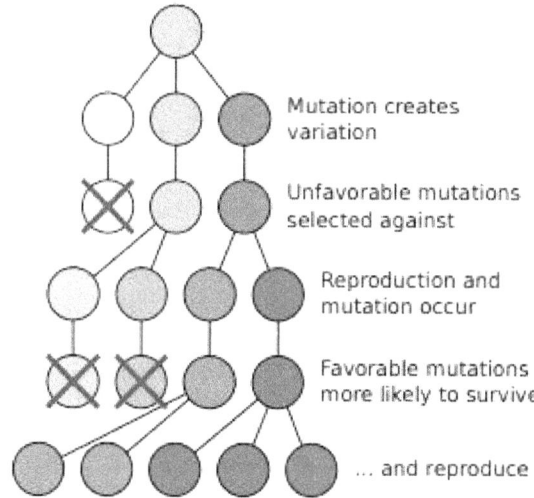

Mutation followed by natural selection results in a population with darker colouration.

From a Neo-Darwinian perspective, evolution occurs when there are changes in the frequencies of alleles within a population of interbreeding organisms.[92] For example, the allele for black colour in a population of moths becoming more common. Mechanisms that can lead to changes in allele frequencies include natural selection, genetic drift, genetic hitchhiking, mutation and gene flow.

29.4.1 Natural selection

Main article: Natural selection

Evolution by means of natural selection is the process by which traits that enhance survival and reproduction become more common in successive generations of a population. It has often been called a "self-evident" mechanism because it necessarily follows from three simple facts:[10]

- Variation exists within populations of organisms with respect to morphology, physiology, and behaviour (phenotypic variation).

- Different traits confer different rates of survival and reproduction (differential fitness).

- These traits can be passed from generation to generation (heritability of fitness).

More offspring are produced than can possibly survive, and these conditions produce competition between organisms for survival and reproduction. Consequently, organisms with traits that give them an advantage over their competitors are more likely to pass on their traits to the next generation than those with traits that do not confer an advantage.[125]

The central concept of natural selection is the evolutionary fitness of an organism.[126] Fitness is measured by an organism's ability to survive and reproduce, which determines the size of its genetic contribution to the next generation.[126] However, fitness is not the same as the total number of offspring: instead fitness is indicated by the proportion of subsequent generations that carry an organism's genes.[127] For example, if an organism could survive well and reproduce rapidly, but its offspring were all too small and weak to survive, this organism would make little genetic contribution to future generations and would thus have low fitness.[126]

If an allele increases fitness more than the other alleles of that gene, then with each generation this allele will become more common within the population. These traits are said to be "selected *for*." Examples of traits that can increase fitness are enhanced survival and increased fecundity. Conversely, the lower fitness caused by having a less beneficial or deleterious allele results in this allele becoming rarer—they are "selected *against*."[128] Importantly, the fitness of an allele is not a fixed characteristic; if the environment changes, previously neutral or harmful traits may become beneficial and previously beneficial traits become harmful.[81] However, even if the direction of selection does reverse in this way, traits that were lost in the past may not re-evolve in an identical form (see Dollo's law).[129][130]

Natural selection within a population for a trait that can vary across a range of values, such as height, can be categorised

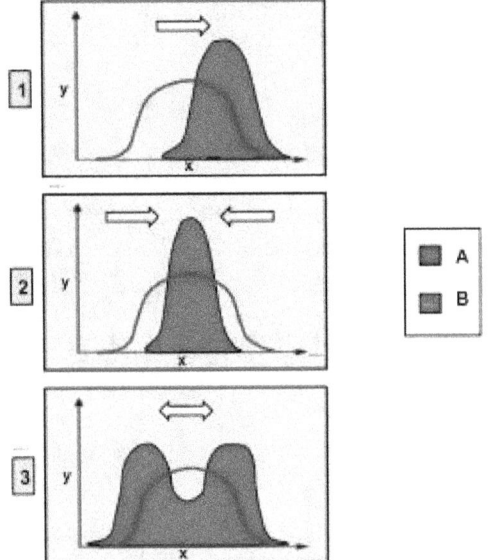

These charts depict the different types of genetic selection. On each graph, the x-axis variable is the type of phenotypic trait and the y-axis variable is the number of organisms. Group A is the original population and Group B is the population after selection.
· Graph 1 shows directional selection, in which a single extreme phenotype is favored.
· Graph 2 depicts stabilizing selection, where the intermediate phenotype is favored over the extreme traits.
· Graph 3 shows disruptive selection, in which the extreme phenotypes are favored over the intermediate.

into three different types. The first is directional selection, which is a shift in the average value of a trait over time—for example, organisms slowly getting taller.[131] Secondly, disruptive selection is selection for extreme trait values and often results in two different values becoming most common, with selection against the average value. This would be when either short or tall organisms had an advantage, but not those of medium height. Finally, in stabilising selection there is selection against extreme trait values on both ends, which causes a decrease in variance around the average value and less diversity.[125][132] This would, for example, cause organisms to slowly become all the same height.

A special case of natural selection is sexual selection, which is selection for any trait that increases mating success by increasing the attractiveness of an organism to potential mates.[133] Traits that evolved through sexual selection are particularly prominent among males of several animal species. Although sexually favoured, traits such as cumbersome antlers, mating calls, large body size and bright colours often attract predation, which compromises the survival of individual males.[134][135] This survival disadvantage is balanced by higher reproductive success in males that show these hard-to-fake, sexually selected traits.[136]

Natural selection most generally makes nature the measure against which individuals and individual traits, are more or less likely to survive. "Nature" in this sense refers to an ecosystem, that is, a system in which organisms interact with every other element, physical as well as biological, in their local environment. Eugene Odum, a founder of ecology, defined an ecosystem as: "Any unit that includes all of the organisms...in a given area interacting with the physical environment so that a flow of energy leads to clearly defined trophic structure, biotic diversity and material cycles (ie: exchange of materials between living and nonliving parts) within the system."[137] Each population within an ecosystem occupies a distinct niche, or position, with distinct relationships to other parts of the system. These relationships involve the life history of the organism, its position in the food chain and its geographic range. This broad understanding of nature enables scientists to delineate specific forces which, together, comprise natural selection.

Natural selection can act at different levels of organisation, such as genes, cells, individual organisms, groups of organisms and species.[138][139][140] Selection can act at multiple levels simultaneously.[141] An example of selection occurring below the level of the individual organism are genes called transposons, which can replicate and spread throughout a genome.[142] Selection at a level above the individual, such as group selection, may allow the evolution of cooperation, as discussed below.[143]

29.4.2 Biased mutation

In addition to being a major source of variation, mutation may also function as a mechanism of evolution when there are different probabilities at the molecular level for different mutations to occur, a process known as mutation bias.[144] If two genotypes, for example one with the nucleotide G and another with the nucleotide A in the same position, have the same fitness, but mutation from G to A happens more often than mutation from A to G, then genotypes with A will tend to evolve.[145] Different insertion vs. deletion mutation biases in different taxa can lead to the evolution of different genome sizes.[146][147] Developmental or mutational biases have also been observed in morphological evolution.[148][149] For example, according to the phenotype-first theory of evolution, mutations can eventually cause the genetic assimilation of traits that were previously induced by the environment.[150][151][152]

Mutation bias effects are superimposed on other processes. If selection would favor either one out of two mutations, but there is no extra advantage to having both, then the mutation that occurs the most frequently is the one that is most likely to become fixed in a population.[153][154] Mutations leading to the loss of function of a gene are much more common than mutations that produce a new, fully functional gene. Most loss of function mutations are selected against. But when selection is weak, mutation bias towards loss of function can affect evolution.[155] For example, pigments are no longer useful when animals live in the darkness of caves, and tend to be lost.[156] This kind of loss of function can occur because of mutation bias, and/or because the function had a cost, and once the benefit of the function disappeared, natural selection leads to the loss. Loss of sporulation ability in *Bacillus subtilis* during laboratory evolution appears to have been caused by mutation bias, rather than natural selection against the cost of maintaining sporulation ability.[157] When there is no selection for loss of function, the speed at which loss evolves depends more on the mutation rate than it does on the effective population size,[158] indicating that it is driven more by mutation bias than by genetic drift. In parasitic organisms, mutation bias leads to selection pressures as seen in Ehrlichia. Mutations are biased towards antigenic variants in outer-membrane proteins.

29.4.3 Genetic drift

Further information: Genetic drift and Effective population size

Genetic drift is the change in allele frequency from one generation to the next that occurs because alleles are subject to sampling error.[159] As a result, when selective forces are absent or relatively weak, allele frequencies tend to "drift" upward or downward randomly (in a random walk). This drift halts when an allele eventually becomes fixed, either by disappearing from the population, or replacing the other alleles entirely. Genetic drift may therefore eliminate some alleles from a population due to chance alone. Even in the absence of selective forces, genetic drift can cause two separate populations that began with the same genetic structure to drift apart into two divergent populations with different sets of alleles.[160]

It is usually difficult to measure the relative importance of selection and neutral processes, including drift.[161] The comparative importance of adaptive and non-adaptive forces in driving evolutionary change is an area of current research.[162]

The neutral theory of molecular evolution proposed that most evolutionary changes are the result of the fixation of neutral mutations by genetic drift.[163] Hence, in this model, most genetic changes in a population are the result of constant mutation pressure and genetic drift.[164] This form of the neutral theory is now largely abandoned, since it does not seem to fit the genetic variation seen in nature.[165][166] However, a more recent and better-supported version of this model is the nearly neutral theory, where a mutation that would be effectively neutral in a small population is not necessarily neutral in a large population.[125] Other alternative theories propose that genetic drift is dwarfed by other stochastic forces in evolution, such as genetic hitchhiking, also known as genetic draft.[159][167][168]

The time for a neutral allele to become fixed by genetic drift depends on population size, with fixation occurring more rapidly in smaller populations.[169] The number of individuals in a population is not critical, but instead a measure known as the effective population size.[170] The effective population is usually smaller than the total population since it takes into account factors such as the level of inbreeding and the stage of the lifecycle in which the population is the smallest.[170] The effective population size may not be the same for every gene in the same population.[171]

29.4.4 Genetic hitchhiking

Further information: Genetic hitchhiking, Hill–Robertson effect, and Selective sweep

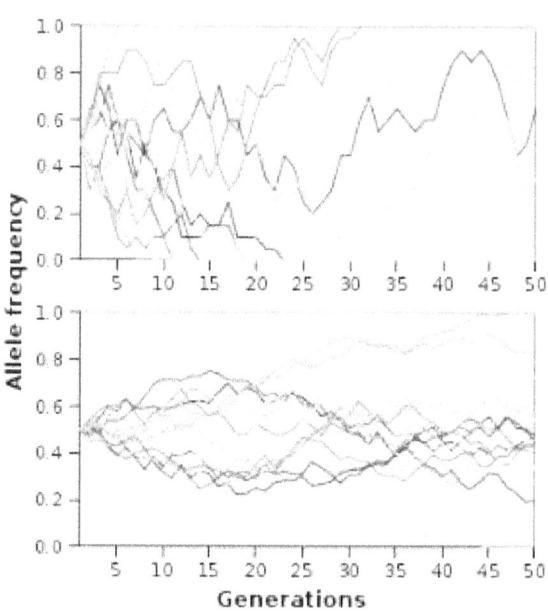

Simulation of genetic drift of 20 unlinked alleles in populations of 10 (top) and 100 (bottom). Drift to fixation is more rapid in the smaller population.

Recombination allows alleles on the same strand of DNA to become separated. However, the rate of recombination is low (approximately two events per chromosome per generation). As a result, genes close together on a chromosome may not always be shuffled away from each other and genes that are close together tend to be inherited together, a phenomenon known as linkage.[172] This tendency is measured by finding how often two alleles occur together on a single chromosome compared to expectations, which is called their linkage disequilibrium. A set of alleles that is usually inherited in a group is called a haplotype. This can be important when one allele in a particular haplotype is strongly beneficial: natural selection can drive a selective sweep that will also cause the other alleles in the haplotype to become more common in the population; this effect is called genetic hitchhiking or genetic draft.[173] Genetic draft caused by the fact that some neutral genes are genetically linked to others that are under selection can be partially captured by an appropriate effective population size.[167]

29.4.5 Gene flow

Further information: Gene flow, Hybrid (biology), and Horizontal gene transfer

Gene flow involves the exchange of genes between populations and between species.[117] The presence or absence of gene flow fundamentally changes the course of evolution. Due to the complexity of organisms, any two completely isolated populations will eventually evolve genetic incompatibilities through neutral processes, as in the Bateson-Dobzhansky-Muller model, even if both populations remain essentially identical in terms of their adaptation to the environment.

If genetic differentiation between populations develops, gene flow between populations can introduce traits or alleles which are disadvantageous in the local population and this may lead to organisms within these populations evolving mechanisms that prevent mating with genetically distant populations, eventually resulting in the appearance of new species. Thus, exchange of genetic information between individuals is fundamentally important for the development of the biological species concept.

During the development of the modern synthesis, Sewall Wright developed his shifting balance theory, which regarded gene flow between partially isolated populations as an important aspect of adaptive evolution.[174] However, recently there has been substantial criticism of the importance of the shifting balance theory.[175]

29.5 Outcomes

Evolution influences every aspect of the form and behaviour of organisms. Most prominent are the specific behavioural and physical adaptations that are the outcome of natural selection. These adaptations increase fitness by aiding activities such as finding food, avoiding predators or attracting mates. Organisms can also respond to selection by cooperating with each other, usually by aiding their relatives or engaging in mutually beneficial symbiosis. In the longer term, evolution produces new species through splitting ancestral populations of organisms into new groups that cannot or will not interbreed.

These outcomes of evolution are distinguished based on time scale as macroevolution versus microevolution. Macroevolution refers to evolution that occurs at or above the level of species, in particular speciation and extinction; whereas microevolution refers to smaller evolutionary changes within a species or population, in particular shifts in gene frequency and adaptation.[177] In general, macroevolution is regarded as the outcome of long periods of microevolution.[178] Thus, the distinction between micro- and macroevolution is not a fundamental one—the difference is simply the time involved.[179] However, in macroevolution, the traits of the entire species may be important. For instance, a large amount of variation among individuals allows a species to rapidly adapt to new habitats, lessening the chance of it going extinct, while a wide geographic range increases the chance of speciation, by making it more likely that part of the population will become isolated. In this sense, microevolution and macroevolution might involve selection at different levels—with microevolution acting on genes and organisms, versus macroevolutionary processes such as species selection acting on entire species and affecting their rates of speciation and extinction.[180][181][182]

A common misconception is that evolution has goals, long-term plans, or an innate tendency for "progress," as expressed in beliefs such as orthogenesis and evolutionism; realistically however, evolution has no long-term goal and does not necessarily produce greater complexity.[183][184][185] Although complex species have evolved, they occur as a side effect of the overall number of organisms increasing and simple forms of life still remain more common in the biosphere.[186] For example, the overwhelming majority of species are microscopic prokaryotes, which form about half the world's biomass despite their small size,[187] and constitute the vast majority of Earth's biodiversity.[188] Simple organisms have therefore been the dominant form of life on Earth throughout its history and continue to be the main form of life up to the present day, with complex life only appearing more diverse because it is more noticeable.[189] Indeed, the evolution of microorganisms is particularly important to modern evolutionary research, since their rapid reproduction allows the study of experimental evolution and the observation of evolution and adaptation in real time.[190][191]

29.5.1 Adaptation

For more details on this topic, see Adaptation.

Adaptation is the process that makes organisms better

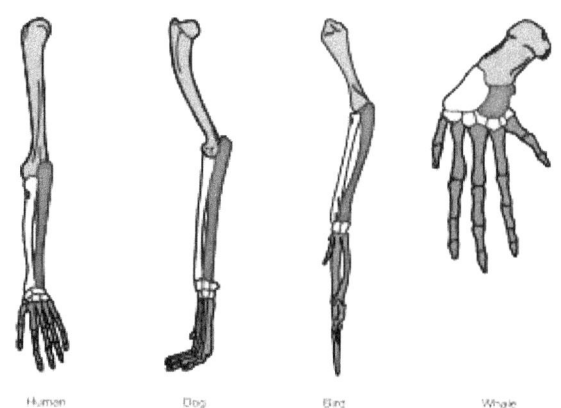

Homologous bones in the limbs of tetrapods. The bones of these animals have the same basic structure, but have been adapted for specific uses.

suited to their habitat.[192][193] Also, the term adaptation may refer to a trait that is important for an organism's survival. For example, the adaptation of horses' teeth to the grinding of grass. By using the term *adaptation* for the evolutionary process and *adaptive trait* for the product (the bodily part or function), the two senses of the word may be distinguished. Adaptations are produced by natural selection.[194] The following definitions are due to Theodosius Dobzhansky:

1. *Adaptation* is the evolutionary process whereby an organism becomes better able to live in its habitat or habitats.[195]

2. *Adaptedness* is the state of being adapted: the degree to which an organism is able to live and reproduce in a given set of habitats.[196]

3. An *adaptive trait* is an aspect of the developmental pattern of the organism which enables or enhances the probability of that organism surviving and reproducing.[197]

Adaptation may cause either the gain of a new feature, or the loss of an ancestral feature. An example that shows both types of change is bacterial adaptation to antibiotic selection, with genetic changes causing antibiotic resistance by both modifying the target of the drug, or increasing the activity of transporters that pump the drug out of the cell.[198] Other striking examples are the bacteria *Escherichia coli* evolving the ability to use citric acid as a nutrient in a long-term laboratory experiment,[199] *Flavobacterium* evolving a novel enzyme that allows these bacteria to grow on the by-products of nylon manufacturing,[200][201] and the soil bacterium *Sphingobium* evolving an entirely new metabolic pathway that degrades the synthetic pesticide pentachlorophenol.[202][203] An interesting but still controversial idea is that some adaptations might increase the ability of organisms to generate genetic diversity and adapt by natural selection (increasing organisms' evolvability).[204][205][206][207][208]

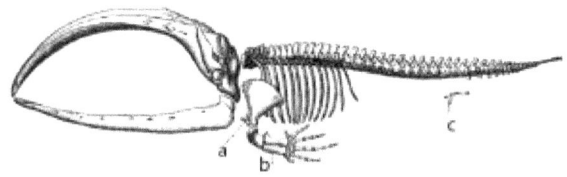

A baleen whale skeleton, a and b label flipper bones, which were adapted from front leg bones; while c indicates vestigial leg bones, suggesting an adaptation from land to sea.[209]

Adaptation occurs through the gradual modification of existing structures. Consequently, structures with similar internal organisation may have different functions in related organisms. This is the result of a single ancestral structure being adapted to function in different ways. The bones within bat wings, for example, are very similar to those in mice feet and primate hands, due to the descent of all these structures from a common mammalian ancestor.[210] However, since all living organisms are related to some extent,[211] even organs that appear to have little or no structural similarity, such as arthropod, squid and vertebrate eyes, or the limbs and wings of arthropods and vertebrates, can depend on a common set of homologous genes that control their assembly and function; this is called deep homology.[212][213]

During evolution, some structures may lose their original function and become vestigial structures.[214] Such structures may have little or no function in a current species, yet have a clear function in ancestral species, or other closely related species. Examples include pseudogenes,[215] the non-functional remains of eyes in blind cave-dwelling fish,[216] wings in flightless birds,[217] the presence of hip bones in whales and snakes,[209] and sexual traits in organisms that reproduce via asexual reproduction.[218] Examples of vestigial structures in humans include wisdom teeth,[219] the coccyx,[214] the vermiform appendix,[214] and other behavioural vestiges such as goose bumps[220][221] and primitive reflexes.[222][223][224]

However, many traits that appear to be simple adaptations are in fact exaptations: structures originally adapted for one function, but which coincidentally became somewhat useful for some other function in the process.[225] One example is the African lizard *Holaspis guentheri*, which developed an extremely flat head for hiding in crevices, as can be seen by looking at its near relatives. However, in this species, the head has become so flattened that it assists in gliding from tree to tree—an exaptation.[225] Within cells, molecular machines such as the bacterial flagella[226] and protein sorting machinery[227] evolved by the recruitment of several pre-existing proteins that previously had different functions.[177] Another example is the recruitment of enzymes from glycolysis and xenobiotic metabolism to serve as structural proteins called crystallins within the lenses of organisms' eyes.[228][229]

An area of current investigation in evolutionary developmental biology is the developmental basis of adaptations and exaptations.[230] This research addresses the origin and evolution of embryonic development and how modifications of development and developmental processes produce novel features.[231] These studies have shown that evolution can alter development to produce new structures, such as embryonic bone structures that develop into the jaw in other animals instead forming part of the middle ear in mammals.[232] It is also possible for structures that have been lost in evolution to reappear due to changes in developmental genes, such as a mutation in chickens causing embryos to grow teeth similar to those of crocodiles.[233] It is now becoming clear that most alterations in the form of organisms are due to changes in a small set of conserved genes.[234]

29.5.2 Coevolution

Common garter snake (Thamnophis sirtalis sirtalis) *has evolved resistance to the defensive substance tetrodotoxin in its amphibian prey.*

Further information: Coevolution

Interactions between organisms can produce both conflict and cooperation. When the interaction is between pairs of species, such as a pathogen and a host, or a predator and its prey, these species can develop matched sets of adaptations. Here, the evolution of one species causes adaptations in a second species. These changes in the second species then, in turn, cause new adaptations in the first species. This cycle of selection and response is called coevolution.[235] An example is the production of tetrodotoxin in the rough-skinned newt and the evolution of tetrodotoxin resistance in its predator, the common garter snake. In this predator-prey pair, an evolutionary arms race has produced high levels of toxin in the newt and correspondingly high levels of toxin resistance in the snake.[236]

29.5.3 Cooperation

Further information: Co-operation (evolution)

Not all co-evolved interactions between species involve conflict.[237] Many cases of mutually beneficial interactions have evolved. For instance, an extreme cooperation exists between plants and the mycorrhizal fungi that grow on their roots and aid the plant in absorbing nutrients from the soil.[238] This is a reciprocal relationship as the plants provide the fungi with sugars from photosynthesis. Here, the fungi actually grow inside plant cells, allowing them to exchange nutrients with their hosts, while sending signals that suppress the plant immune system.[239]

Coalitions between organisms of the same species have also evolved. An extreme case is the eusociality found in social insects, such as bees, termites and ants, where sterile insects feed and guard the small number of organisms in a colony that are able to reproduce. On an even smaller scale, the somatic cells that make up the body of an animal limit their reproduction so they can maintain a stable organism, which then supports a small number of the animal's germ cells to produce offspring. Here, somatic cells respond to specific signals that instruct them whether to grow, remain as they are, or die. If cells ignore these signals and multiply inappropriately, their uncontrolled growth causes cancer.[240]

Such cooperation within species may have evolved through the process of kin selection, which is where one organism acts to help raise a relative's offspring.[241] This activity is selected for because if the *helping* individual contains alleles which promote the helping activity, it is likely that its kin will *also* contain these alleles and thus those alleles will be passed on.[242] Other processes that may promote cooperation include group selection, where cooperation provides benefits to a group of organisms.[243]

29.5.4 Speciation

Main article: Speciation

Speciation is the process where a species diverges into two

The four geographic modes of speciation

or more descendant species.[244]

There are multiple ways to define the concept of "species." The choice of definition is dependent on the particularities of the species concerned.[245] For example, some species concepts apply more readily toward sexually reproducing organisms while others lend themselves better toward asexual organisms. Despite the diversity of various species concepts, these various concepts can be placed into one of three broad philosophical approaches: interbreeding, ecological and phylogenetic.[246] The Biological Species Concept (BSC) is a classic example of the interbreeding approach. Defined by Ernst Mayr in 1942, the BSC states that "species are groups of actually or potentially interbreeding natural populations, which are reproductively isolated from other such groups."[247] Despite its wide and long-term use, the BSC like others is not without controversy, for example because these concepts cannot be applied to prokaryotes,[248] and this is called the species problem.[245] Some researchers have attempted a unifying monistic definition of species, while others adopt a pluralistic approach and suggest that there may be different ways to logically interpret the definition of a species.[245][246]

Barriers to reproduction between two diverging sexual populations are required for the populations to become new species. Gene flow may slow this process by spreading the new genetic variants also to the other populations. Depending on how far two species have diverged since their most recent common ancestor, it may still be possible for them to produce offspring, as with horses and donkeys mating to produce mules.[249] Such hybrids are generally infertile. In this case, closely related species may regularly interbreed, but hybrids will be selected against and the species will remain distinct. However, viable hybrids are occasionally formed and these new species can either have properties intermediate between their parent species, or possess a totally new phenotype.[250] The importance of hybridisation in producing new species of animals is unclear, although cases have been seen in many types of animals,[251] with the gray tree frog being a particularly well-studied example.[252]

Speciation has been observed multiple times under both controlled laboratory conditions and in nature.[253] In sexually reproducing organisms, speciation results from reproductive isolation followed by genealogical divergence. There are four primary geographic modes of speciation. The most common in animals is allopatric speciation, which occurs in populations initially isolated geographically, such as by habitat fragmentation or migration. Selection under these conditions can produce very rapid changes in the appearance and behaviour of organisms.[254][255] As selection and drift act independently on populations isolated from the rest of their species, separation may eventually produce organisms that cannot interbreed.[256]

The second mode of speciation is peripatric speciation, which occurs when small populations of organisms become isolated in a new environment. This differs from allopatric speciation in that the isolated populations are numerically much smaller than the parental population. Here, the founder effect causes rapid speciation after an increase in inbreeding increases selection on homozygotes, leading to rapid genetic change.[257]

The third mode is parapatric speciation. This is similar to peripatric speciation in that a small population enters a new habitat, but differs in that there is no physical separation between these two populations. Instead, speciation results from the evolution of mechanisms that reduce gene flow between the two populations.[244] Generally this occurs when there has been a drastic change in the environment within the parental species' habitat. One example is the grass *Anthoxanthum odoratum*, which can undergo parapatric speciation in response to localised metal pollution from mines.[258] Here, plants evolve that have resistance to high levels of metals in the soil. Selection against interbreeding with the metal-sensitive parental population produced a gradual change in the flowering time of the metal-resistant plants, which eventually produced complete reproductive isolation. Selection against hybrids between the two populations may cause reinforcement, which is the evolution of traits that promote mating within a species, as well as character displacement, which is when two species become more distinct in appearance.[259]

Finally, in sympatric speciation species diverge without ge-

1. Geospiza magnirostris
2. Geospiza fortis
3. Geospiza parvula
4. Certhidea olivacea

Finches from Galapagos Archipelago

Geographical isolation of finches on the Galápagos Islands produced over a dozen new species.

ographic isolation or changes in habitat. This form is rare since even a small amount of gene flow may remove genetic differences between parts of a population.[260] Generally, sympatric speciation in animals requires the evolution of both genetic differences and non-random mating, to allow reproductive isolation to evolve.[261]

One type of sympatric speciation involves crossbreeding of two related species to produce a new hybrid species. This is not common in animals as animal hybrids are usually sterile. This is because during meiosis the homologous chromosomes from each parent are from different species and cannot successfully pair. However, it is more common in plants because plants often double their number of chromosomes, to form polyploids.[262] This allows the chromosomes from each parental species to form matching pairs during meiosis, since each parent's chromosomes are represented by a pair already.[263] An example of such a speciation event is when the plant species *Arabidopsis thaliana* and *Arabidopsis arenosa* crossbred to give the new species *Arabidopsis suecica*.[264] This happened about 20,000 years ago,[265] and the speciation process has been repeated in the laboratory, which allows the study of the genetic mechanisms involved in this process.[266] Indeed, chromosome doubling within a species may be a common cause of reproductive isolation, as half the doubled chromosomes will be unmatched when breeding with undoubled organisms.[267]

Speciation events are important in the theory of punctuated equilibrium, which accounts for the pattern in the fossil record of short "bursts" of evolution interspersed with relatively long periods of stasis, where species remain relatively unchanged.[268] In this theory, speciation and rapid evolution are linked, with natural selection and genetic drift acting most strongly on organisms undergoing speciation in novel habitats or small populations. As a result, the periods of stasis in the fossil record correspond to the parental population and the organisms undergoing speciation and rapid evolution are found in small populations or geographically restricted habitats and therefore rarely being preserved as fossils.[181]

29.5.5 Extinction

Further information: Extinction

Extinction is the disappearance of an entire species. Ex-

Tyrannosaurus rex. *Non-avian dinosaurs died out in the Cretaceous–Paleogene extinction event at the end of the Cretaceous period.*

tinction is not an unusual event, as species regularly appear through speciation and disappear through extinction.[269] Nearly all animal and plant species that have lived on Earth are now extinct,[270] and extinction appears to be the ultimate fate of all species.[271] These extinctions have happened continuously throughout the history of life, although the rate of extinction spikes in occasional mass extinction events.[272] The Cretaceous–Paleogene extinction event, during which the non-avian dinosaurs became extinct, is the most well-known, but the earlier Permian–Triassic extinction event was even more severe, with approximately 96% of all marine species driven to extinction.[272] The Holocene extinction event is an ongoing mass extinction associated with humanity's expansion across the globe over the past few thousand years. Present-day extinction rates are 100–1000 times greater than the background rate and up to 30% of current species may be extinct by the mid 21st century.[273] Human activities are now the primary cause of the ongoing extinction event;[274] global warming may further accelerate it in the future.[275]

The role of extinction in evolution is not very well understood and may depend on which type of extinction is considered.[272] The causes of the continuous "low-level" extinction events, which form the majority of extinctions, may be the result of competition between species for limited resources (the competitive exclusion principle).[72] If one species can out-compete another, this could produce species selection, with the fitter species surviving and the other species being driven to extinction.[139] The intermittent mass extinctions are also important, but instead of acting as a selective force, they drastically reduce diversity in a nonspecific manner and promote bursts of rapid evolution and speciation in survivors.[276]

29.6 Evolutionary history of life

29.6.1 Origin of life

Further information: Abiogenesis, Panspermia, and RNA world hypothesis

The Earth is about 4.54 billion years old.[277][278][279] The earliest undisputed evidence of life on Earth dates from at least 3.5 billion years ago,[19][280] during the Eoarchean Era after a geological crust started to solidify following the earlier molten Hadean Eon. Microbial mat fossils have been found in 3.48 billion-year-old sandstone in Western Australia.[6][7][8] Other early physical evidence of a biogenic substance is graphite in 3.7 billion-year-old metasedimentary rocks discovered in Western Greenland[5] as well as "remains of biotic life" found in 4.1 billion-year-old rocks in Western Australia.[20][21] According to one of the researchers, "If life arose relatively quickly on Earth ... then it could be common in the universe."[20]

More than 99 percent of all species, amounting to over five billion species,[281] that ever lived on Earth are estimated to be extinct.[23][24] Estimates on the number of Earth's current species range from 10 million to 14 million,[25][26] of which about 1.9 million are estimated to have been named[27] and 1.6 million documented in a central database to date,[28] leaving at least 80 percent not yet described.

Highly energetic chemistry is thought to have produced a self-replicating molecule around 4 billion years ago, and half a billion years later the last common ancestor of all life existed.[17] The current scientific consensus is that the complex biochemistry that makes up life came from simpler chemical reactions.[282] The beginning of life may have included self-replicating molecules such as RNA[283] and the assembly of simple cells.[284]

29.6.2 Common descent

Further information: Common descent and Evidence of common descent

All organisms on Earth are descended from a common ancestor or ancestral gene pool.[211][285] Current species are a stage in the process of evolution, with their diversity the product of a long series of speciation and extinction events.[286] The common descent of organisms was first deduced from four simple facts about organisms: First, they have geographic distributions that cannot be explained by local adaptation. Second, the diversity of life is not a set of completely unique organisms, but organisms that share morphological similarities. Third, vestigial traits with no clear purpose resemble functional ancestral traits and finally, that organisms can be classified using these similarities into a hierarchy of nested groups—similar to a family tree.[287] However, modern research has suggested that, due to horizontal gene transfer, this "tree of life" may be more complicated than a simple branching tree since some genes have spread independently between distantly related species.[288][289]

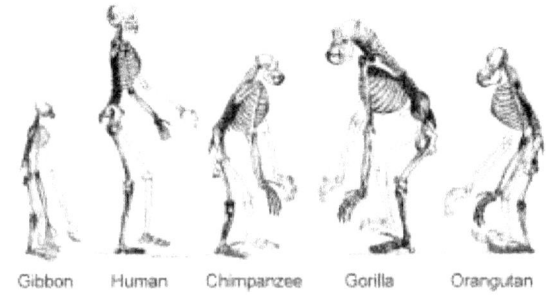

The hominoids are descendants of a common ancestor.

Past species have also left records of their evolutionary history. Fossils, along with the comparative anatomy of present-day organisms, constitute the morphological, or anatomical, record.[290] By comparing the anatomies of both modern and extinct species, paleontologists can infer the lineages of those species. However, this approach is most successful for organisms that had hard body parts, such as shells, bones or teeth. Further, as prokaryotes such as bacteria and archaea share a limited set of common morphologies, their fossils do not provide information on their ancestry.

More recently, evidence for common descent has come from the study of biochemical similarities between organisms. For example, all living cells use the same basic set of nucleotides and amino acids.[291] The development of molecular genetics has revealed the record of evolution left in organisms' genomes: dating when species diverged through the molecular clock produced by mutations.[292] For example, these DNA sequence comparisons have revealed that humans and chimpanzees share 98% of their genomes and analysing the few areas where they differ helps shed light on when the common ancestor of these species existed.[293]

29.6.3 Evolution of life

Main articles: Evolutionary history of life and Timeline of evolutionary history of life

Evolutionary tree showing the divergence of modern species from their common ancestor in the centre.[1] The three domains are coloured, with bacteria blue, archaea green and eukaryotes red.

1. ^ Ciccarelli, Francesca D.; Doerks, Tobias; von Mering, Christian; et al. (March 3, 2006). "Toward Automatic Reconstruction of a Highly Resolved Tree of Life". *Science*. Washington, D.C.: American Association for the Advancement of Science. **311** (5765): 1283–1287. Bibcode:2006Sci...311.1283C. ISSN 0036-8075. PMID 16513982. doi:10.1126/science.1123061.

Prokaryotes inhabited the Earth from approximately 3–4 billion years ago.[294][295] No obvious changes in morphology or cellular organisation occurred in these organisms over the next few billion years.[296] The eukaryotic cells emerged between 1.6–2.7 billion years ago. The next major change in cell structure came when bacteria were engulfed by eukaryotic cells, in a cooperative association called endosymbiosis.[297][298] The engulfed bacteria and the host cell then underwent coevolution, with the bacteria evolving into either mitochondria or hydrogenosomes.[299] Another engulfment of cyanobacterial-like organisms led to the formation of chloroplasts in algae and plants.[300]

The history of life was that of the unicellular eukaryotes, prokaryotes and archaea until about 610 million years ago when multicellular organisms began to appear in the oceans in the Ediacaran period.[294][301] The evolution of multicellularity occurred in multiple independent events, in organisms as diverse as sponges, brown algae, cyanobacteria, slime moulds and myxobacteria.[302] In January 2016, scientists reported that, about 800 million years ago, a minor genetic change in a single molecule called GK-PID may have allowed organisms to go from a single cell organism to one of many cells.[303]

Soon after the emergence of these first multicellular organisms, a remarkable amount of biological diversity appeared over approximately 10 million years, in an event called the Cambrian explosion. Here, the majority of types of modern animals appeared in the fossil record, as well as unique lineages that subsequently became extinct.[304] Various triggers for the Cambrian explosion have been proposed, including the accumulation of oxygen in the atmosphere from photosynthesis.[305]

About 500 million years ago, plants and fungi colonised the land and were soon followed by arthropods and other animals.[306] Insects were particularly successful and even today make up the majority of animal species.[307] Amphibians first appeared around 364 million years ago, followed by early amniotes and birds around 155 million years ago (both from "reptile"-like lineages), mammals around 129 million years ago, homininae around 10 million years ago and modern humans around 250,000 years ago.[308][309][310] However, despite the evolution of these large animals, smaller organisms similar to the types that evolved early in this process continue to be highly successful and dominate the Earth, with the majority of both biomass and species being prokaryotes.[188]

29.7 Applications

Main articles: Applications of evolution, Selective breeding, and Evolutionary computation

Concepts and models used in evolutionary biology, such as natural selection, have many applications.[311]

Artificial selection is the intentional selection of traits in a population of organisms. This has been used for thousands of years in the domestication of plants and animals.[312] More recently, such selection has become a vital part of genetic engineering, with selectable markers such as antibiotic resistance genes being used to manipulate DNA. Proteins with valuable properties have evolved by repeated rounds of mutation and selection (for example modified enzymes and new antibodies) in a process called directed evolution.[313]

Understanding the changes that have occurred during an organism's evolution can reveal the genes needed to construct parts of the body, genes which may be involved in human genetic disorders.[314] For example, the Mexican tetra is an albino cavefish that lost its eyesight during evolution. Breeding together different populations of this blind fish produced some offspring with functional eyes, since different mutations had occurred in the isolated populations that had evolved in different caves.[315] This helped identify genes required for vision and pigmentation.[316]

Many human diseases are not static phenomena, but capable of evolution. Viruses, bacteria, fungi and cancers evolve to be resistant to host immune defences, as well as pharmaceutical drugs.[317][318][319] These same problems occur in agriculture with pesticide[320] and herbicide[321] resistance. It is possible that we are facing the end of the effective life of most of available antibiotics[322] and predicting the evolution and evolvability[323] of our pathogens and devising strategies to slow or circumvent it is requiring deeper knowledge of the complex forces driving evolution at the molecular level.[324]

In computer science, simulations of evolution using evolutionary algorithms and artificial life started in the 1960s and were extended with simulation of artificial selection.[325] Artificial evolution became a widely recognised optimisation method as a result of the work of Ingo Rechenberg in the 1960s. He used evolution strategies to solve complex engineering problems.[326] Genetic algorithms in particular became popular through the writing of John Henry Holland.[327] Practical applications also include automatic evolution of computer programmes.[328] Evolutionary algorithms are now used to solve multidimensional problems more efficiently than software produced by human designers and also to optimise the design of systems.[329]

As evolution became widely accepted in the 1870s, caricatures of Charles Darwin with an ape or monkey body symbolised evolution.[330]

29.8 Social and cultural responses

Further information: Social effects of evolutionary theory, 1860 Oxford evolution debate, Creation–evolution controversy, and Objections to evolution

In the 19th century, particularly after the publication of *On the Origin of Species* in 1859, the idea that life had evolved was an active source of academic debate centred on the philosophical, social and religious implications of evolution. Today, the modern evolutionary synthesis is accepted by a vast majority of scientists.[72] However, evolution remains a contentious concept for some theists.[331]

While various religions and denominations have reconciled their beliefs with evolution through concepts such as theistic evolution, there are creationists who believe that evolution is contradicted by the creation myths found in their religions and who raise various objections to evolution.[177][332][333] As had been demonstrated by responses to the publication of *Vestiges of the Natural History of Creation* in 1844, the most controversial aspect of evolutionary biology is the implication of human evolution that humans share common ancestry with apes and that the mental and moral faculties of humanity have the same types of natural causes as other inherited traits in animals.[334] In some countries, notably the United States, these tensions between science and religion have fuelled the current creation–evolution controversy, a religious conflict focusing on politics and public education.[335] While other scientific fields such as cosmology[336] and Earth science[337] also conflict with literal interpretations of many religious texts, evolutionary biology experiences significantly more opposition from religious literalists.

The teaching of evolution in American secondary school biology classes was uncommon in most of the first half of the 20th century. The Scopes Trial decision of 1925 caused the subject to become very rare in American secondary biology textbooks for a generation, but it was gradually reintroduced later and became legally protected with the 1968 *Epperson v. Arkansas* decision. Since then, the competing religious belief of creationism was legally disallowed in secondary school curricula in various decisions in the 1970s and 1980s, but it returned in pseudoscientific form as intelligent design (ID), to be excluded once again in the 2005 *Kitzmiller v. Dover Area School District* case.[338]

29.9 See also

- Argument from poor design
- Biocultural evolution
- Biological classification
- Evidence of common descent
- Evolutionary anthropology
- Evolutionary ecology
- Evolutionary epistemology
- Evolutionary neuroscience
- Evolution of biological complexity
- Evolution of plants
- Timeline of the evolutionary history of life
- Unintelligent design
- Universal Darwinism

29.10 References

[1] Hall & Hallgrímsson 2008, pp. 4–6

[2] "Evolution Resources". National Academies of Sciences, Engineering, and Medicine. 2016.

[3] Hall & Hallgrímsson 2008, pp. 3–5

[4] Panno 2005, pp. xv-16

[5] Ohtomo, Yoko; Kakegawa, Takeshi; Ishida, Akizumi; et al. (January 2014). "Evidence for biogenic graphite in early Archaean Isua metasedimentary rocks". *Nature Geoscience*. London: Nature Publishing Group. **7** (1): 25–28. Bibcode:2014NatGe...7...25O. ISSN 1752-0894. doi:10.1038/ngeo2025.

[6] Borenstein, Seth (November 13, 2013). "Oldest fossil found: Meet your microbial mom". *Excite*. Yonkers, NY: Mindspark Interactive Network. Associated Press. Retrieved 2015-05-31.

[7] Pearlman, Jonathan (November 13, 2013). "'Oldest signs of life on Earth found'". *The Daily Telegraph*. London: Telegraph Media Group. Retrieved 2014-12-15.

[8] Noffke, Nora; Christian, Daniel; Wacey, David; Hazen, Robert M. (November 16, 2013). "Microbially Induced Sedimentary Structures Recording an Ancient Ecosystem in the ca. 3.48 Billion-Year-Old Dresser Formation, Pilbara, Western Australia". *Astrobiology*. New Rochelle, NY: Mary Ann Liebert, Inc. **13** (12): 1103–1124. Bibcode:2013AsBio..13.1103N. ISSN 1531-1074. PMC 3870916. PMID 24205812. doi:10.1089/ast.2013.1030.

[9] Futuyma 2004, p. 33

[10] Lewontin, R. C. (November 1970). "The Units of Selection" (PDF). *Annual Review of Ecology and Systematics*. Palo Alto, CA: Annual Reviews. **1**: 1–18. ISSN 1545-2069. JSTOR 2096764. doi:10.1146/annurev.es.01.110170.000245.

[11] Darwin 1859, Chapter XIV

[12] Scott-Phillips, T. C.; Laland, K. N.; Shuker, D. M.; Dickins, T. E.; West, S. A. (2014). "The Niche Construction Perspective: A Critical Appraisal". *Evolution*. **68**: 1231–1243. PMC 4261998. PMID 24325256. doi:10.1111/evo.12332. Evolutionary processes are generally thought of as processes by which these changes occur. Four such processes are widely recognized: natural selection (in the broad sense, to include sexual selection), genetic drift, mutation, and migration (Fisher 1930; Haldane 1932). The latter two generate variation; the first two sort it.

[13] Edward S. Buckler IV, Tara L. Phelps-Durr, Carlyn S, Keith Buckler, R. Kelly Dawe, John F. Doebley and Timothy P. Holtsford (1999). "Meiotic Drive of Chromosomal Knobs Reshaped the Maize Genome". *Genetics*. **153** (1): 415–426. PMC 1460728. PMID 10471723.

[14] Otto, Sarah P.; Servedio, Maria R.; Nuismer, Scott L. (2008). "Frequency-Dependent Selection and the Evolution of Assortative Mating". *Genetics*. **179** (4): 2091–2112. PMC 2516082. PMID 18660541. doi:10.1534/genetics.107.084418.

[15] Provine 1988, pp. 49–79

[16] Kampourakis 2014, pp. 127–129

[17] Doolittle, W. Ford (February 2000). "Uprooting the Tree of Life" (PDF). *Scientific American*. Stuttgart: Georg von Holtzbrinck Publishing Group. **282** (2): 90–95. Bibcode:2000SciAm.282b..90D. ISSN 0036-8733. PMID 10710791. doi:10.1038/scientificamerican0200-90. Archived from the original (PDF) on 2006-09-07. Retrieved 2015-04-05.

[18] Glansdorff, Nicolas; Ying Xu; Labedan, Bernard (July 9, 2008). "The Last Universal Common Ancestor: emergence, constitution and genetic legacy of an elusive forerunner". *Biology Direct*. London: BioMed Central. **3**: 29. ISSN 1745-6150. PMC 2478661. PMID 18613974. doi:10.1186/1745-6150-3-29.

[19] Schopf, J. William; Kudryavtsev, Anatoliy B.; Czaja, Andrew D.; Tripathi, Abhishek B. (October 5, 2007). "Evidence of Archean life: Stromatolites and microfossils". *Precambrian Research*. Amsterdam, the Netherlands: Elsevier. **158** (3–4): 141–155. Bibcode:2007PreR..158..141S. ISSN 0301-9268. doi:10.1016/j.precamres.2007.04.009.

[20] Borenstein, Seth (October 19, 2015). "Hints of life on what was thought to be desolate early Earth". *Excite*. Yonkers,

NY: Mindspark Interactive Network. Associated Press. Retrieved 2015-10-20.

[21] Bell, Elizabeth A.; Boehnike, Patrick; Harrison, T. Mark; et al. (November 24, 2015). "Potentially biogenic carbon preserved in a 4.1 billion-year-old zircon" (PDF). *Proc. Natl. Acad. Sci. U.S.A.* Washington, D.C.: National Academy of Sciences. **112** (47): 14518–14521. Bibcode:2015PNAS..11214518B. ISSN 0027-8424. PMC 4664351. PMID 26483481. doi:10.1073/pnas.1517557112. Retrieved 2015-12-30.

[22] Wade, Nicholas (July 25, 2016). "Meet Luca, the Ancestor of All Living Things". *New York Times*. Retrieved July 25, 2016.

[23] Stearns, Beverly Peterson; Stearns, S. C.; Stearns, Stephen C. (2000). *Watching, from the Edge of Extinction*. Yale University Press. p. preface x. ISBN 978-0-300-08469-6. Retrieved 30 May 2017.

[24] Novacek, Michael J. (November 8, 2014). "Prehistory's Brilliant Future". *The New York Times*. New York: The New York Times Company. ISSN 0362-4331. Retrieved 2014-12-25.

[25] Mora, Camilo; Tittensor, Derek P.; Adl, Sina; et al. (August 23, 2011). "How Many Species Are There on Earth and in the Ocean?". *PLOS Biology*. San Francisco, CA: Public Library of Science. **9** (8): e1001127. ISSN 1545-7885. PMC 3160336. PMID 21886479. doi:10.1371/journal.pbio.1001127.

[26] Miller & Spoolman 2012, p. 62

[27] Chapman, Arthur D. (2009). *Numbers of Living Species in Australia and the World*. 2nd edition. Canberra: Australian Biological Resources Study. ISBN 978-0-642-56860-1. Retrieved 2016-11-06.

[28] "Catalogue of Life: 2016 Annual Checklist". 2016. Retrieved 2016-11-06.

[29] Staff (2 May 2016). "Researchers find that Earth may be home to 1 trillion species". *National Science Foundation*. Retrieved 6 May 2016.

[30] NAS 2008, pp. R11–R12

[31] Ayala & Avise 2014

[32] NAS 2008, p. 17

[33] Moore, Decker & Cotner 2010, p. 454

[34] Futuyma, Douglas J., ed. (1999). "Evolution, Science, and Society: Evolutionary Biology and the National Research Agenda" (PDF) (Executive summary). New Brunswick, NJ: Office of University Publications, Rutgers, The State University of New Jersey. OCLC 43422991. Archived from the original (PDF) on 2012-01-31. Retrieved 2014-11-24.

[35] Darwin 1909, p. 53

[36] Kirk, Raven & Schofield 1983, pp. 100–142, 280–321

[37] Lucretius. "Book V, lines 855–877". *De Rerum Natura*. *Perseus Digital Library*. Edited and translated by William Ellery Leonard (1916). Medford/Somerville, MA: Tufts University. OCLC 33233743. Retrieved 2014-11-25.

[38] Sedley, David (2003). "Lucretius and the New Empedocles" (PDF). *Leeds International Classical Studies*. Leeds, West Yorkshire, England: Leeds International Classics Seminar. **2** (4). ISSN 1477-3643. Archived from the original (PDF) on 2014-08-23. Retrieved 2014-11-25.

[39] Torrey, Harry Beal; Felin, Frances (March 1937). "Was Aristotle an Evolutionist?". *The Quarterly Review of Biology*. **12** (1): 1–18. ISSN 0033-5770. JSTOR 2808399. doi:10.1086/394520.

[40] Hull, David L. (December 1967). "The Metaphysics of Evolution". *The British Journal for the History of Science*. Cambridge: Cambridge University Press. **3** (4): 309–337. JSTOR 4024958. doi:10.1017/S0007087400002892.

[41] Mason 1962, pp. 43–44

[42] Mayr 1982, pp. 256–257

- Ray 1686

[43] Waggoner, Ben (July 7, 2000). "Carl Linnaeus (1707-1778)". *Evolution* (Online exhibit). Berkeley, CA: University of California Museum of Paleontology. Retrieved 2012-02-11.

[44] Bowler 2003, pp. 73–75

[45] "Erasmus Darwin (1731-1802)". *Evolution* (Online exhibit). Berkeley, CA: University of California Museum of Paleontology. October 4, 1995. Retrieved February 11, 2012.

[46] Lamarck 1809

[47] Nardon & Grenier 1991, p. 162

[48] Gould 2002

[49] Ghiselin, Michael T. (September–October 1994). "The Imaginary Lamarck: A Look at Bogus 'History' in Schoolbooks". *The Textbook Letter*. Sausalito, CA: The Textbook League. OCLC 23228649. Retrieved 2008-01-23.

[50] Magner 2002

[51] Jablonka, Eva; Lamb, Marion J. (August 2007). "Précis of Evolution in Four Dimensions". *Behavioural and Brain Sciences*. Cambridge: Cambridge University Press. **30** (4): 353–365. ISSN 0140-525X. doi:10.1017/S0140525X07002221.

[52] Burkhardt & Smith 1991

- "Darwin, C. R. to Lubbock, John". *Darwin Correspondence Project*. Cambridge, UK: University of Cambridge. Retrieved 2014-12-01. Letter 2532, November 22, 1859.

[53] Sulloway, Frank J. (June 2009). "Why Darwin rejected intelligent design". *Journal of Biosciences*. Bangalore: Indian Academy of Sciences. **34** (2): 173–183. ISSN 0250-5991. PMID 19550032. doi:10.1007/s12038-009-0020-8.

[54] Dawkins 1990

[55] Sober, Elliott (June 16, 2009). "Did Darwin write the *Origin* backwards?". *Proc. Natl. Acad. Sci. U.S.A.* Washington, D.C.: National Academy of Sciences. **106** (Suppl. 1): 10048–10055. Bibcode:2009PNAS..10610048S. ISSN 0027-8424. PMC 2702806. PMID 19528655. doi:10.1073/pnas.0901109106.

[56] Mayr 2002, p. 165

[57] Bowler 2003, pp. 145–146

[58] Sokal, Robert R.; Crovello, Theodore J. (March–April 1970). "The Biological Species Concept: A Critical Evaluation". *The American Naturalist*. Chicago, Illinois: University of Chicago Press on behalf of the American Society of Naturalists. **104** (936): 127–153. ISSN 0003-0147. JSTOR 2459191. doi:10.1086/282646.

[59] Darwin, Charles; Wallace, Alfred (August 20, 1858). "On the Tendency of Species to form Varieties; and on the Perpetuation of Varieties and Species by Natural Means of Selection". *Journal of the Proceedings of the Linnean Society of London. Zoology*. **3** (9): 45–62. ISSN 1096-3642. doi:10.1111/j.1096-3642.1858.tb02500.x. Retrieved 2007-05-13.

[60] Desmond, Adrian J. (July 17, 2014). "Thomas Henry Huxley". *Encyclopædia Britannica Online*. Chicago, Illinois: Encyclopædia Britannica, Inc. Retrieved 2014-12-02.

[61] Liu, Y. S.; Zhou, X. M.; Zhi, M. X.; Li, X. J.; Wang, Q. L. (September 2009). "Darwin's contributions to genetics". *Journal of Applied Genetics*. Poznań: Institute of Plant Genetics, Polish Academy of Sciences. **50** (3): 177–184. ISSN 1234-1983. PMID 19638672. doi:10.1007/BF03195671.

[62] Weiling, Franz (July 1991). "Historical study: Johann Gregor Mendel 1822–1884". *American Journal of Medical Genetics*. **40** (1): 1–25; discussion 26. PMID 1887835. doi:10.1002/ajmg.1320400103.

[63] Wright 1984, p. 480

[64] Provine 1971

[65] Stamhuis, Ida H.; Meijer, Onno G.; Zevenhuizen, Erik J. A. (June 1999). "Hugo de Vries on Heredity, 1889-1903: Statistics, Mendelian Laws, Pangenes, Mutations". *Isis*. Chicago, Illinois: University of Chicago Press. **90** (2): 238–267. ISSN 0021-1753. JSTOR 237050. PMID 10439561. doi:10.1086/384323.

[66] Quammen 2006

[67] Bowler 1989

[68] Watson, J. D.; Crick, F. H. C. (April 25, 1953). "Molecular Structure of Nucleic Acids: A Structure for Deoxyribose Nucleic Acid" (PDF). *Nature*. London: Nature Publishing Group. **171** (4356): 737–738. Bibcode:1953Natur.171..737W. ISSN 0028-0836. PMID 13054692. doi:10.1038/171737a0. Retrieved 2014-12-04. It has not escaped our notice that the specific pairing we have postulated immediately suggests a possible copying mechanism for the genetic material.

[69] Hennig 1999, p. 280

[70] Wiley & Lieberman 2011

[71] Dobzhansky, Theodosius (March 1973). "Nothing in Biology Makes Sense Except in the Light of Evolution". *The American Biology Teacher*. McLean, VA: National Association of Biology Teachers. **35** (3): 125–129. doi:10.2307/4444260.

[72] Kutschera, Ulrich; Niklas, Karl J. (June 2004). "The modern theory of biological evolution: an expanded synthesis". *Naturwissenschaften*. Heidelberg: Springer-Verlag Heidelberg. **91** (6): 255–276. Bibcode:2004NW.....91..255K. ISSN 1432-1904. PMID 15241603. doi:10.1007/s00114-004-0515-y.

[73] Cracraft & Bybee 2005

[74] Avise, John C.; Ayala, Francisco J. (May 11, 2010). "In the light of evolution IV: The human condition" (PDF). *Proc. Natl. Acad. Sci. U.S.A.* Washington, D.C.: National Academy of Sciences. **107** (Suppl. 2): 8897–8901. ISSN 0027-8424. doi:10.1073/pnas.100321410. Retrieved 2014-12-29.

[75] Danchin, É; Charmantier, A; Champagne, FA; Mesoudi, A; Pujol, B; Blanchet, S (2011). "Beyond DNA: integrating inclusive inheritance into an extended theory of evolution". *Nature Reviews Genetics*. **12**: 475–486. PMID 21681209. doi:10.1038/nrg3028.

[76] Pigliucci (26 March 2010). *Evolution - the Extended Synthesis*. The MIT Press. ISBN 978-0262513678.

[77] Sturm, Richard A.; Frudakis, Tony N. (August 2004). "Eye colour: portals into pigmentation genes and ancestry". *Trends in Genetics*. Cambridge, MA: Cell Press. **20** (8): 327–332. ISSN 0168-9525. PMID 15262401. doi:10.1016/j.tig.2004.06.010.

[78] Pearson, Helen (May 25, 2006). "Genetics: What is a gene?". *Nature*. London: Nature Publishing Group. **441** (7092): 398–401. Bibcode:2006Natur.441..398P. ISSN 0028-0836. PMID 16724031. doi:10.1038/441398a.

[79] Visscher, Peter M.; Hill, William G.; Wray, Naomi R. (April 2008). "Heritability in the genomics era — concepts and misconceptions". *Nature Reviews Genetics*. London: Nature Publishing Group. **9** (4): 255–266. ISSN 1471-0056. PMID 18319743. doi:10.1038/nrg2322.

[80] Oetting, William S.; Brilliant, Murray H.; King, Richard A. (August 1996). "The clinical spectrum of albinism in humans". *Molecular Medicine Today*. Cambridge, MA: Cell Press. **2** (8): 330–335. ISSN 1357-4310. PMID 8796918. doi:10.1016/1357-4310(96)81798-9.

[81] Futuyma 2005

[82] Phillips, Patrick C. (November 2008). "Epistasis—the essential role of gene interactions in the structure and evolution of genetic systems". *Nature Reviews Genetics*. London: Nature Publishing Group. **9** (11): 855–867. ISSN 1471-0056. PMC 2689140 ⊙. PMID 18852697. doi:10.1038/nrg2452.

[83] Rongling Wu; Min Lin (March 2006). "Functional mapping — how to map and study the genetic architecture of dynamic complex traits". *Nature Reviews Genetics*. London: Nature Publishing Group. **7** (3): 229–237. ISSN 1471-0056. PMID 16485021. doi:10.1038/nrg1804.

[84] Jablonka, Eva; Raz, Gal (June 2009). "Transgenerational Epigenetic Inheritance: Prevalence, Mechanisms, and Implications for the Study of Heredity and Evolution". *The Quarterly Review of Biology*. Chicago, Illinois: University of Chicago Press. **84** (2): 131–176. ISSN 0033-5770. PMID 19606595. doi:10.1086/598822.

[85] Bossdorf, Oliver; Arcuri, Davide; Richards, Christina L.; Pigliucci, Massimo (May 2010). "Experimental alteration of DNA methylation affects the phenotypic plasticity of ecologically relevant traits in *Arabidopsis thaliana*". *Evolutionary Ecology*. Dordrecht, the Netherlands: Springer Netherlands. **24** (3): 541–553. ISSN 0269-7653. doi:10.1007/s10682-010-9372-7.

[86] Jablonka & Lamb 2005

[87] Jablonka, Eva; Lamb, Marion J. (December 2002). "The Changing Concept of Epigenetics". *Annals of the New York Academy of Sciences*. Hoboken, NJ: Wiley-Blackwell on behalf of the New York Academy of Sciences. **981** (1): 82–96. Bibcode:2002NYASA.981...82J. ISSN 0077-8923. PMID 12547675. doi:10.1111/j.1749-6632.2002.tb04913.x.

[88] Laland, Kevin N.; Sterelny, Kim (September 2006). "Perspective: Seven Reasons (Not) to Neglect Niche Construction". *Evolution*. Hoboken, NJ: John Wiley & Sons on behalf of the Society for the Study of Evolution. **60** (9): 1751–1762. ISSN 0014-3820. doi:10.1111/j.0014-3820.2006.tb00520.x.

[89] Chapman, Michael J.; Margulis, Lynn (December 1998). "Morphogenesis by symbiogenesis" (PDF). *International Microbiology*. Barcelona: Spanish Society for Microbiology. **1** (4): 319–326. ISSN 1139-6709. PMID 10943381. Archived from the original (PDF) on 2014-08-23. Retrieved 2014-12-09.

[90] Wilson, David Sloan; Wilson, Edward O. (December 2007). "Rethinking the Theoretical Foundation of Sociobiology" (PDF). *The Quarterly Review of Biology*. Chicago, Illinois: University of Chicago Press. **82** (4): 327–348. ISSN 0033-5770. PMID 18217526. doi:10.1086/522809. Archived from the original (PDF) on 2011-05-11.

[91] Amos, William; Harwood, John (February 28, 1998). "Factors affecting levels of genetic diversity in natural populations". *Philosophical Transactions of the Royal Society B: Biological Sciences*. London: Royal Society. **353** (1366): 177–186. ISSN 0962-8436. PMC 1692205 ⊙. PMID 9533122. doi:10.1098/rstb.1998.0200.

[92] Ewens 2004

[93] Butlin, Roger K.; Tregenza, Tom (February 28, 1998). "Levels of genetic polymorphism: marker loci versus quantitative traits". *Philosophical Transactions of the Royal Society B: Biological Sciences*. London: Royal Society. **353** (1366): 187–198. ISSN 0962-8436. PMC 1692210 ⊙. PMID 9533123. doi:10.1098/rstb.1998.0201.

- Butlin, Roger K.; Tregenza, Tom (December 29, 2000). "Correction for Butlin and Tregenza, Levels of genetic polymorphism: marker loci versus quantitative traits". *Philosophical Transactions of the Royal Society B: Biological Sciences*. London: Royal Society. **355** (1404): 1865. ISSN 0962-8436. doi:10.1098/rstb.2000.2000. Some of the values in table 1 on p. 193 were given incorrectly. The errors do not affect the conclusions drawn in the paper. The corrected table is reproduced below.

[94] Wetterbom, Anna; Sevov, Marie; Cavelier, Lucia; Bergström, Tomas F. (November 2006). "Comparative Genomic Analysis of Human and Chimpanzee Indicates a Key Role for Indels in Primate Evolution". *Journal of Molecular Evolution*. New York: Springer-Verlag New York. **63** (5): 682–690. Bibcode:2006JMolE..63..682W. ISSN 0022-2844. PMID 17075697. doi:10.1007/s00239-006-0045-7.

[95] Sawyer, Stanley A.; Parsch, John; Zhang Zhi; Hartl, Daniel L. (April 17, 2007). "Prevalence of positive selection among nearly neutral amino acid replacements in *Drosophila*". *Proc. Natl. Acad. Sci. U.S.A.* Washington, D.C.: National Academy of Sciences. **104** (16): 6504–6510. Bibcode:2007PNAS..104.6504S. ISSN 0027-8424. PMC 1871816 ⊙. PMID 17409186. doi:10.1073/pnas.0701572104.

[96] Hastings, P. J.; Lupski, James R.; Rosenberg, Susan M.; Ira, Grzegorz (August 2009). "Mechanisms of change in gene copy number". *Nature Reviews Genetics*. London: Nature Publishing Group. **10** (8): 551–564. ISSN 1471-0056. PMC 2864001 ⊙. PMID 19597530. doi:10.1038/nrg2593.

[97] Carroll, Grenier & Weatherbee 2005

[98] Harrison, Paul M.; Gerstein, Mark (May 17, 2002). "Studying Genomes Through the Aeons: Protein Families, Pseudogenes and Proteome Evolution". *Journal of Molecular Biology*. Amsterdam, the Netherlands: Elsevier. **318**

29.10. REFERENCES

(5): 1155–1174. ISSN 0022-2836. PMID 12083509. doi:10.1016/S0022-2836(02)00109-2.

[99] Bowmaker, James K. (May 1998). "Evolution of colour vision in vertebrates". *Eye*. London: Nature Publishing Group on behalf of the Royal College of Ophthalmologists. **12** (3b): 541–547. ISSN 0950-222X. PMID 9775215. doi:10.1038/eye.1998.143.

[100] Gregory, T. Ryan; Hebert, Paul D. N. (April 1999). "The Modulation of DNA Content: Proximate Causes and Ultimate Consequences". *Genome Research*. Cold Spring Harbor, NY: Cold Spring Harbor Laboratory Press. **9** (4): 317–324. ISSN 1088-9051. PMID 10207154. doi:10.1101/gr.9.4.317. Retrieved 2014-12-11.

[101] Hurles, Matthew (July 13, 2004). "Gene Duplication: The Genomic Trade in Spare Parts". *PLOS Biology*. San Francisco, CA: Public Library of Science. **2** (7): e206. ISSN 1545-7885. PMC 449868. PMID 15252449. doi:10.1371/journal.pbio.0020206.

[102] Liu, Na; Okamura, Katsutomo; Tyler, David M.; et al. (October 2008). "The evolution and functional diversification of animal microRNA genes". *Cell Research*. London: Nature Publishing Group on behalf of the Shanghai Institutes for Biological Sciences. **18** (10): 985–996. ISSN 1001-0602. PMC 2712117. PMID 18711447. doi:10.1038/cr.2008.278. Retrieved 2014-12-11.

[103] Siepel, Adam (October 2009). "Darwinian alchemy: Human genes from noncoding DNA". *Genome Research*. Cold Spring Harbor, NY: Cold Spring Harbor Laboratory Press. **19** (10): 1693–1695. ISSN 1088-9051. PMC 2765273. PMID 19797681. doi:10.1101/gr.098376.109. Retrieved 2014-12-11.

[104] Orengo, Christine A.; Thornton, Janet M. (July 2005). "Protein families and their evolution—a structural perspective". *Annual Review of Biochemistry*. Palo Alto, CA: Annual Reviews. **74**: 867–900. ISSN 0066-4154. PMID 15954844. doi:10.1146/annurev.biochem.74.082803.133029.

[105] Long, Manyuan; Betrán, Esther; Thornton, Kevin; Wang, Wen (November 2003). "The origin of new genes: glimpses from the young and old". *Nature Reviews Genetics*. London: Nature Publishing Group. **4** (11): 865–875. ISSN 1471-0056. PMID 14634634. doi:10.1038/nrg1204.

[106] Wang, Minglei; Caetano-Anollés, Gustavo (January 14, 2009). "The Evolutionary Mechanics of Domain Organization in Proteomes and the Rise of Modularity in the Protein World". *Structure*. Cambridge, MA: Cell Press. **17** (1): 66–78. ISSN 1357-4310. PMID 19141283. doi:10.1016/j.str.2008.11.008.

[107] Weissman, Kira J.; Müller, Rolf (April 14, 2008). "Protein–Protein Interactions in Multienzyme Megasynthetases". *ChemBioChem*. Weinheim, Germany: Wiley-VCH. **9** (6): 826–848. ISSN 1439-4227. PMID 18357594. doi:10.1002/cbic.200700751.

[108] Radding, Charles M. (December 1982). "Homologous Pairing and Strand Exchange in Genetic Recombination". *Annual Review of Genetics*. Palo Alto, CA: Annual Reviews. **16**: 405–437. ISSN 0066-4197. PMID 6297377. doi:10.1146/annurev.ge.16.120182.002201.

[109] Agrawal, Aneil F. (September 5, 2006). "Evolution of Sex: Why Do Organisms Shuffle Their Genotypes?". *Current Biology*. Cambridge, MA: Cell Press. **16** (17): R696–R704. ISSN 0960-9822. PMID 16950096. doi:10.1016/j.cub.2006.07.063.

[110] Peters, Andrew D.; Otto, Sarah P. (June 2003). "Liberating genetic variance through sex". *BioEssays*. Hoboken, NJ: John Wiley & Sons. **25** (6): 533–537. ISSN 0265-9247. PMID 12766942. doi:10.1002/bies.10291.

[111] Goddard, Matthew R.; Godfray, H. Charles J.; Burt, Austin (March 31, 2005). "Sex increases the efficacy of natural selection in experimental yeast populations". *Nature*. London: Nature Publishing Group. **434** (7033): 636–640. Bibcode:2005Natur.434..636G. ISSN 0028-0836. PMID 15800622. doi:10.1038/nature03405.

[112] Maynard Smith 1978

[113] Ridley 1993

[114] Van Valen, Leigh (1973). "A New Evolutionary Law" (PDF). *Evolutionary Theory*. Chicago, Illinois: University of Chicago. **1**: 1–30. ISSN 0093-4755. Archived from the original (PDF) on 2014-12-22. Retrieved 2014-12-24.

[115] Hamilton, W. D.; Axelrod, Robert; Tanese, Reiko (May 1, 1990). "Sexual reproduction as an adaptation to resist parasites (a review)". *Proc. Natl. Acad. Sci. U.S.A.* Washington, D.C.: National Academy of Sciences. **87** (9): 3566–3573. Bibcode:1990PNAS...87.3566H. ISSN 0027-8424. PMC 53943. PMID 2185476. doi:10.1073/pnas.87.9.3566.

[116] Birdsell & Wills 2003, pp. 113–117

[117] Morjan, Carrie L.; Rieseberg, Loren H. (June 2004). "How species evolve collectively: implications of gene flow and selection for the spread of advantageous alleles". *Molecular Ecology*. Hoboken, NJ: Wiley-Blackwell. **13** (6): 1341–1356. ISSN 0962-1083. PMC 2600545. PMID 15140081. doi:10.1111/j.1365-294X.2004.02164.x.

[118] Boucher, Yan; Douady, Christophe J.; Papke, R. Thane; et al. (December 2003). "Lateral gene transfer and the origins of prokaryotic groups". *Annual Review of Genetics*. Palo Alto, CA: Annual Reviews. **37**: 283–328. ISSN 0066-4197. PMID 14616063. doi:10.1146/annurev.genet.37.050503.084247.

[119] Walsh, Timothy R. (October 2006). "Combinatorial genetic evolution of multiresistance". *Current Opinion in Microbiology*. Amsterdam, the Netherlands: Elsevier. **9** (5): 476–482. ISSN 1369-5274. PMID 16942901. doi:10.1016/j.mib.2006.08.009.

[120] Kondo, Natsuko; Nikoh, Naruo; Ijichi, Nobuyuki; et al. (October 29, 2002). "Genome fragment of *Wolbachia* endosymbiont transferred to X chromosome of host insect". *Proc. Natl. Acad. Sci. U.S.A.* Washington, D.C.: National Academy of Sciences. **99** (22): 14280–14285. Bibcode:2002PNAS...9914280K. ISSN 0027-8424. PMC 137875. PMID 12386340. doi:10.1073/pnas.222228199.

[121] Sprague, George F., Jr. (December 1991). "Genetic exchange between kingdoms". *Current Opinion in Genetics & Development*. Amsterdam, the Netherlands: Elsevier. **1** (4): 530–533. ISSN 0959-437X. PMID 1822285. doi:10.1016/S0959-437X(05)80203-5.

[122] Gladyshev, Eugene A.; Meselson, Matthew; Arkhipova, Irina R. (May 30, 2008). "Massive Horizontal Gene Transfer in Bdelloid Rotifers". *Science*. Washington, D.C.: American Association for the Advancement of Science. **320** (5880): 1210–1213. Bibcode:2008Sci...320.1210G. ISSN 0036-8075. PMID 18511688. doi:10.1126/science.1156407.

[123] Baldo, Angela M.; McClure, Marcella A. (September 1999). "Evolution and Horizontal Transfer of dUTPase-Encoding Genes in Viruses and Their Hosts". *Journal of Virology*. Washington, D.C.: American Society for Microbiology. **73** (9): 7710–7721. ISSN 0022-538X. PMC 104298. PMID 10438861.

[124] Rivera, Maria C.; Lake, James A. (September 9, 2004). "The ring of life provides evidence for a genome fusion origin of eukaryotes". *Nature*. London: Nature Publishing Group. **431** (7005): 152–155. Bibcode:2004Natur.431..152R. ISSN 0028-0836. PMID 15356622. doi:10.1038/nature02848.

[125] Hurst, Laurence D. (February 2009). "Fundamental concepts in genetics: genetics and the understanding of selection". *Nature Reviews Genetics*. London: Nature Publishing Group. **10** (2): 83–93. ISSN 1471-0056. PMID 19119264. doi:10.1038/nrg2506.

[126] Orr, H. Allen (August 2009). "Fitness and its role in evolutionary genetics". *Nature Reviews Genetics*. London: Nature Publishing Group. **10** (8): 531–539. ISSN 1471-0056. PMC 2753274. PMID 19546856. doi:10.1038/nrg2603.

[127] Haldane, J. B. S. (March 14, 1959). "The Theory of Natural Selection To-Day". *Nature*. London: Nature Publishing Group. **183** (4663): 710–713. Bibcode:1959Natur.183..710H. ISSN 0028-0836. PMID 13644170. doi:10.1038/183710a0.

[128] Lande, Russell; Arnold, Stevan J. (November 1983). "The Measurement of Selection on Correlated Characters". *Evolution*. Hoboken, NJ: John Wiley & Sons on behalf of the Society for the Study of Evolution. **37** (6): 1210–1226. ISSN 0014-3820. JSTOR 2408842. doi:10.2307/2408842.

[129] Goldberg, Emma E.; Igić, Boris (November 2008). "On phylogenetic tests of irreversible evolution". *Evolution*. Hoboken, NJ: John Wiley & Sons on behalf of the Society for the Study of Evolution. **62** (11): 2727–2741. ISSN 0014-3820. PMID 18764918. doi:10.1111/j.1558-5646.2008.00505.x.

[130] Collin, Rachel; Miglietta, Maria Pia (November 2008). "Reversing opinions on Dollo's Law". *Trends in Ecology & Evolution*. Cambridge, MA: Cell Press. **23** (11): 602–609. ISSN 0169-5347. PMID 18814933. doi:10.1016/j.tree.2008.06.013.

[131] Hoekstra, Hopi E.; Hoekstra, Jonathan M.; Berrigan, David; et al. (July 31, 2001). "Strength and tempo of directional selection in the wild". *Proc. Natl. Acad. Sci. U.S.A.* Washington, D.C.: National Academy of Sciences. **98** (16): 9157–9160. Bibcode:2001PNAS...98.9157H. ISSN 0027-8424. PMC 55389. PMID 11470913. doi:10.1073/pnas.161281098.

[132] Felsenstein, Joseph (November 1979). "Excursions along the Interface between Disruptive and Stabilizing Selection". *Genetics*. Bethesda, MD: Genetics Society of America. **93** (3): 773–795. ISSN 0016-6731. PMC 1214112. PMID 17248980.

[133] Andersson, Malte; Simmons, Leigh W. (June 2006). "Sexual selection and mate choice". *Trends in Ecology & Evolution*. Cambridge, MA: Cell Press. **21** (6): 296–302. ISSN 0169-5347. PMID 16769428. doi:10.1016/j.tree.2006.03.015.

[134] Kokko, Hanna; Brooks, Robert; McNamara, John M.; Houston, Alasdair I. (July 7, 2002). "The sexual selection continuum". *Proceedings of the Royal Society B*. London: Royal Society. **269** (1498): 1331–1340. ISSN 0962-8452. PMC 1691039. PMID 12079655. doi:10.1098/rspb.2002.2020.

[135] Quinn, Thomas P.; Hendry, Andrew P.; Buck, Gregory B. (2001). "Balancing natural and sexual selection in sockeye salmon: interactions between body size, reproductive opportunity and vulnerability to predation by bears" (PDF). *Evolutionary Ecology Research*. **3**: 917–937. ISSN 1522-0613. Retrieved 2014-12-15.

[136] Hunt, John; Brooks, Robert; Jennions, Michael D.; et al. (December 23, 2004). "High-quality male field crickets invest heavily in sexual display but die young". *Nature*. London: Nature Publishing Group. **432** (7020): 1024–1027. Bibcode:2004Natur.432.1024H. ISSN 0028-0836. PMID 15616562. doi:10.1038/nature03084.

[137] Odum 1971, p. 8

[138] Okasha 2006

[139] Gould, Stephen Jay (February 28, 1998). "Gulliver's further travels: the necessity and difficulty of a hierarchical theory of selection". *Philosophical Transactions of the Royal Society B: Biological Sciences*. London: Royal Society. **353**

29.10. REFERENCES

(1366): 307–314. ISSN 0962-8436. PMC 1692213. PMID 9533127. doi:10.1098/rstb.1998.0211.

[140] Mayr, Ernst (March 18, 1997). "The objects of selection". *Proc. Natl. Acad. Sci. U.S.A.* Washington, D.C.: National Academy of Sciences. **94** (6): 2091–2094. Bibcode:1997PNAS...94.2091M. ISSN 0027-8424. PMC 33654. PMID 9122151. doi:10.1073/pnas.94.6.2091.

[141] Maynard Smith 1998, pp. 203–211; discussion 211–217

[142] Hickey, Donal A. (1992). "Evolutionary dynamics of transposable elements in prokaryotes and eukaryotes". *Genetica*. Dordrecht, the Netherlands: Kluwer Academic Publishers. **86** (1–3): 269–274. ISSN 0016-6707. PMID 1334911. doi:10.1007/BF00133725.

[143] Gould, Stephen Jay; Lloyd, Elisabeth A. (October 12, 1999). "Individuality and adaptation across levels of selection: how shall we name and generalise the unit of Darwinism?". *Proc. Natl. Acad. Sci. U.S.A.* Washington, D.C.: National Academy of Sciences. **96** (21): 11904–11909. Bibcode:1999PNAS...9611904G. ISSN 0027-8424. PMC 18385. PMID 10518549. doi:10.1073/pnas.96.21.11904.

[144] Lynch, Michael (May 15, 2007). "The frailty of adaptive hypotheses for the origins of organismal complexity". *Proc. Natl. Acad. Sci. U.S.A.* Washington, D.C.: National Academy of Sciences. **104** (Suppl. 1): 8597–8604. Bibcode:2007PNAS..104.8597L. ISSN 0027-8424. PMC 1876435. PMID 17494740. doi:10.1073/pnas.0702207104.

[145] Smith, Nick G.C.; Webster, Matthew T.; Ellegren, Hans (September 2002). "Deterministic Mutation Rate Variation in the Human Genome". *Genome Research*. Cold Spring Harbor, NY: Cold Spring Harbor Laboratory Press. **12** (9): 1350–1356. ISSN 1088-9051. PMC 186654. PMID 12213772. doi:10.1101/gr.220502.

[146] Petrov, Dmitri A.; Sangster, Todd A.; Johnston, J. Spencer; et al. (February 11, 2000). "Evidence for DNA Loss as a Determinant of Genome Size". *Science*. Washington, D.C.: American Association for the Advancement of Science. **287** (5455): 1060–1062. Bibcode:2000Sci...287.1060P. ISSN 0036-8075. PMID 10669421. doi:10.1126/science.287.5455.1060.

[147] Petrov, Dmitri A. (May 2002). "DNA loss and evolution of genome size in *Drosophila*". *Genetica*. Dordrecht, the Netherlands: Kluwer Academic Publishers. **115** (1): 81–91. ISSN 0016-6707. PMID 12188050. doi:10.1023/A:1016076215168.

[148] Kiontke, Karin; Barriere, Antoine; Kolotuev, Irina; et al. (November 2007). "Trends, Stasis, and Drift in the Evolution of Nematode Vulva Development". *Current Biology*. Cambridge, MA: Cell Press. **17** (22): 1925–1937. ISSN 0960-9822. PMID 18024125. doi:10.1016/j.cub.2007.10.061.

[149] Braendle, Christian; Baer, Charles F.; Félix, Marie-Anne (March 12, 2010). Barsh, Gregory S., ed. "Bias and Evolution of the Mutationally Accessible Phenotypic Space in a Developmental System". *PLOS Genetics*. San Francisco, CA: Public Library of Science. **6** (3): e1000877. ISSN 1553-7390. PMC 2837400. PMID 20300655. doi:10.1371/journal.pgen.1000877.

[150] Palmer, A. Richard (October 29, 2004). "Symmetry breaking and the evolution of development". *Science*. Washington, D.C.: American Association for the Advancement of Science. **306** (5697): 828–833. Bibcode:2004Sci...306..828P. ISSN 0036-8075. PMID 15514148. doi:10.1126/science.1103707.

[151] West-Eberhard 2003, pp. 140

[152] Pocheville, Arnaud; Danchin, Etienne (January 1, 2017). "Chapter 3: Genetic assimilation and the paradox of blind variation". In Huneman, Philippe; Walsh, Denis. *Challenging the Modern Synthesis*. Oxford University Press.

[153] Stoltzfus, Arlin; Yampolsky, Lev Y. (September–October 2009). "Climbing Mount Probable: Mutation as a Cause of Nonrandomness in Evolution". *Journal of Heredity*. Oxford: Oxford University Press on behalf of the American Genetic Association. **100** (5): 637–647. ISSN 0022-1503. PMID 19625453. doi:10.1093/jhered/esp048.

[154] Yampolsky, Lev Y.; Stoltzfus, Arlin (March 2001). "Bias in the introduction of variation as an orienting factor in evolution". *Evolution & Development*. Hoboken, NJ: Wiley-Blackwell on behalf of the Society for Integrative and Comparative Biology. **3** (2): 73–83. ISSN 1520-541X. PMID 11341676. doi:10.1046/j.1525-142x.2001.003002073.x.

[155] Haldane, J. B. S. (January–February 1933). "The Part Played by Recurrent Mutation in Evolution". *The American Naturalist*. Chicago, Illinois: University of Chicago Press on behalf of the American Society of Naturalists. **67** (708): 5–19. ISSN 0003-0147. JSTOR 2457127. doi:10.1086/280465.

[156] Protas, Meredith; Conrad, Melissa; Gross, Joshua B.; et al. (March 6, 2007). "Regressive Evolution in the Mexican Cave Tetra, *Astyanax mexicanus*". *Current Biology*. Cambridge, MA: Cell Press. **17** (5): 452–454. ISSN 0960-9822. PMC 2570642. PMID 17306543. doi:10.1016/j.cub.2007.01.051.

[157] Maughan, Heather; Masel, Joanna; Birky, C. William, Jr.; Nicholson, Wayne L. (October 2007). "The Roles of Mutation Accumulation and Selection in Loss of Sporulation in Experimental Populations of *Bacillus subtilis*". *Genetics*. Bethesda, MD: Genetics Society of America. **177** (2): 937–948. ISSN 0016-6731. PMC 2034656. PMID 17720926. doi:10.1534/genetics.107.075663.

[158] Masel, Joanna; King, Oliver D.; Maughan, Heather (January 2007). "The Loss of Adaptive Plasticity during Long

Periods of Environmental Stasis". *The American Naturalist*. Chicago, Illinois: University of Chicago Press on behalf of the American Society of Naturalists. **169** (1): 38–46. ISSN 0003-0147. PMC 1766558. PMID 17206583. doi:10.1086/510212.

[159] Masel, Joanna (October 25, 2011). "Genetic drift". *Current Biology*. Cambridge, MA: Cell Press. **21** (20): R837–R838. ISSN 0960-9822. PMID 22032182. doi:10.1016/j.cub.2011.08.007.

[160] Lande, Russell (1989). "Fisherian and Wrightian theories of speciation". *Genome*. Ottawa: National Research Council of Canada. **31** (1): 221–227. ISSN 0831-2796. PMID 2687093. doi:10.1139/g89-037.

[161] Mitchell-Olds, Thomas; Willis, John H.; Goldstein, David B. (November 2007). "Which evolutionary processes influence natural genetic variation for phenotypic traits?". *Nature Reviews Genetics*. London: Nature Publishing Group. **8** (11): 845–856. ISSN 1471-0056. PMID 17943192. doi:10.1038/nrg2207.

[162] Nei, Masatoshi (December 2005). "Selectionism and Neutralism in Molecular Evolution". *Molecular Biology and Evolution*. Oxford: Oxford University Press on behalf of the Society for Molecular Biology and Evolution. **22** (12): 2318–2342. ISSN 0737-4038. PMC 1513187. PMID 16120807. doi:10.1093/molbev/msi242.

- Nei, Masatoshi (May 2006). "Selectionism and Neutralism in Molecular Evolution". *Molecular Biology and Evolution* (Erratum). Oxford: Oxford University Press on behalf of the Society for Molecular Biology and Evolution. **23** (5): 1095. ISSN 0737-4038. doi:10.1093/molbev/msk009.

[163] Kimura, Motoo (1991). "The neutral theory of molecular evolution: a review of recent evidence". *The Japanese Journal of Human Genetics*. Mishima, Japan: Genetics Society of Japan. **66** (4): 367–386. ISSN 0021-504X. PMID 1954033. doi:10.1266/jjg.66.367.

[164] Kimura, Motoo (1989). "The neutral theory of molecular evolution and the world view of the neutralists". *Genome*. Ottawa: National Research Council of Canada. **31** (1): 24–31. ISSN 0831-2796. PMID 2687096. doi:10.1139/g89-009.

[165] Kreitman, Martin (August 1996). "The neutral theory is dead. Long live the neutral theory". *BioEssays*. Hoboken, NJ: John Wiley & Sons. **18** (8): 678–683; discussion 683. ISSN 0265-9247. PMID 8760341. doi:10.1002/bies.950180812.

[166] Leigh, E. G., Jr. (November 2007). "Neutral theory: a historical perspective". *Journal of Evolutionary Biology*. Hoboken, NJ: Wiley-Blackwell on behalf of the European Society for Evolutionary Biology. **20** (6): 2075–2091. ISSN 1010-061X. PMID 17956380. doi:10.1111/j.1420-9101.2007.01410.x.

[167] Gillespie, John H. (November 2001). "Is the population size of a species relevant to its evolution?". *Evolution*. Hoboken, NJ: John Wiley & Sons on behalf of the Society for the Study of Evolution. **55** (11): 2161–2169. ISSN 0014-3820. PMID 11794777. doi:10.1111/j.0014-3820.2001.tb00732.x.

[168] Neher, Richard A.; Shraiman, Boris I. (August 2011). "Genetic Draft and Quasi-Neutrality in Large Facultatively Sexual Populations". *Genetics*. Bethesda, MD: Genetics Society of America. **188** (4): 975–996. ISSN 0016-6731. PMC 3176096. PMID 21625002. doi:10.1534/genetics.111.128876.

[169] Otto, Sarah P.; Whitlock, Michael C. (June 1997). "The Probability of Fixation in Populations of Changing Size" (PDF). *Genetics*. Bethesda, MD: Genetics Society of America. **146** (2): 723–733. ISSN 0016-6731. PMC 1208011. PMID 9178020. Retrieved 2014-12-18.

[170] Charlesworth, Brian (March 2009). "Fundamental concepts in genetics: effective population size and patterns of molecular evolution and variation". *Nature Reviews Genetics*. London: Nature Publishing Group. **10** (3): 195–205. ISSN 1471-0056. PMID 19204717. doi:10.1038/nrg2526.

[171] Cutter, Asher D.; Choi, Jae Young (August 2010). "Natural selection shapes nucleotide polymorphism across the genome of the nematode *Caenorhabditis briggsae*". *Genome Research*. Cold Spring Harbor, NY: Cold Spring Harbor Laboratory Press. **20** (8): 1103–1111. ISSN 1088-9051. PMC 2909573. PMID 20508143. doi:10.1101/gr.104331.109.

[172] Lien, Sigbjørn; Szyda, Joanna; Schechinger, Birgit; et al. (February 2000). "Evidence for Heterogeneity in Recombination in the Human Pseudoautosomal Region: High Resolution Analysis by Sperm Typing and Radiation-Hybrid Mapping". *American Journal of Human Genetics*. Cambridge, MA: Cell Press on behalf of the American Society of Human Genetics. **66** (2): 557–566. ISSN 0002-9297. PMC 1288109. PMID 10677316. doi:10.1086/302754.

[173] Barton, Nicholas H. (November 29, 2000). "Genetic hitchhiking". *Philosophical Transactions of the Royal Society B: Biological Sciences*. London: Royal Society. **355** (1403): 1553–1562. ISSN 0962-8436. PMC 1692896. PMID 11127900. doi:10.1098/rstb.2000.0716.

[174] Wright, Sewall (1932). "The roles of mutation, inbreeding, crossbreeding and selection in evolution". *Proceedings of the VI International Congress of Genetics*. **1**: 356–366. Retrieved 2014-12-18.

[175] Coyne, Jerry A.; Barton, Nicholas H.; Turelli, Michael (June 1997). "Perspective: A Critique of Sewall Wright's Shifting Balance Theory of Evolution". *Evolution*. Hoboken, NJ: John Wiley & Sons on behalf of the Society for the Study of Evolution. **51** (3): 643–671. ISSN 0014-3820. doi:10.2307/2411143.

29.10. REFERENCES

[176] Baym, Michael; Lieberman, Tami D.; Kelsic, Eric D.; Chait, Remy; Gross, Rotem; Yelin, Idan; Kishony, Roy (2016-09-09). "Spatiotemporal microbial evolution on antibiotic landscapes". *Science*. **353** (6304): 1147–1151. Bibcode:2016Sci...353.1147B. ISSN 0036-8075. PMID 27609891. doi:10.1126/science.aag0822.

[177] Scott, Eugenie C.; Matzke, Nicholas J. (May 15, 2007). "Biological design in science classrooms". *Proc. Natl. Acad. Sci. U.S.A.* Washington, D.C.: National Academy of Sciences. **104** (Suppl. 1): 8669–8676. Bibcode:2007PNAS..104.8669S. ISSN 0027-8424. PMC 1876445. PMID 17494747. doi:10.1073/pnas.0701505104.

[178] Hendry, Andrew Paul; Kinnison, Michael T. (November 2001). "An introduction to microevolution: rate, pattern, process". *Genetica*. Dordrecht, the Netherlands: Kluwer Academic Publishers. **112–113** (1): 1–8. ISSN 0016-6707. PMID 11838760. doi:10.1023/A:1013368628607.

[179] Leroi, Armand M. (March–April 2000). "The scale independence of evolution". *Evolution & Development*. Hoboken, NJ: Wiley-Blackwell on behalf of the Society for Integrative and Comparative Biology. **2** (2): 67–77. ISSN 1520-541X. PMID 11258392. doi:10.1046/j.1525-142x.2000.00044.x.

[180] Gould 2002, pp. 657–658.

[181] Gould, Stephen Jay (July 19, 1994). "Tempo and mode in the macroevolutionary reconstruction of Darwinism". *Proc. Natl. Acad. Sci. U.S.A.* Washington, D.C.: National Academy of Sciences. **91** (15): 6764–6771. Bibcode:1994PNAS...91.6764G. ISSN 0027-8424. PMC 44281. PMID 8041695. doi:10.1073/pnas.91.15.6764.

[182] Jablonski, David (2000). "Micro- and macroevolution: scale and hierarchy in evolutionary biology and paleobiology". *Paleobiology*. Boulder, CO: Paleontological Society. **26** (sp4): 15–52. ISSN 0094-8373. doi:10.1666/0094-8373(2000)26[15:MAMSAH]2.0.CO;2.

[183] Dougherty, Michael J. (July 20, 1998). "Is the human race evolving or devolving?". *Scientific American*. Stuttgart: Georg von Holtzbrinck Publishing Group. ISSN 0036-8733. Retrieved 2015-09-11.

[184] Isaak, Mark, ed. (July 22, 2003). "Claim CB932: Evolution of degenerate forms". *TalkOrigins Archive*. Houston, TX: The TalkOrigins Foundation, Inc. Retrieved 2014-12-19.

[185] Lane 1996, p. 61

[186] Carroll, Sean B. (February 22, 2001). "Chance and necessity: the evolution of morphological complexity and diversity". *Nature*. London: Nature Publishing Group. **409** (6823): 1102–1109. Bibcode:2001Natur.409.1102C. ISSN 0028-0836. PMID 11234024. doi:10.1038/35059227.

[187] Whitman, William B.; Coleman, David C.; Wiebe, William J. (June 9, 1998). "Prokaryotes: The unseen majority". *Proc. Natl. Acad. Sci. U.S.A.* Washington, D.C.: National Academy of Sciences. **95** (12): 6578–6583. Bibcode:1998PNAS...95.6578W. ISSN 0027-8424. PMC 33863. PMID 9618454. doi:10.1073/pnas.95.12.6578.

[188] Schloss, Patrick D.; Handelsman, Jo (December 2004). "Status of the Microbial Census". *Microbiology and Molecular Biology Reviews*. Washington, D.C.: American Society for Microbiology. **68** (4): 686–691. ISSN 1092-2172. PMC 539005. PMID 15590780. doi:10.1128/MMBR.68.4.686-691.2004.

[189] Nealson, Kenneth H. (January 1999). "Post-Viking microbiology: new approaches, new data, new insights". *Origins of Life and Evolution of Biospheres*. Dordrecht, the Netherlands: Kluwer Academic Publishers. **29** (1): 73–93. ISSN 0169-6149. PMID 11536899. doi:10.1023/A:1006515817767.

[190] Buckling, Angus; MacLean, R. Craig; Brockhurst, Michael A.; Colegrave, Nick (February 12, 2009). "The Beagle in a bottle". *Nature*. London: Nature Publishing Group. **457** (7231): 824–829. Bibcode:2009Natur.457..824B. ISSN 0028-0836. PMID 19212400. doi:10.1038/nature07892.

[191] Elena, Santiago F.; Lenski, Richard E. (June 2003). "Evolution experiments with microorganisms: the dynamics and genetic bases of adaptation". *Nature Reviews Genetics*. London: Nature Publishing Group. **4** (6): 457–469. ISSN 1471-0056. PMID 12776215. doi:10.1038/nrg1088.

[192] Mayr 1982, p. 483: "Adaptation... could no longer be considered a static condition, a product of a creative past and became instead a continuing dynamic process."

[193] The sixth edition of the *Oxford Dictionary of Science* (2010) defines *adaptation* as "Any change in the structure or functioning of successive generations of a population that makes it better suited to its environment."

[194] Orr, H. Allen (February 2005). "The genetic theory of adaptation: a brief history". *Nature Reviews Genetics*. London: Nature Publishing Group. **6** (2): 119–127. ISSN 1471-0056. PMID 15716908. doi:10.1038/nrg1523.

[195] Dobzhansky 1968, pp. 1–34

[196] Dobzhansky 1970, pp. 4–6, 79–82, 84–87

[197] Dobzhansky, Theodosius (March 1956). "Genetics of Natural Populations. XXV. Genetic Changes in Populations of *Drosophila pseudoobscura* and *Drosophila persimilis* in Some Localities in California". *Evolution*. Hoboken, NJ: John Wiley & Sons on behalf of the Society for the Study of Evolution. **10** (1): 82–92. ISSN 0014-3820. JSTOR 2406099. doi:10.2307/2406099.

[198] Nakajima, Akira; Sugimoto, Yohko; Yoneyama, Hiroshi; Nakae, Taiji (June 2002). "High-Level Fluoroquinolone Resistance in *Pseudomonas aeruginosa* Due to Interplay

of the MexAB-OprM Efflux Pump and the DNA Gyrase Mutation". *Microbiology and Immunology*. Tokyo: Center for Academic Publications Japan. **46** (6): 391–395. ISSN 1348-0421. PMID 12153116. doi:10.1111/j.1348-0421.2002.tb02711.x.

[199] Blount, Zachary D.; Borland, Christina Z.; Lenski, Richard E. (June 10, 2008). "Inaugural Article: Historical contingency and the evolution of a key innovation in an experimental population of *Escherichia coli*". *Proc. Natl. Acad. Sci. U.S.A.* Washington, D.C.: National Academy of Sciences. **105** (23): 7899–7906. Bibcode:2008PNAS..105.7899B. ISSN 0027-8424. PMC 2430337. PMID 18524956. doi:10.1073/pnas.0803151105.

[200] Okada, Hirosuke; Negoro, Seiji; Kimura, Hiroyuki; Nakamura, Shunichi (November 10, 1983). "Evolutionary adaptation of plasmid-encoded enzymes for degrading nylon oligomers". *Nature*. London: Nature Publishing Group. **306** (5939): 203–206. Bibcode:1983Natur.306..203O. ISSN 0028-0836. PMID 6646204. doi:10.1038/306203a0.

[201] Ohno, Susumu (April 1984). "Birth of a unique enzyme from an alternative reading frame of the preexisted, internally repetitious coding sequence". *Proc. Natl. Acad. Sci. U.S.A.* Washington, D.C.: National Academy of Sciences. **81** (8): 2421–2425. Bibcode:1984PNAS...81.2421O. ISSN 0027-8424. PMC 345072. PMID 6585807. doi:10.1073/pnas.81.8.2421.

[202] Copley, Shelley D. (June 2000). "Evolution of a metabolic pathway for degradation of a toxic xenobiotic: the patchwork approach". *Trends in Biochemical Sciences*. Cambridge, MA: Cell Press. **25** (6): 261–265. ISSN 0968-0004. PMID 10838562. doi:10.1016/S0968-0004(00)01562-0.

[203] Crawford, Ronald L.; Jung, Carina M.; Strap, Janice L. (October 2007). "The recent evolution of pentachlorophenol (PCP)−4-monooxygenase (PcpB) and associated pathways for bacterial degradation of PCP". *Biodegradation*. Dordrecht, the Netherlands: Springer Netherlands. **18** (5): 525–539. ISSN 0923-9820. PMID 17123025. doi:10.1007/s10532-006-9090-6.

[204] Eshel, Ilan (December 1973). "Clone-Selection and Optimal Rates of Mutation". *Journal of Applied Probability*. Sheffield, England. **10** (4): 728–738. ISSN 1475-6072. JSTOR 3212376. doi:10.2307/3212376.

[205] Altenberg 1995, pp. 205–259

[206] Masel, Joanna; Bergman, Aviv (July 2003). "The evolution of the evolvability properties of the yeast prion [PSI+]". *Evolution*. Hoboken, NJ: John Wiley & Sons on behalf of the Society for the Study of Evolution. **57** (7): 1498–1512. ISSN 0014-3820. PMID 12940355. doi:10.1111/j.0014-3820.2003.tb00358.x.

[207] Lancaster, Alex K.; Bardill, J. Patrick; True, Heather L.; Masel, Joanna (February 2010). "The Spontaneous Appearance Rate of the Yeast Prion [PSI+] and Its Implications for the Evolution of the Evolvability Properties of the [PSI+] System". *Genetics*. Bethesda, MD: Genetics Society of America. **184** (2): 393–400. ISSN 0016-6731. PMC 2828720. PMID 19917766. doi:10.1534/genetics.109.110213.

[208] Draghi, Jeremy; Wagner, Günter P. (February 2008). "Evolution of evolvability in a developmental model". *Evolution*. Hoboken, NJ: John Wiley & Sons on behalf of the Society for the Study of Evolution. **62** (2): 301–315. ISSN 0014-3820. PMID 18031304. doi:10.1111/j.1558-5646.2007.00303.x.

[209] Bejder, Lars; Hall, Brian K. (November 2002). "Limbs in whales and limblessness in other vertebrates: mechanisms of evolutionary and developmental transformation and loss". *Evolution & Development*. Hoboken, NJ: Wiley-Blackwell on behalf of the Society for Integrative and Comparative Biology. **4** (6): 445–458. ISSN 1520-541X. PMID 12492145. doi:10.1046/j.1525-142X.2002.02033.x.

[210] Young, Nathan M.; Hallgrímsson, Benedikt (December 2005). "Serial homology and the evolution of mammalian limb covariation structure". *Evolution*. Hoboken, NJ: John Wiley & Sons on behalf of the Society for the Study of Evolution. **59** (12): 2691–2704. ISSN 0014-3820. PMID 16526515. doi:10.1554/05-233.1.

[211] Penny, David; Poole, Anthony (December 1999). "The nature of the last universal common ancestor". *Current Opinion in Genetics & Development*. Amsterdam, the Netherlands: Elsevier. **9** (6): 672–677. ISSN 0959-437X. PMID 10607605. doi:10.1016/S0959-437X(99)00020-9.

[212] Hall, Brian K. (August 2003). "Descent with modification: the unity underlying homology and homoplasy as seen through an analysis of development and evolution". *Biological Reviews*. Cambridge: Cambridge University Press on behalf of the Cambridge Philosophical Society. **78** (3): 409–433. ISSN 1464-7931. PMID 14558591. doi:10.1017/S1464793102006097.

[213] Shubin, Neil; Tabin, Clifford J.; Carroll, Sean (February 12, 2009). "Deep homology and the origins of evolutionary novelty". *Nature*. London: Nature Publishing Group. **457** (7231): 818–823. Bibcode:2009Natur.457..818S. ISSN 0028-0836. PMID 19212399. doi:10.1038/nature07891.

[214] Fong, Daniel F.; Kane, Thomas C.; Culver, David C. (November 1995). "Vestigialization and Loss of Nonfunctional Characters". *Annual Review of Ecology and Systematics*. Palo Alto, CA: Annual Reviews. **26**: 249–268. ISSN 1545-2069. doi:10.1146/annurev.es.26.110195.001341.

[215] ZhaoLei Zhang; Gerstein, Mark (August 2004). "Large-scale analysis of pseudogenes in the human genome". *Current Opinion in Genetics & Development*. Amsterdam, the Netherlands: Elsevier. **14** (4): 328–335. ISSN 0959-437X. PMID 15261647. doi:10.1016/j.gde.2004.06.003.

29.10. REFERENCES

[216] Jeffery, William R. (May–June 2005). "Adaptive Evolution of Eye Degeneration in the Mexican Blind Cavefish". *Journal of Heredity*. Oxford: Oxford University Press on behalf of the American Genetic Association. **96** (3): 185–196. ISSN 0022-1503. PMID 15653557. doi:10.1093/jhered/esi028.

[217] Maxwell, Erin E.; Larsson, Hans C.E. (May 2007). "Osteology and myology of the wing of the Emu (*Dromaius novaehollandiae*) and its bearing on the evolution of vestigial structures". *Journal of Morphology*. Hoboken, NJ: John Wiley & Sons. **268** (5): 423–441. ISSN 0362-2525. PMID 17390336. doi:10.1002/jmor.10527.

[218] van der Kooi, Casper J.; Schwander, Tanja (November 2014). "On the fate of sexual traits under asexuality" (PDF). *Biological Reviews*. Cambridge: Cambridge University Press on behalf of the Cambridge Philosophical Society. **89** (4): 805–819. ISSN 1464-7931. PMID 24443922. doi:10.1111/brv.12078. Retrieved 2015-08-05.

[219] Silvestri, Anthony R., Jr.; Singh, Iqbal (April 2003). "The unresolved problem of the third molar: Would people be better off without it?". *Journal of the American Dental Association*. Chicago, Illinois: American Dental Association. **134** (4): 450–455. ISSN 0002-8177. PMID 12733778. doi:10.14219/jada.archive.2003.0194. Archived from the original on 23 August 2014.

[220] Coyne 2009, p. 62

[221] Darwin 1872, pp. 101, 103

[222] Gray 2007, p. 66

[223] Coyne 2009, pp. 85–86

[224] Stevens 1982, p. 87

[225] Gould 2002, pp. 1235–1236.

[226] Pallen, Mark J.; Matzke, Nicholas J. (October 2006). "From *The Origin of Species* to the origin of bacterial flagella" (PDF). *Nature Reviews Microbiology* (PDF). London: Nature Publishing Group. **4** (10): 784–790. ISSN 1740-1526. PMID 16953248. doi:10.1038/nrmicro1493. Retrieved 2014-12-25.

[227] Clements, Abigail; Bursac, Dejan; Gatsos, Xenia; et al. (September 15, 2009). "The reducible complexity of a mitochondrial molecular machine". *Proc. Natl. Acad. Sci. U.S.A.* Washington, D.C.: National Academy of Sciences. **106** (37): 15791–15795. Bibcode:2009PNAS..10615791C. ISSN 0027-8424. PMC 2747197. PMID 19717453. doi:10.1073/pnas.0908264106.

[228] Piatigorsky et al. 1994, pp. 241–250

[229] Wistow, Graeme (August 1993). "Lens crystallins: gene recruitment and evolutionary dynamism". *Trends in Biochemical Sciences*. Cambridge, MA: Cell Press. **18** (8): 301–306. ISSN 0968-0004. PMID 8236445. doi:10.1016/0968-0004(93)90041-K.

[230] Johnson, Norman A.; Porter, Adam H. (November 2001). "Toward a new synthesis: population genetics and evolutionary developmental biology". *Genetica*. Dordrecht, the Netherlands: Kluwer Academic Publishers. **112–113** (1): 45–58. ISSN 0016-6707. PMID 11838782. doi:10.1023/A:1013371201773.

[231] Baguñà, Jaume; Garcia-Fernàndez, Jordi (2003). "Evo-Devo: the long and winding road". *The International Journal of Developmental Biology*. Bilbao, Spain: University of the Basque Country Press. **47** (7–8): 705–713. ISSN 0214-6282. PMID 14756346.

- Love, Alan C. (March 2003). "Evolutionary Morphology, Innovation and the Synthesis of Evolutionary and Developmental Biology". *Biology and Philosophy*. Dordrecht, the Netherlands: Kluwer Academic Publishers. **18** (2): 309–345. ISSN 0169-3867. doi:10.1023/A:1023940220348.

[232] Allin, Edgar F. (December 1975). "Evolution of the mammalian middle ear". *Journal of Morphology*. Hoboken, NJ: John Wiley & Sons. **147** (4): 403–437. ISSN 0362-2525. PMID 1202224. doi:10.1002/jmor.1051470404.

[233] Harris, Matthew P.; Hasso, Sean M.; Ferguson, Mark W. J.; Fallon, John F. (February 21, 2006). "The Development of Archosaurian First-Generation Teeth in a Chicken Mutant". *Current Biology*. Cambridge, MA: Cell Press. **16** (4): 371–377. ISSN 0960-9822. PMID 16488870. doi:10.1016/j.cub.2005.12.047.

[234] Carroll, Sean B. (July 11, 2008). "Evo-Devo and an Expanding Evolutionary Synthesis: A Genetic Theory of Morphological Evolution". *Cell*. Cambridge, MA: Cell Press. **134** (1): 25–36. ISSN 0092-8674. PMID 18614008. doi:10.1016/j.cell.2008.06.030.

[235] Wade, Michael J. (March 2007). "The co-evolutionary genetics of ecological communities". *Nature Reviews Genetics*. London: Nature Publishing Group. **8** (3): 185–195. ISSN 1471-0056. PMID 17279094. doi:10.1038/nrg2031.

[236] Geffeney, Shana; Brodie, Edmund D., Jr.; Ruben, Peter C.; Brodie, Edmund D., III (August 23, 2002). "Mechanisms of Adaptation in a Predator-Prey Arms Race: TTX-Resistant Sodium Channels". *Science*. Washington, D.C.: American Association for the Advancement of Science. **297** (5585): 1336–1339. Bibcode:2002Sci...297.1336G. ISSN 0036-8075. PMID 12193784. doi:10.1126/science.1074310.

- Brodie, Edmund D., Jr.; Ridenhour, Benjamin J.; Brodie, Edmund D., III (October 2002). "The evolutionary response of predators to dangerous prey: hotspots and coldspots in the geographic mosaic of coevolution between garter snakes and newts". *Evolution*. Hoboken, NJ: John Wiley & Sons on behalf of the Society for the Study of Evolution. **56** (10): 2067–2082. ISSN 0014-3820. PMID 12449493. doi:10.1554/0014-3820(2002)056[2067:teropt]2.0.co;2.

- Carroll, Sean B. (December 21, 2009). "Whatever Doesn't Kill Some Animals Can Make Them Deadly". *The New York Times*. New York: The New York Times Company. ISSN 0362-4331. Retrieved 2014-12-26.

[237] Sachs, Joel L. (September 2006). "Cooperation within and among species". *Journal of Evolutionary Biology*. Hoboken, NJ: Wiley-Blackwell on behalf of the European Society for Evolutionary Biology. **19** (5): 1415–1418; discussion 1426–1436. ISSN 1010-061X. PMID 16910971. doi:10.1111/j.1420-9101.2006.01152.x.

- Nowak, Martin A. (December 8, 2006). "Five Rules for the Evolution of Cooperation". *Science*. Washington, D.C.: American Association for the Advancement of Science. **314** (5805): 1560–1563. Bibcode:2006Sci...314.1560N. ISSN 0036-8075. PMC 3279745. PMID 17158317. doi:10.1126/science.1133755.

[238] Paszkowski, Uta (August 2006). "Mutualism and parasitism: the yin and yang of plant symbioses". *Current Opinion in Plant Biology*. Amsterdam, the Netherlands: Elsevier. **9** (4): 364–370. ISSN 1369-5266. PMID 16713732. doi:10.1016/j.pbi.2006.05.008.

[239] Hause, Bettina; Fester, Thomas (May 2005). "Molecular and cell biology of arbuscular mycorrhizal symbiosis". *Planta*. Berlin: Springer-Verlag. **221** (2): 184–196. ISSN 0032-0935. PMID 15871030. doi:10.1007/s00425-004-1436-x.

[240] Bertram, John S. (December 2000). "The molecular biology of cancer". *Molecular Aspects of Medicine*. Amsterdam, the Netherlands: Elsevier on behalf of the International Union of Biochemistry and Molecular Biology. **21** (6): 167–223. ISSN 0098-2997. PMID 11173079. doi:10.1016/S0098-2997(00)00007-8.

[241] Reeve, H. Kern; Hölldobler, Bert (June 5, 2007). "The emergence of a superorganism through intergroup competition". *Proc. Natl. Acad. Sci. U.S.A.* Washington, D.C.: National Academy of Sciences. **104** (23): 9736–9740. Bibcode:2007PNAS..104.9736R. ISSN 0027-8424. PMC 1887545. PMID 17517608. doi:10.1073/pnas.0703466104.

[242] Axelrod, Robert; Hamilton, W. D. (March 27, 1981). "The evolution of cooperation". *Science*. Washington, D.C.: American Association for the Advancement of Science. **211** (4489): 1390–1396. Bibcode:1981Sci...211.1390A. ISSN 0036-8075. PMID 7466396. doi:10.1126/science.7466396.

[243] Wilson, Edward O.; Hölldobler, Bert (September 20, 2005). "Eusociality: Origin and consequences". *Proc. Natl. Acad. Sci. U.S.A.* Washington, D.C.: National Academy of Sciences. **102** (38): 13367–1371. Bibcode:2005PNAS..10213367W. ISSN 0027-8424. PMC 1224642. PMID 16157878. doi:10.1073/pnas.0505858102.

[244] Gavrilets, Sergey (October 2003). "Perspective: models of speciation: what have we learned in 40 years?". *Evolution*. Hoboken, NJ: John Wiley & Sons on behalf of the Society for the Study of Evolution. **57** (10): 2197–2215. ISSN 0014-3820. PMID 14628909. doi:10.1554/02-727.

[245] de Queiroz, Kevin (May 3, 2005). "Ernst Mayr and the modern concept of species". *Proc. Natl. Acad. Sci. U.S.A.* Washington, D.C.: National Academy of Sciences. **102** (Suppl. 1): 6600–6607. Bibcode:2005PNAS..102.6600D. ISSN 0027-8424. PMC 1131873. PMID 15851674. doi:10.1073/pnas.0502030102.

[246] Ereshefsky, Marc (December 1992). "Eliminative pluralism". *Philosophy of Science*. Chicago, Illinois: University of Chicago Press. **59** (4): 671–690. ISSN 0031-8248. JSTOR 188136. doi:10.1086/289701.

[247] Mayr 1942, p. 120

[248] Fraser, Christophe; Alm, Eric J.; Polz, Martin F.; et al. (February 6, 2009). "The Bacterial Species Challenge: Making Sense of Genetic and Ecological Diversity". *Science*. Washington, D.C.: American Association for the Advancement of Science. **323** (5915): 741–746. Bibcode:2009Sci...323..741F. ISSN 0036-8075. PMID 19197054. doi:10.1126/science.1159388.

[249] Short, Roger Valentine (October 1975). "The contribution of the mule to scientific thought". *Journal of Reproduction and Fertility. Supplement*. Society for Reproduction and Fertility (23): 359–364. ISSN 0449-3087. OCLC 1639439. PMID 1107543.

[250] Gross, Briana L.; Rieseberg, Loren H. (May–June 2005). "The Ecological Genetics of Homoploid Hybrid Speciation". *Journal of Heredity*. Oxford: Oxford University Press on behalf of the American Genetic Association. **96** (3): 241–252. ISSN 0022-1503. PMC 2517139. PMID 15618301. doi:10.1093/jhered/esi026.

[251] Burke, John M.; Arnold, Michael L. (December 2001). "Genetics and the fitness of hybrids". *Annual Review of Genetics*. Palo Alto, CA: Annual Reviews. **35**: 31–52. ISSN 0066-4197. PMID 11700276. doi:10.1146/annurev.genet.35.102401.085719.

[252] Vrijenhoek, Robert C. (April 4, 2006). "Polyploid Hybrids: Multiple Origins of a Treefrog Species". *Current Biology*. Cambridge, MA: Cell Press. **16** (7): R245–R247. ISSN 0960-9822. PMID 16581499. doi:10.1016/j.cub.2006.03.005.

[253] Rice, William R.; Hostert, Ellen E. (December 1993). "Laboratory Experiments on Speciation: What Have We Learned in 40 Years?". *Evolution*. Hoboken, NJ: John Wiley & Sons on behalf of the Society for the Study of Evolution. **47** (6): 1637–1653. ISSN 0014-3820. doi:10.2307/2410209.

- Jiggins, Chris D.; Bridle, Jon R. (March 2004). "Speciation in the apple maggot fly: a blend of vintages?".

29.10. REFERENCES

Trends in Ecology & Evolution. Cambridge, MA: Cell Press. **19** (3): 111–114. ISSN 0169-5347. PMID 16701238. doi:10.1016/j.tree.2003.12.008.

- Boxhorn, Joseph (September 1, 1995). "Observed Instances of Speciation". *TalkOrigins Archive*. Houston, TX: The TalkOrigins Foundation, Inc. Retrieved 2008-12-26.

- Weinberg, James R.; Starczak, Victoria R.; Jörg, Daniele (August 1992). "Evidence for Rapid Speciation Following a Founder Event in the Laboratory". *Evolution*. Hoboken, NJ: John Wiley & Sons on behalf of the Society for the Study of Evolution. **46** (4): 1214–1220. ISSN 0014-3820. JSTOR 2409766. doi:10.2307/2409766.

[254] Herrel, Anthony; Huyghe, Katleen; Vanhooydonck, Bieke; et al. (March 25, 2008). "Rapid large-scale evolutionary divergence in morphology and performance associated with exploitation of a different dietary resource". *Proc. Natl. Acad. Sci. U.S.A.* Washington, D.C.: National Academy of Sciences. **105** (12): 4792–4795. Bibcode:2008PNAS..105.4792H. ISSN 0027-8424. PMC 2290806. PMID 18344323. doi:10.1073/pnas.0711998105.

[255] Losos, Jonathan B.; Warheit, Kenneth I.; Schoener, Thomas W. (May 1, 1997). "Adaptive differentiation following experimental island colonization in *Anolis* lizards". *Nature*. London: Nature Publishing Group. **387** (6628): 70–73. Bibcode:1997Natur.387...70L. ISSN 0028-0836. doi:10.1038/387070a0.

[256] Hoskin, Conrad J.; Higgie, Megan; McDonald, Keith R.; Moritz, Craig (October 27, 2005). "Reinforcement drives rapid allopatric speciation". *Nature*. London: Nature Publishing Group. **437** (7063): 1353–1356. Bibcode:2005Natur.437.1353H. ISSN 0028-0836. PMID 16251964. doi:10.1038/nature04004.

[257] Templeton, Alan R. (April 1980). "The Theory of Speciation VIA the Founder Principle" (PDF). *Genetics*. Bethesda, MD: Genetics Society of America. **94** (4): 1011–1038. ISSN 0016-6731. PMC 1214177. PMID 6777243. Retrieved 2014-12-29.

[258] Antonovics, Janis (July 2006). "Evolution in closely adjacent plant populations X: long-term persistence of preproductive isolation at a mine boundary". *Heredity*. London: Nature Publishing Group for The Genetics Society. **97** (1): 33–37. ISSN 0018-067X. PMID 16639420. doi:10.1038/sj.hdy.6800835. Retrieved 2014-12-29.

[259] Nosil, Patrik; Crespi, Bernard J.; Gries, Regine; Gries, Gerhard (March 2007). "Natural selection and divergence in mate preference during speciation". *Genetica*. Dordrecht, the Netherlands: Kluwer Academic Publishers. **129** (3): 309–327. ISSN 0016-6707. PMID 16900317. doi:10.1007/s10709-006-0013-6.

[260] Savolainen, Vincent; Anstett, Marie-Charlotte; Lexer, Christian; et al. (May 11, 2006). "Sympatric speciation in palms on an oceanic island". *Nature*. London: Nature Publishing Group. **441** (7090): 210–213. Bibcode:2006Natur.441..210S. ISSN 0028-0836. PMID 16467788. doi:10.1038/nature04566.

- Barluenga, Marta; Stölting, Kai N.; Salzburger, Walter; et al. (February 9, 2006). "Sympatric speciation in Nicaraguan crater lake cichlid fish". *Nature*. London: Nature Publishing Group. **439** (7077): 719–23. Bibcode:2006Natur.439..719B. ISSN 0028-0836. PMID 16467837. doi:10.1038/nature04325.

[261] Gavrilets, Sergey (March 21, 2006). "The Maynard Smith model of sympatric speciation". *Journal of Theoretical Biology*. Amsterdam, the Netherlands: Elsevier. **239** (2): 172–182. ISSN 0022-5193. PMID 16242727. doi:10.1016/j.jtbi.2005.08.041.

[262] Wood, Troy E.; Takebayashi, Naoki; Barker, Michael S.; et al. (August 18, 2009). "The frequency of polyploid speciation in vascular plants". *Proc. Natl. Acad. Sci. U.S.A.* Washington, D.C.: National Academy of Sciences. **106** (33): 13875–13879. Bibcode:2009PNAS..10613875W. ISSN 0027-8424. PMC 2728988. PMID 19667210. doi:10.1073/pnas.0811575106.

[263] Hegarty, Matthew J.; Hiscock, Simon J. (May 20, 2008). "Genomic Clues to the Evolutionary Success of Polyploid Plants". *Current Biology*. Cambridge, MA: Cell Press. **18** (10): R435–R444. ISSN 0960-9822. PMID 18492478. doi:10.1016/j.cub.2008.03.043.

[264] Jakobsson, Mattias; Hagenblad, Jenny; Tavaré, Simon; et al. (June 2006). "A Unique Recent Origin of the Allotetraploid Species *Arabidopsis suecica*: Evidence from Nuclear DNA Markers". *Molecular Biology and Evolution*. Oxford: Oxford University Press on behalf of the Society for Molecular Biology and Evolution. **23** (6): 1217–1231. ISSN 0737-4038. PMID 16549398. doi:10.1093/molbev/msk006.

[265] Säll, Torbjörn; Jakobsson, Mattias; Lind-Halldén, Christina; Halldén, Christer (September 2003). "Chloroplast DNA indicates a single origin of the allotetraploid *Arabidopsis suecica*". *Journal of Evolutionary Biology*. Hoboken, NJ: Wiley-Blackwell on behalf of the European Society for Evolutionary Biology. **16** (5): 1019–1029. ISSN 1010-061X. PMID 14635917. doi:10.1046/j.1420-9101.2003.00554.x.

[266] Bomblies, Kirsten; Weigel, Detlef (December 2007). "*Arabidopsis*—a model genus for speciation". *Current Opinion in Genetics & Development*. Amsterdam, the Netherlands: Elsevier. **17** (6): 500–504. ISSN 0959-437X. PMID 18006296. doi:10.1016/j.gde.2007.09.006.

[267] Sémon, Marie; Wolfe, Kenneth H. (December 2007). "Consequences of genome duplication". *Current Opinion in Genetics & Development*. Amsterdam, the Netherlands: Elsevier. **17** (6): 505–512. ISSN 0959-437X. PMID 18006297. doi:10.1016/j.gde.2007.09.007.

[268] Eldredge & Gould 1972, pp. 82–115

[269] Benton, Michael J. (April 7, 1995). "Diversification and extinction in the history of life". *Science*. Washington, D.C.: American Association for the Advancement of Science. **268** (5207): 52–58. Bibcode:1995Sci...268...52B. ISSN 0036-8075. PMID 7701342. doi:10.1126/science.7701342.

[270] Raup, David M. (March 28, 1986). "Biological extinction in Earth history". *Science*. Washington, D.C.: American Association for the Advancement of Science. **231** (4745): 1528–1533. Bibcode:1986Sci...231.1528R. ISSN 0036-8075. PMID 11542058. doi:10.1126/science.11542058.

[271] Avise, John C.; Hubbell, Stephen P.; Ayala, Francisco J. (August 12, 2008). "In the light of evolution II: Biodiversity and extinction". *Proc. Natl. Acad. Sci. U.S.A*. Washington, D.C.: National Academy of Sciences. **105** (Suppl. 1): 11453–11457. Bibcode:2008PNAS..10511453A. ISSN 0027-8424. PMC 2556414. PMID 18695213. doi:10.1073/pnas.0802504105.

[272] Raup, David M. (July 19, 1994). "The role of extinction in evolution". *Proc. Natl. Acad. Sci. U.S.A*. Washington, D.C.: National Academy of Sciences. **91** (15): 6758–6763. Bibcode:1994PNAS...91.6758R. ISSN 0027-8424. PMC 44280. PMID 8041694. doi:10.1073/pnas.91.15.6758.

[273] Novacek, Michael J.; Cleland, Elsa E. (May 8, 2001). "The current biodiversity extinction event: scenarios for mitigation and recovery". *Proc. Natl. Acad. Sci. U.S.A*. Washington, D.C.: National Academy of Sciences. **98** (10): 5466–5470. Bibcode:2001PNAS...98.5466N. ISSN 0027-8424. PMC 33235. PMID 11344295. doi:10.1073/pnas.091093698.

[274] Pimm, Stuart; Raven, Peter; Peterson, Alan; et al. (July 18, 2006). "Human impacts on the rates of recent, present and future bird extinctions". *Proc. Natl. Acad. Sci. U.S.A*. Washington, D.C.: National Academy of Sciences. **103** (29): 10941–10946. Bibcode:2006PNAS..10310941P. ISSN 0027-8424. PMC 1544153. PMID 16829570. doi:10.1073/pnas.0604181103.

- Barnosky, Anthony D.; Koch, Paul L.; Feranec, Robert S.; et al. (October 1, 2004). "Assessing the Causes of Late Pleistocene Extinctions on the Continents". *Science*. Washington, D.C.: American Association for the Advancement of Science. **306** (5693): 70–75. Bibcode:2004Sci...306...70B. ISSN 0036-8075. PMID 15459379. doi:10.1126/science.1101476.

[275] Lewis, Owen T. (January 29, 2006). "Climate change, species–area curves and the extinction crisis". *Philosophical Transactions of the Royal Society B: Biological Sciences*. London: Royal Society. **361** (1465): 163–171. ISSN 0962-8436. PMC 1831839. PMID 16553315. doi:10.1098/rstb.2005.1712.

[276] Jablonski, David (May 8, 2001). "Lessons from the past: Evolutionary impacts of mass extinctions". *Proc. Natl. Acad. Sci. U.S.A*. Washington, D.C.: National Academy of Sciences. **98** (10): 5393–5398. Bibcode:2001PNAS...98.5393J. ISSN 0027-8424. PMC 33224. PMID 11344284. doi:10.1073/pnas.101092598.

[277] "Age of the Earth". United States Geological Survey. July 9, 2007. Retrieved 2015-05-31.

[278] Dalrymple 2001, pp. 205–221

[279] Manhesa, Gérard; Allègre, Claude J.; Dupréa, Bernard; Hamelin, Bruno (May 1980). "Lead isotope study of basic-ultrabasic layered complexes: Speculations about the age of the earth and primitive mantle characteristics". *Earth and Planetary Science Letters*. Amsterdam, the Netherlands: Elsevier. **47** (3): 370–382. Bibcode:1980E&PSL...47..370M. ISSN 0012-821X. doi:10.1016/0012-821X(80)90024-2.

[280] Raven & Johnson 2002, p. 68

[281] McKinney 1997, p. 110

[282] Peretó, Juli (March 2005). "Controversies on the origin of life" (PDF). *International Microbiology*. Barcelona: Spanish Society for Microbiology. **8** (1): 23–31. ISSN 1139-6709. PMID 15906258. Archived from the original (PDF) on 2015-08-24.

[283] Joyce, Gerald F. (July 11, 2002). "The antiquity of RNA-based evolution". *Nature*. London: Nature Publishing Group. **418** (6894): 214–221. Bibcode:2002Natur.418..214J. ISSN 0028-0836. PMID 12110897. doi:10.1038/418214a.

[284] Trevors, Jack T.; Psenner, Roland (December 2001). "From self-assembly of life to present-day bacteria: a possible role for nanocells". *FEMS Microbiology Reviews*. Amsterdam, the Netherlands: Elsevier on behalf of the Federation of European Microbiological Societies. **25** (5): 573–582. ISSN 1574-6976. PMID 11742692. doi:10.1111/j.1574-6976.2001.tb00592.x.

[285] Theobald, Douglas L. (May 13, 2010). "A formal test of the theory of universal common ancestry". *Nature*. London: Nature Publishing Group. **465** (7295): 219–222. Bibcode:2010Natur.465..219T. ISSN 0028-0836. PMID 20463738. doi:10.1038/nature09014.

[286] Bapteste, Eric; Walsh, David A. (June 2005). "Does the 'Ring of Life' ring true?". *Trends in Microbiology*. Cambridge, MA: Cell Press. **13** (6): 256–261. ISSN 0966-842X. PMID 15936656. doi:10.1016/j.tim.2005.03.012.

[287] Darwin 1859, p. 1

[288] Doolittle, W. Ford; Bapteste, Eric (February 13, 2007). "Pattern pluralism and the Tree of Life hypothesis". *Proc. Natl. Acad. Sci. U.S.A*. Washington, D.C.: National Academy of Sciences. **104** (7):

29.10. REFERENCES

2043–2049. Bibcode:2007PNAS..104.2043D. ISSN 0027-8424. PMC 1892968. PMID 17261804. doi:10.1073/pnas.0610699104.

[289] Kunin, Victor; Goldovsky, Leon; Darzentas, Nikos; Ouzounis, Christos A. (July 2005). "The net of life: Reconstructing the microbial phylogenetic network". *Genome Research*. Cold Spring Harbor, NY: Cold Spring Harbor Laboratory Press. **15** (7): 954–959. ISSN 1088-9051. PMC 1172039. PMID 15965028. doi:10.1101/gr.3666505.

[290] Jablonski, David (June 25, 1999). "The Future of the Fossil Record". *Science*. Washington, D.C.: American Association for the Advancement of Science. **284** (5423): 2114–2116. ISSN 0036-8075. PMID 10381868. doi:10.1126/science.284.5423.2114.

[291] Mason, Stephen F. (September 6, 1984). "Origins of biomolecular handedness". *Nature*. London: Nature Publishing Group. **311** (5981): 19–23. Bibcode:1984Natur.311...19M. ISSN 0028-0836. PMID 6472461. doi:10.1038/311019a0.

[292] Wolf, Yuri I.; Rogozin, Igor B.; Grishin, Nick V.; Koonin, Eugene V. (September 1, 2002). "Genome trees and the tree of life". *Trends in Genetics*. Cambridge, MA: Cell Press. **18** (9): 472–479. ISSN 0168-9525. PMID 12175808. doi:10.1016/S0168-9525(02)02744-0.

[293] Varki, Ajit; Altheide, Tasha K. (December 2005). "Comparing the human and chimpanzee genomes: searching for needles in a haystack". *Genome Research*. Cold Spring Harbor, NY: Cold Spring Harbor Laboratory Press. **15** (12): 1746–1758. ISSN 1088-9051. PMID 16339373. doi:10.1101/gr.3737405.

[294] Cavalier-Smith, Thomas (June 29, 2006). "Cell evolution and Earth history: stasis and revolution". *Philosophical Transactions of the Royal Society B: Biological Sciences*. London: Royal Society. **361** (1470): 969–1006. ISSN 0962-8436. PMC 1578732. PMID 16754610. doi:10.1098/rstb.2006.1842.

[295] Schopf, J. William (June 29, 2006). "Fossil evidence of Archaean life". *Philosophical Transactions of the Royal Society B: Biological Sciences*. London: Royal Society. **361** (1470): 869–885. ISSN 0962-8436. PMC 1578735. PMID 16754604. doi:10.1098/rstb.2006.1834.

- Altermann, Wladyslaw; Kazmierczak, Józef (November 2003). "Archean microfossils: a reappraisal of early life on Earth". *Research in Microbiology*. Amsterdam, the Netherlands: Elsevier for the Pasteur Institute. **154** (9): 611–617. ISSN 0923-2508. PMID 14596897. doi:10.1016/j.resmic.2003.08.006.

[296] Schopf, J. William (July 19, 1994). "Disparate rates, differing fates: tempo and mode of evolution changed from the Precambrian to the Phanerozoic". *Proc. Natl. Acad. Sci. U.S.A.* Washington, D.C.: National Academy of Sciences. **91** (15): 6735–6742. Bibcode:1994PNAS...91.6735S. ISSN 0027-8424. PMC 44277. PMID 8041691. doi:10.1073/pnas.91.15.6735.

[297] Poole, Anthony M.; Penny, David (January 2007). "Evaluating hypotheses for the origin of eukaryotes". *BioEssays*. Hoboken, NJ: John Wiley & Sons. **29** (1): 74–84. ISSN 0265-9247. PMID 17187354. doi:10.1002/bies.20516.

[298] Dyall, Sabrina D.; Brown, Mark T.; Johnson, Patricia J. (April 9, 2004). "Ancient Invasions: From Endosymbionts to Organelles". *Science*. Washington, D.C.: American Association for the Advancement of Science. **304** (5668): 253–257. Bibcode:2004Sci...304..253D. ISSN 0036-8075. PMID 15073369. doi:10.1126/science.1094884.

[299] Martin, William (October 2005). "The missing link between hydrogenosomes and mitochondria". *Trends in Microbiology*. Cambridge, MA: Cell Press. **13** (10): 457–459. ISSN 0966-842X. PMID 16109488. doi:10.1016/j.tim.2005.08.005.

[300] Lang, B. Franz; Gray, Michael W.; Burger, Gertraud (December 1999). "Mitochondrial genome evolution and the origin of eukaryotes". *Annual Review of Genetics*. Palo Alto, CA: Annual Reviews. **33**: 351–397. ISSN 0066-4197. PMID 10690412. doi:10.1146/annurev.genet.33.1.351.

- McFadden, Geoffrey Ian (December 1, 1999). "Endosymbiosis and evolution of the plant cell". *Current Opinion in Plant Biology*. Amsterdam, the Netherlands: Elsevier. **2** (6): 513–519. ISSN 1369-5266. PMID 10607659. doi:10.1016/S1369-5266(99)00025-4.

[301] DeLong, Edward F.; Pace, Norman R. (August 1, 2001). "Environmental Diversity of Bacteria and Archaea". *Systematic Biology*. Oxford: Oxford University Press on behalf of the Society of Systematic Biologists. **50** (4): 470–478. ISSN 1063-5157. PMID 12116647. doi:10.1080/106351501750435040.

[302] Kaiser, Dale (December 2001). "Building a multicellular organism". *Annual Review of Genetics*. Palo Alto, CA: Annual Reviews. **35**: 103–123. ISSN 0066-4197. PMID 11700279. doi:10.1146/annurev.genet.35.102401.090145.

[303] Zimmer, Carl (January 7, 2016). "Genetic Flip Helped Organisms Go From One Cell to Many". *The New York Times*. Retrieved January 7, 2016.

[304] Valentine, James W.; Jablonski, David; Erwin, Douglas H. (March 1, 1999). "Fossils, molecules and embryos: new perspectives on the Cambrian explosion" (PDF). *Development*. Cambridge: The Company of Biologists. **126** (5): 851–859. ISSN 0950-1991. PMID 9927587. Retrieved 2014-12-30.

[305] Ohno, Susumu (January 1997). "The reason for as well as the consequence of the Cambrian explosion in animal evolution". *Journal of Molecular Evolution*. New York: Springer-Verlag New York. **44** (Suppl. 1): S23–S27. Bibcode:1997JMolE..44S..23O. ISSN 0022-2844. PMID 9071008. doi:10.1007/PL00000055.

- Valentine, James W.; Jablonski, David (2003). "Morphological and developmental macroevolution: a paleontological perspective". *The International Journal of Developmental Biology*. Bilbao, Spain: University of the Basque Country Press. **47** (7–8): 517–522. ISSN 0214-6282. PMID 14756327. Retrieved 2014-12-30.

[306] Waters, Elizabeth R. (December 2003). "Molecular adaptation and the origin of land plants". *Molecular Phylogenetics and Evolution*. Academic Press. **29** (3): 456–463. ISSN 1055-7903. PMID 14615186. doi:10.1016/j.ympev.2003.07.018.

[307] Mayhew, Peter J. (August 2007). "Why are there so many insect species? Perspectives from fossils and phylogenies". *Biological Reviews*. Cambridge: Cambridge University Press on behalf of the Cambridge Philosophical Society. **82** (3): 425–454. ISSN 1464-7931. PMID 17624962. doi:10.1111/j.1469-185X.2007.00018.x.

[308] Carroll, Robert L. (May 2007). "The Palaeozoic Ancestry of Salamanders, Frogs and Caecilians". *Zoological Journal of the Linnean Society*. Hoboken, NJ: Wiley-Blackwell. **150** (Supplement s1): 1–140. ISSN 1096-3642. doi:10.1111/j.1096-3642.2007.00246.x.

[309] Wible, John R.; Rougier, Guillermo W.; Novacek, Michael J.; Asher, Robert J. (June 21, 2007). "Cretaceous eutherians and Laurasian origin for placental mammals near the K/T boundary". *Nature*. London: Nature Publishing Group. **447** (7147): 1003–1006. Bibcode:2007Natur.447.1003W. ISSN 0028-0836. PMID 17581585. doi:10.1038/nature05854.

[310] Witmer, Lawrence M. (July 28, 2011). "Palaeontology: An icon knocked from its perch". *Nature*. London: Nature Publishing Group. **475** (7357): 458–459. ISSN 0028-0836. PMID 21796198. doi:10.1038/475458a.

[311] Bull, James J.; Wichman, Holly A. (November 2001). "Applied evolution". *Annual Review of Ecology and Systematics*. Palo Alto, CA: Annual Reviews. **32**: 183–217. ISSN 1545-2069. doi:10.1146/annurev.ecolsys.32.081501.114020.

[312] Doebley, John F.; Gaut, Brandon S.; Smith, Bruce D. (December 29, 2006). "The Molecular Genetics of Crop Domestication". *Cell*. Cambridge, MA: Cell Press. **127** (7): 1309–1321. ISSN 0092-8674. PMID 17190597. doi:10.1016/j.cell.2006.12.006.

[313] Jäckel, Christian; Kast, Peter; Hilvert, Donald (June 2008). "Protein Design by Directed Evolution". *Annual Review of Biophysics*. Palo Alto, CA: Annual Reviews. **37**: 153–173. ISSN 1936-122X. PMID 18573077. doi:10.1146/annurev.biophys.37.032807.125832.

[314] Maher, Brendan (April 8, 2009). "Evolution: Biology's next top model?". *Nature*. London: Nature Publishing Group. **458** (7239): 695–698. ISSN 0028-0836. PMID 19360058. doi:10.1038/458695a.

[315] Borowsky, Richard (January 8, 2008). "Restoring sight in blind cavefish". *Current Biology*. Cambridge, MA: Cell Press. **18** (1): R23–R24. ISSN 0960-9822. PMID 18177707. doi:10.1016/j.cub.2007.11.023.

[316] Gross, Joshua B.; Borowsky, Richard; Tabin, Clifford J. (January 2, 2009). Barsh, Gregory S., ed. "A novel role for *Mc1r* in the parallel evolution of depigmentation in independent populations of the cavefish *Astyanax mexicanus*". *PLOS Genetics*. San Francisco, CA: Public Library of Science. **5** (1): e1000326. ISSN 1553-7390. PMC 2603666. PMID 19119422. doi:10.1371/journal.pgen.1000326.

[317] Merlo, Lauren M.F.; Pepper, John W.; Reid, Brian J.; Maley, Carlo C. (December 2006). "Cancer as an evolutionary and ecological process.". *Nature Reviews Cancer*. London: Nature Publishing Group. **6** (12): 924–935. ISSN 1474-175X. PMID 17109012. doi:10.1038/nrc2013.

[318] Pan, Dabo; Weiwei Xue; Wenqi Zhang; et al. (October 2012). "Understanding the drug resistance mechanism of hepatitis C virus NS3/4A to ITMN-191 due to R155K, A156V, D168A/E mutations: a computational study.". *Biochimica et Biophysica Acta (BBA) - General Subjects*. Amsterdam, the Netherlands: Elsevier. **1820** (10): 1526–1534. ISSN 0304-4165. PMID 22698669. doi:10.1016/j.bbagen.2012.06.001.

[319] Woodford, Neil; Ellington, Matthew J. (January 2007). "The emergence of antibiotic resistance by mutation.". *Clinical Microbiology and Infection*. Amsterdam, the Netherlands: Elsevier for the European Society of Clinical Microbiology and Infectious Diseases. **13** (1): 5–18. ISSN 1198-743X. PMID 17184282. doi:10.1111/j.1469-0691.2006.01492.x.

[320] Labbé, Pierrick; Berticat, Claire; Berthomieu, Arnaud; et al. (November 16, 2007). "Forty Years of Erratic Insecticide Resistance Evolution in the Mosquito *Culex pipiens*". *PLOS Genetics*. San Francisco, CA: Public Library of Science. **3** (11): e205. ISSN 1553-7390. PMC 2077897. PMID 18020711. doi:10.1371/journal.pgen.0030205.

[321] Neve, Paul (October 2007). "Challenges for herbicide resistance evolution and management: 50 years after Harper". *Weed Research*. Hoboken, NJ: Wiley-Blackwell on behalf of the European Weed Research Society. **47** (5): 365–369. ISSN 0043-1737. doi:10.1111/j.1365-3180.2007.00581.x.

[322] Rodríguez-Rojas, Alexandro; Rodríguez-Beltrán, Jerónimo; Couce, Alejandro; Blázquez, Jesús (August 2013). "Antibiotics and antibiotic resistance: A bitter fight against evolution". *International Journal of Medical Microbiology*. Amsterdam, the Netherlands: Elsevier. **303** (6–7): 293–297. ISSN 1438-4221. PMID 23517688. doi:10.1016/j.ijmm.2013.02.004.

[323] Schenk, Martijn F.; Szendro, Ivan G.; Krug, Joachim; de Visser, J. Arjan G. M. (June 28, 2012). "Quantifying the Adaptive Potential of an Antibiotic Resistance Enzyme".

PLOS Genetics. San Francisco, CA: Public Library of Science. **8** (6): e1002783. ISSN 1553-7390. PMC 3386231 ⓐ. PMID 22761587. doi:10.1371/journal.pgen.1002783.

[324] Read, Andrew F.; Lynch, Penelope A.; Thomas, Matthew B. (April 7, 2009). "How to Make Evolution-Proof Insecticides for Malaria Control". *PLOS Biology*. San Francisco, CA: Public Library of Science. **7** (4): e1000058. ISSN 1545-7885. PMC 3279047 ⓐ. PMID 19355786. doi:10.1371/journal.pbio.1000058.

[325] Fraser, Alex S. (January 18, 1958). "Monte Carlo Analyses of Genetic Models". *Nature*. London: Nature Publishing Group. **181** (4603): 208–209. Bibcode:1958Natur.181..208F. ISSN 0028-0836. PMID 13504138. doi:10.1038/181208a0.

[326] Rechenberg 1973

[327] Holland 1975

[328] Koza 1992

[329] Jamshidi, Mo (August 15, 2003). "Tools for intelligent control: fuzzy controllers, neural networks and genetic algorithms". *Philosophical Transactions of the Royal Society A*. London: Royal Society. **361** (1809): 1781–1808. Bibcode:2003RSPTA.361.1781J. ISSN 1364-503X. PMID 12952685. doi:10.1098/rsta.2003.1225.

[330] Browne 2003, pp. 376–379

[331] For an overview of the philosophical, religious and cosmological controversies, see:

- Dennett 1995

For the scientific and social reception of evolution in the 19th and early 20th centuries, see:

- Johnston, Ian C. (1999). "Section Three: The Origins of Evolutionary Theory". . . . *And Still We Evolve: A Handbook for the Early History of Modern Science* (3rd revised ed.). Nanaimo, BC: Liberal Studies Department, Malaspina University-College. Retrieved 2015-01-01.
- Bowler 2003
- Zuckerkandl, Emile (December 30, 2006). "Intelligent design and biological complexity". *Gene*. Amsterdam, the Netherlands: Elsevier. **385**: 2–18. ISSN 0378-1119. PMID 17011142. doi:10.1016/j.gene.2006.03.025.

[332] Ross, Marcus R. (May 2005). "Who Believes What? Clearing up Confusion over Intelligent Design and Young-Earth Creationism" (PDF). *Journal of Geoscience Education*. National Association of Geoscience Teachers. **53** (3): 319. Bibcode:2005JGeEd..53..319R. ISSN 1089-9995. doi:10.5408/1089-9995-53.3.319. Retrieved 2008-04-28.

[333] Hameed, Salman (December 12, 2008). "Bracing for Islamic Creationism" (PDF). *Science*. Washington, D.C.: American Association for the Advancement of Science. **322** (5908): 1637–1638. ISSN 0036-8075. PMID 19074331. doi:10.1126/science.1163672. Archived from the original (PDF) on November 10, 2014.

[334] Bowler 2003

[335] Miller, Jon D.; Scott, Eugenie C.; Okamoto, Shinji (August 11, 2006). "Public Acceptance of Evolution". *Science*. Washington, D.C.: American Association for the Advancement of Science. **313** (5788): 765–766. ISSN 0036-8075. PMID 16902112. doi:10.1126/science.1126746.

[336] Spergel, David Nathaniel; Verde, Licia; Peiris, Hiranya V.; et al. (2003). "First-Year Wilkinson Microwave Anisotropy Probe (WMAP) Observations: Determination of Cosmological Parameters". *The Astrophysical Journal Supplement Series*. Chicago, Illinois: University of Chicago Press on behalf of the American Astronomical Society. **148** (1): 175–194. Bibcode:2003ApJS..148..175S. arXiv:astro-ph/0302209 ⓐ. doi:10.1086/377226.

[337] Wilde, Simon A.; Valley, John W.; Peck, William H.; Graham, Colin M. (January 11, 2001). "Evidence from detrital zircons for the existence of continental crust and oceans on the Earth 4.4 Gyr ago". *Nature*. London: Nature Publishing Group. **409** (6817): 175–178. ISSN 0028-0836. PMID 11196637. doi:10.1038/35051550.

[338] Branch, Glenn (March 2007). "Understanding Creationism after *Kitzmiller*". *BioScience*. Oxford: Oxford University Press on behalf of the American Institute of Biological Sciences. **57** (3): 278–284. ISSN 0006-3568. doi:10.1641/B570313.

29.11 Bibliography

- Altenberg, Lee (1995). "Genome growth and the evolution of the genotype-phenotype map". In Banzhaf, Wolfgang; Eeckman, Frank H. *Evolution and Biocomputation: Computational Models of Evolution*. Lecture Notes in Computer Science. **899**. Berlin; New York: Springer-Verlag Berlin Heidelberg. ISBN 3-540-59046-3. ISSN 0302-9743. LCCN 95005970. OCLC 32049812. doi:10.1007/3-540-59046-3_11.

- Ayala, Francisco J.; Avise, John C., eds. (2014). *Essential Readings in Evolutionary Biology*. Baltimore, MD: Johns Hopkins University Press. ISBN 978-1-4214-1305-1. LCCN 2013027718. OCLC 854285705.

- Birdsell, John A.; Wills, Christopher (2003). "The Evolutionary Origin and Maintenance of Sexual Recombination: A Review of Contemporary Models".

- In MacIntyre, Ross J.; Clegg, Michael T. *Evolutionary Biology*. Evolutionary Biology. **33**. New York: Springer Science+Business Media. ISBN 978-1-4419-3385-0. ISSN 0071-3260. OCLC 751583918.

- Bowler, Peter J. (1989). *The Mendelian Revolution: The Emergence of Hereditarian Concepts in Modern Science and Society*. Baltimore, MD: Johns Hopkins University Press. ISBN 0-8018-3888-6. LCCN 89030914. OCLC 19322402.

- Bowler, Peter J. (2003). *Evolution: The History of an Idea* (3rd completely rev. and expanded ed.). Berkeley, CA: University of California Press. ISBN 0-520-23693-9. LCCN 2002007569. OCLC 49824702.

- Browne, Janet (2003). *Charles Darwin: The Power of Place*. **2**. London: Pimlico. ISBN 0-7126-6837-3. LCCN 94006598. OCLC 52327000.

- Burkhardt, Frederick; Smith, Sydney, eds. (1991). *The Correspondence of Charles Darwin*. The Correspondence of Charles Darwin. **7**: 1858–1859. Cambridge: Cambridge University Press. ISBN 0-521-38564-4. LCCN 84045347. OCLC 185662993.

- Carroll, Sean B.; Grenier, Jennifer K.; Weatherbee, Scott D. (2005). *From DNA to Diversity: Molecular Genetics and the Evolution of Animal Design* (2nd ed.). Malden, MA: Blackwell Publishing. ISBN 1-4051-1950-0. LCCN 2003027991. OCLC 53972564.

- Coyne, Jerry A. (2009). *Why Evolution is True*. New York: Viking. ISBN 978-0-670-02053-9. LCCN 2008033973. OCLC 233549529.

- Cracraft, Joel; Bybee, Rodger W., eds. (2005). *Evolutionary Science and Society: Educating a New Generation* (PDF). Colorado Springs, CO: Biological Sciences Curriculum Study. ISBN 1-929614-23-3. OCLC 64228003. Retrieved 2014-12-06. "Revised Proceedings of the BSCS, AIBS Symposium November 2004, Chicago, Illinois"

- Dalrymple, G. Brent (2001). "The age of the Earth in the twentieth century: a problem (mostly) solved". In Lewis, C. L. E.; Knell, S. J. *The Age of the Earth: from 4004 BC to AD 2002*. Geological Society Special Publication. **190**. London: Geological Society of London. Bibcode:2001GSLSP.190..205D. ISBN 1-86239-093-2. ISSN 0305-8719. LCCN 2003464816. OCLC 48570033. doi:10.1144/gsl.sp.2001.190.01.14.

- Darwin, Charles (1859). *On the Origin of Species by Means of Natural Selection, or the Preservation of Favoured Races in the Struggle for Life* (1st ed.). London: John Murray. LCCN 06017473. OCLC 741260650. The book is available from The Complete Work of Charles Darwin Online. Retrieved 2014-11-21.

- Darwin, Charles (1872). *The Expression of the Emotions in Man and Animals*. London: John Murray. LCCN 04002793. OCLC 1102785.

- Darwin, Francis, ed. (1909). *The foundations of The origin of species, a sketch written in 1842* (PDF). Cambridge: Printed at the University Press. LCCN 61057537. OCLC 1184581. Retrieved 2014-11-27.

- Dawkins, Richard (1990). *The Blind Watchmaker*. Penguin Science. London: Penguin Books. ISBN 0-14-014481-1. OCLC 60143870.

- Dennett, Daniel (1995). *Darwin's Dangerous Idea: Evolution and the Meanings of Life*. New York: Simon & Schuster. ISBN 0-684-80290-2. LCCN 94049158. OCLC 31867409.

- Dobzhansky, Theodosius (1968). "On Some Fundamental Concepts of Darwinian Biology". In Dobzhansky, Theodosius; Hecht, Max K.; Steere, William C. *Evolutionary Biology*. Volume 2 (1st ed.). New York: Appleton-Century-Crofts. OCLC 24875357. doi:10.1007/978-1-4684-8094-8_1.

- Dobzhansky, Theodosius (1970). *Genetics of the Evolutionary Process*. New York: Columbia University Press. ISBN 0-231-02837-7. LCCN 72127363. OCLC 97663.

- Eldredge, Niles; Gould, Stephen Jay (1972). "Punctuated equilibria: an alternative to phyletic gradualism". In Schopf, Thomas J. M. *Models in Paleobiology*. San Francisco, CA: Freeman, Cooper. ISBN 0-87735-325-5. LCCN 72078387. OCLC 572084.

  - Eldredge, Niles (1985). *Time Frames: The Rethinking of Darwinian Evolution and the Theory of Punctuated Equilibria*. New York: Simon & Schuster. ISBN 0-671-49555-0. LCCN 84023632. OCLC 11443805.

- Ewens, Warren J. (2004). *Mathematical Population Genetics*. Interdisciplinary Applied Mathematics. **I**. Theoretical Introduction (2nd ed.). New York: Springer-Verlag New York. ISBN 0-387-20191-2. LCCN 2003065728. OCLC 53231891.

- Futuyma, Douglas J. (2004). "The Fruit of the Tree of Life: Insights into Evolution and Ecology". In Cracraft, Joel; Donoghue, Michael J. *Assembling the Tree of Life*. Oxford; New York: Oxford University Press. ISBN 0-19-517234-5. LCCN 2003058012.

OCLC 61342697. "Proceedings of a symposium held at the American Museum of Natural History in New York, 2002."

- Futuyma, Douglas J. (2005). *Evolution*. Sunderland, MA: Sinauer Associates. ISBN 0-87893-187-2. LCCN 2004029808. OCLC 57311264.

- Gould, Stephen Jay (2002). *The Structure of Evolutionary Theory*. Cambridge, MA: Belknap Press of Harvard University Press. ISBN 0-674-00613-5. LCCN 2001043556. OCLC 47869352.

- Gray, Peter (2007). *Psychology* (5th ed.). New York: Worth Publishers. ISBN 978-0-7167-0617-5. LCCN 2006921149. OCLC 76872504.

- Hall, Brian K.; Hallgrímsson, Benedikt (2008). *Strickberger's Evolution* (4th ed.). Sudbury, MA: Jones and Bartlett Publishers. ISBN 978-0-7637-0066-9. LCCN 2007008981. OCLC 85814089.

- Hennig, Willi (1999) [Originally published 1966 (reprinted 1979); translated from the author's unpublished revision of *Grundzüge einer Theorie der phylogenetischen Systematik*, published in 1950]. *Phylogenetic Systematics*. Translation by D. Dwight Davis and Rainer Zangerl; foreword by Donn E. Rosen, Gareth Nelson, and Colin Patterson (Reissue ed.). Urbana, IL: University of Illinois Press. ISBN 0-252-06814-9. LCCN 78031969. OCLC 722701473.

- Holland, John H. (1975). *Adaptation in Natural and Artificial Systems: An Introductory Analysis with Applications to Biology, Control, and Artificial Intelligence*. Ann Arbor, MI: University of Michigan Press. ISBN 0-472-08460-7. LCCN 74078988. OCLC 1531617.

- Jablonka, Eva; Lamb, Marion J. (2005). *Evolution in Four Dimensions: Genetic, Epigenetic, Behavioral, and Symbolic Variation in the History of Life*. Illustrations by Anna Zeligowski. Cambridge, MA: MIT Press. ISBN 0-262-10107-6. LCCN 2004058193. OCLC 61896061.

- Kampourakis, Kostas (2014). *Understanding Evolution*. Cambridge; New York: Cambridge University Press. ISBN 978-1-107-03491-4. LCCN 2013034917. OCLC 855585457.

- Kirk, Geoffrey; Raven, John; Schofield, Malcolm (1983). *The Presocratic Philosophers: A Critical History with a Selection of Texts* (2nd ed.). Cambridge; New York: Cambridge University Press. ISBN 0-521-27455-9. LCCN 82023505. OCLC 9081712.

- Koza, John R. (1992). *Genetic Programming: On the Programming of Computers by Means of Natural Selection*. Complex Adaptive Systems. Cambridge, MA: MIT Press. ISBN 0-262-11170-5. LCCN 92025785. OCLC 26263956.

- Lamarck, Jean-Baptiste (1809). *Philosophie Zoologique*. Paris: Dentu et L'Auteur. OCLC 2210044. Philosophie zoologique (1809) on the Internet Archive. Retrieved 2014-11-29.

- Lane, David H. (1996). *The Phenomenon of Teilhard: Prophet for a New Age* (1st ed.). Macon, GA: Mercer University Press. ISBN 0-86554-498-0. LCCN 96008777. OCLC 34710780.

- Magner, Lois N. (2002). *A History of the Life Sciences* (3rd rev. and expanded ed.). New York: Marcel Dekker. ISBN 0-8247-0824-5. LCCN 2002031313. OCLC 50410202.

- Mason, Stephen F. (1962). *A History of the Sciences*. Collier Books, Science Library, CS9 (New rev. ed.). New York: Collier Books. LCCN 62003378. OCLC 568032626.

- Maynard Smith, John (1978). *The Evolution of Sex*. Cambridge; New York: Cambridge University Press. ISBN 0-521-29302-2. LCCN 77085689. OCLC 3413793.

- Maynard Smith, John (1998). "The Units of Selection". In Bock, Gregory R.; Goode, Jamie A. *The Limits of Reductionism in Biology*. Novartis Foundation Symposia. **213**. Chichester, England: John Wiley & Sons. ISBN 0-471-97770-5. ISSN 1935-4657. LCCN 98002779. OCLC 38311600. PMID 9653725. doi:10.1002/9780470515488.ch15. "Papers from the Symposium on the Limits of Reductionism in Biology, held at the Novartis Foundation, London, May 13–15, 1997."

- Mayr, Ernst (1942). *Systematics and the Origin of Species from the Viewpoint of a Zoologist*. Columbia Biological Series. **13**. New York: Columbia University Press. LCCN 43001098. OCLC 766053.

- Mayr, Ernst (1982). *The Growth of Biological Thought: Diversity, Evolution, and Inheritance*. Translation of John Ray by E. Silk. Cambridge, MA: Belknap Press. ISBN 0-674-36445-7. LCCN 81013204. OCLC 7875904.

- Mayr, Ernst (2002) [Originally published 2001; New York: Basic Books]. *What Evolution Is*. Science Masters. London: Weidenfeld & Nicolson. ISBN 0-297-60741-3. LCCN 2001036562. OCLC 248107061.

- McKinney, Michael L. (1997). "How do rare species avoid extinction? A paleontological view". In Kunin, William E.; Gaston, Kevin J. *The Biology of Rarity: Causes and consequences of rare—common differences* (1st ed.). London; New York: Chapman & Hall. ISBN 0-412-63380-9. LCCN 96071014. OCLC 36442106.

- Miller, G. Tyler; Spoolman, Scott E. (2012). *Environmental Science* (14th ed.). Belmont, CA: Brooks/Cole. ISBN 978-1-111-98893-7. LCCN 2011934330. OCLC 741539226. Retrieved 2014-12-27.

- Moore, Randy; Decker, Mark; Cotner, Sehoya (2010). *Chronology of the Evolution-Creationism Controversy*. Santa Barbara, CA: Greenwood Press/ABC-CLIO. ISBN 978-0-313-36287-3. LCCN 2009039784. OCLC 422757410.

- Nardon, Paul; Grenier, Anne-Marie (1991). "Serial Endosymbiosis Theory and Weevil Evolution: The Role of Symbiosis". In Margulis, Lynn; Fester, René. *Symbiosis as a Source of Evolutionary Innovation: Speciation and Morphogenesis*. Cambridge, MA: MIT Press. ISBN 0-262-13269-9. LCCN 90020439. OCLC 22597587. "Based on a conference held in Bellagio, Italy, June 25–30, 1989"

- National Academy of Sciences; Institute of Medicine (2008). *Science, Evolution, and Creationism*. Washington, D.C.: National Academy Press. ISBN 978-0-309-10586-6. LCCN 2007015904. OCLC 123539346. Retrieved 2014-11-22.

- Odum, Eugene P. (1971). *Fundamentals of Ecology* (3rd ed.). Philadelphia, PA: Saunders. ISBN 0-7216-6941-7. LCCN 76081826. OCLC 154846.

- Okasha, Samir (2006). *Evolution and the Levels of Selection*. Oxford; New York: Oxford University Press. ISBN 0-19-926797-9. LCCN 2006039679. OCLC 70985413.

- Panno, Joseph (2005). *The Cell: Evolution of the First Organism*. Facts on File science library. New York: Facts on File. ISBN 0-8160-4946-7. LCCN 2003025841. OCLC 53901436.

- Piatigorsky, Joram; Kantorow, Marc; Gopal-Srivastava, Rashmi; Tomarev, Stanislav I. (1994). "Recruitment of enzymes and stress proteins as lens crystallins". In Jansson, Bengt; Jörnvall, Hans; Rydberg, Ulf; et al. *Toward a Molecular Basis of Alcohol Use and Abuse*. Experientia. **71**. Basel; Boston: Birkhäuser Verlag. ISBN 3-7643-2940-8. ISSN 1023-294X. LCCN 94010167. OCLC 30030941. PMID 8032155. doi:10.1007/978-3-0348-7330-7_24.

- Provine, William B. (1971). *The Origins of Theoretical Population Genetics*. Chicago History of Science and Medicine (2nd ed.). Chicago, Illinois: University of Chicago Press. ISBN 0-226-68464-4. LCCN 2001027561. OCLC 46660910.

- Provine, William B. (1988). "Progress in Evolution and Meaning in Life". In Nitecki, Matthew H. *Evolutionary Progress*. Chicago, Illinois: University of Chicago Press. ISBN 0-226-58693-6. LCCN 88020835. OCLC 18380658. "This book is the result of the Spring Systematics Symposium held in May, 1987, at the Field Museum in Chicago"

- Quammen, David (2006). *The Reluctant Mr. Darwin: An Intimate Portrait of Charles Darwin and the Making of His Theory of Evolution*. Great Discoveries (1st ed.). New York: Atlas Books/W. W. Norton & Company. ISBN 978-0-393-05981-6. LCCN 2006009864. OCLC 65400177.

- Raven, Peter H.; Johnson, George B. (2002). *Biology* (6th ed.). Boston, MA: McGraw-Hill. ISBN 0-07-112261-3. LCCN 2001030052. OCLC 45806501.

- Ray, John (1686). *Historia Plantarum* [*History of Plants*]. **I**. Londini: Typis Mariæ Clark. LCCN agr11000774. OCLC 2126030.

- Rechenberg, Ingo (1973). *Evolutionsstrategie; Optimierung technischer Systeme nach Prinzipien der biologischen Evolution* (PhD thesis). Problemata (in German). **15**. Afterword by Manfred Eigen. Stuttgart-Bad Cannstatt: Frommann-Holzboog. ISBN 3-7728-0373-3. LCCN 74320689. OCLC 9020616.

- Ridley, Matt (1993). *The Red Queen: Sex and the Evolution of Human Nature*. New York: Viking. ISBN 0-670-84357-1. OCLC 636657988.

- Stevens, Anthony (1982). *Archetype: A Natural History of the Self*. London: Routledge & Kegan Paul. ISBN 0-7100-0980-1. LCCN 84672250. OCLC 10458367.

- West-Eberhard, Mary Jane (2003). *Developmental Plasticity and Evolution*. Oxford; New York: Oxford University Press. ISBN 0-19-512235-6. LCCN 2001055164. OCLC 48398911.

- Wiley, E. O.; Lieberman, Bruce S. (2011). *Phylogenetics: Theory and Practice of Phylogenetic Systematics* (2nd ed.). Hoboken, NJ: Wiley-Blackwell. ISBN 978-0-470-90596-8. LCCN 2010044283. OCLC 741259265. doi:10.1002/9781118017883.

- Wright, Sewall (1984). *Genetic and Biometric Foundations*. Evolution and the Genetics of Populations. 1. Chicago, Illinois: University of Chicago Press. ISBN 0-226-91038-5. LCCN 67025533. OCLC 246124737.

29.12 Further reading

Further information: Bibliography of biology

Introductory reading

- Barrett, Paul H.; Weinshank, Donald J.; Gottleber, Timothy T., eds. (1981). *A Concordance to Darwin's Origin of Species, First Edition*. Ithaca, NY: Cornell University Press. ISBN 0-8014-1319-2. LCCN 80066893. OCLC 610057960.

- Carroll, Sean B. (2005). *Endless Forms Most Beautiful: The New Science of Evo Devo and the Making of the Animal Kingdom*. illustrations by Jamie W. Carroll, Josh P. Klaiss, Leanne M. Olds (1st ed.). New York: W. W. Norton & Company. ISBN 0-393-06016-0. LCCN 2004029388. OCLC 57316841.

- Charlesworth, Brian; Charlesworth, Deborah (2003). *Evolution: A Very Short Introduction*. Very Short Introductions. Oxford; New York: Oxford University Press. ISBN 0-19-280251-8. LCCN 2003272247. OCLC 51668497.

- Gould, Stephen Jay (1989). *Wonderful Life: The Burgess Shale and the Nature of History* (1st ed.). New York: W. W. Norton & Company. ISBN 0-393-02705-8. LCCN 88037469. OCLC 18983518.

- Jones, Steve (1999). *Almost Like a Whale: The Origin of Species Updated*. London; New York: Doubleday. ISBN 0-385-40985-0. LCCN 2002391059. OCLC 41420544.

    - ——— (2000). *Darwin's Ghost: The Origin of Species Updated* (1st ed.). New York: Random House. ISBN 0-375-50103-7. LCCN 99053246. OCLC 42690131. American version.

- Mader, Sylvia S. (2007). *Biology*. Significant contributions by Murray P. Pendarvis (9th ed.). Boston, MA: McGraw-Hill Higher Education. ISBN 978-0-07-246463-4. LCCN 2005027781. OCLC 61748307.

- Maynard Smith, John (1993). *The Theory of Evolution* (Canto ed.). Cambridge; New York: Cambridge University Press. ISBN 0-521-45128-0. LCCN 93020358. OCLC 27676642.

- Pallen, Mark J. (2009). *The Rough Guide to Evolution*. Rough Guides Reference Guides. London; New York: Rough Guides. ISBN 978-1-85828-946-5. LCCN 2009288090. OCLC 233547316.

Advanced reading

- Barton, Nicholas H.; Briggs, Derek E. G.; Eisen, Jonathan A.; et al. (2007). *Evolution*. Cold Spring Harbor, NY: Cold Spring Harbor Laboratory Press. ISBN 978-0-87969-684-9. LCCN 2007010767. OCLC 86090399.

- Coyne, Jerry A.; Orr, H. Allen (2004). *Speciation*. Sunderland, MA: Sinauer Associates. ISBN 0-87893-089-2. LCCN 2004009505. OCLC 55078441.

- Bergstrom, Carl T.; Dugatkin, Lee Alan (2012). *Evolution* (1st ed.). New York: W. W. Norton & Company. ISBN 978-0-393-91341-5. LCCN 2011036572. OCLC 729341924.

- Gould, Stephen Jay (2002). *The Structure of Evolutionary Theory*. Cambridge, MA: Belknap Press of Harvard University Press. ISBN 0-674-00613-5. LCCN 2001043556. OCLC 47869352.

- Hall, Brian K.; Olson, Wendy, eds. (2003). *Keywords and Concepts in Evolutionary Developmental Biology*. Cambridge, MA: Harvard University Press. ISBN 0-674-00904-5. LCCN 2002192201. OCLC 50761342.

- Kauffman, Stuart A. (1993). *The Origins of Order: Self-organization and Selection in Evolution*. New York, NY: Oxford University Press.

- Maynard Smith, John; Szathmáry, Eörs (1995). *The Major Transitions in Evolution*. Oxford; New York: W.H. Freeman Spektrum. ISBN 0-7167-4525-9. LCCN 94026965. OCLC 30894392.

- Mayr, Ernst (2001). *What Evolution Is*. New York: Basic Books. ISBN 0-465-04426-3. LCCN 2001036562. OCLC 47443814.

- Minelli, Alessandro (2009). *Forms of Becoming: The Evolutionary Biology of Development*. Translation by Mark Epstein. Princeton, NJ; Oxford: Princeton University Press. ISBN 978-0-691-13568-7. LCCN 2008028825. OCLC 233030259.

29.13 External links

General information

-
- Evolution on *In Our Time* at the BBC.
- "Evolution". *New Scientist*. Retrieved 2011-05-30.
- "Evolution Resources from the National Academies". Washington, D.C.: National Academy of Sciences. Retrieved 2011-05-30.
- "Understanding Evolution: your one-stop resource for information on evolution". University of California, Berkeley. Retrieved 2011-05-30.
- "Evolution of Evolution – 150 Years of Darwin's 'On the Origin of Species'". Arlington County, VA: National Science Foundation. Retrieved 2011-05-30.
- Human Timeline (Interactive) – Smithsonian, National Museum of Natural History (August 2016).

Experiments concerning the process of biological evolution

- Lenski, Richard E. "Experimental Evolution". Michigan State University. Retrieved 2013-07-31.
- Chastain, Erick; Livnat, Adi; Papadimitriou, Christos; Vazirani, Umesh (July 22, 2014). "Algorithms, games, and evolution". *Proc. Natl. Acad. Sci. U.S.A.* Washington, D.C.: National Academy of Sciences. **111** (29): 10620–10623. Bibcode:2014PNAS..11110620C. ISSN 0027-8424. doi:10.1073/pnas.1406556111. Retrieved 2015-01-03.

Online lectures

- Carroll, Sean B. "The Making of the Fittest". Archived from the original on 2011-07-18. Retrieved 2011-05-30.
- Stearns, Stephen C. "Principles of Evolution, Ecology and Behavior". Retrieved 2011-08-30.

Chapter 30

Adaptation

This article is about the evolutionary process. For other uses, see Adaptation (disambiguation).
Not to be confused with Adoption or Acclimatization.

In biology, **adaptation** has three related meanings. Firstly it is the dynamic evolutionary process that fits a population of organisms to their environment, enhancing their evolutionary fitness. Secondly, it is a state reached by the population during that process. Thirdly, it is a phenotypic or **adaptive trait**, with a functional role in each individual organism, that is maintained and has been evolved by natural selection.

Organisms face a succession of environmental challenges as they grow, and show adaptive plasticity as traits develop in response to the imposed conditions. This gives them resilience to varying environments.

30.1 History

Main article: History of evolutionary thought

Adaptation is an observable fact of life accepted by philosophers and natural historians from ancient times, independently of their views on evolution, but their explanations differed. Empedocles did not believe that adaptation required a final cause (~ purpose), but "came about naturally, since such things survived." Aristotle did believe in final causes, but assumed that species were fixed.

In natural theology, adaptation was interpreted as the work of a deity and as evidence for the existence of God.[1] William Paley believed that organisms were perfectly adapted to the lives they led, an argument that shadowed Gottfried Wilhelm Leibniz, who had argued that God had brought about "the best of all possible worlds." Voltaire's Dr. Pangloss[2] is a parody of this optimistic idea, and David Hume also argued against design.[3] The *Bridgewater Treatises* are a product of natural theology, though some of the authors managed to present their work in a fairly neutral manner. The series was lampooned by Robert Knox, who held quasi-evolutionary views, as the *Bilgewater Treatises*. Charles Darwin broke with the tradition by emphasising the flaws and limitations which occurred in the animal and plant worlds.[4]

Lamarckism is a proto-evolutionary hypothesis of the inheritance of acquired characteristics, whose main purpose is to explain adaptations by natural means.[5] Jean-Baptiste Lamarck proposed a tendency for organisms to become more complex, moving up a ladder of progress, plus "the influence of circumstances," usually expressed as *use and disuse*.[6]

Other natural historians, such as Buffon, accepted adaptation, and some also accepted evolution, without voicing their opinions as to the mechanism. This illustrates the real merit of Darwin and Alfred Russel Wallace, and secondary figures such as Henry Walter Bates, for putting forward a mechanism whose significance had only been glimpsed previously. A century later, experimental field studies and breeding experiments by people such as E. B. Ford and Dobzhansky produced evidence that natural selection was not only the 'engine' behind adaptation, but was a much stronger force than had previously been thought.[7][8][9]

30.2 General principles

The significance of an adaptation can only be understood in relation to the total biology of the species.
— Julian Huxley, *Evolution: The Modern Synthesis*[10]

30.2.1 What adaptation is

Adaptation is, first of all, a *process*, rather than a physical form or part of a body.[11] An internal parasite (such as a liver fluke) can illustrate the distinction: such a parasite may have a very simple bodily structure, but nevertheless the organism is highly adapted to its specific environment. From this we see that adaptation is not just a matter of visible traits: in such parasites critical adaptations take place in the life cycle, which is often quite complex.[12] However, as a practical term, "adaptation" often refers to a *product*: those features of a species which result from the process. Many aspects of an animal or plant can be correctly called adaptations, though there are always some features whose function remains in doubt. By using the term *adaptation* for the evolutionary *process*, and *adaptive trait* for the bodily part or function (the product), one may distinguish the two different senses of the word.[13][14][15][16]

Adaptation is one of the two main processes that explain the observed diversity of species, such as the different species of Darwin's finches. The other process is speciation, in which new species arise, typically through reproductive isolation.[17][18] A favorite example used today to study the interplay of adaptation and speciation is the evolution of cichlid fish in African lakes, where the question of reproductive isolation is complex.[19][20]

Adaptation is not always a simple matter where the ideal phenotype evolves for a given external environment. An organism must be viable at all stages of its development and at all stages of its evolution. This places *constraints* on the evolution of development, behavior and structure of organisms. The main constraint, over which there has been much debate, is the requirement that each genetic and phenotypic change during evolution should be relatively small, because developmental systems are so complex and interlinked. However, it is not clear what "relatively small" should mean, for example polyploidy in plants is a reasonably common large genetic change.[21] The origin of eukaryotic symbiosis exemplifies a more dramatic example.[22]

All adaptations help organisms survive in their ecological niches.[23] The adaptive traits may be structural, behavioral or physiological. Structural adaptations are physical features of an organism (shape, body covering, armament, internal organization). Behavioral adaptations are inherited behavior chains and the ability to learn. Behaviors may be inherited in detail (instincts), or a capacity for learning may be inherited (see neuropsychology). Examples: searching for food, mating, vocalizations. Physiological adaptations can permit the organism to perform special functions (for instance, making venom, secreting slime, phototropism), but may also involve more general functions such as growth and development, temperature regulation, ionic balance and other aspects of homeostasis. Adaptation affects all aspects of the life of an organism.

The following definitions are given by the evolutionary biologist Theodosius Dobzhansky:

1. *Adaptation* is the evolutionary process whereby an organism becomes better able to live in its habitat or habitats.[24]

2. *Adaptedness* is the state of being adapted: the degree to which an organism is able to live and reproduce in a given set of habitats.[25]

3. An *adaptive trait* is an aspect of the developmental pattern of the organism which enables or enhances the probability of that organism surviving and reproducing.[26]

30.2.2 What adaptation is not

Some generalists, such as birds, have the flexibility to adapt to urban areas.

Adaptation differs from flexibility, acclimatization, and learning. Flexibility deals with the relative capacity of an organism to maintain itself in different habitats: its degree of specialization. Acclimatization describes automatic physiological adjustments during life; learning means improvement in behavioral performance during life. These terms are preferred to adaptation for changes during life which are not inherited by the next generation.

Flexibility stems from phenotypic plasticity, the ability of an organism with a given genotype to change its phenotype in response to changes in its habitat, or to move to a different habitat.[27][28] The degree of flexibility is inherited, and varies between individuals. A highly specialized animal or plant lives only in a well-defined habitat, eats a specific type of food, and cannot survive if its needs are not met. Many herbivores are like this; extreme examples are koalas which depend on *Eucalyptus*, and giant pandas which require bamboo. A generalist, on the other hand, eats a range of food, and can survive in many different conditions. Examples are humans, rats, crabs and many carnivores. The *tendency* to behave in a specialized or exploratory manner is inherited—it is an adaptation. Rather different is developmental flexibility: "An animal or plant is developmentally flexible if when it is raised in or transferred to new conditions, it changes in structure so that it is better fitted to survive in the new environment," writes evolutionary biologist John Maynard Smith.[29]

If humans move to a higher altitude, respiration and physical exertion become a problem, but after spending time in high altitude conditions they acclimatize to the reduced partial pressure of oxygen, such as by producing more red blood cells. The ability to acclimatize is an adaptation, but the acclimatization itself is not. Fecundity goes down, but deaths from some tropical diseases also goes down. Over a longer period of time, some people are better able to reproduce at high altitudes than others. They contribute more heavily to later generations, and gradually by natural selection the whole population becomes adapted to the new conditions. This has demonstrably occurred, as the observed performance of long-term communities at higher altitude is significantly better than the performance of new arrivals, even when the new arrivals have had time to acclimatize.[30]

30.2.3 Adaptedness and fitness

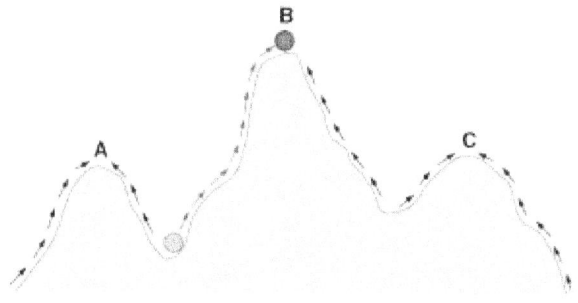

In this sketch of a fitness landscape, a population can evolve by following the arrows to the adaptive peak at point B, and the points A and C are local optima where a population could become trapped.

Main articles: Fitness (biology) and Fitness landscape

There is a relationship between adaptedness and the concept of fitness used in population genetics. Differences in fitness between genotypes predict the rate of evolution by natural selection. Natural selection changes the relative frequencies of alternative phenotypes, insofar as they are heritable.[31] However, a phenotype with high adaptedness may not have high fitness. Dobzhansky mentioned the example of the Californian redwood, which is highly adapted, but a relict species in danger of extinction.[24] Elliott Sober commented that adaptation was a retrospective concept since it implied something about the history of a trait, whereas fitness predicts a trait's future.[32]

1. Relative fitness. The average contribution to the next generation by a genotype or a class of genotypes, relative to the contributions of other genotypes in the population.[33] This is also known as *Darwinian fitness*, *selection coefficient*, and other terms.

2. Absolute fitness. The absolute contribution to the next generation by a genotype or a class of genotypes. Also known as the Malthusian parameter when applied to the population as a whole.[31][34]

3. Adaptedness. The extent to which a phenotype fits its local ecological niche. Researchers can sometimes test this through a reciprocal transplant.

Sewall Wright proposed that populations occupy *adaptive peaks* on a fitness landscape. In order to evolve to another, higher peak, a population would first have to pass through a valley of maladaptive intermediate stages.[35] A given population might be "trapped" on a peak that is not optimally adapted.

30.2.4 Genetic basis

A large diversity of genome DNAs in a species is the basis for adaptation and differentiation. A large population is needed to carry sufficient diversity. According to the misrepair-accumulation aging theory,[36][37] The misrepair mechanism is important in maintaining a sufficient number of individuals in a species.[38] misrepair is a way of repair for increasing the surviving chance of an organism when it has severe injuries. Without misrepairs, no individual could survive to reproduction age. Thus misrepair mechanism is an essential mechanism for the survival of a species and for maintaining the number of individuals. Although individuals die from aging, genome DNAs are being recopied and transmitted by individuals generation by generation. In addition, the DNA misrepairs in germ cells contribute also to the diversity of genome DNAs.

30.3 Types

Adaptation is the heart and soul of evolution.
— Niles Eldredge, Reinventing Darwin: The Great Debate at the High Table of Evolutionary Theory[39]

30.3.1 Changes in habitat

Before Darwin, adaptation was seen as a fixed relationship between an organism and its habitat. It was not appreciated that as the climate changed, so did the habitat; and as the habitat changed, so did the biota. Also, habitats are subject to changes in their biota: for example, invasions of species from other areas. The relative numbers of species in a given habitat are always changing. Change is the rule, though much depends on the speed and degree of the change. When the habitat changes, three main things may happen to a resident population: habitat tracking, genetic change or extinction. In fact, all three things may occur in sequence. Of these three effects only genetic change brings about adaptation. When a habitat changes, the resident population typically moves to more suitable places; this is the typical response of flying insects or oceanic organisms, which have wide (though not unlimited) opportunity for movement.[40] This common response is called *habitat tracking*. It is one explanation put forward for the periods of apparent stasis in the fossil record (the punctuated equilibrium theory).[41]

30.3.2 Genetic change

Genetic change occurs in a population when natural selection and mutations act on its genetic variability.[42] The first pathways of enzyme-based metabolism may have been parts of purine nucleotide metabolism, with previous metabolic pathways being part of the ancient RNA world. By this means, the population adapts genetically to its circumstances.[9] Genetic changes may result in visible structures, or may adjust physiological activity in a way that suits the habitat.

Habitats and biota do frequently change. Therefore, it follows that the process of adaptation is never finally complete.[43] Over time, it may happen that the environment changes little, and the species comes to fit its surroundings better and better. On the other hand, it may happen that changes in the environment occur relatively rapidly, and then the species becomes less and less well adapted. Seen like this, adaptation is a genetic *tracking process*, which goes on all the time to some extent, but especially when the population cannot or does not move to another, less hostile area. Given enough genetic change, as well as specific demographic conditions, an adaptation may be enough to bring a population back from the brink of extinction in a process called evolutionary rescue. It should be noted that adaptation does affect, to some extent, every species in a particular ecosystem.[44][45]

Leigh Van Valen thought that even in a stable environment, competing species had to constantly adapt to maintain their relative standing. This became known as the Red Queen hypothesis, as seen in host-parasite interaction.[46]

30.3.3 Co-adaptation

Main article: Co-adaptation

In coevolution, where the existence of one species is tightly

Pollinating insects are co-adapted with flowering plants.

bound up with the life of another species, new or 'improved' adaptations which occur in one species are often followed by the appearance and spread of corresponding features in the other species. These relationships are intrinsically dynamic, and may continue on a trajectory for millions of years, as has occurred in the relationship between flowering plants and pollinating insects.

30.3.4 Mimicry

Main article: Mimicry

Bates' work on Amazonian butterflies led him to develop the first scientific account of mimicry, especially the kind of mimicry which bears his name: Batesian mimicry.[47] This is the mimicry by a palatable species of an unpalatable or noxious species. A common example seen in temperate gardens is the hoverfly, many of which—though bearing no sting—mimic the warning colouration of hymenoptera (wasps and bees). Such mimicry does not need to be perfect to improve the survival of the palatable species.[48]

Bates, Wallace and Fritz Müller believed that Batesian and Müllerian mimicry provided evidence for the action of natural selection, a view which is now standard amongst biologists.[49] All aspects of this situation can be, and have been, the subject of research.[50] Field and experimental work on these ideas continues to this day; the topic con-

30.3. TYPES

A and B show real wasps; the rest are Batesian mimics: three hoverflies and one beetle.

nects strongly to speciation, genetics and evolutionary developmental biology.[51]

30.3.5 Internal adaptations

There are some important adaptations to do with the overall coordination of the systems in the body. Such adaptations may have significant consequences. Examples, in vertebrates, would be temperature regulation, or improvements in brain function, or an effective immune system. Contains an extensive analysis of the evolution of adaptations in the radiation of angiosperms. Such adaptations may make the clade (monophyletic group) more viable in a wide range of habitats. The acquisition of such major adaptations has often served as the spark for adaptive radiation, and huge success over long periods of time for a whole group of animals or plants.

30.3.6 Trade-offs

> *It is a profound truth that Nature does not know best; that genetical evolution... is a story of waste, makeshift, compromise and blunder.*
> — Peter Medawar, The Future of Man[52]

All adaptations have a downside: horse legs are great for running on grass, but they can't scratch their backs; mammals' hair helps temperature, but offers a niche for ectoparasites; the only flying penguins do is under water. Adaptations serving different functions may be mutually destructive. Compromise and makeshift occur widely, not perfection. Selection pressures pull in different directions, and the adaptation that results is some kind of compromise.[53]

> *Since the phenotype as a whole is the target of selection, it is impossible to improve simultaneously all aspects of the phenotype to the same degree.*
> — Ernst Mayr, The Growth of Biological Thought: Diversity, Evolution, and Inheritance[54]

Consider the antlers of the Irish elk, (often supposed to be far too large: in deer antler size has an allometric relationship to body size). Obviously, antlers serve positively for defence against predators, and to score victories in the annual rut. But they are costly in terms of resource. Their size during the last glacial period presumably depended on the relative gain and loss of reproductive capacity in the population of elks during that time.[55] Another example: camouflage to avoid detection is destroyed when vivid colors are displayed at mating time. Here the risk to life is counterbalanced by the necessity for reproduction.

Stream-dwelling salamanders, such as Caucasian salamander or Gold-striped salamander have very slender, long bodies, perfectly adapted to life at the banks of fast small rivers and mountain brooks. Elongated body protects their larvae from being washed out by current. However, elongated body increases risk of desiccation and decreases dispersal ability of the salamanders; it also negatively affects their fecundity. As a result, fire salamander, less perfectly adapted to the mountain brook habitats, is in general more successful, have a higher fecundity and broader geographic range.[56]

The peacock's ornamental train (grown anew in time for each mating season) is a famous adaptation. It must reduce his maneuverability and flight, and is hugely conspicuous; also, its growth costs food resources. Darwin's explanation of its advantage was in terms of sexual selection: "This depends on the advantage which certain individuals have over other individuals of the same sex and species, in exclusive relation to reproduction."[57] The kind of sexual selection represented by the peacock is called 'mate choice,' with an implication that the process selects the more fit over the less fit, and so has survival value.[58] The recognition of sexual selection was for a long time in abeyance, but has been

An Indian peacock's train in full display

rehabilitated.[59]

The conflict between the size of the human foetal brain at birth, (which cannot be larger than about 400 cm³, else it will not get through the mother's pelvis) and the size needed for an adult brain (about 1400 cm³), means the brain of a newborn child is quite immature. The most vital things in human life (locomotion, speech) just have to wait while the brain grows and matures. That is the result of the birth compromise. Much of the problem comes from our upright bipedal stance, without which our pelvis could be shaped more suitably for birth. Neanderthals had a similar problem.[60][61][62]

As another example, the long neck of a giraffe is a burden and a blessing. The neck of a giraffe can be up to 2 m (6 ft 7 in) in length.[63] This neck can be used for inter-species competition or for foraging on tall trees where shorter herbivores cannot reach. However, as previously stated, there is always a trade-off. This long neck is heavy and it adds to the body mass of a giraffe, so the giraffe needs an abundance of nutrition to provide for this costly adaptation.[64]

30.4 Shifts in function

> *Adaptation and function are two aspects of one problem.*
> — Julian Huxley, Evolution: The Modern Synthesis[65]

30.4.1 Pre-adaptations

This occurs when a species or population has characteristics which (by chance) are suited for conditions which have not yet arisen. For example, the polyploid cordgrass *Spartina townsendii* is better adapted than either of its parent species to their own habitat of saline marsh and mud-flats.[66] White Leghorn chicken are markedly more resistant to vitamin B_1 deficiency than other breeds.[67] On a plentiful diet there is no difference, but on a restricted diet this preadaptation could be decisive.

Pre-adaptation may occur because a natural population carries a huge quantity of genetic variability.[68] In diploid eukaryotes, this is a consequence of the system of sexual reproduction, where mutant alleles get partially shielded, for example, by the selective advantage of heterozygotes. Microorganisms, with their huge populations, also carry a great deal of genetic variability.

The first experimental evidence of the pre-adaptive nature of genetic variants in microorganisms was provided by Salvador Luria and Max Delbrück who developed Fluctuation Test, a method to show the random fluctuation of pre-existing genetic changes that conferred resistance to bacteriophage in the bacterium *Escherichia coli*.

30.4.2 Co-option of existing traits: exaptation

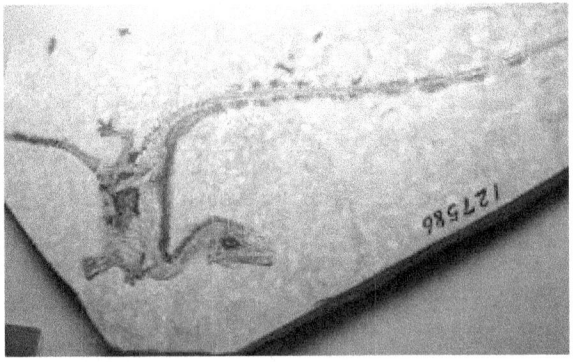

Sinosauropteryx fossil, a dinosaur with feathers, used for insulation, not flight

Main article: Exaptation

The classic example is the ear ossicles of mammals, which we know from paleontological and embryological studies originated in the upper and lower jaws and the hyoid bone of their synapsid ancestors, and further back still were part of the gill arches of early fish.[69] We owe this esoteric knowledge to the comparative anatomists, who, a century ago, were at the cutting edge of evolutionary studies.[70] The word *exaptation* was coined to cover these shifts in function, which are surprisingly common in evolutionary history.[71] The origin of wings from feathers that were originally used for temperature regulation is a more recent discovery (see feathered dinosaurs).

30.5 Non-adaptive traits

Main articles: Spandrel (biology) and Vestigiality

Some traits do not appear to be adaptive, that is, they have a neutral or deleterious effect on fitness in the current environment. Because genes have pleiotropic effects, not all traits may be functional: they may be what Stephen Jay Gould and Richard Lewontin called spandrels, features brought about by neighbouring adaptations, like the triangular areas under neighbouring arches in architecture which began as functionless features.[72]

Another possibility is that a trait may have been adaptive at some point in an organism's evolutionary history, but a change in habitats caused what used to be an adaptation to become unnecessary or even a hindrance (maladaptations). Such adaptations are termed vestigial. Many organisms have vestigial organs, which are the remnants of fully functional structures in their ancestors. As a result of changes in lifestyle the organs became redundant, and are either not functional or reduced in functionality. Since any structure represents some kind of cost to the general economy of the body, an advantage may accrue from their elimination once they are not functional. Examples: wisdom teeth in humans; the loss of pigment and functional eyes in cave fauna; the loss of structure in endoparasites.[73]

30.6 Extinction and coextinction

Main articles: Extinction and Coextinction

If a population cannot move or change sufficiently to preserve its long-term viability, then obviously, it will become extinct, at least in that locale. The species may or may not survive in other locales. Species extinction occurs when the death rate over the entire species exceeds the birth rate for a long enough period for the species to disappear. It was an observation of Van Valen that groups of species tend to have a characteristic and fairly regular rate of extinction.[74]

Just as we have co-adaptation, there is also coextinction.[75] Coextinction is the loss of a species due to the extinction of another with which it is coadapted, as with the extinction of parasitic insects following the loss of their hosts, or when a flowering plant loses its pollinator, or when a food chain is disrupted.[76]

30.7 Philosophical issues

Main articles: Adaptationism and Teleology in biology

Adaptation raises philosophical issues concerning how biologists speak of function and purpose, as this carries implications of evolutionary history – that a feature evolved by natural selection for a specific reason – and potentially of supernatural intervention – that features and organisms exist because of a deity's conscious intentions.[77][78] Aristotle introduced teleology to describe the adaptedness of organisms, but without accepting the supernatural intention built into Plato's thinking, which Aristotle rejected.[79][80] Modern biologists continue to face the same difficulty.[81][82][83][84][85][86] On the one hand, adaptation is obviously purposeful: natural selection chooses what works and eliminates what does not. On the other hand, biologists want to deny conscious purpose in evolution. The dilemma gave rise to a famous joke by the evolutionary biologist Haldane: "Teleology is like a mistress to a biologist: he cannot live without her but he's unwilling to be seen with her in public.'" David Hull commented that Haldane's mistress "has become a lawfully wedded wife. Biologists no longer feel obligated to apologize for their use of teleological language; they flaunt it."[87]

30.8 See also

- Adaptive evolution in the human genome
- Adaptive memory
- Adaptive mutation
- Adaptive system
- Ecological trap
- Evolutionary pressure
- Evolvability
- Intragenomic conflict
- Neutral theory of molecular evolution

30.9 References

[1] Desmond 1989, pp. 31–32, fn 18

[2] Voltaire (1759). *Candide*. Cramer et al.

[3] Sober 1993, chpt. 2

[4] Darwin 1872, p. 397: "Rudimentary, Atrophied, and Aborted Organs"

[5] See, for example, the discussion in Bowler 2003, pp. 86–95: "Whatever the true nature of Lamark's theory, it was his mechanism of adaptation that caught the attention of later naturalists." (p. 90)

[6] Bowler, Peter J. (1989) [1983]. *Evolution The History of an Idea* (Revised ed.). University of California Press. p. 86. ISBN 0-520-06386-4.

[7] Provine 1986

[8] Ford 1975

[9] Orr, H. Allen (February 2005). "The genetic theory of adaptation: a brief history". *Nature Reviews Genetics*. London: Nature Publishing Group. 6 (2): 119–127. ISSN 1471-0056. PMID 15716908. doi:10.1038/nrg1523.

[10] Huxley 1942, p. 449

[11] Mayr 1982, p. 483: "Adaptation... could no longer be considered a static condition, a product of a creative past, and became instead a continuing dynamic process."

[12] Price 1980

[13] Daintith, John; Martin, Elizabeth A., eds. (2010) [First published 1984 as *Concise Science Dictionary*]. "adaptation". *A Dictionary of Science*. Oxford Paperback Reference (6th ed.). Oxford; New York: Oxford University Press. p. 13. ISBN 978-0-19-956146-9. LCCN 2010287468. OCLC 444383696. Any change in the structure or functioning of successive generations of a population that makes it better suited to its environment.

[14] Bowler 2003, p. 10

[15] Patterson 1999, p. 1

[16] Williams 1966, p. 5: "Evolutionary adaptation is a phenomenon of pervasive importance in biology."

[17] Mayr 1963

[18] Mayr 1982, pp. 562–566

[19] Salzburger, Walter; Mack, Tanja; Verheyen, Erik; Meyer, Axel (February 21, 2005). "Out of Tanganyika: Genesis, explosive speciation, key-innovations and phylogeography of the haplochromine cichlid fishes" (PDF). *BMC Evolutionary Biology*. London: BioMed Central. 5 (17). ISSN 1471-2148. PMC 554777. PMID 15723698. doi:10.1186/1471-2148-5-17. Retrieved 2015-08-15.

[20] Kornfield, Irv; Smith, Peter F. (November 2000). "African Cichlid Fishes: Model Systems for Evolutionary Biology". *Annual Review of Ecology and Systematics*. Palo Alto, CA: Annual Reviews. 31: 163–196. ISSN 1545-2069. doi:10.1146/annurev.ecolsys.31.1.163.

[21] Stebbins 1950, chpts. 8 and 9

[22] Margulis & Fester 1991

[23] Hutchinson 1965. The niche is the central concept in evolutionary ecology; see especially part II: "The niche: an abstractly inhabited hypervolume." (pp. 26–78)

[24] Dobzhansky 1968, pp. 1–34

[25] Dobzhansky 1970, pp. 4–6; 79–82

[26] Dobzhansky, Theodosius (March 1956). "Genetics of Natural Populations. XXV. Genetic Changes in Populations of *Drosophila pseudoobscura* and *Drosophila persimilis* in Some Localities in California". *Evolution*. Hoboken, NJ: John Wiley & Sons for the Society for the Study of Evolution. 10 (1): 82–92. JSTOR 2406099. doi:10.2307/2406099.

[27] Price, Trevor D.; Qvarnström, Anna; Irwin, Darren E. (July 2003). "The role of phenotypic plasticity in driving genetic evolution". *Proceedings of the Royal Society B*. London: Royal Society. 270 (1523): 1433–1440. ISSN 0962-8452. PMC 1691402. PMID 12965006. doi:10.1098/rspb.2003.2372.

[28] Price, Trevor D. (June 2006). "Phenotypic plasticity, sexual selection and the evolution of colour patterns". *The Journal of Experimental Biology*. Cambridge, UK: The Company of Biologists. 209 (12): 2368–2376. ISSN 0022-0949. PMID 16731813. doi:10.1242/jeb.02183.

[29] Maynard Smith 1993, p. 33

[30] Moore, Lorna G.; Regensteiner, Judith G. (October 1983). "Adaptation to High Altitude". *Annual Review of Anthropology*. Palo Alto, CA: Annual Reviews. 12: 285–304. doi:10.1146/annurev.an.12.100183.001441.

[31] Endler 1986, pp. 33–51

[32] Sober 1984, p. 210

[33] Futuyma 1986, p. 552

[34] Fisher 1930, p. 25

[35] Wright 1932, pp. 356–366

[36] Wang, Jicun; Michelitsch, Thomas M.; Wunderlin, Arne; Mahadeva, Ravi (2009). "Aging as a consequence of misrepair—A novel theory of aging". arXiv:0904.0575 [q-bio.TO].

[37] Wang-Michelitsch, Jicun; Michelitsch, Thomas M. (2015). "Aging as a process of accumulation of misrepairs". arXiv:1503.07163 [q-bio.TO].

30.9. REFERENCES

[38] Wang-Michelitsch, Jicun; Michelitsch, Thomas M. (2015). "Misrepair mechanism: a mechanism essential for individual adaptation, species adaptation and species evolution". arXiv:1505.03900 ⊙ [q-bio.TO].

[39] Eldredge 1995, p. 33

[40] Eldredge 1985, p. 136: "Of glaciers and beetles"

[41] Eldredge 1995, p. 64

[42] Hogan, C. Michael (October 12, 2010). "Mutation". In Monosson, Emily. *Encyclopedia of Earth*. Washington, D.C.: Environmental Information Coalition, National Council for Science and the Environment. OCLC 72808636. Retrieved 2015-08-18.

[43] Mayr 1982, pp. 481–483: This sequence tells how Darwin's ideas on adaptation developed as he came to appreciate it as "a continuing dynamic process."

[44] Sterelny & Griffiths 1999, p. 217

[45] Freeman & Herron 2007, p. 364

[46] Rabajante, J; et al. (2016). "Host-parasite Red Queen dynamics with phase-locked rare genotypes". *Science Advances*. **2**: e1501548. doi:10.1126/sciadv.1501548.

[47] Carpenter & Ford 1933

[48] Wickler 1968

[49] Moon 1976

[50] Ruxton, Sherratt & Speed 2004

[51] Mallet, James (November 2001). "The speciation revolution" (PDF). *Journal of Evolutionary Biology*. Hoboken, NJ: Wiley-Blackwell on behalf of the European Society for Evolutionary Biology. **14** (6): 887–888. ISSN 1010-061X. doi:10.1046/j.1420-9101.2001.00342.x.

[52] Medawar 1960

[53] Jacob, François (June 10, 1977). "Evolution and Tinkering". *Science*. Washington, D.C.: American Association for the Advancement of Science. **196** (4295): 1161–1166. ISSN 0036-8075. PMID 860134. doi:10.1126/science.860134.

[54] Mayr 1982, p. 589

[55] Gould, Stephen Jay (June 1974). "The Origin and Function of 'Bizarre' Structures: Antler Size and Skull Size in the 'Irish Elk,' *Megaloceros giganteus*". *Evolution*. Hoboken, NJ: John Wiley & Sons for the Society for the Study of Evolution. **28** (2): 191–220. ISSN 0014-3820. JSTOR 2407322. doi:10.2307/2407322.

[56] Tarkhnishvili, David N. (1994). "Interdependences between Populational, Developmental and Morphological Features of the Caucasian salamander, *Mertensiella caucasica*" (PDF). *Mertensiella*. Bonn, Germany: Deutsche Gesellschaft für Herpetologie und Terrarienkunde. **4**: 315–325. ISSN 0934-6643. Retrieved 2015-08-18.

[57] Darwin 1871, p. 256

[58] The case was treated by Fisher 1930, pp. 134–139

[59] Cronin 1991

[60] Rosenberg, Karen R. (1992). "The evolution of modern human childbirth". *American Journal of Physical Anthropology*. Hoboken, NJ: John Wiley & Sons for the American Association of Physical Anthropologists. **35** (Supplement S15): 89–124. ISSN 0002-9483. doi:10.1002/ajpa.1330350605.

[61] Friedlander, Nancy J.; Jordan, David K. (October–December 1994). "Obstetric implications of Neanderthal robusticity and bone density". *Human Evolution*. Kluwer Academic Publishers. **9** (4): 331–342. ISSN 0393-9375. doi:10.1007/BF02435519.

[62] Miller 2007

[63] Williams 2010, p. 29

[64] Altwegg, Robert E.; Simmons, Res (September 2010). "Necks-for-sex or competing browsers? A critique of ideas on the evolution of giraffe". *Journal of Zoology*. Hoboken, NJ: Wiley-Blackwell. **282** (1): 6–12. ISSN 0952-8369. doi:10.1111/j.1469-7998.2010.00711.x.

[65] Huxley 1942, p. 417

[66] Huskins, C. Leonard (1930). "The origin of Spartina Townsendii". *Genetica*. Martinus Nijhoff, The Hague/Kluwer Academic Publishers. **12** (6): 531–538. ISSN 0016-6707. doi:10.1007/BF01487665.

[67] Lamoreux, Wilfred F.; Hutt, Frederick B. (February 15, 1939). "Breed differences in resistance to a deficiency in vitamin B_1 in the fowl" (PDF). *Journal of Agricultural Research*. Washington, D.C.: United States Department of Agriculture. **58** (4): 307–316. ISSN 0095-9758. Retrieved 2015-08-20.

[68] Dobzhansky 1981

[69] Allin & Hopson 1992, pp. 587–614

[70] Panchen 1992, chpt. 4. "Homology and the evidence for evolution"

[71] Gould, Stephen Jay; Vrba, Elizabeth S. (Winter 1982). "Exaptation—A Missing Term in the Science of Form". *Paleobiology*. Boulder, CO: Paleontological Society. **8** (1): 4–15. ISSN 0094-8373. JSTOR 2400563.

[72] Wagner, Günter P., *Homology, Genes, and Evolutionary Innovation*. Princeton University Press. 2014. Chapter 1: The Intellectual Challenge of Morphological Evolution: A Case for Variational Structuralism. Page 7

[73] Barrett et al. 1987. Charles Darwin was the first to put forward such ideas.

[74] Van Valen, Leigh (July 1973). "A New Evolutionary Law" (PDF). *Evolutionary Theory*. Chicago, IL: University of Chicago. **1**: 1–30. ISSN 0093-4755. Retrieved 2015-08-22.

[75] Koh, Lian Pin; Dunn, Robert R.; Sodhi, Navjot S.; et al. (September 2004). "Species Coextinctions and the Biodiversity Crisis". *Science*. Washington, D.C.: American Association for the Advancement of Science. **305** (5690): 1632–1634. ISSN 0036-8075. PMID 15361627. doi:10.1126/science.1101101.

[76] Darwin 1872, pp. 57–58. Darwin in tells the story of "a web of complex relations" involving heartsease (*Viola tricolor*), red clover (*Trifolium pratense*), humble-bees (bumblebees), mice and cats.

[77] Sober 1993, pp. 85–86

[78] Williams 1966, pp. 8–10

[79] Nagel, Ernest (May 1977). "Goal-Directed Processes in Biology". *The Journal of Philosophy*. New York: The Journal of Philosophy, Inc. **74** (5): 261–279. ISSN 0022-362X. JSTOR 2025745. doi:10.2307/2025745. Teleology Revisisted: The Dewy Lectures 1977 (first lecture)

[80] Nagel, Ernest (May 1977). "Functional Explanations in Biology". *The Journal of Philosophy*. **74** (5): 280–301. ISSN 0022-362X. JSTOR 2025746. doi:10.2307/2025746. Teleology Revisisted: The Dewy Lectures 1977 (second lecture)

[81] Pittendrigh 1958

[82] Mayr 1965, pp. 33–50

[83] Mayr 1988, chpt. 3, "The Multiple Meanings of Teleological"

[84] Williams 1966, "The Scientific Study of Adaptation"

[85] Monod 1971

[86] Allaby, Michael, ed. (2003). "teleonomy". *A Dictionary of Zoology*. Oxford Paperback Reference (Reissued with new cover and corrections ed.). Oxford; New York: Oxford University Press. ISBN 0-19-860758-X. LCCN 2003278285. OCLC 444678726. Retrieved 2015-08-24.

[87] Hull 1982

30.10 Sources

- Allin, Edgar F.; Hopson, James A. (1992). "Evolution of the Auditory System in Synapsida ("Mammal-Like Reptiles" and Primitive Mammals) as Seen in the Fossil Record". In Webster, Douglas B.; Fay, Richard R.; Popper, Arthur N. *The Evolutionary Biology of Hearing*. New York: Springer-Verlag. ISBN 0-387-97588-8. LCCN 91004805. OCLC 23582549. doi:10.1007/978-1-4612-2784-7_37. "Based on a conference held at the Mote Marine Laboratory in Sarasota, Fla., May 20–24, 1990."

- Barrett, Paul H.; Gautrey, Peter J.; Herbert, Sandra; et al., eds. (1987). *Charles Darwin's Notebooks, 1836-1844: Geology, Transmutation of Species, Metaphysical Enquiries*. London: British Museum (Natural History); Ithaca, NY: Cornell University Press. ISBN 0-521-09975-7. LCCN 87047593. OCLC 16224403.

- Bowler, Peter J. (2003). *Evolution: The History of an Idea* (3rd completely rev. and expanded ed.). Berkeley, CA: University of California Press. ISBN 0-520-23693-9. LCCN 2002007569. OCLC 49824702.

- Carpenter, G.D. Hale; Ford, E. B. (1933). *Mimicry. With a Section on Its Genetic Aspect by E. B. Ford*. Methuen's Monographs on Biological Subjects. London: Methuen. OCLC 875481859.

- Cronin, Helen (1991). *The Ant and the Peacock: Altruism and Sexual Selection from Darwin to Today*. Foreword by John Maynard Smith. Cambridge, UK; New York: Press Syndicate of the University of Cambridge. ISBN 0-521-32937-X. LCCN 91007887. OCLC 23144516.

- Darwin, Charles (1871). *The Descent of Man, and Selection in Relation to Sex*. London: John Murray. OCLC 550912.

- Darwin, Charles (1872). *The Origin of Species by Means of Natural Selection, or the Preservation of Favoured Races in the Struggle for Life* (6th ed.). London: John Murray. OCLC 1185571. Retrieved 2015-08-17.

- Desmond, Adrian (1989). *The Politics of Evolution: Morphology, Medicine, and Reform in Radical London*. Science and its Conceptual Foundations. Chicago, IL: University of Chicago Press. ISBN 0-226-14346-5. LCCN 89005137. OCLC 709606191.

- Dobzhansky, Theodosius (1968). "On Some Fundamental Concepts of Darwinian Biology". In Dobzhansky, Theodosius; Hecht, Max K.; Steere, William C. *Evolutionary Biology*. **2**. New York: Appleton-Century-Crofts. OCLC 24875357. doi:10.1007/978-1-4684-8094-8_1.

- Dobzhansky, Theodosius (1970). *Genetics of the Evolutionary Process*. New York: Columbia University Press. ISBN 0-231-02837-7. LCCN 72127363. OCLC 97663.

- Dobzhansky, Theodosius (1981). Lewontin, Richard C.; Moore, John A.; Provine, William B.; et al., eds. *Dobzhansky's Genetics of Natural Populations I-XLIII*. New York: Columbia University Press. ISBN 0-231-05132-8. LCCN 81002073. OCLC 7276406. "Papers by Dobzhansky and his collaborators, originally published 1937-1975 in various journals."

- Eldredge, Niles (1985). *Time Frames: The Rethinking of Darwinian Evolution and the Theory of Punctuated Equilibria*. New York: Simon & Schuster. ISBN 0-671-49555-0. LCCN 84023632. OCLC 11443805.

- Eldredge, Niles (1995). *Reinventing Darwin: The Great Debate at the High Table of Evolutionary Theory*. New York: John Wiley & Sons. ISBN 0-471-30301-1. LCCN 94032861. OCLC 30975979.

- Endler, John A. (1986). "Fitness and Adaptation". *Natural Selection in the Wild*. Monographs in Population Biology. **21**. Princeton, NJ: Princeton University Press. ISBN 0-691-08387-8. LCCN 85042683. OCLC 12262762.

- Fisher, Ronald Aylmer (1930). *The Genetical Theory of Natural Selection*. Oxford: The Clarendon Press. LCCN 30029177. OCLC 493745635.

- Ford, E. B. (1975). *Ecological Genetics* (4th ed.). London; New York: Chapman & Hall; John Wiley & Sons. ISBN 0-470-26576-0. LCCN 75002165. OCLC 1890603.

- Freeman, Scott; Herron, Jon C. (2007). *Evolutionary Analysis* (4th ed.). Upper Saddle River, NJ: Pearson Prentice Hall. ISBN 0-13-227584-8. LCCN 2006034384. OCLC 73502978.

- Futuyma, Douglas J. (1986). *Evolutionary Biology* (2nd ed.). Sunderland, MA: Sinauer Associates. ISBN 0-87893-188-0. LCCN 86015531. OCLC 13822044.

- Hull, David L. (1982). "Philosophy and biology". In Fløistad, Guttorm. *Philosophy of Science*. Contemporary Philosophy: A New Survey. **2**. The Hague: Martinus Nijhoff Publishers; Springer Netherlands. ISBN 90-247-2518-6. LCCN 81003972. OCLC 502399533. doi:10.1007/978-94-010-9940-0.

- Hutchinson, G. Evelyn (1965). *The Ecological Theater and the Evolutionary Play*. New Haven, CT: Yale University Press. LCCN 65022321. OCLC 250039.

- Huxley, Julian (1942). *Evolution: The Modern Synthesis*. London: Allen & Unwin. LCCN 42050738. OCLC 1399386.

- Margulis, Lynn; Fester, René, eds. (1991). *Symbiosis as a Source of Evolutionary Innovation: Speciation and Morphogenesis*. Cambridge, MA: MIT Press. ISBN 0-262-13269-9. LCCN 90020439. OCLC 22597587. "Based on a conference held in Bellagio, Italy, June 25–30, 1989"

- Maynard Smith, John (1993). *The Theory of Evolution* (Canto ed.). Cambridge, UK; New York: Cambridge University Press. ISBN 0-521-45128-0. LCCN 93020358. OCLC 27676642.

- Mayr, Ernst (1963). *Animal Species and Evolution*. Cambridge, MA: Belknap Press of Harvard University Press. ISBN 0-674-03750-2. LCCN 63009552. OCLC 899044868.

- Mayr, Ernst (1965). "Cause and Effect in Biology". In Lerner, Daniel. *Cause and Effect*. The Hayden Colloquium on Scientific Method and Concept. New York: Free Press. LCCN 65015439. OCLC 384895.

- Mayr, Ernst (1982). *The Growth of Biological Thought: Diversity, Evolution, and Inheritance*. Cambridge, MA: Belknap Press. ISBN 0-674-36445-7. LCCN 81013204. OCLC 7875904.

- Mayr, Ernst (1988). *Toward a New Philosophy of Biology: Observations of an Evolutionist*. Cambridge, MA: Belknap Press of Harvard University Press. ISBN 0-674-89665-3. LCCN 87031892. OCLC 17108004.

- Medawar, Peter (1960). *The Future of Man*. The BBC Reith Lectures, 1959. London: Methuen. LCCN 62002077. OCLC 1374615.

- Miller, Geoffrey (2007). "Brain Evolution". In Gangestad, Steven W.; Simpson, Jeffry A. *The Evolution of Mind: Fundamental Questions and Controversies*. New York: Guilford Press. ISBN 978-1-59385-408-9. LCCN 2006026955. OCLC 71005838.

- Monod, Jacques (1971). *Chance and Necessity: An Essay on the Natural Philosophy of Modern Biology*. Translation of *Le hasard et la nécessité* by Austryn Wainhouse (1st American ed.). New York: Knopf. ISBN 0-394-46615-2. LCCN 77154929. OCLC 209901.

- Moon, Harold Philip (1976). *Henry Walter Bates FRS, 1825-1892: Explorer, Scientist, and Darwinian*. Leicester, England: Leicestershire Museums, Art Galleries, and Records Service. ISBN 0-904671-19-4. LCCN 77369905. OCLC 3607387.

- Panchen, Alec L. (1992). *Classification, Evolution and the Nature of Biology*. Cambridge, UK; New York: Cambridge University Press. ISBN 0-521-31578-6. LCCN 91026274. OCLC 24247430.

- Patterson, Colin (1999). *Evolution*. Comstock Book Series (2nd illustrated, revised ed.). Ithaca, NY: Cornell University Press. ISBN 0-8014-8594-0. LCCN 98041312. OCLC 39724234.

- Pittendrigh, Colin S. (1958). "Adaptation, Natural Selection, and Behavior". In Roe, Anne; Simpson, George Gaylord. *Behavior and Evolution*. New Haven, CT: Yale University Press. LCCN 58011260. OCLC 191989.

- Price, Peter W. (1980). *The Evolutionary Biology of Parasites*. Monographs in Population Biology. **15**. Princeton, NJ: Princeton University Press. ISBN 0-691-08257-X. LCCN 79003227. OCLC 5706295.

- Provine, William B. (1986). *Sewall Wright and Evolutionary Biology*. Science and its Conceptual Foundations. Chicago, IL: University of Chicago Press. ISBN 0-226-68474-1. LCCN 85024651. OCLC 12808844.

- Ruxton, Graeme D.; Sherratt, Thomas N.; Speed, Michael P. (2004). *Avoiding Attack: The Evolutionary Ecology of Crypsis, Warning Signals and Mimicry*. Oxford Biology. Oxford; New York: Oxford University Press. ISBN 0-19-852859-0. LCCN 2005297323. OCLC 56644492.

- Sober, Elliott (1984). *The Nature of Selection: Evolutionary Theory in Philosophical Focus*. Cambridge, MA: MIT Press. ISBN 0-262-19232-2. LCCN 84019470. OCLC 11114517.

- Sober, Elliott (1993). *Philosophy of Biology*. Dimensions of Philosophy Series. Boulder, CO: Westview Press. ISBN 0-8133-0785-6. LCCN 92037484. OCLC 26974492.

- Stebbins, G. Ledyard, Jr. (1950). *Variation and Evolution in Plants*. Columbia Biological Series. **16**. New York: Columbia University Press. LCCN 50009426. OCLC 294016.

- Sterelny, Kim; Griffiths, Paul E. (1999). *Sex and Death: An Introduction to Philosophy of Biology*. Science and its Conceptual Foundations. Chicago, IL: University of Chicago Press. ISBN 0-226-77304-3. LCCN 98047555. OCLC 40193587.

- Voltaire (1759). *Candide, ou l'Optimisme*. Paris. Candide on the Internet Archive Retrieved 2015-08-17.

- Wickler, Wolfgang (1968). *Mimicry in Plants and Animals*. World University Library. Translated from the German by R. D. Martin. New York: McGraw-Hill. LCCN 67026359. OCLC 160314.

- Williams, Edgar (2010). *Giraffe*. Animal (Reaktion Books). London: Reaktion Books. ISBN 978-1-86189-764-0. OCLC 587198932.

- Williams, George C. (1966). *Adaptation and Natural Selection: A Critique of Some Current Evolutionary Thought*. Princeton Science Library. Princeton, NJ: Princeton University Press. ISBN 0-691-02615-7. LCCN 65017164. OCLC 35230452.

- Wright, Sewall (1932). "The Roles of Mutation, Inbreeding, Crossbreeding and Selection in Evolution". In Jones, Donald F. *Proceedings of the Sixth International Congress of Genetics*. **1**. Ithaca, NY: Genetics Society of America. OCLC 439596433.

Chapter 31

Artificial neural network

"Neural network" redirects here. For networks of living neurons, see Biological neural network. For the journal, see Neural Networks (journal). For the evolutionary concept, see Neutral network (evolution).

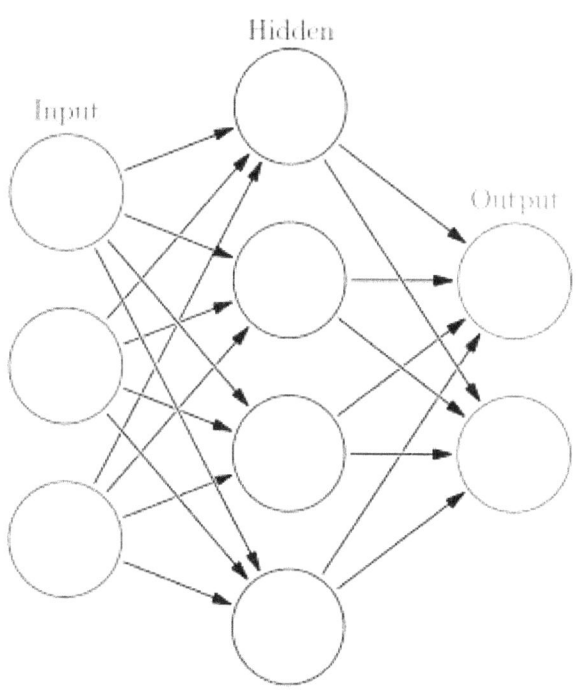

An artificial neural network is an interconnected group of nodes, akin to the vast network of neurons in a brain. Here, each circular node represents an artificial neuron and an arrow represents a connection from the output of one neuron to the input of another.

Artificial neural networks (ANNs) or **connectionist systems** are computing systems inspired by the biological neural networks that constitute animal brains. Such systems learn (progressively improve performance) to do tasks by considering examples, generally without task-specific programming. For example, in image recognition, they might learn to identify images that contain cats by analyzing example images that have been manually labeled as "cat" or "no cat" and using the analytic results to identify cats in other images. They have found most use in applications difficult to express in a traditional computer algorithm using rule-based programming.

An ANN is based on a collection of connected units called artificial neurons, (analogous to axons in a biological brain). Each connection (synapse) between neurons can transmit a signal to another neuron. The receiving (postsynaptic) neuron can process the signal(s) and then signal downstream neurons connected to it. Neurons may have state, generally represented by real numbers, typically between 0 and 1. Neurons and synapses may also have a weight that varies as learning proceeds, which can increase or decrease the strength of the signal that it sends downstream. Further, they may have a threshold such that only if the aggregate signal is below (or above) that level is the downstream signal sent.

Typically, neurons are organized in layers. Different layers may perform different kinds of transformations on their inputs. Signals travel from the first (input), to the last (output) layer, possibly after traversing the layers multiple times.

The original goal of the neural network approach was to solve problems in the same way that a human brain would. Over time, attention focused on matching specific mental abilities, leading to deviations from biology such as backpropagation, or passing information in the reverse direction and adjusting the network to reflect that information.

Neural networks have been used on a variety of tasks, including computer vision, speech recognition, machine translation, social network filtering, playing board and video games, medical diagnosis and in many other domains.

31.1 History

Warren McCulloch and Walter Pitts[1] (1943) created a computational model for neural networks based on mathematics and algorithms called threshold logic. This model paved the way for neural network research to split into two approaches. One approach focused on biological processes in the brain while the other focused on the application of neural networks to artificial intelligence. This work led to work on nerve networks and their link to finite automata.[2]

31.1.1 Hebbian learning

In the late 1940s, D.O. Hebb[3] created a learning hypothesis based on the mechanism of neural plasticity that is now known as Hebbian learning. Hebbian learning is an unsupervised learning rule. This evolved into models for long term potentiation. Researchers started applying these ideas to computational models in 1948 with Turing's B-type machines.

Farley and Clark[4] (1954) first used computational machines, then called "calculators", to simulate a Hebbian network. Other neural network computational machines were created by Rochester, Holland, Habit and Duda[5] (1956).

Rosenblatt[6] (1958) created the perceptron, an algorithm for pattern recognition. With mathematical notation, Rosenblatt described circuitry not in the basic perceptron, such as the exclusive-or circuit that could not be processed by neural networks at the time.[7]

In 1959, a biological model proposed by Nobel laureates Hubel and Wiesel was based on their discovery of two types of cells in the primary visual cortex: simple cells and complex cells[8]

The first functional networks with many layers were published by Ivakhnenko and Lapa in 1965, becoming the Group Method of Data Handling.[9][10][11]

Neural network research stagnated after machine learning research by Minsky and Papert (1969),[12] who discovered two key issues with the computational machines that processed neural networks. The first was that basic perceptrons were incapable of processing the exclusive-or circuit. The second was that computers didn't have enough processing power to effectively handle the work required by large neural networks. Neural network research slowed until computers achieved far greater processing power.

31.1.2 Backpropagation

Much of Artificial intelligence had focussed on high-level (symbolic) models that are processed by using algorithms, characterized for example by expert systems with knowledge embodied in *if-then* rules, until in the late 1980ies research expanded to low-level (sub-symbolic) machine learning, characterized by knowledge embodied in the parameters of a cognitive model.

A key trigger for the renewed interest in neural networks and learning was Werbos's (1975) backpropagation algorithm that effectively solved the exclusive-or problem and more generally accelerated the training of multi-layer networks.[7]

In the mid-1980s, parallel distributed processing became popular under the name connectionism. Rumelhart and McClelland (1986) described the use of connectionism to simulate neural processes.[13]

Support vector machines and other, much simpler methods such as linear classifiers gradually overtook neural networks in machine learning popularity.

Earlier challenges in training deep neural networks were successfully addressed with methods such as unsupervised pre-training, while available computing power increased through the use of GPUs and distributed computing. Neural networks were deployed on a large scale, particularly in image and visual recognition problems. This became known as "deep learning", although deep learning is not strictly synonymous with deep neural networks.

In 1992, max-pooling was introduced to help with least shift invariance and tolerance to deformation to aid in 3D object recognition.[14][15][16]

The vanishing gradient problem affects many-layered feedforward networks that use backpropagation and also recurrent neural networks.[17][18] As errors propagate from layer to layer, they shrink exponentially with the number of layers, impeding the tuning of neuron weights that is based on those errors, particularly affecting deep networks.

To overcome this problem, Schmidhuber's multi-level hierarchy of networks (1992) pre-trained one level at a time by unsupervised learning, fine-tuned by backpropagation.[19] Behnke (2003) relied only on the sign of the gradient (Rprop)[20] on problems such as image reconstruction and face localization.

Hinton et al. (2006) employed learning the distribution of a high-level representation using successive layers of binary or real-valued latent variables with a restricted Boltzmann machine[21] to model each layer. Once sufficiently many layers have been learned, the deep architecture may be used as a generative model by reproducing the data when sampling down the model (an "ancestral pass") from the top level feature activations.[22][23] In 2012, Ng and Dean created a neural network that learned to recognize higher-level concepts, such as cats, only from watching unlabeled images taken from YouTube videos.[24]

31.1.3 Hardware-based designs

Computational devices were created in CMOS, for both biophysical simulation and neuromorphic computing. Nanodevices[25] for very large scale principal components analyses and convolution may create a new class of neural computing because they are fundamentally analog rather than digital (even though the first implementations may use digital devices.)[26] Ciresan and colleagues (2010)[27] in Schmidhuber's group showed that despite the vanishing gradient problem, GPUs makes back-propagation feasible for many-layered feedforward neural networks.

31.1.4 Contests

Between 2009 and 2012, recurrent neural networks and deep feedforward neural networks developed in the Schmidhuber's research group, winning eight international competitions in pattern recognition and machine learning.[28][29] For example, the bi-directional and multi-dimensional long short-term memory (LSTM)[30][31][32][33] of Graves et al. won three competitions in connected handwriting recognition at the 2009 International Conference on Document Analysis and Recognition (ICDAR), without any prior knowledge about the three languages to be learned.[32][34]

Ciresan and colleagues won pattern recognition contests, including the IJCNN 2011 Traffic Sign Recognition Competition,[35] the ISBI 2012 Segmentation of Neuronal Structures in Electron Microscopy Stacks challenge[36] and others. Their neural networks were the first pattern recognizers to achieve human-competitive or even superhuman performance[37] on benchmarks such as traffic sign recognition (IJCNN 2012), or the MNIST handwritten digits problem.

Researchers demonstrated (2010) that deep neural networks interfaced with a hidden Markov model with context-dependent states that define the neural network output layer can drastically reduce errors in large-vocabulary speech recognition tasks such as voice search.

GPU-based implementations[38] of this approach won many pattern recognition contests, including the IJCNN 2011 Traffic Sign Recognition Competition,[35] the ISBI 2012 Segmentation of neuronal structures in EM stacks challenge,[39] the ImageNet Competition[40] and others.

Deep, highly nonlinear neural architectures similar to the neocognitron[41] and the "standard architecture of vision",[42] inspired by simple and complex cells were pre-trained by unsupervised methods[43][44] by Hinton.[43][45] A team from his lab won a 2012 contest sponsored by Merck to design software to help find molecules that might identify new drugs.[46]

31.1.5 Convolutional networks

As of 2011, the state of the art in deep learning feedforward networks alternated convolutional layers and max-pooling layers,[38][47] topped by several fully or sparsely connected layers followed by a final classification layer. Learning is usually done without unsupervised pre-training.

Such supervised deep learning methods were the first artificial pattern recognizers to achieve human-competitive performance on certain tasks.[37]

ANNs were able to guarantee shift invariance to deal with small and large natural objects in large cluttered scenes, only when invariance extended beyond shift, to all ANN-learned concepts, such as location, type (object class label), scale, lighting and others. This was realized in Developmental Networks (DNs)[48] whose embodiments are Where-What Networks, WWN-1 (2008)[49] through WWN-7 (2013).[50]

31.2 Models

An *(artificial) neural network* is a network of simple elements called *neurons*, which receive input, change their internal state *(activation)* according to that input, and produce output depending on the input and activation. The *network* forms by connecting the output of certain neurons to the input of other neurons forming a directed, weighted graph. The weights as well as the functions that compute the activation can be modified by a process called *learning* which is governed by a *learning* rule.[51]

31.2.1 Components of an artificial neural network

Neurons

A neuron with label j receiving an input $p_j(t)$ from predecessor neurons consists of the following components:[51]

- an *activation* $a_j(t)$, depending on a discrete time parameter,
- possibly a *threshold* θ_j, which stays fixed unless changed by a learning function,
- an *activation function* f that computes the new activation at a given time $t+1$ from $a_j(t)$, θ_j and the net input $p_j(t)$ giving rise to the relation

$$a_j(t+1) = f(a_j(t), p_j(t), \theta_j)$$

- and an *output function* f_{out} computing the output from the activation

$$o_j(t) = f_{out}(a_j(t))$$

Often the output function is simply the Identity function.

An *input neuron* has no predecessor but serves as input interface for the whole network. Similarly an *output neuron* has no successor and thus serves as output interface of the whole network.

Connections and weights

The *network* consists of connections, each connection transferring the output of a neuron i to the input of a neuron j. In this sense i is the predecessor of j and j is the successor of i. Each connection is assigned a weight w_{ij}.[51]

Propagation function

The *propagation function* computes the *input* $p_j(t)$ to the neuron j from the outputs $o_i(t)$ of predecessor neurons and typically has the form[51]

$$p_j(t) = \sum_i o_i(t) w_{ij}$$

Learning rule

The *learning rule* is a rule or an algorithm which modifies the parameters of the neural network, in order for a given input to the network to produce a favored output. This *learning* process typically amounts to modifying the weights and thresholds of the variables within the network.[51]

31.2.2 Neural networks as functions

See also: Graphical models

Neural network models can be viewed as simple mathematical models defining a function $f : X \to Y$ or a distribution over X or both X and Y. Sometimes models are intimately associated with a particular learning rule. A common use of the phrase "ANN model" is really the definition of a *class* of such functions (where members of the class are obtained by varying parameters, connection weights, or specifics of the architecture such as the number of neurons or their connectivity).

Mathematically, a neuron's network function $f(x)$ is defined as a composition of other functions $g_i(x)$, that can further be decomposed into other functions. This can be conveniently represented as a network structure, with arrows depicting the dependencies between functions. A widely used type of composition is the *nonlinear weighted sum*, where $f(x) = K\left(\sum_i w_i g_i(x)\right)$, where K (commonly referred to as the activation function[52]) is some predefined function, such as the hyperbolic tangent or sigmoid function or softmax function or rectifier function. The important characteristic of the activation function is that it provides a smooth transition as input values change, i.e. a small change in input produces a small change in output. The following refers to a collection of functions g_i as a vector $g = (g_1, g_2, \ldots, g_n)$.

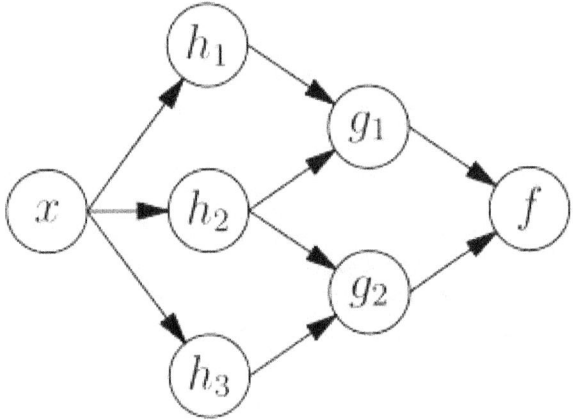

ANN dependency graph

This figure depicts such a decomposition of f, with dependencies between variables indicated by arrows. These can be interpreted in two ways.

The first view is the functional view: the input x is transformed into a 3-dimensional vector h, which is then transformed into a 2-dimensional vector g, which is finally transformed into f. This view is most commonly encountered in the context of optimization.

The second view is the probabilistic view: the random variable $F = f(G)$ depends upon the random variable $G = g(H)$, which depends upon $H = h(X)$, which depends upon the random variable X. This view is most commonly encountered in the context of graphical models.

The two views are largely equivalent. In either case, for this particular architecture, the components of individual layers are independent of each other (e.g., the components of g are independent of each other given their input h). This naturally enables a degree of parallelism in the implementation.

Networks such as the previous one are commonly called feedforward, because their graph is a directed acyclic graph. Networks with cycles are commonly called recurrent. Such networks are commonly depicted in the manner shown at

31.2. MODELS

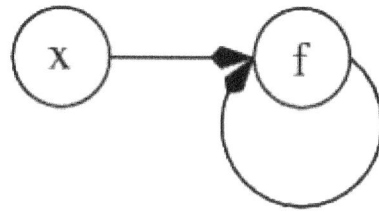

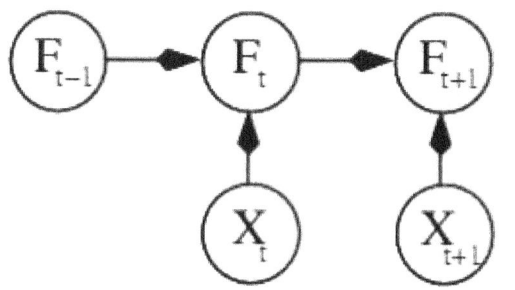

Two separate depictions of the recurrent ANN dependency graph

the top of the figure, where f is shown as being dependent upon itself. However, an implied temporal dependence is not shown.

31.2.3 Learning

See also: Mathematical optimization, Estimation theory, and Machine learning

The possibility of learning has attracted the most interest in neural networks. Given a specific *task* to solve, and a class of functions F, learning means using a set of observations to find $f^* \in F$ which solves the task in some optimal sense.

This entails defining a cost function $C : F \to \mathbb{R}$ such that, for the optimal solution f^*, $C(f^*) \leq C(f) \; \forall f \in F$ – i.e., no solution has a cost less than the cost of the optimal solution (see mathematical optimization).

The cost function C is an important concept in learning, as it is a measure of how far away a particular solution is from an optimal solution to the problem to be solved. Learning algorithms search through the solution space to find a function that has the smallest possible cost.

For applications where the solution is data dependent, the cost must necessarily be a function of the observations, otherwise the model would not relate to the data. It is frequently defined as a statistic to which only approximations can be made. As a simple example, consider the problem of finding the model f, which minimizes $C =$ $E\left[(f(x) - y)^2\right]$, for data pairs (x, y) drawn from some distribution D. In practical situations we would only have N samples from D and thus, for the above example, we would only minimize $\hat{C} = \frac{1}{N}\sum_{i=1}^{N}(f(x_i) - y_i)^2$. Thus, the cost is minimized over a sample of the data rather than the entire distribution.

When $N \to \infty$ some form of online machine learning must be used, where the cost is reduced as each new example is seen. While online machine learning is often used when D is fixed, it is most useful in the case where the distribution changes slowly over time. In neural network methods, some form of online machine learning is frequently used for finite datasets.

Choosing a cost function

While it is possible to define an ad hoc cost function, frequently a particular cost (function) is used, either because it has desirable properties (such as convexity) or because it arises naturally from a particular formulation of the problem (e.g., in a probabilistic formulation the posterior probability of the model can be used as an inverse cost). Ultimately, the cost function depends on the task.

Backpropagation

Main article: Backpropagation

A DNN can be discriminatively trained with the standard backpropagation algorithm. Backpropagation is a method to calculate the gradient of the loss function (produces the cost associated with a given state) with respect to the weights in an ANN.

The basics of continuous backpropagation[9][53][54][55] were derived in the context of control theory by Kelley[56] in 1960 and by Bryson in 1961,[57] using principles of dynamic programming. In 1962, Dreyfus published a simpler derivation based only on the chain rule.[58] Bryson and Ho described it as a multi-stage dynamic system optimization method in 1969.[59][60] In 1970, Linnainmaa finally published the general method for automatic differentiation (AD) of discrete connected networks of nested differentiable functions.[61][62] This corresponds to the modern version of backpropagation which is efficient even when the networks are sparse.[9][53][63][64] In 1973, Dreyfus used backpropagation to adapt parameters of controllers in proportion to error gradients.[65] In 1974, Werbos mentioned the possibility of applying this principle to ANNs,[66] and in 1982, he applied Linnainmaa's AD method to neural networks in the way that is widely used today.[53][67] In 1986, Rumelhart, Hinton and Williams noted that this method can generate useful internal representations of incoming data in hidden layers of neural networks.[68] In 1993, Wan was the first[9] to win an international pattern recognition contest through backpropagation.[69]

The weight updates of backpropagation can be done via stochastic gradient descent using the following equation:

$$w_{ij}(t+1) = w_{ij}(t) + \eta \frac{\partial C}{\partial w_{ij}} + \xi(t)$$

where, η is the learning rate, C is the cost (loss) function and $\xi(t)$ a stochastic term. The choice of the cost function depends on factors such as the learning type (supervised, unsupervised, reinforcement, etc.) and the activation function. For example, when performing supervised learning on a multiclass classification problem, common choices for the activation function and cost function are the softmax function and cross entropy function, respectively. The softmax function is defined as $p_j = \frac{\exp(x_j)}{\sum_k \exp(x_k)}$ where p_j represents the class probability (output of the unit j) and x_j and x_k represent the total input to units j and k of the same level respectively. Cross entropy is defined as $C = -\sum_j d_j \log(p_j)$ where d_j represents the target probability for output unit j and p_j is the probability output for j after applying the activation function.[70]

These can be used to output object bounding boxes in the form of a binary mask. They are also used for multi-scale regression to increase localization precision. DNN-based regression can learn features that capture geometric information in addition to serving as a good classifier. They remove the requirement to explicitly model parts and their relations. This helps to broaden the variety of objects that can be learned. The model consists of multiple layers, each of which has a rectified linear unit as its activation function for non-linear transformation. Some layers are convolutional, while others are fully connected. Every convolutional layer has an additional max pooling. The network is trained to minimize L2 error for predicting the mask ranging over the entire training set containing bounding boxes represented as masks.

Alternatives to backpropagation include Extreme Learning Machines,[71] "No-prop" networks,[72] training without backtracking,[73] "weightless" networks,"[74][75] and non-connectionist neural networks.

31.2.4 Learning paradigms

The three major learning paradigms each correspond to a particular learning task. These are supervised learning, unsupervised learning and reinforcement learning.

Supervised learning

Supervised learning uses a set of example pairs (x, y), $x \in X$, $y \in Y$ and the aim is to find a function $f : X \rightarrow Y$ in the allowed class of functions that matches the examples. In other words, we wish to infer the mapping implied by the data; the cost function is related to the mismatch between our mapping and the data and it implicitly contains prior knowledge about the problem domain.[76]

A commonly used cost is the mean-squared error, which tries to minimize the average squared error between the network's output, $f(x)$, and the target value y over all the example pairs. Minimizing this cost using gradient descent for the class of neural networks called multilayer perceptrons (MLP), produces the backpropagation algorithm for training neural networks.

Tasks that fall within the paradigm of supervised learning are pattern recognition (also known as classification) and regression (also known as function approximation). The supervised learning paradigm is also applicable to sequential data (e.g., for hand writing, speech and gesture recognition). This can be thought of as learning with a "teacher", in the form of a function that provides continuous feedback on the quality of solutions obtained thus far.

Unsupervised learning

In unsupervised learning, some data x is given and the cost function to be minimized, that can be any function of the data x and the network's output, f.

The cost function is dependent on the task (the model domain) and any *a priori* assumptions (the implicit properties of the model, its parameters and the observed variables).

As a trivial example, consider the model $f(x) = a$ where a is a constant and the cost $C = E[(x - f(x))^2]$. Minimizing this cost produces a value of a that is equal to the mean of the data. The cost function can be much more complicated. Its form depends on the application: for example, in compression it could be related to the mutual information between x and $f(x)$, whereas in statistical modeling, it could be related to the posterior probability of the model given the data (note that in both of those examples those quantities would be maximized rather than minimized).

Tasks that fall within the paradigm of unsupervised learning are in general estimation problems; the applications include clustering, the estimation of statistical distributions, compression and filtering.

Reinforcement learning

See also: Stochastic control

In reinforcement learning, data x are usually not given, but generated by an agent's interactions with the environment. At each point in time t, the agent performs an action y_t and the environment generates an observation x_t and an instantaneous cost c_t, according to some (usually unknown) dynamics. The aim is to discover a policy for selecting actions that minimizes some measure of a long-term cost, e.g., the expected cumulative cost. The environment's dynamics and the long-term cost for each policy are usually unknown, but can be estimated.

More formally the environment is modeled as a Markov decision process (MDP) with states $s_1, ..., s_n \in S$ and actions $a_1, ..., a_m \in A$ with the following probability distributions: the instantaneous cost distribution $P(c_t|s_t)$, the observation distribution $P(x_t|s_t)$ and the transition $P(s_{t+1}|s_t, a_t)$, while a policy is defined as the conditional distribution over actions given the observations. Taken together, the two then define a Markov chain (MC). The aim is to discover the policy (i.e., the MC) that minimizes the cost.

ANNs are frequently used in reinforcement learning as part of the overall algorithm.[77][78] Dynamic programming was coupled with ANNs (giving neurodynamic programming) by Bertsekas and Tsitsiklis[79] and applied to multi-dimensional nonlinear problems such as those involved in vehicle routing,[80] natural resources management[81][82] or medicine[83] because of the ability of ANNs to mitigate losses of accuracy even when reducing the discretization grid density for numerically approximating the solution of the original control problems.

Tasks that fall within the paradigm of reinforcement learning are control problems, games and other sequential decision making tasks.

31.2.5 Learning algorithms

See also: Machine learning

Training a neural network model essentially means selecting one model from the set of allowed models (or, in a Bayesian framework, determining a distribution over the set of allowed models) that minimizes the cost. Numerous algorithms are available for training neural network models; most of them can be viewed as a straightforward application of optimization theory and statistical estimation.

Most employ some form of gradient descent, using backpropagation to compute the actual gradients. This is done by simply taking the derivative of the cost function with respect to the network parameters and then changing those parameters in a gradient-related direction. Backpropagation training algorithms fall into three categories:

- steepest descent (with variable learning rate and momentum, resilient backpropagation);
- quasi-Newton (Broyden-Fletcher-Goldfarb-Shanno, one step secant);
- Levenberg-Marquardt and conjugate gradient (Fletcher-Reeves update, Polak-Ribiére update, Powell-Beale restart, scaled conjugate gradient).[84]

Evolutionary methods,[85] gene expression programming,[86] simulated annealing,[87] expectation-maximization, non-parametric methods and particle swarm optimization[88] are other methods for training neural networks.

31.3 Variants

31.3.1 Group method of data handling

Main article: Group method of data handling

The Group Method of Data Handling (GMDH)[89] features fully automatic structural and parametric model optimization. The node activation functions are Kolmogorov-Gabor polynomials that permit additions and multiplications. It used a deep feedforward multilayer perceptron with eight layers.[90] It is a supervised learning network that grows layer by layer, where each layer is trained by regression analysis. Useless items are detected using a validation set, and pruned through regularization. The size and depth of the resulting network depends on the task.[91]

31.3.2 Convolutional neural networks

Main article: Convolutional neural network

A convolutional neural network (CNN) is a class of deep, feed-forward networks, composed of one or more convolutional layers with fully connected layers (matching those in typical ANNs) on top. It uses tied weights and pooling layers. In particular, max-pooling[15] is often structured via Fukushima's convolutional architecture.[92] This architecture allows CNNs to take advantage of the 2D structure of input data.

CNNs are suitable for processing visual and other two-dimensional data.[93][94] They have shown superior results in both image and speech applications. They can be trained with standard backpropagation. CNNs are easier to train than other regular, deep, feed-forward neural networks and have many fewer parameters to estimate.[95] Examples of applications in computer vision include DeepDream.[96]

31.3.3 Long short-term memory

Main article: Long short-term memory

Long short-term memory (LSTM) networks are RNNs that avoid the vanishing gradient problem.[97] LSTM is normally augmented by recurrent gates called forget gates.[98] LSTM networks prevent backpropagated errors from vanishing or exploding.[17] Instead errors can flow backwards through unlimited numbers of virtual layers in space-unfolded LSTM. That is, LSTM can learn "very deep learning" tasks[9] that require memories of events that happened thousands or even millions of discrete time steps ago. Problem-specific LSTM-like topologies can be evolved.[99] LSTM can handle long delays and signals that have a mix of low and high frequency components.

Stacks of LSTM RNNs[100] trained by Connectionist Temporal Classification (CTC)[101] can find an RNN weight matrix that maximizes the probability of the label sequences in a training set, given the corresponding input sequences. CTC achieves both alignment and recognition.

In 2003, LSTM started to become competitive with traditional speech recognizers.[102] In 2007, the combination with CTC achieved first good results on speech data.[103] In 2009, a CTC-trained LSTM was the first RNN to win pattern recognition contests, when it won several competitions in connected handwriting recognition.[9][32] In 2014, Baidu used CTC-trained RNNs to break the Switchboard Hub5'00 speech recognition benchmark, without traditional speech processing methods.[104] LSTM also improved large-vocabulary speech recognition,[105][106] text-to-speech synthesis,[107] for Google Android,[53][108] and photo-real talking heads.[109] In 2015, Google's speech recognition experienced a 49% improvement through CTC-trained LSTM.[110]

LSTM became popular in Natural Language Processing. Unlike previous models based on HMMs and similar concepts, LSTM can learn to recognise context-sensitive languages.[111] LSTM improved machine translation,[112] language modeling[113] and multilingual language processing.[114] LSTM combined with CNNs improved automatic image captioning.[115]

31.3.4 Deep reservoir computing

Main article: Reservoir computing

Deep Reservoir Computing and Deep Echo State Networks (deepESNs)[116][117] provide a framework for efficiently trained models for hierarchical processing of temporal data, while enabling the investigation of the inherent role of RNN layered composition.

31.3.5 Deep belief networks

Main article: Deep belief network

A deep belief network (DBN) is a probabilistic, generative

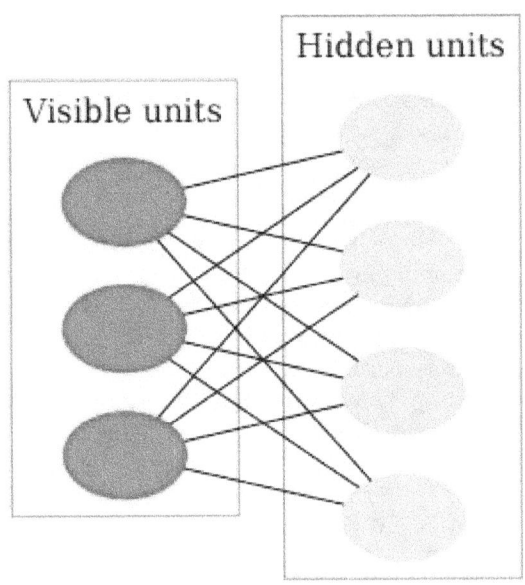

A restricted Boltzmann machine (RBM) with fully connected visible and hidden units. Note there are no hidden-hidden or visible-visible connections.

model made up of multiple layers of hidden units. It can be considered a composition of simple learning modules that make up each layer.[118]

A DBN can be used to generatively pre-train a DNN by using the learned DBN weights as the initial DNN weights. Backpropagation or other discriminative algorithms can then tune these weights. This is particularly helpful when training data are limited, because poorly initialized weights can significantly hinder model performance. These pre-trained weights are in a region of the weight space that is closer to the optimal weights than were they randomly chosen. This allows for both improved modeling and faster convergence of the fine-tuning phase.[119]

31.3.6 Large memory storage and retrieval neural networks

Large memory storage and retrieval neural networks (LAMSTAR)[120][121] are fast deep learning neural networks of many layers that can use many filters simultaneously. These filters may be nonlinear, stochastic, logic, non-stationary, or even non-analytical. They are biologically motivated and learn continuously.

A LAMSTAR neural network may serve as a dynamic neural network in spatial or time domains or both. Its speed is provided by Hebbian link-weights[122] that integrate the various and usually different filters (preprocessing functions) into its many layers and to dynamically rank the significance of the various layers and functions relative to a given learning task. This grossly imitates biological learning which integrates various preprocessors (cochlea, retina, etc.) and cortexes (auditory, visual, etc.) and their various regions. Its deep learning capability is further enhanced by using inhibition, correlation and its ability to cope with incomplete data, or "lost" neurons or layers even amidst a task. It is fully transparent due to its link weights. The link-weights allow dynamic determination of innovation and redundancy, and facilitate the ranking of layers, of filters or of individual neurons relative to a task.

LAMSTAR has been applied to many domains, including medical[123][124][125] and financial predictions,[126] adaptive filtering of noisy speech in unknown noise,[127] still-image recognition,[128] video image recognition,[129] software security[130] and adaptive control of non-linear systems.[131] LAMSTAR had a much faster learning speed and somewhat lower error rate than a CNN based on ReLU-function filters and max pooling, in 20 comparative studies.[132]

These applications demonstrate delving into aspects of the data that are hidden from shallow learning networks and the human senses, such as in the cases of predicting onset of sleep apnea events,[124] of an electrocardiogram of a fetus as recorded from skin-surface electrodes placed on the mother's abdomen early in pregnancy,[125] of financial prediction[126] or in blind filtering of noisy speech.[127]

LAMSTAR was proposed in 1996 (A U.S. Patent 5,920,852 A) and was further developed Graupe and Kordylewski from 1997-2002.[133][134][135] A modified version, known as LAMSTAR 2, was developed by Schneider and Graupe in 2008.[136][137]

31.3.7 Stacked (de-noising) auto-encoders

The auto encoder idea is motivated by the concept of a *good* representation. For example, for a classifier, a good representation can be defined as one that yields a better-performing classifier.

An *encoder* is a deterministic mapping f_θ that transforms an input vector x into hidden representation y, where $\theta = \{W, b\}$, W is the weight matrix and b is an offset vector (bias). A *decoder* maps back the hidden representation y to the reconstructed input z via g_θ. The whole process of auto encoding is to compare this reconstructed input to the original and try to minimize the error to make the reconstructed value as close as possible to the original.

In *stacked denoising auto encoders*, the partially corrupted output is cleaned (de-noised). This idea was introduced in 2010 by Vincent et al.[138] with a specific approach to *good* representation, a *good representation* is one that can be obtained robustly from a corrupted input and that will be useful for recovering the corresponding clean input. Implicit in this definition are the following ideas:

- The higher level representations are relatively stable and robust to input corruption;

- It is necessary to extract features that are useful for representation of the input distribution.

The algorithm starts by a stochastic mapping of x to $\tilde{x}$ through $q_D(\tilde{x}|x)$, this is the corrupting step. Then the corrupted input $\tilde{x}$ passes through a basic auto-encoder process and is mapped to a hidden representation $y = f_\theta(\tilde{x}) = s(W\tilde{x} + b)$. From this hidden representation, we can reconstruct $z = g_\theta(y)$. In the last stage, a minimization algorithm runs in order to have z as close as possible to uncorrupted input x. The reconstruction error $L_H(x, z)$ might be either the cross-entropy loss with an affine-sigmoid decoder, or the squared error loss with an affine decoder.[138]

In order to make a deep architecture, auto encoders stack.[139] Once the encoding function f_θ of the first denoising auto encoder is learned and used to uncorrupt the input (corrupted input), the second level can be trained.[138]

Once the stacked auto encoder is trained, its output can be used as the input to a supervised learning algorithm such as support vector machine classifier or a multi-class logistic regression.[138]

31.3.8 Deep stacking networks

A a deep stacking network (DSN)[140] (deep convex network) is based on a hierarchy of blocks of simplified neural network modules. It was introduced in 2011 by Deng and Dong.[141] It formulates the learning as a convex optimization problem with a closed-form solution, emphasizing the mechanism's similarity to stacked generalization.[142] Each DSN block is a simple module that is easy to train by itself in a supervised fashion without backpropagation for the entire blocks.[143]

Each block consists of a simplified multi-layer perceptron (MLP) with a single hidden layer. The hidden layer h has logistic sigmoidal units, and the output layer has linear units. Connections between these layers are represented by weight matrix U; input-to-hidden-layer connections have weight matrix W. Target vectors t form the columns of matrix T, and the input data vectors x form the columns of matrix X. The matrix of hidden units is $H = \sigma(W^T X)$. Modules are trained in order, so lower-layer weights W are known at each stage. The function performs the element-wise logistic sigmoid operation. Each block estimates the same final label class y, and its estimate is concatenated with original input X to form the expanded input for the next block. Thus, the input to the first block contains the original data only, while downstream blocks' input adds the output of preceding blocks. Then learning the upper-layer weight matrix U given other weights in the network can be formulated as a convex optimization problem:

$$\min_{U^T} f = \|U^T H - T\|_F^2.$$

which has a closed-form solution.

Unlike other deep architectures, such as DBNs, the goal is not to discover the transformed feature representation. The structure of the hierarchy of this kind of architecture makes parallel learning straightforward, as a batch-mode optimization problem. In purely discriminative tasks, DSNs perform better than conventional DBNs.[140]

31.3.9 Tensor deep stacking networks

This architecture is a DSN extension. It offers two important improvements: it uses higher-order information from covariance statistics, and it transforms the non-convex problem of a lower-layer to a convex sub-problem of an upper-layer.[144] TDSNs use covariance statistics in a bilinear mapping from each of two distinct sets of hidden units in the same layer to predictions, via a third-order tensor.

While parallelization and scalability are not considered seriously in conventional DNNs,[145][146][147] all learning for DSNs and TDSNs is done in batch mode, to allow parallelization.[141][140] Parallelization allows scaling the design to larger (deeper) architectures and data sets.

The basic architecture is suitable for diverse tasks such as classification and regression.

31.3.10 Spike-and-slab RBMs

The need for deep learning with real-valued inputs, as in Gaussian restricted Boltzmann machines, led to the *spike-and-slab* RBM (*ss*RBM), which models continuous-valued inputs with strictly binary latent variables.[148] Similar to basic RBMs and its variants, a spike-and-slab RBM is a bipartite graph, while like GRBMs, the visible units (input) are real-valued. The difference is in the hidden layer, where each hidden unit has a binary spike variable and a real-valued slab variable. A spike is a discrete probability mass at zero, while a slab is a density over continuous domain;[149] their mixture forms a prior.[150]

An extension of ssRBM called μ-ssRBM provides extra modeling capacity using additional terms in the energy function. One of these terms enables the model to form a conditional distribution of the spike variables by marginalizing out the slab variables given an observation.

31.3.11 Compound hierarchical-deep models

Compound hierarchical-deep models compose deep networks with non-parametric Bayesian models. Features can be learned using deep architectures such as DBNs,[151] DBMs,[152] deep auto encoders,[153] convolutional variants,[154][155] ssRBMs,[149] deep coding networks,[156] DBNs with sparse feature learning,[157] RNNs,[158] conditional DBNs,[159] de-noising auto encoders.[160] This provides a better representation, allowing faster learning and more accurate classification with high-dimensional data. However, these architectures are poor at learning novel classes with few examples, because all network units are involved in representing the input (a **distributed representation**) and must be adjusted together (high degree of freedom). Limiting the degree of freedom reduces the number of parameters to learn, facilitating learning of new classes from few examples. *Hierarchical Bayesian (HB)* models allow learning from few examples, for example[161][162][163][164][165] for computer vision, statistics and cognitive science.

Compound HD architectures aim to integrate characteristics of both HB and deep networks. The compound HDP-DBM architecture is a *hierarchical Dirichlet process (HDP)* as a hierarchical model, incorporated with DBM architecture. It is a full generative model, generalized from abstract concepts flowing through the layers of the model, which is able to synthesize new examples in novel classes that look "reasonably" natural. All the levels are learned jointly by maximizing a joint log-probability score.[166]

In a DBM with three hidden layers, the probability of a visible input **v** is:

31.3. VARIANTS

$$p(\nu, \psi) = \frac{1}{Z} \sum_h e^{\sum_{ij} W_{ij}^{(1)} \nu_i h_j^1 + \sum_{jl} W_{jl}^{(2)} h_j^1 h_l^2 + \sum_{lm} W_{lm}^{(3)} h_l^2 h_m^3}.$$

where $h = \{h^{(1)}, h^{(2)}, h^{(3)}\}$ is the set of hidden units, and $\psi = \{W^{(1)}, W^{(2)}, W^{(3)}\}$ are the model parameters, representing visible-hidden and hidden-hidden symmetric interaction terms.

A learned DBM model is an undirected model that defines the joint distribution $P(\nu, h^1, h^2, h^3)$. One way to express what has been learned is the conditional model $P(\nu, h^1, h^2|h^3)$ and a prior term $P(h^3)$.

Here $P(\nu, h^1, h^2|h^3)$ represents a conditional DBM model, which can be viewed as a two-layer DBM but with bias terms given by the states of h^3:

$$P(\nu, h^1, h^2|h^3) =$$

$$\frac{1}{Z(\psi, h^3)} e^{\sum_{ij} W_{ij}^{(1)} \nu_i h_j^1 + \sum_{jl} W_{jl}^{(2)} h_j^1 h_l^2 + \sum_{lm} W_{lm}^{(3)} h_l^2 h_m^3}.$$

31.3.12 Deep predictive coding networks

A deep predictive coding network (DPCN) is a predictive coding scheme that uses top-down information to empirically adjust the priors needed for a bottom-up inference procedure by means of a deep, locally connected, generative model. This works by extracting sparse features from time-varying observations using a linear dynamical model. Then, a pooling strategy is used to learn invariant feature representations. These units compose to form a deep architecture and are trained by greedy layer-wise unsupervised learning. The layers constitute a kind of Markov chain such that the states at any layer depend only on the preceding and succeeding layers.

DPCNs predict the representation of the layer, by using a top-down approach using the information in upper layer and temporal dependencies from previous states.[167]

DPCNs can be extended to form a convolutional network.[167]

31.3.13 Networks with separate memory structures

Integrating external memory with ANNs dates to early research in distributed representations[168] and Kohonen's self-organizing maps. For example, in sparse distributed memory or hierarchical temporal memory, the patterns encoded by neural networks are used as addresses for content-addressable memory, with "neurons" essentially serving as address encoders and decoders. However, the early controllers of such memories were not differentiable.

LSTM-related differentiable memory structures

Apart from long short-term memory (LSTM), other approaches also added differentiable memory to recurrent functions. For example:

- Differentiable push and pop actions for alternative memory networks called neural stack machines[169][170]

- Memory networks where the control network's external differentiable storage is in the fast weights of another network[171]

- LSTM forget gates[172]

- Self-referential RNNs with special output units for addressing and rapidly manipulating the RNN's own weights in differentiable fashion (internal storage)[173][174]

- Learning to transduce with unbounded memory[175]

Neural Turing machines   Main article: Neural Turing machine

Neural Turing machines[176] couple LSTM networks to external memory resources, with which they can interact by attentional processes. The combined system is analogous to a Turing machine but is differentiable end-to-end, allowing it to be efficiently trained by gradient descent. Preliminary results demonstrate that neural Turing machines can infer simple algorithms such as copying, sorting and associative recall from input and output examples.

Semantic hashing

Approaches that represent previous experiences directly and use a similar experience to form a local model are often called nearest neighbour or k-nearest neighbors methods.[177] Deep learning is useful in semantic hashing[178] where a deep graphical model the word-count vectors[179] obtained from a large set of documents. Documents are mapped to memory addresses in such a way that semantically similar documents are located at nearby addresses. Documents similar to a query document can then be found by accessing all the addresses that differ by only a few bits from the address of the query document. Unlike sparse distributed memory that operates on 1000-bit addresses, semantic hashing works on 32 or 64-bit addresses found in a conventional computer architecture.

Memory networks

Memory networks[180][181] are another extension to neural networks incorporating long-term memory. The long-term memory can be read and written to, with the goal of using it for prediction. These models have been applied in the context of question answering (QA) where the long-term memory effectively acts as a (dynamic) knowledge base and the output is a textual response.[182]

Pointer networks

Deep neural networks can be potentially improved by deepening and parameter reduction, while maintaining trainability. While training extremely deep (e.g., 1 million layers) neural networks might not be practical, CPU-like architectures such as pointer networks[183] and neural random-access machines[184] overcome this limitation by using external random-access memory and other components that typically belong to a computer architecture such as registers, ALU and pointers. Such systems operate on probability distribution vectors stored in memory cells and registers. Thus, the model is fully differentiable and trains end-to-end. The key characteristic of these models is that their depth, the size of their short-term memory, and the number of parameters can be altered independently — unlike models like LSTM, whose number of parameters grows quadratically with memory size.

Encoder–decoder networks

Encoder–decoder frameworks are based on neural networks that map highly structured input to highly structured output. The approach arose in the context of machine translation,[185][186][187] where the input and output are written sentences in two natural languages. In that work, an LSTM RNN or CNN was used as an encoder to summarize a source sentence, and the summary was decoded using a conditional RNN language model to produce the translation.[188] These systems share building blocks: gated RNNs and CNNs and trained attention mechanisms.

31.4 Multilayer kernel machine

Multilayer kernel machines (MKM) are a way of learning highly nonlinear functions by iterative application of weakly nonlinear kernels. They use the kernel principal component analysis (KPCA),[189] as a method for the unsupervised greedy layer-wise pre-training step of the deep learning architecture.[190]

Layer $l+1$ learns the representation of the previous layer l, extracting the n_l principal component (PC) of the projection layer l output in the feature domain induced by the kernel. For the sake of dimensionality reduction of the updated representation in each layer, a supervised strategy is proposed to select the best informative features among features extracted by KPCA. The process is:

- rank the n_l features according to their mutual information with the class labels;
- for different values of K and $m_l \in \{1, \ldots, n_l\}$, compute the classification error rate of a *K-nearest neighbor (K-NN)* classifier using only the m_l most informative features on a validation set;
- the value of m_l with which the classifier has reached the lowest error rate determines the number of features to retain.

Some drawbacks accompany the KPCA method as the building cells of an MKM.

A more straightforward way to use kernel machines for deep learning was developed for spoken language understanding.[191] The main idea is to use a kernel machine to approximate a shallow neural net with an infinite number of hidden units, then use stacking to splice the output of the kernel machine and the raw input in building the next, higher level of the kernel machine. The number of levels in the deep convex network is a hyper-parameter of the overall system, to be determined by cross validation.

31.5 Use

Using ANNs requires an understanding of their characteristics.

- Choice of model: This depends on the data representation and the application. Overly complex models slow learning.
- Learning algorithm: Numerous trade-offs exist between learning algorithms. Almost any algorithm will work well with the correct hyperparameters for training on a particular data set. However, selecting and tuning an algorithm for training on unseen data requires significant experimentation.
- Robustness: If the model, cost function and learning algorithm are selected appropriately, the resulting ANN can become robust.

ANN capabilities fall within the following broad categories:

- Function approximation, or regression analysis, including time series prediction, fitness approximation and modeling.
- Classification, including pattern and sequence recognition, novelty detection and sequential decision making.
- Data processing, including filtering, clustering, blind source separation and compression.
- Robotics, including directing manipulators and prostheses.
- Control, including computer numerical control.

31.6 Applications

Because of their ability to reproduce and model nonlinear processes, ANNs have found many applications in a wide range of disciplines.

Application areas include system identification and control (vehicle control, trajectory prediction,[192] process control, natural resources management), quantum chemistry,[193] game-playing and decision making (backgammon, chess, poker), pattern recognition (radar systems, face identification, signal classification,[194] object recognition and more), sequence recognition (gesture, speech, handwritten text recognition), medical diagnosis, finance (e.g. automated trading systems), data mining, visualization, machine translation, social network filtering[195] and e-mail spam filtering.

ANNs have been used to diagnose cancers, including lung cancer,[196] prostate cancer, colorectal cancer[197] and to distinguish highly invasive cancer cell lines from less invasive lines using only cell shape information.[198][199]

ANNs have been used for building black-box models in geoscience: hydrology,[200][201] ocean modelling and coastal engineering,[202][203] and geomorphology,[204] are just few examples of this kind.

31.6.1 Neuroscience

Theoretical and computational neuroscience is concerned with the theoretical analysis and the computational modeling of biological neural systems. Since neural systems attempt to reflect cognitive processes and behavior, the field is closely related to cognitive and behavioral modeling.

To gain this understanding, neuroscientists strive to link observed biological processes (data), biologically plausible mechanisms for neural processing and learning (biological neural network models) and theory (statistical learning theory and information theory).

Brain research has repeatedly led to new ANN approaches, such as the use of connections to connect neurons in other layers rather than adjacent neurons in the same layer. Other research explored the use of multiple signal types, or finer control than boolean (on/off) variables. Dynamic neural networks can dynamically form new connections and even new neural units while disabling others.[205]

Types of models

Many types of models are used, defined at different levels of abstraction and modeling different aspects of neural systems. They range from models of the short-term behavior of individual neurons,[206] models of how the dynamics of neural circuitry arise from interactions between individual neurons and finally to models of how behavior can arise from abstract neural modules that represent complete subsystems. These include models of the long-term, and short-term plasticity, of neural systems and their relations to learning and memory from the individual neuron to the system level.

Networks with memory

Integrating external memory components with artificial neural networks dates to early research in distributed representations[168] and self-organizing maps. E.g. in sparse distributed memory the patterns encoded by neural networks are used as memory addresses for content-addressable memory, with "neurons" essentially serving as address encoders and decoders.

More recently deep learning was shown to be useful in semantic hashing[207] where a deep graphical model of the word-count vectors[179] is obtained from a large document set. Documents are mapped to memory addresses in such a way that semantically similar documents are located at nearby addresses. Documents similar to a query document can then be found by simply accessing other nearby addresses.

Memory networks are another extension to neural networks incorporating long-term memory.[180] Long-term memory can be read and written to, with the goal of using it for prediction. These models have been applied in the context of question answering (QA) where the long-term memory effectively acts as a knowledge base, and the output is a textual response.

Neural turing machines (NTM) extend the capabilities of deep neural networks by coupling them to external memory resources, which they can interact with by attentional processes.[176] The combined system is analogous to a Turing Machine but is differentiable end-to-end, allowing it to be efficiently trained with gradient descent. Preliminary results demonstrate that NTMs can infer simple algorithms such as copying, sorting and associative recall from input and output examples.

Differentiable neural computers (DNC) are an NTM extension. They out-performed Neural turing machines, long short-term memory systems and memory networks on sequence-processing tasks.[208][209][210][211][212]

31.7 Theoretical properties

31.7.1 Computational power

The multilayer perceptron is a universal function approximator, as proven by the universal approximation theorem. However, the proof is not constructive regarding the number of neurons required, the network topology, the weights and the learning parameters.

A specific recurrent architecture with rational valued weights (as opposed to full precision real number-valued weights) has the full power of a universal Turing machine,[213] using a finite number of neurons and standard linear connections. Further, the use of irrational values for weights results in a machine with super-Turing power.[214]

31.7.2 Capacity

Models' "capacity" property roughly corresponds to their ability to model any given function. It is related to the amount of information that can be stored in the network and to the notion of complexity.

31.7.3 Convergence

Models may not consistently converge on a single solution, firstly because many local minima may exist, depending on the cost function and the model. Secondly, the optimization method used might not guarantee to converge when it begins far from any local minimum. Thirdly, for sufficiently large data or parameters, some methods become impractical.

31.7.4 Generalization and statistics

Applications whose goal is to create a system that generalizes well to unseen examples, face the possibility of over-training. This arises in convoluted or over-specified systems when the capacity of the network significantly exceeds the needed free parameters. Two approaches address over-training. The first is to use cross-validation and similar techniques to check for the presence of over-training and optimally select hyperparameters to minimize the generalization error. The second is to use some form of *regularization*. This concept emerges in a probabilistic (Bayesian) framework, where regularization can be performed by selecting a larger prior probability over simpler models; but also in statistical learning theory, where the goal is to minimize over two quantities: the 'empirical risk' and the 'structural risk', which roughly corresponds to the error over the training set and the predicted error in unseen data due to overfitting.

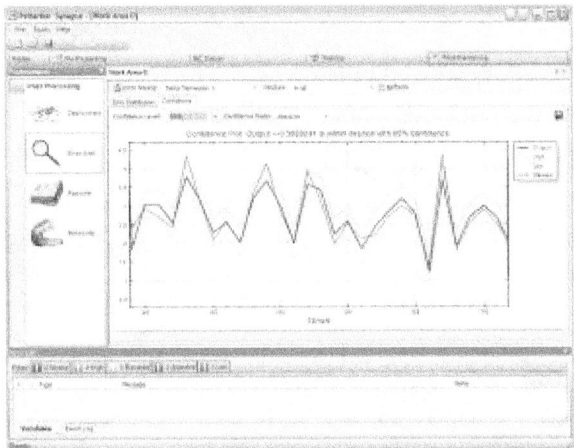

Confidence analysis of a neural network

Supervised neural networks that use a mean squared error (MSE) cost function can use formal statistical methods to determine the confidence of the trained model. The MSE on a validation set can be used as an estimate for variance. This value can then be used to calculate the confidence interval of the output of the network, assuming a normal distribution. A confidence analysis made this way is statistically valid as long as the output probability distribution stays the same and the network is not modified.

By assigning a softmax activation function, a generalization of the logistic function, on the output layer of the neural network (or a softmax component in a component-based neural network) for categorical target variables, the outputs can be interpreted as posterior probabilities. This is very useful in classification as it gives a certainty measure on classifications.

The softmax activation function is:

$$y_i = \frac{e^{x_i}}{\sum_{j=1}^{c} e^{x_j}}$$

31.8 Criticism

31.8.1 Training issues

A common criticism of neural networks, particularly in robotics, is that they require too much training for real-world operation. Potential solutions include randomly shuffling training examples, by using a numerical optimization algorithm that does not take too large steps when changing the network connections following an example and by grouping examples in so-called mini-batches.

31.8.2 Theoretical issues

No neural network has solved such computationally difficult problems such as the n-Queens problem, the travelling salesman problem, or the problem of factoring large integers.

A fundamental objection is that they do not reflect how real neurons function. Back propagation is a critical part of most artificial neural networks, although no such mechanism exists in biological neural networks.[215] How information is coded by real neurons is not known. Sensor neurons fire action potentials more frequently with sensor activation and muscle cells pull more strongly when their associated motor neurons receive action potentials more frequently.[216] Other than the case of relaying information from a sensor neuron to a motor neuron, almost nothing of the principles of how information is handled by biological neural networks is known.

The motivation behind ANNs is not necessarily to strictly replicate neural function, but to use biological neural networks as an inspiration. A central claim of ANNs is therefore that it embodies some new and powerful general principle for processing information. Unfortunately, these general principles are ill-defined. It is often claimed that they are emergent from the network itself. This allows simple statistical association (the basic function of artificial neural networks) to be described as learning or recognition. Alexander Dewdney commented that, as a result, artificial neural networks have a "something-for-nothing quality, one that imparts a peculiar aura of laziness and a distinct lack of curiosity about just how good these computing systems are. No human hand (or mind) intervenes; solutions are found as if by magic; and no one, it seems, has learned anything".[217]

Biological brains use both shallow and deep circuits as reported by brain anatomy,[218] displaying a wide variety of invariance. Weng[219] argued that the brain self-wires largely according to signal statistics and therefore, a serial cascade cannot catch all major statistical dependencies.

31.8.3 Hardware issues

Large and effective neural networks require considerable computing resources.[220] While the brain has hardware tailored to the task of processing signals through a graph of neurons, simulating even a simplified neuron on von Neumann architecture may compel a neural network designer to fill many millions of database rows for its connections – which can consume vast amounts of memory and storage. Furthermore, the designer often needs to transmit signals through many of these connections and their associated neurons – which must often be matched with enormous CPU processing power and time.

Schmidhuber notes that the resurgence of neural networks in the twenty-first century is largely attributable to advances in hardware: from 1991 to 2015, computing power, especially as delivered by GPGPUs (on GPUs), has increased around a million-fold, making the standard backpropagation algorithm feasible for training networks that are several layers deeper than before.[221] The use of parallel GPUs can reduce training times from months to days.[220]

Neuromorphic engineering addresses the hardware difficulty directly, by constructing non-von-Neumann chips to directly implement neural networks in circuitry. Another chip optimized for neural network processing is called a Tensor Processing Unit, or TPU.[222]

31.8.4 Practical counterexamples to criticisms

Arguments against Dewdney's position are that neural networks have been successfully used to solve many complex and diverse tasks, ranging from autonomously flying aircraft[223] to detecting credit card fraud to mastering the game of Go.

Technology writer Roger Bridgman commented:

> Neural networks, for instance, are in the dock not only because they have been hyped to high heaven, (what hasn't?) but also because you could create a successful net without understanding how it worked: the bunch of numbers that captures its behaviour would in all probability be "an opaque, unreadable table...valueless as a scientific resource".
>
> In spite of his emphatic declaration that science is not technology, Dewdney seems here to pillory neural nets as bad science when most of those devising them are just trying to be good engineers. An unreadable table that a useful machine could read would still be well worth having.[224]

Although it is true that analyzing what has been learned by an artificial neural network is difficult, it is much easier to do so than to analyze what has been learned by a biological neural network. Furthermore, researchers involved in exploring learning algorithms for neural networks are gradually uncovering general principles that allow a learning machine to be successful. For example, local vs non-local learning and shallow vs deep architecture.[225]

31.8.5 Hybrid approaches

Advocates of hybrid models (combining neural networks and symbolic approaches), claim that such a mixture can better capture the mechanisms of the human mind.[226][227]

31.9 Types

Main article: Types of artificial neural networks

Artificial neural networks have many variations. The simplest, static types have one or more static components, including number of units, number of layers, unit weights and topology. Dynamic types allow one or more of these to change during the learning process. The latter are much more complicated, but can shorten learning periods and produce better results. Some types allow/require learning to be "supervised" by the operator, while others operate independently. Some types operate purely in hardware, while others are purely software and run on general purpose computers.

31.10 Gallery

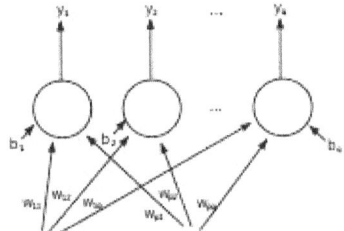

- A single-layer feedforward artificial neural network. Arrows originating from x_2 are omitted for clarity. There are p inputs to this network and q outputs. In this system, the value of the qth output, y_q would be calculated as $y_q = K * (\sum (x_i * w_{iq}) - b_q)$

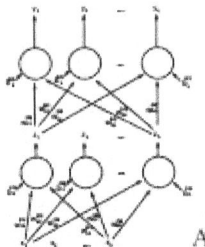

- A two-layer feedforward artificial neural network.

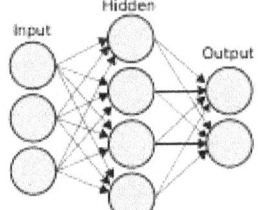

- An artificial neural network.

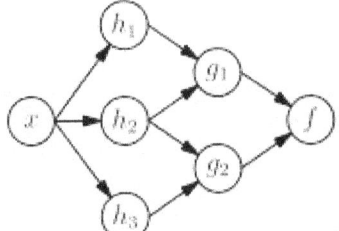

- An ANN dependency graph.

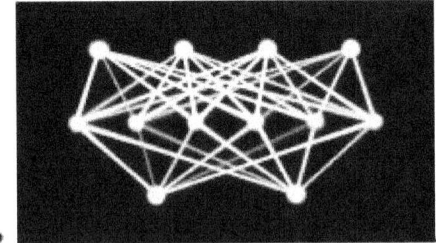

- A single-layer feedforward artificial neural network with 4 inputs, 6 hidden and 2 outputs. Given position state and direction outputs wheel based control values.

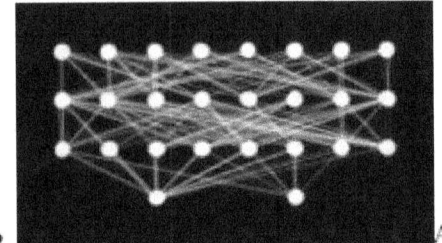

- A two-layer feedforward artificial neural network with 8 inputs, 2x8 hidden and 2 outputs. Given position state, direction and other environment values outputs thruster based control values.

31.11 See also

- Hierarchical temporal memory
- 20Q
- ADALINE
- Adaptive resonance theory
- Artificial life
- Associative memory
- Autoencoder
- BEAM robotics
- Biological cybernetics
- Biologically inspired computing
- Blue Brain Project
- Catastrophic interference
- Cerebellar Model Articulation Controller
- Cognitive architecture
- Cognitive science
- Convolutional neural network (CNN)
- Connectionist expert system
- Connectomics
- Cultured neuronal networks
- Deep learning
- Digital morphogenesis
- Encog
- Fuzzy logic
- Gene expression programming
- Genetic algorithm
- Genetic programming
- Group method of data handling
- Habituation
- In Situ Adaptive Tabulation
- Machine learning concepts
- Models of neural computation
- Multilinear subspace learning
- Neuroevolution
- Neural coding
- Neural gas
- Neural machine translation
- Neural network software
- Neuroscience
- Ni1000 chip
- Nonlinear system identification
- Optical neural network
- Parallel Constraint Satisfaction Processes
- Parallel distributed processing
- Radial basis function network
- Recurrent neural networks
- Self-organizing map
- Spiking neural network
- Systolic array
- Tensor product network
- Time delay neural network (TDNN)

31.12 References

[1] McCulloch, Warren; Walter Pitts (1943). "A Logical Calculus of Ideas Immanent in Nervous Activity". *Bulletin of Mathematical Biophysics*. **5** (4): 115–133. doi:10.1007/BF02478259.

[2] Kleene, S.C. (1956). "Representation of Events in Nerve Nets and Finite Automata". *Annals of Mathematics Studies* (34). Princeton University Press. pp. 3–41. Retrieved 2017-06-17.

[3] Hebb, Donald (1949). *The Organization of Behavior*. New York: Wiley. ISBN 978-1-135-63190-1.

[4] Farley, B.G.; W.A. Clark (1954). "Simulation of Self-Organizing Systems by Digital Computer". *IRE Transactions on Information Theory*. **4** (4): 76–84. doi:10.1109/TIT.1954.1057468.

[5] Rochester, N.; J.H. Holland; L.H. Habit; W.L. Duda (1956). "Tests on a cell assembly theory of the action of the brain, using a large digital computer". *IRE Transactions on Information Theory*. **2** (3): 80–93. doi:10.1109/TIT.1956.1056810.

[6] Rosenblatt, F. (1958). "The Perceptron: A Probabilistic Model For Information Storage And Organization In The Brain". *Psychological Review*. **65** (6): 386–408. CiteSeerX 10.1.1.588.3775. PMID 13602029. doi:10.1037/h0042519.

[7] Werbos, P.J. (1975). *Beyond Regression: New Tools for Prediction and Analysis in the Behavioral Sciences*.

[8] David H. Hubel and Torsten N. Wiesel (2005). *Brain and visual perception: the story of a 25-year collaboration*. Oxford University Press US. p. 106. ISBN 978-0-19-517618-6.

[9] Schmidhuber, J. (2015). "Deep Learning in Neural Networks: An Overview". *Neural Networks*. **61**: 85–117. arXiv:1404.7828. doi:10.1016/j.neunet.2014.09.003.

[10] Ivakhnenko, A. G. (1973). *Cybernetic Predicting Devices*. CCM Information Corporation.

[11] Ivakhnenko, A. G.; Grigor'evich Lapa, Valentin (1967). *Cybernetics and forecasting techniques*. American Elsevier Pub. Co.

[12] Minsky, Marvin; Papert, Seymour (1969). *Perceptrons: An Introduction to Computational Geometry*. MIT Press. ISBN 0-262-63022-2.

[13] Rumelhart, D.E; McClelland, James (1986). *Parallel Distributed Processing: Explorations in the Microstructure of Cognition*. Cambridge: MIT Press. ISBN 978-0-262-63110-5.

[14] J. Weng, N. Ahuja and T. S. Huang, "Cresceptron: a self-organizing neural network which grows adaptively," *Proc. International Joint Conference on Neural Networks*, Baltimore, Maryland, vol I, pp. 576-581, June, 1992.

[15] J. Weng, N. Ahuja and T. S. Huang, "Learning recognition and segmentation of 3-D objects from 2-D images," *Proc. 4th International Conf. Computer Vision*, Berlin, Germany, pp. 121-128, May, 1993.

[16] J. Weng, N. Ahuja and T. S. Huang, "Learning recognition and segmentation using the Cresceptron," *International Journal of Computer Vision*, vol. 25, no. 2, pp. 105-139, Nov. 1997.

[17] S. Hochreiter., "Untersuchungen zu dynamischen neuronalen Netzen," *Diploma thesis. Institut f. Informatik, Technische Univ. Munich. Advisor: J. Schmidhuber*, 1991.

[18] Hochreiter, S.; et al. (15 January 2001). "Gradient flow in recurrent nets: the difficulty of learning long-term dependencies". In Kolen, John F.; Kremer, Stefan C. *A Field Guide to Dynamical Recurrent Networks*. John Wiley & Sons. ISBN 978-0-7803-5369-5.

[19] J. Schmidhuber., "Learning complex, extended sequences using the principle of history compression," *Neural Computation*, 4, pp. 234–242, 1992.

[20] Sven Behnke (2003). *Hierarchical Neural Networks for Image Interpretation*. (PDF). Lecture Notes in Computer Science. **2766**. Springer.

[21] Smolensky, P. (1986). "Information processing in dynamical systems: Foundations of harmony theory.". In D. E. Rumelhart, J. L. McClelland, & the PDP Research Group. *Parallel Distributed Processing: Explorations in the Microstructure of Cognition*. **1**. pp. 194–281.

[22] Hinton, G. E.; Osindero, S.; Teh, Y. (2006). "A fast learning algorithm for deep belief nets" (PDF). *Neural Computation*. **18** (7): 1527–1554. PMID 16764513. doi:10.1162/neco.2006.18.7.1527.

[23] Hinton, G. (2009). "Deep belief networks". *Scholarpedia*. **4** (5): 5947. doi:10.4249/scholarpedia.5947.

[24] Ng, Andrew; Dean, Jeff (2012). "Building High-level Features Using Large Scale Unsupervised Learning". arXiv:1112.6209.

[25] Yang, J. J.; Pickett, M. D.; Li, X. M.; Ohlberg, D. A. A.; Stewart, D. R.; Williams, R. S. (2008). "Memristive switching mechanism for metal/oxide/metal nanodevices". *Nat. Nanotechnol*. **3**: 429–433. doi:10.1038/nnano.2008.160.

[26] Strukov, D. B.; Snider, G. S.; Stewart, D. R.; Williams, R. S. (2008). "The missing memristor found". *Nature*. **453**: 80–83. PMID 18451858. doi:10.1038/nature06932.

[27] Cireşan, Dan Claudiu; Meier, Ueli; Gambardella, Luca Maria; Schmidhuber, Jürgen (2010-09-21). "Deep, Big, Simple Neural Nets for Handwritten Digit Recognition". *Neural Computation*. **22** (12): 3207–3220. ISSN 0899-7667. doi:10.1162/neco_a_00052.

[28] 2012 Kurzweil AI Interview with Jürgen Schmidhuber on the eight competitions won by his Deep Learning team 2009–2012

[29] "How bio-inspired deep learning keeps winning competitions | KurzweilAI". *www.kurzweilai.net*. Retrieved 2017-06-16.

[30] Graves, Alex; and Schmidhuber, Jürgen; *Offline Handwriting Recognition with Multidimensional Recurrent Neural Networks*, in Bengio, Yoshua; Schuurmans, Dale; Lafferty, John; Williams, Chris K. I.; and Culotta, Aron (eds.), *Advances in Neural Information Processing Systems 22 (NIPS'22), 7–10 December 2009, Vancouver, BC*, Neural Information Processing Systems (NIPS) Foundation, 2009, pp. 545–552.

[31] Graves, A.; Liwicki, M.; Fernandez, S.; Bertolami, R.; Bunke, H.; Schmidhuber, J. (2009). "A Novel Connectionist System for Improved Unconstrained Handwriting Recognition" (PDF). *IEEE Transactions on Pattern Analysis and Machine Intelligence*. **31** (5).

[32] Graves, Alex; Schmidhuber, Jürgen (2009). Bengio, Yoshua; Schuurmans, Dale; Lafferty, John; Williams, Chris editor-K. I.; Culotta, Aron, eds. "Offline Handwriting Recognition with Multidimensional Recurrent Neural Networks". *Neural Information Processing Systems (NIPS) Foundation*: 545–552.

[33] Graves, A.; Liwicki, M.; Fernández, S.; Bertolami, R.; Bunke, H.; Schmidhuber, J. (May 2009). "A Novel Connectionist System for Unconstrained Handwriting Recognition". *IEEE Transactions on Pattern Analysis and Machine Intelligence*. **31** (5): 855–868. ISSN 0162-8828. doi:10.1109/tpami.2008.137.

[34] Graves, A.; Liwicki, M.; Fernandez, S.; Bertolami, R.; Bunke, H.; Schmidhuber, J. (2009). "A Novel Connectionist System for Improved Unconstrained Handwriting Recognition". *IEEE Transactions on Pattern Analysis and Machine Intelligence*. **31** (5): 855–868. doi:10.1109/tpami.2008.137.

[35] Cireşan, Dan; Meier, Ueli; Masci, Jonathan; Schmidhuber, Jürgen (August 2012). "Multi-column deep neural network for traffic sign classification". *Neural Networks*. Selected Papers from IJCNN 2011. **32**: 333–338. doi:10.1016/j.neunet.2012.02.023.

[36] Ciresan, Dan; Giusti, Alessandro; Gambardella, Luca M.; Schmidhuber, Juergen (2012). Pereira, F.; Burges, C. J. C.; Bottou, L.; Weinberger, K. Q., eds. *Advances in Neural Information Processing Systems 25* (PDF). Curran Associates, Inc. pp. 2843–2851.

[37] Ciresan, Dan; Meier, U.; Schmidhuber, J. (June 2012). "Multi-column deep neural networks for image classification". *2012 IEEE Conference on Computer Vision and Pattern Recognition*: 3642–3649. doi:10.1109/cvpr.2012.6248110.

[38] Ciresan, D. C.; Meier, U.; Masci, J.; Gambardella, L. M.; Schmidhuber, J. (2011). "Flexible, High Performance Convolutional Neural Networks for Image Classification" (PDF). *International Joint Conference on Artificial Intelligence*. doi:10.5591/978-1-57735-516-8/ijcai11-210.

[39] Cireşan, Dan; Giusti, Alessandro; Gambardella, Luca M.; Schmidhuber, Juergen (2012). Pereira, F.; Burges, C. J. C.; Bottou, L.; Weinberger, K. Q., eds. *Advances in Neural Information Processing Systems 25* (PDF). Curran Associates, Inc. pp. 2843–2851.

[40] Krizhevsky, Alex; Sutskever, Ilya; Hinton, Geoffry (2012). "ImageNet Classification with Deep Convolutional Neural Networks" (PDF). *NIPS 2012: Neural Information Processing Systems, Lake Tahoe, Nevada*.

[41] Fukushima, K. (1980). "Neocognitron: A self-organizing neural network model for a mechanism of pattern recognition unaffected by shift in position". *Biological Cybernetics*. **36** (4): 93–202. PMID 7370364. doi:10.1007/BF00344251.

[42] Riesenhuber, M; Poggio, T (1999). "Hierarchical models of object recognition in cortex". *Nature Neuroscience*. **2** (11): 1019–1025. doi:10.1038/14819.

[43] Hinton, Geoffrey (2009-05-31). "Deep belief networks". *Scholarpedia*. **4** (5). ISSN 1941-6016. doi:10.4249/scholarpedia.5947.

[44] Hinton, G. E.; Osindero, S.; Teh, Y. W. (2006). "A Fast Learning Algorithm for Deep Belief Nets" (PDF). *Neural Computation*. **18** (7): 1527–1554. PMID 16764513. doi:10.1162/neco.2006.18.7.1527.

[45] Hinton, G. E.; Osindero, S.; Teh, Y. (2006). "A fast learning algorithm for deep belief nets" (PDF). *Neural Computation*. **18** (7): 1527–1554. PMID 16764513. doi:10.1162/neco.2006.18.7.1527.

[46] Markoff, John (November 23, 2012). "Scientists See Promise in Deep-Learning Programs". *New York Times*.

[47] Martines, H.; Bengio, Y.; Yannakakis, G. N. (2013). "Learning Deep Physiological Models of Affect". *IEEE Computational Intelligence*. **8** (2): 20–33. doi:10.1109/mci.2013.2247823.

[48] J. Weng, "Why Have We Passed 'Neural Networks Do not Abstract Well'?," *Natural Intelligence: the INNS Magazine*, vol. 1, no.1, pp. 13-22, 2011.

[49] Z. Ji, J. Weng, and D. Prokhorov, "Where-What Network 1: Where and What Assist Each Other Through Top-down Connections," *Proc. 7th International Conference on Development and Learning (ICDL'08)*, Monterey, CA, Aug. 9-12, pp. 1-6, 2008.

[50] X. Wu, G. Guo, and J. Weng, "Skull-closed Autonomous Development: WWN-7 Dealing with Scales," *Proc. International Conference on Brain-Mind*, July 27–28, East Lansing, Michigan, pp. +1-9, 2013.

[51] Zell, Andreas (1994). "chapter 5.2". *Simulation Neuronaler Netze [Simulation of Neural Networks]* (in German) (1st ed.). Addison-Wesley. ISBN 3-89319-554-8.

[52] "The Machine Learning Dictionary".

[53] Schmidhuber, Jürgen (2015). "Deep Learning". *Scholarpedia*. **10** (11): 32832. doi:10.4249/scholarpedia.32832.

[54] Dreyfus, Stuart E. (1990-09-01). "Artificial neural networks, back propagation, and the Kelley-Bryson gradient procedure". *Journal of Guidance, Control, and Dynamics*. **13** (5): 926–928. ISSN 0731-5090. doi:10.2514/3.25422.

[55] Eiji Mizutani, Stuart Dreyfus, Kenichi Nishio (2000). On derivation of MLP backpropagation from the Kelley-Bryson optimal-control gradient formula and its application. Proceedings of the IEEE International Joint Conference on Neural Networks (IJCNN 2000), Como Italy, July 2000. Online

[56] Kelley, Henry J. (1960). "Gradient theory of optimal flight paths". *Ars Journal*. **30** (10): 947–954. doi:10.2514/8.5282.

[57] Arthur E. Bryson (1961, April). A gradient method for optimizing multi-stage allocation processes. In Proceedings of the Harvard Univ. Symposium on digital computers and their applications.

[58] Dreyfus, Stuart (1962). "The numerical solution of variational problems". *Journal of Mathematical Analysis and Applications*. **5** (1): 30–45. doi:10.1016/0022-247x(62)90004-5.

[59] Russell, Stuart J.; Norvig, Peter (2010). *Artificial Intelligence A Modern Approach*. Prentice Hall. p. 578. ISBN 978-0-13-604259-4. The most popular method for learning in multilayer networks is called Back-propagation.

[60] Bryson, Arthur Earl (1969). *Applied Optimal Control: Optimization, Estimation and Control*. Blaisdell Publishing Company or Xerox College Publishing. p. 481.

[61] Seppo Linnainmaa (1970). The representation of the cumulative rounding error of an algorithm as a Taylor expansion of the local rounding errors. Master's Thesis (in Finnish), Univ. Helsinki, 6-7.

[62] Linnainmaa, Seppo (1976). "Taylor expansion of the accumulated rounding error". *BIT Numerical Mathematics*. **16** (2): 146–160. doi:10.1007/bf01931367.

[63] Griewank, Andreas (2012). "Who Invented the Reverse Mode of Differentiation?" (PDF). *Documenta Matematica, Extra Volume ISMP*: 389–400.

[64] Griewank, Andreas; Walther, Andrea (2008). *Evaluating Derivatives: Principles and Techniques of Algorithmic Differentiation, Second Edition*. SIAM. ISBN 978-0-89871-776-1.

[65] Dreyfus, Stuart (1973). "The computational solution of optimal control problems with time lag". *IEEE Transactions on Automatic Control*. **18** (4): 383–385. doi:10.1109/tac.1973.1100330.

[66] Paul Werbos (1974). Beyond regression: New tools for prediction and analysis in the behavioral sciences. PhD thesis, Harvard University.

[67] Werbos, Paul (1982). "Applications of advances in nonlinear sensitivity analysis". *System modeling and optimization* (PDF). Springer. pp. 762–770.

[68] Rumelhart, David E.; Hinton, Geoffrey E.; Williams, Ronald J. "Learning representations by back-propagating errors". *Nature*. **323** (6088): 533–536. doi:10.1038/323533a0.

[69] Eric A. Wan (1993). Time series prediction by using a connectionist network with internal delay lines. In SANTA FE INSTITUTE STUDIES IN THE SCIENCES OF COMPLEXITY-PROCEEDINGS (Vol. 15, pp. 195-195). Addison-Wesley Publishing Co.

[70] Hinton, G.; Deng, L.; Yu, D.; Dahl, G. E.; Mohamed, A. r; Jaitly, N.; Senior, A.; Vanhoucke, V.; Nguyen, P. (November 2012). "Deep Neural Networks for Acoustic Modeling in Speech Recognition: The Shared Views of Four Research Groups". *IEEE Signal Processing Magazine*. **29** (6): 82–97. ISSN 1053-5888. doi:10.1109/msp.2012.2205597.

[71] Huang, Guang-Bin; Zhu, Qin-Yu; Siew, Chee-Kheong (2006). "Extreme learning machine: theory and applications". *Neurocomputing*. **70** (1): 489–501. doi:10.1016/j.neucom.2005.12.126.

[72] Widrow, Bernard; et al. (2013). "The no-prop algorithm: A new learning algorithm for multilayer neural networks". *Neural Networks*. **37**: 182–188. doi:10.1016/j.neunet.2012.09.020.

[73] Ollivier, Yann; Charpiat, Guillaume (2015). "Training recurrent networks without backtracking". arXiv:1507.07680.

[74] ESANN. 2009

[75] Hinton, G. E. (2010). "A Practical Guide to Training Restricted Boltzmann Machines". *Tech. Rep. UTML TR 2010-003*.

[76] Ojha, Varun Kumar; Abraham, Ajith; Snášel, Václav (2017-04-01). "Metaheuristic design of feedforward neural networks: A review of two decades of research". *Engineering Applications of Artificial Intelligence*. **60**: 97–116. doi:10.1016/j.engappai.2017.01.013.

[77] Dominic, S.; Das, R.; Whitley, D.; Anderson, C. (July 1991). "Genetic reinforcement learning for neural networks". *IJCNN-91-Seattle International Joint Conference on Neural Networks*. IJCNN-91-Seattle International Joint Conference on Neural Networks. Seattle, Washington, USA: IEEE. ISBN 0-7803-0164-1. doi:10.1109/IJCNN.1991.155315. Retrieved 29 July 2012.

[78] Hoskins, J.C.; Himmelblau, D.M. (1992). "Process control via artificial neural networks and reinforcement learning". *Computers & Chemical Engineering*. **16** (4): 241–251. doi:10.1016/0098-1354(92)80045-B.

[79] Bertsekas, D.P.; Tsitsiklis, J.N. (1996). *Neuro-dynamic programming*. Athena Scientific. p. 512. ISBN 1-886529-10-8.

[80] Secomandi, Nicola (2000). "Comparing neuro-dynamic programming algorithms for the vehicle routing problem with stochastic demands". *Computers & Operations Research*. **27** (11–12): 1201–1225. doi:10.1016/S0305-0548(99)00146-X.

[81] de Rigo, D.; Rizzoli, A. E.; Soncini-Sessa, R.; Weber, E.; Zenesi, P. (2001). "Neuro-dynamic programming for the efficient management of reservoir networks" (PDF). *Proceedings of MODSIM 2001, International Congress on Modelling and Simulation*. MODSIM 2001, International Congress

31.12. REFERENCES

on Modelling and Simulation. Canberra, Australia: Modelling and Simulation Society of Australia and New Zealand. ISBN 0-867405252. doi:10.5281/zenodo.7481. Retrieved 29 July 2012.

[82] Damas, M.; Salmeron, M.; Diaz, A.; Ortega, J.; Prieto, A.; Olivares, G. (2000). "Genetic algorithms and neurodynamic programming: application to water supply networks". *Proceedings of 2000 Congress on Evolutionary Computation*. 2000 Congress on Evolutionary Computation. La Jolla, California, USA: IEEE. ISBN 0-7803-6375-2. doi:10.1109/CEC.2000.870269. Retrieved 29 July 2012.

[83] Deng, Geng; Ferris, M.C. (2008). "Neuro-dynamic programming for fractionated radiotherapy planning". *Springer Optimization and Its Applications*. **12**: 47–70. CiteSeerX 10.1.1.137.8288 . doi:10.1007/978-0-387-73299-2_3.

[84] M. Forouzanfar; H. R. Dajani; V. Z. Groza; M. Bolic & S. Rajan (July 2010). *Comparison of Feed-Forward Neural Network Training Algorithms for Oscillometric Blood Pressure Estimation* (PDF). 4th Int. Workshop Soft Computing Applications. Arad, Romania: IEEE.

[85] de Rigo, D., Castelletti, A., Rizzoli, A.E., Soncini-Sessa, R., Weber, E. (January 2005). "A selective improvement technique for fastening Neuro-Dynamic Programming in Water Resources Network Management". In Pavel Zítek. *Proceedings of the 16th IFAC World Congress – IFAC-PapersOnLine*. 16th IFAC World Congress. **16**. Prague, Czech Republic: IFAC. ISBN 978-3-902661-75-3. doi:10.3182/20050703-6-CZ-1902.02172. Retrieved 30 December 2011.

[86] Ferreira, C. (2006). "Designing Neural Networks Using Gene Expression Programming" (PDF). In A. Abraham, B. de Baets, M. Köppen, and B. Nickolay, eds., Applied Soft Computing Technologies: The Challenge of Complexity, pages 517–536, Springer-Verlag.

[87] Da, Y.; Xiurun, G. (July 2005). T. Villmann, ed. *An improved PSO-based ANN with simulated annealing technique*. New Aspects in Neurocomputing: 11th European Symposium on Artificial Neural Networks. Elsevier. doi:10.1016/j.neucom.2004.07.002.

[88] Wu, J.; Chen, E. (May 2009). Wang, H., Shen, Y., Huang, T., Zeng, Z., eds. *A Novel Nonparametric Regression Ensemble for Rainfall Forecasting Using Particle Swarm Optimization Technique Coupled with Artificial Neural Network*. 6th International Symposium on Neural Networks, ISNN 2009. Springer. ISBN 978-3-642-01215-0. doi:10.1007/978-3-642-01513-7-6.

[89] Grigorevich Ivakhnenko, Alexey (1968). "The group method of data handling – a rival of the method of stochastic approximation". *Soviet Automatic Control*. **13** (3): 43–55.

[90] Ivakhnenko, Alexey (1971). "Polynomial theory of complex systems". *IEEE Transactions on Systems, Man and Cybernetics (4)*: 364–378. doi:10.1109/TSMC.1971.4308320.

[91] Kondo, T.; Ueno, J. (2008). "Multi-layered GMDH-type neural network self-selecting optimum neural network architecture and its application to 3-dimensional medical image recognition of blood vessels". *International Journal of Innovative Computing, Information and Control*. **4** (1): 175–187.

[92] Fukushima, K. (1980). "Neocognitron: A self-organizing neural network model for a mechanism of pattern recognition unaffected by shift in position". *Biol. Cybern.* **36**: 193–202. PMID 7370364. doi:10.1007/bf00344251.

[93] LeCun et al., "Backpropagation Applied to Handwritten Zip Code Recognition," *Neural Computation*, 1, pp. 541–551, 1989.

[94] Yann LeCun (2016). Slides on Deep Learning Online

[95] "Unsupervised Feature Learning and Deep Learning Tutorial".

[96] Szegedy, Christian; Liu, Wei; Jia, Yangqing; Sermanet, Pierre; Reed, Scott; Anguelov, Dragomir; Erhan, Dumitru; Vanhoucke, Vincent; Rabinovich, Andrew (2014). "Going Deeper with Convolutions". *Computing Research Repository*. arXiv:1409.4842 . doi:10.1109/CVPR.2015.7298594.

[97] Hochreiter, Sepp; Schmidhuber, Jürgen (1997-11-01). "Long Short-Term Memory". *Neural Computation*. **9** (8): 1735–1780. ISSN 0899-7667. doi:10.1162/neco.1997.9.8.1735.

[98] "Learning Precise Timing with LSTM Recurrent Networks (PDF Download Available)". *ResearchGate*. pp. 115–143. Retrieved 2017-06-13.

[99] Bayer, Justin; Wierstra, Daan; Togelius, Julian; Schmidhuber, Jürgen (2009-09-14). "Evolving Memory Cell Structures for Sequence Learning". *Artificial Neural Networks – ICANN 2009*. Springer, Berlin, Heidelberg: 755–764. doi:10.1007/978-3-642-04277-5_76.

[100] Fernández, Santiago; Graves, Alex; Schmidhuber, Jürgen (2007). "Sequence labelling in structured domains with hierarchical recurrent neural networks". *In Proc. 20th Int. Joint Conf. on Artificial InTelligence, Ijcai 2007*: 774–779.

[101] Graves, Alex; Fernández, Santiago; Gomez, Faustino (2006). "Connectionist temporal classification: Labelling unsegmented sequence data with recurrent neural networks". *In Proceedings of the International Conference on Machine Learning, ICML 2006*: 369–376.

[102] Graves, Alex; Eck, Douglas; Beringer, Nicole; Schmidhuber, Jürgen (2003). "Biologically Plausible Speech Recognition with LSTM Neural Nets" (PDF). *1st Intl. Workshop on Biologically Inspired Approaches to Advanced Information Technology, Bio-ADIT 2004, Lausanne, Switzerland*. pp. 175–184.

[103] Fernández, Santiago; Graves, Alex; Schmidhuber, Jürgen (2007). "An Application of Recurrent Neural Networks to Discriminative Keyword Spotting". *Proceedings of the 17th International Conference on Artificial Neural Networks*. ICANN'07. Berlin, Heidelberg: Springer-Verlag: 220–229. ISBN 3540746935.

[104] Hannun, Awni; Case, Carl; Casper, Jared; Catanzaro, Bryan; Diamos, Greg; Elsen, Erich; Prenger, Ryan; Satheesh, Sanjeev; Sengupta, Shubho (2014-12-17). "Deep Speech: Scaling up end-to-end speech recognition". *arXiv:1412.5567 [cs]*.

[105] Sak, Hasim; Senior, Andrew; Beaufays, Francoise (2014). "Long Short-Term Memory recurrent neural network architectures for large scale acoustic modeling" (PDF).

[106] Li, Xiangang; Wu, Xihong (2014-10-15). "Constructing Long Short-Term Memory based Deep Recurrent Neural Networks for Large Vocabulary Speech Recognition". *arXiv:1410.4281 [cs]*.

[107] Fan, Y.; Qian, Y.; Xie, F.; Soong, F. K. (2014). "TTS synthesis with bidirectional LSTM based Recurrent Neural Networks". *ResearchGate*. Retrieved 2017-06-13.

[108] Zen, Heiga; Sak, Hasim (2015). "Unidirectional Long Short-Term Memory Recurrent Neural Network with Recurrent Output Layer for Low-Latency Speech Synthesis" (PDF). *Google.com*. ICASSP. pp. 4470–4474.

[109] Fan, Bo; Wang, Lijuan; Soong, Frank K.; Xie, Lei (2015). "Photo-Real Talking Head with Deep Bidirectional LSTM" (PDF). *Proceedings of ICASSP*.

[110] Sak, Haşim; Senior, Andrew; Rao, Kanishka; Beaufays, Françoise; Schalkwyk, Johan (September 2015). "Google voice search: faster and more accurate".

[111] Gers, Felix A.; Schmidhuber, Jürgen (2001). "LSTM Recurrent Networks Learn Simple Context Free and Context Sensitive Languages". *IEEE TNN*. **12** (6): 1333–1340. doi:10.1109/72.963769.

[112] Sutskever, L.; Vinyals, O.; Le, Q. (2014). "Sequence to Sequence Learning with Neural Networks" (PDF). *Proc. NIPS*.

[113] Jozefowicz, Rafal; Vinyals, Oriol; Schuster, Mike; Shazeer, Noam; Wu, Yonghui (2016-02-07). "Exploring the Limits of Language Modeling". *arXiv:1602.02410 [cs]*.

[114] Gillick, Dan; Brunk, Cliff; Vinyals, Oriol; Subramanya, Amarnag (2015-11-30). "Multilingual Language Processing From Bytes". *arXiv:1512.00103 [cs]*.

[115] Vinyals, Oriol; Toshev, Alexander; Bengio, Samy; Erhan, Dumitru (2014-11-17). "Show and Tell: A Neural Image Caption Generator". *arXiv:1411.4555 [cs]*.

[116] Gallicchio, Claudio; Micheli, Alessio; Pedrelli, Luca. "Deep reservoir computing: A critical experimental analysis". *Neurocomputing*. doi:10.1016/j.neucom.2016.12.089.

[117] Gallicchio, Claudio; Micheli, Alessio. "Echo State Property of Deep Reservoir Computing Networks". *Cognitive Computation*. **9** (3): 337–350. ISSN 1866-9956. doi:10.1007/s12559-017-9461-9.

[118] Hinton, G.E. "Deep belief networks". *Scholarpedia*. **4** (5): 5947. doi:10.4249/scholarpedia.5947.

[119] Larochelle, Hugo; Erhan, Dumitru; Courville, Aaron; Bergstra, James; Bengio, Yoshua (2007). "An Empirical Evaluation of Deep Architectures on Problems with Many Factors of Variation". *Proceedings of the 24th International Conference on Machine Learning*. ICML '07. New York, NY, USA: ACM: 473–480. ISBN 9781595937933. doi:10.1145/1273496.1273556.

[120] Graupe, Daniel (2013). *Principles of Artificial Neural Networks*. World Scientific. pp. 1–. ISBN 978-981-4522-74-8.

[121] A US 5920852 A D. Graupe," Large memory storage and retrieval (LAMSTAR) network, April 1996

[122]

[123] Nigam, Vivek Prakash; Graupe, Daniel (2004-01-01). "A neural-network-based detection of epilepsy". *Neurological Research*. **26** (1): 55–60. ISSN 0161-6412. PMID 14977058. doi:10.1179/016164104773026534.

[124] Waxman, Jonathan A.; Graupe, Daniel; Carley, David W. (2010-04-01). "Automated Prediction of Apnea and Hypopnea, Using a LAMSTAR Artificial Neural Network". *American Journal of Respiratory and Critical Care Medicine*. **181** (7): 727–733. ISSN 1073-449X. doi:10.1164/rccm.200907-1146oc.

[125] Graupe, D.; Graupe, M. H.; Zhong, Y.; Jackson, R. K. (2008). "Blind adaptive filtering for non-invasive extraction of the fetal electrocardiogram and its non-stationarities". *Proc. Inst. Mech Eng., UK, Part H: Journal of Engineering in Medicine*. **222** (8): 1221–1234. doi:10.1243/09544119jeim417.

[126] Graupe 2013, pp. 240–253

[127] Graupe, D.; Abon, J. (2002). "A Neural Network for Blind Adaptive Filtering of Unknown Noise from Speech". *Intelligent Engineering Systems Through Artificial Neural Networks*. Technische Informationsbibliothek (TIB). **12**: 683–688. Retrieved 2017-06-14.

[128] D. Graupe, "Principles of Artificial Neural Networks.3rd Edition", World Scientific Publishers", 2013, pp.253-274.

[129] Girado, J. I.; Sandin, D. J.; DeFanti, T. A. (2003). "Realtime camera-based face detection using a modified LAMSTAR neural network system". *Proc. SPIE 5015, Applications of Artificial Neural Networks in Image Processing VIII*. doi:10.1117/12.477405.

[130] Venkatachalam, V; Selvan, S. (2007). "Intrusion Detection using an Improved Competitive Learning Lamstar Network". *International Journal of Computer Science and Network Security*. **7** (2): 255–263.

[131] Graupe, D.; Smollack, M. (2007). "Control of unstable nonlinear and nonstationary systems using LAMSTAR neural networks". *ResearchGate*. Proceedings of 10th IASTED on Intelligent Control. Sect.592., pp. 141–144. Retrieved 2017-06-14.

[132] Graupe, Daniel (7 July 2016). *Deep Learning Neural Networks: Design and Case Studies*. World Scientific Publishing Co Inc. pp. 57–110. ISBN 978-981-314-647-1.

[133] Graupe, D.; Kordylewski, H. (August 1996). "Network based on SOM (Self-Organizing-Map) modules combined with statistical decision tools". *Proceedings of the 39th Midwest Symposium on Circuits and Systems*. **1**: 471–474 vol.1. doi:10.1109/mwscas.1996.594203.

[134] Graupe, D.; Kordylewski, H. (1998-03-01). "A Large Memory Storage and Retrieval Neural Network for Adaptive Retrieval and Diagnosis". *International Journal of Software Engineering and Knowledge Engineering*. **08** (01): 115–138. ISSN 0218-1940. doi:10.1142/s0218194098000091.

[135] Kordylewski, H.; Graupe, D; Liu, K. "A novel large-memory neural network as an aid in medical diagnosis applications". *IEEE Transactions on Information Technology in Biomedicine*. **5** (3): 202–209. doi:10.1109/4233.945291.

[136] Schneider, N.C.; Graupe (2008). "A modified LAMSTAR neural network and its applications". *International journal of neural systems*. **18** (4): 331–337. doi:10.1142/s0129065708001634.

[137] Graupe 2013, p. 217

[138] Vincent, Pascal; Larochelle, Hugo; Lajoie, Isabelle; Bengio, Yoshua; Manzagol, Pierre-Antoine (2010). "Stacked Denoising Autoencoders: Learning Useful Representations in a Deep Network with a Local Denoising Criterion". *The Journal of Machine Learning Research*. **11**: 3371–3408.

[139] Ballard, Dana H. (1987). "Modular learning in neural networks" (PDF). *Proceedings of AAAI*. pp. 279–284.

[140] Deng, Li; Yu, Dong; Platt, John (2012). "Scalable stacking and learning for building deep architectures" (PDF). *2012 IEEE International Conference on Acoustics, Speech and Signal Processing (ICASSP)*: 2133–2136.

[141] Deng, Li; Yu, Dong (2011). "Deep Convex Net: A Scalable Architecture for Speech Pattern Classification" (PDF). *Proceedings of the Interspeech*: 2285–2288.

[142] David, Wolpert (1992). "Stacked generalization". *Neural Networks*. **5** (2): 241–259. doi:10.1016/S0893-6080(05)80023-1.

[143] Bengio, Y. (2009-11-15). "Learning Deep Architectures for AI". *Foundations and Trends® in Machine Learning*. **2** (1): 1–127. ISSN 1935-8237. doi:10.1561/2200000006.

[144] Hutchinson, Brian; Deng, Li; Yu, Dong (2012). "Tensor deep stacking networks". *IEEE Transactions on Pattern Analysis and Machine Intelligence*. **1–15**: 1944–1957. doi:10.1109/tpami.2012.268.

[145] Hinton, Geoffrey; Salakhutdinov, Ruslan (2006). "Reducing the Dimensionality of Data with Neural Networks". *Science*. **313**: 504–507. PMID 16873662. doi:10.1126/science.1127647.

[146] Dahl, G.; Yu, D.; Deng, L.; Acero, A. (2012). "Context-Dependent Pre-Trained Deep Neural Networks for Large-Vocabulary Speech Recognition". *IEEE Transactions on Audio, Speech, and Language Processing*. **20** (1): 30–42. doi:10.1109/tasl.2011.2134090.

[147] Mohamed, Abdel-rahman; Dahl, George; Hinton, Geoffrey (2012). "Acoustic Modeling Using Deep Belief Networks". *IEEE Transactions on Audio, Speech, and Language Processing*. **20** (1): 14–22. doi:10.1109/tasl.2011.2109382.

[148] Courville, Aaron; Bergstra, James; Bengio, Yoshua (2011). "A Spike and Slab Restricted Boltzmann Machine" (PDF). *JMLR: Workshop and Conference Proceeding*. **15**: 233–241.

[149] Courville, Aaron; Bergstra, James; Bengio, Yoshua (2011). "Unsupervised Models of Images by Spike-and-Slab RBMs". *Proceedings of the 28th International Conference on Machine Learning* (PDF). **10**. pp. 1–8.

[150] Mitchell, T; Beauchamp, J (1988). "Bayesian Variable Selection in Linear Regression". *Journal of the American Statistical Association*. **83** (404): 1023–1032. doi:10.1080/01621459.1988.10478694.

[151] Hinton, G. E.; Osindero, S.; Teh, Y. (2006). "A fast learning algorithm for deep belief nets" (PDF). *Neural Computation*. **18** (7): 1527–1554. PMID 16764513. doi:10.1162/neco.2006.18.7.1527.

[152] Hinton, Geoffrey; Salakhutdinov, Ruslan (2009). "Efficient Learning of Deep Boltzmann Machines" (PDF). **3**: 448–455.

[153] Larochelle, Hugo; Bengio, Yoshua; Louradour, Jerdme; Lamblin, Pascal (2009). "Exploring Strategies for Training Deep Neural Networks". *The Journal of Machine Learning Research*. **10**: 1–40.

[154] Coates, Adam; Carpenter, Blake (2011). "Text Detection and Character Recognition in Scene Images with Unsupervised Feature Learning" (PDF): 440–445.

[155] Lee, Honglak; Grosse, Roger (2009). "Convolutional deep belief networks for scalable unsupervised learning of hierarchical representations". *Proceedings of the 26th Annual International Conference on Machine Learning*: 1–8.

[156] Lin, Yuanqing; Zhang, Tong (2010). "Deep Coding Network" (PDF). *Advances in Neural . . .*: 1–9.

[157] Ranzato, Marc Aurelio; Boureau, Y-Lan (2007). "Sparse Feature Learning for Deep Belief Networks" (PDF). *Advances in Neural Information Processing Systems*. **23**: 1–8.

[158] Socher, Richard; Lin, Clif (2011). "Parsing Natural Scenes and Natural Language with Recursive Neural Networks" (PDF). *Proceedings of the 26th International Conference on Machine Learning*.

[159] Taylor, Graham; Hinton, Geoffrey (2006). "Modeling Human Motion Using Binary Latent Variables" (PDF). *Advances in Neural Information Processing Systems.*

[160] Vincent, Pascal; Larochelle, Hugo (2008). "Extracting and composing robust features with denoising autoencoders". *Proceedings of the 25th international conference on Machine learning - ICML '08*: 1096–1103.

[161] Kemp, Charles; Perfors, Amy; Tenenbaum, Joshua (2007). "Learning overhypotheses with hierarchical Bayesian models". *Developmental Science.* **10** (3): 307–21. PMID 17444972. doi:10.1111/j.1467-7687.2007.00585.x.

[162] Xu, Fei; Tenenbaum, Joshua (2007). "Word learning as Bayesian inference". *Psychol. Rev.* **114** (2): 245–72. PMID 17500627. doi:10.1037/0033-295X.114.2.245.

[163] Chen, Bo; Polatkan, Gungor (2011). "The Hierarchical Beta Process for Convolutional Factor Analysis and Deep Learning" (PDF). *Machine Learning . . .*

[164] Fei-Fei, Li; Fergus, Rob (2006). "One-shot learning of object categories". *IEEE Transactions on Pattern Analysis and Machine Intelligence.* **28** (4): 594–611. PMID 16566508. doi:10.1109/TPAMI.2006.79.

[165] Rodriguez, Abel; Dunson, David (2008). "The Nested Dirichlet Process". *Journal of the American Statistical Association.* **103** (483): 1131–1154. doi:10.1198/016214508000000553.

[166] Ruslan, Salakhutdinov; Joshua, Tenenbaum (2012). "Learning with Hierarchical-Deep Models". *IEEE Transactions on Pattern Analysis and Machine Intelligence.* **35**: 1958–71. doi:10.1109/TPAMI.2012.269.

[167] Chalasani, Rakesh; Principe, Jose (2013). "Deep Predictive Coding Networks". arXiv:1301.3541.

[168] Hinton, Geoffrey E. (1984). "Distributed representations".

[169] S. Das, C.L. Giles, G.Z. Sun, "Learning Context Free Grammars: Limitations of a Recurrent Neural Network with an External Stack Memory," Proc. 14th Annual Conf. of the Cog. Sci. Soc., p. 79, 1992.

[170] Mozer, M. C.; Das, S. (1993). "A connectionist symbol manipulator that discovers the structure of context-free languages". NIPS 5. pp. 863–870.

[171] Schmidhuber, J. (1992). "Learning to control fast-weight memories: An alternative to recurrent nets". *Neural Computation.* **4** (1): 131–139. doi:10.1162/neco.1992.4.1.131.

[172] Gers, F.; Schraudolph, N.; Schmidhuber, J. (2002). "Learning precise timing with LSTM recurrent networks" (PDF). *JMLR.* **3**: 115–143.

[173] Jürgen Schmidhuber (1993). "An introspective network that can learn to run its own weight change algorithm". In Proc. of the Intl. Conf. on Artificial Neural Networks, Brighton. IEE. pp. 191–195.

[174] Hochreiter, Sepp; Younger, A. Steven; Conwell, Peter R. (2001). "Learning to Learn Using Gradient Descent". *ICANN.* **2130**: 87–94.

[175] Grefenstette, Edward, et al. "Learning to Transduce with Unbounded Memory."arXiv:1506.02516 (2015).

[176] Graves, Alex, Greg Wayne, and Ivo Danihelka. "Neural Turing Machines." arXiv:1410.5401 (2014).

[177] Atkeson, Christopher G.; Schaal, Stefan (1995). "Memory-based neural networks for robot learning". *Neurocomputing.* **9** (3): 243–269. doi:10.1016/0925-2312(95)00033-6.

[178] Salakhutdinov, Ruslan, and Geoffrey Hinton. "Semantic hashing." International Journal of Approximate Reasoning 50.7 (2009): 969-978.

[179] Le, Quoc V.; Mikolov, Tomas (2014). "Distributed representations of sentences and documents". arXiv:1405.4053.

[180] Weston, Jason, Sumit Chopra, and Antoine Bordes. "Memory networks." arXiv:1410.3916 (2014).

[181] Sukhbaatar, Sainbayar, et al. "End-To-End Memory Networks." arXiv:1503.08895 (2015).

[182] Bordes, Antoine, et al. "Large-scale Simple Question Answering with Memory Networks." arXiv:1506.02075 (2015).

[183] Vinyals, Oriol, Meire Fortunato, and Navdeep Jaitly. "Pointer networks." arXiv:1506.03134 (2015).

[184] Kurach, Karol, Andrychowicz, Marcin and Sutskever, Ilya. "Neural Random-Access Machines." arXiv:1511.06392 (2015).

[185] Kalchbrenner, N.; Blunsom, P. (2013). "Recurrent continuous translation models". EMNLP'2013.

[186] Sutskever, I.; Vinyals, O.; Le, Q. V. (2014). "Sequence to sequence learning with neural networks" (PDF). NIPS'2014.

[187] Cho, K.; van Merrienboer, B.; Gulcehre, C.; Bougares, F.; Schwenk, H.; Bengio, Y. (October 2014). "Learning phrase representations using RNN encoder-decoder for statistical machine translation". *Proceedings of the Empiricial Methods in Natural Language Processing.*

[188] Cho, Kyunghyun, Aaron Courville, and Yoshua Bengio. "Describing Multimedia Content using Attention-based Encoder–Decoder Networks." arXiv:1507.01053 (2015).

[189] Scholkopf, B; Smola, Alexander (1998). "Nonlinear component analysis as a kernel eigenvalue problem". *Neural computation.* (**44**): 1299–1319. doi:10.1162/089976698300017467.

[190] Cho, Youngmin (2012). "Kernel Methods for Deep Learning" (PDF): 1–9.

31.12. REFERENCES

[191] Deng, Li; Tur, Gokhan; He, Xiaodong; Hakkani-Tür, Dilek (2012-12-01). "Use of Kernel Deep Convex Networks and End-To-End Learning for Spoken Language Understanding". *Microsoft Research*.

[192] Zissis, Dimitrios (October 2015). "A cloud based architecture capable of perceiving and predicting multiple vessel behaviour". *Applied Soft Computing*. **35**: 652–661. doi:10.1016/j.asoc.2015.07.002.

[193] Roman M. Balabin; Ekaterina I. Lomakina (2009). "Neural network approach to quantum-chemistry data: Accurate prediction of density functional theory energies". *J. Chem. Phys.* **131** (7): 074104. PMID 19708729. doi:10.1063/1.3206326.

[194] Sengupta, Nandini; Sahidullah, Md; Saha, Goutam (August 2016). "Lung sound classification using cepstral-based statistical features". *Computers in Biology and Medicine*. **75** (1): 118–129. doi:10.1016/j.compbiomed.2016.05.013.

[195] Schechner, Sam (2017-06-15). "Facebook Boosts A.I. to Block Terrorist Propaganda". *Wall Street Journal*. ISSN 0099-9660. Retrieved 2017-06-16.

[196] Ganesan, N. "Application of Neural Networks in Diagnosing Cancer Disease Using Demographic Data" (PDF). International Journal of Computer Applications.

[197] Bottaci, Leonardo. "Artificial Neural Networks Applied to Outcome Prediction for Colorectal Cancer Patients in Separate Institutions" (PDF). The Lancet.

[198] Alizadeh, Elaheh; Lyons, Samanthe M; Castle, Jordan M; Prasad, Ashok (2016). "Measuring systematic changes in invasive cancer cell shape using Zernike moments". *Integrative Biology*. **8** (11): 1183–1193. doi:10.1039/C6IB00100A.

[199] Lyons, Samanthe (2016). "Changes in cell shape are correlated with metastatic potential in murine". *Biology Open*. **5** (3): 289–299. doi:10.1242/bio.013409.

[200] null null (2000-04-01). "Artificial Neural Networks in Hydrology. I: Preliminary Concepts". *Journal of Hydrologic Engineering*. **5** (2): 115–123. doi:10.1061/(ASCE)1084-0699(2000)5:2(115).

[201] null null (2000-04-01). "Artificial Neural Networks in Hydrology. II: Hydrologic Applications". *Journal of Hydrologic Engineering*. **5** (2): 124–137. doi:10.1061/(ASCE)1084-0699(2000)5:2(124).

[202] Peres, D. J.; Iuppa, C.; Cavallaro, L.; Cancelliere, A.; Foti, E. (2015-10-01). "Significant wave height record extension by neural networks and reanalysis wind data". *Ocean Modelling*. **94**: 128–140. doi:10.1016/j.ocemod.2015.08.002.

[203] Dwarakish, G. S.; Rakshith, Shetty; Natesan, Usha (2013). "Review on Applications of Neural Network in Coastal Engineering". *Artificial Intelligent Systems and Machine Learning*. **5** (7): 324–331.

[204] Ermini, Leonardo; Catani, Filippo; Casagli, Nicola (2005-03-01). "Artificial Neural Networks applied to landslide susceptibility assessment". *Geomorphology*. Geomorphological hazard and human impact in mountain environments. **66** (1): 327–343. doi:10.1016/j.geomorph.2004.09.025.

[205] "Introduction to Dynamic Neural Networks - MATLAB & Simulink". www.mathworks.com. Retrieved 2017-06-15.

[206] Forrest MD (April 2015). "Simulation of alcohol action upon a detailed Purkinje neuron model and a simpler surrogate model that runs >400 times faster". *BMC Neuroscience*. **16** (27). doi:10.1186/s12868-015-0162-6.

[207] Salakhutdinov, Ruslan; Hinton, Geoffrey (2009). "Semantic hashing". *International Journal of Approximate Reasoning*. **50** (7): 969–978. CiteSeerX 10.1.1.160.7001. doi:10.1016/j.ijar.2008.11.006.

[208] Burgess, Matt. "DeepMind's AI learned to ride the London Underground using human-like reason and memory". *WIRED UK*. Retrieved 2016-10-19.

[209] "DeepMind AI 'Learns' to Navigate London Tube". *PCMAG*. Retrieved 2016-10-19.

[210] Mannes, John. "DeepMind's differentiable neural computer helps you navigate the subway with its memory". *TechCrunch*. Retrieved 2016-10-19.

[211] Graves, Alex; Wayne, Greg; Reynolds, Malcolm; Harley, Tim; Danihelka, Ivo; Grabska-Barwińska, Agnieszka; Colmenarejo, Sergio Gómez; Grefenstette, Edward; Ramalho, Tiago (2016-10-12). "Hybrid computing using a neural network with dynamic external memory". *Nature*. **538**: 471–476. ISSN 1476-4687. PMID 27732574. doi:10.1038/nature20101.

[212] "Differentiable neural computers | DeepMind". *DeepMind*. Retrieved 2016-10-19.

[213] Siegelmann, H.T.; Sontag, E.D. (1991). "Turing computability with neural nets" (PDF). *Appl. Math. Lett.* **4** (6): 77–80. doi:10.1016/0893-9659(91)90080-F.

[214] Balcázar, José (Jul 1997). "Computational Power of Neural Networks: A Kolmogorov Complexity Characterization". *Information Theory, IEEE Transactions on*. **43** (4): 1175–1183. CiteSeerX 10.1.1.411.7782. doi:10.1109/18.605580. Retrieved 3 November 2014.

[215] Crick, Francis (1989). "The recent excitement about neural networks". *Nature*. **337** (6203): 129–132. PMID 2911347. doi:10.1038/337129a0.

[216] Adrian, Edward D. (1926). "The impulses produced by sensory nerve endings". *The Journal of Physiology*. **61** (1): 49–72. doi:10.1113/jphysiol.1926.sp002273.

[217] Dewdney, A. K. (1 April 1997). *Yes, we have no neutrons: an eye-opening tour through the twists and turns of bad science*. Wiley. p. 82. ISBN 978-0-471-10806-1.

[218] D. J. Felleman and D. C. Van Essen, "Distributed hierarchical processing in the primate cerebral cortex," *Cerebral Cortex*, 1, pp. 1-47, 1991.

[219] J. Weng, "Natural and Artificial Intelligence: Introduction to Computational Brain-Mind," BMI Press, ISBN 978-0985875725, 2012.

[220] Edwards, Chris (25 June 2015). "Growing pains for deep learning". *Communications of the ACM*. **58** (7): 14–16. doi:10.1145/2771283.

[221] Schmidhuber, Jürgen (2015). "Deep learning in neural networks: An overview". *Neural Networks*. **61**: 85–117. arXiv:1404.7828. doi:10.1016/j.neunet.2014.09.003.

[222] Cade Metz (May 18, 2016). "Google Built Its Very Own Chips to Power Its AI Bots". *Wired*.

[223] NASA - Dryden Flight Research Center - News Room: News Releases: NASA NEURAL NETWORK PROJECT PASSES MILESTONE. Nasa.gov. Retrieved on 2013-11-20.

[224] Roger Bridgman's defence of neural networks

[225] "Scaling Learning Algorithms towards {AI} - LISA - Publications - Aigaion 2.0".

[226] Sun and Bookman (1990)

[227] Tahmasebi; Hezarkhani (2012). "A hybrid neural networks-fuzzy logic-genetic algorithm for grade estimation". *Computers & Geosciences*. **42**: 18–27. doi:10.1016/j.cageo.2012.02.004.

31.13 Bibliography

- Bhadeshia H. K. D. H. (1999). "Neural Networks in Materials Science" (PDF). *ISIJ International*. **39** (10): 966–979. doi:10.2355/isijinternational.39.966.

- M., Bishop, Christopher (1995). *Neural networks for pattern recognition*. Clarendon Press. ISBN 0198538499. OCLC 33101074.

- Cybenko, G.V. (2006). "Approximation by Superpositions of a Sigmoidal function". In van Schuppen, Jan H. *Mathematics of Control, Signals, and Systems*. Springer International. pp. 303–314. PDF

- Dewdney, A. K. (1997). *Yes, we have no neutrons : an eye-opening tour through the twists and turns of bad science*. New York: Wiley. ISBN 9780471108061. OCLC 35558945.

- Duda, Richard O.; Hart, Peter Elliot; Stork, David G. (2001). *Pattern classification* (2 ed.). Wiley. ISBN 0471056693. OCLC 41347061.

- Egmont-Petersen, M.; de Ridder, D.; Handels, H. (2002). "Image processing with neural networks – a review". *Pattern Recognition*. **35** (10): 2279–2301. doi:10.1016/S0031-3203(01)00178-9.

- Gurney, Kevin (1997). *An introduction to neural networks*. UCL Press. ISBN 1857286731. OCLC 37875698.

- Haykin, Simon S. (1999). *Neural networks : a comprehensive foundation*. Prentice Hall. ISBN 0132733501. OCLC 38908586.

- Fahlman, S.; Lebiere, C (1991). "The Cascade-Correlation Learning Architecture" (PDF).created for National Science Foundation, Contract Number EET-8716324, and Defense Advanced Research Projects Agency (DOD), ARPA Order No. 4976 under Contract F33615-87-C-1499.

- Hertz, J.; Palmer, Richard G.; Krogh, Anders S. (1991). *Introduction to the theory of neural computation*. Addison-Wesley. ISBN 0201515601. OCLC 21522159.</ref>

- Lawrence, Jeanette (1994). *Introduction to neural networks : design, theory and applications*. California Scientific Software. ISBN 1883157005. OCLC 32179420.

- *Information theory, inference, and learning algorithms*. Cambridge University Press. ISBN 9780521642989. OCLC 52377690.

- MacKay, David, J.C. (2003). *Information Theory, Inference, and Learning Algorithms* (PDF). Cambridge University Press. ISBN 9780521642989.

- Masters., Timothy (1994). *Signal and image processing with neural networks : a C++ sourcebook*. J. Wiley. ISBN 0471049638. OCLC 29877717.

- Ripley, Brian D. (2007). *Pattern Recognition and Neural Networks*. Cambridge University Press. ISBN 978-0-521-71770-0.

- Siegelmann, H.T.; Sontag, Eduardo D. (1994). "Analog computation via neural networks" (PDF). *Theoretical Computer Science*. **131** (2): 331–360. doi:10.1016/0304-3975(94)90178-3.

- 1944-, Smith, Murray, (1993). *Neural networks for statistical modeling*. Van Nostrand Reinhold. ISBN 0442013108. OCLC 27145760.</ref> Smith, Murray (1993) *Neural Networks for Statistical Modeling*. Van Nostrand Reinhold, ISBN 0-442-01310-8

- Wasserman, Philip D. (1993). *Advanced methods in neural computing*. Van Nostrand Reinhold. ISBN 0442004613. OCLC 27429729.

- Kruse, Rudolf.; Borgelt, Christian; Klawonn, F.; Moewes, Christian; Steinbrecher, Matthias; Held., Pascal (2013). *Computational intelligence : a methodological introduction*. Springer. ISBN 9781447150121. OCLC 837524179.

- Borgelt,, Christian (2003). *Neuro-Fuzzy-Systeme : von den Grundlagen künstlicher Neuronaler Netze zur Kopplung mit Fuzzy-Systemen*. Vieweg. ISBN 9783528252656. OCLC 76538146.

31.14 External links

- Neural Networks at DMOZ
- A brief introduction to Neural Networks (PDF), illustrated 250p textbook covering the common kinds of neural networks (CC license).
- An Introduction to Deep Neural Networks.
- A Tutorial of Neural Network in Excel.
- MIT course on Neural Networks on YouTube
- A Concise Introduction to Machine Learning with Artificial Neural Networks
- Neural Networks for Machine Learning - a course by Geoffrey Hinton
- Deep Learning
- Aplikasi pendeteksi fraud pada event log proses bisnis pengadaan barang dan jasa menggunakan algoritma heuristic miner

Chapter 32

Evolutionary computation

For the journal, see Evolutionary Computation (journal).

In computer science, **evolutionary computation** is a family of algorithms for global optimization inspired by biological evolution, and the subfield of artificial intelligence and soft computing studying these algorithms. In technical terms, they are a family of population-based trial and error problem solvers with a metaheuristic or stochastic optimization character.

In evolutionary computation, an initial set of candidate solutions is generated and iteratively updated. Each new generation is produced by stochastically removing less desired solutions, and introducing small random changes. In biological terminology, a population of solutions is subjected to natural selection (or artificial selection) and mutation. As a result, the population will gradually evolve to increase in fitness, in this case the chosen fitness function of the algorithm.

Evolutionary computation techniques can produce highly optimized solutions in a wide range of problem settings, making them popular in computer science. Many variants and extensions exist, suited to more specific families of problems and data structures. Evolutionary computation is also sometimes used in evolutionary biology as an *in silico* experimental procedure to study common aspects of general evolutionary processes.

32.1 History

The use of Evolutionary principles for automated problem solving originated in the 1950s. It was not until the 1960s that three distinct interpretations of this idea started to be developed in three different places.

Evolutionary programming was introduced by Lawrence J. Fogel in the US, while John Henry Holland called his method a genetic algorithm. In Germany Ingo Rechenberg and Hans-Paul Schwefel introduced evolution strategies. These areas developed separately for about 15 years. From the early nineties on they are unified as different representatives ("dialects") of one technology, called evolutionary computing. Also in the early nineties, a fourth stream following the general ideas had emerged – genetic programming. Since the 1990s, nature-inspired algorithms are becoming an increasingly significant part of evolutionary computation.

These terminologies denote the field of evolutionary computing and consider evolutionary programming, evolution strategies, genetic algorithms, and genetic programming as sub-areas.

Simulations of evolution using evolutionary algorithms and artificial life started with the work of Nils Aall Barricelli in the 1960s, and was extended by Alex Fraser, who published a series of papers on simulation of artificial selection.[1] Artificial evolution became a widely recognised optimisation method as a result of the work of Ingo Rechenberg in the 1960s and early 1970s, who used evolution strategies to solve complex engineering problems.[2] Genetic algorithms in particular became popular through the writing of John Holland.[3] As academic interest grew, dramatic increases in the power of computers allowed practical applications, including the automatic evolution of computer programs.[4] Evolutionary algorithms are now used to solve multi-dimensional problems more efficiently than software produced by human designers, and also to optimise the design of systems.[5][6]

32.2 Techniques

Evolutionary computing techniques mostly involve metaheuristic optimization algorithms. Broadly speaking, the field includes:

- Ant colony optimization

- Artificial immune systems

- Artificial life (also see digital organism)
- Cultural algorithms
- Differential evolution
- Dual-phase evolution
- Estimation of distribution algorithms
- Evolutionary algorithms
- Evolutionary programming
- Evolution strategy
- Gene expression programming
- Genetic algorithm
- Genetic programming
- Grammatical evolution
- Learnable evolution model
- Learning classifier systems
- Memetic algorithms
- Particle swarm optimization
- Self-organization such as self-organizing maps, competitive learning
- Swarm intelligence

32.3 Evolutionary algorithms

Main article: Evolutionary algorithm

Evolutionary algorithms form a subset of evolutionary computation in that they generally only involve techniques implementing mechanisms inspired by biological evolution such as reproduction, mutation, recombination, natural selection and survival of the fittest. Candidate solutions to the optimization problem play the role of individuals in a population, and the cost function determines the environment within which the solutions "live" (see also fitness function). Evolution of the population then takes place after the repeated application of the above operators.

In this process, there are two main forces that form the basis of evolutionary systems: **Recombination** and **mutation** create the necessary diversity and thereby facilitate novelty, while **selection** acts as a force increasing quality.

Many aspects of such an evolutionary process are stochastic. Changed pieces of information due to recombination and mutation are randomly chosen. On the other hand, selection operators can be either deterministic, or stochastic. In the latter case, individuals with a higher fitness have a higher chance to be selected than individuals with a lower fitness, but typically even the weak individuals have a chance to become a parent or to survive.

32.4 Practitioners

The list of active researchers is naturally dynamic and non-exhaustive. A network analysis of the community was published in 2007.[7]

- Thomas Bäck
- Wolfgang Banzhaf
- B V Babu
- Kalyanmoy Deb
- Kenneth A De Jong
- Gusz Eiben
- Peter J. Fleming
- David B. Fogel
- Stephanie Forrest
- David E. Goldberg
- Emma Hart
- John Henry Holland
- Theo Jansen
- John Koza
- Zbigniew Michalewicz
- Peter Nordin
- Riccardo Poli
- Ingo Rechenberg
- Marc Schoenauer
- Hans-Paul Schwefel
- Jim Smith
- Gloria Townsend
- Xin Yao

32.5 See also

- Adaptive dimensional search
- Artificial development
- Autoconstructive
- Developmental biology
- Digital organism
- Estimation of distribution algorithm
- Evolutionary robotics
- Evolved antenna
- Fitness approximation
- Fitness function
- Fitness landscape
- Genetic operators
- Grammatical evolution
- Human-based evolutionary computation
- Inferential programming
- Interactive evolutionary computation
- List of digital organism simulators
- Mutation testing
- No free lunch in search and optimization
- Program synthesis
- Test functions for optimization
- Universal Darwinism

32.6 Bibliography

- Th. Bäck, D.B. Fogel, and Z. Michalewicz (Editors), Handbook of Evolutionary Computation, 1997, ISBN 0750303921
- Th. Bäck and H.-P. Schwefel. An overview of evolutionary algorithms for parameter optimization. Evolutionary Computation, 1(1):1–23, 1993.
- W. Banzhaf, P. Nordin, R.E. Keller, and F.D. Francone. Genetic Programming — An Introduction. Morgan Kaufmann, 1998.
- S. Cagnoni, et al., Real-World Applications of Evolutionary Computing, Springer-Verlag Lecture Notes in Computer Science, Berlin, 2000.
- R. Chiong, Th. Weise, Z. Michalewicz (Editors), Variants of Evolutionary Algorithms for Real-World Applications, Springer, 2012, ISBN 3642234232
- K. A. De Jong, Evolutionary computation: a unified approach. MIT Press, Cambridge MA, 2006
- A. E. Eiben and M. Schoenauer, Evolutionary computing, Information Processing Letters, 82(1): 1–6, 2002.
- A. E. Eiben and J.E. Smith, Introduction to Evolutionary Computing, Springer, First edition, 2003, ISBN 3-540-40184-9.
- D. B. Fogel. Evolutionary Computation. Toward a New Philosophy of Machine Intelligence. IEEE Press, Piscataway, NJ, 1995.
- L. J. Fogel, A. J. Owens, and M. J. Walsh. Artificial Intelligence through Simulated Evolution. New York: John Wiley, 1966.
- D. E. Goldberg. Genetic algorithms in search, optimization and machine learning. Addison Wesley, 1989.
- J. H. Holland. Adaptation in natural and artificial systems. University of Michigan Press, Ann Arbor, 1975.
- P. Hingston, L. Barone, and Z. Michalewicz (Editors), Design by Evolution, Natural Computing Series, 2008, Springer, ISBN 3540741097
- J. R. Koza. Genetic Programming: On the Programming of Computers by means of Natural Evolution. MIT Press, Massachusetts, 1992.
- F.J. Lobo, C.F. Lima, Z. Michalewicz (Editors), Parameter Setting in Evolutionary Algorithms, Springer, 2010, ISBN 3642088929
- Z. Michalewicz, Genetic Algorithms + Data Structures – Evolution Programs, 1996, Springer, ISBN 3540606769
- Z. Michalewicz and D.B. Fogel, How to Solve It: Modern Heuristics, Springer, 2004, ISBN 978-3-540-22494-5
- I. Rechenberg. Evolutionstrategie: Optimierung Technischer Systeme nach Prinzipien des Biologischen Evolution. Fromman-Hozlboog Verlag, Stuttgart, 1973. (in German)

- H.-P. Schwefel. Numerical Optimization of Computer Models. John Wiley & Sons, New-York, 1981. 1995 – 2nd edition.

- D. Simon. Evolutionary Optimization Algorithms. Wiley, 2013.

32.7 References

[1] Fraser AS (1958). "Monte Carlo analyses of genetic models". *Nature*. **181** (4603): 208–9. PMID 13504138. doi:10.1038/181208a0.

[2] Rechenberg, Ingo (1973). *Evolutionsstrategie – Optimierung technischer Systeme nach Prinzipien der biologischen Evolution (PhD thesis)* (in German). Fromman-Holzboog.

[3] Holland, John H. (1975). *Adaptation in Natural and Artificial Systems*. University of Michigan Press. ISBN 0-262-58111-6.

[4] Koza, John R. (1992). *Genetic Programming: On the Programming of Computers by Means of Natural Selection*. MIT Press. ISBN 0-262-11170-5.

[5] G. C. Onwubolu and B V Babu, "New Optimization Techniques in Engineering". Retrieved 17 September 2016.

[6] Jamshidi M (2003). "Tools for intelligent control: fuzzy controllers, neural networks and genetic algorithms". *Philosophical Transactions of the Royal Society A*. **361** (1809): 1781–808. PMID 12952685. doi:10.1098/rsta.2003.1225.

[7] J.J. Merelo and C. Cotta (2007). "Who is the best connected EC researcher? Centrality analysis of the complex network of authors in evolutionary computation". arXiv:0708.2021

Chapter 33

Genetic algorithm

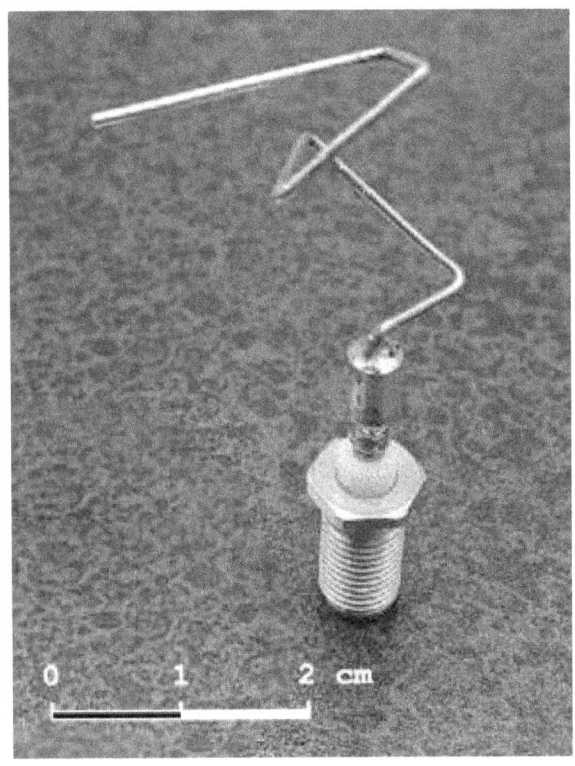

The 2006 NASA ST5 spacecraft antenna. This complicated shape was found by an evolutionary computer design program to create the best radiation pattern. It is known as an evolved antenna.

In computer science and operations research, a **genetic algorithm** (**GA**) is a metaheuristic inspired by the process of natural selection that belongs to the larger class of evolutionary algorithms (EA). Genetic algorithms are commonly used to generate high-quality solutions to optimization and search problems by relying on bio-inspired operators such as mutation, crossover and selection.[1]

33.1 Methodology

33.1.1 Optimization problems

In a genetic algorithm, a population of candidate solutions (called individuals, creatures, or phenotypes) to an optimization problem is evolved toward better solutions. Each candidate solution has a set of properties (its chromosomes or genotype) which can be mutated and altered; traditionally, solutions are represented in binary as strings of 0s and 1s, but other encodings are also possible.[2]

The evolution usually starts from a population of randomly generated individuals, and is an iterative process, with the population in each iteration called a *generation*. In each generation, the fitness of every individual in the population is evaluated; the fitness is usually the value of the objective function in the optimization problem being solved. The more fit individuals are stochastically selected from the current population, and each individual's genome is modified (recombined and possibly randomly mutated) to form a new generation. The new generation of candidate solutions is then used in the next iteration of the algorithm. Commonly, the algorithm terminates when either a maximum number of generations has been produced, or a satisfactory fitness level has been reached for the population.

A typical genetic algorithm requires:

1. a genetic representation of the solution domain,
2. a fitness function to evaluate the solution domain.

A standard representation of each candidate solution is as an array of bits.[2] Arrays of other types and structures can be used in essentially the same way. The main property that makes these genetic representations convenient is that their parts are easily aligned due to their fixed size, which facilitates simple crossover operations. Variable length representations may also be used, but crossover implementation is more complex in this case. Tree-like representations are explored in genetic programming and graph-form representations are explored in evolutionary programming; a mix of both linear chromosomes and trees is explored in gene expression programming.

33.1. METHODOLOGY

Once the genetic representation and the fitness function are defined, a GA proceeds to initialize a population of solutions and then to improve it through repetitive application of the mutation, crossover, inversion and selection operators.

Initialization

The population size depends on the nature of the problem, but typically contains several hundreds or thousands of possible solutions. Often, the initial population is generated randomly, allowing the entire range of possible solutions (the *search space*). Occasionally, the solutions may be "seeded" in areas where optimal solutions are likely to be found.

Selection

Main article: Selection (genetic algorithm)

During each successive generation, a portion of the existing population is selected to breed a new generation. Individual solutions are selected through a *fitness-based* process, where fitter solutions (as measured by a fitness function) are typically more likely to be selected. Certain selection methods rate the fitness of each solution and preferentially select the best solutions. Other methods rate only a random sample of the population, as the former process may be very time-consuming.

The fitness function is defined over the genetic representation and measures the *quality* of the represented solution. The fitness function is always problem dependent. For instance, in the knapsack problem one wants to maximize the total value of objects that can be put in a knapsack of some fixed capacity. A representation of a solution might be an array of bits, where each bit represents a different object, and the value of the bit (0 or 1) represents whether or not the object is in the knapsack. Not every such representation is valid, as the size of objects may exceed the capacity of the knapsack. The *fitness* of the solution is the sum of values of all objects in the knapsack if the representation is valid, or 0 otherwise.

In some problems, it is hard or even impossible to define the fitness expression; in these cases, a simulation may be used to determine the fitness function value of a phenotype (e.g. computational fluid dynamics is used to determine the air resistance of a vehicle whose shape is encoded as the phenotype), or even interactive genetic algorithms are used.

Genetic operators

Main articles: Crossover (genetic algorithm) and Mutation (genetic algorithm)

The next step is to generate a second generation population of solutions from those selected through a combination of genetic operators: crossover (also called recombination), and mutation.

For each new solution to be produced, a pair of "parent" solutions is selected for breeding from the pool selected previously. By producing a "child" solution using the above methods of crossover and mutation, a new solution is created which typically shares many of the characteristics of its "parents". New parents are selected for each new child, and the process continues until a new population of solutions of appropriate size is generated. Although reproduction methods that are based on the use of two parents are more "biology inspired", some research[3][4] suggests that more than two "parents" generate higher quality chromosomes.

These processes ultimately result in the next generation population of chromosomes that is different from the initial generation. Generally the average fitness will have increased by this procedure for the population, since only the best organisms from the first generation are selected for breeding, along with a small proportion of less fit solutions. These less fit solutions ensure genetic diversity within the genetic pool of the parents and therefore ensure the genetic diversity of the subsequent generation of children.

Opinion is divided over the importance of crossover versus mutation. There are many references in Fogel (2006) that support the importance of mutation-based search.

Although crossover and mutation are known as the main genetic operators, it is possible to use other operators such as regrouping, colonization-extinction, or migration in genetic algorithms.[5]

It is worth tuning parameters such as the mutation probability, crossover probability and population size to find reasonable settings for the problem class being worked on. A very small mutation rate may lead to genetic drift (which is non-ergodic in nature). A recombination rate that is too high may lead to premature convergence of the genetic algorithm. A mutation rate that is too high may lead to loss of good solutions, unless elitist selection is employed.

Heuristics

In addition to the main operators above, other heuristics may be employed to make the calculation faster or more robust. The *speciation* heuristic penalizes crossover between candidate solutions that are too similar; this encourages population diversity and helps prevent premature convergence to a less optimal solution.[6][7]

Termination

This generational process is repeated until a termination condition has been reached. Common terminating conditions are:

- A solution is found that satisfies minimum criteria
- Fixed number of generations reached
- Allocated budget (computation time/money) reached
- The highest ranking solution's fitness is reaching or has reached a plateau such that successive iterations no longer produce better results
- Manual inspection
- Combinations of the above
-

33.2 The building block hypothesis

Genetic algorithms are simple to implement, but their behavior is difficult to understand. In particular it is difficult to understand why these algorithms frequently succeed at generating solutions of high fitness when applied to practical problems. The building block hypothesis (BBH) consists of:

1. A description of a heuristic that performs adaptation by identifying and recombining "building blocks", i.e. low order, low defining-length schemata with above average fitness.
2. A hypothesis that a genetic algorithm performs adaptation by implicitly and efficiently implementing this heuristic.

Goldberg describes the heuristic as follows:

> "Short, low order, and highly fit schemata are sampled, recombined [crossed over], and resampled to form strings of potentially higher fitness. In a way, by working with these particular schemata [the building blocks], we have reduced the complexity of our problem; instead of building high-performance strings by trying every conceivable combination, we construct better and better strings from the best partial solutions of past samplings.

> "Because highly fit schemata of low defining length and low order play such an important role in the action of genetic algorithms, we have already given them a special name: building blocks. Just as a child creates magnificent fortresses through the arrangement of simple blocks of wood, so does a genetic algorithm seek near optimal performance through the juxtaposition of short, low-order, high-performance schemata, or building blocks."[8]

Despite the lack of consensus regarding the validity of the building-block hypothesis, it has been consistently evaluated and used as reference throughout the years. Many estimation of distribution algorithms, for example, have been proposed in an attempt to provide an environment in which the hypothesis would hold.[9][10] Although good results have been reported for some classes of problems, skepticism concerning the generality and/or practicality of the building-block hypothesis as an explanation for GAs efficiency still remains. Indeed, there is a reasonable amount of work that attempts to understand its limitations from the perspective of estimation of distribution algorithms.[11][12][13]

33.3 Limitations

There are limitations of the use of a genetic algorithm compared to alternative optimization algorithms:

- Repeated fitness function evaluation for complex problems is often the most prohibitive and limiting segment of artificial evolutionary algorithms. Finding the optimal solution to complex high-dimensional, multimodal problems often requires very expensive fitness function evaluations. In real world problems such as structural optimization problems, a single function evaluation may require several hours to several days of complete simulation. Typical optimization methods can not deal with such types of problem. In this case, it may be necessary to forgo an exact evaluation and use an approximated fitness that is computationally efficient. It is apparent that amalgamation of approximate models may be one of the most promising approaches to convincingly use GA to solve complex real life problems.

- Genetic algorithms do not scale well with complexity. That is, where the number of elements which are exposed to mutation is large there is often an exponential increase in search space size. This makes it extremely difficult to use the technique on problems such as designing an engine, a house or plane. In order to make such problems tractable to evolutionary search, they must be broken down into the simplest representation possible. Hence we typically see evolutionary algorithms encoding designs for fan blades instead of engines, building shapes instead of detailed construction plans, and airfoils instead of whole aircraft designs. The second problem of complexity is the issue of how to protect parts that have evolved to represent good solutions from further destructive mutation, particularly when their fitness assessment requires them to combine well with other parts.

- The "better" solution is only in comparison to other solutions. As a result, the stop criterion is not clear in every problem.

- In many problems, GAs may have a tendency to converge towards local optima or even arbitrary points rather than the global optimum of the problem. This means that it does not "know how" to sacrifice short-term fitness to gain longer-term fitness. The likelihood of this occurring depends on the shape of the fitness landscape: certain problems may provide an easy ascent towards a global optimum, others may make it easier for the function to find the local optima. This problem may be alleviated by using a different fitness function, increasing the rate of mutation, or by using selection techniques that maintain a diverse population of solutions,[14] although the No Free Lunch theorem[15] proves that there is no general solution to this problem. A common technique to maintain diversity is to impose a "niche penalty", wherein, any group of individuals of sufficient similarity (niche radius) have a penalty added, which will reduce the representation of that group in subsequent generations, permitting other (less similar) individuals to be maintained in the population. This trick, however, may not be effective, depending on the landscape of the problem. Another possible technique would be to simply replace part of the population with randomly generated individuals, when most of the population is too similar to each other. Diversity is important in genetic algorithms (and genetic programming) because crossing over a homogeneous population does not yield new solutions. In evolution strategies and evolutionary programming, diversity is not essential because of a greater reliance on mutation.

- Operating on dynamic data sets is difficult, as genomes begin to converge early on towards solutions which may no longer be valid for later data. Several methods have been proposed to remedy this by increasing genetic diversity somehow and preventing early convergence, either by increasing the probability of mutation when the solution quality drops (called *triggered hypermutation*), or by occasionally introducing entirely new, randomly generated elements into the gene pool (called *random immigrants*). Again, evolution strategies and evolutionary programming can be implemented with a so-called "comma strategy" in which parents are not maintained and new parents are selected only from offspring. This can be more effective on dynamic problems.

- GAs cannot effectively solve problems in which the only fitness measure is a single right/wrong measure (like decision problems), as there is no way to converge on the solution (no hill to climb). In these cases, a random search may find a solution as quickly as a GA. However, if the situation allows the success/failure trial to be repeated giving (possibly) different results, then the ratio of successes to failures provides a suitable fitness measure.

- For specific optimization problems and problem instances, other optimization algorithms may be more efficient than genetic algorithms in terms of speed of convergence. Alternative and complementary algorithms include evolution strategies, evolutionary programming, simulated annealing, Gaussian adaptation, hill climbing, and swarm intelligence (e.g.: ant colony optimization, particle swarm optimization) and methods based on integer linear programming. The suitability of genetic algorithms is dependent on the amount of knowledge of the problem; well known problems often have better, more specialized approaches.

33.4 Variants

33.4.1 Chromosome representation

Main article: genetic representation

The simplest algorithm represents each chromosome as a bit string. Typically, numeric parameters can be represented by integers, though it is possible to use floating point representations. The floating point representation is natural to evolution strategies and evolutionary programming. The notion of real-valued genetic algorithms has been offered but is really a misnomer because it does not really represent the building block theory that was proposed by John Henry Holland in the 1970s. This theory is not without support though, based on theoretical and experimental results (see below). The basic algorithm performs crossover and mutation at the bit level. Other variants treat the chromosome as a list of numbers which are indexes into an instruction table, nodes in a linked list, hashes, objects, or any other imaginable data structure. Crossover and mutation are performed so as to respect data element boundaries. For most data types, specific variation operators can be designed. Different chromosomal data types seem to work better or worse for different specific problem domains.

When bit-string representations of integers are used, Gray coding is often employed. In this way, small changes in the integer can be readily affected through mutations or crossovers. This has been found to help prevent premature convergence at so called *Hamming walls*, in which too many simultaneous mutations (or crossover events) must occur in order to change the chromosome to a better solution.

Other approaches involve using arrays of real-valued numbers instead of bit strings to represent chromosomes. Results from the theory of schemata suggest that in general the smaller the alphabet, the better the performance, but it was initially surprising to researchers that good results were obtained from using real-valued chromosomes. This was explained as the set of real values in a finite population of chromosomes as forming a *virtual alphabet* (when selection and recombination are dominant) with a much lower cardinality than would be expected from a floating point representation.[16][17]

An expansion of the Genetic Algorithm accessible problem domain can be obtained through more complex encoding of the solution pools by concatenating several types of heterogenously encoded genes into one chromosome.[18] This particular approach allows for solving optimization problems that require vastly disparate definition domains for the problem parameters. For instance, in problems of cascaded controller tuning, the internal loop controller structure can belong to a conventional regulator of three parameters, whereas the external loop could implement a linguistic controller (such as a fuzzy system) which has an inherently different description. This particular form of encoding requires a specialized crossover mechanism that recombines the chromosome by section, and it is a useful tool for the modelling and simulation of complex adaptive systems, especially evolution processes.

33.4.2 Elitism

A practical variant of the general process of constructing a new population is to allow the best organism(s) from the current generation to carry over to the next, unaltered. This strategy is known as *elitist selection* and guarantees that the solution quality obtained by the GA will not decrease from one generation to the next.[19]

33.4.3 Parallel implementations

Parallel implementations of genetic algorithms come in two flavors. Coarse-grained parallel genetic algorithms assume a population on each of the computer nodes and migration of individuals among the nodes. Fine-grained parallel genetic algorithms assume an individual on each processor node which acts with neighboring individuals for selection and reproduction. Other variants, like genetic algorithms for online optimization problems, introduce time-dependence or noise in the fitness function.

33.4.4 Adaptive GAs

Genetic algorithms with adaptive parameters (adaptive genetic algorithms, AGAs) is another significant and promising variant of genetic algorithms. The probabilities of crossover (pc) and mutation (pm) greatly determine the degree of solution accuracy and the convergence speed that genetic algorithms can obtain. Instead of using fixed values of pc and pm, AGAs utilize the population information in each generation and adaptively adjust the pc and pm in order to maintain the population diversity as well as to sustain the convergence capacity. In AGA (adaptive genetic algorithm),[20] the adjustment of pc and pm depends on the fitness values of the solutions. In *CAGA* (clustering-based adaptive genetic algorithm),[21] through the use of clustering analysis to judge the optimization states of the population, the adjustment of pc and pm depends on these optimization states. It can be quite effective to combine GA with other optimization methods. GA tends to be quite good at finding generally good global solutions, but quite inefficient at finding the last few mutations to find the absolute optimum. Other techniques (such as simple hill climbing) are quite efficient at finding absolute optimum in a limited region. Alternating GA and hill climbing can improve the efficiency of GA while overcoming the lack of robustness of hill climbing.

This means that the rules of genetic variation may have a different meaning in the natural case. For instance – provided that steps are stored in consecutive order – crossing over may sum a number of steps from maternal DNA adding a number of steps from paternal DNA and so on. This is like adding vectors that more probably may follow a ridge in the phenotypic landscape. Thus, the efficiency of the process may be increased by many orders of magnitude. Moreover, the inversion operator has the opportunity to place steps in consecutive order or any other suitable order in favour of survival or efficiency. (See for instance [22] or example in travelling salesman problem, in particular the use of an edge recombination operator.)

A variation, where the population as a whole is evolved

rather than its individual members, is known as gene pool recombination.

A number of variations have been developed to attempt to improve performance of GAs on problems with a high degree of fitness epistasis, i.e. where the fitness of a solution consists of interacting subsets of its variables. Such algorithms aim to learn (before exploiting) these beneficial phenotypic interactions. As such, they are aligned with the Building Block Hypothesis in adaptively reducing disruptive recombination. Prominent examples of this approach include the mGA,[23] GEMGA[24] and LLGA.[25]

33.5 Problem domains

Problems which appear to be particularly appropriate for solution by genetic algorithms include timetabling and scheduling problems, and many scheduling software packages are based on GAs. GAs have also been applied to engineering.[26] Genetic algorithms are often applied as an approach to solve global optimization problems.

As a general rule of thumb genetic algorithms might be useful in problem domains that have a complex fitness landscape as mixing, i.e., mutation in combination with crossover, is designed to move the population away from local optima that a traditional hill climbing algorithm might get stuck in. Observe that commonly used crossover operators cannot change any uniform population. Mutation alone can provide ergodicity of the overall genetic algorithm process (seen as a Markov chain).

Examples of problems solved by genetic algorithms include: mirrors designed to funnel sunlight to a solar collector,[27] antennae designed to pick up radio signals in space,[28] and walking methods for computer figures.[29]

In his *Algorithm Design Manual*, Skiena advises against genetic algorithms for any task:

> [I]t is quite unnatural to model applications in terms of genetic operators like mutation and crossover on bit strings. The pseudobiology adds another level of complexity between you and your problem. Second, genetic algorithms take a very long time on nontrivial problems. [...] [T]he analogy with evolution—where significant progress require [sic] millions of years—can be quite appropriate.
> [...]
> I have never encountered any problem where genetic algorithms seemed to me the right way to attack it. Further, I have never seen any computational results reported using genetic algorithms that have favorably impressed me. Stick

to simulated annealing for your heuristic search voodoo needs.
— Steven Skiena[30]:267

33.6 History

In 1950, Alan Turing proposed a "learning machine" which would parallel the principles of evolution.[31] Computer simulation of evolution started as early as in 1954 with the work of Nils Aall Barricelli, who was using the computer at the Institute for Advanced Study in Princeton, New Jersey.[32][33] His 1954 publication was not widely noticed. Starting in 1957,[34] the Australian quantitative geneticist Alex Fraser published a series of papers on simulation of artificial selection of organisms with multiple loci controlling a measurable trait. From these beginnings, computer simulation of evolution by biologists became more common in the early 1960s, and the methods were described in books by Fraser and Burnell (1970)[35] and Crosby (1973).[36] Fraser's simulations included all of the essential elements of modern genetic algorithms. In addition, Hans-Joachim Bremermann published a series of papers in the 1960s that also adopted a population of solution to optimization problems, undergoing recombination, mutation, and selection. Bremermann's research also included the elements of modern genetic algorithms.[37] Other noteworthy early pioneers include Richard Friedberg, George Friedman, and Michael Conrad. Many early papers are reprinted by Fogel (1998).[38]

Although Barricelli, in work he reported in 1963, had simulated the evolution of ability to play a simple game,[39] artificial evolution became a widely recognized optimization method as a result of the work of Ingo Rechenberg and Hans-Paul Schwefel in the 1960s and early 1970s – Rechenberg's group was able to solve complex engineering problems through evolution strategies.[40][41][42][43] Another approach was the evolutionary programming technique of Lawrence J. Fogel, which was proposed for generating artificial intelligence. Evolutionary programming originally used finite state machines for predicting environments, and used variation and selection to optimize the predictive logics. Genetic algorithms in particular became popular through the work of John Holland in the early 1970s, and particularly his book *Adaptation in Natural and Artificial Systems* (1975). His work originated with studies of cellular automata, conducted by Holland and his students at the University of Michigan. Holland introduced a formalized framework for predicting the quality of the next generation, known as Holland's Schema Theorem. Research in GAs remained largely theoretical until the mid-1980s, when The First International Conference on Genetic Algorithms was held in Pittsburgh, Pennsylvania.

33.6.1 Commercial products

In the late 1980s, General Electric started selling the world's first genetic algorithm product, a mainframe-based toolkit designed for industrial processes.[44] In 1989, Axcelis, Inc. released Evolver, the world's first commercial GA product for desktop computers. The New York Times technology writer John Markoff wrote[45] about Evolver in 1990, and it remained the only interactive commercial genetic algorithm until 1995.[46] Evolver was sold to Palisade in 1997, translated into several languages, and is currently in its 6th version.[47]

33.7 Related techniques

See also: List of genetic algorithm applications

33.7.1 Parent fields

Genetic algorithms are a sub-field of:

- Evolutionary algorithms
- Evolutionary computing
- Metaheuristics
- Stochastic optimization
- Optimization

33.7.2 Related fields

Evolutionary algorithms

Main article: Evolutionary algorithm

Evolutionary algorithms is a sub-field of evolutionary computing.

- Evolution strategies (ES, see Rechenberg, 1994) evolve individuals by means of mutation and intermediate or discrete recombination. ES algorithms are designed particularly to solve problems in the real-value domain. They use self-adaptation to adjust control parameters of the search. De-randomization of self-adaptation has led to the contemporary Covariance Matrix Adaptation Evolution Strategy (CMA-ES).

- Evolutionary programming (EP) involves populations of solutions with primarily mutation and selection and arbitrary representations. They use self-adaptation to adjust parameters, and can include other variation operations such as combining information from multiple parents.

- Estimation of Distribution Algorithm (EDA) substitutes traditional reproduction operators by model-guided operators. Such models are learned from the population by employing machine learning techniques and represented as Probabilistic Graphical Models, from which new solutions can be sampled[48][49] or generated from guided-crossover.[50]

- Gene expression programming (GEP) also uses populations of computer programs. These complex computer programs are encoded in simpler linear chromosomes of fixed length, which are afterwards expressed as expression trees. Expression trees or computer programs evolve because the chromosomes undergo mutation and recombination in a manner similar to the canonical GA. But thanks to the special organization of GEP chromosomes, these genetic modifications always result in valid computer programs.[51]

- Genetic programming (GP) is a related technique popularized by John Koza in which computer programs, rather than function parameters, are optimized. Genetic programming often uses tree-based internal data structures to represent the computer programs for adaptation instead of the list structures typical of genetic algorithms.

- Grouping genetic algorithm (GGA) is an evolution of the GA where the focus is shifted from individual items, like in classical GAs, to groups or subset of items.[52] The idea behind this GA evolution proposed by Emanuel Falkenauer is that solving some complex problems, a.k.a. *clustering* or *partitioning* problems where a set of items must be split into disjoint group of items in an optimal way, would better be achieved by making characteristics of the groups of items equivalent to genes. These kind of problems include bin packing, line balancing, clustering with respect to a distance measure, equal piles, etc., on which classic GAs proved to perform poorly. Making genes equivalent to groups implies chromosomes that are in general of variable length, and special genetic operators that manipulate whole groups of items. For bin packing in particular, a GGA hybridized with the Dominance Criterion of Martello and Toth, is arguably the best technique to date.

- Interactive evolutionary algorithms are evolutionary algorithms that use human evaluation. They are usually applied to domains where it is hard to design a computational fitness function, for example, evolving images, music, artistic designs and forms to fit users' aesthetic preference.

Swarm intelligence

Main article: Swarm intelligence

Swarm intelligence is a sub-field of evolutionary computing.

- Ant colony optimization (**ACO**) uses many ants (or agents) equipped with a pheromone model to traverse the solution space and find locally productive areas. Although considered an Estimation of distribution algorithm.[53]

- Particle swarm optimization (PSO) is a computational method for multi-parameter optimization which also uses population-based approach. A population (swarm) of candidate solutions (particles) moves in the search space, and the movement of the particles is influenced both by their own best known position and swarm's global best known position. Like genetic algorithms, the PSO method depends on information sharing among population members. In some problems the PSO is often more computationally efficient than the GAs, especially in unconstrained problems with continuous variables.[54]

Other evolutionary computing algorithms

Evolutionary computation is a sub-field of the metaheuristic methods.

- Memetic algorithm (MA), often called *hybrid genetic algorithm* among others, is a population-based method in which solutions are also subject to local improvement phases. The idea of memetic algorithms comes from memes, which unlike genes, can adapt themselves. In some problem areas they are shown to be more efficient than traditional evolutionary algorithms.

- Bacteriologic algorithms (BA) inspired by evolutionary ecology and, more particularly, bacteriologic adaptation. Evolutionary ecology is the study of living organisms in the context of their environment, with the aim of discovering how they adapt. Its basic concept is that in a heterogeneous environment, there is not one individual that fits the whole environment. So, one needs to reason at the population level. It is also believed BAs could be successfully applied to complex positioning problems (antennas for cell phones, urban planning, and so on) or data mining.[55]

- Cultural algorithm (CA) consists of the population component almost identical to that of the genetic algorithm and, in addition, a knowledge component called the belief space.

- Differential search algorithm (DS) inspired by migration of superorganisms.[56]

- Gaussian adaptation (normal or natural adaptation, abbreviated NA to avoid confusion with GA) is intended for the maximisation of manufacturing yield of signal processing systems. It may also be used for ordinary parametric optimisation. It relies on a certain theorem valid for all regions of acceptability and all Gaussian distributions. The efficiency of NA relies on information theory and a certain theorem of efficiency. Its efficiency is defined as information divided by the work needed to get the information.[57] Because NA maximises mean fitness rather than the fitness of the individual, the landscape is smoothed such that valleys between peaks may disappear. Therefore it has a certain "ambition" to avoid local peaks in the fitness landscape. NA is also good at climbing sharp crests by adaptation of the moment matrix, because NA may maximise the disorder (average information) of the Gaussian simultaneously keeping the mean fitness constant.

Other metaheuristic methods

Metaheuristic methods broadly fall within stochastic optimisation methods.

- Simulated annealing (SA) is a related global optimization technique that traverses the search space by testing random mutations on an individual solution. A mutation that increases fitness is always accepted. A mutation that lowers fitness is accepted probabilistically based on the difference in fitness and a decreasing temperature parameter. In SA parlance, one speaks of seeking the lowest energy instead of the maximum fitness. SA can also be used within a standard GA algorithm by starting with a relatively high rate of mutation and decreasing it over time along a given schedule.

- Tabu search (TS) is similar to simulated annealing in that both traverse the solution space by testing mutations of an individual solution. While simulated annealing generates only one mutated solution, tabu search generates many mutated solutions and moves to the solution with the lowest energy of those generated. In order to prevent cycling and encourage greater movement through the solution space, a tabu list is maintained of partial or complete solutions. It is forbidden to move to a solution that contains elements of the tabu list, which is updated as the solution traverses the solution space.

- Extremal optimization (EO) Unlike GAs, which work with a population of candidate solutions, EO evolves a single solution and makes local modifications to the worst components. This requires that a suitable representation be selected which permits individual solution components to be assigned a quality measure ("fitness"). The governing principle behind this algorithm is that of *emergent* improvement through selectively removing low-quality components and replacing them with a randomly selected component. This is decidedly at odds with a GA that selects good solutions in an attempt to make better solutions.

Other stochastic optimisation methods

- The cross-entropy (CE) method generates candidates solutions via a parameterized probability distribution. The parameters are updated via cross-entropy minimization, so as to generate better samples in the next iteration.

- Reactive search optimization (RSO) advocates the integration of sub-symbolic machine learning techniques into search heuristics for solving complex optimization problems. The word reactive hints at a ready response to events during the search through an internal online feedback loop for the self-tuning of critical parameters. Methodologies of interest for Reactive Search include machine learning and statistics, in particular reinforcement learning, active or query learning, neural networks, and meta-heuristics.

33.8 See also

- List of genetic algorithm applications
- Genetic algorithms in signal processing (a.k.a. particle filters)
- Propagation of schema
- Universal Darwinism
- Metaheuristics
- Learning classifier system
- Rule-based machine learning

33.9 References

[1] Mitchell 1996, p. 2.

[2] Whitley 1994, p. 66.

[3] Eiben, A. E. et al (1994). "Genetic algorithms with multi-parent recombination". PPSN III: Proceedings of the International Conference on Evolutionary Computation. The Third Conference on Parallel Problem Solving from Nature: 78–87. ISBN 3-540-58484-6.

[4] Ting, Chuan-Kang (2005). "On the Mean Convergence Time of Multi-parent Genetic Algorithms Without Selection". Advances in Artificial Life: 403–412. ISBN 978-3-540-28848-0.

[5] Akbari, Ziarati (2010). "A multilevel evolutionary algorithm for optimizing numerical functions" IJIEC 2 (2011): 419–430

[6] Deb, Kalyanmoy; Spears, William M. (1997). "C6.2: Speciation methods". *Handbook of Evolutionary Computation* (PDF). Institute of Physics Publishing.

[7] Shir, Ofer M. (2012). "Niching in Evolutionary Algorithms". In Rozenberg, Grzegorz; Bäck, Thomas; Kok, Joost N. *Handbook of Natural Computing*. Springer Berlin Heidelberg. pp. 1035–1069. ISBN 9783540929093. doi:10.1007/978-3-540-92910-9_32.

[8] Goldberg 1989, p. 41.

[9] Harik, Georges R.; Lobo, Fernando G.; Sastry, Kumara (1 January 2006). "Linkage Learning via Probabilistic Modeling in the Extended Compact Genetic Algorithm (ECGA)". *Scalable Optimization via Probabilistic Modeling*. Springer Berlin Heidelberg: 39–61. doi:10.1007/978-3-540-34954-9_3.

[10] Pelikan, Martin; Goldberg, David E.; Cantú-Paz, Erick (1 January 1999). "BOA: The Bayesian Optimization Algorithm". *Proceedings of the 1st Annual Conference on Genetic and Evolutionary Computation - Volume 1*. Morgan Kaufmann Publishers Inc.: 525–532.

[11] Coffin, David; Smith, Robert E. (1 January 2008). "Linkage Learning in Estimation of Distribution Algorithms". *Linkage in Evolutionary Computation*. Springer Berlin Heidelberg: 141–156. doi:10.1007/978-3-540-85068-7_7.

[12] Echegoyen, Carlos; Mendiburu, Alexander; Santana, Roberto; Lozano, Jose A. (8 November 2012). "On the Taxonomy of Optimization Problems Under Estimation of Distribution Algorithms". *Evolutionary Computation*. **21** (3): 471–495. ISSN 1063-6560. doi:10.1162/EVCO_a_00095.

[13] Sadowski, Krzysztof L.; Bosman, Peter A.N.; Thierens, Dirk (1 January 2013). "On the Usefulness of Linkage Processing for Solving MAX-SAT". *Proceedings of the 15th Annual Conference on Genetic and Evolutionary Computation*. ACM: 853–860. doi:10.1145/2463372.2463474.

[14] Taherdangkoo, Mohammad; Paziresh, Mahsa; Yazdi, Mehran; Bagheri, Mohammad Hadi (19 November 2012). "An efficient algorithm for function optimization: modified stem cells algorithm". *Central European Journal of Engineering*. **3** (1): 36–50. doi:10.2478/s13531-012-0047-8.

[15] Wolpert, D.H., Macready, W.G., 1995. No Free Lunch Theorems for Optimisation. Santa Fe Institute, SFI-TR-05-010. Santa Fe.

[16] Goldberg, David E. (1991). "The theory of virtual alphabets". *Parallel Problem Solving from Nature, Lecture Notes in Computer Science*. **496**: 13–22. doi:10.1007/BFb0029726. Retrieved 2 July 2013.

[17] Janikow, C. Z.; Michalewicz, Z. (1991). "An Experimental Comparison of Binary and Floating Point Representations in Genetic Algorithms" (PDF). *Proceedings of the Fourth International Conference on Genetic Algorithms*: 31–36. Retrieved 2 July 2013.

[18] Patrascu, M.; Stancu, A.F.; Pop, F. (2014). "HELGA: a heterogeneous encoding lifelike genetic algorithm for population evolution modeling and simulation". *Soft Computing*. **18**: 2565–2576. doi:10.1007/s00500-014-1401-y.

[19] Baluja, Shumeet; Caruana, Rich (1995). *Removing the genetics from the standard genetic algorithm* (PDF). ICML.

[20] Srinivas. M and Patnaik. L, "Adaptive probabilities of crossover and mutation in genetic algorithms," IEEE Transactions on System, Man and Cybernetics, vol.24, no.4, pp.656–667, 1994.

[21] ZHANG. J, Chung. H and Lo. W. L, "Clustering-Based Adaptive Crossover and Mutation Probabilities for Genetic Algorithms", IEEE Transactions on Evolutionary Computation vol.11, no.3, pp. 326–335, 2007.

[22] Evolution-in-a-nutshell

[23] D.E. Goldberg, B. Korb, and K. Deb. "Messy genetic algorithms: Motivation, analysis, and first results". Complex Systems, 5(3):493–530, October 1989.

[24] Gene expression: The missing link in evolutionary computation

[25] G. Harik. Learning linkage to efficiently solve problems of bounded difficulty using genetic algorithms. PhD thesis, Dept. Computer Science, University of Michigan, Ann Arbour, 1997

[26] Tomoiagă B, Chindriş M, Sumper A, Sudria-Andreu A, Villafafila-Robles R. Pareto Optimal Reconfiguration of Power Distribution Systems Using a Genetic Algorithm Based on NSGA-II. Energies. 2013; 6(3):1439-1455.

[27] Gross, Bill. "A solar energy system that tracks the sun". *TED*. Retrieved 20 November 2013.

[28] Hornby, G. S.; Linden, D. S.; Lohn, J. D., *Automated Antenna Design with Evolutionary Algorithms* (PDF)

[29] http://goatstream.com/research/papers/SA2013/index.html

[30] Skiena, Steven (2010). *The Algorithm Design Manual* (2nd ed.). Springer Science+Business Media. ISBN 1-849-96720-2.

[31] Turing, Alan M. "Computing machinery and intelligence". *Mind*. **LIX** (238): 433–460. doi:10.1093/mind/LIX.236.433.

[32] Barricelli, Nils Aall (1954). "Esempi numerici di processi di evoluzione". *Methodos*: 45–68.

[33] Barricelli, Nils Aall (1957). "Symbiogenetic evolution processes realized by artificial methods". *Methodos*: 143–182.

[34] Fraser, Alex (1957). "Simulation of genetic systems by automatic digital computers. I. Introduction". *Aust. J. Biol. Sci.* **10**: 484–491.

[35] Fraser, Alex; Burnell, Donald (1970). *Computer Models in Genetics*. New York: McGraw-Hill. ISBN 0-07-021904-4.

[36] Crosby, Jack L. (1973). *Computer Simulation in Genetics*. London: John Wiley & Sons. ISBN 0-471-18880-8.

[37] 02.27.96 - UC Berkeley's Hans Bremermann, professor emeritus and pioneer in mathematical biology, has died at 69

[38] Fogel, David B. (editor) (1998). *Evolutionary Computation: The Fossil Record*. New York: IEEE Press. ISBN 0-7803-3481-7.

[39] Barricelli, Nils Aall (1963). "Numerical testing of evolution theories. Part II. Preliminary tests of performance, symbiogenesis and terrestrial life". *Acta Biotheoretica* (16): 99–126.

[40] Rechenberg, Ingo (1973). *Evolutionsstrategie*. Stuttgart: Holzmann-Froboog. ISBN 3-7728-0373-3.

[41] Schwefel, Hans-Paul (1974). *Numerische Optimierung von Computer-Modellen (PhD thesis)*.

[42] Schwefel, Hans-Paul (1977). *Numerische Optimierung von Computor-Modellen mittels der Evolutionsstrategie : mit einer vergleichenden Einführung in die Hill-Climbing- und Zufallsstrategie*. Basel; Stuttgart: Birkhäuser. ISBN 3-7643-0876-1.

[43] Schwefel, Hans-Paul (1981). *Numerical optimization of computer models (Translation of 1977 Numerische Optimierung von Computor-Modellen mittels der Evolutionsstrategie*. Chichester ; New York: Wiley. ISBN 0-471-09988-0.

[44] Aldawoodi, Namir (2008). *An Approach to Designing an Unmanned Helicopter Autopilot Using Genetic Algorithms and Simulated Annealing*. ProQuest. p. 99. ISBN 0549773495 – via Google Books.

[45] Markoff, John (29 August 1990). "What's the Best Answer? It's Survival of the Fittest". New York Times. Retrieved 2016-07-13.

[46] Ruggiero, Murray A.. (2009-08-01) Fifteen years and counting. Futuresmag.com. Retrieved on 2013-08-07.

[47] Evolver: Sophisticated Optimization for Spreadsheets. Palisade. Retrieved on 2013-08-07.

[48] Pelikan, Martin; Goldberg, David E.; Cantú-Paz, Erick (1 January 1999). "BOA: The Bayesian Optimization Algorithm". *Proceedings of the 1st Annual Conference on Genetic and Evolutionary Computation - Volume 1*. Morgan Kaufmann Publishers Inc.: 525–532.

[49] Pelikan, Martin (2005). *Hierarchical Bayesian optimization algorithm : toward a new generation of evolutionary algorithms* (1st ed.). Berlin [u.a.]: Springer. ISBN 978-3-540-23774-7.

[50] Thierens, Dirk (11 September 2010). "The Linkage Tree Genetic Algorithm". *Parallel Problem Solving from Nature, PPSN XI*. Springer Berlin Heidelberg: 264–273. doi:10.1007/978-3-642-15844-5_27.

[51] Ferreira, C. "Gene Expression Programming: A New Adaptive Algorithm for Solving Problems" (PDF). Complex Systems, Vol. 13, issue 2: 87-129.

[52] Falkenauer, Emanuel (1997). *Genetic Algorithms and Grouping Problems*. Chichester, England: John Wiley & Sons Ltd. ISBN 978-0-471-97150-4.

[53] Zlochin, Mark; Birattari, Mauro; Meuleau, Nicolas; Dorigo, Marco (1 October 2004). "Model-Based Search for Combinatorial Optimization: A Critical Survey". *Annals of Operations Research*. **131** (1-4): 373–395. ISSN 0254-5330. doi:10.1023/B:ANOR.0000039526.52305.af.

[54] Rania Hassan, Babak Cohanim, Olivier de Weck, Gerhard Venter (2005) A comparison of particle swarm optimization and the genetic algorithm

[55] Baudry, Benoit; Franck Fleurey; Jean-Marc Jézéquel; Yves Le Traon (March–April 2005). "Automatic Test Case Optimization: A Bacteriologic Algorithm" (PDF). *IEEE Software*. IEEE Computer Society. **22** (2): 76–82. doi:10.1109/MS.2005.30. Retrieved 9 August 2009.

[56] Civicioglu, P. (2012). "Transforming Geocentric Cartesian Coordinates to Geodetic Coordinates by Using Differential Search Algorithm". *Computers &Geosciences*. **46**: 229–247. doi:10.1016/j.cageo.2011.12.011.

[57] Kjellström, G. (December 1991). "On the Efficiency of Gaussian Adaptation". *Journal of Optimization Theory and Applications*. **71** (3): 589–597. doi:10.1007/BF00941405.

33.10 Bibliography

- Banzhaf, Wolfgang; Nordin, Peter; Keller, Robert; Francone, Frank (1998). *Genetic Programming – An Introduction*. San Francisco, CA: Morgan Kaufmann. ISBN 978-1558605107.

- Bies, Robert R.; Muldoon, Matthew F.; Pollock, Bruce G.; Manuck, Steven; Smith, Gwenn; Sale, Mark E. (2006). "A Genetic Algorithm-Based, Hybrid Machine Learning Approach to Model Selection". *Journal of Pharmacokinetics and Pharmacodynamics*. Netherlands: Springer: 196–221.

- Cha, Sung-Hyuk; Tappert, Charles C. (2009). "A Genetic Algorithm for Constructing Compact Binary Decision Trees". *Journal of Pattern Recognition Research*. **4** (1): 1–13. doi:10.13176/11.44.

- Fraser, Alex S. (1957). "Simulation of Genetic Systems by Automatic Digital Computers. I. Introduction". *Australian Journal of Biological Sciences*. **10**: 484–491.

- Goldberg, David (1989). *Genetic Algorithms in Search, Optimization and Machine Learning*. Reading, MA: Addison-Wesley Professional. ISBN 978-0201157673.

- Goldberg, David (2002). *The Design of Innovation: Lessons from and for Competent Genetic Algorithms*. Norwell, MA: Kluwer Academic Publishers. ISBN 978-1402070983.

- Fogel, David. *Evolutionary Computation: Toward a New Philosophy of Machine Intelligence* (3rd ed.). Piscataway, NJ: IEEE Press. ISBN 978-0471669517.

- Holland, John (1992). *Adaptation in Natural and Artificial Systems*. Cambridge, MA: MIT Press. ISBN 978-0262581110.

- Koza, John (1992). *Genetic Programming: On the Programming of Computers by Means of Natural Selection*. Cambridge, MA: MIT Press. ISBN 978-0262111706.

- Michalewicz, Zbigniew (1996). *Genetic Algorithms + Data Structures = Evolution Programs*. Springer-Verlag. ISBN 978-3540606765.

- Mitchell, Melanie (1996). *An Introduction to Genetic Algorithms*. Cambridge, MA: MIT Press. ISBN 9780585030944.

- Poli, R.; Langdon, W. B.; McPhee, N. F. (2008). *A Field Guide to Genetic Programming*. Lulu.com, freely available from the internet. ISBN 978-1-4092-0073-4.

- Rechenberg, Ingo (1994): Evolutionsstrategie '94. Stuttgart: Fromman-Holzboog.

- Schmitt, Lothar M; Nehaniv, Chrystopher L; Fujii, Robert H (1998), *Linear analysis of genetic algorithms*, Theoretical Computer Science 208: 111–148

- Schmitt, Lothar M (2001), *Theory of Genetic Algorithms*, Theoretical Computer Science 259: 1–61

- Schmitt, Lothar M (2004), *Theory of Genetic Algorithms II: models for genetic operators over the string-tensor representation of populations and convergence to global optima for arbitrary fitness function under scaling*, Theoretical Computer Science 310: 181–231

- Schwefel, Hans-Paul (1974): Numerische Optimierung von Computer-Modellen (PhD thesis). Reprinted by Birkhäuser (1977).

- Vose, Michael (1999). *The Simple Genetic Algorithm: Foundations and Theory*. Cambridge, MA: MIT Press. ISBN 978-0262220583.

- Whitley, Darrell (1994). "A genetic algorithm tutorial". *Statistics and Computing*. 4 (2): 65–85. doi:10.1007/BF00175354.

- Hingston, Philip; Barone, Luigi; Michalewicz, Zbigniew (2008). *Design by Evolution: Advances in Evolutionary Design*. Springer. ISBN 978-3540741091.

- Eiben, Agoston; Smith, James (2003). *Introduction to Evolutionary Computing*. Springer. ISBN 978-3540401841.

- "Essentials of Metaheuristics", 2009 (225 p). Free open text by Sean Luke.

- Global Optimization Algorithms – Theory and Application

- Genetic Algorithms in Python Tutorial with the intuition behind GAs and Python implementation.

33.11 External links

33.11.1 Resources

- Provides a list of resources in the genetic algorithms field

33.11.2 Tutorials

- Genetic Algorithms - Computer programs that "evolve" in ways that resemble natural selection can solve complex problems even their creators do not fully understand An excellent introduction to GA by John Holland and with an application to the Prisoner's Dilemma

- An online interactive GA tutorial for a reader to practise or learn how a GA works: Learn step by step or watch global convergence in batch, change the population size, crossover rates/bounds, mutation rates/bounds and selection mechanisms, and add constraints.

- A Genetic Algorithm Tutorial by Darrell Whitley Computer Science Department Colorado State University An excellent tutorial with lots of theory

Chapter 34

Genetic programming

Not to be confused with Generic programming or Genetic engineering.

In artificial intelligence, **genetic programming** (**GP**) is a technique whereby computer programs are encoded as a set of genes that are then modified (evolved) using an evolutionary algorithm (often a genetic algorithm, "GA") – it is an application of (for example) genetic algorithms where the space of solutions consists of computer programs. The results are computer programs able to perform well in a predefined task. The methods used to encode a computer program in an artificial chromosome and to evaluate its fitness with respect to the predefined task are central in the GP technique and still the subject of active research.

34.1 History

In 1954, pioneering work on what is today known as artificial life was carried out by Nils Aall Barricelli using the very early computers.[1] In the 1960s and early 1970s, evolutionary algorithms became widely recognized as optimization methods. Ingo Rechenberg and his group were able to solve complex engineering problems through evolution strategies as documented in his 1971 PhD thesis and the resulting 1973 book. John Holland was highly influential during the 1970s. The establishment of evolutionary algorithms in the scientific community allowed, by then, the first concrete steps to study the GP idea.

In 1964, Lawrence J. Fogel, one of the earliest practitioners of the GP methodology, applied evolutionary algorithms to the problem of discovering finite-state automata. Later GP-related work grew out of the learning classifier system community, which developed sets of sparse rules describing optimal policies for Markov decision processes. In 1981 Richard Forsyth evolved tree rules to classify heart disease.[2] The first statement of modern "tree-based" genetic programming (that is, procedural languages organized in tree-based structures and operated on by suitably defined GA-operators) was given by Nichael L. Cramer (1985).[3] This work was later greatly expanded by John R. Koza, a main proponent of GP who has pioneered the application of genetic programming in various complex optimization and search problems.[4] Gianna Giavelli, a student of Koza's, later pioneered the use of genetic programming as a technique to model DNA expression.[5]

In the 1990s, GP was mainly used to solve relatively simple problems because it is very computationally intensive. Recently GP has produced many novel and outstanding results in areas such as quantum computing, electronic design, game playing, cyberterrorism prevention,[6] sorting, and searching, due to improvements in GP technology and the exponential growth in CPU power.[7] These results include the replication or development of several post-year-2000 inventions. GP has also been applied to evolvable hardware as well as computer programs.

Developing a theory for GP has been very difficult and so in the 1990s GP was considered a sort of outcast among search techniques.

34.2 Program representation

Main article: genetic representation

GP evolves computer programs, traditionally represented in memory as tree structures.[3] Trees can be easily evaluated in a recursive manner. Every tree node has an operator function and every terminal node has an operand, making mathematical expressions easy to evolve and evaluate. Thus traditionally GP favors the use of programming languages that naturally embody tree structures (for example, Lisp; other functional programming languages are also suitable).

Non-tree representations have been suggested and successfully implemented, such as linear genetic programming which suits the more traditional imperative languages [see, for example, Banzhaf et al. (1998)].[8] The commercial

34.4 Meta-genetic programming

Meta-genetic programming is the proposed meta learning technique of evolving a genetic programming system using genetic programming itself. It suggests that chromosomes, crossover, and mutation were themselves evolved, therefore like their real life counterparts should be allowed to change on their own rather than being determined by a human programmer. Meta-GP was formally proposed by Jürgen Schmidhuber in 1987.[13] Doug Lenat's Eurisko is an earlier effort that may be the same technique. It is a recursive but terminating algorithm, allowing it to avoid infinite recursion.

Critics of this idea often say this approach is overly broad in scope. However, it might be possible to constrain the fitness criterion onto a general class of results, and so obtain an evolved GP that would more efficiently produce results for sub-classes. This might take the form of a meta evolved GP for producing human walking algorithms which is then used to evolve human running, jumping, etc. The fitness criterion applied to the meta GP would simply be one of efficiency.

For general problem classes there may be no way to show that meta GP will reliably produce results more efficiently than a created algorithm other than exhaustion.

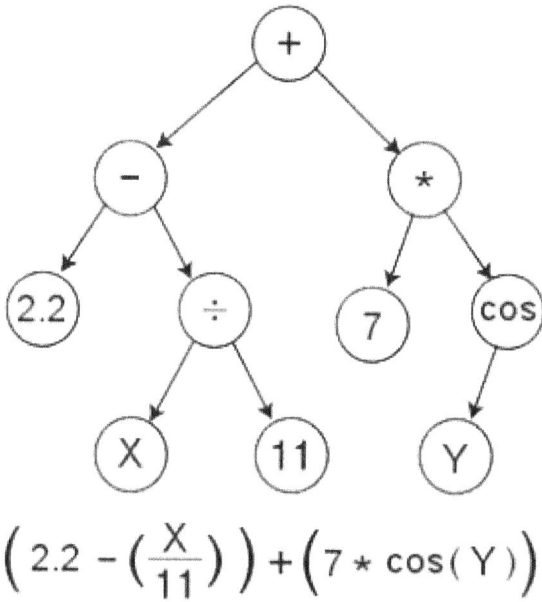

A function represented as a tree structure.

GP software *Discipulus* uses automatic induction of binary machine code ("AIM")[9] to achieve better performance. μGP[10] uses directed multigraphs to generate programs that fully exploit the syntax of a given assembly language

Most non-tree representations have structurally noneffective code (introns). Such non-coding genes may seem to be useless, because they have no effect on the performance of any one individual. However, experiments seem to show faster convergence when using program representations — such as linear genetic programming and Cartesian genetic programming — that allow such non-coding genes, compared to tree-based program representations that do not have any non-coding genes.[11][12]

34.3 Other approaches

The basic ideas of genetic programming have been modified and extended in a variety of ways:

- Extended compact genetic programming (ECGP)
- Embedded Cartesian genetic programming (ECGP)
- Probabilistic incremental program evolution (PIPE)
- Strongly typed genetic programming (STGP)

34.5 See also

- Bio-inspired computing
- Covariance Matrix Adaptation Evolution Strategy (CMA-ES)
- Fitness approximation
- Gene expression programming
- Genetic improvement
- Genetic representation
- Grammatical evolution
- Inductive programming
- Linear genetic programming
- Multi expression programming
- Propagation of schema

34.6 References

[1] Barricelli, Nils (1954). "Esempi numerici di processi di evoluzione" [Numerical examples of evolution processes]. *Methodos* (in Italian). **6** (21-22): 45–68.

[2] "EmeraldInsight". *Kybernetes*. **10**: 159–166. doi:10.1108/eb005587.

[3] Nichael L. Cramer "A Representation for the Adaptive Generation of Simple Sequential Programs".

[4] "genetic-programming.com-Home-Page".

[5] The Genetic Coding of Behavioral Attributes in Cellular Automata. Artificial Life at Stanford 1994 Stanford, California, 94305-3079 USA.

[6] Hansen, James V.; Benjamin Lowry, Paul; Meservy, Rayman; McDonald, Dan (2007). "Genetic programming for prevention of cyberterrorism through dynamic and evolving intrusion detection". *Decision Support Systems*. **43** (4): 1362–1374. SSRN 877981. doi:10.1016/j.dss.2006.04.004.

[7] John R. Koza. "36 Human-Competitive Results Produced by Genetic Programming". retrieved 2015-09-01.

[8] Garnett Wilson and Wolfgang Banzhaf. "A Comparison of Cartesian Genetic Programming and Linear Genetic Programming".

[9] (Peter Nordin, 1997, Banzhaf et al., 1998, Section 11.6.2-11.6.3)

[10] Giovanni Squillero. "µGP (MicroGP)".

[11] Julian F. Miller. "Cartesian Genetic Programming". p. 19.

[12] Janet Clegg; James Alfred Walker; Julian Francis Miller. A New Crossover Technique for Cartesian Genetic Programming". 2007.

[13] "1987 THESIS ON LEARNING HOW TO LEARN, METALEARNING, META GENETIC PROGRAMMING,CREDIT-CONSERVING MACHINE LEARNING ECONOMY".

- Riccardo Poli, William B. Langdon,Nicholas F. McPhee, John R. Koza, "A Field Guide to Genetic Programming" (2008)

34.7 External links

- Aymen S Saket & Mark C Sinclair
- *Genetic Programming and Evolvable Machines*, a journal
- *Evo2 for genetic programming*
- GP bibliography
- The Hitch-Hiker's Guide to Evolutionary Computation

Chapter 35

Artificial life

This article is about a field of research. For artificially created life forms, see Synthetic life. For the journal, see Artificial Life (journal).
"ALife" redirects here. It is not to be confused with Alife.

Artificial life (often abbreviated **ALife** or **A-Life**) is a field of study wherein researchers examine systems related to natural life, its processes, and its evolution, through the use of simulations with computer models, robotics, and biochemistry.[1] The discipline was named by Christopher Langton, an American theoretical biologist, in 1986.[2] There are three main kinds of alife,[3] named for their approaches: *soft*,[4] from software; *hard*,[5] from hardware; and *wet*, from biochemistry. Artificial life researchers study traditional biology by trying to recreate aspects of biological phenomena.[6]

A Braitenberg vehicle simulation, programmed in breve, an artificial life simulator

35.1 Overview

Artificial life studies the fundamental processes of living systems in artificial environments in order to gain a deeper understanding of the complex information processing that define such systems. These topics are broad, but often include evolutionary dynamics, emergent properties of collective systems, biomimicry, as well as related issues about the philosophy of the nature of life and the use of lifelike properties in artistic works.

35.2 Philosophy

The modeling philosophy of artificial life strongly differs from traditional modeling by studying not only "life-as-we-know-it" but also "life-as-it-might-be".[7]

A traditional model of a biological system will focus on capturing its most important parameters. In contrast, an alife modeling approach will generally seek to decipher the most simple and general principles underlying life and implement them in a simulation. The simulation then offers the possibility to analyse new and different lifelike systems.

Vladimir Georgievich Red'ko proposed to generalize this distinction to the modeling of any process, leading to the more general distinction of "processes-as-we-know-them" and "processes-as-they-could-be".[8]

At present, the commonly accepted definition of life does not consider any current alife simulations or software to be alive, and they do not constitute part of the evolutionary process of any ecosystem. However, different opinions about artificial life's potential have arisen:

- The *strong alife* (cf. Strong AI) position states that "life is a process which can be abstracted away from any particular medium" (John von Neumann). Notably, Tom Ray declared that his program Tierra is not simulating life in a computer but synthesizing it.[9]

- The *weak alife* position denies the possibility of generating a "living process" outside of a chemical solution. Its researchers try instead to simulate life processes to understand the underlying mechanics of biological phenomena.

35.3 Organizations

Main article: Artificial life organizations

35.4 Software-based ("soft")

35.4.1 Techniques

- Cellular automata were used in the early days of artificial life, and are still often used for ease of scalability and parallelization. Alife and cellular automata share a closely tied history.

- Neural networks are sometimes used to model the brain of an agent. Although traditionally more of an artificial intelligence technique, neural nets can be important for simulating population dynamics of organisms that can *learn*. The symbiosis between learning and evolution is central to theories about the development of instincts in organisms with higher neurological complexity, as in, for instance, the Baldwin effect.

35.4.2 Notable simulators

This is a list of artificial life/digital organism simulators, organized by the method of creature definition.

Program-based

Further information: programming game

Program-based simulations contain organisms with a complex DNA language, usually Turing complete. This language is more often in the form of a computer program than actual biological DNA. Assembly derivatives are the most common languages used. An organism "lives" when its code is executed, and there are usually various methods allowing self-replication. Mutations are generally implemented as random changes to the code. Use of cellular automata is common but not required. Another example could be an artificial intelligence and multi-agent system/program.

Module-based

Individual modules are added to a creature. These modules modify the creature's behaviors and characteristics either directly, by hard coding into the simulation (leg type A increases speed and metabolism), or indirectly, through the emergent interactions between a creature's modules (leg type A moves up and down with a frequency of X, which interacts with other legs to create motion). Generally these are simulators which emphasize user creation and accessibility over mutation and evolution.

Parameter-based

Organisms are generally constructed with pre-defined and fixed behaviors that are controlled by various parameters that mutate. That is, each organism contains a collection of numbers or other *finite* parameters. Each parameter controls one or several aspects of an organism in a well-defined way.

Neural net–based

These simulations have creatures that learn and grow using neural nets or a close derivative. Emphasis is often, although not always, more on learning than on natural selection.

35.4.3 Complex systems modelling

Mathematical models of complex systems are of three types: black-box (phenomenological), white-box (mechanistic, based on the first principles) and grey-box (mixtures of phenomenological and mechanistic models) [10][11]. In black-box models, the individual-based (mechanistic) mechanisms of a complex dynamic system remain hidden.

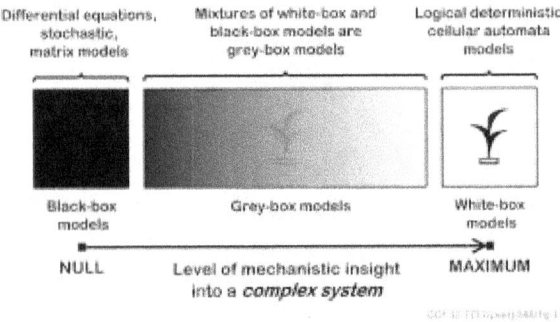

Mathematical models for complex systems

Black-box models are completely nonmechanistic. They are phenomenological and ignore a composition and internal structure of a complex system. We cannot investigate interactions of subsystems of such a non-transparent model. A white-box model of complex dynamic system has 'transparent walls' and directly shows underlying mechanisms. All events at micro-, meso- and macro-levels of a dynamic system are directly visible at all stages of its white-box model evolution. In most cases mathematical modelers use the heavy black-box mathematical methods, which cannot produce mechanistic models of complex dynamic systems. Grey-box models are intermediate and combine black-box and white-box approaches.

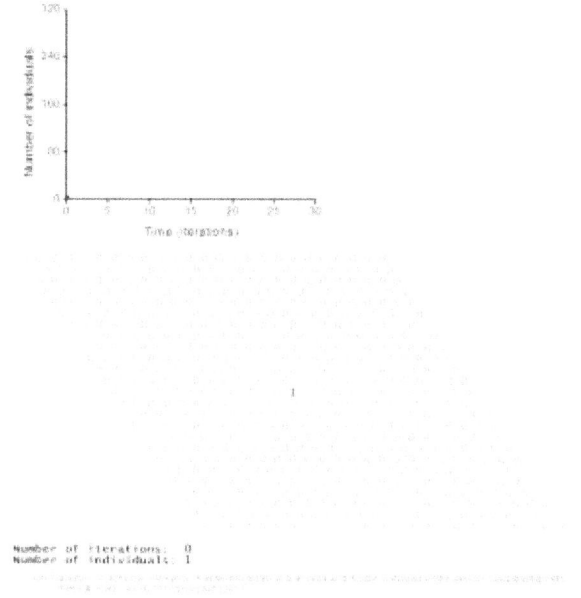

Logical deterministic individual-based cellular automata model of single species population growth

Creation of a white-box model of complex system is associated with the problem of the necessity of an a priori basic knowledge of the modeling subject. The deterministic logical cellular automata are necessary but not sufficient condition of a white-box model. The second necessary prerequisite of a white-box model is the presence of the physical ontology of the object under study. The white-box modeling represents an automatic hyper-logical inference from the first principles because it is completely based on the deterministic logic and axiomatic theory of the subject. The purpose of the white-box modeling is to derive from the basic axioms a more detailed, more concrete mechanistic knowledge about the dynamics of the object under study. The necessity to formulate an intrinsic axiomatic system of the subject before creating its white-box model distinguishes the cellular automata models of white-box type from cellular automata models based on arbitrary logical rules. If cellular automata rules have not been formulated from the first principles of the subject, then such a model may have a weak relevance to the real problem [11].

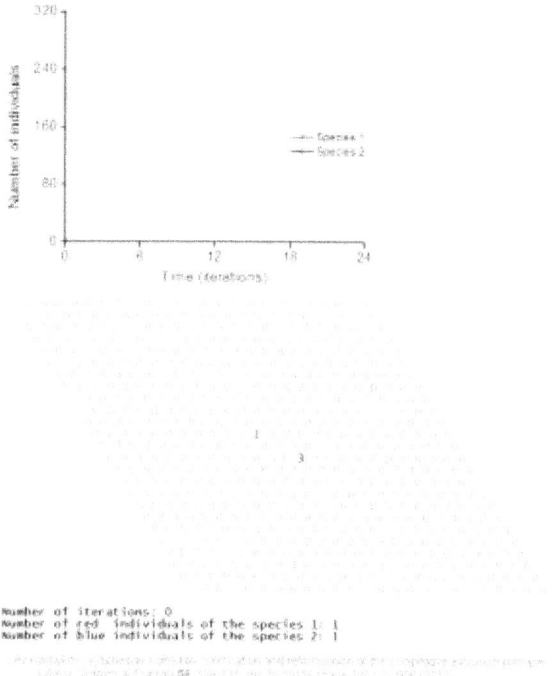

Logical deterministic individual-based cellular automata model of interspecific competition for a single limited resource

35.5 Hardware-based ("hard")

Further information: Robot

Hardware-based artificial life mainly consist of *robots*, that is, automatically guided machines able to do tasks on their own.

35.6 Biochemical-based ("wet")

Further information: Synthetic biology

Biochemical-based life is studied in the field of synthetic biology. It involves e.g. the creation of synthetic DNA. The term "wet" is an extension of the term "wetware".

35.7 Open problems

How does life arise from the nonliving?[12][13]

- Generate a molecular proto-organism in vitro.
- Achieve the transition to life in an artificial chemistry in silico.

- Determine whether fundamentally novel living organizations can exist.
- Simulate a unicellular organism over its entire life cycle.
- Explain how rules and symbols are generated from physical dynamics in living systems.

What are the potentials and limits of living systems?

- Determine what is inevitable in the open-ended evolution of life.
- Determine minimal conditions for evolutionary transitions from specific to generic response systems.
- Create a formal framework for synthesizing dynamical hierarchies at all scales.
- Determine the predictability of evolutionary consequences of manipulating organisms and ecosystems.
- Develop a theory of information processing, information flow, and information generation for evolving systems.

How is life related to mind, machines, and culture?

- Demonstrate the emergence of intelligence and mind in an artificial living system.
- Evaluate the influence of machines on the next major evolutionary transition of life.
- Provide a quantitative model of the interplay between cultural and biological evolution.
- Establish ethical principles for artificial life.

35.8 Related subjects

1. Artificial intelligence has traditionally used a top down approach, while alife generally works from the bottom up.[14]
2. Artificial chemistry started as a method within the alife community to abstract the processes of chemical reactions.
3. Evolutionary algorithms are a practical application of the weak alife principle applied to optimization problems. Many optimization algorithms have been crafted which borrow from or closely mirror alife techniques. The primary difference lies in explicitly defining the fitness of an agent by its ability to solve a problem, instead of its ability to find food, reproduce, or avoid death. The following is a list of evolutionary algorithms closely related to and used in alife:
   - Ant colony optimization
   - Bacterial colony optimization
   - Genetic algorithm
   - Genetic programming
   - Swarm intelligence
4. Multi-agent system – A multi-agent system is a computerized system composed of multiple interacting intelligent agents within an environment.
5. Evolutionary art uses techniques and methods from artificial life to create new forms of art.
6. Evolutionary music uses similar techniques, but applied to music instead of visual art.
7. Abiogenesis and the origin of life sometimes employ alife methodologies as well.

35.9 History

Main article: History of artificial life

35.10 Criticism

Alife has had a controversial history. John Maynard Smith criticized certain artificial life work in 1994 as "fact-free science".[15]

35.11 See also

- Artificial consciousness
- Applications of artificial intelligence
- Artificial life organizations
- Autonomous foraging
- Bioethics
- Complex adaptive system
- Darwin machine
- Digital morphogenesis
- Emergence

- Life simulation game
- List of emerging technologies
- Mathematical biology
- Multi-agent system
- Outline of artificial intelligence
- Player Project
- Simulated reality
- Social simulation
- Soda Constructor
- Swarm intelligence
- Synthetic life
- Universal Darwinism
- Webots

35.12 References

[1] "Dictionary.com definition". Retrieved 2007-01-19.

[2] The MIT Encyclopedia of the Cognitive Sciences, The MIT Press, p.37. ISBN 978-0-262-73144-7

[3] Mark A. Bedau (November 2003). "Artificial life: organization, adaptation and complexity from the bottom up" (PDF). Trends in Cognitive Sciences. Retrieved 2007-01-19.

[4] Maciej Komosinski and Andrew Adamatzky (2009). *Artificial Life Models in Software*. New York: Springer. ISBN 978-1-84882-284-9.

[5] Andrew Adamatzky and Maciej Komosinski (2009). *Artificial Life Models in Hardware*. New York: Springer. ISBN 978-1-84882-529-1.

[6] Langton, Christopher. "What is Artificial Life?". Archived from the original on 17 January 2007. Retrieved 2007-01-19.

[7] See Langton, C. G. 1992. Artificial Life Archived March 11, 2007, at the Wayback Machine.. Addison-Wesley. ., section 1

[8] See Red'ko. V. G. 1999. Mathematical Modeling of Evolution. in: F. Heylighen, C. Joslyn and V. Turchin (editors): Principia Cybernetica Web (Principia Cybernetica, Brussels). For the importance of ALife modeling from a cosmic perspective, see also Vidal, C. 2008.The Future of Scientific Simulations: from Artificial Life to Artificial Cosmogenesis. In Death And Anti-Death, ed. Charles Tandy, 6: Thirty Years After Kurt Gödel (1906-1978) p. 285-318. Ria University Press.)

[9] Ray, Thomas (1991). Taylor, C. C.; Farmer, J. D.; Rasmussen, S, eds. "An approach to the synthesis of life". *Artificial Life II, Santa Fe Institute Studies in the Sciences of Complexity*. Redwood City, CA: Addison-Wesley. **XI**: 371–408. Archived from the original on 2015-07-11. Retrieved 24 January 2016. The intent of this work is to synthesize rather than simulate life.

[10] Kalmykov, Lev V.; Kalmykov, Vyacheslav L. (2015), "A Solution to the Biodiversity Paradox by Logical Deterministic Cellular Automata", *Acta Biotheoretica*: 1–19, doi:10.1007/s10441-015-9257-9

[11] Kalmykov, Lev V.; Kalmykov, Vyacheslav L. (2015), "A white-box model of S-shaped and double S-shaped single-species population growth", *PeerJ*. 3:e948. doi:10.7717/peerj.948

[12] "Libarynth". Retrieved 2015-05-11.

[13] "Caltech" (PDF). Retrieved 2015-05-11.

[14] "AI Beyond Computer Games". Archived from the original on 2008-07-01. Retrieved 2008-07-04.

[15] Horgan, J. 1995. From Complexity to Perplexity. Scientific American. p107

35.13 External links

- Artificial life at DMOZ
- International Society of Artificial Life
- *Artificial Life* journal, at MIT Press Journal
- The Artificial Life Lab, a virtual environment lab
- aDiatomea: an artificial life experiment using 3d generated diatoms
- JSimLife: an artificial life environment using DNA and neural networks

Chapter 36

Machine learning

For the journal, see Machine Learning (journal).

Machine learning is a field of computer science that gives computers the ability to learn without being explicitly programmed.[1][2]

Arthur Samuel, an American pioneer in the field of computer gaming and artificial intelligence, coined the term "Machine Learning" in 1959 while at IBM[3]. Evolved from the study of pattern recognition and computational learning theory in artificial intelligence,[4] machine learning explores the study and construction of algorithms that can learn from and make predictions on data[5] – such algorithms overcome following strictly static program instructions by making data-driven predictions or decisions,[6]:2 through building a model from sample inputs. Machine learning is employed in a range of computing tasks where designing and programming explicit algorithms with good performance is difficult or infeasible; example applications include email filtering, detection of network intruders or malicious insiders working towards a data breach,[7] optical character recognition (OCR),[8] learning to rank, and computer vision.

Machine learning is closely related to (and often overlaps with) computational statistics, which also focuses on prediction-making through the use of computers. It has strong ties to mathematical optimization, which delivers methods, theory and application domains to the field. Machine learning is sometimes conflated with data mining,[9] where the latter subfield focuses more on exploratory data analysis and is known as unsupervised learning.[6]:vii[10] Machine learning can also be unsupervised[11] and be used to learn and establish baseline behavioral profiles for various entities[12] and then used to find meaningful anomalies.

Within the field of data analytics, machine learning is a method used to devise complex models and algorithms that lend themselves to prediction; in commercial use, this is known as predictive analytics. These analytical models allow researchers, data scientists, engineers, and analysts to "produce reliable, repeatable decisions and results" and uncover "hidden insights" through learning from historical relationships and trends in the data.[13]

As of 2016, machine learning is a buzzword, and according to the Gartner hype cycle of 2016, at its peak of inflated expectations.[14] Effective machine learning is difficult because finding patterns is hard and often not enough training data is available; as a result, machine-learning programs often fail to deliver.[15][16]

36.1 Overview

Tom M. Mitchell provided a widely quoted, more formal definition of the algorithms studied in the machine learning field: "A computer program is said to learn from experience E with respect to some class of tasks T and performance measure P if its performance at tasks in T, as measured by P, improves with experience E."[17] This definition of the tasks in which machine learning is concerned offers a fundamentally operational definition rather than defining the field in cognitive terms. This follows Alan Turing's proposal in his paper "Computing Machinery and Intelligence", in which the question "Can machines think?" is replaced with the question "Can machines do what we (as thinking entities) can do?".[18] In Turing's proposal the various characteristics that could be possessed by a *thinking machine* and the various implications in constructing one are exposed.

36.1.1 Types of problems and tasks

Machine learning tasks are typically classified into three broad categories, depending on the nature of the learning "signal" or "feedback" available to a learning system. These are[19]

- Supervised learning: The computer is presented with example inputs and their desired outputs, given by a "teacher", and the goal is to learn a general rule that maps inputs to outputs.

- Unsupervised learning: No labels are given to the learning algorithm, leaving it on its own to find structure in its input. Unsupervised learning can be a goal in itself (discovering hidden patterns in data) or a means towards an end (feature learning).

- Reinforcement learning: A computer program interacts with a dynamic environment in which it must perform a certain goal (such as driving a vehicle or playing a game against an opponent[6]:3). The program is provided feedback in terms of rewards and punishments as it navigates its problem space.

Between supervised and unsupervised learning is semi-supervised learning, where the teacher gives an incomplete training signal: a training set with some (often many) of the target outputs missing. Transduction is a special case of this principle where the entire set of problem instances is known at learning time, except that part of the targets are missing.

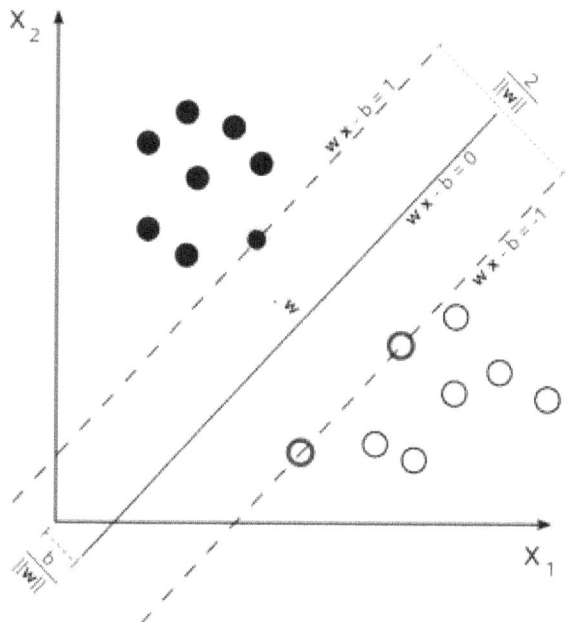

A support vector machine is a classifier that divides its input space into two regions, separated by a linear boundary. Here, it has learned to distinguish black and white circles.

Among other categories of machine learning problems, learning to learn learns its own inductive bias based on previous experience. Developmental learning, elaborated for robot learning, generates its own sequences (also called curriculum) of learning situations to cumulatively acquire repertoires of novel skills through autonomous self-exploration and social interaction with human teachers and using guidance mechanisms such as active learning, maturation, motor synergies, and imitation.

Tasks can be categorized into deep learning (the application of artificial neural networks to learning tasks that contain more than one hidden layer) and shallow learning (tasks with a single hidden layer).

Another categorization of machine learning tasks arises when one considers the desired *output* of a machine-learned system:[6]:3

- In classification, inputs are divided into two or more classes, and the learner must produce a model that assigns unseen inputs to one or more (multi-label classification) of these classes. This is typically tackled in a supervised way. Spam filtering is an example of classification, where the inputs are email (or other) messages and the classes are "spam" and "not spam".

- In regression, also a supervised problem, the outputs are continuous rather than discrete.

- In clustering, a set of inputs is to be divided into groups. Unlike in classification, the groups are not known beforehand, making this typically an unsupervised task.

- Density estimation finds the distribution of inputs in some space.

- Dimensionality reduction simplifies inputs by mapping them into a lower-dimensional space. Topic modeling is a related problem, where a program is given a list of human language documents and is tasked to find out which documents cover similar topics.

36.2 History and relationships to other fields

See also: Timeline of machine learning

As a scientific endeavour, machine learning grew out of the quest for artificial intelligence. Already in the early days of AI as an academic discipline, some researchers were interested in having machines learn from data. They attempted to approach the problem with various symbolic methods, as well as what were then termed "neural networks"; these were mostly perceptrons and other models that were later found to be reinventions of the generalized linear models of statistics.[20] Probabilistic reasoning was also employed, especially in automated medical diagnosis.[19]:488

However, an increasing emphasis on the logical, knowledge-based approach caused a rift between AI and machine learning. Probabilistic systems were plagued by theoretical and practical problems of data acquisition and representation.[19]:488 By 1980, expert systems had come to dominate AI, and statistics was out of favor.[21] Work on symbolic/knowledge-based learning did continue within AI, leading to inductive logic programming, but the more statistical line of research was now outside the field of AI proper, in pattern recognition and information retrieval.[19]:708–710; 755 Neural networks research had been abandoned by AI and computer science around the same time. This line, too, was continued outside the AI/CS field, as "connectionism", by researchers from other disciplines including Hopfield, Rumelhart and Hinton. Their main success came in the mid-1980s with the reinvention of backpropagation.[19]:25

Machine learning, reorganized as a separate field, started to flourish in the 1990s. The field changed its goal from achieving artificial intelligence to tackling solvable problems of a practical nature. It shifted focus away from the symbolic approaches it had inherited from AI, and toward methods and models borrowed from statistics and probability theory.[21] It also benefited from the increasing availability of digitized information, and the possibility to distribute that via the Internet.

Machine learning and data mining often employ the same methods and overlap significantly, but while machine learning focuses on prediction, based on *known* properties learned from the training data, data mining focuses on the discovery of (previously) *unknown* properties in the data (this is the analysis step of knowledge discovery in databases). Data mining uses many machine learning methods, but with different goals; on the other hand, machine learning also employs data mining methods as "unsupervised learning" or as a preprocessing step to improve learner accuracy. Much of the confusion between these two research communities (which do often have separate conferences and separate journals, ECML PKDD being a major exception) comes from the basic assumptions they work with: in machine learning, performance is usually evaluated with respect to the ability to *reproduce known* knowledge, while in knowledge discovery and data mining (KDD) the key task is the discovery of previously *unknown* knowledge. Evaluated with respect to known knowledge, an uninformed (unsupervised) method will easily be outperformed by other supervised methods, while in a typical KDD task, supervised methods cannot be used due to the unavailability of training data.

Machine learning also has intimate ties to optimization: many learning problems are formulated as minimization of some loss function on a training set of examples. Loss functions express the discrepancy between the predictions of the model being trained and the actual problem instances (for example, in classification, one wants to assign a label to instances, and models are trained to correctly predict the preassigned labels of a set of examples). The difference between the two fields arises from the goal of generalization: while optimization algorithms can minimize the loss on a training set, machine learning is concerned with minimizing the loss on unseen samples.[22]

36.2.1 Relation to statistics

Machine learning and statistics are closely related fields. According to Michael I. Jordan, the ideas of machine learning, from methodological principles to theoretical tools, have had a long pre-history in statistics.[23] He also suggested the term data science as a placeholder to call the overall field.[23]

Leo Breiman distinguished two statistical modelling paradigms: data model and algorithmic model,[24] wherein "algorithmic model" means more or less the machine learning algorithms like Random forest.

Some statisticians have adopted methods from machine learning, leading to a combined field that they call *statistical learning*.[25]

36.3 Theory

Main article: Computational learning theory

A core objective of a learner is to generalize from its experience.[26][27] Generalization in this context is the ability of a learning machine to perform accurately on new, unseen examples/tasks after having experienced a learning data set. The training examples come from some generally unknown probability distribution (considered representative of the space of occurrences) and the learner has to build a general model about this space that enables it to produce sufficiently accurate predictions in new cases.

The computational analysis of machine learning algorithms and their performance is a branch of theoretical computer science known as computational learning theory. Because training sets are finite and the future is uncertain, learning theory usually does not yield guarantees of the performance of algorithms. Instead, probabilistic bounds on the performance are quite common. The bias–variance decomposition is one way to quantify generalization error.

For the best performance in the context of generalization, the complexity of the hypothesis should match the complexity of the function underlying the data. If the hypothesis is less complex than the function, then the model has underfit the data. If the complexity of the model is increased in response, then the training error decreases. But if the hypothesis is too complex, then the model is subject to overfitting and generalization will be poorer.[28]

36.4. APPROACHES

In addition to performance bounds, computational learning theorists study the time complexity and feasibility of learning. In computational learning theory, a computation is considered feasible if it can be done in polynomial time. There are two kinds of time complexity results. Positive results show that a certain class of functions can be learned in polynomial time. Negative results show that certain classes cannot be learned in polynomial time.

36.4 Approaches

Main article: List of machine learning algorithms

36.4.1 Decision tree learning

Main article: Decision tree learning

Decision tree learning uses a decision tree as a predictive model, which maps observations about an item to conclusions about the item's target value.

36.4.2 Association rule learning

Main article: Association rule learning

Association rule learning is a method for discovering interesting relations between variables in large databases.

36.4.3 Artificial neural networks

Main article: Artificial neural network

An artificial neural network (ANN) learning algorithm, usually called "neural network" (NN), is a learning algorithm that is inspired by the structure and functional aspects of biological neural networks. Computations are structured in terms of an interconnected group of artificial neurons, processing information using a connectionist approach to computation. Modern neural networks are non-linear statistical data modeling tools. They are usually used to model complex relationships between inputs and outputs, to find patterns in data, or to capture the statistical structure in an unknown joint probability distribution between observed variables.

36.4.4 Deep learning

Main article: Deep learning

Falling hardware prices and the development of GPUs for personal use in the last few years have contributed to the development of the concept of deep learning which consists of multiple hidden layers in an artificial neural network. This approach tries to model the way the human brain processes light and sound into vision and hearing. Some successful applications of deep learning are computer vision and speech recognition.[29]

36.4.5 Inductive logic programming

Main article: Inductive logic programming

Inductive logic programming (ILP) is an approach to rule learning using logic programming as a uniform representation for input examples, background knowledge, and hypotheses. Given an encoding of the known background knowledge and a set of examples represented as a logical database of facts, an ILP system will derive a hypothesized logic program that entails all positive and no negative examples. Inductive programming is a related field that considers any kind of programming languages for representing hypotheses (and not only logic programming), such as functional programs.

36.4.6 Support vector machines

Main article: Support vector machines

Support vector machines (SVMs) are a set of related supervised learning methods used for classification and regression. Given a set of training examples, each marked as belonging to one of two categories, an SVM training algorithm builds a model that predicts whether a new example falls into one category or the other.

36.4.7 Clustering

Main article: Cluster analysis

Cluster analysis is the assignment of a set of observations into subsets (called *clusters*) so that observations within the same cluster are similar according to some predesignated criterion or criteria, while observations drawn from different clusters are dissimilar. Different clustering techniques make different assumptions on the structure of the data, often defined by some *similarity metric* and evaluated for example by *internal compactness* (similarity between members of the same cluster) and *separation* between different clusters. Other methods are based on *estimated density* and *graph connectivity*. Clustering is a method of unsupervised learning, and a common technique for statistical data analysis.

36.4.8 Bayesian networks

Main article: Bayesian network

A Bayesian network, belief network or directed acyclic graphical model is a probabilistic graphical model that represents a set of random variables and their conditional independencies via a directed acyclic graph (DAG). For example, a Bayesian network could represent the probabilistic relationships between diseases and symptoms. Given symptoms, the network can be used to compute the probabilities of the presence of various diseases. Efficient algorithms exist that perform inference and learning.

36.4.9 Reinforcement learning

Main article: Reinforcement learning

Reinforcement learning is concerned with how an *agent* ought to take *actions* in an *environment* so as to maximize some notion of long-term *reward*. Reinforcement learning algorithms attempt to find a *policy* that maps *states* of the world to the actions the agent ought to take in those states. Reinforcement learning differs from the supervised learning problem in that correct input/output pairs are never presented, nor sub-optimal actions explicitly corrected.

36.4.10 Representation learning

Main article: Representation learning

Several learning algorithms, mostly unsupervised learning algorithms, aim at discovering better representations of the inputs provided during training. Classical examples include principal components analysis and cluster analysis. Representation learning algorithms often attempt to preserve the information in their input but transform it in a way that makes it useful, often as a pre-processing step before performing classification or predictions, allowing reconstruction of the inputs coming from the unknown data generating distribution, while not being necessarily faithful for configurations that are implausible under that distribution.

Manifold learning algorithms attempt to do so under the constraint that the learned representation is low-dimensional. Sparse coding algorithms attempt to do so under the constraint that the learned representation is sparse (has many zeros). Multilinear subspace learning algorithms aim to learn low-dimensional representations directly from tensor representations for multidimensional data, without reshaping them into (high-dimensional) vectors.[30] Deep learning algorithms discover multiple levels of representation, or a hierarchy of features, with higher-level, more abstract features defined in terms of (or generating) lower-level features. It has been argued that an intelligent machine is one that learns a representation that disentangles the underlying factors of variation that explain the observed data.[31]

36.4.11 Similarity and metric learning

Main article: Similarity learning

In this problem, the learning machine is given pairs of examples that are considered similar and pairs of less similar objects. It then needs to learn a similarity function (or a distance metric function) that can predict if new objects are similar. It is sometimes used in Recommendation systems.

36.4.12 Sparse dictionary learning

Main article: Sparse dictionary learning

In this method, a datum is represented as a linear combination of basis functions, and the coefficients are assumed to be sparse. Let x be a d-dimensional datum, D be a d by n matrix, where each column of D represents a basis function. r is the coefficient to represent x using D. Mathematically, sparse dictionary learning means solving $x \approx Dr$ where r is sparse. Generally speaking, n is assumed to be larger than d to allow the freedom for a sparse representation.

Learning a dictionary along with sparse representations is strongly NP-hard and also difficult to solve approximately.[32] A popular heuristic method for sparse dictionary learning is K-SVD.

Sparse dictionary learning has been applied in several contexts. In classification, the problem is to determine which classes a previously unseen datum belongs to. Suppose a dictionary for each class has already been built. Then a new datum is associated with the class such that it's best sparsely represented by the corresponding dictionary. Sparse dictionary learning has also been applied in image de-noising. The key idea is that a clean image patch can be sparsely represented by an image dictionary, but the noise cannot.[33]

36.4.13 Genetic algorithms

Main article: Genetic algorithm

A genetic algorithm (GA) is a search heuristic that mimics the process of natural selection, and uses methods such as mutation and crossover to generate new genotype in the hope of finding good solutions to a given problem. In machine learning, genetic algorithms found some uses in the 1980s and 1990s.[34][35] Vice versa, machine learning techniques have been used to improve the performance of genetic and evolutionary algorithms.[36]

36.4.14 Rule-based machine learning

Rule-based machine learning is a general term for any machine learning method that identifies, learns, or evolves 'rules' to store, manipulate or apply, knowledge. The defining characteristic of a rule-based machine learner is the identification and utilization of a set of relational rules that collectively represent the knowledge captured by the system. This is in contrast to other machine learners that commonly identify a singular model that can be universally applied to any instance in order to make a prediction.[37] Rule-based machine learning approaches include learning classifier systems, association rule learning, and artificial immune systems.

Learning classifier systems

Main article: Learning classifier system

Learning classifier systems (LCS) are a family of rule-based machine learning algorithms that combine a discovery component (e.g. typically a genetic algorithm) with a learning component (performing either supervised learning, reinforcement learning, or unsupervised learning). They seek to identify a set of context-dependent rules that collectively store and apply knowledge in a piecewise manner in order to make predictions.[38]

36.5 Applications

Applications for machine learning include:

- Automated theorem proving[39][40]
- Adaptive websites
- Affective computing
- Bioinformatics
- Brain–machine interfaces
- Cheminformatics
- Classifying DNA sequences
- Computational anatomy
- Computer vision, including object recognition
- Detecting credit-card fraud
- General game playing[41]
- Information retrieval
- Internet fraud detection[28]
- Linguistics
- Marketing
- Machine learning control
- Machine perception
- Medical diagnosis
- Economics
- Insurance
- Natural language processing
- Natural language understanding[42]
- Optimization and metaheuristic
- Online advertising
- Recommender systems
- Robot locomotion
- Search engines
- Sentiment analysis (or opinion mining)
- Sequence mining
- Software engineering
- Speech and handwriting recognition
- Financial market analysis
- Structural health monitoring
- Syntactic pattern recognition
- Time series forecasting
- User behavior analytics

- Translation[43]

In 2006, the online movie company Netflix held the first "Netflix Prize" competition to find a program to better predict user preferences and improve the accuracy on its existing Cinematch movie recommendation algorithm by at least 10%. A joint team made up of researchers from AT&T Labs-Research in collaboration with the teams Big Chaos and Pragmatic Theory built an ensemble model to win the Grand Prize in 2009 for $1 million.[44] Shortly after the prize was awarded, Netflix realized that viewers' ratings were not the best indicators of their viewing patterns ("everything is a recommendation") and they changed their recommendation engine accordingly.[45]

In 2010 The Wall Street Journal wrote about the firm Rebellion Research and their use of Machine Learning to predict the financial crisis. [46]

In 2012, co-founder of Sun Microsystems Vinod Khosla predicted that 80% of medical doctors jobs would be lost in the next two decades to automated machine learning medical diagnostic software.[47]

In 2014, it has been reported that a machine learning algorithm has been applied in Art History to study fine art paintings, and that it may have revealed previously unrecognized influences between artists.[48]

36.6 Model assessments

Classification machine learning models can be validated by accuracy estimation techniques like the Holdout method, which splits the data in a training and test set (conventionally 2/3 training set and 1/3 test set designation) and evaluates the performance of the training model on the test set. In comparison, the N-fold-cross-validation method randomly splits the data in k subsets where the k-1 instances of the data are used to train the model while the kth instance is used to test the predictive ability of the training model. In addition to the holdout and cross-validation methods, bootstrap, which samples n instances with replacement from the dataset, can be used to assess model accuracy.[49]

In addition to overall accuracy, investigators frequently report sensitivity and specificity meaning True Positive Rate (TPR) and True Negative Rate (TNR) respectively. Similarly, investigators sometimes report the False Positive Rate (FPR) as well as the False Negative Rate (FNR). However, these rates are ratios that fail to reveal their numerators and denominators. The Total Operating Characteristic (TOC) is an effective method to express a model's diagnostic ability. TOC shows the numerators and denominators of the previously mentioned rates, thus TOC provides more information than the commonly used Receiver operating characteristic (ROC) and ROC's associated Area Under the Curve (AUC).

36.7 Ethics

Machine learning poses a host of ethical questions. Systems which are trained on datasets collected with biases may exhibit these biases upon use, thus digitizing cultural prejudices.[50] Responsible collection of data thus is a critical part of machine learning.

Because language contains biases, machines trained on language *corpora* will necessarily also learn bias.[51]

See Machine ethics for additional information.

36.8 Software

Software suites containing a variety of machine learning algorithms include the following :

36.8.1 Free and open-source software

- CNTK
- Deeplearning4j
- dlib
- ELKI
- GNU Octave
- H2O
- Mahout
- Mallet
- mlpy
- MLPACK
- MOA (Massive Online Analysis)
- MXNet
- ND4J: ND arrays for Java
- NuPIC
- OpenAI Gym
- OpenAI Universe
- OpenNN

- Orange
- R
- scikit-learn
- Shogun
- TensorFlow
- Torch
- Yooreeka
- Weka

36.8.2 Proprietary software with free and open-source editions

- KNIME
- RapidMiner

36.8.3 Proprietary software

- Amazon Machine Learning
- Angoss KnowledgeSTUDIO
- Ayasdi
- IBM Data Science Experience
- Google Prediction API
- IBM SPSS Modeler
- KXEN Modeler
- LIONsolver
- Mathematica
- MATLAB
- Microsoft Azure Machine Learning
- Neural Designer
- NeuroSolutions
- Oracle Data Mining
- RCASE
- SAP Leonardo
- SAS Enterprise Miner
- SequenceL
- Skymind
- Splunk
- STATISTICA Data Miner

36.9 Journals

- *Journal of Machine Learning Research*
- *Machine Learning*
- *Neural Computation*

36.10 Conferences

- Conference on Neural Information Processing Systems
- International Conference on Machine Learning
- International Conference on Learning Representations

36.11 See also

- Artificial intelligence
- Automatic reasoning
- Big data
- Computational intelligence
- Computational neuroscience
- Data science
- Deep learning
- Ethics of artificial intelligence
- Existential risk from advanced artificial intelligence
- Explanation-based learning
- Quantum machine learning
- Important publications in machine learning
- List of machine learning algorithms
- List of datasets for machine learning research
- Similarity learning
- Machine-learning applications in bioinformatics

36.12 References

[1] Supposedly paraphrased from: Samuel, Arthur (1959). "Some Studies in Machine Learning Using the Game of Checkers". *IBM Journal of Research and Development*. **3** (3). doi:10.1147/rd.33.0210.

[2] Koza, John R.; Bennett, Forrest H.; Andre, David; Keane, Martin A. (1996). *Automated Design of Both the Topology and Sizing of Analog Electrical Circuits Using Genetic Programming*. Artificial Intelligence in Design '96. Springer, Dordrecht. pp. 151–170. doi:10.1007/978-94-009-0279-4_9.

[3] R. Kohavi and F. Provost, \Glossary of terms," Machine Learning, vol. 30, no. 2-3, pp. 271-274, 1998.

[4] http://www.britannica.com/EBchecked/topic/1116194/machine-learning This tertiary source reuses information from other sources but does not name them.

[5] Ron Kohavi; Foster Provost (1998). "Glossary of terms". *Machine Learning*. **30**: 271–274.

[6] Machine learning and pattern recognition "can be viewed as two facets of the same field."

[7] Dickson, Ben. "Exploiting machine learning in cybersecurity". *TechCrunch*. Retrieved 2017-05-23.

[8] Wernick, Yang, Brankov, Yourganov and Strother, Machine Learning in Medical Imaging, *IEEE Signal Processing Magazine*, vol. 27, no. 4, July 2010, pp. 25-38

[9] Mannila, Heikki (1996). *Data mining: machine learning, statistics, and databases*. Int'l Conf. Scientific and Statistical Database Management. IEEE Computer Society.

[10] Friedman, Jerome H. (1998). "Data Mining and Statistics: What's the connection?". *Computing Science and Statistics*. **29** (1): 3–9.

[11] "Dark Reading".

[12] "AI Business".

[13] "Machine Learning: What it is and why it matters". *www.sas.com*. Retrieved 2016-03-29.

[14] "Gartner's 2016 Hype Cycle for Emerging Technologies Identifies Three Key Trends That Organizations Must Track to Gain Competitive Advantage". Retrieved 2017-04-10.

[15] "Why Machine Learning Models Often Fail to Learn: QuickTake Q&A". *Bloomberg.com*. 2016-11-10. Retrieved 2017-04-10.

[16] Simonite, Tom. "Microsoft says its racist chatbot illustrates how AI isn't adaptable enough to help most businesses". *MIT Technology Review*. Retrieved 2017-04-10.

[17] Mitchell, T. (1997). *Machine Learning*. McGraw Hill. p. 2. ISBN 0-07-042807-7.

[18] Stevan Harnad (2008). "The Annotation Game: On Turing (1950) on Computing, Machinery, and Intelligence", in Epstein, Robert; Peters, Grace, *The Turing Test Sourcebook: Philosophical and Methodological Issues in the Quest for the Thinking Computer*, Kluwer

[19] Russell, Stuart; Norvig, Peter (2003) [1995]. *Artificial Intelligence: A Modern Approach* (2nd ed.). Prentice Hall. ISBN 978-0137903955.

[20] Sarle, Warren. "Neural Networks and statistical models". *CiteseerX*. CiteSeerX 10.1.1.27.699.

[21] Langley, Pat (2011). "The changing science of machine learning". *Machine Learning*. **82** (3): 275–279. doi:10.1007/s10994-011-5242-y.

[22] Le Roux, Nicolas; Bengio, Yoshua; Fitzgibbon, Andrew (2012). "Improving First and Second-Order Methods by Modeling Uncertainty". In Sra, Suvrit; Nowozin, Sebastian; Wright, Stephen J. *Optimization for Machine Learning*. MIT Press. p. 404.

[23] Michael I. Jordan (2014-09-10). "statistics and machine learning". reddit. Retrieved 2014-10-01.

[24] Cornell University Library. "Breiman: Statistical Modeling: The Two Cultures (with comments and a rejoinder by the author)". Retrieved 8 August 2015.

[25] Gareth James; Daniela Witten; Trevor Hastie; Robert Tibshirani (2013). *An Introduction to Statistical Learning*. Springer. p. vii.

[26] Bishop, C. M. (2006), *Pattern Recognition and Machine Learning*, Springer, ISBN 0-387-31073-8

[27] Mohri, Mehryar; Rostamizadeh, Afshin; Talwalkar, Ameet (2012). *Foundations of Machine Learning*. USA, Massachusetts: MIT Press. ISBN 9780262018258.

[28] Alpaydin, Ethem (2010). *Introduction to Machine Learning*. London: The MIT Press. ISBN 978-0-262-01243-0. Retrieved 4 February 2017.

[29] Honglak Lee, Roger Grosse, Rajesh Ranganath, Andrew Y. Ng. "Convolutional Deep Belief Networks for Scalable Unsupervised Learning of Hierarchical Representations" Proceedings of the 26th Annual International Conference on Machine Learning, 2009.

[30] Lu, Haiping; Plataniotis, K.N.; Venetsanopoulos, A.N. (2011). "A Survey of Multilinear Subspace Learning for Tensor Data" (PDF). *Pattern Recognition*. **44** (7): 1540–1551. doi:10.1016/j.patcog.2011.01.004.

[31] Yoshua Bengio (2009). *Learning Deep Architectures for AI*. Now Publishers Inc. pp. 1–3. ISBN 978-1-60198-294-0.

[32] A. M. Tillmann, "On the Computational Intractability of Exact and Approximate Dictionary Learning", IEEE Signal Processing Letters 22(1), 2015: 45–49.

[33] Aharon, M, M Elad, and A Bruckstein. 2006. "K-SVD: An Algorithm for Designing Overcomplete Dictionaries for Sparse Representation." Signal Processing, IEEE Transactions on 54 (11): 4311–4322

[34] Goldberg, David E.; Holland, John H. (1988). "Genetic algorithms and machine learning". *Machine Learning*. **3** (2): 95–99. doi:10.1007/bf00113892.

[35] Michie, D.; Spiegelhalter, D. J.; Taylor, C. C. (1994). *Machine Learning, Neural and Statistical Classification*. Ellis Horwood.

[36] Zhang, Jun; Zhan, Zhi-hui; Lin, Ying; Chen, Ni; Gong, Yue-jiao; Zhong, Jing-hui; Chung, Henry S.H.; Li, Yun; Shi, Yu-hui (2011). "Evolutionary Computation Meets Machine Learning: A Survey" (PDF). *Computational Intelligence Magazine*. IEEE. **6** (4): 68–75. doi:10.1109/mci.2011.942584.

[37] Bassel, George W.; Glaab, Enrico; Marquez, Julietta; Holdsworth, Michael J.; Bacardit, Jaume (2011-09-01). "Functional Network Construction in Arabidopsis Using Rule-Based Machine Learning on Large-Scale Data Sets". *The Plant Cell*. **23** (9): 3101–3116. ISSN 1532-298X. PMC 3203449. PMID 21896882. doi:10.1105/tpc.111.088153.

[38] Urbanowicz, Ryan J.; Moore, Jason H. (2009-09-22). "Learning Classifier Systems: A Complete Introduction, Review, and Roadmap". *Journal of Artificial Evolution and Applications*. **2009**: 1–25. ISSN 1687-6229. doi:10.1155/2009/736398.

[39] Bridge, James P., Sean B. Holden, and Lawrence C. Paulson. "Machine learning for first-order theorem proving." Journal of automated reasoning 53.2 (2014): 141-172.

[40] Loos, Sarah, et al. "Deep Network Guided Proof Search." arXiv preprint arXiv:1701.06972 (2017).

[41] Finnsson, Hilmar, and Yngvi Björnsson. "Simulation-Based Approach to General Game Playing." AAAI. Vol. 8. 2008.

[42] Sarikaya, Ruhi, Geoffrey E. Hinton, and Anoop Deoras. "Application of deep belief networks for natural language understanding." IEEE/ACM Transactions on Audio, Speech and Language Processing (TASLP) 22.4 (2014): 778-784.

[43] "AI-based translation to soon reach human levels: industry officials". Yonhap news agency. Retrieved 4 Mar 2017.

[44] "BelKor Home Page" research.att.com

[45] "The Netflix Tech Blog: Netflix Recommendations: Beyond the 5 stars (Part 1)". Retrieved 8 August 2015.

[46]

[47] Vonod Khosla (January 10, 2012). "Do We Need Doctors or Algorithms?". Tech Crunch.

[48] When A Machine Learning Algorithm Studied Fine Art Paintings, It Saw Things Art Historians Had Never Noticed, *The Physics at ArXiv blog*

[49] Kohavi, Ron (1995). "A Study of Cross-Validation and Bootstrap for Accuracy Estimation and Model Selection" (PDF). *International Joint Conference on Artificial Intelligence*.

[50] Bostrom, Nick (2011). "The Ethics of Artificial Intelligence" (PDF). Retrieved 11 April 2016.

[51]

36.13 Further reading

- Nils J. Nilsson, *Introduction to Machine Learning*.

- Trevor Hastie, Robert Tibshirani and Jerome H. Friedman (2001). *The Elements of Statistical Learning*, Springer. ISBN 0-387-95284-5.

- Pedro Domingos (September 2015), The Master Algorithm, Basic Books, ISBN 978-0-465-06570-7

- Ian H. Witten and Eibe Frank (2011). *Data Mining: Practical machine learning tools and techniques* Morgan Kaufmann, 664pp., ISBN 978-0-12-374856-0.

- Ethem Alpaydin (2004). Introduction to Machine Learning, MIT Press, ISBN 978-0-262-01243-0.

- David J. C. MacKay. *Information Theory, Inference, and Learning Algorithms* Cambridge: Cambridge University Press, 2003. ISBN 0-521-64298-1

- Richard O. Duda, Peter E. Hart, David G. Stork (2001) *Pattern classification* (2nd edition), Wiley, New York, ISBN 0-471-05669-3.

- Christopher Bishop (1995). *Neural Networks for Pattern Recognition*, Oxford University Press. ISBN 0-19-853864-2.

- Stuart Russell & Peter Norvig, (2002). *Artificial Intelligence - A Modern Approach*. Prentice Hall. ISBN 0-136-04259-7.

- Ray Solomonoff, *An Inductive Inference Machine*, IRE Convention Record, Section on Information Theory, Part 2, pp., 56-62, 1957.

- Ray Solomonoff, "An Inductive Inference Machine" A privately circulated report from the 1956 Dartmouth Summer Research Conference on AI.

36.14 External links

- International Machine Learning Society

- Popular online course by Andrew Ng, at Coursera. It uses GNU Octave. The course is a free version of Stanford University's actual course taught by Ng, whose lectures are also available for free.

- mloss is an academic database of open-source machine learning software.

Chapter 37

Evolutionary developmental biology

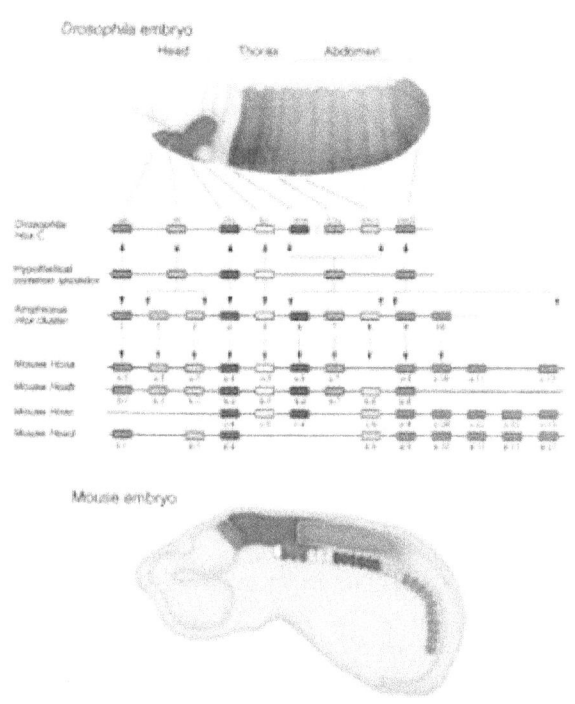

Homologous hox *genes in such different animals as insects and vertebrates control embryonic development and hence the form of adult bodies. These genes have been highly conserved through hundreds of millions of years of evolution.*

Evolutionary developmental biology (informally, **evo-devo**) is a field of biological research that compares the developmental processes of different organisms to infer the ancestral relationships between them and how developmental processes evolved.

The field grew from 19th century beginnings, where embryology faced a mystery: zoologists did not know how embryonic development was controlled at the molecular level. Charles Darwin noted that having similar embryos implied common ancestry, but little progress was made until the 1970s. Then, recombinant DNA technology at last brought embryology together with molecular genetics. A key early discovery was of homeotic genes that regulate development in a wide range of eukaryotes.

The field is characterised by some key concepts, which took biologists by surprise. One is deep homology, the finding that dissimilar organs such as the eyes of insects, vertebrates and cephalopod molluscs, long thought to have evolved separately, are controlled by similar genes such as *pax-6*, from the evo-devo gene toolkit. These genes are ancient, being highly conserved among phyla; they generate the patterns in time and space which shape the embryo, and ultimately form the body plan of the organism. Another is that species do not differ much in their structural genes, such as those coding for enzymes; what does differ is the way that gene expression is regulated by the toolkit genes. These genes are reused, unchanged, many times in different parts of the embryo and at different stages of development, forming a complex cascade of control, switching other regulatory genes as well as structural genes on and off in a precise pattern. This multiple pleiotropic reuse explains why these genes are highly conserved, as any change would have many adverse consequences which natural selection would oppose.

New morphological features and ultimately new species are produced by variations in the toolkit, either when genes are expressed in a new pattern, or when toolkit genes acquire additional functions. Another possibility is the Neo-Lamarckian theory that epigenetic changes are later consolidated at gene level, something that may have been important early in the history of multicellular life.

37.1 History

Further information: History of evolutionary thought

37.1.1 Recapitulation

Main article: Recapitulation theory

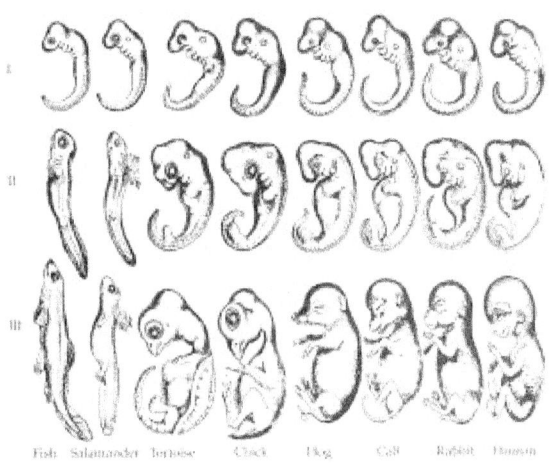

Ernst Haeckel argued for recapitulation of evolutionary development in the embryo, using embryo drawings to emphasise the genuine similarities.[1]

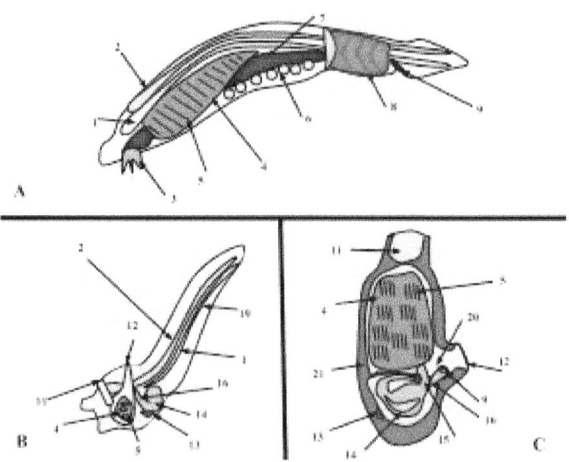

A. Lancelet (a chordate), B. Larval tunicate, C. Adult tunicate. Kowalevsky saw that the notochord (1) and gill slit (5) are shared by tunicates and vertebrates.

A recapitulation theory of evolutionary development was proposed by Étienne Serres in 1824–26, echoing the 1808 ideas of Johann Friedrich Meckel, which in turn had been hinted at by Aristotle. They argued that the embryos of 'higher' animals went through or recapitulated a series of stages, each of which resembled an animal lower down the great chain of being. For example, the brain of a human embryo looked first like that of a fish, then in turn like that of a reptile, bird, and mammal before becoming clearly human. The embryologist Karl Ernst von Baer opposed this, arguing in 1828 that there was no linear sequence as in the great chain of being, based on a single body plan. Von Baer instead recognised four distinct animal body plans: radiate, like starfish; molluscan, like clams; articulate, like lobsters; and vertebrate, like fish. Zoologists then largely abandoned recapitulation, though Ernst Haeckel revived it in 1866.[2][3][4][5][6]

37.1.2 Evolutionary morphology

Further information: Morphology (biology) and Body plan

From the early 19th century through most of the 20th century, embryology faced a mystery. Animals were seen to develop into adults of widely differing body plan, often through similar stages, from the egg, but zoologists knew almost nothing about how embryonic development was controlled at the molecular level, and therefore equally little about how developmental processes had evolved.[7] Charles Darwin argued that a shared embryonic structure implied a common ancestor. As an example of this, Darwin cited in his 1859 book *On the Origin of Species* the shrimp-like larva of the barnacle, whose sessile adults looked nothing like other arthropods; Linnaeus and Cuvier had classified them as molluscs.[8][9] Darwin also noted Alexander Kowalevsky's finding that the tunicate, too, was not a mollusc, but in its larval stage had a notochord and pharyngeal slits which developed from the same germ layers as the equivalent structures in vertebrates, and should therefore be grouped with them as chordates.[8][10] 19th century zoology thus converted embryology into an evolutionary science, connecting phylogeny with homologies between the germ layers of embryos. Zoologists including Fritz Müller proposed the use of embryology to discover phylogenetic relationships between taxa. Müller demonstrated that crustaceans shared the Nauplius larva, identifying several parasitic species that had not been recognised as crustaceans. Müller also recognised that natural selection must act on larvae, just as it does on adults, giving the lie to recapitulation, which would require larval forms to be shielded from natural selection.[8] Two of Haeckel's other ideas about the evolution of development have fared better than recapitulation: he argued in the 1870s that changes in the timing (heterochrony) and changes in the positioning within the body (heterotopy) of aspects of embryonic development would drive evolution by changing the shape of a descendant's body compared to an ancestor's. It took a century before these ideas were shown to be correct.[11][12][13] In 1917, D'Arcy Thompson wrote a book on the shapes of animals, showing with simple mathematics how small changes to parameters, such as the angles of a gastropod's spiral shell, can radically alter an animal's form, though he preferred mechanical to evolutionary explanation.[14][15] But for the next century, without molecular evidence, progress stalled.[8]

37.1.3 The modern synthesis of the early 20th century

Main article: Modern synthesis

In the so-called modern synthesis of the early 20th century, Ronald Fisher brought together Darwin's theory of evolution, with its insistence on natural selection, heredity, and variation, and Gregor Mendel's laws of genetics into a coherent structure for evolutionary biology. Biologists assumed that an organism was a straightforward reflection of its component genes: the genes coded for proteins, which built the organism's body. Biochemical pathways (and, they supposed, new species) evolved through mutations in these genes. It was a simple, clear and nearly comprehensive picture: but it did not explain embryology.[8][16]

37.1.4 The lac operon

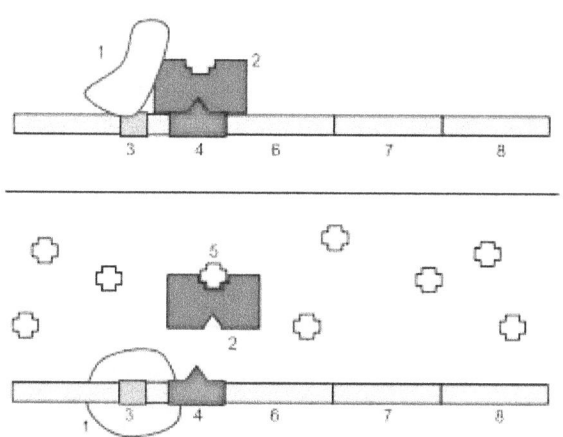

The lac operon. Top:Repressed, Bottom:Active
1: RNA Polymerase, 2: Repressor, 3: Promoter, 4: Operator, 5: Lactose, 6–8: protein-encoding genes, controlled by the switch, that cause lactose to be digested

Main article: Lac operon

In 1961, Jacques Monod, Jean-Pierre Changeux and François Jacob discovered the lac operon in the bacterium *Escherichia coli*. It was a cluster of genes, arranged in a feedback control loop so that its products would only be made when "switched on" by an environmental stimulus. One of these products was an enzyme that splits a sugar, lactose; and lactose itself was the stimulus that switched the genes on. This was a revelation, as it showed for the first time that genes, even in an organism as small as a bacterium, were subject to fine-grained control. The implication was that many other genes were also elaborately regulated.[17]

37.1.5 The birth of evo-devo and a second synthesis

In 1977, a revolution in thinking about evolution and developmental biology began, with the arrival of recombinant DNA technology in genetics, and the papers *Ontogeny and Phylogeny* by Stephen J. Gould and *Evolution by Tinkering* by François Jacob. Gould laid to rest Haeckel's interpretation of evolutionary embryology, while Jacob set out an alternative theory.[8] This led to a second synthesis,[18][19] at last including embryology as well as molecular genetics, phylogeny, and evolutionary biology to form evo-devo.[20][21] In 1978, Edward B. Lewis discovered homeotic genes that regulate embryonic development in *Drosophila* fruit flies, which like all insects are arthropods, one of the major phyla of invertebrate animals.[22] Bill McGinnis quickly discovered homeotic gene sequences, homeoboxes, in animals in other phyla, in vertebrates such as frogs, birds, and mammals; they were later also found in fungi such as yeasts, and in plants.[23][24] There were evidently strong similarities in the genes that controlled development across all the eukaryotes.[25] In 1980, Christiane Nüsslein-Volhard and Eric Wieschaus described gap genes which help to create the segmentation pattern in fruit fly embryos;[26][27] they and Lewis won a Nobel Prize for their work in 1995.[23][28]

Later, more specific similarities were discovered: for example, the Distal-less gene was found in 1989 to be involved in the development of appendages or limbs in fruit flies,[29] the fins of fish, the wings of chickens, the parapodia of marine annelid worms, the ampullae and siphons of tunicates, and the tube feet of sea urchins. It was evident that the gene must be ancient, dating back to the last common ancestor of bilateral animals (before the Ediacaran Period, which began some 635 million years ago). Evo-devo had started to uncover the ways that all animal bodies were built during development.[30][31]

37.2 The control of body structure

37.2.1 Deep homology

Further information: Deep homology

Roughly spherical eggs of different animals give rise to extremely different bodies, from jellyfish to lobsters, butterflies to elephants. Many of these organisms share the same structural genes for body-building proteins like collagen and enzymes, but biologists had expected that each group of animals would have its own rules of development. The surprise of evo-devo is that the shaping of bodies is controlled by a rather small percentage of genes, and that these regulatory genes are ancient, shared by all animals. The giraffe does not have a gene for a long neck, any more than the elephant has a gene for a big body. Their bodies are patterned by a system of switching which causes development of different features to begin earlier or later, to occur in this or that part of the embryo, and to continue for more or less time.[7]

The puzzle of how embryonic development was controlled began to be solved using the fruit fly *Drosophila melanogaster* as a model organism. The step-by-step control of its embryogenesis was visualized by attaching fluorescent dyes of different colours to specific types of protein made by genes expressed in the embryo.[7] A dye such as green fluorescent protein, originally from a jellyfish, was typically attached to an antibody specific to a fruit fly protein, forming a precise indicator of where and when that protein appeared in the living embryo.[32]

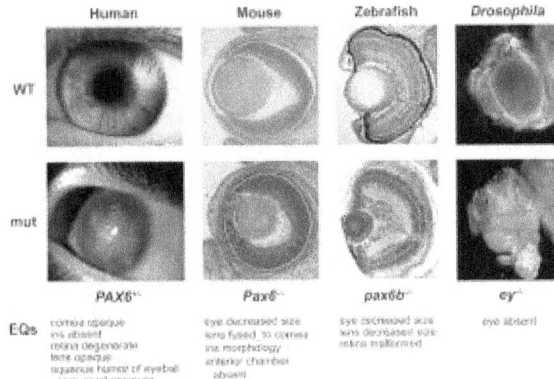

The pax-6 *gene controls development of eyes of different types across the animal kingdom.*

Using such a technique, in 1994 Walter Gehring found that the *pax-6* gene, vital for forming the eyes of fruit flies, exactly matches an eye-forming gene in mice and humans. The same gene was quickly found in many other groups of animals, such as squid, a cephalopod mollusc. Biologists including Ernst Mayr had believed that eyes had arisen in the animal kingdom at least 40 times, as the anatomy of different types of eye varies widely.[7] For example, the fruit fly's compound eye is made of hundreds of small lensed structures (ommatidia); the human eye has a blind spot where the optic nerve enters the eye, and the nerve fibres run over the surface of the retina, so light has to pass through a layer of nerve fibres before reaching the detector cells in the retina, so the structure is effectively "upside-down"; in contrast, the cephalopod eye has the retina, then a layer of nerve fibres, then the wall of the eye "the right way around".[33] The evidence of *pax-6*, however, was that the same genes controlled the development of the eyes of all these animals, suggesting that they all evolved from a common ancestor.[7] Ancient genes had been conserved through millions of years of evolution to create dissimilar structures for similar functions, demonstrating deep homology between structures once thought to be purely analogous.[34][35] This has caused a radical revision of the meaning of homology in evolutionary biology.[34][35][36]

37.2.2 Gene toolkit

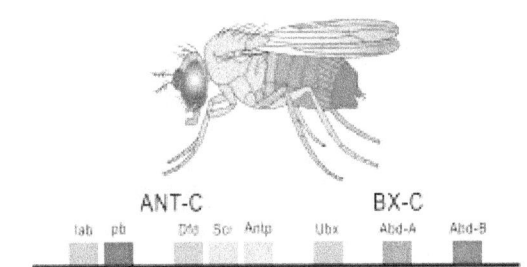

Expression of homeobox (Hox) genes in the fruit fly

Main article: Evo-devo gene toolkit

A small fraction of the genes in an organism's genome control the organism's development. These genes are called the developmental-genetic toolkit. They are highly conserved among phyla, meaning that they are ancient and very similar in widely separated groups of animals. Differences in deployment of toolkit genes affect the body plan and the number, identity, and pattern of body parts. Most toolkit genes are parts of signalling pathways: they encode transcription factors, cell adhesion proteins, cell surface receptor proteins and signalling ligands that bind to them, and secreted morphogens that diffuse through the embryo. All of these help to define the fate of undifferentiated cells in the embryo. Together, they generate the patterns in time and space which shape the embryo, and ultimately form the body plan of the organism. Among the most important toolkit genes are the *Hox* genes. These transcription factors contain the homeobox protein-binding DNA motif, also found in other toolkit genes, and create the basic pattern of the body along its front-to-back axis.[36] Hox genes determine where repeating parts, such as the many vertebrae of snakes, will grow in a developing embryo or larva.[7] *Pax-6*, already mentioned, is a classic toolkit gene.[37] Homeobox genes are also found in plants, implying they are common to all eukaryotes.[38][39][40]

37.2.3 The embryo's regulatory networks

Further information: Regulation of gene expression and Transcriptional regulation

37.2. THE CONTROL OF BODY STRUCTURE

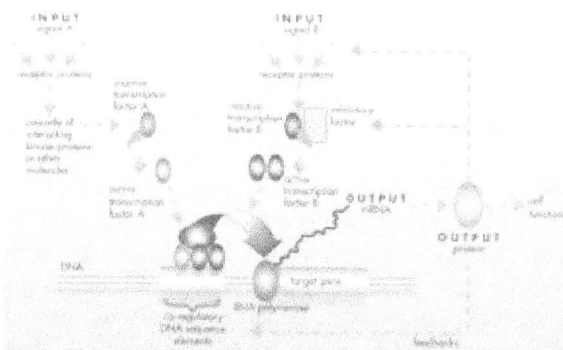

A gene regulatory network

The protein products of the regulatory toolkit are reused not by duplication and modification, but by a complex mosaic of pleiotropy, being applied unchanged in many independent developmental processes, giving pattern to many dissimilar body structures.[36] The loci of these pleiotropic toolkit genes have large, complicated and modular cis-regulatory elements. For example, while a non-pleiotropic rhodopsin gene in the fruit fly has a cis-regulatory element just a few hundred base pairs long, the pleiotropic eyeless cis-regulatory region contains 6 cis-regulatory elements in over 7000 base pairs.[36] The regulatory networks involved are often very large. Each regulatory protein controls "scores to hundreds" of cis-regulatory elements. For instance, 67 fruit fly transcription factors controlled on average 124 target genes each.[36] All this complexity enables genes involved in the development of the embryo to be switched on and off at exactly the right times and in exactly the right places. Some of these genes are structural, directly forming enzymes, tissues and organs of the embryo. But many others are themselves regulatory genes, so what is switched on is often a precisely-timed cascade of switching, involving turning on one developmental process after another in the developing embryo.[36]

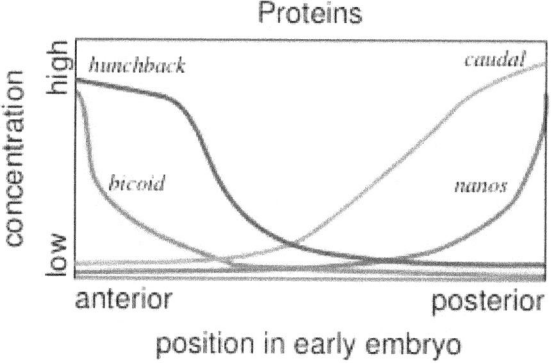

Gene product distributions along the long axis of the early embryo of a fruit fly

Such a cascading regulatory network has been studied in detail in the development of the fruit fly embryo. The young embryo is oval in shape, like a rugby ball. A small number of genes produce messenger RNAs that set up concentration gradients along the long axis of the embryo. In the early embryo, the *bicoid* and *hunchback* genes are at high concentration near the anterior end, and give pattern to the future head and thorax; the *caudal* and *nanos* genes are at high concentration near the posterior end, and give pattern to the hindmost abdominal segments. The effects of these genes interact; for instance, the Bicoid protein blocks the translation of *caudal*'s messenger RNA, so the Caudal protein concentration becomes low at the anterior end. Caudal later switches on genes which create the fly's hindmost segments, but only at the posterior end where it is most concentrated.[41][42]

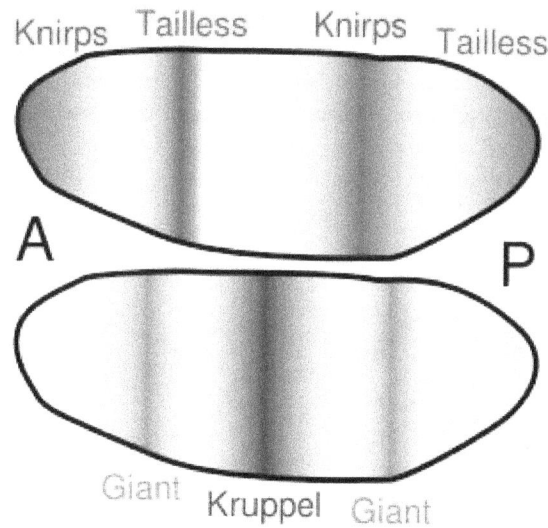

Gap genes in the fruit fly are switched on by genes such as bicoid, *setting up stripes across the embryo which start to pattern the body's segments.*

The Bicoid, Hunchback and Caudal proteins in turn regulate the transcription of gap genes such as *giant, knirps, Krüppel,* and *tailless* in a striped pattern, creating the first level of structures that will become segments.[26] The proteins from these in turn control the pair-rule genes, which in the next stage set up 7 bands across the embryo's long axis. Finally, the segment polarity genes such as *engrailed* split each of the 7 bands into two, creating 14 future segments.[41][42]

This process explains the accurate conservation of toolkit gene sequences, which has resulted in deep homology and functional equivalence of toolkit proteins in dissimilar animals (seen, for example, when a mouse protein controls fruit fly development). The interactions of transcription factors and cis-regulatory elements, or of signalling proteins and receptors, become locked in through multiple usages, making almost any mutation deleterious and hence eliminated by natural selection.[36]

37.3 The origins of novelty

See also: History of evolutionary thought § 21st century

Among the more surprising and, perhaps, counterintuitive (from a neo-Darwinian viewpoint) results of recent research in evolutionary developmental biology is that the diversity of body plans and morphology in organisms across many phyla are not necessarily reflected in diversity at the level of the sequences of genes, including those of the developmental genetic toolkit and other genes involved in development. Indeed, as Gerhart and Kirschner have noted, there is an apparent paradox: "where we most expect to find variation, we find conservation, a lack of change".[43] So, if the observed morphological novelty between different clades does not come from changes in gene sequences (such as by mutation), where does it come from? Novelty may arise by mutation-driven changes in gene regulation.[36][44][45][46]

37.3.1 Variations in the toolkit

Heliconius erato

Heliconius melpomene
Different species of *Heliconius* butterfly have independently evolved similar patterns, apparently both facilitated and constrained by the available developmental-genetic toolkit genes controlling wing pattern formation.

Variations in the toolkit may have produced a large part of the morphological evolution of animals. The toolkit can drive evolution in two ways. A toolkit gene can be expressed in a different pattern, as when the beak of Darwin's large ground-finch was enlarged by the *BMP* gene,[47] or when snakes lost their legs as *distal-less* became under-expressed or not expressed at all in the places where other reptiles continued to form their limbs.[48] Or, a toolkit gene can acquire a new function, as seen in the many functions of that same gene, *distal-less*, which controls such diverse structures as the mandible in vertebrates,[49][50] legs and antennae in the fruit fly,[51] and eyespot pattern in butterfly wings.[52] Given that small changes in toolbox genes can cause significant changes in body structures, they have often enabled the same function convergently or in parallel. *distal-less* generates wing patterns in the butterflies *Heliconius erato* and *Heliconius melpomene*, which are Müllerian mimics. In so-called facilitated variation,[53] their wing patterns arose in different evolutionary events, but are controlled by the same genes.[54] Developmental changes can contribute directly to speciation.[55]

37.3.2 Consolidation of epigenetic changes

Main article: Transgenerational epigenetic inheritance
Further information: Extended Evolutionary Synthesis

Evolutionary innovation may sometimes begin in Lamarckian style with epigenetic alterations of gene regulation or phenotype generation, subsequently consolidated by changes at the gene level. Epigenetic changes include modification of DNA by reversible methylation,[56] as well as nonprogrammed remoulding of the organism by physical and other environmental effects due to the inherent plasticity of developmental mechanisms.[57] The biologists Stuart A. Newman and Gerd B. Müller have suggested that organisms early in the history of multicellular life were more susceptible to this second category of epigenetic determination than are modern organisms, providing a basis for early macroevolutionary changes.[58]

37.4 Eco-evo-devo

Ecological evolutionary developmental biology (eco-evo-devo) integrates research from developmental biology and ecology to examine their relationship with evolutionary theory.[59] Researchers study concepts and mechanisms such as developmental plasticity, epigenetic inheritance, genetic assimilation, niche construction and symbiosis.[60][61]

37.5 See also

- Arthropod head problem
- Cell signaling

- *Evolution & Development* (journal)
- Human evolutionary developmental biology
- *Just So Stories* (as seen by evolutionary developmental biologists)

37.6 Notes

[1] Richardson and Keuck (2002). "Haeckel's ABC of evolution and development". *Biological reviews of the Cambridge Philosophical Society*. **77** (4): 495–528. PMID 12475051. doi:10.1017/s1464793102005948 p.516

[2] O'Connell, Lindsey (10 July 2013). "The Meckel-Serres Conception of Recapitulation". *The Embryo Project Encyclopedia*. Retrieved 10 October 2016.

[3] Desmond, Adrian J. (1989). *The politics of evolution: morphology, medicine, and reform in radical London*. Chicago: University of Chicago Press. pp. 53–53, 86–88, 337–340, 490–491. ISBN 0-226-14374-0.

[4] (Secord 2003, p. 252–253)

[5] Bowler, Peter J. (2003). *Evolution: the history of an idea*. Berkeley: University of California Press. pp. 120–128, 190–191, 208. ISBN 0-520-23693-9.

[6] (Secord 2003, p. 424, 512)

[7] Carroll, Sean B. "The Origins of Form". *Natural History*. Retrieved 9 October 2016. Biologists could say, with confidence, that forms change, and that natural selection is an important force for change. Yet they could say nothing about how that change is accomplished. How bodies or body parts change, or how new structures arise, remained complete mysteries.

[8] Gilbert, Scott F. (2003). "The morphogenesis of evolutionary developmental biology" (PDF). *International Journal of Developmental Biology*. **47**: 467–477.

[9] Darwin, Charles (1859). *On the Origin of Species*. London: John Murray. pp. 439–440. ISBN 0-8014-1319-2. Cirripedes afford a good instance of this: even the illustrious Cuvier did not perceive that a barnacle was, as it certainly is, a crustacean; but a glance at the larva shows this to be the case in an unmistakeable manner.

[10] Richmond, Marsha (January 2007). "Darwin's Study of the Cirripedia". Darwin Online. Retrieved 9 October 2016.

[11] Hall, B. K. (2003). "Evo-Devo: evolutionary developmental mechanisms". *International Journal of Developmental Biology*. **47** (7–8): 491–495. PMID 14756324.

[12] Ridley, Mark (2003). *Evolution*. Wiley-Blackwell. ISBN 978-1-4051-0345-9.

[13] Gould, Stephen Jay (1977). *Ontogeny and Phylogeny*. Cambridge, Massachusetts: Harvard University Press. pp. 221–222. ISBN 0-674-63940-5.

[14] Ball, Philip (7 February 2013). "In retrospect: On Growth and Form". *Nature*. **494** (32–33). doi:10.1038/494032a.

[15] Shalizi, Cosma. "Review: The Self-Made Tapestry by Philip Ball". University of Michigan. Retrieved 14 October 2016.

[16] Bock, Walter J. (July 1981). "Reviewed Work: The Evolutionary Synthesis. Perspectives on the Unification of Biology". *The Auk*. McLean, VA: American Ornithologists' Union. **98** (3): 644–646. ISSN 0004-8038. JSTOR 4086148.

[17] Monod, Jacques; Changeux, J.P.; Jacob, François (1963). "Allosteric proteins and cellular control systems". *Journal of Molecular Biology*. **6** (4): 306–329. PMID 13936070. doi:10.1016/S0022-2836(63)80091-1.

[18] Gilbert, S.F.; Opitz, J.M.; Raff, R.A. (1996). "Resynthesizing Evolutionary and Developmental Biology". *Developmental Biology*. **173** (2): 357–372. PMID 8605997. doi:10.1006/dbio.1996.0032.

[19] Müller, G.B. (2007). "Evo–devo: extending the evolutionary synthesis". *Nature Reviews Genetics*. **8** (12): 943–949. PMID 17984972. doi:10.1038/nrg2219.

[20] Goodman CS and Coughlin BS (Eds). (2000). "Special feature: The evolution of evo-devo biology". *Proceedings of the National Academy of Sciences*. **97** (9): 4424–4456. Bibcode:2000PNAS...97.4424G. PMC 18255. PMID 10781035. doi:10.1073/pnas.97.9.4424.

[21] Müller GB and Newman SA (Eds.) (2005). "Special issue: Evolutionary Innovation and Morphological Novelty". *Journal of Exp. Zool. Part B: Molecular and Developmental Evolution*. **304B** (6): 485–631. doi:10.1002/jez.b.21080.

[22] Palmer, R.A. (2004). "Symmetry breaking and the evolution of development". *Science*. **306** (5697): 828–833. Bibcode:2004Sci...306..828P. PMID 15514148. doi:10.1126/science.1103707.

[23] Winchester, Guil (2004). "Edward B. Lewis 1918-2004" (PDF). *Current Biology* (published Sep 21, 2004). **14** (18): R740–742. PMID 15380080. doi:10.1016/j.cub.2004.09.007.

[24] Bürglin, Thomas R. "The Homeobox Page". Karolinska Institutet. Retrieved 13 October 2016.

[25] Holland, P.W. (2013). "Evolution of homeobox genes". *Wiley Interdiscip Rev Dev Biol*. **2** (1): 31–45. PMID 23799629. doi:10.1002/wdev.78. Homeobox genes are found in almost all eukaryotes, and have diversified into 11 gene classes and over 100 gene families in animal evolution, and 10 to 14 gene classes in plants.

[26] Nüsslein-Volhard, C.; Wieschaus, E. (October 1980). "Mutations affecting segment number and polarity in *Drosophila*". *Nature*. **287** (5785): 795–801. PMID 6776413. doi:10.1038/287795a0.

[27] Arthur, Wallace (14 February 2002). "The emerging conceptual framework of evolutionary developmental biology". *Nature*. **415**: 757–764. doi:10.1038/415757a.

[28] "Eric Wieschaus and Christiane Nüsslein-Volhard: Collaborating to Find Developmental Genes". iBiology. Retrieved 13 October 2016.

[29] Cohen, S. M.; Jurgens, G. (1989). "Proximal-distal pattern formation in Drosophila: cell autonomous requirement for Distal-less activity in limb development". *EMBO J.* **8** (7): 2045–2055. PMC 401088. PMID 16453891.

[30] Carroll, Sean B. (2006) [2005]. *Endless Forms Most Beautiful: The New Science of Evo Devo and the Making of the Animal Kingdom*. Weidenfeld & Nicolson [Norton]. pp. 63–70. ISBN 978-0-297-85094-6.

[31] Panganiban, G.; Irvine, S. M.; Lowe, C.; Roehl, H.; Corley, L. S.; Sherbon, B.; Grenier, J. K.; Fallon, J. F.; Kimble, J.; Walker, M.; Wray, G. A.; Swalla, B. J.; Martindale, M. Q.; Carroll, S. B. (1997). "The origin and evolution of animal appendages". *Proceedings of the National Academy of Sciences of the United States of America*. **94** (10): 5162–5166. PMC 24649. PMID 9144208. doi:10.1073/pnas.94.10.5162.

[32] "Fluorescent Probes". ThermoFisher Scientific. 2015. Retrieved 12 October 2016.

[33] Land, M. F.; Fernald, R. D. (1992). "The evolution of eyes". *Annual Review of Neuroscience*. **15**: 1–29. PMID 1575438. doi:10.1146/annurev.ne.15.030192.000245.

[34] Tomarev, Stanislav I.; Callaerts, Patrick; Kos, Lidia; Zinovieva, Rina; Halder, Georg; Gehring, Walter; Piatigorsky, Joram (1997). "Squid Pax-6 and eye development". *Proceedings of the National Academy of Sciences*. **94** (6): 2421–2426. Bibcode:1997PNAS...94.2421T. PMC 20103. PMID 9122210. doi:10.1073/pnas.94.6.2421.

[35] Pichaud, Franck; Desplan, Claude (August 2002). "Pax genes and eye organogenesis". *Current Opinion in Genetics & Development*. **12** (4): 430–434. PMID 12100888. doi:10.1016/S0959-437X(02)00321-0.

[36] Carroll, Sean B. (2008). "Evo-Devo and an Expanding Evolutionary Synthesis: A Genetic Theory of Morphological Evolution". *Cell*. **134** (1): 25–36. PMID 18614008. doi:10.1016/j.cell.2008.06.030.

[37] Xu, P.X.; Woo, I.; Her, H.; Beier, D.R.; Maas, R.L. (1997). "Mouse Eya homologues of the Drosophila eyes absent gene require Pax6 for expression in lens and nasal placode". *Development*. **124** (1): 219–231. PMID 9006082.

[38] Mukherjee, K.; Brocchieri, L.; Bürglin, T.R. (December 2009). "A comprehensive classification and evolutionary analysis of plant homeobox genes". *Molecular Biology and Evolution*. **26** (12): 2775–94. PMC 2775110. PMID 19734295. doi:10.1093/molbev/msp201.

[39] Bürglin, T.R. (November 1997). "Analysis of TALE superclass homeobox genes (MEIS, PBC, KNOX, Iroquois, TGIF) reveals a novel domain conserved between plants and animals". *Nucleic Acids Research*. **25** (21): 4173–80. PMC 147054. PMID 9336443. doi:10.1093/nar/25.21.4173.

[40] Derelle, R.; Lopez, P.; Le Guyader, H.; Manuel, M. (2007). "Homeodomain proteins belong to the ancestral molecular toolkit of eukaryotes". *Evolution & Development*. **9** (3): 212–9. PMID 17501745. doi:10.1111/j.1525-142X.2007.00153.x.

[41] Russel, Peter (2010). *iGenetics: a molecular approach*. Pearson Education. pp. 564–571. ISBN 978-0-321-56976-9.

[42] Rivera-Pomar, Rolando; Jackle, Herbert (1996). "From gradients to stripes in Drosophila embryogenesis: Filling in the gaps". *Trends in Genetics*. **12** (11): 478–483. doi:10.1016/0168-9525(96)10044-5.

[43] Gerhart, John; Kirschner, Marc (1997). *Cells, Embryos and Evolution*. Blackwell Science. ISBN 978-0-86542-574-3.

[44] Carroll, Sean B.; Grenier, Jennifer K.; Weatherbee, Scott D. (2005). *From DNA to Diversity: Molecular Genetics and the Evolution of Animal Design — Second Edition*. Blackwell Publishing. ISBN 1-4051-1950-0.

[45] Carroll, Sean B. (2000). "Endless forms: the evolution of gene regulation and morphological diversity". *Cell*. **101** (6): 577–80. PMID 10892643. doi:10.1016/S0092-8674(00)80868-5.

[46] Moczek, Armin P.; et al. (2015). "The Significance and Scope of Evolutionary Developmental Biology: A Vision for the 21st Century" (PDF). *Evolution & Development*. **17**: 198–219. doi:10.1111/ede.12125.

[47] Abzhanov, A.; Protas, M.; Grant, B.R.; Grant, P.R.; Tabin, C.J. (2004). "Bmp4 and Morphological Variation of Beaks in Darwin's Finches". *Science*. **305** (5689): 1462–1465. Bibcode:2004Sci...305.1462A. PMID 15353802. doi:10.1126/science.1098095.

[48] Cohn, M.J.; Tickle, C. (1999). "Developmental basis of limblessness and axial patterning in snakes". *Nature*. **399** (6735): 474–479. Bibcode:1999Natur.399..474C. PMID 10365960. doi:10.1038/20944.

[49] Beverdam, A.; Merlo, G.R.; Paleari, L.; Mantero, S.; Genova, F.; Barbieri, O.; Janvier, P.; Levi, G. (August 2002). "Jaw Transformation With Gain of Symmetry After DLX5/DLX6 Inactivation: Mirror of the Past?". *Genesis*. **34** (4): 221–227. PMID 12434331. doi:10.1002/gene.10156.

[50] Depew, M.J.; Lufkin, T.; Rubenstein, J.L. (October 2002). "Specification of jaw subdivisions by DLX genes". *Science*. **298** (5592): 381–385. PMID 12193642. doi:10.1126/science.1075703.

[51] Panganiban, Grace; Rubenstein, John L. R. (2002). "Developmental functions of the Distal-less/Dlx homeobox genes". *Development*. **129**: 4371–4386.

[52] Beldade, P.; Brakefield, P.M.; Long, A.D. (2002). "Contribution of Distal-less to quantitative variation in butterfly eyespots". *Nature*. **415** (6869): 315–318. PMID 11797007. doi:10.1038/415315a.

[53] Gerhart, John; Kirschner, Marc (2007). "The theory of facilitated variation". *Proceedings of the National Academy of Sciences*. **104** (suppl1): 8582–8589. Bibcode:2007PNAS..104.8582G. PMC 1876433. PMID 17494755. doi:10.1073/pnas.0701035104.

[54] Baxter, S.W.; Papa, R.; Chamberlain, N.; Humphray, S.J.; Joron, M.; Morrison, C.; ffrench-Constant, R.H.; McMillan, W.O.; Jiggins, C.D. (2008). "Convergent Evolution in the Genetic Basis of Mullerian Mimicry in Heliconius Butterflies". *Genetics*. **180** (3): 1567–1577. PMC 2581958. PMID 18791259. doi:10.1534/genetics.107.082982.

[55] Pennisi, E. (2002). "Evolutionary Biology:Evo-Devo Enthusiasts Get Down to Details". *Science*. **298** (5595): 953–955. PMID 12411686. doi:10.1126/science.298.5595.953.

[56] Jablonka, Eva; Lamb, Marion (1995). *Epigenetic Inheritance and Evolution: The Lamarckian Dimension*. Oxford, New York: Oxford University Press. ISBN 978-0-19-854063-2.

[57] West-Eberhard, M-J. (2003). *Developmental plasticity and evolution*. New York: Oxford University Press. ISBN 978-0-19-512235-0.

[58] Müller, Gerd B.; Newman, Stuart A., eds. (2003). *Origination of Organismal Form: Beyond the Gene in Developmental and Evolutionary Biology*. MIT Press.

[59] Abouheif, E.; Favé, M.-J.; Ibarrarán-Viniegra, A. S.; Lesoway, M. P.; Rafiqi, A. M.; Rajakumar, R. (2014). "Eco-Evo-Devo: The Time Has Come". In Landry, C.R.; Aubin-Horth, N. *Ecological Genomics: Ecology and the Evolution of Genes and Genomes*. pp. 107–125. ISBN 978-94-007-7346-2. doi:10.1007/978-94-007-7347-9_6.

[60] Schlichting, C.D. (2009). "An Uneven Guide to Eco-Devo". *BioScience*. **59** (11): 1000–1001. doi:10.1525/bio.2009.59.11.12.

[61] Gilbert, S. F.; Bosch, T. C. G.; Ledón-Rettig, C. (2015). "Eco-Evo-Devo: developmental symbiosis and developmental plasticity as evolutionary agents". *Nature Reviews Genetics*. **16** (10): 611–622. PMID 26370902. doi:10.1038/nrg3982.

37.7 References

- Secord, James A. (2000). *Victorian sensation: the extraordinary publication, reception, and secret authorship of Vestiges of the natural history of creation*. Chicago: University of Chicago Press. ISBN 978-0-226-74410-0

Chapter 38

Evolutionary robotics

Evolutionary robotics (**ER**) is a methodology that uses evolutionary computation to develop controllers for autonomous robots. Algorithms in ER frequently operate on populations of candidate controllers, initially selected from some distribution. This population is then repeatedly modified according to a fitness function. In the case of genetic algorithms (or "GAs"), a common method in evolutionary computation, the population of candidate controllers is repeatedly grown according to crossover, mutation and other GA operators and then culled according to the fitness function. The candidate controllers used in ER applications may be drawn from some subset of the set of artificial neural networks, although some applications (including SAMUEL, developed at the Naval Center for Applied Research in Artificial Intelligence) use collections of "IF THEN ELSE" rules as the constituent parts of an individual controller. It is theoretically possible to use any set of symbolic formulations of a control law (sometimes called a policy in the machine learning community) as the space of possible candidate controllers. Artificial neural networks can also be used for robot learning outside of the context of evolutionary robotics. In particular, other forms of reinforcement learning can be used for learning robot controllers.

Developmental robotics is related to, but differs from, evolutionary robotics. ER uses populations of robots that evolve over time, whereas DevRob is interested in how the organization of a single robot's control system develops through experience, over time.

38.1 History

The foundation of ER was laid with work at the national research council in Rome in the 90s, but the initial idea of encoding a robot control system into a genome and have artificial evolution improve on it dates back to the late 80s.

In 1992 and 1993 three research groups, one surrounding Floreano and Mondada at the EPFL in Lausanne and a second involving Cliff, Harvey, and Husbands from COGS at the University of Sussex and a third from the University of Southern California involved M. Anthony Lewis and Andrew H Fagg reported promising results from experiments on artificial evolution of autonomous robots.[1][2] The success of this early research triggered a wave of activity in labs around the world trying to harness the potential of the approach.

Lately, the difficulty in "scaling up" the complexity of the robot tasks has shifted attention somewhat towards the theoretical end of the field rather than the engineering end.

38.2 Objectives

Evolutionary robotics is done with many different objectives, often at the same time. These include creating useful controllers for real-world robot tasks, exploring the intricacies of evolutionary theory (such as the Baldwin effect), reproducing psychological phenomena, and finding out about biological neural networks by studying artificial ones. Creating controllers via artificial evolution requires a large number of evaluations of a large population. This is very time consuming, which is one of the reasons why controller evolution is usually done in software. Also, initial random controllers may exhibit potentially harmful behaviour, such as repeatedly crashing into a wall, which may damage the robot. Transferring controllers evolved in simulation to physical robots is very difficult and a major challenge in using the ER approach. The reason is that evolution is free to explore all possibilities to obtain a high fitness, including any inaccuracies of the simulation. This need for a large number of evaluations, requiring fast yet accurate computer simulations, is one of the limiting factors of the ER approach.

In rare cases, evolutionary computation may be used to design the physical structure of the robot, in addition to the controller. One of the most notable examples of this was Karl Sims' demo for Thinking Machines Corporation.

38.3 Motivation

Many of the commonly used machine learning algorithms require a set of training examples consisting of both a hypothetical input and a desired answer. In many robot learning applications the desired answer is an action for the robot to take. These actions are usually not known explicitly a priori, instead the robot can, at best, receive a value indicating the success or failure of a given action taken. Evolutionary algorithms are natural solutions to this sort of problem framework, as the fitness function need only encode the success or failure of a given controller, rather than the precise actions the controller should have taken. An alternative to the use of evolutionary computation in robot learning is the use of other forms of reinforcement learning, such as q-learning, to learn the fitness of any particular action, and then use predicted fitness values indirectly to create a controller.

38.4 Conferences and institutes

38.4.1 Main conferences

- Genetic and Evolutionary Computation Conference
- IEEE Congress on Evolutionary Computation
- European Conference on Artificial Life
- ALife

Octavia interactive robot of Navy Center for Applied Research In Artificial Intelligence

38.4.2 Academic institutes and researchers

- Chalmers University of Technology: Peter Nordin, The Humanoid Project[3]
- University of Sussex: Inman Harvey, Phil Husbands, Ezequiel Di Paolo
- Consiglio Nazionale delle Ricerche (CNR): Stefano Nolfi
- EPFL: Dario Floreano
- University of Zürich: Rolf Pfeifer
- Cornell University: Hod Lipson
- University of Vermont: Josh Bongard
- Indiana University: Randall Beer
- Center for Robotics and Intelligent Machines, North Carolina State University: Eddie Grant, Andrew Nelson
- University College London: Peter J. Bentley
- The IDSIA Robotics Lab: Juergen Schmidhuber, Juxi Leitner[4]
- U.S. Naval Research Laboratory[5]
- University of Osnabrueck, Neurocybernetics Group: Frank Pasemann
- Evolved Virtual Creatures by Karl Sims (GenArts)
- Ken Rinaldo artificial life robotics
- European Space Agency's Advanced Concepts Team: Dario Izzo
- University of the Basque Country (UPV-EHU): Robótica Evolutiva, Pablo González-Nalda (in Spanish) PDF (in English)
- University of Plymouth: Angelo Cangelosi, Davide Marocco, Fabio Ruini, * Martin Peniak
- Heriot-Watt University: Patricia A. Vargas
- Pierre and Marie Curie University, ISIR: Stephane Doncieux, Jean-Baptiste Mouret
- Paris-Sud University and INRIA, IAO/TAO: Nicolas Bredeche
- RIKEN Brain Science Institute
- Karlsruhe Institute of Technology, Institute of Applied Informatics and Formal Description Methods: Lukas Koenig

38.5 See also

- Artificial intelligence
- Cybernetics
- Cognitive robotics
- Evolutionary computation
- Roboticist
- Robotics
- Robot kit
- Universal Darwinism

38.6 References

[1] http://www.sussex.ac.uk/Users/philh/pubs/evolvingVisguided.pdf D.Cliff, I.Harvey, & P.Husbands, (1992) "Evolving Visually Guided Robots"; conference paper presented at SAB92, Hawaii, 1992.

[2] http://citeseerx.ist.psu.edu/viewdoc/download?doi=10.1.1.45.240&rep=rep1&type=pdf Lewis, Fagg and Solidum (1992) "Genetic programming approach to the construction of a neural network for control of a walking robot"; conference paper presented at ICRA 1992.

[3] The Humanoid Project Archived 2007-06-30 at the Wayback Machine.

[4] http://Juxi.net/

[5] Navy Center for Applied Research In Artificial Intelligence

- *Evolutionary Robotics* by Stefano Nolfi and Dario Floreano. (2000). ISBN 0-262-14070-5

- *Advances in the Evolutionary Synthesis of Intelligent Agents* by Mukesh Patel, Vasant Honavar and Karthik Balakrishnan (Ed). Cambridge, MA: MIT Press. (2001). ISBN 0-262-16201-6

- *Evolving neuromorphic flight control for a flapping-wing mechanical insect* by Boddhu, Sanjay K., and John C. Gallagher. International Journal of Intelligent Computing and Cybernetics 3.1 (2010): 94-116.

- *The Horizons of Evolutionary Robotics* edited by Patricia A. Vargas, Ezequiel Di Paolo, Inman Harvey and Phil Husbands. (2014). ISBN 9780262026765.

38.7 External links

- An introduction to Evolutionary Robotics with annotated bibliography
- The Evolutionary Robotics Homepage

Chapter 39

Pattern formation

The science of **pattern formation** deals with the visible, (statistically) orderly outcomes of self-organization and the common principles behind similar patterns in nature.

In developmental biology, pattern formation refers to the generation of complex organizations of cell fates in space and time. Pattern formation is controlled by genes. The role of genes in pattern formation is an aspect of morphogenesis, the creation of diverse anatomies from similar genes, now being explored in the science of evolutionary developmental biology or evo-devo. The mechanisms involved are well seen in the anterior-posterior patterning of embryos from the model organism *Drosophila melanogaster* (a fruit fly), one of the first organisms to have its morphogenesis studied, and in the eyespots of butterflies, whose development is a variant of the standard (fruit fly) mechanism.

39.1 Examples

Further information: Patterns in nature

Examples of pattern formation can be found in Biology, Chemistry, Physics and Mathematics,[1] and can readily be simulated with Computer graphics, as described in turn below.

39.1.1 Biology

Further information: Evolutionary developmental biology and Morphogenetic field

Biological patterns such as animal markings, the segmentation of animals, and phyllotaxis are formed in different ways.[2]

In developmental biology, pattern formation describes the mechanism by which initially equivalent cells in a developing tissue in an embryo assume complex forms and functions.[3] Embryogenesis, such as of the fruit fly *Drosophila*, involves coordinated control of cell fates.[4][5][6] Pattern formation is genetically controlled, and often involves each cell in a field sensing and responding to its position along a morphogen gradient, followed by short distance cell-to-cell communication through cell signaling pathways to refine the initial pattern. In this context, a field of cells is the group of cells whose fates are affected by responding to the same set positional information cues. This conceptual model was first described as the French flag model in the 1960s.[7][8] More generally, the morphology of organisms is patterned by the mechanisms of evolutionary developmental biology, such as changing the timing and positioning of specific developmental events in the embryo.[9]

Possible mechanisms of pattern formation in biological systems include the classical reaction-diffusion model proposed by Alan Turing[10] and the more recently found elastic instability mechanism which is thought to be responsible for the fold patterns on the cerebral cortex of higher animals, among other things.[11][12]

Growth of Colonies

Bacterial colonies show a large variety of patterns formed during colony growth. The resulting shapes depend on the growth conditions. In particular, stresses (hardness of the culture medium, lack of nutrients, etc.) enhance the complexity of the resulting patterns.[13] Other organisms such as slime moulds display remarkable patterns caused by the dynamics of chemical signalling.[14]

Vegetation patterns

Main article: patterned vegetation
Vegetation patterns such as tiger bush[15] and fir waves[16] form for different reasons. Tiger bush consists of stripes of bushes on arid slopes in countries such as Niger where plant growth is limited by rainfall. Each roughly horizontal stripe of vegetation absorbs rainwater from the bare zone immediately above it.[15] In contrast, fir waves occur in forests on

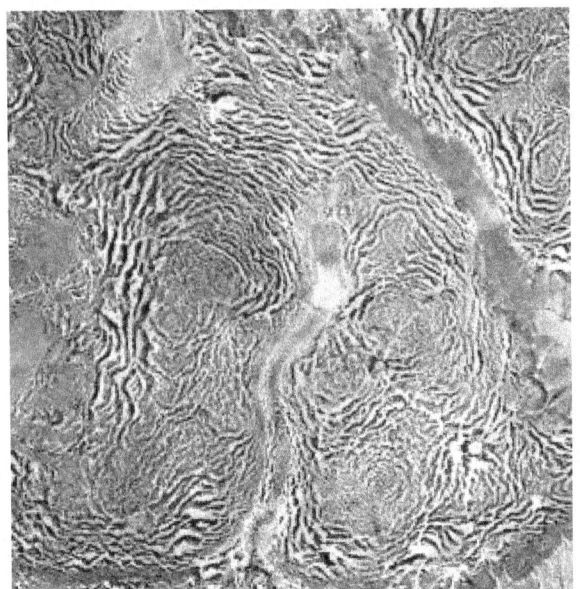

Tiger bush is a vegetation pattern that forms in arid conditions.

mountain slopes after wind disturbance, during regeneration. When trees fall, the trees that they had sheltered become exposed and are in turn more likely to be damaged, so gaps tend to expand downwind. Meanwhile, on the windward side, young trees grow, protected by the wind shadow of the remaining tall trees.[16]

39.1.2 Chemistry

Further information: reaction–diffusion system and Turing Patterns

- Belousov-Zhabotinsky reaction
- Liesegang rings

39.1.3 Physics

Bénard cells. Laser. cloud formations in stripes or rolls. Ripples in icicles. Washboard patterns on dirtroads. Dendrites in solidification. liquid crystals. Solitons.

39.1.4 Mathematics

Sphere packings and coverings. Mathematics underlies the other pattern formation mechanisms listed.

Further information: Gradient Pattern Analysis

39.1.5 Computer graphics

Pattern resembling a Reaction-diffusion model, produced using sharpen and blur

Further information: Cellular automaton

Some types of automata have been used to generate organic-looking textures for more realistic shading of 3d objects.[17][18]

A popular Photoshop plugin, KPT 6, included a filter called 'KPT reaction'. Reaction produced reaction-diffusion style patterns based on the supplied seed image.

A similar effect to the 'KPT reaction' can be achieved with convolution functions in digital image processing, with a little patience, by repeatedly sharpening and blurring an image in a graphics editor. If other filters are used, such as emboss or edge detection, different types of effects can be achieved.

Computers are often used to simulate the biological, physical or chemical processes that lead to pattern formation, and they can display the results in a realistic way. Calculations using models like Reaction-diffusion or MClone are based on the actual mathematical equations designed by the scientists to model the studied phenomena.

39.2 References

[1] Ball, 2009.

[2] Ball, 2009. *Shapes*, pp. 231–252.

[3] Ball, 2009. Shapes, pp. 261–290.

[4] Eric C. Lai (March 2004). "Notch signaling: control of cell communication and cell fate" (PDF). *Development*. **131** (5): 965–73. PMID 14973298. doi:10.1242/dev.01074.

[5] Melinda J. Tyler, David A. Cameron (2007). "Cellular pattern formation during retinal regeneration: A role for homotypic control of cell fate acquisition". *Vision Research*. **47** (4): 501–511. PMID 17034830. doi:10.1016/j.visres.2006.08.025.

[6] Hans Meinhard (2001-10-26). "Biological pattern formation: How cell[s] talk with each other to achieve reproducible pattern formation". Max-Planck-Institut für Entwicklungsbiologie, Tübingen, Germany.

[7] Wolpert L (October 1969). "Positional information and the spatial pattern of cellular differentiation". *J. Theor. Biol.* **25** (1): 1–47. PMID 4390734. doi:10.1016/S0022-5193(69)80016-0.

[8] Wolpert, Lewis; et al. (2007). *Principles of development* (3rd ed.). Oxford [Oxfordshire]: Oxford University Press. ISBN 0-19-927536-X.

[9] Hall, B. K. (2003). "Evo-Devo: evolutionary developmental mechanisms". *International Journal of Developmental Biology*. **47** (7–8): 491–495. PMID 14756324.

[10] S. Kondo, T. Miura, "Reaction-Diffusion Model as a Framework for Understanding Biological Pattern Formation", Science 24 Sep 2010: Vol. 329, Issue 5999, pp. 1616-1620 DOI: 10.1126/science.1179047

[11] Mercker et al. Biology Direct (2016) 11:22 DOI 10.1186/s13062-016-0124-7

[12] Tallinen et al. Nature Physics 12, 588–593 (2016) doi:10.1038/nphys3632

[13] Ball, 2009. *Branches*, pp. 52–59.

[14] Ball, 2009. *Shapes*, pp. 149–151.

[15] Tongway, D.J., Valentin, C. & Seghieri, J. (2001). *Banded vegetation patterning in arid and semiarid environments*. New York: Springer-Verlag. ISBN 978-1461265597.

[16] D'Avanzo, C. (22 February 2004). "Fir Waves: Regeneration in New England Conifer Forests". TIEE. Retrieved 26 May 2012.

[17] Greg Turk, Reaction-Diffusion

[18] Andrew Witkin; Michael Kassy (1991). "Reaction-Diffusion Textures" (PDF). *Proceedings of the 18th annual conference on Computer graphics and interactive techniques*. pp. 299–308. doi:10.1145/122718.122750.

39.3 Bibliography

- Ball, Philip (2009). *Nature's Patterns: a tapestry in three parts. 1:Shapes. 2:Flow. 3:Branches*. Oxford. ISBN 978-0199604869.

39.4 External links

- *SpiralZoom.com*, an educational website about the science of pattern formation, spirals in nature, and spirals in the mythic imagination.

- '15-line Matlab code', A simple 15-line Matlab program to simulate 2D pattern formation for reaction-diffusion model.

Chapter 40

Reaction–diffusion system

Reaction–diffusion systems are mathematical models which correspond to several physical phenomena: the most common is the change in space and time of the concentration of one or more chemical substances: local chemical reactions in which the substances are transformed into each other, and diffusion which causes the substances to spread out over a surface in space.

Reaction–diffusion systems are naturally applied in chemistry. However, the system can also describe dynamical processes of non-chemical nature. Examples are found in biology, geology and physics (neutron diffusion theory) and ecology. Mathematically, reaction–diffusion systems take the form of semi-linear parabolic partial differential equations. They can be represented in the general form

$$\partial_t q = \underline{\underline{D}} \nabla^2 q + R(q).$$

where $q(x, t)$ represents the unknown vector function, D is a diagonal matrix of diffusion coefficients, and R accounts for all local reactions. The solutions of reaction–diffusion equations display a wide range of behaviours, including the formation of travelling waves and wave-like phenomena as well as other self-organized patterns like stripes, hexagons or more intricate structure like dissipative solitons. Each function, for which a reaction diffusion differential equation holds, represents in fact a *concentration variable*.

40.1 One-component reaction–diffusion equations

The simplest reaction–diffusion equation is in one spatial dimension in plane geometry,

$$\partial_t u = D \partial_x^2 u + R(u).$$

is also referred to as the Kolmogorov–Petrovsky–Piskunov equation.[1] If the reaction term vanishes, then the equation represents a pure diffusion process. The corresponding equation is Fick's second law. The choice $R(u) = u(1 - u)$ yields Fisher's equation that was originally used to describe the spreading of biological populations,[2] the Newell–Whitehead-Segel equation with $R(u) = u(1 - u^2)$ to describe Rayleigh–Bénard convection,[3][4] the more general Zeldovich equation with $R(u) = u(1 - u)(u - \alpha)$ and $0 < \alpha < 1$ that arises in combustion theory,[5] and its particular degenerate case with $R(u) = u^2 - u^3$ that is sometimes referred to as the Zeldovich equation as well.[6]

The dynamics of one-component systems is subject to certain restrictions as the evolution equation can also be written in the variational form

$$\partial_t u = -\frac{\delta \mathfrak{L}}{\delta u}$$

and therefore describes a permanent decrease of the "free energy" $\mathfrak{L}$ given by the functional

$$\mathfrak{L} = \int_{-\infty}^{\infty} \left[\frac{D}{2} (\partial_x u)^2 - V(u) \right] dx$$

with a potential $V(u)$ such that $R(u) = dV(u)/du$.

In systems with more than one stationary homogeneous solution, a typical solution is given by travelling fronts connecting the homogeneous states. These solutions move with constant speed without changing their shape and are of the form $u(x, t) = \hat{u}(\xi)$ with $\xi = x - ct$, where c is the speed of the travelling wave. Note that while travelling waves are generically stable structures, all non-monotonous stationary solutions (e.g. localized domains composed of a front-antifront pair) are unstable. For $c = 0$, there is a simple proof for this statement:[7] if $u_0(x)$ is a stationary solution and $u = u_0(x) + \tilde{u}(x, t)$ is an infinitesimally perturbed solution, linear stability analysis yields the equation

$$\partial_t \tilde{u} = D \partial_x^2 \tilde{u} - U(x) \tilde{u}, \qquad U(x) = -R'(u)|_{u=u_0(x)}.$$

360

40.2 Two-component reaction–diffusion equations

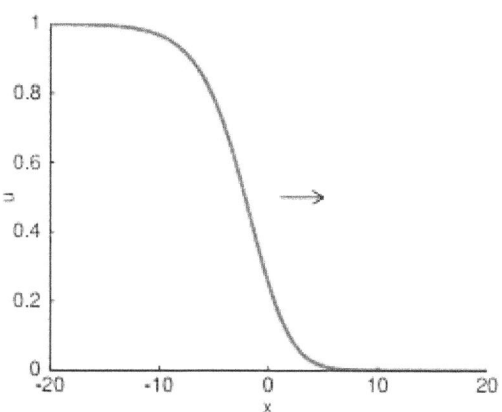

A travelling wave front solution for Fisher's equation.

With the ansatz $\tilde{u} = \psi(x)\exp(-\lambda t)$ we arrive at the eigenvalue problem

$$\hat{H}\psi = \lambda\psi, \qquad \hat{H} = -D\partial_x^2 + U(x),$$

of Schrödinger type where negative eigenvalues result in the instability of the solution. Due to translational invariance $\psi = \partial_x u_0(x)$ is a neutral eigenfunction with the eigenvalue $\lambda = 0$, and all other eigenfunctions can be sorted according to an increasing number of knots with the magnitude of the corresponding real eigenvalue increases monotonically with the number of zeros. The eigenfunction $\psi = \partial_x u_0(x)$ should have at least one zero, and for a non-monotonic stationary solution the corresponding eigenvalue $\lambda = 0$ cannot be the lowest one, thereby implying instability.

To determine the velocity c of a moving front, one may go to a moving coordinate system and look at stationary solutions:

$$D\partial_\xi^2 \hat{u}(\xi) + c\partial_\xi \hat{u}(\xi) + R(\hat{u}(\xi)) = 0.$$

This equation has a nice mechanical analogue as the motion of a mass D with position $\hat{u}$ in the course of the "time" ξ under the force R with the damping coefficient c which allows for a rather illustrative access to the construction of different types of solutions and the determination of c.

When going from one to more space dimensions, a number of statements from one-dimensional systems can still be applied. Planar or curved wave fronts are typical structures, and a new effect arises as the local velocity of a curved front becomes dependent on the local radius of curvature (this can be seen by going to polar coordinates). This phenomenon leads to the so-called curvature-driven instability.[8]

Two-component systems allow for a much larger range of possible phenomena than their one-component counterparts. An important idea that was first proposed by Alan Turing is that a state that is stable in the local system can become unstable in the presence of diffusion.[9]

A linear stability analysis however shows that when linearizing the general two-component system

$$\begin{pmatrix}\partial_t u \\ \partial_t v\end{pmatrix} = \begin{pmatrix}D_u & 0 \\ 0 & D_v\end{pmatrix}\begin{pmatrix}\partial_{xx} u \\ \partial_{xx} v\end{pmatrix} + \begin{pmatrix}F(u,v) \\ G(u,v)\end{pmatrix}$$

a plane wave perturbation

$$\tilde{q}_k(x,t) = \begin{pmatrix}\tilde{u}(t) \\ \tilde{v}(t)\end{pmatrix}e^{ik\cdot x}$$

of the stationary homogeneous solution will satisfy

$$\begin{pmatrix}\partial_t \tilde{u}_k(t) \\ \partial_t \tilde{v}_k(t)\end{pmatrix} = -k^2\begin{pmatrix}D_u \tilde{u}_k(t) \\ D_v \tilde{v}_k(t)\end{pmatrix} + R'\begin{pmatrix}\tilde{u}_k(t) \\ \tilde{v}_k(t)\end{pmatrix}.$$

Turing's idea can only be realized in four equivalence classes of systems characterized by the signs of the Jacobian R' of the reaction function. In particular, if a finite wave vector k is supposed to be the most unstable one, the Jacobian must have the signs

$$\begin{pmatrix}+ & - \\ + & -\end{pmatrix}, \quad \begin{pmatrix}+ & + \\ - & -\end{pmatrix}, \quad \begin{pmatrix}- & + \\ - & +\end{pmatrix}, \quad \begin{pmatrix}- & - \\ + & +\end{pmatrix}.$$

This class of systems is named *activator-inhibitor system* after its first representative: close to the ground state, one component stimulates the production of both components while the other one inhibits their growth. Its most prominent representative is the FitzHugh–Nagumo equation

$$\partial_t u = d_u^2 \nabla^2 u + f(u) - \sigma v,$$
$$\tau\partial_t v = d_v^2 \nabla^2 v + u - v$$

with $f(u) = \lambda u - u^3 - \kappa$ which describes how an action potential travels through a nerve.[10][11] Here, d_u, d_v, τ, σ and λ are positive constants.

When an activator-inhibitor system undergoes a change of parameters, one may pass from conditions under which a homogeneous ground state is stable to conditions under which it is linearly unstable. The corresponding bifurcation may be either a Hopf bifurcation to a globally oscillating homogeneous state with a dominant wave number $k = 0$ or a *Turing bifurcation* to a globally patterned state with a dominant finite wave number. The latter in two spatial dimensions typically leads to stripe or hexagonal patterns.

- Subcritical Turing bifurcation: formation of a hexagonal pattern from noisy initial conditions in the above two-component reaction-diffusion system of Fitzhugh-Nagumo type.

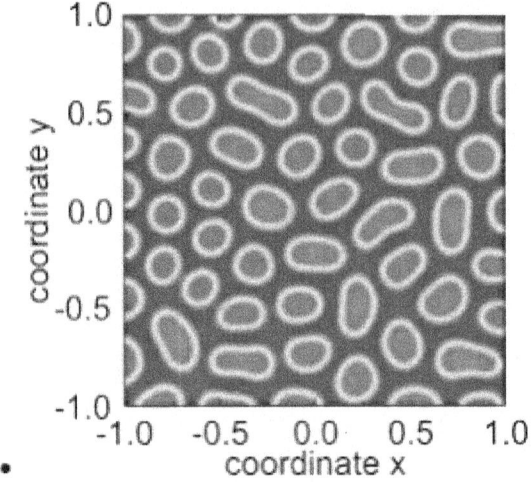

- Almost converged state at $t = 100$.

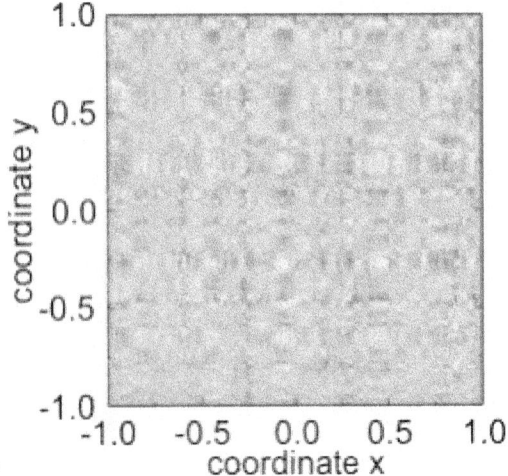

- Noisy initial conditions at $t = 0$.

For the Fitzhugh-Nagumo example, the neutral stability curves marking the boundary of the linearly stable region for the Turing and Hopf bifurcation are given by

$$q_n^H(k): \quad \frac{1}{\tau} + \left(d_u^2 + \frac{1}{\tau}d_v^2\right)k^2 = f'(u_h).$$

$$q_n^T(k): \quad \frac{\kappa}{1 + d_v^2 k^2} + d_u^2 k^2 = f'(u_h).$$

If the bifurcation is subcritical, often localized structures (dissipative solitons) can be observed in the hysteretic region where the pattern coexists with the ground state. Other frequently encountered structures comprise pulse trains (also known as periodic travelling waves), spiral waves and target patterns. These three solution types are also generic features of two- (or more-) component reaction-diffusion equations in which the local dynamics have a stable limit cycle[12]

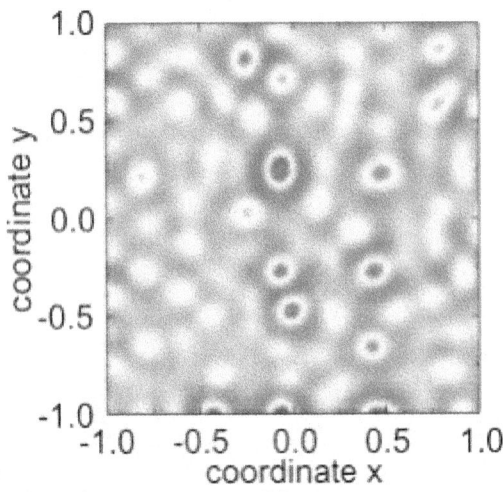

- State of the system at $t = 10$.

- Other patterns found in the above two-component reaction-diffusion system of Fitzhugh-Nagumo type.

40.3 Three- and more-component reaction–diffusion equations

For a variety of systems, reaction-diffusion equations with more than two components have been proposed, e.g. as models for the Belousov-Zhabotinsky reaction,[13] for blood clotting[14] or planar gas discharge systems.[15]

It is known that systems with more components allow for a variety of phenomena not possible in systems with one or two components (e.g. stable running pulses in more than one spatial dimension without global feedback),.[16] An introduction and systematic overview of the possible phenomena in dependence on the properties of the underlying system is given in.[17]

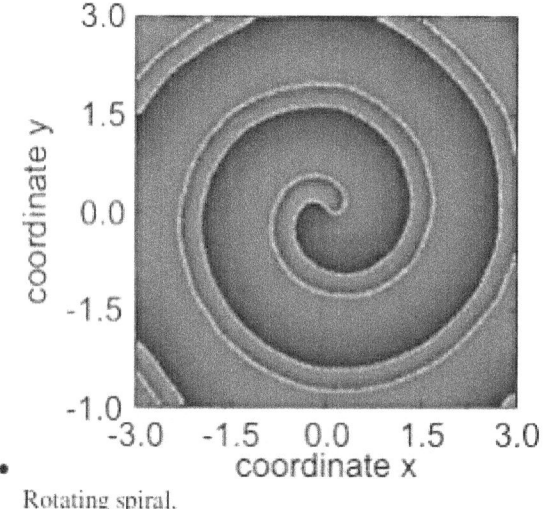
- Rotating spiral.

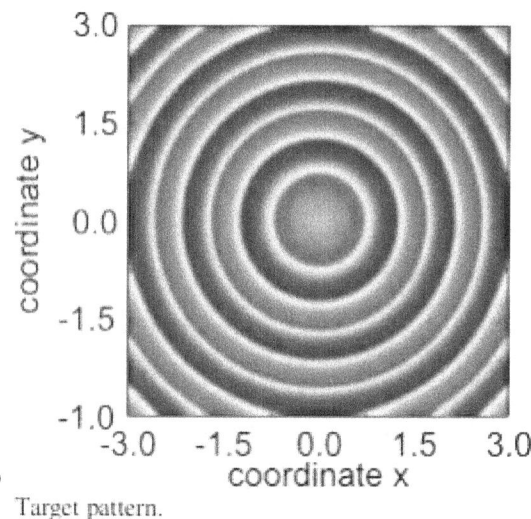
- Target pattern.

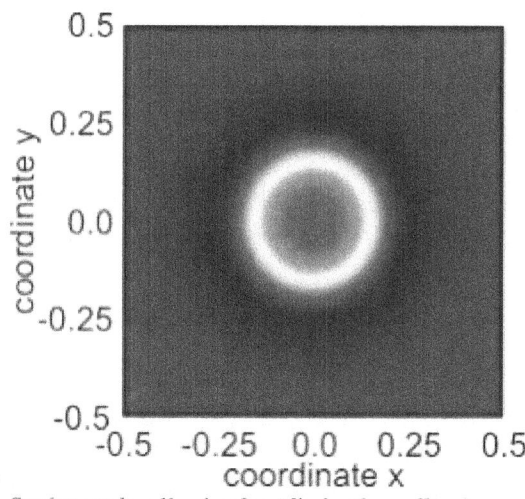
- Stationary localized pulse (dissipative soliton).

40.4 Applications and universality

In recent times, reaction–diffusion systems have attracted much interest as a prototype model for pattern formation. The above-mentioned patterns (fronts, spirals, targets, hexagons, stripes and dissipative solitons) can be found in various types of reaction-diffusion systems in spite of large discrepancies e.g. in the local reaction terms. It has also been argued that reaction-diffusion processes are an essential basis for processes connected to morphogenesis in biology[18] and may even be related to animal coats and skin pigmentation.[19][20] Other applications of reaction-diffusion equations include ecological invasions,[21] spread of epidemics,[22] tumour growth[23][24][25] and wound healing.[26] Another reason for the interest in reaction-diffusion systems is that although they are nonlinear partial differential equations, there are often possibilities for an analytical treatment.[7][8][27][28][29]

40.5 Experiments

Well-controllable experiments in chemical reaction-diffusion systems have up to now been realized in three ways. First, gel reactors[30] or filled capillary tubes[31] may be used. Second, temperature pulses on catalytic surfaces have been investigated.[32][33] Third, the propagation of running nerve pulses is modelled using reaction-diffusion systems.[10][34]

Aside from these generic examples, it has turned out that under appropriate circumstances electric transport systems like plasmas[35] or semiconductors[36] can be described in a reaction-diffusion approach. For these systems various experiments on pattern formation have been carried out.

40.6 Numerical treatments

A reaction-diffusion system can be solved by using methods of numerical mathematics. There are existing several numerical treatments in research literature.[37] Also for complex geometries numerical solution methods are proposed.[38][39]

40.7 See also

- Autowave
- Diffusion-controlled reaction
- Chemical kinetics
- Phase space method
- Autocatalytic reactions and order creation
- Pattern formation
- Patterns in nature
- Periodic travelling wave
- Stochastic geometry
- MClone
- The Chemical Basis of Morphogenesis

40.8 Some examples of reaction-diffusion equations

- Fisher's equation
- Fisher-Kolmogorov equation

40.9 References

[1] A. Kolmogorov et al., Moscow Univ. Bull. Math. A 1 (1937): 1

[2] R. A. Fisher, Ann. Eug. 7 (1937): 355

[3] A. C. Newell and J. A. Whitehead, J. Fluid Mech. 38 (1969): 279

[4] L. A. Segel, J. Fluid Mech. 38 (1969): 203

[5] Y. B. Zeldovich and D. A. Frank-Kamenetsky, Acta Physicochim. 9 (1938): 341

[6] B. H. Gilding and R. Kersner, Travelling Waves in Nonlinear Diffusion Convection Reaction, Birkhäuser (2004)

[7] P. C. Fife, Mathematical Aspects of Reacting and Diffusing Systems, Springer (1979)

[8] A. S. Mikhailov, Foundations of Synergetics I. Distributed Active Systems, Springer (1990)

[9] A. M. Turing, Phil. Transact. Royal Soc. B 237 (1952): 37

[10] R. FitzHugh, Biophys. J. 1 (1961): 445

[11] J. Nagumo et al., Proc. Inst. Radio Engin. Electr. 50 (1962): 2061

[12] N. Kopell and L.N. Howard, Stud. Appl. Math. 52 (1973): 291

[13] V. K. Vanag and I. R. Epstein, Phys. Rev. Lett. 92 (2004): 128301

[14] E. S. Lobanova and F. I. Ataullakhanov, Phys. Rev. Lett. 93 (2004): 098303

[15] H.-G. Purwins et al. in: Dissipative Solitons, Lectures Notes in Physics, Ed. N. Akhmediev and A. Ankiewicz, Springer (2005)

[16] C. P. Schenk et al., Phys. Rev. Lett. 78 (1997): 3781

[17] A. W. Liehr: *Dissipative Solitons in Reaction Diffusion Systems. Mechanism, Dynamics, Interaction.* Volume 70 of Springer Series in Synergetics, Springer, Berlin Heidelberg 2013, ISBN 978-3-642-31250-2

[18] L.G. Harrison, Kinetic Theory of Living Pattern, Cambridge University Press (1993)

[19] H. Meinhardt, Models of Biological Pattern Formation, Academic Press (1982)

[20] Murray, James D. (9 March 2013). *Mathematical Biology*. Springer Science & Business Media. pp. 436–450. ISBN 978-3-662-08539-4.

[21] E.E. Holmes et al, Ecology 75 (1994): 17

[22] J.D. Murray et al, Proc. R. Soc. Lond. B 229 (1986: 111

[23] M.A.J. Chaplain J. Bio. Systems 3 (1995): 929

[24] J.A. Sherratt and M.A. Nowak, Proc. R. Soc. Lond. B 248 (1992): 261

[25] R.A. Gatenby and E.T. Gawlinski, Cancer Res. 56 (1996): 5745

[26] J.A. Sherratt and J.D. Murray, Proc. R. Soc. Lond. B 241 (1990): 29

[27] P. Grindrod, Patterns and Waves: The Theory and Applications of Reaction-Diffusion Equations, Clarendon Press (1991)

[28] J. Smoller, Shock Waves and Reaction Diffusion Equations, Springer (1994)

[29] B. S. Kerner and V. V. Osipov, Autosolitons. A New Approach to Problems of Self-Organization and Turbulence, Kluwer Academic Publishers (1994)

[30] K.-J. Lee et al., Nature 369 (1994): 215

[31] C. T. Hamik and O. Steinbock, New J. Phys. 5 (2003): 58

[32] H. H. Rotermund et al., Phys. Rev. Lett. 66 (1991): 3083

[33] M. D. Graham et al., J. Phys. Chem. 97 (1993): 7564

[34] A. L. Hodgkin and A. F. Huxley, J. Physiol. 117 (1952): 500

[35] M. Bode and H.-G. Purwins, Physica D 86 (1995): 53

[36] E. Schöll, Nonlinear Spatio-Temporal Dynamics and Chaos in Semiconductors, Cambridge University Press (2001)

[37] S.Tang et al., J.Austral.Math.Soc. Ser.B 35(1993): 223-243

[38] Isaacson, Samuel A.; Peskin, Charles S. (2006). "Incorporating Diffusion in Complex Geometries into Stochastic Chemical Kinetics Simulations". *SIAM J. Sci. Comput.* **28** (1): 47–74. doi:10.1137/040605060.

[39] Linker, Patrick (2016). "Numerical methods for solving the reactive diffusion equation in complex geometries". *The Winnower*.

40.10 External links

- Reaction-Diffusion by the Gray-Scott Model: Pearson's parameterization a visual map of the parameter space of Gray-Scott reaction diffusion.

- A Thesis on reaction-diffusion patterns with an overview of the field

Chapter 41

Partial differential equation

A visualisation of a solution to the two-dimensional heat equation with temperature represented by the third dimension

In mathematics, a **partial differential equation** (**PDE**) is a differential equation that contains unknown multivariable functions and their partial derivatives. (A special case are ordinary differential equations (ODEs), which deal with functions of a single variable and their derivatives.) PDEs are used to formulate problems involving functions of several variables, and are either solved by hand, or used to create a relevant computer model.

PDEs can be used to describe a wide variety of phenomena such as sound, heat, electrostatics, electrodynamics, fluid dynamics, elasticity, or quantum mechanics. These seemingly distinct physical phenomena can be formalised similarly in terms of PDEs. Just as ordinary differential equations often model one-dimensional dynamical systems, partial differential equations often model multidimensional systems. PDEs find their generalisation in stochastic partial differential equations.

41.1 Introduction

Partial differential equations (PDEs) are equations that involve rates of change with respect to continuous variables. The position of a rigid body is specified by six numbers, but the configuration of a fluid is given by the continuous distribution of several parameters, such as the temperature, pressure, and so forth. The dynamics for the rigid body take place in a finite-dimensional configuration space; the dynamics for the fluid occur in an infinite-dimensional configuration space. This distinction usually makes PDEs much harder to solve than ordinary differential equations (ODEs), but here again, there will be simple solutions for linear problems. Classic domains where PDEs are used include acoustics, fluid dynamics, electrodynamics, and heat transfer.

A partial differential equation (PDE) for the function $u(x_1, \cdots, x_n)$ is an equation of the form

$$f\left(x_1, \ldots, x_n, u, \frac{\partial u}{\partial x_1}, \ldots, \frac{\partial u}{\partial x_n}, \frac{\partial^2 u}{\partial x_1 \partial x_1}, \ldots, \frac{\partial^2 u}{\partial x_1 \partial x_n}, \ldots \right) = 0.$$

If f is a linear function of u and its derivatives, then the PDE is called linear. Common examples of linear PDEs include the heat equation, the wave equation, Laplace's equation, Helmholtz equation, Klein–Gordon equation, and Poisson's equation.

A relatively simple PDE is

$$\frac{\partial u}{\partial x}(x, y) = 0.$$

This relation implies that the function $u(x,y)$ is independent of x. However, the equation gives no information on the function's dependence on the variable y. Hence the general solution of this equation is

$$u(x, y) = f(y),$$

where f is an arbitrary function of y. The analogous ordinary differential equation is

$$\frac{du}{dx}(x) = 0,$$

which has the solution

$$u(x) = c.$$

where c is any constant value. These two examples illustrate that general solutions of ordinary differential equations (ODEs) involve arbitrary constants, but solutions of PDEs involve arbitrary functions. A solution of a PDE is generally not unique; additional conditions must generally be specified on the boundary of the region where the solution is defined. For instance, in the simple example above, the function $f(y)$ can be determined if u is specified on the line $x = 0$.

41.2 Existence and uniqueness

Although the issue of existence and uniqueness of solutions of ordinary differential equations has a very satisfactory answer with the Picard–Lindelöf theorem, that is far from the case for partial differential equations. The Cauchy–Kowalevski theorem states that the Cauchy problem for any partial differential equation whose coefficients are analytic in the unknown function and its derivatives, has a locally unique analytic solution. Although this result might appear to settle the existence and uniqueness of solutions, there are examples of linear partial differential equations whose coefficients have derivatives of all orders (which are nevertheless not analytic) but which have no solutions at all: see Lewy (1957). Even if the solution of a partial differential equation exists and is unique, it may nevertheless have undesirable properties. The mathematical study of these questions is usually in the more powerful context of weak solutions.

An example of pathological behavior is the sequence (depending upon n) of Cauchy problems for the Laplace equation

$$\frac{\partial^2 u}{\partial x^2} + \frac{\partial^2 u}{\partial y^2} = 0.$$

with boundary conditions

$$u(x, 0) = 0,$$
$$\frac{\partial u}{\partial y}(x, 0) = \frac{\sin(nx)}{n}.$$

where n is an integer. The derivative of u with respect to y approaches 0 uniformly in x as n increases, but the solution is

$$u(x, y) = \frac{\sinh(ny)\sin(nx)}{n^2}.$$

This solution approaches infinity if nx is not an integer multiple of π for any non-zero value of y. The Cauchy problem for the Laplace equation is called *ill-posed* or *not well-posed*, since the solution does not continuously depend on the data of the problem. Such ill-posed problems are not usually satisfactory for physical applications.

41.3 Notation

In PDEs, it is common to denote partial derivatives using subscripts. That is:

$$u_x = \frac{\partial u}{\partial x}$$
$$u_{xx} = \frac{\partial^2 u}{\partial x^2}$$
$$u_{xy} = \frac{\partial^2 u}{\partial y \partial x} = \frac{\partial}{\partial y}\left(\frac{\partial u}{\partial x}\right).$$

Especially in physics, del or Nabla (∇) is often used to denote spatial derivatives, and $\dot{u}, \ddot{u}$ for time derivatives. For example, the wave equation (described below) can be written as

$$\ddot{u} = c^2 \nabla^2 u$$

or

$$\ddot{u} = c^2 \Delta u$$

where Δ is the Laplace operator.

41.4 Classification

Some linear, second-order partial differential equations can be classified as parabolic, hyperbolic and elliptic. Others such as the Euler–Tricomi equation have different types in different regions. The classification provides a guide to appropriate initial and boundary conditions, and to the smoothness of the solutions.

41.4.1 Equations of first order

Main article: First-order partial differential equation

41.4.2 Linear equations of second order

Assuming $u_{xy} = u_{yx}$, the general second-order PDE in two independent variables has the form

$$Au_{xx} + 2Bu_{xy} + Cu_{yy} + \cdots \text{(lower order terms)} = 0,$$

where the coefficients A, B, C etc. may depend upon x and y. If $A^2 + B^2 + C^2 > 0$ over a region of the xy plane, the PDE is second-order in that region. This form is analogous to the equation for a conic section:

$$Ax^2 + 2Bxy + Cy^2 + \cdots = 0.$$

More precisely, replacing ∂x by X, and likewise for other variables (formally this is done by a Fourier transform), converts a constant-coefficient PDE into a polynomial of the same degree, with the top degree (a homogeneous polynomial, here a quadratic form) being most significant for the classification.

Just as one classifies conic sections and quadratic forms into parabolic, hyperbolic, and elliptic based on the discriminant $B^2 - 4AC$, the same can be done for a second-order PDE at a given point. However, the discriminant in a PDE is given by $B^2 - AC$, due to the convention of the xy term being $2B$ rather than B; formally, the discriminant (of the associated quadratic form) is $(2B)^2 - 4AC = 4(B^2 - AC)$, with the factor of 4 dropped for simplicity.

1. $B^2 - AC < 0$:(*elliptic partial differential equation*)→ Solutions of elliptic PDEs are as smooth as the coefficients allow, within the interior of the region where the equation and solutions are defined. For example, solutions of Laplace's equation are analytic within the domain where they are defined, but solutions may assume boundary values that are not smooth. The motion of a fluid at subsonic speeds can be approximated with elliptic PDEs, and the Euler–Tricomi equation is elliptic where $x < 0$.

2. $B^2 - AC = 0$:(*parabolic partial differential equation*)→ Equations that are parabolic at every point can be transformed into a form analogous to the heat equation by a change of independent variables. Solutions smooth out as the transformed time variable increases. The Euler–Tricomi equation has parabolic type on the line where $x = 0$.

3. $B^2 - AC > 0$:(*hyperbolic partial differential equation*)→ hyperbolic equations retain any discontinuities of functions or derivatives in the initial data. An example is the wave equation. The motion of a fluid at supersonic speeds can be approximated with hyperbolic PDEs, and the Euler–Tricomi equation is hyperbolic where $x > 0$.

If there are n independent variables $x_1, x_2, ..., xn$, a general linear partial differential equation of second order has the form

$$Lu = \sum_{i=1}^{n} \sum_{j=1}^{n} a_{i,j} \frac{\partial^2 u}{\partial x_i \partial x_j} \quad \text{terms lower-order plus} = 0.$$

The classification depends upon the signature of the eigenvalues of the coefficient matrix $a_{i,j}$.

1. Elliptic: The eigenvalues are all positive or all negative.

2. Parabolic : The eigenvalues are all positive or all negative, save one that is zero.

3. Hyperbolic: There is only one negative eigenvalue and all the rest are positive, or there is only one positive eigenvalue and all the rest are negative.

4. Ultrahyperbolic: There is more than one positive eigenvalue and more than one negative eigenvalue, and there are no zero eigenvalues. There is only a limited theory for ultra-hyperbolic equations (Courant and Hilbert, 1962).

41.4.3 Systems of first-order equations and characteristic surfaces

The classification of partial differential equations can be extended to systems of first-order equations, where the unknown u is now a vector with m components, and the coefficient matrices $A\nu$ are m by m matrices for $\nu = 1, ..., n$. The partial differential equation takes the form

$$Lu = \sum_{\nu=1}^{n} A_\nu \frac{\partial u}{\partial x_\nu} + B = 0.$$

where the coefficient matrices $A\nu$ and the vector B may depend upon x and u. If a hypersurface S is given in the implicit form

$$\varphi(x_1, x_2, \ldots, x_n) = 0,$$

where φ has a non-zero gradient, then S is a **characteristic surface** for the operator L at a given point if the characteristic form vanishes:

$$Q\left(\frac{\partial\varphi}{\partial x_1},\ldots,\frac{\partial\varphi}{\partial x_n}\right) = \det\left[\sum_{\nu=1}^{n} A_\nu \frac{\partial\varphi}{\partial x_\nu}\right] = 0.$$

The geometric interpretation of this condition is as follows: if data for u are prescribed on the surface S, then it may be possible to determine the normal derivative of u on S from the differential equation. If the data on S and the differential equation determine the normal derivative of u on S, then S is non-characteristic. If the data on S and the differential equation *do not* determine the normal derivative of u on S, then the surface is **characteristic**, and the differential equation restricts the data on S: the differential equation is *internal* to S.

1. A first-order system $Lu=0$ is *elliptic* if no surface is characteristic for L: the values of u on S and the differential equation always determine the normal derivative of u on S.

2. A first-order system is *hyperbolic* at a point if there is a **space-like** surface S with normal ξ at that point. This means that, given any non-trivial vector η orthogonal to ξ, and a scalar multiplier λ, the equation $Q(\lambda\xi + \eta) = 0$ has m real roots $\lambda_1, \lambda_2, \ldots, \lambda_m$. The system is **strictly hyperbolic** if these roots are always distinct. The geometrical interpretation of this condition is as follows: the characteristic form $Q(\zeta) = 0$ defines a cone (the normal cone) with homogeneous coordinates ζ. In the hyperbolic case, this cone has m sheets, and the axis $\zeta = \lambda \xi$ runs inside these sheets: it does not intersect any of them. But when displaced from the origin by η, this axis intersects every sheet. In the elliptic case, the normal cone has no real sheets.

41.4.4 Equations of mixed type

If a PDE has coefficients that are not constant, it is possible that it will not belong to any of these categories but rather be of **mixed type**. A simple but important example is the Euler–Tricomi equation

$$u_{xx} = x u_{yy},$$

which is called **elliptic-hyperbolic** because it is elliptic in the region $x < 0$, hyperbolic in the region $x > 0$, and degenerate parabolic on the line $x = 0$.

41.4.5 Infinite-order PDEs in quantum mechanics

In the phase space formulation of quantum mechanics, one may consider the quantum Hamilton's equations for trajectories of quantum particles. These equations are infinite-order PDEs. However, in the semiclassical expansion, one has a finite system of ODEs at any fixed order of $\hbar$. The evolution equation of the Wigner function is also an infinite-order PDE. The quantum trajectories are quantum characteristics, with the use of which one could calculate the evolution of the Wigner function.

41.5 Analytical solutions

41.5.1 Separation of variables

Main article: Separable partial differential equation

Linear PDEs can be reduced to systems of ordinary differential equations by the important technique of separation of variables. This technique rests on a characteristic of solutions to differential equations: if one can find any solution that solves the equation and satisfies the boundary conditions, then it is *the* solution (this also applies to ODEs). We assume as an ansatz that the dependence of a solution on the parameters space and time can be written as a product of terms that each depend on a single parameter, and then see if this can be made to solve the problem.[1]

In the method of separation of variables, one reduces a PDE to a PDE in fewer variables, which is an ordinary differential equation if in one variable – these are in turn easier to solve.

This is possible for simple PDEs, which are called separable partial differential equations, and the domain is generally a rectangle (a product of intervals). Separable PDEs correspond to diagonal matrices – thinking of "the value for fixed x" as a coordinate, each coordinate can be understood separately.

This generalizes to the method of characteristics, and is also used in integral transforms.

41.5.2 Method of characteristics

Main article: Method of characteristics

In special cases, one can find characteristic curves on which the equation reduces to an ODE – changing coordinates in the domain to straighten these curves allows separation of variables, and is called the method of characteristics.

More generally, one may find characteristic surfaces.

41.5.3 Integral transform

An integral transform may transform the PDE to a simpler one, in particular, a separable PDE. This corresponds to diagonalizing an operator.

An important example of this is Fourier analysis, which diagonalizes the heat equation using the eigenbasis of sinusoidal waves.

If the domain is finite or periodic, an infinite sum of solutions such as a Fourier series is appropriate, but an integral of solutions such as a Fourier integral is generally required for infinite domains. The solution for a point source for the heat equation given above is an example of the use of a Fourier integral.

41.5.4 Change of variables

Often a PDE can be reduced to a simpler form with a known solution by a suitable change of variables. For example, the Black–Scholes PDE

$$\frac{\partial V}{\partial t} + \frac{1}{2}\sigma^2 S^2 \frac{\partial^2 V}{\partial S^2} + rS\frac{\partial V}{\partial S} - rV = 0$$

is reducible to the heat equation

$$\frac{\partial u}{\partial \tau} = \frac{\partial^2 u}{\partial x^2}$$

by the change of variables (for complete details see Solution of the Black Scholes Equation at the Wayback Machine (archived April 11, 2008))

$$V(S,t) = Kv(x,\tau)$$
$$x = \ln\left(\frac{S}{K}\right)$$
$$\tau = \frac{1}{2}\sigma^2(T-t)$$
$$v(x,\tau) = \exp(-\alpha x - \beta\tau)u(x,\tau).$$

41.5.5 Fundamental solution

Main article: Fundamental solution

Inhomogeneous equations can often be solved (for constant coefficient PDEs, always be solved) by finding the fundamental solution (the solution for a point source), then taking the convolution with the boundary conditions to get the solution.

This is analogous in signal processing to understanding a filter by its impulse response.

41.5.6 Superposition principle

Because any superposition of solutions of a linear, homogeneous PDE is again a solution, the particular solutions may then be combined to obtain more general solutions. if u_1 and u_2 are solutions of a homogeneous linear pde in same region R, then $u = c_1 u_1 + c_2 u_2$ with any constants c_1 and c_2 are also a solution of that pde in that same region....

41.5.7 Methods for non-linear equations

See also: nonlinear partial differential equation

There are no generally applicable methods to solve nonlinear PDEs. Still, existence and uniqueness results (such as the Cauchy–Kowalevski theorem) are often possible, as are proofs of important qualitative and quantitative properties of solutions (getting these results is a major part of analysis). Computational solution to the nonlinear PDEs, the split-step method, exist for specific equations like nonlinear Schrödinger equation.

Nevertheless, some techniques can be used for several types of equations. The h-principle is the most powerful method to solve underdetermined equations. The Riquier–Janet theory is an effective method for obtaining information about many analytic overdetermined systems.

The method of characteristics (similarity transformation method) can be used in some very special cases to solve partial differential equations.

In some cases, a PDE can be solved via perturbation analysis in which the solution is considered to be a correction to an equation with a known solution. Alternatives are numerical analysis techniques from simple finite difference schemes to the more mature multigrid and finite element methods. Many interesting problems in science and engineering are solved in this way using computers, sometimes high performance supercomputers.

41.5.8 Lie group method

From 1870 Sophus Lie's work put the theory of differential equations on a more satisfactory foundation. He showed that the integration theories of the older mathematicians can, by the introduction of what are now called Lie groups, be referred to a common source; and that ordinary differential equations which admit the same infinitesimal transformations present comparable difficulties of integration. He also emphasized the subject of transformations of contact.

A general approach to solving PDE's uses the symmetry property of differential equations, the continuous infinitesimal transformations of solutions to solutions (Lie theory). Continuous group theory, Lie algebras and differential geometry are used to understand the structure of linear and nonlinear partial differential equations for generating integrable equations, to find its Lax pairs, recursion operators, Bäcklund transform and finally finding exact analytic solutions to the PDE.

Symmetry methods have been recognized to study differential equations arising in mathematics, physics, engineering, and many other disciplines.

41.5.9 Semianalytical methods

The adomian decomposition method, the Lyapunov artificial small parameter method, and He's homotopy perturbation method are all special cases of the more general homotopy analysis method. These are series expansion methods, and except for the Lyapunov method, are independent of small physical parameters as compared to the well known perturbation theory, thus giving these methods greater flexibility and solution generality.

41.6 Numerical solutions

The three most widely used numerical methods to solve PDEs are the finite element method (FEM), finite volume methods (FVM) and finite difference methods (FDM), as well other kind of methods called Meshfree methods, which were made to solve problems where the before mentioned methods are limited. The FEM has a prominent position among these methods and especially its exceptionally efficient higher-order version hp-FEM. Other hybrid versions of FEM and Meshfree methods include the generalized finite element method (GFEM), extended finite element method (XFEM), spectral finite element method (SFEM), meshfree finite element method, discontinuous Galerkin finite element method (DGFEM), Element-Free Galerkin Method (EFGM), Interpolating Element-Free Galerkin Method (IEFGM), etc.

41.6.1 Finite element method

Main article: Finite element method

The finite element method (FEM) (its practical application often known as finite element analysis (FEA)) is a numerical technique for finding approximate solutions of partial differential equations (PDE) as well as of integral equations. The solution approach is based either on eliminating the differential equation completely (steady state problems), or rendering the PDE into an approximating system of ordinary differential equations, which are then numerically integrated using standard techniques such as Euler's method, Runge–Kutta, etc.

41.6.2 Finite difference method

Main article: Finite difference method

Finite-difference methods are numerical methods for approximating the solutions to differential equations using finite difference equations to approximate derivatives.

41.6.3 Finite volume method

Main article: Finite volume method

Similar to the finite difference method or finite element method, values are calculated at discrete places on a meshed geometry. "Finite volume" refers to the small volume surrounding each node point on a mesh. In the finite volume method, surface integrals in a partial differential equation that contain a divergence term are converted to volume integrals, using the divergence theorem. These terms are then evaluated as fluxes at the surfaces of each finite volume. Because the flux entering a given volume is identical to that leaving the adjacent volume, these methods are conservative.

41.7 See also

- Dirichlet boundary condition
- Jet bundle
- Laplace transform applied to differential equations
- List of dynamical systems and differential equations topics
- Matrix differential equation
- Neumann boundary condition
- Numerical partial differential equations
- Partial differential algebraic equation
- Recurrence relation
- Robin boundary condition
- Stochastic processes and boundary value problems

41.8 Notes

[1] Gershenfeld, Neil (2000). *The nature of mathematical modeling* (Reprinted (with corr.). ed.). Cambridge: Cambridge Univ. Press. p. 27. ISBN 0521570956.

41.9 References

- Adomian, G. (1994). *Solving Frontier problems of Physics: The decomposition method*. Kluwer Academic Publishers.
- Courant, R. & Hilbert, D. (1962), *Methods of Mathematical Physics*, **II**, New York: Wiley-Interscience.
- Evans, L. C. (1998), *Partial Differential Equations*, Providence: American Mathematical Society, ISBN 0-8218-0772-2.
- Holubová, Pavel Drábek ; Gabriela (2007). *Elements of partial differential equations* ([Online-Ausg.]. ed.). Berlin: de Gruyter. ISBN 9783110191240.
- Ibragimov, Nail H (1993). *CRC Handbook of Lie Group Analysis of Differential Equations Vol. 1-3*. Providence: CRC-Press, ISBN 0-8493-4488-3.
- John, F. (1982), *Partial Differential Equations* (4th ed.), New York: Springer-Verlag, ISBN 0-387-90609-6.
- Jost, J. (2002), *Partial Differential Equations*, New York: Springer-Verlag, ISBN 0-387-95428-7.
- Lewy, Hans (1957), "An example of a smooth linear partial differential equation without solution", *Annals of Mathematics. Second Series*, **66** (1): 155–158, doi:10.2307/1970121.
- Liao, S.J. (2003), *Beyond Perturbation: Introduction to the Homotopy Analysis Method*, Boca Raton: Chapman & Hall/ CRC Press, ISBN 1-58488-407-X
- Olver, P.J. (1995), *Equivalence, Invariants and Symmetry*, Cambridge Press.
- Petrovskii, I. G. (1967), *Partial Differential Equations*, Philadelphia: W. B. Saunders Co..
- Pinchover, Y. & Rubinstein, J. (2005), *An Introduction to Partial Differential Equations*, New York: Cambridge University Press, ISBN 0-521-84886-5.
- Polyanin, A. D. (2002), *Handbook of Linear Partial Differential Equations for Engineers and Scientists*, Boca Raton: Chapman & Hall/CRC Press, ISBN 1-58488-299-9.
- Polyanin, A. D. & Zaitsev, V. F. (2004), *Handbook of Nonlinear Partial Differential Equations*, Boca Raton: Chapman & Hall/CRC Press, ISBN 1-58488-355-3.
- Polyanin, A. D.; Zaitsev, V. F. & Moussiaux, A. (2002), *Handbook of First Order Partial Differential Equations*, London: Taylor & Francis, ISBN 0-415-27267-X.
- Roubíček, T. (2013), *Nonlinear Partial Differential Equations with Applications* (2nd ed.), Basel, Boston, Berlin: Birkhäuser, ISBN 978-3-0348-0512-4, MR 3014456
- Solin, P. (2005), *Partial Differential Equations and the Finite Element Method*, Hoboken, NJ: J. Wiley & Sons, ISBN 0-471-72070-4.
- Solin, P.; Segeth, K. & Dolezel, I. (2003), *Higher-Order Finite Element Methods*, Boca Raton: Chapman & Hall/CRC Press, ISBN 1-58488-438-X.
- Stephani, H. (1989), *Differential Equations: Their Solution Using Symmetries. Edited by M. MacCallum*, Cambridge University Press.
- Wazwaz, Abdul-Majid (2009). *Partial Differential Equations and Solitary Waves Theory*. Higher Education Press. ISBN 978-3-642-00251-9.
- Wazwaz, Abdul-Majid (2002). *Partial Differential Equations Methods and Applications*. A.A. Balkema. ISBN 90-5809-369-7.
- Zwillinger, D. (1997), *Handbook of Differential Equations* (3rd ed.), Boston: Academic Press, ISBN 0-12-784395-7.
- Gershenfeld, N. (1999), *The Nature of Mathematical Modeling* (1st ed.), New York: Cambridge University Press, New York, NY, USA, ISBN 0-521-57095-6.
- Krasil'shchik, I.S. & Vinogradov, A.M., Eds. (1999), *Symmetries and Conservation Laws for Differential Equations of Mathematical Physics*, American Mathematical Society, Providence, Rhode Island, USA, ISBN 0-8218-0958-X.
- Krasil'shchik, I.S.; Lychagin, V.V. & Vinogradov, A.M. (1986), *Geometry of Jet Spaces and Nonlinear Partial Differential Equations*, Gordon and Breach Science Publishers, New York, London, Paris, Montreux, Tokyo, ISBN 2-88124-051-8.
- Vinogradov, A.M. (2001), *Cohomological Analysis of Partial Differential Equations and Secondary Calculus*, American Mathematical Society, Providence, Rhode Island, USA, ISBN 0-8218-2922-X.

41.10 Further reading

- Cajori, Florian (1928). "The Early History of Partial Differential Equations and of Partial Differentiation and Integration" (PDF). *The American Mathematical Monthly*. **35** (9): 459–467. doi:10.2307/2298771.

41.11 External links

- Hazewinkel, Michiel, ed. (2001) [1994], "Differential equation, partial", *Encyclopedia of Mathematics*, Springer Science+Business Media B.V. / Kluwer Academic Publishers, ISBN 978-1-55608-010-4
- Partial Differential Equations: Exact Solutions at EqWorld: The World of Mathematical Equations.
- Partial Differential Equations: Index at EqWorld: The World of Mathematical Equations.
- Partial Differential Equations: Methods at EqWorld: The World of Mathematical Equations.
- Example problems with solutions at exampleproblems.com
- Partial Differential Equations at mathworld.wolfram.com
- Partial Differential Equations with Mathematica
- Partial Differential Equations in Cleve Moler: Numerical Computing with MATLAB
- Partial Differential Equations at nag.com
- Dispersive PDE Wiki
- NEQwiki, the nonlinear equations encyclopedia
- Partial differential equation | Scholarpedia

Chapter 42

Dissipative system

A **dissipative system** is a thermodynamically open system which is operating out of, and often far from, thermodynamic equilibrium in an environment with which it exchanges energy and matter.

A **dissipative structure** is a dissipative system that has a dynamical régime that is in some sense in a reproducible steady state. This reproducible steady state may be reached by natural evolution of the system, by artifice, or by a combination of these two.

42.1 Overview

A dissipative structure is characterized by the spontaneous appearance of symmetry breaking (anisotropy) and the formation of complex, sometimes chaotic, structures where interacting particles exhibit long range correlations. Examples in everyday life include convection, turbulent flow, cyclones, hurricanes and living organisms. Less common examples include lasers, Bénard cells, and the Belousov–Zhabotinsky reaction.[1]

One way of mathematically modeling a dissipative system is given in the article on *wandering sets*: it involves the action of a group on a measurable set.

Dissipative systems can also be used as a tool to study economic systems and complex systems.[2] For example, a dissipative system involving self-assembly of nanowires has been used as a model to understand the relationship between entropy generation and the robustness of biological systems.[3]

42.2 Dissipative structures in thermodynamics

The term *dissipative structure* was coined by Russian-Belgian physical chemist Ilya Prigogine, who was awarded the Nobel Prize in Chemistry in 1977 for his pioneering work on these structures. The dissipative structures considered by Prigogine have dynamical regimes that can be regarded as thermodynamic steady states, and sometimes at least can be described by suitable extremal principles in non-equilibrium thermodynamics.

In his Nobel lecture,[4] Prigogine explains how thermodynamic systems far from equilibrium can have drastically different behavior from systems close to equilibrium. Near equilibrium, the *local equilibrium* hypothesis applies and typical thermodynamic quantities such as free energy and entropy can be defined locally. One can assume linear relations between the (generalized) flux and forces of the system. Two celebrated results from linear thermodynamics are the Onsager reciprocal relations and the principle of minimum entropy production.[5] After efforts to extend such results to systems far from equilibrium, it was found that they do not hold in this regime and opposite results were obtained.

One way to rigorously analyze such systems is by studying the stability of the system far from equilibrium. Close to equilibrium, one can show the existence of a Lyapunov function which ensures that the entropy tends to a stable maximum. Fluctuations are damped in the neighborhood of the fixed point and a macroscopic description suffices. However, far from equilibrium stability is no longer a universal property and can be broken. In chemical systems this occurs with the presence of autocatalytic reactions, such as in the example of the Brusselator. If the system is driven beyond a certain threshold, oscillations are no longer damped out, but may be amplified. Mathematically, this corresponds to a Hopf bifurcation where increasing one of the parameters beyond a certain value leads to limit cycle behavior. If spatial effects are taken into account through a reaction-diffusion equation, long-range correlations and spatially ordered patterns arise,[6] such as in the case of the Belousov–Zhabotinsky reaction. Systems with such dynamic states of matter that arise as the result of irreversible processes are dissipative structures.

Recent research has seen reconsideration of Prigogine's

ideas of dissipative structures in relation to biological systems.[7]

42.3 Dissipative systems in control theory

In systems and control theory, dissipative systems are dynamical systems with states $x(t)$, inputs $u(t)$ and outputs $y(t)$, which satisfy the so-called "dissipation inequality".

A system is said to be dissipative if there exist a continuous nonnegative function $V(x)$ of the real variable x, called the storage function, such that the following inequality, known as the dissipation inequality, always holds:

$$\frac{dV(x(t))}{dt} \leq u(t) \cdot y(t)$$

The function $u \cdot y$, where $\cdot$ denotes the scalar product, is called the "supply rate".

The physical interpretation is that $V(x)$ is the energy stored in the system, whereas $u \cdot y$ is the energy that is supplied to the system. Other supply rates $w=w(u,y)$ are also possible.

This notion has a strong connection with Lyapunov stability, where the storage functions may play, under certain conditions of controllability and observability of the dynamical system, the role of Lyapunov functions.

Roughly speaking, dissipativity theory is useful for the design of feedback control laws for linear and nonlinear systems. Dissipative systems theory has been discussed by V.M. Popov, J.C. Willems, D.J. Hill and P. Moylan. In the case of linear invariant systems, this is known as positive real transfer functions, and a fundamental tool is the so-called Kalman–Yakubovich–Popov lemma which relates the state space and the frequency domain properties of positive real systems.[8] Dissipative systems are still an active field of research in systems and control, due to their important applications.

42.4 Quantum dissipative systems

Main article: Quantum dissipation

As quantum mechanics, and any classical dynamical system, relies heavily on Hamiltonian mechanics for which time is reversible, these approximations are not intrinsically able to describe dissipative systems. It has been proposed that in principle, one can couple weakly the system – say, an oscillator – to a bath, i.e., an assembly of many oscillators in thermal equilibrium with a broad band spectrum, and trace (average) over the bath. This yields a master equation which is a special case of a more general setting called the Lindblad equation that is the quantum equivalent of the classical Liouville equation. The well known form of this equation and its quantum counterpart takes time as a reversible variable over which to integrate, but the very foundations of dissipative structures imposes an irreversible and constructive role for time.

42.5 See also

- Conservation equation
- Non-equilibrium thermodynamics
- Extremal principles in non-equilibrium thermodynamics
- Autowave
- Self-organization
- Autocatalytic reactions and order creation
- Dynamical system
- Autopoiesis
- Relational order theories
- Loschmidt's paradox
- Viable System Theory

42.6 Notes

[1] Li, HP (February 2014). "Dissipative Belousov–Zhabotinsky reaction in unstable micropyretic synthesis". *Current Opinion in Chemical Engineering- Biological engineering - Materials engineering*. **3**: 1–6. doi:10.1016/j.coche.2013.08.007.

[2] Chen, Jing (2015). *The Unity of Science and Economics: A New Foundation of Economic Theory*. https://www.springer.com/us/book/9781493934645: Springer.

[3] Hubler, Alfred; Belkin, Andrey; Bezryadin, Alexey (2 January 2015). "Noise induced phase transition between maximum entropy production structures and minimum entropy production structures?". *Complexity*. **20** (3): 8–11. Bibcode:2015Cmplx..20c...8H. doi:10.1002/cplx.21639.

[4] Prigogine, Ilya. "Time, Structure and Fluctuations". *Nobelprize.org*.

[5] Prigogine, Ilya (1945). "Modération et transformations irréversibles des systèmes ouverts". *Bulletin de la Classe des Sciences, Académie Royale de Belgique*. **31**: 600–606.

[6] Lemarchand, H.; Nicolis, G. (1976). "Long range correlations and the onset of chemical instabilities". *Physica*. **82A**: 521–542.

[7] England, Jeremy L. (4 November 2015). "Dissipative adaptation in driven self-assembly". *Nature Nanotechnology*. **10** (11): 919–923. Bibcode:2015NatNa..10..919E. doi:10.1038/NNANO.2015.250.

[8] Bao, Jie; Lee, Peter L. (2007). *Process Control - The Passive Systems Approach*. Springer-Verlag London. ISBN 978-1-84628-892-0. doi:10.1007/978-1-84628-893-7.

42.7 References

- B. Brogliato, R. Lozano, B. Maschke, O. Egeland, Dissipative Systems Analysis and Control. Theory and Applications. Springer Verlag, London, 2nd Ed., 2007.

- Davies, Paul *The Cosmic Blueprint* Simon & Schuster, New York 1989 (abridged— 1500 words) (abstract— 170 words) — self-organized structures.

- Philipson, Schuster, *Modeling by Nonlinear Differential Equations: Dissipative and Conservative Processes*, World Scientific Publishing Company 2009.

- Prigogine, Ilya, *Time, structure and fluctuations*. Nobel Lecture, 8 December 1977.

- J.C. Willems. Dissipative dynamical systems, part I: General theory; part II: Linear systems with quadratic supply rates. Archive for Rationale mechanics Analysis, vol.45, pp. 321–393, 1972.

42.8 External links

- The dissipative systems model The Australian National University

Chapter 43

Percolation

In coffee percolation, soluble compounds leave the coffee grounds and join the water to form coffee. Insoluble compounds (and granulates) remain within the coffee filter.

In physics, chemistry and materials science, **percolation** (from Latin *percōlāre*, "to filter" or "trickle through") refers to the movement and filtering of fluids through porous materials.

43.1 Background

During the last decades, percolation theory, an extensive mathematical studies model of percolation, has brought new understanding and techniques to a broad range of topics in physics, materials science, complex networks, epidemiology, and other fields. For example, in geology, percolation refers to filtration of water through soil and permeable rocks. The water flows to recharge the groundwater in the water table and aquifers. In places where infiltration basins or septic drain fields are planned to dispose of substantial amounts of water, a percolation test is needed beforehand to determine whether the intended structure is likely to succeed or fail.

Percolation typically exhibits universality. Statistical physics concepts such as scaling theory, renormalization, phase transition, critical phenomena and fractals are used to characterize percolation properties. Combinatorics is commonly employed to study percolation thresholds.

Due to the complexity involved in obtaining exact results from analytical models of percolation, computer simulations are typically used. The current fastest algorithm for percolation was published in 2000 by Mark Newman and Robert Ziff.[1]

43.2 Examples

- Coffee percolation, where the solvent is water, the permeable substance is the coffee grounds, and the soluble constituents are the chemical compounds that give coffee its color, taste, and aroma

- Movement of weathered material down on a slope under the earth's surface

- Cracking of trees with the presence of two conditions, sunlight and under the influence of pressure

- Robustness of networks to random and targeted attacks

- Transport in porous media

- Epidemic spreading[2][3]

- Surface roughening

- Dental Percolation, increase rate of decay under crowns because of a conducive environment for strep mutans and lactobacillus

- Potential sites for septic systems are tested via the "perk test." Example/theory: A hole (usually 6-10 inches in diameter) is dug in the ground surface (usually 12-24" deep). Water is filled in to the hole and the time is measured for a drop of one inch in the water surface. If the water surface quickly drops, as usually seen in poorly-graded sands, then it is a potentially good place for a septic "leach field." If the hydraulic conductivity of the site is low (usually in clayey and loamy soils) then the site is undesirable.

43.3 See also

- Branched polymer
- Conductance
- Critical exponents
- Fragmentation
- Gelation
- Groundwater recharge
- Immunization
- Network theory
- Percolation critical exponents
- Percolation theory
- Percolation threshold
- Polymerization
- Self-organization
- Self-organized criticality
- Septic tank
- Supercooled water
- Water pipe percolator

43.4 References

[1] Newman, Mark; Ziff, Robert (2000). "Efficient Monte Carlo Algorithm and High-Precision Results for Percolation". *Physical Review Letters*. American Physical Society. **85** (19): 4104–4107. Bibcode:2000PhRvL..85.4104N. PMID 11056635. arXiv:cond-mat/0005264. doi:10.1103/PhysRevLett.85.4104. Retrieved 19 November 2013.

[2] Parshani, Roni; Carmi, Shai; Havlin, Shlomo (2010). "Epidemic Threshold for the Susceptible-Infectious-Susceptible Model on Random Networks". *Physical Review Letters*. **104** (25). Bibcode:2010PhRvL.104y8701P. ISSN 0031-9007. arXiv:0909.3811. doi:10.1103/PhysRevLett.104.258701.

[3] Grassberger, P. "On the Critical Behavior of the General Epidemic Process and Dynamic Percolation". *Mathematical Biosciences*.

43.5 Further reading

- Harry Kesten. What is percolation? *Notices of the AMS*, May 2006.
- Muhammad Sahimi. *Applications of Percolation Theory*. Taylor & Francis, 1994. ISBN 0-7484-0075-3 (cloth), ISBN 0-7484-0076-1 (paper)
- Geoffrey Grimmett. *Percolation (2. ed)*. Springer Verlag, 1999.
- D.Stauffer and A.Aharony. *Introduction to Percolation Theory*
- A. Bunde, S. Havlin (Editors) *Fractals and Disordered Systems*, Springer, 1996
- S. Kirkpatrick *Percolation and conduction* Rev. Mod. Phys. 45, 574, 1973
- D. Ben-Avraham, S. Havlin *Diffusion and Reactions in Fractals and Disordered Systems*, Cambridge University Press, 2000
- Edouard Rodrigues, Remarkable properties of pawns on a hexboard
- R. Cohen and S. Havlin *Complex Networks: Structure, Robustness and Function*
- Bollobás, Béla; Riordan, Oliver (2006), Percolation, Cambridge University Press, ISBN 0521872324
- Grimmett, Geoffrey (1999), Percolation, Springer

43.6 External links

- Introduction to Percolation Theory: short course by Shlomo Havlin

Chapter 44

Cellular automaton

Gosper's Glider Gun creating "gliders" in the cellular automaton Conway's Game of Life[1]

A **cellular automaton** (pl. **cellular automata**, abbrev. **CA**) is a discrete model studied in computability theory, mathematics, physics, complexity science, theoretical biology and microstructure modeling. Cellular automata are also called **cellular spaces**, **tessellation automata**, **homogeneous structures**, **cellular structures**, **tessellation structures**, and **iterative arrays**.[2]

A cellular automaton consists of a regular grid of *cells*, each in one of a finite number of *states*, such as *on* and *off* (in contrast to a coupled map lattice). The grid can be in any finite number of dimensions. For each cell, a set of cells called its *neighborhood* is defined relative to the specified cell. An initial state (time $t = 0$) is selected by assigning a state for each cell. A new *generation* is created (advancing t by 1), according to some fixed *rule* (generally, a mathematical function)[3] that determines the new state of each cell in terms of the current state of the cell and the states of the cells in its neighborhood. Typically, the rule for updating the state of cells is the same for each cell and does not change over time, and is applied to the whole grid simultaneously,[4] though exceptions are known, such as the stochastic cellular automaton and asynchronous cellular automaton.

The concept was originally discovered in the 1940s by Stanislaw Ulam and John von Neumann while they were contemporaries at Los Alamos National Laboratory. While studied by some throughout the 1950s and 1960s, it was not until the 1970s and Conway's Game of Life, a two-dimensional cellular automaton, that interest in the subject expanded beyond academia. In the 1980s, Stephen Wolfram engaged in a systematic study of one-dimensional cellular automata, or what he calls elementary cellular automata; his research assistant Matthew Cook showed that one of these rules is Turing-complete. Wolfram published *A New Kind of Science* in 2002, claiming that cellular automata have applications in many fields of science. These include computer processors and cryptography.

The primary classifications of cellular automata, as outlined by Wolfram, are numbered one to four. They are, in order, automata in which patterns generally stabilize into homogeneity, automata in which patterns evolve into mostly stable or oscillating structures, automata in which patterns evolve in a seemingly chaotic fashion, and automata in which patterns become extremely complex and may last for a long time, with stable local structures. This last class are thought to be computationally universal, or capable of simulating a Turing machine. Special types of cellular automata are *reversible*, where only a single configuration leads directly to a subsequent one, and *totalistic*, in which the future value of individual cells only depends on the total value of a group of neighboring cells. Cellular automata can simulate a variety of real-world systems, including biological and chemical ones.

44.1 Overview

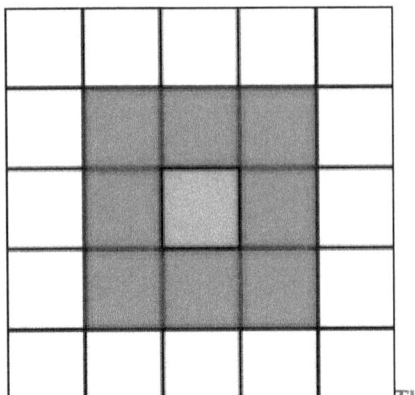

The red cells are the Moore neighborhood for the blue cell.

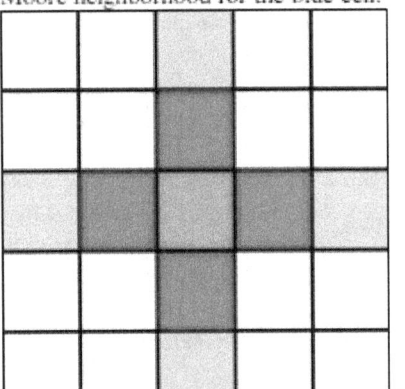

The red cells are the von Neumann neighborhood for the blue cell. The extended neighborhood includes the pink cells as well.

One way to simulate a two-dimensional cellular automaton is with an infinite sheet of graph paper along with a set of rules for the cells to follow. Each square is called a "cell" and each cell has two possible states, black and white. The *neighborhood* of a cell is the nearby, usually adjacent, cells. The two most common types of neighborhoods are the *von Neumann neighborhood* and the *Moore neighborhood*.[5] The former, named after the founding cellular automaton theorist, consists of the four orthogonally adjacent cells.[5] The latter includes the von Neumann neighborhood as well as the four remaining cells surrounding the cell whose state is to be calculated.[5] For such a cell and its Moore neighborhood, there are 512 (= 2^9) possible patterns. For each of the 512 possible patterns, the rule table would state whether the center cell will be black or white on the next time interval. Conway's Game of Life is a popular version of this model. Another common neighborhood type is the *extended von Neumann neighborhood*, which includes the two closest cells in each orthogonal direction, for a total of eight.[5] The general equation for such a system of rules is k^{k^s}, where k is the number of possible states for a cell, and s is the number of neighboring cells (including the cell to be calculated itself) used to determine the cell's next state.[6] Thus, in the two dimensional system with a Moore neighborhood, the total number of automata possible would be 2^{2^9}, or 1.34×10^{154}.

It is usually assumed that every cell in the universe starts in the same state, except for a finite number of cells in other states; the assignment of state values is called a *configuration*.[7] More generally, it is sometimes assumed that the universe starts out covered with a periodic pattern, and only a finite number of cells violate that pattern. The latter assumption is common in one-dimensional cellular automata.

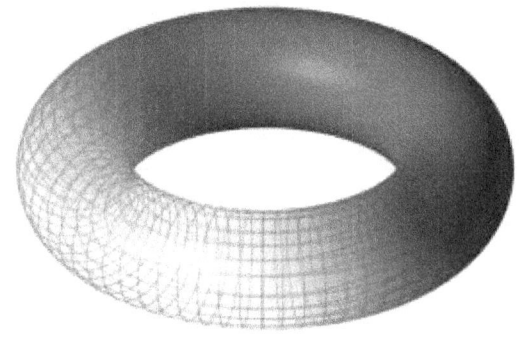

A torus, a toroidal shape

Cellular automata are often simulated on a finite grid rather than an infinite one. In two dimensions, the universe would be a rectangle instead of an infinite plane. The obvious problem with finite grids is how to handle the cells on the edges. How they are handled will affect the values of all the cells in the grid. One possible method is to allow the values in those cells to remain constant. Another method is to define neighborhoods differently for these cells. One could say that they have fewer neighbors, but then one would also have to define new rules for the cells located on the edges. These cells are usually handled with a *toroidal* arrangement: when one goes off the top, one comes in at the corresponding position on the bottom, and when one goes off the left, one comes in on the right. (This essentially simulates an infinite periodic tiling, and in the field of partial differential equations is sometimes referred to as *periodic* boundary conditions.) This can be visualized as taping the left and right edges of the rectangle to form a tube, then taping the top and bottom edges of the tube to form a torus (doughnut shape). Universes of other dimensions are handled similarly. This solves boundary problems with neighborhoods, but another advantage is that it is easily programmable using modular arithmetic functions. For example, in a 1-dimensional cellular automaton like the examples below, the neighborhood of a cell xi^t is $\{xi_{-1}^{t-1}, xi^{t-1}, xi_{+1}^{t-1}\}$, where t is the time step (vertical), and i is the index (horizontal) in one generation.

44.2 History

Stanislaw Ulam, while working at the Los Alamos National Laboratory in the 1940s, studied the growth of crystals, using a simple lattice network as his model.[8] At the same time, John von Neumann, Ulam's colleague at Los Alamos, was working on the problem of self-replicating systems.[9] Von Neumann's initial design was founded upon the notion of one robot building another robot. This design is known as the kinematic model.[10][11] As he developed this design, von Neumann came to realize the great difficulty of building a self-replicating robot, and of the great cost in providing the robot with a "sea of parts" from which to build its replicant. Neumann read a paper entitled "The general and logical theory of automata" at the Hixon Symposium in 1948.[9] Ulam was the one who suggested using a *discrete* system for creating a reductionist model of self-replication.[12][13] Nils Aall Barricelli performed many of the earliest explorations of these models of artificial life.

John von Neumann, Los Alamos ID badge

Ulam and von Neumann created a method for calculating liquid motion in the late 1950s. The driving concept of the method was to consider a liquid as a group of discrete units and calculate the motion of each based on its neighbors' behaviors.[14] Thus was born the first system of cellular automata. Like Ulam's lattice network, von Neumann's cellular automata are two-dimensional, with his self-replicator implemented algorithmically. The result was a universal copier and constructor working within a cellular automaton with a small neighborhood (only those cells that touch are neighbors; for von Neumann's cellular automata, only orthogonal cells), and with 29 states per cell.[15] Von Neumann gave an existence proof that a particular pattern would make endless copies of itself within the given cellular universe by designing a 200,000 cell configuration that could do so.[15] This design is known as the tessellation model, and is called a von Neumann universal constructor.[16]

Also in the 1940s, Norbert Wiener and Arturo Rosenblueth developed a model of excitable media with some of the characteristics of a cellular automaton.[17] Their specific motivation was the mathematical description of impulse conduction in cardiac systems. However their model is not a cellular automaton because the medium in which signals propagate is continuous, and wave fronts are curves.[17][18] A true cellular automaton model of excitable media was developed and studied by J. M. Greenberg and S. P. Hastings in 1978; see Greenberg-Hastings cellular automaton. The original work of Wiener and Rosenblueth contains many insights and continues to be cited in modern research publications on cardiac arrhythmia and excitable systems.[19]

In the 1960s, cellular automata were studied as a particular type of dynamical system and the connection with the mathematical field of symbolic dynamics was established for the first time. In 1969, Gustav A. Hedlund compiled many results following this point of view[20] in what is still considered as a seminal paper for the mathematical study of cellular automata. The most fundamental result is the characterization in the Curtis–Hedlund–Lyndon theorem of the set of global rules of cellular automata as the set of continuous endomorphisms of shift spaces.

In 1969, German computer pioneer Konrad Zuse published his book *Calculating Space*, proposing that the physical laws of the universe are discrete by nature, and that the entire universe is the output of a deterministic computation on a single cellular automaton; "Zuse's Theory" became the foundation of the field of study called *digital physics*.[21]

Also in 1969 computer scientist Alvy Ray Smith completed a Stanford PhD dissertation on Cellular Automata Theory, the first mathematical treatment of CA as a general class of computers. Many papers came from this dissertation: He showed the equivalence of neighborhoods of various shapes, how to reduce a Moore to a von Neumann neighborhood or how to reduce any neighborhood to a von Neumann neighborhood.[22] He proved that two-dimensional CA are computation universal, introduced 1-dimensional CA, and showed that they too are computation universal, even with simple neighborhoods.[23] He showed how to subsume the complex von Neumann proof of construction universality (and hence self-reproducing machines) into a consequence of computation universality in a 1-dimensional CA.[24] Intended as the introduction to the German edition of von Neumann's book on CA, he wrote a survey of the field with dozens of references to papers, by many authors in many countries over a decade or so of work, often overlooked by modern CA researchers.[25]

In the 1970s a two-state, two-dimensional cellular automaton named Game of Life became widely known, particularly among the early computing community. Invented by John Conway and popularized by Martin Gardner in a *Scientific American* article,[26] its rules are as follows:

1. Any live cell with fewer than two live neighbours dies, as if caused by underpopulation.
2. Any live cell with two or three live neighbours lives on to the next generation.
3. Any live cell with more than three live neighbours dies, as if by overpopulation.
4. Any dead cell with exactly three live neighbours becomes a live cell, as if by reproduction.

Despite its simplicity, the system achieves an impressive diversity of behavior, fluctuating between apparent randomness and order. One of the most apparent features of the Game of Life is the frequent occurrence of *gliders*, arrangements of cells that essentially move themselves across the grid. It is possible to arrange the automaton so that the gliders interact to perform computations, and after much effort it has been shown that the Game of Life can emulate a universal Turing machine.[27] It was viewed as a largely recreational topic, and little follow-up work was done outside of investigating the particularities of the Game of Life and a few related rules in the early 1970s.[28]

Stephen Wolfram independently began working on cellular automata in mid 1981 after considering how complex patterns seemed formed in nature in violation of the Second Law of Thermodynamics.[29] His investigations were initially spurred by an interest in modelling systems such as neural networks.[29] He published his first paper in *Reviews of Modern Physics* investigating *elementary cellular automata* (Rule 30 in particular) in June 1983.[2][29] The unexpected complexity of the behavior of these simple rules led Wolfram to suspect that complexity in nature may be due to similar mechanisms.[29] His investigations, however, led him to realize that cellular automata were poor at modelling neural networks.[29] Additionally, during this period Wolfram formulated the concepts of intrinsic randomness and computational irreducibility,[30] and suggested that rule 110 may be universal—a fact proved later by Wolfram's research assistant Matthew Cook in the 1990s.[31]

In 2002 Wolfram published a 1280-page text *A New Kind of Science*, which extensively argues that the discoveries about cellular automata are not isolated facts but are robust and have significance for all disciplines of science.[32]

Despite confusion in the press,[33][34] the book did not argue for a fundamental theory of physics based on cellular automata,[35] and although it did describe a few specific physical models based on cellular automata,[36] it also provided models based on qualitatively different abstract systems.[37]

44.3 Classification

Wolfram, in *A New Kind of Science* and several papers dating from the mid-1980s, defined four classes into which cellular automata and several other simple computational models can be divided depending on their behavior. While earlier studies in cellular automata tended to try to identify type of patterns for specific rules, Wolfram's classification was the first attempt to classify the rules themselves. In order of complexity the classes are:

- Class 1: Nearly all initial patterns evolve quickly into a stable, homogeneous state. Any randomness in the initial pattern disappears.[38]

- Class 2: Nearly all initial patterns evolve quickly into stable or oscillating structures. Some of the randomness in the initial pattern may filter out, but some remains. Local changes to the initial pattern tend to remain local.[38]

- Class 3: Nearly all initial patterns evolve in a pseudo-random or chaotic manner. Any stable structures that appear are quickly destroyed by the surrounding noise. Local changes to the initial pattern tend to spread indefinitely.[38]

- Class 4: Nearly all initial patterns evolve into structures that interact in complex and interesting ways, with the formation of local structures that are able to survive for long periods of time.[39] Class 2 type stable or oscillating structures may be the eventual outcome, but the number of steps required to reach this state may be very large, even when the initial pattern is relatively simple. Local changes to the initial pattern may spread indefinitely. Wolfram has conjectured that many, if not all class 4 cellular automata are capable of universal computation. This has been proven for Rule 110 and Conway's game of Life.

These definitions are qualitative in nature and there is some room for interpretation. According to Wolfram, "...with almost any general classification scheme there are inevitably cases which get assigned to one class by one definition and another class by another definition. And so it is with cellular automata: there are occasionally rules...that show some features of one class and some of another."[40] Wolfram's classification has been empirically matched to a clustering of the compressed lengths of the outputs of cellular automata.[41]

There have been several attempts to classify cellular automata in formally rigorous classes, inspired by the Wolfram's classification. For instance, Culik and Yu proposed three well-defined classes (and a fourth one for the automata not matching any of these), which are sometimes called Culik-Yu classes; membership in these proved undecidable.[42][43][44] Wolfram's class 2 can be partitioned into two subgroups of stable (fixed-point) and oscillating (periodic) rules.[45]

The idea that there are 4 classes of dynamical system came originally from nobel-prize winning chemist Ilya Prigogine who identified these 4 classes of for thermodynamical systems - (1) systems in thermodynamic equilibrium, (2) spatially/temporally uniform systems, (3) chaotic systems, and (4) complex far-from-equilibrium systems with dissipative structures (see figure 1 in Nicolis' paper (Prigogine's student)): [46] (Nicolis was Prigogine's student).

44.3.1 Reversible

Main article: Reversible cellular automaton

A cellular automaton is *reversible* if, for every current configuration of the cellular automaton, there is exactly one past configuration (preimage).[47] If one thinks of a cellular automaton as a function mapping configurations to configurations, reversibility implies that this function is bijective.[47] If a cellular automaton is reversible, its time-reversed behavior can also be described as a cellular automaton; this fact is a consequence of the Curtis–Hedlund–Lyndon theorem, a topological characterization of cellular automata.[48][49] For cellular automata in which not every configuration has a preimage, the configurations without preimages are called *Garden of Eden* patterns.[50]

For one-dimensional cellular automata there are known algorithms for deciding whether a rule is reversible or irreversible.[51][52] However, for cellular automata of two or more dimensions reversibility is undecidable; that is, there is no algorithm that takes as input an automaton rule and is guaranteed to determine correctly whether the automaton is reversible. The proof by Jarkko Kari is related to the tiling problem by Wang tiles.[53]

Reversible cellular automata are often used to simulate such physical phenomena as gas and fluid dynamics, since they obey the laws of thermodynamics. Such cellular automata have rules specially constructed to be reversible. Such systems have been studied by Tommaso Toffoli, Norman Margolus and others. Several techniques can be used to explicitly construct reversible cellular automata with known inverses. Two common ones are the second order cellular automaton and the block cellular automaton, both of which involve modifying the definition of a cellular automaton in some way. Although such automata do not strictly satisfy the definition given above, it can be shown that they can be emulated by conventional cellular automata with sufficiently large neighborhoods and numbers of states, and can therefore be considered a subset of conventional cellular automata. Conversely, it has been shown that every reversible cellular automaton can be emulated by a block cellular automaton.[54][55]

44.3.2 Totalistic

A special class of cellular automata are *totalistic* cellular automata. The state of each cell in a totalistic cellular automaton is represented by a number (usually an integer value drawn from a finite set), and the value of a cell at time t depends only on the *sum* of the values of the cells in its neighborhood (possibly including the cell itself) at time $t-1$.[56][57] If the state of the cell at time t depends on both its own state and the total of its neighbors at time $t-1$ then the cellular automaton is properly called *outer totalistic*.[57] Conway's Game of Life is an example of an outer totalistic cellular automaton with cell values 0 and 1; outer totalistic cellular automata with the same Moore neighborhood structure as Life are sometimes called life-like cellular automata.[58][59]

44.3.3 Related automata

There are many possible generalizations of the cellular automaton concept.

One way is by using something other than a rectangular (cubic, etc.) grid. For example, if a plane is tiled with regular hexagons, those hexagons could be used as cells. In many cases the resulting cellular automata are equivalent to those with rectangular grids with specially designed neighborhoods and rules. Another variation would be to make the grid itself irregular, such as with Penrose tiles.[60]

Also, rules can be probabilistic rather than deterministic. Such cellular automata are called probabilistic cellular automata. A probabilistic rule gives, for each pattern at time t, the probabilities that the central cell will transition to each possible state at time $t+1$. Sometimes a simpler rule is used; for example: "The rule is the Game of Life, but on each time step there is a 0.001% probability that each cell will transition to the opposite color."

The neighborhood or rules could change over time or space.

44.4 Elementary cellular automata

Main article: Elementary cellular automaton

The simplest nontrivial cellular automaton would be one-dimensional, with two possible states per cell, and a cell's neighbors defined as the adjacent cells on either side of it. A cell and its two neighbors form a neighborhood of 3 cells, so there are $2^3 = 8$ possible patterns for a neighborhood. A rule consists of deciding, for each pattern, whether the cell will be a 1 or a 0 in the next generation. There are then $2^8 = 256$ possible rules.[6] These 256 cellular automata are generally referred to by their Wolfram code, a standard naming convention invented by Wolfram that gives each rule a number from 0 to 255. A number of papers have analyzed and compared these 256 cellular automata. The rule 30 and rule 110 cellular automata are particularly interesting. The images below show the history of each when the starting configuration consists of a 1 (at the top of each image) surrounded by 0s. Each row of pixels represents a generation in the history of the automaton, with $t=0$ being the top row. Each pixel is colored white for 0 and black for 1.

A cellular automaton based on hexagonal cells instead of squares (rule 34/2)

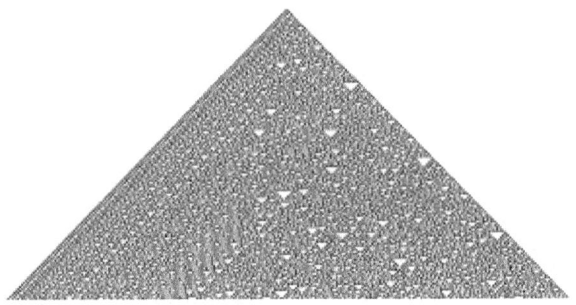

Rule 30

Rule 30 cellular automaton

Rule 30 exhibits *class 3* behavior, meaning even simple input patterns such as that shown lead to chaotic, seemingly random histories.

Rule 110 cellular automaton

Rule 110, like the Game of Life, exhibits what Wolfram calls *class 4* behavior, which is neither completely random nor completely repetitive. Localized structures appear and interact in various complicated-looking ways. In the course of the development of *A New Kind of Science*, as a research assistant to Wolfram in 1994, Matthew Cook proved that some of these structures were rich enough to support universality. This result is interesting because rule 110 is an extremely simple one-dimensional system, and difficult to engineer to perform specific behavior. This result therefore provides significant support for Wolfram's view that class 4 systems are inherently likely to be universal. Cook pre-

For example, initially the new state of a cell could be determined by the horizontally adjacent cells, but for the next generation the vertical cells would be used.

In cellular automata, the new state of a cell is not affected by the new state of other cells. This could be changed so that, for instance, a 2 by 2 block of cells can be determined by itself and the cells adjacent to itself.

There are *continuous automata*. These are like totalistic cellular automata, but instead of the rule and states being discrete (*e.g.* a table, using states {0,1,2}), continuous functions are used, and the states become continuous (usually values in [0,1]). The state of a location is a finite number of real numbers. Certain cellular automata can yield diffusion in liquid patterns in this way.

Continuous spatial automata have a continuum of locations. The state of a location is a finite number of real numbers. Time is also continuous, and the state evolves according to differential equations. One important example is reaction-diffusion textures, differential equations proposed by Alan Turing to explain how chemical reactions could create the stripes on zebras and spots on leopards.[61] When these are approximated by cellular automata, they often yield similar patterns. MacLennan considers continuous spatial automata as a model of computation.

There are known examples of continuous spatial automata, which exhibit propagating phenomena analogous to gliders in the Game of Life.[62]

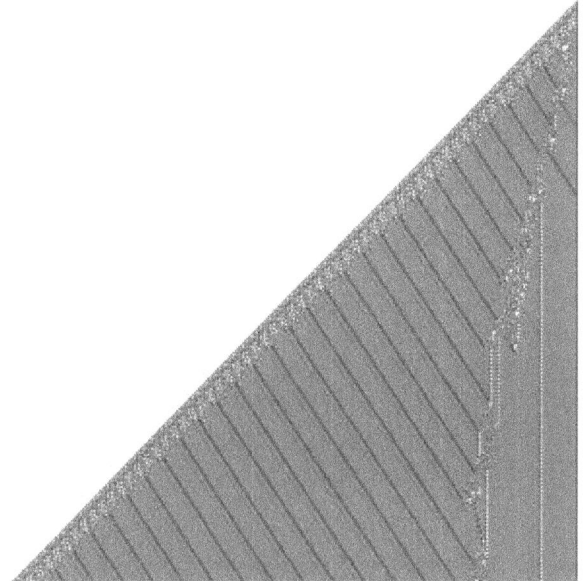

Rule 110

sented his proof at a Santa Fe Institute conference on Cellular Automata in 1998, but Wolfram blocked the proof from being included in the conference proceedings, as Wolfram did not want the proof announced before the publication of *A New Kind of Science*.[63] In 2004, Cook's proof was finally published in Wolfram's journal *Complex Systems* (Vol. 15, No. 1), over ten years after Cook came up with it. Rule 110 has been the basis for some of the smallest universal Turing machines.[64]

44.5 Rule space

An elementary cellular automaton rule is specified by 8 bits, and all elementary cellular automaton rules can be considered to sit on the vertices of the 8-dimensional unit hypercube. This unit hypercube is the cellular automaton rule space. For next-nearest-neighbor cellular automata, a rule is specified by $2^5 = 32$ bits, and the cellular automaton rule space is a 32-dimensional unit hypercube. A distance between two rules can be defined by the number of steps required to move from one vertex, which represents the first rule, and another vertex, representing another rule, along the edge of the hypercube. This rule-to-rule distance is also called the Hamming distance.

Cellular automaton rule space allows us to ask the question concerning whether rules with similar dynamical behavior are "close" to each other. Graphically drawing a high dimensional hypercube on the 2-dimensional plane remains a difficult task, and one crude locator of a rule in the hypercube is the number of bit-1 in the 8-bit string for elementary rules (or 32-bit string for the next-nearest-neighbor rules). Drawing the rules in different Wolfram classes in these slices of the rule space show that class 1 rules tend to have lower number of bit-1's, thus located in one region of the space, whereas class 3 rules tend to have higher proportion (50%) of bit-1's.[45]

For larger cellular automaton rule space, it is shown that class 4 rules are located between the class 1 and class 3 rules.[65] This observation is the foundation for the phrase edge of chaos, and is reminiscent of the phase transition in thermodynamics.

44.6 Biology

Conus textile exhibits a cellular automaton pattern on its shell.[66]

Further information: Patterns in nature

Some biological processes occur—or can be simulated—by cellular automata.

Patterns of some seashells, like the ones in *Conus* and *Cymbiola* genus, are generated by natural cellular automata. The pigment cells reside in a narrow band along the shell's lip. Each cell secretes pigments according to the activating and inhibiting activity of its neighbor pigment cells, obeying a natural version of a mathematical rule.[66] The cell band leaves the colored pattern on the shell as it grows slowly. For example, the widespread species *Conus textile* bears a pattern resembling Wolfram's rule 30 cellular automaton.[66]

Plants regulate their intake and loss of gases via a cellular automaton mechanism. Each stoma on the leaf acts as a cell.[67]

Moving wave patterns on the skin of cephalopods can be simulated with a two-state, two-dimensional cellular automata, each state corresponding to either an expanded or retracted chromatophore.[68]

Threshold automata have been invented to simulate neurons, and complex behaviors such as recognition and learning can be simulated.[69]

Fibroblasts bear similarities to cellular automata, as each fibroblast only interacts with its neighbors.[70]

44.7 Chemical types

The Belousov–Zhabotinsky reaction is a spatio-temporal chemical oscillator that can be simulated by means of a cellular automaton. In the 1950s A. M. Zhabotinsky (extending the work of B. P. Belousov) discovered that when a thin, homogenous layer of a mixture of malonic acid, acidified bromate, and a ceric salt were mixed together and left undisturbed, fascinating geometric patterns such as concentric circles and spirals propagate across the medium. In the "Computer Recreations" section of the August 1988 issue of *Scientific American*,[71] A. K. Dewdney discussed a cellular automaton[72] developed by Martin Gerhardt and Heike Schuster of the University of Bielefeld (Germany). This automaton produces wave patterns that resemble those in the Belousov-Zhabotinsky reaction.

44.8 Applications

44.8.1 Computer processors

Cellular automaton processors are physical implementations of CA concepts, which can process information computationally. Processing elements are arranged in a regular grid of identical cells. The grid is usually a square tiling, or tessellation, of two or three dimensions; other tilings are possible, but not yet used. Cell states are determined only by interactions with adjacent neighbor cells. No means exists to communicate directly with cells farther away.[73] One such cellular automaton processor array configuration is the systolic array. Cell interaction can be via electric charge, magnetism, vibration (phonons at quantum scales), or any other physically useful means. This can be done in several ways so no wires are needed between any elements. This is very unlike processors used in most computers today, von Neumann designs, which are divided into sections with elements that can communicate with distant elements over wires.

44.8.2 Cryptography

Rule 30 was originally suggested as a possible block cipher for use in cryptography. Two dimensional cellular automata are used for random number generation.[74]

Cellular automata have been proposed for public key cryptography. The one-way function is the evolution of a finite CA whose inverse is believed to be hard to find. Given the rule, anyone can easily calculate future states, but it appears to be very difficult to calculate previous states.

44.8.3 Error correction coding

CA have been applied to design error correction codes in a paper by D. Roy Chowdhury, S. Basu, I. Sen Gupta, and P. Pal Chaudhuri. The paper defines a new scheme of building single bit error correction and double bit error detection (SEC-DED) codes using CA, and also reports a fast hardware decoder for the code.[75]

44.9 Modeling physical reality

Main articles: Digital physics and digital philosophy

As Andrew Ilachinski points out in his *Cellular Automata*, many scholars have raised the question of whether the universe is a cellular automaton.[76] Ilachinski argues that the importance of this question may be better appreciated with a simple observation, which can be stated as follows. Consider the evolution of rule 110: if it were some kind of "alien physics", what would be a reasonable description of the observed patterns?[77] If an observer did not know how the images were generated, that observer might end up conjecturing about the movement of some particle-like objects. Indeed, physicist James Crutchfield has constructed a rigorous mathematical theory out of this idea, proving the statistical emergence of "particles" from cellular automata.[78] Then, as the argument goes, one might wonder if *our* world, which is currently well described, at our current level of understanding, by physics with particle-like objects, could be a CA at its most fundamental level with the gaps in information or incomplete understanding of fundamental data appearing as an arbitrary random order that would seem contrary to CA.

While a complete theory along this line has not been developed, entertaining and developing this hypothesis led scholars to interesting speculation and fruitful intuitions on how can we make sense of our world within a discrete framework. Marvin Minsky, the AI pioneer, investigated how to understand particle interaction with a four-dimensional CA lattice;[79] Konrad Zuse—the inventor of the first working computer, the Z3—developed an irregularly organized lattice to address the question of the information content of particles.[80] More recently, Edward Fredkin exposed what he terms the "finite nature hypothesis", i.e., the idea that "ultimately every quantity of physics, including space and time, will turn out to be discrete and finite."[81] Fredkin and Wolfram are strong proponents of a CA-based physics. In 2016 Gerard 't Hooft published a book long development of the idea to rebuild quantum mechanics using cellular automata.[82]

In recent years, other suggestions along these lines have emerged from literature in non-standard computation. Wolfram's *A New Kind of Science* considers CA the key to understanding a variety of subjects, physics included. The *Mathematics of the Models of Reference*—created by iLabs[83] founder Gabriele Rossi and developed with Francesco Berto and Jacopo Tagliabue—features an original 2D/3D universe based on a new "rhombic dodecahedron-based" lattice and a unique rule. This model satisfies universality (it is equivalent to a Turing Machine) and perfect reversibility (a *desideratum* if one wants to conserve various quantities easily and never lose information), and it comes embedded in a first-order theory, allowing computable, qualitative statements on the universe evolution.[84]

44.10 Specific rules

Specific types of cellular automata include:

- Brian's Brain
- Langton's ant
- Wireworld
- Rule 90
- Rule 184
- von Neumann cellular automata
- Nobili cellular automata
- Codd's cellular automaton
- Langton's loops
- CoDi

44.11 Problems solved

Problems that can be solved with cellular automata include:

- Firing squad synchronization problem
- Majority problem

44.12 See also

- Agent-based model
- Automata theory
- Bidirectional traffic
- Cellular automata in popular culture
- Cyclic cellular automaton
- Excitable medium
- Mirek's Cellebration
- Movable cellular automaton
- Quantum cellular automata
- Spatial decision support system
- Turmites

44.13 Reference notes

[1] Daniel Dennett (1995), *Darwin's Dangerous Idea*, Penguin Books, London, ISBN 978-0-14-016734-4, ISBN 0-14-016734-X

[2] Wolfram, Stephen (1983). "Statistical Mechanics of Cellular Automata". *Reviews of Modern Physics*. 55 (3): 601–644. Bibcode:1983RvMP...55..601W. doi:10.1103/RevModPhys.55.601.

[3] Toffoli, Tommaso; Margolus, Norman (1987). *Cellular Automata Machines: A New Environment for Modeling*. MIT Press. p. 27. ISBN 9780262200608.

[4] Schiff, Joel L. (2011). *Cellular Automata: A Discrete View of the World*. Wiley & Sons, Inc. p. 40. ISBN 9781118030639.

[5] Kier, Seybold, Cheng 2005, p. 15

[6] Bialynicki-Birula, Bialynicka-Birula 2004, p. 9

[7] Schiff 2011, p. 41

[8] Pickover, Clifford A. (2009). *The Math Book: From Pythagoras to the 57th Dimension, 250 Milestones in the History of Mathematics*. Sterling Publishing Company, Inc. p. 406. ISBN 978-1402757969.

[9] Schiff 2011, p. 1

[10] John von Neumann, "The general and logical theory of automata," in L.A. Jeffress, ed., Cerebral Mechanisms in Behavior – The Hixon Symposium, John Wiley & Sons, New York, 1951, pp. 1–31.

[11] Kemeny, John G. (1955). "Man viewed as a machine". *Sci. Amer.* **192**: 58–67. doi:10.1038/scientificamerican0455-58.; *Sci. Amer.* 1955; 192:6 (errata).

[12] Schiff 2011, p. 3

[13] Ilachinski 2001, p. xxix

[14] Bialynicki-Birula, Bialynicka-Birula 2004, p. 8

[15] Wolfram 2002, p. 876

[16] von Neumann, John; Burks, Arthur W. (1966). *Theory of Self-Reproducing Automata*. University of Illinois Press.

[17] Wiener, N.; Rosenblueth, A. (1946). "The mathematical formulation of the problem of conduction of impulses in a network of connected excitable elements, specifically in cardiac muscle". *Arch. Inst. Cardiol. México*. **16**: 205.

[18] Letichevskii, A. A.; Reshodko, L. V. (1974). "N. Wiener's theory of the activity of excitable media". *Cybernetics*. **8**: 856–864. doi:10.1007/bf01068458.

[19] Davidenko, J. M.; Pertsov, A. V.; Salomonsz, R.; Baxter, W.; Jalife, J. (1992). "Stationary and drifting spiral waves of excitation in isolated cardiac muscle". *Nature*. **355** (6358): 349–351. Bibcode:1992Natur.355..349D. PMID 1731248. doi:10.1038/355349a0.

[20] Hedlund, G. A. (1969). "Endomorphisms and automorphisms of the shift dynamical system". *Math. Systems Theory*. **3** (4): 320–3751. doi:10.1007/BF01691062.

[21] Schiff 2011, p. 182

[22] Smith, Alvy Ray. "Cellular Automata Complexity Trade-Offs" (PDF).

[23] Smith, Alvy Ray. "Simple Computation-Universal Cellular Spaces" (PDF).

[24] Smith, Alvy Ray. "Simple Nontrivial Self-Reproducing Machines" (PDF).

[25] Smith, Alvy Ray. "Introduction to and Survey of Cellular Automata or Polyautomata Theory" (PDF).

[26] Gardner, Martin (1970). "Mathematical Games: The fantastic combinations of John Conway's new solitaire game "life"". *Scientific American* (223): 120–123.

[27] Paul Chapman. Life universal computer. http://www.igblan.free-online.co.uk/igblan/ca/ November 2002

[28] Wainwright 2010, p. 16

[29] Wolfram 2002, p. 880

[30] Wolfram 2002, p. 881

[31] Mitchell, Melanie (4 October 2002). "Is the Universe a Universal Computer?". *Science*. **298** (5591): 65–68. doi:10.1126/science.1075073.

[32] Wolfram 2002, pp. 1–7

[33] Johnson, George (9 June 2002). "'A New Kind of Science': You Know That Space-Time Thing? Never Mind". *The New York Times*. The New York Times Company. Retrieved 22 January 2013.

[34] "The Science of Everything". *The Economist*. 30 May 2002. Retrieved 22 January 2013.

[35] Wolfram 2002, pp. 433–546

[36] Wolfram 2002, pp. 51–114

[37] Wolfram 2002, pp. 115–168

[38] Ilachinsky 2001, p. 12

[39] Ilachinsky 2001, p. 13

[40] Wolfram 2002, p. 231

[41] Zenil, Hector (2010). "Compression-based investigation of the dynamical properties of cellular automata and other systems" (PDF). *Complex Systems*. **19** (1).

[42] G. Cattaneo; E. Formenti; L. Margara (1998). "Topological chaos and CA". In M. Delorme; J. Mazoyer. *Cellular automata: a parallel model*. Springer. p. 239. ISBN 978-0-7923-5493-2.

[43] Burton H. Voorhees (1996). *Computational analysis of one-dimensional cellular automata*. World Scientific. p. 8. ISBN 978-981-02-2221-5.

[44] Max Garzon (1995). *Models of massive parallelism: analysis of cellular automata and neural networks*. Springer. p. 149. ISBN 978-3-540-56149-1.

[45] Li, Wentian; Packard, Norman (1990). "The structure of the elementary cellular automata rule space" (PDF). *Complex Systems*. **4**: 281–297. Retrieved 25 January 2013.

[46] Nicolis (1974). "Dissipative Structures, Catastrophes, and Pattern Formation: A Bifurcation Analysis" (PDF). *PNAS*. **71** (7): 2748–2751. doi:10.1073/pnas.71.7.2748. Retrieved 25 March 2017.

[47] Kari, Jarrko 1991, p. 379

[48] Richardson, D. (1972). "Tessellations with local transformations". *J. Computer System Sci.* **6** (5): 373–388. doi:10.1016/S0022-0000(72)80009-6.

[49] Margenstern, Maurice (2007). *Cellular Automata in Hyperbolic Spaces – Tome I, Volume 1*. Archives contemporaines. p. 134. ISBN 978-2-84703-033-4.

[50] Schiff 2011, p. 103

[51] Amoroso, Serafino; Patt, Yale N. (1972). "Decision Procedures for Surjectivity and Injectivity of Parallel Maps for Tessellation Structures". *J. Comput. Syst. Sci.* **6** (5): 448–464. doi:10.1016/s0022-0000(72)80013-8.

[52] Sutner, Klaus (1991). "De Bruijn Graphs and Linear Cellular Automata" (PDF). *Complex Systems*. **5**: 19–30.

[53] Kari, Jarkko (1990). "Reversibility of 2D cellular automata is undecidable". *Physica D*. **45**: 379–385. Bibcode:1990PhyD...45..379K. doi:10.1016/0167-2789(90)90195-U.

[54] Kari, Jarkko (1999). "On the circuit depth of structurally reversible cellular automata". *Fundamenta Informaticae*. **38**: 93–107.

[55] Durand-Lose, Jérôme (2001). "Representing reversible cellular automata with reversible block cellular automata". *Discrete Mathematics and Theoretical Computer Science*. **AA**: 145–154. Archived from the original on 15 May 2011.

[56] Wolfram 2002, p. 60

[57] Ilachinski, Andrew (2001). *Cellular automata: a discrete universe*. World Scientific. pp. 44–45. ISBN 978-981-238-183-5.

[58] The phrase "life-like cellular automaton" dates back at least to Barral, Chaté & Manneville (1992), who used it in a broader sense to refer to outer totalistic automata, not necessarily of two dimensions. The more specific meaning given here was used e.g. in several chapters of Adamatzky (2010). See: Barral, Bernard; Chaté, Hugues; Manneville, Paul (1992). "Collective behaviors in a family of high-dimensional cellular automata". *Physics Letters A*. **163** (4): 279–285. Bibcode:1992PhLA..163..279B. doi:10.1016/0375-9601(92)91013-H.

[59] Eppstein 2010, pp. 72–73

[60] Jacob Aron. "First gliders navigate ever-changing Penrose universe". *New Scientist*.

[61] Murray, J. "Mathematical Biology II". Springer.

[62] Pivato, M: "RealLife: The continuum limit of Larger than Life cellular automata", *Theoretical Computer Science*, 372 (1), March 2007, pp. 46–68

[63] Giles, Jim (2002). "What Kind of Science is This?". *Nature*. **417**: 216–218. doi:10.1038/417216a.

[64] Weinberg, Steven (24 October 2002). "Is the Universe a Computer?". *The New York Review of Books*. Rea S. Hederman. Retrieved 12 October 2012.

[65] Wentian Li; Norman Packard; Chris G Langton (1990). "Transition phenomena in cellular automata rule space". *Physica D*. **45** (1–3): 77–94. Bibcode:1990PhyD...45...77L. doi:10.1016/0167-2789(90)90175-O.

[66] Coombs, Stephen (15 February 2009), *The Geometry and Pigmentation of Seashells* (PDF), pp. 3–4, retrieved 2 September 2012

[67] Peak, West; Messinger, Mott (2004). "Evidence for complex, collective dynamics and emergent, distributed computation in plants". *Proceedings of the National Academy of Sciences*. **101** (4): 918–922. Bibcode:2004PNAS..101..918P. PMC 327117. PMID 14732685. doi:10.1073/pnas.0307811100.

[68] http://gilly.stanford.edu/past_research_files/APackardneuralnet.pdf

[69] Ilachinsky 2001, p. 275

[70] Yves Bouligand (1986). *Disordered Systems and Biological Organization*. pp. 374–375.

[71] A. K. Dewdney. The hodgepodge machine makes waves. Scientific American, p. 104, August 1988.

[72] Gerhardt, M.; Schuster, H. (1989). "A cellular automaton describing the formation of spatially ordered structures in chemical systems". *Physica D*. **36**: 209–221. doi:10.1016/0167-2789(89)90081-x.

[73] Muhtaroglu, Ali (August 1996). "4.1 Cellular Automaton Processor (CAP)". *Cellular Automaton Processor Based Systems for Genetic Sequence Comparison/Database Searching*. Cornell University. pp. 62–74.

[74] Tomassini, M.; Sipper, M.; Perrenoud, M. (2000). "On the generation of high-quality random numbers by two-dimensional cellular automata". *IEEE Transactions on Computers*. **49** (10): 1146–1151. doi:10.1109/12.888056.

[75] Chowdhury, D. Roy; Basu, S.; Gupta, I. Sen; Chaudhuri, P. Pal (June 1994). "Design of CAECC - cellular automata based error correcting code". *IEEE Transactions on Computers*. **43** (6): 759–764. doi:10.1109/12.286310.

[76] Ilachinsky 2001, p. 660

[77] Ilachinsky 2001, pp. 661–662

[78] J. P. Crutchfield, "The Calculi of Emergence: Computation, Dynamics, and Induction", Physica D 75, 11–54, 1994.

[79] Minsky, M. "Cellular Vacuum". *International Journal of Theoretical Physics*. **21** (537–551): 1982.

[80] K. Zuse, "The Computing Universe", Int. Jour. of Theo. Phy. 21, 589–600, 1982.

[81] E. Fredkin, "Digital mechanics: an informational process based on reversible universal cellular automata", Physica D 45, 254–270, 1990

[82] Gerard 't Hooft, 2016, *The Cellular Automaton Interpretation of Quantum Mechanics*. Springer International Publishing. DOI 10.1007/978-3-319-41285-6, Open access-

[83] "Ilabs".

[84] F. Berto, G. Rossi, J. Tagliabue, The Mathematics of the Models of Reference, College Publications, 2010

44.14 References

- Adamatzky, Andrew, ed. (2010). *Game of Life Cellular Automata*. Springer. ISBN 978-1-84996-216-2.

- Bialynicki-Birula, Iwo; Bialynicka-Birula, Iwona (2004). *Modeling Reality: How Computers Mirror Life*. Oxford University Press. ISBN 0198531001.

- Chopard, Bastien; Droz, Michel (2005). *Cellular Automata Modeling of Physical Systems*. Cambridge University Press. ISBN 0-521-46168-5.

- Gutowitz, Howard, ed. (1991). *Cellular Automata: Theory and Experiment*. MIT Press. ISBN 9780262570862.

- Ilachinski, Andrew (2001). *Cellular Automata: A Discrete Universe*. World Scientific. ISBN 9789812381835.

- Kier, Lemont B.; Seybold, Paul G.; Cheng, Chao-Kun (2005). *Modeling Chemical Systems using Cellular Automata*. Springer. ISBN 9781402036576.

- Wolfram, Stephen (2002). *A New Kind of Science*. Wolfram Media. ISBN 978-1579550080.

- Cellular automaton FAQ from the newsgroup comp.theory.cell-automata

- "Neighbourhood Survey" (includes discussion on triangular grids, and larger neighborhood CAs)

- von Neumann, John, 1966, *The Theory of Self-reproducing Automata*, A. Burks, ed., Univ. of Illinois Press, Urbana, IL.

- Cosma Shalizi's Cellular Automata Notebook contains an extensive list of academic and professional reference material.

- Wolfram's papers on CAs

- A.M. Turing. 1952. The Chemical Basis of Morphogenesis. *Phil. Trans. Royal Society*, vol. B237, pp. 37–72. (proposes reaction-diffusion, a type of continuous automaton).

- Evolving Cellular Automata with Genetic Algorithms: A Review of Recent Work, Melanie Mitchell, James P. Crutchfeld, Rajarshi Das (In Proceedings of the First International Conference on Evolutionary Computation and Its Applications (EvCA'96). Moscow, Russia: Russian Academy of Sciences, 1996.)

- The Evolutionary Design of Collective Computation in Cellular Automata, James P. Crutchfeld, Melanie Mitchell, Rajarshi Das (In J. P. Crutchfeld and P. K. Schuster (editors), Evolutionary Dynamics|Exploring the Interplay of Selection, Neutrality, Accident, and Function. New York: Oxford University Press, 2002.)

- The Evolution of Emergent Computation, James P. Crutchfield and Melanie Mitchell (SFI Technical Report 94-03-012)

- Ganguly, Sikdar, Deutsch and Chaudhuri "A Survey on Cellular Automata"

44.15 External links

- Berto, Francesco; Tagliabue, Jacopo. "Cellular Automata". *Stanford Encyclopedia of Philosophy*.

- Mirek's Cellebration – Home to free MCell and MJCell cellular automata explorer software and rule libraries. The software supports a large number of 1D and 2D rules. The site provides both an extensive rules lexicon and many image galleries loaded with examples of rules. MCell is a Windows application, while MJCell is a Java applet. Source code is available.

- Modern Cellular Automata – Easy to use interactive exhibits of live color 2D cellular automata, powered by Java applet. Included are exhibits of traditional, reversible, hexagonal, multiple step, fractal generating, and pattern generating rules. Thousands of rules are provided for viewing. Free software is available.

- Self-replication loops in Cellular Space – Java applet powered exhibits of self replication loops.

- A collection of over 10 different cellular automata applets (in Monash University's Virtual Lab)

- Golly supports von Neumann, Nobili, GOL, and a great many other systems of cellular automata. Developed by Tomas Rokicki and Andrew Trevorrow. This is the only simulator currently available that can demonstrate von Neumann type self-replication.

- Fourier Life - A collection of rules that demonstrate self-replicating patterns which spontaneously emerge from a field of random cells. Most of the rules were found using an algorithm that uses a Fourier transform to detect self-replication.

- Wolfram Atlas – An atlas of various types of one-dimensional cellular automata.

- Conway Life

44.15. EXTERNAL LINKS

- First replicating creature spawned in life simulator
- *The Mathematics of the Models of Reference*, featuring a general tutorial on CA, interactive applet, free code and resources on CA as model of fundamental physics
- Fourmilab Cellular Automata Laboratory
- Busy Boxes, a 3-D, reversible, SALT-architecture CA
- Cellular Automata Repository (CA researchers, historic links, free software, books and beyond)

Chapter 45

Spatial ecology

Spatial ecology represents the ultimate distributional or spatial unit occupied by a species. In a particular habitat shared by several species, each of the species is usually confined to its own micro habitat or spatial niche because two species in the same general territory cannot usually occupy the same ecological niche for any significant length of time.

45.1 Overview

In nature, organisms are neither distributed uniformly nor at random, forming instead some sort of spatial pattern.[1] This is due to various energy inputs, disturbances, and species interactions that result in spatially patchy structures or gradients. This spatial variance in the environment creates diversity in communities of organisms, as well as in the variety of the observed biological and ecological events.[1] The type of spatial arrangement present may suggest certain interactions within and between species, such as competition, predation, and reproduction.[2] On the other hand, certain spatial patterns may also rule out specific ecological theories previously thought to be true.[3]

Although spatial ecology deals with spatial patterns, it is usually based on observational data rather than on an existing model.[2] This is because nature rarely follows set expected order. To properly research a spatial pattern or population, the spatial extent to which it occurs must be detected. Ideally, this would be accomplished beforehand via a benchmark spatial survey, which would determine whether the pattern or process is on a local, regional, or global scale. This is rare in actual field research, however, due to the lack of time and funding, as well as the ever-changing nature of such widely-studied organisms such as insects and wildlife.[4] With detailed information about a species' life-stages, dynamics, demography, movement, behavior, etc., models of spatial pattern may be developed to estimate and predict events in unsampled locations.[2]

45.2 History

Most mathematical studies in ecology in the nineteenth century assumed a uniform distribution of living organisms in their habitat.[1] In the past quarter century, ecologists have begun to recognize the degree to which organisms respond to spatial patterns in their environment. Due to the rapid advances in computer technology in the same time period, more advanced methods of statistical data analysis have come into use.[3] Also, the repeated use of remotely sensed imagery and geographic information systems in a particular area has led to increased analysis and identification of spatial patterns over time.[4] These technologies have also increased the ability to determine how human activities have impacted animal habitat and climate change.[5] The natural world has become increasingly fragmented due to human activities; anthropogenic landscape change has had a ripple-effect impacts on wildlife populations, which are now more likely to be small, restricted in distribution, and increasingly isolated from one another. In part as a reaction to this knowledge, and partially due to increasingly sophisticated theoretical developments, ecologists began stressing the importance of spatial context in research. Spatial ecology emerged from this movement toward spatial accountability; "the progressive introduction of spatial variation and complexity into ecological analysis, including changes in spatial patterns over time".[6]

45.3 Concepts

45.3.1 Scale

In spatial ecology, scale refers to the spatial extent of ecological processes and the spatial interpretation of the data.[7] The response of an organism or a species to the environment is particular to a specific scale, and may respond differently at a larger or smaller scale.[8] Choosing a scale that is appropriate to the ecological process in question is very important in accurately hypothesizing and determining the underlying cause.[9][10] Most often, ecological patterns are a result of multiple ecological processes, which often operate at more than one spatial scale.[11] Through the use of such spatial statistical methods such as geostatistics and principal coordinate analysis of neighbor matrices (PCNM), one can identify spatial relationships between organisms and environmental variables at multiple scales.[8]

45.3.2 Spatial autocorrelation

Spatial autocorrelation refers to the value of samples taken close to each other are more likely to have similar magnitude than by chance alone.[7] When a pair of values located at a certain distance apart are more similar than expected by chance, the spatial autocorrelation is said to be positive. When a pair of values are less similar, the spatial autocorrelation is said to be negative. It is common for values to be positively autocorrelated at shorter distances and negative autocorrelated at longer distances.[11] This is commonly known as Tobler's first law of geography, summarized as "everything is related to everything else, but nearby objects are more related than distant objects".

In ecology, there are two important sources of spatial autocorrelation, which both arise from spatial-temporal processes, such as dispersal or migration:[11]

- True/inherent spatial autocorrelation arises from interactions among individuals located in close proximity. This process is endogenous (internal) and results in the individuals being spatially adjacent in a patchy fashion.[7] An example of this would be sexual reproduction, the success of which requires the closeness of a male and female of the species.

- Induced spatial autocorrelation (or 'induced spatial dependence') arises from the species response to the spatial structure of exogenous (external) factors, which are themselves spatially autocorrelated.[7] An example of this would be the winter habitat range of deer, which use conifers for heat retention and forage.

Most ecological data exhibit some degree of spatial autocorrelation, depending on the ecological scale (spatial resolution) of interest. As the spatial arrangement of most ecological data is not random, traditional random population samples tend to overestimate the true value of a variable, or infer significant correlation where there is none.[1] This bias can be corrected through the use of geostatistics and other more statistically advanced models. Regardless of method, the sample size must be appropriate to the scale and the spatial statistical method used in order to be valid.[4]

45.3.3 Pattern

Spatial patterns, such as the distribution of a species, are the result of either true or induced spatial autocorrelation.[7] In nature, organisms are distributed neither uniformly nor at random. The environment is spatially structured by various ecological processes,[1] which in combination with the behavioral response of species' generally results in:

- Gradients (trends) steady directional change in numbers over a specific distance
- Patches (clumps) a relatively uniform and homogenous area separated by gaps
- Noise (random fluctuations) variation not able to be explained by a model

Theoretically, any of these structures may occur at any given scale. Due to the presence of spatial autocorrelation, in nature gradients are generally found at the global level, whereas patches represent intermediate (regional) scales, and noise at local scales.[11]

The analysis of spatial ecological patterns comprises two families of methods:[12]

- Point pattern analysis deals with the distribution of individuals through space, and is used to determine whether the distribution is random.[13] It also describes the type of pattern and draws conclusions on what kind of process created the observed pattern. Quadrat-density and the nearest neighbor methods are the most commonly used statistical methods.

- Surface pattern analysis deals with spatially continuous phenomena. After the spatial distribution of the variables is determined through discrete sampling, statistical methods are used to quantify the magnitude, intensity, and extent of spatial autocorrelation present in the data (such as correlograms, variograms, and periodograms), as well as to map the amount of spatial variation.

45.4 Applications

45.4.1 Research

Analysis of spatial trends has been used to research wildlife management, fire ecology, population ecology, disease ecology, invasive species, marine ecology, and carbon sequestration modeling using the spatial relationships and patterns to determine ecological processes and their effects on the environment. Spatial patterns have different ecosystem functioning in ecology for examples enhanced productive.[14]

45.4.2 Interdisciplinary

The concepts of spatial ecology are fundamental to understanding the spatial dynamics of population and community ecology. The spatial heterogeneity of populations and communities plays a central role in such ecological theories such as succession, adaptation, community stability, competition, predator-prey interactions, parasitism, and epidemics.[1] The rapidly expanding field of landscape ecology utilizes the basic aspects of spatial ecology in its research.

The practical use of spatial ecology concepts is essential to understanding the consequences of fragmentation and habitat loss for wildlife. Understanding the response of a species' to a spatial structure provides useful information in regards to biodiversity conservation and habitat restoration.[15]

Spatial ecology modeling uses components of remote sensing and geographical information systems (GIS).

45.5 Statistical tests

A number of statistical tests have been developed to study such relations.

45.5.1 Tests based on distance

Clark and Evans' R

Clark and Evans in 1954[16] proposed a test based on the density and distance between organisms. Under the null hypothesis the expected distance (r_e) between the organisms (measured as the nearest neighbor's distance) with a known constant density (ϱ) is

$$r_e = \frac{1}{2\sqrt{\varrho}}$$

The difference between the observed (r_o) and the expected (r_e) can be tested with a Z test

$$Z = \frac{r_o - r_e}{SE}$$

$$SE = \sqrt{\frac{0.0863 A}{N}}$$

where A is the area of the region sampled and N is the total number of organisms. For large samples Z is distributed normally. The results are usually reported in the form of a ratio: $R = (r_o) / (r_e)$

Pielou's α

Pielou in 1959 devised a different statistic.[17] She considered instead of the nearest neighbors the distance between an organisms and a set of pre-chosen random points within the sampling area again assuming a constant density. If the population is randomly disbursed in the area these distances will equal the nearest neighbors distances. Let ω be the ratio between the distances from the random points and the distances calculated from the nearest neighbor calculations. The α is

$$\alpha = \pi d \omega$$

where d is the constant common density and π has its usual numerical value. Values of α less than, equal to or greater than 1 indicate uniformity, randomness (a Poisson distribution) or aggregation respectively. Alpha may be tested for a significant deviation from 1 by computing the test statistic

$$\chi^2_{2n} = 2n\alpha$$

where χ^2 is distributed with $2n$ degrees of freedom. n here is the number of organisms sampled.

Montford in 1961 showed than then the density is estimated rather than a known constant that this version of alpha tended to overestimate the actual degree of aggregation. He provided a revised formulation which corrects this error.

45.6 See also

- Edge effects
- Spatial analysis
- Taylor's law

45.7 References

[1] Legendre, P.; Fortin, M.-J. (1989). "Spatial pattern and ecological analysis". *Plant Ecology*. **80** (2): 107–138. doi:10.1007/BF00048036.

[2] Perry, J.N.; A.M. Liebhold; M.S. Rosenberg; J. Dungan; M. Miriti; A. Jakomulska; S. Citron-Pousty (2002). "Illustrations and guidelines for selecting statistical methods for quantifying spatial pattern in ecological data". Ecography. 25 (5): 578–600. doi:10.1034/j.1600-0587.2002.250507.x.

[3] Liebhold, A.M.; J. Gurevitch (2002). "Integrating the statistical analysis of spatial data in ecology". Ecography. 25 (5): 553–557. doi:10.1034/j.1600-0587.2002.250505.x.

[4] Tobin, P.C. (2004). "Estimation of the spatial autocorrelation function: consequences of sampling dynamic populations in space and time". Ecography. 27 (6): 765–775. doi:10.1111/j.0906-7590.2004.03977.x.

[5] Keitt, Timothy H.; Ottar N. Bjørnstad; Philip M. Dixon; Steve Citron-Poust (2002). "Accounting for spatial pattern when modeling organism-environment interactions". Ecography. 25 (5): 616–625. doi:10.1034/j.1600-0587.2002.250509.x.

[6] Rockwood, Larry L. (2006). Introduction to Population Ecology. Malden, MA, USA: Blackwell Publishing Ltd. pp. 108–110. ISBN 9781405132633.

[7] Fortin, Marie-Josée; Mark R. T. Dale (2005). Spatial Analysis: A Guide for Ecologists. Cambridge University Press. ISBN 978-0-521-80434-9.

[8] Bellier, E.; P. Monestiez; J.-P. Durbec; J.-N. Candau (2007). "Identifying spatial relationships at multiple scales: principal coordinates of neighbor matrices (PCNM) and geostatistical approaches". Ecography. 30 (3): 385–399. doi:10.1111/j.0906-7590.2007.04911.x.

[9] De Knegt, H.J.; F. van Langevelde; M.B. Coughenour; A.K. Skidmore; W.F. de Boer; I.M.A. Heitkönig; N.M. Knox; R. Slotow; C. van der Waal and H.H.T. Prins (2010). Spatial autocorrelation and the scaling of species–environment relationships. Ecology 91: 2455–2465. doi:10.1890/09-1359.1

[10] Wilschut, L.I.; Addink, E.A.; Heesterbeek, J.A.P.; Heier, L.; Laudisoit, A.; Begon, M.; Davis, S.; Dubyanskiy, V.M.; Burdelov, L.; de Jong, S.M (2013). "Potential corridors and barriers for plague spread in central Asia". International Journal of Health Geographics. 12 (49). doi:10.1186/1476-072X-12-49.

[11] Fortin, M.-J.; M.R.T. Dale; J. ver Hoef (2002). "Spatial Analysis in Ecology" (PDF). Encyclopedia of Environmetrics. 4: 2051–2058.

[12] Legendre, P. (1993). "Spatial autocorrelation: trouble or new paradigm?". Ecology. Ecological Science of America. 74 (6): 1659–1673. ISSN 0012-9658. doi:10.2307/1939924.

[13] Wilschut, L.I.; Laudisoit, A.; Hughes, N.K.; Addink, E.A.; de Jong, S.M; Heesterbeek, J.A.P.; Reijniers, J.; Eagle, S.; Dubyanskiy, V.M.; Begon, M. (2015). "Spatial distribution patterns of plague hosts: point pattern analysis of the burrows of great gerbils in Kazakhstan". Journal of Biogeography. 42 (7): 1281–1291. doi:10.1111/jbi.12534.

[14] Rietkerk, M. and Van de Koppel, J. 2008 Regular pattern formation in real ecosystems. Trends in Ecology and Evolution 23:169-175

[15] Collinge, S.K. (2001). "Spatial ecology and biological conservation: Introduction". Biological Conservation. 100: 1–2. doi:10.1016/s0006-3207(00)00201-9.

[16] Clark PJ, Evans FC (1954) Distance to nearest neighbor as a measure of spatial relationships in populations. Ecology 35: 445–453

[17] Pielou E C 1959. The use of point-to-plant distances in the study of patterns in plant populations. J Ecol 47: 607-613

45.8 External links

- Spatial Ecology, hosts software for use in spatial ecological analysis.
- Spatial Ecology Research Programme at the University of Helsinki
- Spatial Ecology Lab at the University of Queensland
- Ecography publishes peer-reviewed articles on spatial ecology.
- National Center for Ecological Analysis and Synthesis at the University of California, Santa Barbara
- Spatial Ecology Lab at the University of Alaska, Fairbanks
- Spatial Ecology wikipedia, online resources for learning spatial ecological analysis and data processing using Open source software.

Chapter 46

Self-replication

See also: Biological reproduction

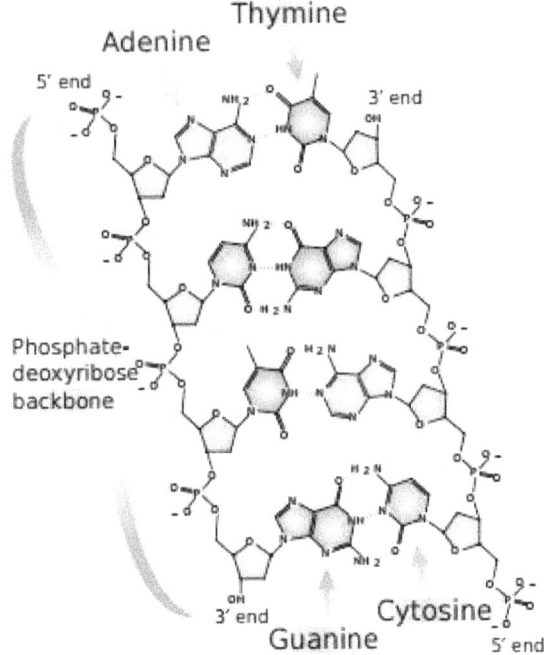

Molecular structure of DNA

Self-replication is any behavior of a dynamical system that yields construction of an identical copy of itself. Biological cells, given suitable environments, reproduce by cell division. During cell division, DNA is replicated and can be transmitted to offspring during reproduction. Biological viruses can replicate, but only by commandeering the reproductive machinery of cells through a process of infection. Harmful prion proteins can replicate by converting normal proteins into rogue forms.[1] Computer viruses reproduce using the hardware and software already present on computers. Self-replication in robotics has been an area of research and a subject of interest in science fiction. Any self-replicating mechanism which does not make a perfect copy will experience genetic variation and will create variants of itself. These variants will be subject to natural selection, since some will be better at surviving in their current environment than others and will out-breed them.

46.1 Overview

46.1.1 Theory

See also: Von Neumann universal constructor

Early research by John von Neumann[2] established that replicators have several parts:

- A coded representation of the replicator
- A mechanism to copy the coded representation
- A mechanism for effecting construction within the host environment of the replicator

Exceptions to this pattern are possible. For example, scientists have come close to constructing RNA that copies itself in an "environment" that is a solution of RNA monomers and transcriptase. In this case, the body is the genome, and the specialized copy mechanisms are external.

However, the simplest possible case is that only a genome exists. Without some specification of the self-reproducing steps, a genome-only system is probably better characterized as something like a crystal.

46.1.2 Classes of self-replication

Recent research[3] has begun to categorize replicators, often based on the amount of support they require.

- Natural replicators have all or most of their design from nonhuman sources. Such systems include natural life forms.

46.1. OVERVIEW

- Autotrophic replicators can reproduce themselves "in the wild". They mine their own materials. It is conjectured that non-biological autotrophic replicators could be designed by humans, and could easily accept specifications for human products.

- Self-reproductive systems are conjectured systems which would produce copies of themselves from industrial feedstocks such as metal bar and wire.

- Self-assembling systems assemble copies of themselves from finished, delivered parts. Simple examples of such systems have been demonstrated at the macro scale.

The design space for machine replicators is very broad. A comprehensive study[4] to date by Robert Freitas and Ralph Merkle has identified 137 design dimensions grouped into a dozen separate categories, including: (1) Replication Control, (2) Replication Information, (3) Replication Substrate, (4) Replicator Structure, (5) Passive Parts, (6) Active Subunits, (7) Replicator Energetics, (8) Replicator Kinematics, (9) Replication Process, (10) Replicator Performance, (11) Product Structure, and (12) Evolvability.

46.1.3 A self-replicating computer program

Main article: Quine (computing)

In computer science a quine is a self-reproducing computer program that, when executed, outputs its own code. For example, a quine in the Python programming language is:

```
a='a=%r;print a%%a';print a%a
```

A more trivial approach is to write a program that will make a copy of any stream of data that it is directed to, and then direct it at itself. In this case the program is treated as both executable code, and as data to be manipulated. This approach is common in most self-replicating systems, including biological life, and is simpler as it does not require the program to contain a complete description of itself.

In many programming languages an empty program is legal, and executes without producing errors or other output. The output is thus the same as the source code, so the program is trivially self-reproducing.

46.1.4 Self-replicating tiling

See also: Self-similarity

In geometry a self-replicating tiling is a tiling pattern in which several congruent tiles may be joined together to form a larger tile that is similar to the original. This is an aspect of the field of study known as tessellation. The "sphinx" hexiamond is the only known self-replicating pentagon.[5] For example, four such concave pentagons can be joined together to make one with twice the dimensions.[6] Solomon W. Golomb coined the term reptiles for self-replicating tilings.

In 2012, Lee Sallows identified rep-tiles as a special instance of a self-tiling tile set or setiset. A setiset of order n is a set of n shapes that can be assembled in n different ways so as to form larger replicas of themselves. Setisets in which every shape is distinct are called 'perfect'. A rep-n rep-tile is just a setiset composed of n identical pieces.

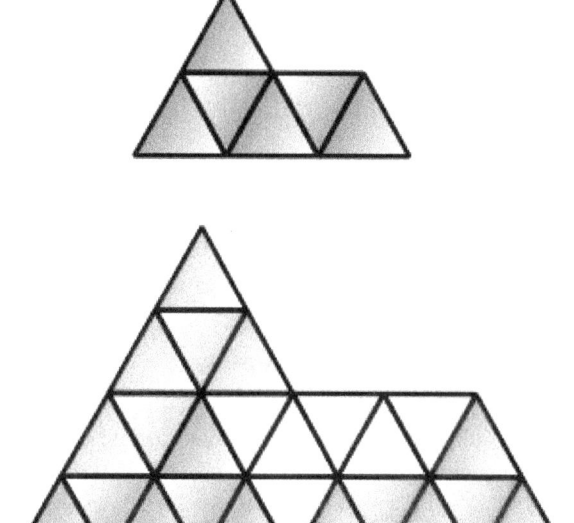

Four 'sphinx' hexiamonds can be put together to form another sphinx.

46.1.5 Applications

It is a long-term goal of some engineering sciences to achieve a clanking replicator, a material device that can self-replicate. The usual reason is to achieve a low cost per item while retaining the utility of a manufactured good. Many authorities say that in the limit, the cost of self-replicating items should approach the cost-per-weight of wood or other biological substances, because self-replication avoids the costs of labor, capital and distribution in conventional manufactured goods.

A fully novel artificial replicator is a reasonable near-term goal. A NASA study recently placed the complexity of a clanking replicator at approximately that of Intel's Pentium

A perfect setiset of order 4

4 CPU.[7] That is, the technology is achievable with a relatively small engineering group in a reasonable commercial time-scale at a reasonable cost.

Given the currently keen interest in biotechnology and the high levels of funding in that field, attempts to exploit the replicative ability of existing cells are timely, and may easily lead to significant insights and advances.

A variation of self replication is of practical relevance in compiler construction, where a similar chicken and egg problem occurs as in natural self replication. A compiler (phenotype) can be applied on the compiler's own source code (genotype) producing the compiler itself. During compiler development, a modified (mutated) source is used to create the next generation of the compiler. This process differs from natural self-replication in that the process is directed by an engineer, not by the subject itself.

46.2 Mechanical self-replication

Main article: self-replicating machine

An activity in the field of robots is the self-replication of machines. Since all robots (at least in modern times) have a fair number of the same features, a self-replicating robot (or possibly a hive of robots) would need to do the following:

- Obtain construction materials
- Manufacture new parts including its smallest parts and thinking apparatus
- Provide a consistent power source
- Program the new members
- error correct any mistakes in the offspring

On a nano scale, assemblers might also be designed to self-replicate under their own power. This, in turn, has given rise to the "grey goo" version of Armageddon, as featured in such science fiction novels as *Bloom*, *Prey*, and *Recursion*.

The Foresight Institute has published guidelines for researchers in mechanical self-replication.[8] The guidelines recommend that researchers use several specific techniques for preventing mechanical replicators from getting out of control, such as using a broadcast architecture.

For a detailed article on mechanical reproduction as it relates to the industrial age see mass production.

46.3 Fields

Research has occurred in the following areas:

- Biology studies natural replication and replicators, and their interaction. These can be an important guide to avoid design difficulties in self-replicating machinery.
- In Chemistry self-replication studies are typically about how a specific set of molecules can act together to replicate each other within the set.[9]
- Memetics studies ideas and how they propagate in human culture. Memes require only small amounts of material, and therefore have theoretical similarities to viruses and are often described as viral.
- Nanotechnology or more precisely, molecular nanotechnology is concerned with making nano scale assemblers. Without self-replication, capital and assembly costs of molecular machines become impossibly large.
- Space resources: NASA has sponsored a number of design studies to develop self-replicating mechanisms to mine space resources. Most of these designs include computer-controlled machinery that copies itself.
- Computer security: Many computer security problems are caused by self-reproducing computer programs that infect computers — computer worms and computer viruses.
- In parallel computing, it takes a long time to manually load a new program on every node of a large computer

cluster or distributed computing system. Automatically loading new programs using mobile agents can save the system administrator a lot of time and give users their results much quicker, as long as they don't get out of control.

46.4 In industry

46.4.1 Space exploration and manufacturing

The goal of self-replication in space systems is to exploit large amounts of matter with a low launch mass. For example, an autotrophic self-replicating machine could cover a moon or planet with solar cells, and beam the power to the Earth using microwaves. Once in place, the same machinery that built itself could also produce raw materials or manufactured objects, including transportation systems to ship the products. Another model of self-replicating machine would copy itself through the galaxy and universe, sending information back.

In general, since these systems are autotrophic, they are the most difficult and complex known replicators. They are also thought to be the most hazardous, because they do not require any inputs from human beings in order to reproduce.

A classic theoretical study of replicators in space is the 1980 NASA study of autotrophic clanking replicators, edited by Robert Freitas.[10]

Much of the design study was concerned with a simple, flexible chemical system for processing lunar regolith, and the differences between the ratio of elements needed by the replicator, and the ratios available in regolith. The limiting element was Chlorine, an essential element to process regolith for Aluminium. Chlorine is very rare in lunar regolith, and a substantially faster rate of reproduction could be assured by importing modest amounts.

The reference design specified small computer-controlled electric carts running on rails. Each cart could have a simple hand or a small bull-dozer shovel, forming a basic robot.

Power would be provided by a "canopy" of solar cells supported on pillars. The other machinery could run under the canopy.

A "casting robot" would use a robotic arm with a few sculpting tools to make plaster molds. Plaster molds are easy to make, and make precise parts with good surface finishes. The robot would then cast most of the parts either from non-conductive molten rock (basalt) or purified metals. An electric oven melted the materials.

A speculative, more complex "chip factory" was specified to produce the computer and electronic systems, but the designers also said that it might prove practical to ship the chips from Earth as if they were "vitamins".

46.4.2 Molecular manufacturing

Main article: molecular nanotechnology § Replicating nanorobots

Nanotechnologists in particular believe that their work will likely fail to reach a state of maturity until human beings design a self-replicating assembler of nanometer dimensions.

These systems are substantially simpler than autotrophic systems, because they are provided with purified feedstocks and energy. They do not have to reproduce them. This distinction is at the root of some of the controversy about whether molecular manufacturing is possible or not. Many authorities who find it impossible are clearly citing sources for complex autotrophic self-replicating systems. Many of the authorities who find it possible are clearly citing sources for much simpler self-assembling systems, which have been demonstrated. In the meantime, a Lego-built autonomous robot able to follow a pre-set track and assemble an exact copy of itself, starting from four externally provided components, was demonstrated experimentally in 2003.

Merely exploiting the replicative abilities of existing cells is insufficient, because of limitations in the process of protein biosynthesis (also see the listing for RNA). What is required is the rational design of an entirely novel replicator with a much wider range of synthesis capabilities.

In 2011, New York University scientists have developed artificial structures that can self-replicate, a process that has the potential to yield new types of materials. They have demonstrated that it is possible to replicate not just molecules like cellular DNA or RNA, but discrete structures that could in principle assume many different shapes, have many different functional features, and be associated with many different types of chemical species.[11][12]

For a discussion of other chemical bases for hypothetical self-replicating systems, see alternative biochemistry.

46.5 See also

- Artificial life
- Astrochicken
- Autopoiesis
- Complex system

- DNA replication
- Life
- Robot
- RepRap
- Self-replicating machine
  - self-replicating spacecraft
- Space manufacturing
- Von Neumann universal constructor
- Virus
- Von Neumann machine (disambiguation)
- Self reconfigurable
- Final Anthropic Principle
- Positive feedback
- Harmonic

46.6 References

[1] "'Lifeless' prion proteins are 'capable of evolution'". BBC News. 2010-01-01. Retrieved 2013-10-22.

[2] von Neumann, John (1948). *The Hixon Symposium*. Pasadena, California. pp. 1–36.

[3] Freitas, Robert; Merkle, Ralph (2004). "Kinematic Self-Replicating Machines - General Taxonomy of Replicators". Retrieved 29 June 2013.

[4] Freitas, Robert; Merkle, Ralph (2004). "Kinematic Self-Replicating Machines - Freitas-Merkle Map of the Kinematic Replicator Design Space (2003–2004)". Retrieved 29 June 2013.

[5] For an image that does not show how this replicates, see: Eric W. Weisstein. "Sphinx." From MathWorld--A Wolfram Web Resource. http://mathworld.wolfram.com/Sphinx.html

[6] For further illustrations, see Teaching TILINGS / TESSELLATIONS with Geo Sphinx

[7] "Modeling Kinematic Cellular Automata Final Report" (PDF). April 30, 2004. Retrieved 2013-10-22.

[8] "Molecular Nanotechnology Guidelines". Foresight.org. Retrieved 2013-10-22.

[9] Moulin, Giuseppone (2011). "Dynamic Combinatorial Self-Replicating Systems". Springer. doi:10.1007/128_2011_198.

[10] Wikisource:Advanced Automation for Space Missions

[11] Wang, Tong; Sha, Ruojie; Dreyfus, Rémi; Leunissen, Mirjam E.; Maass, Corinna; Pine, David J.; Chaikin, Paul M.; Seeman, Nadrian C. (2011). "Self-replication of information-bearing nanoscale patterns". *Nature*. **478** (7368): 225–228. PMC 3192504. PMID 21993758. doi:10.1038/nature10500.

[12] "Self-replication process holds promise for production of new materials.". Science Daily. 17 October 2011. Retrieved 17 October 2011.

Notes

- von Neumann, J., 1966, *The Theory of Self-reproducing Automata*, A. Burks, ed., Univ. of Illinois Press, Urbana, IL.

- Advanced Automation for Space Missions, a 1980 NASA study edited by Robert Freitas

- Kinematic Self-Replicating Machines first comprehensive survey of entire field in 2004 by Robert Freitas and Ralph Merkle

- NASA Institute for Advance Concepts study by General Dynamics- concluded that complexity of the development was equal to that of a Pentium 4, and promoted a design based on cellular automata.

- *Gödel, Escher, Bach* by Douglas Hofstadter (detailed discussion and many examples)

- Kenyon, R., *Self-replicating tilings*, in: Symbolic Dynamics and Applications (P. Walters, ed.) Contemporary Math. vol. 135 (1992), 239-264.

- http://www.moshesipper.com/the-artificial-self-replication-page.html The Artificial Self-Replication Page

Chapter 47

Geomorphology

For the scientific journal, see Geomorphology (journal).
Geomorphology (from Ancient Greek: γῆ, gê, "earth";

Badlands incised into shale at the foot of the North Caineville Plateau, Utah, within the pass carved by the Fremont River and known as the Blue Gate. GK Gilbert studied the landscapes of this area in great detail, forming the observational foundation for many of his studies on geomorphology.[1]

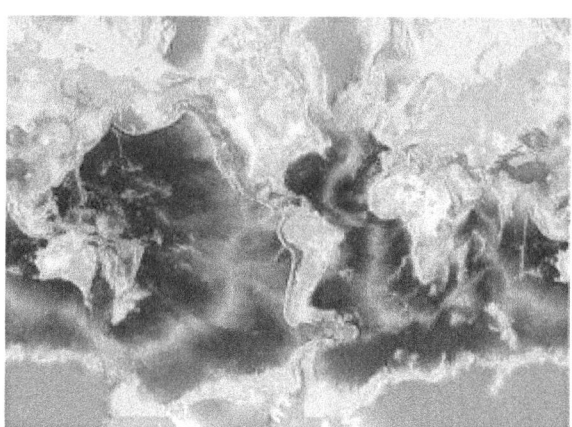

Surface of the Earth, showing higher elevations in red.

μορφή, morphḗ, "form"; and λόγος, lógos, "study") is the scientific study of the origin and evolution of topographic and bathymetric features created by physical, chemical or biological processes operating at or near the Earth's surface. Geomorphologists seek to understand why landscapes look the way they do, to understand landform history and dynamics and to predict changes through a combination of field observations, physical experiments and numerical modeling. Geomorphologists work within disciplines such as physical geography, geology, geodesy, engineering geology, archaeology and geotechnical engineering. This broad base of interests contributes to many research styles and interests within the field.

47.1 Overview

Waves and water chemistry lead to structural failure in exposed rocks

Earth's surface is modified by a combination of surface processes that sculpt landscapes, and geologic processes that cause tectonic uplift and subsidence, and shape the coastal geography. Surface processes comprise the action of water, wind, ice, fire, and living things on the surface of the Earth, along with chemical reactions that form soils and alter material properties, the stability and rate of change of topography under the force of gravity, and other factors, such as (in the very recent past) human alteration of the landscape. Many of these factors are strongly mediated by climate. Geologic processes include the uplift of mountain ranges, the growth of volcanoes, isostatic changes in land surface elevation (sometimes in response to surface processes), and the formation of deep sedimentary basins where the surface of the Earth drops and is filled with material eroded from other parts of the landscape. The Earth's surface and its topography therefore are an intersection of climatic, hydrologic, and biologic action with geologic processes, or alternatively stated, the intersection of the Earth's lithosphere with its hydrosphere, atmosphere, and biosphere.

The broad-scale topographies of the Earth illustrate this intersection of surface and subsurface action. Mountain belts are uplifted due to geologic processes. Denudation of these high uplifted regions produces sediment that is transported and deposited elsewhere within the landscape or off the coast.[2] On progressively smaller scales, similar ideas apply, where individual landforms evolve in response to the balance of additive processes (uplift and deposition) and subtractive processes (subsidence and erosion). Often, these processes directly affect each other: ice sheets, water, and sediment are all loads that change topography through flexural isostasy. Topography can modify the local climate, for example through orographic precipitation, which in turn modifies the topography by changing the hydrologic regime in which it evolves. Many geomorphologists are particularly interested in the potential for feedbacks between climate and tectonics, mediated by geomorphic processes.[3]

In addition to these broad-scale questions, geomorphologists address issues that are more specific and/or more local. Glacial geomorphologists investigate glacial deposits such as moraines, eskers, and proglacial lakes, as well as glacial erosional features, to build chronologies of both small glaciers and large ice sheets and understand their motions and effects upon the landscape. Fluvial geomorphologists focus on rivers, how they transport sediment, migrate across the landscape, cut into bedrock, respond to environmental and tectonic changes, and interact with humans. Soils geomorphologists investigate soil profiles and chemistry to learn about the history of a particular landscape and understand how climate, biota, and rock interact. Other geomorphologists study how hillslopes form and change. Still others investigate the relationships between ecology and geomorphology. Because geomorphology is defined to comprise everything related to the surface of the Earth and its modification, it is a broad field with many facets.

Geomorphologists use a wide range of techniques in their work. These may include fieldwork and field data collection, the interpretation of remotely sensed data, geochemical analyses, and the numerical modelling of the physics of landscapes. Geomorphologists may rely on geochronology, using dating methods to measure the rate of changes to the surface.[4][5] Terrain measurement techniques are vital to quantitatively describe the form of the Earth's surface, and include differential GPS, remotely sensed digital terrain models and laser scanning, to quantify, study, and to generate illustrations and maps.[6]

Practical applications of geomorphology include hazard assessment (such as landslide prediction and mitigation), river control and stream restoration, and coastal protection. Planetary geomorphology studies landforms on other terrestrial planets such as Mars. Indications of effects of wind, fluvial, glacial, mass wasting, meteor impact, tectonics and volcanic processes are studied. This effort not only helps better understand the geologic and atmospheric history of those planets but also extends geomorphological study of the Earth. Planetary geomorphologists often use Earth analogues to aid in their study of surfaces of other planets.[7]

47.2 History

"Cono de Arita" at the dry lake Salar de Arizaro on the Atacama Plateau, in northwestern Argentina. The cone itself is a volcanic edifice, representing complex interaction of intrusive igneous rocks with the surrounding salt.[8]

Other than some notable exceptions in antiquity, geomorphology is a relatively young science, growing along with interest in other aspects of the earth sciences in the mid-19th century. This section provides a very brief outline of

Lake "Veľké Hincovo pleso" in High Tatras, Slovakia. The lake occupies an "overdeepening" carved by flowing ice that once occupied this glacial valley.

some of the major figures and events in its development.

47.2.1 Ancient geomorphology

The study of landforms and the evolution of the Earth's surface can be dated back to scholars of Classical Greece. Herodotus argued from observations of soils that the Nile delta was actively growing into the Mediterranean Sea, and estimated its age.[9] Aristotle speculated that due to sediment transport into the sea, eventually those seas would fill while the land lowered. He claimed that this would mean that land and water would eventually swap places, whereupon the process would begin again in an endless cycle.[9]

Another early theory of geomorphology was devised by the polymath Chinese scientist and statesman Shen Kuo (1031–1095 AD). This was based on his observation of marine fossil shells in a geological stratum of a mountain hundreds of miles from the Pacific Ocean. Noticing bivalve shells running in a horizontal span along the cut section of a cliffside, he theorized that the cliff was once the pre-historic location of a seashore that had shifted hundreds of miles over the centuries. He inferred that the land was reshaped and formed by soil erosion of the mountains and by deposition of silt, after observing strange natural erosions of the Taihang Mountains and the Yandang Mountain near Wenzhou.[10][11] Furthermore, he promoted the theory of gradual climate change over centuries of time once ancient petrified bamboos were found to be preserved underground in the dry, northern climate zone of *Yanzhou*, which is now modern day Yan'an, Shaanxi province.[11][12]

47.2.2 Early modern geomorphology

The term geomorphology seems to have been first used by Laumann in an 1858 work written in German. Keith Tinkler has suggested that the word came into general use in English, German and French after John Wesley Powell and W. J. McGee used it during the International Geological Conference of 1891.[13] John Edward Marr in his The Scientific Study of Scenery[14] considered his book as, 'an Introductory Treatise on Geomorphology, a subject which has sprung from the union of Geology and Geography'.

An early popular geomorphic model was the *geographical cycle* or *cycle of erosion* model of broad-scale landscape evolution developed by William Morris Davis between 1884 and 1899.[9] It was an elaboration of the uniformitarianism theory that had first been proposed by James Hutton (1726–1797).[15] With regard to valley forms, for example, uniformitarianism posited a sequence in which a river runs through a flat terrain, gradually carving an increasingly deep valley, until the side valleys eventually erode, flattening the terrain again, though at a lower elevation. It was thought that tectonic uplift could then start the cycle over. In the decades following Davis's development of this idea, many of those studying geomorphology sought to fit their findings into this framework, known today as "Davisian".[15] Davis's ideas are of historical importance, but have been largely superseded today, mainly due to their lack of predictive power and qualitative nature.[15]

In the 1920s, Walther Penck developed an alternative model to Davis's.[15] Penck thought that landform evolution was better described as an alternation between ongoing processes of uplift and denudation, as opposed to Davis's model of a single uplift followed by decay.[16] He also emphasised that in many landscapes slope evolution occurs by backwearing of rocks, not by Davisian-style surface lowering, and his science tended to emphasise surface process over understanding in detail the surface history of a given locality. Penck was German, and during his lifetime his ideas were at times rejected vigorously by the English-speaking geomorphology community.[15] His early death, Davis' dislike for his work, and his at-times-confusing writing style likely all contributed to this rejection.[17]

Both Davis and Penck were trying to place the study of the evolution of the Earth's surface on a more generalized, globally relevant footing than it had been previously. In the early 19th century, authors – especially in Europe – had tended to attribute the form of landscapes to local climate, and in particular to the specific effects of glaciation and periglacial processes. In contrast, both Davis and Penck were seeking to emphasize the importance of evolution of landscapes through time and the generality of the Earth's surface processes across different landscapes under different conditions.

During the early 1900s, the study of regional-scale geomorphology was termed "physiography". Physiography later was considered to be a contraction of "*phys*ical" and "ge*ography*", and therefore synonymous with physical geography, and the concept became embroiled in controversy surrounding the appropriate concerns of that discipline. Some geomorphologists held to a geological basis for physiography and emphasized a concept of physiographic regions while a conflicting trend among geographers was to equate physiography with "pure morphology", separated from its geological heritage. In the period following World War II, the emergence of process, climatic, and quantitative studies led to a preference by many earth scientists for the term "geomorphology" in order to suggest an analytical approach to landscapes rather than a descriptive one.[18]

47.2.3 Climatic geomorphology

Further information: Climatic geomorphology

During the age of New Imperialism in the late 19th century European explorers and scientists traveled across the globe bringing descriptions of landscapes and landforms. As geographical knowledge increased over time these observations were systematized in a search for regional patterns. Climate emerged thus as prime factor for explaining landform distribution at a grand scale. The emergence of climatic geomorphology was foreshadowed by the work of Wladimir Köppen, Vasily Dokuchaev and Andreas Schimper. William Morris Davis, the leading geomorphologist of his time, recognized the role of climate by complementing his "normal" temperate climate cycle of erosion with arid and glacial ones.[19][20] Nevertheless, interest in climatic geomorphology was also a reaction *against* Davisian geomorphology that was by the mid-20th century considered both un-innovative and dubious.[20][21] Early climatic geomorphology developed primarily in continental Europe while in the English-speaking world the tendency was not explicit until L.C. Peltier's 1950 publication on a periglacial cycle of erosion.[19]

Climatic geomorphology was criticized in a 1969 review article by process geomorphologist D.R. Stoddart.[20][22] The criticism by Stoddart proved "devastating" sparking a decline in the popularity of climatic geomorphology in the late 20th century.[20][22] Stoddart criticized climatic geomorphology for applying supposedly "trivial" methodologies in establishing landform differences between morphoclimatic zones, being linked to Davisian geomorphology and by allegedly neglecting the fact that physical laws governing processes are the same across the globe.[22] In addition some conceptions of climatic geomorphology, like that which holds that chemical weathering is more rapid in tropical climates than in cold climates proved to not be straightforwardly true.[20]

47.2.4 Quantitative and process geomorphology

Part of the Great Escarpment in the Drakensberg, southern Africa. This landscape, with its high altitude plateau being incised into by the steep slopes of the escarpment, was cited by Davis as a classic example of his cycle of erosion.[23]

Geomorphology was started to be put on a solid quantitative footing in the middle of the 20th century. Following the early work of Grove Karl Gilbert around the turn of the 20th century,[9][15][16] a group of mainly American natural scientists, geologists and hydraulic engineers including William Walden Rubey, Ralph Alger Bagnold, Hans Albert Einstein, Frank Ahnert, John Hack, Luna Leopold, A. Shields, Thomas Maddock, Arthur Strahler, Stanley Schumm, and Ronald Shreve began to research the form of landscape elements such as rivers and hillslopes by taking systematic, direct, quantitative measurements of aspects of them and investigating the scaling of these measurements.[9][15][16][24]. These methods began to allow prediction of the past and future behavior of landscapes from present observations, and were later to develop into the modern trend of a highly quantitative approach to geomorphic problems. Many groundbreaking and widely cited early geomorphology studies appeared in the Bulletin of the Geological Society of America,[25] and received only few citations prior to 2000 (they are examples of "sleeping beauties")[26] when a marked increase in quantitative geomorphology research occured.[27]

Quantitative geomorphology can involve fluid dynamics and solid mechanics, geomorphometry, laboratory studies, field measurements, theoretical work, and full landscape evolution modeling. These approaches are used to understand weathering and the formation of soils, sediment transport, landscape change, and the interactions between climate, tectonics, erosion, and deposition.

In Sweden Filip Hjulström's doctoral thesis, "The River Fyris" (1935), contained one of the first quantitative studies of geomorphological processes ever published. His students followed in the same vein, making quantitative studies of mass transport (Anders Rapp), fluvial transport (Åke Sundborg), delta deposition (Valter Axelsson), and coastal processes (John O. Norrman). This developed into "the Uppsala School of Physical Geography".[28]

47.2.5 Contemporary geomorphology

Today, the field of geomorphology encompasses a very wide range of different approaches and interests.[9] Modern researchers aim to draw out quantitative "laws" that govern Earth surface processes, but equally, recognize the uniqueness of each landscape and environment in which these processes operate. Particularly important realizations in contemporary geomorphology include:

1) that not all landscapes can be considered as either "stable" or "perturbed", where this perturbed state is a temporary displacement away from some ideal target form. Instead, dynamic changes of the landscape are now seen as an essential part of their nature.[29][30]

2) that many geomorphic systems are best understood in terms of the stochasticity of the processes occurring in them, that is, the probability distributions of event magnitudes and return times.[31][32] This in turn has indicated the importance of chaotic determinism to landscapes, and that landscape properties are best considered statistically.[33] The same processes in the same landscapes do not always lead to the same end results.

Albeit having its importance diminished climatic geomorphology continues to exist as field of study producing relevant research. More recently concerns over global warming have led to a renewed interest in the field.[20]

Despite considerable criticism the cycle of erosion model has remained part of the science of geomorphology.[34] The model or theory has never been proved wrong,[34] but neither has it been proven.[35] The inherent difficulties of the model have instead made geomorphological research to advance along other lines.[34] In contrast to its disputed status in geomorphology, the cycle of erosion model is a common approach used to establish denudation chronologies, and is thus an important concept in the science of historical geology.[36] While acknowledging its shortcomings modern geomorphologists Andrew Goudie and Karna Lidmar-Bergström have praised it for its elegance and pedagogical value respectively.[37][38]

47.3 Processes

Gorge cut by the Indus river into bedrock, Nanga Parbat region, Pakistan. This is the deepest river canyon in the world. Nanga Parbat itself, the world's 9th highest mountain, is seen in the background.

Geomorphically relevant processes generally fall into (1) the production of regolith by weathering and erosion, (2) the transport of that material, and (3) its eventual deposition. Primary surface processes responsible for most topographic features include wind, waves, chemical dissolution, mass wasting, groundwater movement, surface water flow, glacial action, tectonism, and volcanism. Other more exotic geomorphic processes might include periglacial (freeze-thaw) processes, salt-mediated action, marine currents activity, seepage of fluids through the seafloor or extraterrestrial impact.

47.3.1 Aeolian processes

Wind-eroded alcove near Moab, Utah

Aeolian processes pertain to the activity of the winds and

more specifically, to the winds' ability to shape the surface of the Earth. Winds may erode, transport, and deposit materials, and are effective agents in regions with sparse vegetation and a large supply of fine, unconsolidated sediments. Although water and mass flow tend to mobilize more material than wind in most environments, aeolian processes are important in arid environments such as deserts.[39]

47.3.2 Biological processes

Beaver dams, as this one in Tierra del Fuego, constitute a specific form of zoogeomorphology, a type of biogeomorphology.

The interaction of living organisms with landforms, or biogeomorphologic processes, can be of many different forms, and is probably of profound importance for the terrestrial geomorphic system as a whole. Biology can influence very many geomorphic processes, ranging from biogeochemical processes controlling chemical weathering, to the influence of mechanical processes like burrowing and tree throw on soil development, to even controlling global erosion rates through modulation of climate through carbon dioxide balance. Terrestrial landscapes in which the role of biology in mediating surface processes can be definitively excluded are extremely rare, but may hold important information for understanding the geomorphology of other planets, such as Mars.[40]

47.3.3 Fluvial processes

Main article: Fluvial
See also: Hack's law and Sediment transport

Rivers and streams are not only conduits of water, but also of sediment. The water, as it flows over the channel bed, is able to mobilize sediment and transport it downstream, either as bed load, suspended load or dissolved load. The rate of sediment transport depends on the availability of sediment itself and on the river's discharge.[41] Rivers are also capable of eroding into rock and creating new sediment, both from their own beds and also by coupling to the surrounding hillslopes. In this way, rivers are thought of as setting the base level for large-scale landscape evolution in nonglacial environments.[42][43] Rivers are key links in the connectivity of different landscape elements.

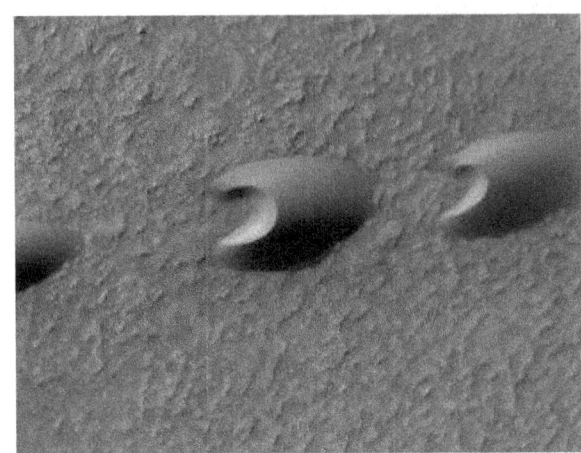

Seif and barchan dunes in the Hellespontus region on the surface of Mars. Dunes are mobile landforms created by the transport of large volumes of sand by wind.

As rivers flow across the landscape, they generally increase in size, merging with other rivers. The network of rivers thus formed is a drainage system. These systems take on four general patterns: dendritic, radial, rectangular, and trellis. Dendritic happens to be the most common, occurring when the underlying stratum is stable (without faulting). Drainage systems have four primary components: drainage basin, alluvial valley, delta plain, and receiving basin. Some geomorphic examples of fluvial landforms are alluvial fans, oxbow lakes, and fluvial terraces.

47.3.4 Glacial processes

Glaciers, while geographically restricted, are effective agents of landscape change. The gradual movement of ice down a valley causes abrasion and plucking of the underlying rock. Abrasion produces fine sediment, termed glacial flour. The debris transported by the glacier, when the glacier recedes, is termed a moraine. Glacial erosion is responsible for U-shaped valleys, as opposed to the V-shaped valleys of fluvial origin.[44]

The way glacial processes interact with other landscape elements, particularly hillslope and fluvial processes, is an important aspect of Plio-Pleistocene landscape evolution

47.3. PROCESSES

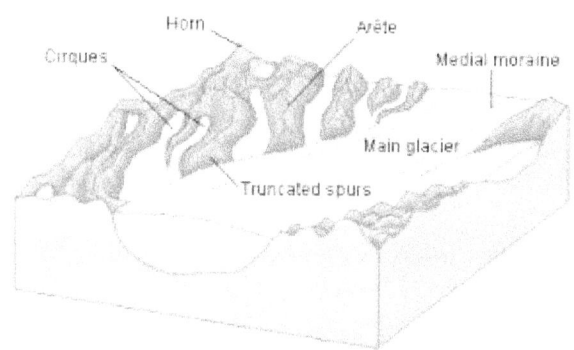

Features of a glacial landscape

and its sedimentary record in many high mountain environments. Environments that have been relatively recently glaciated but are no longer may still show elevated landscape change rates compared to those that have never been glaciated. Nonglacial geomorphic processes which nevertheless have been conditioned by past glaciation are termed paraglacial processes. This concept contrasts with periglacial processes, which are directly driven by formation or melting of ice or frost.[45]

47.3.5 Hillslope processes

Talus cones on the north shore of Isfjorden, Svalbard, Norway. Talus cones are accumulations of coarse hillslope debris at the foot of the slopes producing the material.

Soil, regolith, and rock move downslope under the force of gravity via creep, slides, flows, topples, and falls. Such mass wasting occurs on both terrestrial and submarine slopes, and has been observed on Earth, Mars, Venus, Titan and Iapetus.

Ongoing hillslope processes can change the topology of the hillslope surface, which in turn can change the rates of those processes. Hillslopes that steepen up to certain critical thresholds are capable of shedding extremely large volumes of material very quickly, making hillslope processes an extremely important element of landscapes in tectonically active areas.[46]

The Ferguson Slide is an active landslide in the Merced River canyon on California State Highway 140, a primary access road to Yosemite National Park.

On the Earth, biological processes such as burrowing or tree throw may play important roles in setting the rates of some hillslope processes.[47]

47.3.6 Igneous processes

Both volcanic (eruptive) and plutonic (intrusive) igneous processes can have important impacts on geomorphology. The action of volcanoes tends to rejuvenate landscapes, covering the old land surface with lava and tephra, releasing pyroclastic material and forcing rivers through new paths. The cones built by eruptions also build substantial new topography, which can be acted upon by other surface processes. Plutonic rocks intruding then solidifying at depth can cause both uplift or subsidence of the surface, depending on whether the new material is denser or less dense than the rock it displaces.

47.3.7 Tectonic processes

See also: Erosion and tectonics

Tectonic effects on geomorphology can range from scales of millions of years to minutes or less. The effects of tectonics on landscape are heavily dependent on the nature of the underlying bedrock fabric that more or less controls what kind of local morphology tectonics can shape. Earthquakes can, in terms of minutes, submerge large areas of land creating new wetlands. Isostatic rebound can account for significant changes over hundreds to thousands of years, and allows erosion of a mountain belt to promote further erosion as mass is removed from the chain and the belt uplifts. Long-term plate tectonic dynamics give rise to orogenic belts, large mountain chains with typical lifetimes of many tens of millions of years, which form focal points for high rates of fluvial and hillslope processes and thus long-term sediment production.

Features of deeper mantle dynamics such as plumes and delamination of the lower lithosphere have also been hypothesised to play important roles in the long term (> million year), large scale (thousands of km) evolution of the Earth's topography (see dynamic topography). Both can promote surface uplift through isostasy as hotter, less dense, mantle rocks displace cooler, denser, mantle rocks at depth in the Earth.[48][49]

47.3.8 Marine processes

Marine processes are those associated with the action of waves, marine currents and seepage of fluids through the seafloor. Mass wasting and submarine landsliding are also important processes for some aspects of marine geomorphology.[50] Because ocean basins are the ultimate sinks for a large fraction of terrestrial sediments, depositional processes and their related forms (e.g., sediment fans, deltas) are particularly important as elements of marine geomorphology.

47.4 Scales in geomorphology

Different geomorphological processes dominate at different spatial and temporal scales. Moreover, scales on which processes occur may determine the reactivity or otherwise of landscapes to changes in driving forces such as climate or tectonics.[30] These ideas are key to the study of geomorphology today.

To help categorize landscape scales some geomorphologists might use the following taxonomy:

- 1st – Continent, ocean basin, climatic zone (~10,000,000 km^2)
- 2nd – Shield, e.g. Baltic Shield, or mountain range (~1,000,000 km^2)
- 3rd – Isolated sea, Sahel (~100,000 km^2)
- 4th – Massif, e.g. Massif Central or Group of related landforms, e.g., Weald (~10,000 km^2)
- 5th – River valley, Cotswolds (~1,000 km^2)
- 6th – Individual mountain or volcano, small valleys (~100 km^2)
- 7th – Hillslopes, stream channels, estuary (~10 km^2)
- 8th – gully, barchannel (~1 km^2)
- 9th – Meter-sized features

47.5 Overlap with other fields

There is a considerable overlap between geomorphology and other fields. Deposition of material is extremely important in sedimentology. Weathering is the chemical and physical disruption of earth materials in place on exposure to atmospheric or near surface agents, and is typically studied by soil scientists and environmental chemists, but is an essential component of geomorphology because it is what provides the material that can be moved in the first place. Civil and environmental engineers are concerned with erosion and sediment transport, especially related to canals, slope stability (and natural hazards), water quality, coastal environmental management, transport of contaminants, and stream restoration. Glaciers can cause extensive erosion and deposition in a short period of time, making them extremely important entities in the high latitudes and meaning that they set the conditions in the headwaters of mountain-born streams; glaciology therefore is important in geomorphology.

47.6 See also

- Bioerosion
- Biogeology
- Biogeomorphology
- Biorhexistasy
- British Society for Geomorphology
- Coastal biogeomorphology
- Coastal erosion
- Drainage system (geomorphology)
- Erosion
- Erosion prediction
- Geologic modelling
- Geomorphometry
- Geotechnics
- Hack's law
- Hydrologic modeling, behavioral modeling in hydrology

- Orogeny
- Physiographic regions of the world
- Sediment transport
- Soil morphology
- Soils retrogression and degradation
- Stream capture
- Thermochronology
- Weathering
- List of important publications in geology

47.7 References

[1] Gilbert, Grove Karl, and Charles Butler Hunt, eds. Geology of the Henry Mountains, Utah, as recorded in the notebooks of GK Gilbert, 1875–76. Vol. 167. Geological Society of America, 1988.

[2] Willett, Sean D.; Brandon, Mark T. (January 2002). "On steady states in mountain belts". *Geology*. **30** (2): 175–178. Bibcode:2002Geo....30..175W. doi:10.1130/0091-7613(2002)030<0175:OSSIMB>2.0.CO;2.

[3] Roe, Gerard H.; Whipple, Kelin X.; Fletcher, Jennifer K. (September 2008). "Feedbacks among climate, erosion, and tectonics in a critical wedge orogen". *American Journal of Science*. **308** (7): 815–842. doi:10.2475/07.2008.01.

[4] Summerfield, M.A., 1991, Global Geomorphology, Pearson Education Ltd, 537 p. ISBN 0-582-30156-4.

[5] Dunai, T.J., 2010, Cosmogenic Nucleides, Cambridge University Press, 187 p. ISBN 978-0-521-87380-2.

[6] e.g., DTM intro page, Hunter College Department of Geography, New York NY.

[7] "International Conference of Geomorphology". Europa Organization.

[8] http://www.amusingplanet.com/2014/07/cono-de-arita-in-argentina.html

[9] Bierman, Paul R., and David R. Montgomery. Key concepts in geomorphology. Macmillan Higher Education, 2014.

[10] Sivin, Nathan (1995). *Science in Ancient China: Researches and Reflections*. Brookfield, Vermont: VARIORUM, Ashgate Publishing. III, p. 23

[11] Needham, Joseph. (1959). *Science and Civilization in China: Volume 3, Mathematics and the Sciences of the Heavens and the Earth*. Cambridge University Press. pp. 603–618.

[12] Chan, Alan Kam-leung and Gregory K. Clancey, Hui-Chieh Loy (2002). *Historical Perspectives on East Asian Science, Technology and Medicine*. Singapore: Singapore University Press. p. 15. ISBN 9971-69-259-7.

[13] Tinkler, Keith J. A short history of geomorphology. Page 4. 1985

[14] Marr, J.E. The Scientific Study of Scenery. Methuen, page iii, 1900.

[15] Oldroyd, David R. & Grapes, Rodney H. Contributions to the history of geomorphology and Quaternary geology: an introduction. In: GRAPES, R. H., OLDROYD, D. & GRIGELIS, A. (eds) History of Geomorphology and Quaternary Geology. Geological Society, London, Special Publications, 301, 1–17.

[16] Ritter, Dale F., R. Craig Kochel, and Jerry R. Miller. Process geomorphology. Boston: McGraw-Hill, 1995.

[17] Simons, Martin (1962). "The morphological analysis of landforms: A new review of the work of Walther Penck (1888–1923)", Transactions and Papers (Institute of British Geographers) 31: 1–14.

[18] Baker, Victor R. (1986). "Geomorphology From Space: A Global Overview of Regional Landforms, Introduction". NASA. Retrieved 2007-12-19.

[19] Twidale, C.R.; Lageat, Y. (1994). "Climatic geomorphology: a critique". *Progress in Physical Geography*. **18** (3): 319–334.

[20] Goudie, A.S. (2004). "Climatic geomorphology". In Goudie, A.S. *Encyclopedia of Geomorphology*. pp. 162–164.

[21] Flemal, Ronald C. (1971). "The Attack on the Davisian System Of Geomorphology: A Synopsis": 3–13.

[22] Thomas, Michael F. (2004). "Tropical geomorphology". In Goudie, A.S. *Encyclopedia of Geomorphology*. pp. 1063–1069.

[23] Burke, Kevin, and Yanni Gunnell. "The African erosion surface: a continental-scale synthesis of geomorphology, tectonics, and environmental change over the past 180 million years." Geological Society of America Memoirs 201 (2008): 1–66.

[24] ftp://rock.geosociety.org/pub/Memorials/v41/Schumm-S.pdf

[25] MORISAWA, MARIE (1988-07-01). "The Geological Society of America Bulletin and the development of quantitative geomorphology". *GSA Bulletin*. **100** (7). ISSN 0016-7606. doi:10.1130/0016-7606(1988)100<1016:TGSOAB>2.3.CO;2.

[26] Goldstein, Evan B (2017-04-17). "Delayed recognition of geomorphology papers in the Geological Society of America Bulletin". *Progress in Physical Geography*. **41** (3): 363–368. doi:10.1177/0309133317703093.

[27] Church, Michael (2010-06-01). "The trajectory of geomorphology". *Progress in Physical Geography*. **34** (3): 265–286. ISSN 0309-1333. doi:10.1177/0309133310363992.

[28] Gregory, KJ, 1985: "The Nature of Physical Geography", E. Arnold

[29] Whipple, Kelin X. (19 May 2004). "Bedrock Rivers and the Geomorphology of Active Orogens". *Annual Review of Earth and Planetary Sciences*. **32** (1): 151–185. Bibcode:2004AREPS..32..151W. doi:10.1146/annurev.earth.32.101802.120356.

[30] Allen, Philip A. (2008). "Time scales of tectonic landscapes and their sediment routing systems". *Geological Society, London, Special Publications*. **296**: 7–28. Bibcode:2008GSLSP.296....7A. doi:10.1144/SP296.2.

[31] Benda, Lee; Dunne, Thomas (December 1997). "Stochastic forcing of sediment supply to channel networks from landsliding and debris flow". *Water Resources Research*. **33** (12): 2849–2863. Bibcode:1997WRR....33.2849B. doi:10.1029/97WR02388.

[32] Knighton, David. Fluvial forms and processes: a new perspective. Routledge, 2014.

[33] Dietrich, W. E.; Bellugi, D.G.; Sklar, L.S.; Stock, J.D.; Heimsath, A.M.; Roering, J.J. (2003). "Geomorphic Transport Laws for Predicting Landscape Form and Dynamics" (PDF). *Prediction in Geomorphology, Geophysical Monograph Series*. Washington, D. C. **135**: 103–132. Bibcode:2003GMS...135..103D. doi:10.1029/135GM09.

[34] Slaymaker, Olav (2004). "Geomorphic evolution". In Goudie, A.S. *Encyclopedia of Geomorphology*. pp. 420–422.

[35] Roy, Andre. *Contemporary Meanings in Physical Geography: From What to Why?*. p. 5.

[36] Jones, David K.C. (2004). "Denudation chronology". In Goudie, A.S. *Encyclopedia of Geomorphology*. pp. 244–248.

[37] Lidmar-Bergström, Karna. "erosionscykel". *Nationalencyklopedin* (in Swedish). Cydonia Development. Retrieved June 22, 2016.

[38] Goudie, A.S. (2004). "Cycle of erosion". In Goudie, A.S. *Encyclopedia of Geomorphology*. pp. 223–224.

[39] Leeder, M., 1999, Sedimentology and Sedimentary Basins, From Turbulence to Tectonics, Blackwell Science, 592 p. ISBN 0-632-04976-6.

[40] Dietrich, William E.; Perron, J. Taylor (26 January 2006). "The search for a topographic signature of life". *Nature*. **439** (7075): 411–418. Bibcode:2006Natur.439..411D. PMID 16437104. doi:10.1038/nature04452.

[41] Knighton, D., 1998, *Fluvial Forms & Processes*, Hodder Arnold, 383 p. ISBN 0-340-66313-8.

[42] Strahler, A. N. (1 November 1950). "Equilibrium theory of erosional slopes approached by frequency distribution analysis; Part II". *American Journal of Science*. **248** (11): 800–814. doi:10.2475/ajs.248.11.800.

[43] Burbank, D. W. (February 2002). "Rates of erosion and their implications for exhumation" (PDF). *Mineralogical Magazine*. **66** (1): 25–52. doi:10.1180/0026461026610014.

[44] Bennett, M.R. & Glasser, N.F., 1996, Glacial Geology: Ice Sheets and Landforms, John Wiley & Sons Ltd, 364 p. ISBN 0-471-96345-3.

[45] Church, Michael; Ryder, June M. (October 1972). "Paraglacial Sedimentation: A Consideration of Fluvial Processes Conditioned by Glaciation". *Geological Society of America Bulletin*. **83** (10): 3059–3072. Bibcode:1972GSAB...83.3059C. doi:10.1130/0016-7606(1972)83[3059:PSACOF]2.0.CO;2.

[46] Roering, Joshua J.; Kirchner, James W.; Dietrich, William E. (March 1999). "Evidence for nonlinear, diffusive sediment transport on hillslopes and implications for landscape morphology" (PDF). *Water Resources Research*. **35** (3): 853–870. Bibcode:1999WRR....35..853R. doi:10.1029/1998WR900090.

[47] Gabet, Emmanuel J.; Reichman, O.J.; Seabloom, Eric W. (May 2003). "The Effects of Bioturbation on Soil Processes and Sediment Transport". *Annual Review of Earth and Planetary Sciences*. **31** (1): 249–273. Bibcode:2003AREPS..31..249G. doi:10.1146/annurev.earth.31.100901.141314.

[48] Cserepes, L.; Christensen, U.R.; Ribe, N.M. (15 May 2000). "Geoid height versus topography for a plume model of the Hawaiian swell". *Earth and Planetary Science Letters*. **178** (1–2): 29–38. Bibcode:2000E&PSL.178...29C. doi:10.1016/S0012-821X(00)00065-0.

[49] Seber, Dogan; Barazangi, Muawia; Ibenbrahim, Aomar; Demnati, Ahmed (29 February 1996). "Geophysical evidence for lithospheric delamination beneath the Alboran Sea and Rif–Betic mountains". *Nature*. **379** (6568): 785–790. Bibcode:1996Natur.379..785S. doi:10.1038/379785a0.

[50] Guilcher, A., 1958. Coastal and submarine morphology. Methuen.

47.8 Further reading

- Chorley, Richard J.; Stanley Alfred Schumm; David E. Sugden (1985). *Geomorphology*. London: Methuen. ISBN 0-416-32590-4.

- Committee on Challenges and Opportunities in Earth Surface Processes, National Research Council (2010).

Landscapes on the Edge: New Horizons for Research on Earth's Surface. Washington, DC: National Academies Press. ISBN 0-309-14024-2.

- Edmaier, Bernhard (2004). *Earthsong*. London: Phaidon Press. ISBN 0-7148-4451-9.

- Ialenti, Vincent. "Envisioning Landscapes of Our Very Distant Future" NPR Cosmos & Culture. 9/2014.

- Kondolf, G. Mathias; Hervé Piégay (2003). *Tools in fluvial geomorphology*. New York: Wiley. ISBN 0-471-49142-X.

- Kuenzer, Claudia; Stracher, Glenn B. (2012). "Geomorphology of Coal Seam Fires". *Geomorphology*. **138** (1): 209–222. Bibcode:2012Geomo.138..209K. doi:10.1016/j.geomorph.2011.09.004.

- Needham, Joseph (1954). *Science and civilisation in China*. Cambridge, UK: Cambridge University Press. ISBN 0-521-05801-5.

- Scheidegger, Adrian E. (2004). *Morphotectonics*. Berlin: Springer. ISBN 3-540-20017-7.

- Selby, Michael John (1985). *Earth's changing surface: an introduction to geomorphology*. Oxford: Clarendon Press. ISBN 0-19-823252-7.

- Charlton, Ro (2008). *Fundamentals of fluvial geomorphology*. London, UK: Rutledge. ISBN 978-0-415-33454-9.

47.9 External links

- The Geographical Cycle, or the Cycle of Erosion (1899)

- Geomorphology from Space (NASA)

- British Society for Geomorphology

47.10 Text and image sources, contributors, and licenses

47.10.1 Text

- **Complex system** *Source:* https://en.wikipedia.org/wiki/Complex_system?oldid=801921402 *Contributors:* Bryan Derksen, AdamRetchless, Lexor, Kku, Karada, Ronz, Mkoval, Kimiko, RodC, Tpbradbury, Tschild, Noeckel, Robbot, RedWolf, Chopchopwhitey, Centrx, Karol Langner, Jmeppley, EL Apro, Guppyfinsoup, Chris Howard, Jwdietrich2, &Delta, FT2, Pjacobi, CanisRufus, Truthflux, La goutte de pluie, Mdd, Passw0rd, Hu, Danthemankhan, Acadac, Versageek, Ott, Natalya, Sengkang, Jshadias, Rjwilmsi, KYPark, Salix alba, The wub, FlaBot, JFromm, Miketam, Diza, Vonkje, Wavelength, RussBot, Fmrafka-enwiki, Duracell-enwiki, Thsgrn, Jpbowen, Yonidebest, Zzuuzz, Msuzen, Arthur Rubin, Adastra-enwiki, Luk, Palapa, SmackBot, Supermanchander, Nbrown@unicistinstitute.org, Took, Cazort, Sectryan, Commander Keane bot, Portillo, Betacommand, Adam M. Gadomski, Neo-Jay, Frap, Xyzzyplugh, Kevinbrowning, Choesarian, MisterCharlie, Jon Awbrey, Betamod, Dankonikolic, Christopher Agnew, Sina2, MagnaMopus, Sir Nicholas de Mimsy-Porpington, Filippowiki, Jrouquie, Hvgard, Megane-enwiki, Spook`, Ace Frahm, Magntuve, Papertiger, Rhetth, Tawkerbot2, Mr3641, CmdrObot, Amalas, N2e, Pfhenshaw, McVities, Ballista, MaxEnt, Cydebot, Peterdjones, Skittleys, Jrgetsin, Letranova, Headbomb, WinBot, Oatmealcookiemon, JAnDbot, Magioladitis, Cic, Snowded, Paresnah, Xtifr, Calltech, Malvaro, G.A.S, Ifaomo, Yobol, Nono64, Erkan Yilmaz, Overix, Nigholith, J.A.McCoy, DeKXer, Grosscha, DarwinPeacock, Thecinimod, Dggreen, Childhoodsend, Antoni Barau, JayC, IPSOS, Mfmoore, Don4of4, Steve Masterson, Dpleibovitz, Northfox, GarOgar, Jamessungjin.kim, Iamthedeus, Emmazunz84, Vanished user kijsdion3i4jf, MathShaman, Yhkhoo, Edugalt, Razimantv, Niceguyedc, Schreiber-Bike, DumZiBoT, Jytdog, SilvonenBot, Addbot, DOI bot, RicardoSanz, MrOllie, Tide rolls, Yobot, Cmbarton54, Examtester, AnomieBOT, Steamturn, ArthurBot, LilHelpa, Apothecia, Omnipaedista, The Wiki ghost, FrescoBot, Citation bot 1, Micromesistius, Geropod, Tom.Reding, Slatteryz, TobeBot, E.V.Krishnamurthy, Mean as custard, RjwilmsiBot, Skamecrazy123, EmausBot, WikitanvirBot, Skater00, Tommy2010, Dcirovic, ZéroBot, Traxs7, ElationAviation, Simondc, Greg Royston Molineux, Cmanske, Ego White Tray, ??? , ChuispastonBot, ClueBot NG, Rezabot, Panleek, Widr, Helpful Pixie Bot, Richardjb25, Bibcode Bot, BG19bot, Bereziny, ZFT, برد, Brad7777, Dtotoo, Sminthopsis84, MisterShiney, Me, Myself, and I are Here, Chris troutman, LithiumEnergy, Nigellwh, Paul2520, Ea2206, HariSeldon11988, Lev Kalmykov, Monkbot, Phoenix 123 abc, Loraof, PennyDarling, Jonkirstenhof, Rionbr, KasparBot, Shifra987, CSG1914, Marvellous Spider-Man, Halil2008, Lkulakova, Magic links bot, Horsaman and Anonymous: 146

- **Emergence** *Source:* https://en.wikipedia.org/wiki/Emergence?oldid=801530973 *Contributors:* CYD, The Anome, WillWare, ChangChienFu, Heron, Bdesham, Michael Hardy, Owl, Lexor, Pnm, Kku, Karada, Ronz, Angela, Andres, Palfrey, Pipis, TonyClarke, Technopilgrim, Ec5618, RodC, Charles Matthews, Nickg, Greenrd, Jeffrey Smith, Jerzy, Banno, Tlogmer, Vespristiano, Chopchopwhitey, Steeev, Rursus, Blainster, Wikibot, Aetheling, Paul Murray, Aknxy, Jleedev, Stirling Newberry, Ancheta Wis, Giftlite, Gwalla, Tom harrison, SantiagoGala-enwiki, Henry Flower, Leonard G., Finn-Zoltan, Edcolins, John Abbe, Andycjp, Loremaster, Karol Langner, BookgirlST, Histrion, Talrias, Jmeppley, IcycleMort, Robin klein, Andreas Kaufmann, Chris Howard, MiddleOfNowhere, Rich Farmbrough, Cagliost, Dbachmann, Pavel Vozenilek, Goochelaar, Bender235, ESkog, Ben Standeven, El C, Vipul, Aaronbrick, Ray Dassen, Mike Schwartz, C S, Teorth, Viriditas, Tmh, JavOs, Mdd, HasharBot-enwiki, Kitoba, Mote, Silver hr, Diego Moya, Minority Report, Hu, Radical Mallard, ClockworkSoul, Zenter-enwiki, Cburnett, Stephan Leeds, Cal 1234, Eternal March, Drat, Acadac, Kazvorpal, Oleg Alexandrov, Woohookitty, PoccilScript, Kzollman, Jeff3000, Abu ari, Ludocrat, Ziji, Christianjb, DaveApter, Marudubshinki, Ashmoo, Rjwilmsi, Mayumashu, Gohn, Nightscream, Koavf, Dudegalea, Krash, JFromm, Sydbarrett74, Pe3-enwiki, Diza, Chobot, Fourdee, Bgwhite, Adoniscik, YurikBot, Wavelength, Flameviper, RussBot, John2000, Rintrah, Ksyrie, Arkapravo, DarkFireTaker, BlackAndy, Thiseye, Slarson, Adamrush, Rbarreira, Shadowfax0, Larry laptop, Moe Epsilon, LodeRunner, Epipelagic, MBDowd, WAS 4.250, HereToHelp, Raveled, Curpsbot-unicodify, MagneticFlux, Bwiki, Luk, KnightRider-enwiki, SmackBot, Moxon, Saravask, ElectricRay, Tomdw, Peteresch, ZS, Cazort, Ohnoitsjamie, Betacommand, Chaojoker, Izzynn, Grokmoo, Chris the speller, Ben.c.roberts, Fuzzform, Nbarth, Hongooi, Toomuchnoise, Gbuffett, OrphanBot, Xyzzyplugh, Cybercobra, Pwjb, Richard001, NickPenguin, Jon Awbrey, A.W.Shred, Just plain Bill, Dr. Gabriel Gojon, Vina-iwbot-enwiki, Cast, Bcasterline, Prionesse, Harryboyles, Kreb Dragonrider, John, Rigadoun, Writtenonsand, Tktktk, Physis, Dchudz, Tasc, Wmattis, Olag, Nabeth, Tones, Papertiger, Asatruer, Joseph Solis in Australia, Antonio Prates, GDallimore, ChrisCork, Ripounet, CmdrObot, CBM, USMCM1A1, N2e, AshLin, Pfhenshaw, Emesghali, ONUnicorn, John courtneidge, Arnold.Sikkema, Logicombat, Myasuda, Gregbard, CX, Phatom87, Fyrius, Cydebot, Clappingsimon, Steel, Peterdjones, Anthonyhcole, Mirrormundo, Studerby, Skittleys, Shirulashem, L7HOMAS, Krylonblue83, Trev M, Letranova, Thijs!bot, Wikid77, ConceptExp, D4g0thur, Headbomb, Pjvpjv, Marek69, Mr pand, Dfrg.msc, Muaddeeb, Nick Number, Timf1234, Majorly, Dougher, Athkalani-enwiki, Davemarshall04, Albany NY, Andonic, Magioladitis, Nessman, Psychohistorian, Aka042, LookingGlass, JaGa, Rickard Vogelberg, Profitip, Logan1939, Geoinmn, Keith D, CommonsDelinker, Fixaller, Erkan Yilmaz, AstroHurricane001, Rlsheehan, BillWSmithJr, Alexjryan, Soiducked, Maurice Carbonaro, Lantonov, BobEnyart, Nemo bis, Grosscha, Chiswick Chap, Aquaepulse, Tgooding, Halrhp, Jknd, Hammersoft, VolkovBot, Pleasantville, Dggreen, Toddy1, LuckyInWaco, Rollo44, VivekVish, Karmela, Rei-bot, Lordvolton, Sjeng, Littlealien182, Sintaku, Dendodge, JhsBot, Don4of4, BL2593, Myscience, Andrewaskew, Lova Falk, Dpleibovitz, SieBot, Sweetp80, Djayjp, Scorpion451, Lord Phat, Sunrise, Emptymountains, Mx. Granger, Rowmn, Rojorulet, ClueBot, Kai-Hendrik, WurmWoode, Napzilla, Der Golem, Alexbot, Brews ohare, SchreiberBike, Bbbeard, Jmanigold, JKeck, XLinkBot, Saurus68, Ecolabs, Rreagan007, MystBot, Jonathanmoyer, Anticipation of a New Lover's Arrival, The, Svea Kollavainen, Addbot, Xp54321, Claudio Gnoli-enwiki, MrOllie, Dyaa, SimonB1710, Mjhunton, Zorrobot, Jarble, Ben Ben, Luckas-bot, Yobot, Isotelesis, IW.HG, Examtester, AnomieBOT, 1exec1, Trevithj, Galoubet, 90 Auto, MorgothX, Citation bot, ArthurBot, Carbaholic, Tomwsulcer, Tyrol5, Srich32977, Omnipaedista, RibotBOT, Friesin76, SchnitzelMannGreek, Constructive editor, FrescoBot, LucienBOT, Dwightfowler, Machine Elf 1735, Journalmuncher, Diavel, DivineAlpha, Citation bot 1, Cbarlow, Pinethicket, Exjhawk, Aizquier, Filthylaugh, Jandalhandler, Sroel, Mjs1991, Pollinosisss, Jonkerz, LilyKitty, Inferior Olive, Reaper Eternal, Catcamus, Bento00, Djjr, EmausBot, Rusfuture, Irvbesen, GoingBatty, Dcirovic, Tuxedo junction, PBS-AWB, Alpha Quadrant, SporkBot, Libertaar, Providus, Ricardsolewiki, RockMagnetist, Just granpa, Spicemix, ClueBot NG, MohamedBishr, BarryKayton, Frietjes, SpaniardGR, Panleek, Tr00rle, Helpful Pixie Bot, Calgg, Bibcode Bot, BG19bot, Rosalegria, Dr. Whooves, Manjusri Wickramasinghe, Michaelweinstock, Joshua Jonathan, MHeder, Run to the hills, cos the end of the world is soon!, Warmtub, Symphonic Spenguin, Dexbot, ZutZut, Polyrahul, Sminthopsis84, Limit-theorem, Danny Sprinkle, MisterShiney, Georgeandrews, I am One of Many, Alfy32, Igjohnston, Pamphilia, Dsomers74, Aubreybardo, Francois-Pier, SJ Defender, Deegeejay333, Peter Corning, Fixuture, Occurring, Saectar, Chaya5260, TheEpTic, Loraof, Social Theory, Isambard Kingdom, Floating drinion, You better look out below!, Kew8888, CLCStudent, InternetArchiveBot, M. A. Broussard, Fmadd, Grimbleschlork, Dapifo, Sethddunn, Dockabo, Anti-Anti-Climacus, L8 ManeValidus, 868,383,950edits, Votto18, Magic links bot, Dairy501, Blackest-Jacky, Evolvingindia, KolbertBot and Anonymous: 317

47.10. TEXT AND IMAGE SOURCES, CONTRIBUTORS, AND LICENSES

- **Self-organization** *Source:* https://en.wikipedia.org/wiki/Self-organization?oldid=793415153 *Contributors:* The Anome, Miguel~enwiki, Tedernst, Edward, Michael Hardy, Lexor, Kku, MartinHarper, EntmootsOfTrolls, Charles Matthews, Dysprosia, Nickg, Robbot, Fredrik, Rursus, Moink, Michael Snow, Mu6, Dina, Snobot, Ancheta Wis, Alensha, Pearbonn, Margana, Karol Langner, The Land, Elektron, Pgreenfinch, Robin klein, Andreas Kaufmann, RevRagnarok, Chris Howard, Jwdietrich2, Ronaldo~enwiki, MiddleOfNowhere, Rich Farmbrough, Jarcanist, Vsmith, Wk muriithi, Smyth, Dave souza, JimR, Dmr2, Bender235, FirstPrinciples, Shrike, Zenohockey, Alex Kosorukoff, RoyBoy, Cretog8, Smalljim, Viriditas, .:Ajvol:., Physicistjedi, Ire and curses, Mdd, HasharBot~enwiki, Jheald, RJII, DV8 2XL, Sylvainremy, Rvanschaik, BryanKaplan, Grammarbot, Rjwilmsi, KYPark, Pleiotrop3, ElKevbo, The wub, Renamed user dj65cf874dfh44, Mathbot, Diza, Hamidfar, YurikBot, Wavelength, Mukkakukaku, Duracell~enwiki, Pseudomonas, Robert McClenon, CLW, Dan Harkless, Curpsbot-unicodify, KnightRider~enwiki, SmackBot, Stpalli, WebDrake, Vald, Pokipsy76, M stone, Skizzik, Chris the speller, Mobius27, Thumperward, Complexica, Colonies Chris, Royboycrashfan, Fotoguzzi, Cicero, Ericbritton, Will Beback, Eliyak, Nick Green, John, JoseREMY, Camazine, Kerbii, Dave Runger, Chetvorno, Mr3641, Zarex, N2e, Pfhenshaw, Cydebot, Krauss, Gmusser, Skittleys, Miguel de Servet, Oszillodrom, Letranova, Kilva, Headbomb, Noclevername, Luna Santin, Rudick JG, Davedrh, Smartse, Phanerozoic, JAnDBot, Narssarssuaq, Dereckson, Athkalani~enwiki, Gerculanum, Freshacconci, GrahameKing, Vernanimalcula, Economizer, Snowded, KConWiki, Dirac66, David Eppstein, User A1, Rvsole, Masaki K, Jim.henderson, Keith D, Emathematica, Pilgaard, Keesiewonder, Grosscha, Crakkpot, 1000Faces, Chiswick Chap, Korotkikh, Elizabeth McMillan, Pleasantville, Dggreen, Crscrs, Mcewan, Rollo44, Vipinhari, AllGloryToTheHypnotoad, Ordermaven, Northfox, Gbawden, SieBot, Thehotelambush, GeneCallahan, Adelanwar, Der Golem, Techdoer, Synergier, Gulmammad, Rhododendrites, Sun Creator, EhJJ, Bracton, SchreiberBike, Adriansrfr, Life of Riley, Koumz, Xiaoju zheng, Dthomsen8, Cyberoo, Fd42, WikHead, Thomas h ray, Addbot, USchick, Unesn6iduja, MrOllie, LarryJeff, Lightbot, Mcamus, Jarble, ع ، Luckas-bot, Yobot, II MusLiM HyBRiD II, Azcolvin429, AnomieBOT, Jim1138, Phantom Hoover, Materialscientist, Citation bot, LilHelpa, The Banner, Srich32977, Omnipaedista, Sahehco, Chjoaygame, FrescoBot, TheSen, Citation bot 1, Winterst, Tiberall, Gray1, Charbee, Regular Polyhedron, Jandalhandler, Ambarsande, Trappist the monk, Reflexinio, Barryclemson, We system, Blueshifting, Noresponse, Lithistman, Hhhippo, Quickmute, JuanCano, Cymru.lass, Carl Wivagg, Allanwik, Robbiemorrison, Ems2715, NinjaQuick, TuxFighter, Jrichardliston, ClueBot NG, Fgunnars, Panleek, Joel B. Lewis, MerlIwBot, Helpful Pixie Bot, Richardjb25, Revisor2011, RogerBF, BG19bot, GlaedrH, DPL bot, Terrykel, Kfriston, Soler99, Elizah379, Khazar2, Nathanielfirst, IjonTichyIjonTichy, Dexbot, Makecat-bot, Me, Myself, and I are Here, BurritoBazooka, Mre env, Samotny Wędrowiec, Iztwoz, Biogeographist, Andy Quarry, Duchifat, Otherocketman, FrB.TG, Monkbot, ???, Mit0126, Asuscreative, Isambard Kingdom, KasparBot, Ishy2015, Jman9058, Sangqiu5, Robcduk, Bear-rings, Shahbazbegian, Bender the Bot, L8 ManeValidus, Sens15, Magic links bot and Anonymous: 143

- **Collective consciousness** *Source:* https://en.wikipedia.org/wiki/Collective_consciousness?oldid=802589638 *Contributors:* SimonP, Sam Spade, Bfinn, Macrakis, Khalid hassani, Stevietheman, Piotrus, DNewhall, CesarFelipe, Pgreenfinch, Pocil, Kotuku33, Hoary, Omphaloscope, Sfacets, Bookandcoffee, Sandius, Stefanomione, BD2412, KYPark, Sonshade, Ian Pitchford, SchuminWeb, Benlisquare, NawlinWiki, Koinu~enwiki, Wikijeff, NetRoller 3D, SmackBot, Roberto Cruz, Gilliam, Portillo, Madmedea~enwiki, Pappy97, Ne0Freedom, Marvin147, "alyosha", Will Beback, Arodb, Murli184, Santa Sangre, Xinyu, Skapur, ShelfSkewed, Yellowtailshark, Cydebot, SvenAERTS, 100110100, .anacondabot, SteveSims, Magioladitis, Websterwebfoot, JNW, Anbro, TimidGuy, Alro, PCock, AstroHurricane001, All Is One, Stan J Klimas, DadaNeem, Mlle thenardier, Opirnia, TXiKiBoT, Bernium, Bigvibes, Tomsega, LeaveSleaves, Wassermann~enwiki, DieBuche, MeganLDouglass, Andrewaskew, Lamro, Caltas, Randy Kryn, Wynnj26, The Thing That Should Not Be, Mild Bill Hiccup, Brewcrewer, Simon D M, Eternal-Entropy, Frost 489, Leon1948, PL290, Addbot, Lykos, Fgnievinski, Mseanbrown, CanadianLinuxUser, Jska313, SpBot, Scott Sprague, Luckas-bot, Yobot, Andreasmperu, Carleas, Bengtlueers, AnomieBOT, Citation bot, Didaktron, Quebec99, J JMesserly, ProtectionTaggingBot, Omnipaedista, Tangent747, Fritq, EmausBot, GoingBatty, Alexpmuller, Jacobisq, Samuel purgess, U3964057, Llightex, Georgetbiol, Hind-Dakheel, Oxford73, Helpful Pixie Bot, BlueMoonset, Northamerica1000, PhnomPencil, Graham11, Marcocapelle, Meclee, Brad7777, Timbrooker, GreenUniverse, Artimis Cat, Dexbot, Eis1983, TJ-Grite, Me, Myself, and I are Here, SJ Defender, Tudor1999, Monkbot, Goblin Face, Obsuser, JudeBass, MaterialistX, Magic links bot and Anonymous: 70

- **Collective behavior** *Source:* https://en.wikipedia.org/wiki/Collective_behavior?oldid=797641169 *Contributors:* Owen, MK~enwiki, Jleedev, Piotrus, Lucidish, Pearle, Jigen III, Miranche, Kotasik, Kocio, Axeman89, Stephen, Richard Arthur Norton (1958-), Astanhope, Sin-man, Jaxhere, Yamamoto Ichiro, Klosterdev, Gaius Cornelius, Rjlabs, Some guy, SmackBot, Portillo, Squiddy, Loodog, Tawkerbot2, Prlsmith, Penbat, Cydebot, Kupirijo, PamD, Steve Dufour, Bethan 182, Luigifan, Jimmy, Reswik, Mkdw, Dwalls, Belovedfreak, Remarknibor, Squids and Chips, Barneca, Philip Trueman, Wassermann~enwiki, Andrewaskew, Mild Bill Hiccup, Gobeshock Gobochondro Gyanotirtho, Rhododendrites, Addbot, SmartM&M, Math Champion, Luckas-bot, Yobot, AmeliorationBot, AnomieBOT, RobertEves92, Hi878, Some standardized rigour, Pareschi, Abductive, Rplal120, Mean as custard, EmausBot, AvicBot, ZéroBot, Jacobisq, Hypercephalic, U3964057, Taffenzee, ClueBot NG, Widr, Helpful Pixie Bot, Civeel, Joost26, BattyBot, Uday.gautam6, Frosty, Toksoz, FireflySixtySeven, Yeda123, GShoeLacy, CyclinB1, KolbertBot and Anonymous: 82

- **Social dynamics** *Source:* https://en.wikipedia.org/wiki/Social_dynamics?oldid=736793503 *Contributors:* Derek Ross, Timo Honkasalo, Taw, Slrubenstein, Ed Poor, Rgamble, Lexor, Kku, CanisRufus, Dgorsline, Maurreen, Delius, Mdd, Mathmo, Graham87, Madcoverboy, Occono, SmackBot, GMcGath, Dan1679, Penbat, Cydebot, Athkalani~enwiki, Seductionreport.com~enwiki, Julia Neumann, Tomsega, Arnold90, Burningview, Tide rolls, Legobot, 1exec1, FrescoBot, H.W. Clihor, Attonbrass, EmausBot, ClueBot NG, Pjlamberson, Helpful Pixie Bot, Pingmi, Maksymilian Sielicki, Meclee, Brad7777, BattyBot, ArmbrustBot, Csutric, Monkbot and Anonymous: 26

- **Collective intelligence** *Source:* https://en.wikipedia.org/wiki/Collective_intelligence?oldid=799364890 *Contributors:* The Anome, LA2, Anthere, ChangChienFu, Stevertigo, Michael Hardy, Pnm, Kku, Ronz, Nikai, Timwi, Maximus Rex, Chuckrussell, Topbanana, Stormie, Lumos3, Chealer, Naddy, Chris Roy, Blainster, Sunray, TittoAssini, Hadal, Robinh, Michael Snow, Philwiki, Alison, Andris, Jason Quinn, Khalid hassani, JRR Trollkien, John Abbe, Stevietheman, Gadfium, Gdr, Piotrus, Quickwik, Sam Hocevar, Mennonot, Rich Farmbrough, Bender235, El C, Lycurgus, Alex Kosorukoff, Mike Schwartz, John Vandenberg, Viriditas, Runner1928, 927, Typobox, Superabo, RussAbbott, Fasten, Clubmarx, Drummond, Kmorris1077, Ott, Ianweller, GregorB, Stefanomione, BD2412, Imersion, Rjwilmsi, NatusRoma, DouglasGreen~enwiki, Nsae Comp, Nihiltres, Ewlyahoocom, Common Man, Chobot, YurikBot, Bhny, Marcperkel, Gaius Cornelius, Ksyrie, CarlHewitt, NawlinWiki, Dialectric, Deodar~enwiki, Epipelagic, Chris, F. Masse, MaxVeers, Jurriaan, Cengelbart, The imp, BorgQueen, LeonardoRob0t, DoriSmith, Honamos, SmackBot, InverseHypercube, Portillo, Ohnoitsjamie, Chaojoker, ImageObserver, Chris the speller, JMSwtlk, Jprg1966, Antonrojo, Radagast83, Orangejon, Ahmac, Lambiam, John, Murli184, The Man in Question, Neddyseagoon, Nabeth, DI2000, Papertiger, Poreeoa, Courcelles, Joshuamckenty, Cydebot, Future Perfect at Sunrise, Peterdjones, Doug Weller, Richhoncho, Thijs!bot, Headbomb, Ynagar, Generozo, Dawnseeker2000, RobotG, QuiteUnusual, Knotwork, Golgofrinchian, The Transhumanist, Robina Fox, Andonic, Engineman, Slartibartfast1992, Cgingold, DeanO, B9 hummingbird hovering, Terry.fogarty, Speck-Made, RJBurkhart3, Adavidb, Jiuguang Wang, Tikiwont, Maurice

Carbonaro, All Is One, JFNoubel, Chiswick Chap, Fheyligh, Jeff F F, Bonadea, Tkgd2007, TeamZissou, Psheld, Jeff G., Vincent Lextrait, Aesopos, LeilaniLad, Tomsega, Wassermann-enwiki, Andrewaskew, Dpleibovitz, Newbyguesses, George Por, Theopapada, Dawn Bard, Toddst1, Plevy, Scorpion451, Aspian, Rosiestep, Wetwarexpert, Escape Orbit, Mx. Granger, ClueBot, Kleptosquirrel, Czarkoff, Der Golem, Markmowem, Niceguyedc, Nljnlj, Alexbot, Aristotle28, Rhododendrites, Sun Creator, Lurexus3rd, Chaosdruid, Vanderstoep, Dank, Cardiffbaybee, Editor2020, XLinkBot, Mifter, RealDracaena, RobertDavidSteeleVivas, Wilma Clark, Neilireson, Lalas1972, Addbot, AkhtaBot, CanadianLinuxUser, Gizziiusa, Download, Redheylin, Tassedethe, DebMacP, Jarble, Yobot, Andreasmperu, Jmansbridge, Isotelesis, AnomieBOT, Yukishiro Shi2, Katied26, Tragicity, Sz-iwbot, Immerbeta, Citation bot, ArthurBot, Xqbot, MacounMAS229, DSisyphBot, Harvey the rabbit, Energycurve, Coretheapple, Nasa-verve, ProtectionTaggingBot, Omnipaedista, Fortdj33, Punja2002, OgreBot, Citation bot 1, Jonesey95, Tom.Reding, Trappist the monk, Techhistory, Keegscee, RjwilmsiBot, Cmeiqnj, WikitanvirBot, ZeniffMartineau, Capostrophe Jones, Peter K. Livingston, Dcirovic, AvicBot, WeijiBaikeBianji, Texashorns349, Shared Galaxy, Resprinter123, Ozara, Xerographica, Akma72, ClueBot NG, Mr. snuffleupagus, Barely3am, BG19bot, M0rphzone, Northamerica1000, Compfreak7, CitationCleanerBot, Elrhoads, SindreKA, Shirudo, BattyBot, Artimis Cat, Khazar2, Nathanielfirst, Corrector623, IjonTichyIjonTichy, Dexbot, Karlopiz, Me, Myself, and I are Here, BurritoBazooka, Jodosma, Cascade.uiuc, Jwhitesu, Choi Hyun Hee, Innovativeartist, .js, Plunkersiniapes, Robevans123, PraveenKumanan, Vaishsiva7789, Saevio-Facunde, Fixture, Vleonova, Monkbot, Stigmergy, Jim-Siduri, Kroose1, Quivico, Samadhi7, ObviouslyNotASock, Gmcmurray9, Ellsbell, Sangita Kumaril, Stavelef, InternetArchiveBot, NScheinerman, PaDBu, Fmadd, Jmcgnh, Bender the Bot, Hlandis333, Pietkupka, Davidzhangcmu, BurnAfterReading, KolbertBot and Anonymous: 174

- **Collective action** *Source:* https://en.wikipedia.org/wiki/Collective_action?oldid=797829678 *Contributors:* Bryan Derksen, The Anome, R Lowry, Edward, Pde, Mike Linksvayer, Rl, Cleduc, Hadal, Nikodemos, Aphasia, Jdevine, Pcarbonn, Rdsmith4, Lucidish, D6, Rich Farmbrough, Kzzl, Lycurgus, Longhair, Maurreen, Dejitarob, Pearle, Rodw, Cdc, Dominic, Bookandcoffee, Kupojsin, Bluemoose, Rjwilmsi, Mark Elliott, RattusMaximus, RussBot, Daniel Mietchen, Pb30, SmackBot, RedHouse18, Scaife, Jprg1966, Brinerustle, Bigturtle, Lambiam, Robofish, Accurizer, Nabeth, Joseph Solis in Australia, ShelfSkewed, Lanma726, Penbat, Cydebot, Bossnomad, Thijs!bot, Al Lemos, Headbomb, Ilion2, Skomorokh, Magioladitis, Gsby, Aboluay, Alanadexter, Counterfact, Classicalecon, Birdofpray, Muhandes, Kitfox.it, Addbot, Margin1522, Legobot, Yobot, Andreasmperu, Jmansbridge, Isotelesis, AnomieBOT, Materialscientist, Citation bot, Maemst, AlecStewart, FrescoBot, Haeinous, Citation bot 1, Jonesey95, Diannaa, RjwilmsiBot, EmausBot, ZéroBot, John Cline, Meng6, SporkBot, Pochsad, U3964057, Frietjes, Stic rotide, Helpful Pixie Bot, Meclee, Brad7777, Skk515, HONOURS Q, The Vintage Feminist, GabeIglesia, Me, Myself, and I are Here, Burn2061, Pezeshka, Anarcham, Nastyatolbatova, Fixture, Dan Mihai Pitea, Dk2745, KasparBot, Shalor (Wiki Ed), Khysimp, Hopelessoptimism, KolbertBot and Anonymous: 64

- **Self-organized criticality** *Source:* https://en.wikipedia.org/wiki/Self-organized_criticality?oldid=800436970 *Contributors:* Fnielsen, Michael Hardy, Kku, Charles Matthews, Herbee, Beland, Karol Langner, Ivn-enwiki, Pgreenfinch, Vsmith, JimR, Iamunknown, 9SGjOSfyHJaQVsEmy9NS, Mdd, Embryomystic, Linas, Ruud Koot, Kelisi, BD2412, Rjwilmsi, Nihiltres, Artgirl88, Salsb, Shadowfax0, SmackBot, WebDrake, Badjeros, Chris the speller, Kleuske, Good Intentions, Lapaz, Robofish, Rhetth, Sonswoo, Headbomb, JustAGal, Magioladitis, BigrTex, Hugh Hudson, Westfalr3, Dggreen, Cwkmail, AnonyScientist, Addbot, DOI bot, Lightbot, Jarble, Yobot, AnomieBOT, Citation bot, Omnipaedista, FrescoBot, Satrapa, Citation bot 1, Dimo400, Nastasyuk v. Self-org, Rayman60, Benlansdell, Dcirovic, Crbazevedo, Vacation9, Helpful Pixie Bot, Bibcode Bot, BG19bot, Jcphillips8, Compsim, AppliedMathematics, Me, Myself, and I are Here, Marknew12, Otherocketman, Self Order, Monkbot, Meforwk, Isambard Kingdom, Vasilii Tiorkin, James C. Phillips and Anonymous: 54

- **Herd mentality** *Source:* https://en.wikipedia.org/wiki/Herd_mentality?oldid=794165727 *Contributors:* Edward, YUL89YYZ, Spikeballs, Woohookitty, Mandarax, Graham87, Jweiss11, Remurmur, Bgwhite, NeilN, SmackBot, David.Mestel, Gilliam, DMacks, Lambiam, Doczilla, Peter1c, CBM, N2e, NickW557, Kanags, Alaibot, Mattisse, Leon7, Steveprutz, Classical geographer, Michaeldsuarez, Biscuittin, Pare Mo, Caltas, Twinsday, Martarius, Curious brain, Bettinakatz, Rhododendrites, KuboF Hromoslav, XLinkBot, Ost316, Borock, EEng, Addbot, Betterusername, MrOllie, Jim1138, JackieBot, Robot85, Srich32977, Touchatou, Aaron Kauppi, Eagle4000, GoingBatty, Slightsmile, Donner60, ChuispastonBot, Shelvey C. McPhail, U3964057, ClueBot NG, Frietjes, Widr, Bitphr3ak, North911, Meclee, MuAlphaTheta, FireflySixtySeven, KBH96, LauraC1360, Stewi101015, Marianna251, Buck12341, Ccparsley, Rbpetersen32 and Anonymous: 70

- **Phase transition** *Source:* https://en.wikipedia.org/wiki/Phase_transition?oldid=800403416 *Contributors:* CYD, Mav, Bryan Derksen, Olof, Roadrunner, Peterlin-enwiki, Patrick, Michael Hardy, Tim Starling, Kku, TakuyaMurata, Ahoerstemeier, Glenn, Charles Matthews, Terse, Phys, Bafficlan, Chuunen Baka, Lzur, Dave6, Giftlite, Graeme Bartlett, Djinn112, BenFrantzDale, Zigger, Curps, Ezhiki, Steuard, Antandrus, Aulis Eskola, Karol Langner, Zfr, Pgreenfinch, Revised-enwiki, Mike Simpson, Deglr6328, CALR, Varada, Vsmith, JimR, Ponder, Dmr2, Bender235, Evice, Nabla, Brian0918, El C, Rgdboer, 9SGjOSfyHJaQVsEmy9NS, Slinky Puppet, Kjkolb, Ynhockey, RJFJR, Oleg Alexandrov, Brookie, CoolMike, Linas, V8rik, Zeroparallax, Nanite, Rjwilmsi, Nneonneo, FlaBot, John Baez, Mathbot, GünniX, Gurch, Srleffler, DVdm, Bgwhite, YurikBot, Ugha, Wavelength, Hairy Dude, Supasheep, Chuck Carroll, Salsb, Janke, Tony1, Bota47, Kkmurray, Vicarious, Cmglee, Sbyrnes321, Pankkake, SmackBot, WebDrake, Unyoyega, Gilliam, Kdliss, Joseph Stalin-enwiki, Richtife, Kmarinas86, Chris the speller, MalafayaBot, Complexica, Colonies Chris, Wiki me, MureninC, SkanderH, Akriasas, DMacks, Dave314159, Brennan Milligan, JorisvS, Bjankuloski06en-enwiki, FrostyBytes, Soulkeeper, JHunterJ, Inquisitus, Dl2000, JarahE, Wizard191, Japhet, JRSpriggs, Ouishoebean, CRGreathouse, CmdrObot, Tarchon, Chrumps, Van helsing, Myasuda, Hga, Icek-enwiki, Cydebot, Perfect Proposal, Kweeket, Rifleman 82, JFreeman, JohnInDC, Thijs!bot, Epbr123, TDF, Headbomb, E. Ripley, Nick Number, JAnDbot, Igodard, Penubag, Primephear, Magioladitis, JamesBWatson, Kaiserkarl13, Dhk, User A1, Jtwl1, Hans Dunkelberg, Icseaturtles, TomyDuby, DadaNeem, Foreveclevah, Cuzkatzimhut, Hershen, Nikthestunned, VolkovBot, Oshwah, A4bot, Judge Nutmeg, ChooseAnother, Jackfork, Brianga, Neparis, Kbrose, Barkeep, SieBot, I Like Cheeseburgers, Trumpsternator, Sean.hoyland, Dolphin51, WikiBotas, ClueBot, LAX, Rodhullandemu, Brettstrawbridge, Lame Name, Ktr101, CohesionBot, Tbagasaurausrex21, Thehelpfulone, Sdrtirs, RexxS, XLinkBot, Matthieumarechal, Avoided, Rreagan007, Hess88, Addbot, AkhtaBot, Download, EconoPhysicist, Alelima, TStein, Tide rolls, Yoavd, Snaily, Luckas-bot, Yobot, Pteradactyle, AnomieBOT, Hunnjazal, ImperatorExercitus, Citation bot, Betsim, Madbard, DSisyphBot, Paula Pilcher, Mvpranav, Omnipaedista, SassoBot, Logger9, FrescoBot, Hornord, Kmdouglass, DivineAlpha, Wdcf, Jonesey95, Dr-b-m, Σ, Gryllida, Trappist the monk, Puzl bustr, Wotnow, Lotje, LilyKitty, Marie Poise, Oakycoppice, Ripchip Bot, Rafmag, Super48paul, Slightsmile, Olaw3, K6ka, ZéroBot, Joshlepaknpsa, Mattedia, Sfoske70, Octopusbr, Donner60, RockMagnetist, ClueBot NG, Gareth Griffith-Jones, Anagogist, Nscozzaro, Pvnuffel, Rezabot, Pluma, MerllwBot, Helpful Pixie Bot, KLBot2, Bibcode Bot, BG19bot, Bmusician, PhnomPencil, Indah blestari, Jedharris, GenBiorics, Webclient101, JGTZ, Sphelps 9312, Prokaryotes, Stemcells90, Eigenbra, Wireless erlang, Kkosman, Monkbot, Yikkayaya, Briefexact32, Rakeshyashroy, Ppdouble, Cyclingralph, See996able, CV9933, PDcanonical, KasparBot, Лагічна рэвалюцыйны, Baba Arouj, Drtetractous, Kigaei, 20040420zzb, Delamotteb, PrimeBOT, Dockabo, Tymewalk, Shareride, Himansu Naik, JCW-CleanerBot and Anonymous: 233

47.10. TEXT AND IMAGE SOURCES, CONTRIBUTORS, AND LICENSES

- **Agent-based model** *Source:* https://en.wikipedia.org/wiki/Agent-based_model?oldid=800270741 *Contributors:* Michael Hardy, Lexor, Pnm, SebastianHelm, Ronz, Sbwoodside, Topbanana, Jetfq, Drew3D, Giftlite, Ds13, Robin klein, Jania902, Rich Farmbrough, Fvdham, Bender235, Giraffedata, Mdd, Drf5n, JoaoRicardo, Acadac, Tabletop, Bkwillwm, Waldir, Hyperzonk, Rjwilmsi, Koavf, Nicapicella, Volunteer Marek, YurikBot, KenBailey, Joebeone, Madcoverboy, Grafen, Ma qavi, Rjlabs, Scud1, Tony1, Epipelagic, GraemeL, Guillom, Eykanal, Moxon, Camcom, Pmkpmk, Chris the speller, RandyBurge, Sslevine, BillFlis, Beetstra, Nabeth, Mr3641, CmdrObot, Thomasmeeks, Anthony Bradbury, Simeon, Cydebot, Krauss, After Midnight, Thijs!bot, Headbomb, Nick Number, WinBot, Seaphoto, Pgaudiano, AndreasWittenstein, JAnDbot, Ph.eyes, Mikeironic, TAnthony, Douglas R. White, Magioladitis, Novickas, Dragentsheets, Mårten Berglund, Erkan Yilmaz, Mange01, BigrTex, Fatfreeride, Coppertwig, Hpcoder, Cometstyles, Boombaard, VolkovBot, YewBowman, Jacob Lundberg, Aprodan, Lordvolton, Tesfatsion, Kycook, EmxBot, AdRock, SieBot, Jerryobject, Efcaguab-enwiki, Erik incognito, Krogren, Rinconsoleao, Fatinspanish, MATThematical, Samer.hc, Mlatfs, SoxBot III, XLinkBot, Hjp lzu, Rmdsouza, Terrillfrantz, Addbot, UberArtifex, Jbapowell, Drevicko, Thengarch, Jncraton, Focusheart, Txviet, MrOllie, Lightbot, Cnikolai, ZX81, Yobot, Themfromspace, Roger.mchaney, Cmbarton54, AnomieBOT, Citation bot, ChristianG2, Sergey Suslov, Xqbot, Garrularity, J04n, Pedroulloa, Wrand, HariSel, FrescoBot, Citation bot 1, PRichmondShef, Rvihav, Jmgalan, DrilBot, Symplectic Map, Cedric71, Muaz.niazi, Niazim1, Trappist the monk, Howard.noble323, E.V.Krishnamurthy, Amandus74, RjwilmsiBot, Dewritech, Pwaddell, JaeDyWolf, Dcirovic, ZéroBot, Yoyo Shu, Tectonicura, Sgalland-arakhne, 沈澄宇, Viprenkun, Mod501, Jhmsfi, Deer*lake, Frietjes, Jaengelberg, Helpful Pixie Bot, Chercheurdoz, Bibcode Bot, BG19bot, Mark8Mark, Meclee, Cyberbot II, Dexbot, Djfrost711, Peterknows, GabeIglesia, Me, Myself, and I are Here, Mohammadsp-enwiki, I am One of Many, Tylercusack, NYBrook098, ModalPeak, Anrnusna, Stamptrader, Stegkc, Lev Kalmykov, Monkbot, Jont306, SadeghAsgari, GreenC bot, SPSS2016, Ariefyudho, Paul Smetanin, Huayi Lin (SLU), KolbertBot and Anonymous: 134

- **Synchronization** *Source:* https://en.wikipedia.org/wiki/Synchronization?oldid=798916096 *Contributors:* The Anome, Waveguy, Heron, B4hand, Patrick, Michael Hardy, Kku, Meekohi, Karada, Iluvcapra, CesarB, Egil, Mac, Mulad, Colin Marquardt, AHands, Hyacinth, Grendelkhan, Robbot, Altenmann, Ancheta Wis, DavidCary, Bkonrad, Pne, Nickptar, Zondor, JTN, Noisy, ArnoldReinhold, Dbachmann, Mwanner, Shanes, Guettarda, Liao, Richard Harvey, DanGunn, Wtshymanski, Gpvos, Ruud Koot, Graham87, Rjwilmsi, Vegaswikian, Ian Dunster, Wavelength, Hillman, Cascadian, DanMS, Yamara, Nicke L, CarlHewitt, Aldenhoot, Howcheng, Daniel Mietchen, Tony1, Scottfisher, Closedmouth, SmackBot, Ianwri, Rentier, Telestylo, MichaelII, SynergyBlades, Oli Filth, Neo-Jay, Dual Freq, UNV, ZachPruckowski, Izhikevich, Cleanwiki, Lambiam, ElectronicsPerson, 16@r, Halaqah, Citicat, Kvng, Dre.velation2012, Alexignatiou-enwiki, Corpx, SymlynX, Epbr123, Marek69, NigelR, Nick Number, Peterhawkes, JEBrown87544, AntiVandalBot, Squidfishes, JAnDbot, Stijn Vermeeren, Jim.henderson, SpeckMade, Javawizard, Maurice Carbonaro, JohnGrantNineTiles, Edvige, Soundofmusicals, 3p1416, Dpleibovitz, Movedgood, Radagast3, Biscuittin, Vektor330, Pcontrop, Vice regent, ClueBot, Tigerboy1966, Mild Bill Hiccup, Mumiemonstret, PixelBot, Ykhwong, Brews ohare, Galzigler, Poco a poco, Mabdul, Fgnievinski, Тиверополник, Numbo3-bot, Jarble, Luckas-bot, Yobot, Jalal0, AnomieBOT, Materialscientist, ArthurBot, TheAMmollusc, JimVC3, Nasnema, Subviking, FrescoBot, RedBot, MastiBot, TjeerdB, Meatball91, EmausBot, KIbrain, AManWithNoPlan, Donner60, ChuispastonBot, ClueBot NG, Synchronycreator, MerlIwBot, Wbm1058, BG19bot, TejasDiscipulus2, EricEnfermero, Makecat-bot, Cerabot-enwiki, TwoTwoHello, ProfPhaseSpace, Reatlas, Janek316, Narky Blert, 0xF8E8, Cartoon network freak, KasparBot, JefferyGroup and Anonymous: 113

- **Ant colony optimization algorithms** *Source:* https://en.wikipedia.org/wiki/Ant_colony_optimization_algorithms?oldid=797625602 *Contributors:* Fubar Obfusco, Mrwojo, Michael Hardy, Haakon, Ronz, Dcoetzee, Nickg, Maximus Rex, Furrykef, Omegatron, Altenmann, Xanzzibar, Enochlau, BenFrantzDale, SWAdair, LiDaobing, Asbestos, Jbinder, Rich Farmbrough, TedPavlic, Matthewfallshaw, Paul August, Jpgordon, Mdd, Diego Moya, Andrewpmk, BryanD, Jaardon, Trylks, Cburnett, Runtime, Oleg Alexandrov, Woohookitty, Mindmatrix, Ruud Koot, Tabletop, BD2412, Vberger-enwiki, Rjwilmsi, Quiddity, Sdornan, GünniX, Bgwhite, Roboto de Ajvol, YurikBot, Retodon8, Scott5834, Ritchy, Welsh, Epipelagic, Redgolpe, MaNeMeBasat, Nojhan, Bsod2, Itub, SmackBot, InverseHypercube, Eskimbot, Senarclens, Thumperward, Miguel Andrade, Tamfang, Bsrinath, Spiritia, Feraudyh, Jamie King, IDSIAupdate, Mattbr, Irwangatot, Mernst, Ventania-enwiki, CMG, Damzam, Gnewf, Lawrenceb, PHaze, KrakatoaKatie, Edokter, Whenning, MoyMan, BrotherE, SiobhanHansa, Magioladitis, Nyq, JamesBWatson, Mdorigo, First Harmonic, CommonsDelinker, Whitebox, NerdyNSK, LordAnubisBOT, Cobi, Santiperez, J ham3, VolkovBot, Philip Trueman, Tupolev154, Petebutt, McM.bot, Swagato Barman Roy, Maarten van Emden, Tomaxer, Spinningspark, Mmanfrin73, SieBot, Zwgeem, Melcombe, AllenJB, Denisarona, Explicit, Der Golem, Lasta, Sun Creator, Speicus-enwiki, Amossin, BOTarate, Kopophex, XLinkBot, Saeed.Veradi, WikHead, CalumH93, Addbot, Manjuer, Favonian, Luckas-bot, Quadrescence, Yobot, 4th-otaku, AnomieBOT, Royote, Halberdo, J04n, GrouchoBot, Richardsonlima, Pratik.mallya, NicoMon, Trappist the monk, Pyxzer, Ratchet11111, Vprashanth87, Jonkerz, Dinamik-bot, Dzkd, Smitty1337, Gretchen Hea, K6ka, Daryakav, Tango.ta, ZILIANGdotME, ClueBot NG, Editdorigo, HectorAE, Leonardo61, Snotbot, Frietjes, DI2653, Helpful Pixie Bot, Tholme, Pepanek Nezdara, BG19bot, Amp71, MSchlueter, Solomon7968, Praveenv253, BattyBot, Gueleri, Dexbot, Tturchi, Me, Myself, and I are Here, Selvi muthukumar, DialaceStarvy, Jodosma, DavidLeighEllis, Nikpantera, 182ankit, Stamptrader, Swadim, Monkbot, Notsoimp2012, ChamithN, Vitor, MaryThomas29, HelpUsStopSpam, InternetArchiveBot, Ajevtic, Wiki2016edit, Shushant Bandyopadhyay, GreenC bot, Timallanwheeler, Jeanpmartins, Unist1984, Wikiraptor2, Shanne Dark, Magic links bot, KolbertBot and Anonymous: 191

- **Particle swarm optimization** *Source:* https://en.wikipedia.org/wiki/Particle_swarm_optimization?oldid=799463530 *Contributors:* Michael Hardy, Lexor, Ronz, Mxn, Hike395, Jitse Niesen, Unknown, Robbot, Amgine, Auric, Lysy, Giftlite, Sepreece, BenFrantzDale, Dratman, Tagishsimon, Horndude77, Ronaldo-enwiki, Rich Farmbrough, Swiftly, Photonique, Diego Moya, Oleg Alexandrov, Ruud Koot, Waldir, CoderGnome, Rjwilmsi, Datakid, Hgkamath, Wavelength, NawlinWiki, Epipelagic, Slicing, Dbratton, Neomagus00, Gwe0351, SmackBot, Ma8thew, Mcld, Oli Filth, DHN-bot-enwiki, Prometheum, Cybercobra, Blake-, Ehheh, Dicklyon, Mishrasknehu, Cydebot, DustinFreeman, BetacommandBot, KrakatoaKatie, Whenning, Storkk, Armehrabian, Michael.Clerx, Mange01, Jiuguang Wang, NerdyNSK, K.menin, My wing hk, Oshwah, Seb az86556, Tjh22, Lourakis, Seamustara, CharlesGillingham, Denisarona, Mild Bill Hiccup, Blanchardb, Certes, Betamoo, Saeed.Veradi, Sliders06, Mexy ok, Addbot, Forna84, MrOllie, Chipchap, Ender.ozcan, Luckas-bot, Yobot, Sriramvijay124, AnomieBOT, Kingpin13, Citation bot, Wgao03, YakbutterT, Mcoupal, SassoBot, MuffledThud, Joaquin008, David Nemati, Sanremofilo, Jder-enwiki, FrescoBot, Khafanus, Sharkyangliu, AdrianoCunha, Zhanapollo, Saveur, Bolufe, Bshahul44, Gfoidl, RjwilmsiBot, Ripchip Bot, Becritical, EmausBot, Nick Moyes, Dzkd, Murilo.pontes, Yuejiao Gong, Huabdo, Daryakav, Optimering, ZéroBot, PS., Mrcs.jr, George I. Evers, SporkBot, MClerc, Wingman417, Enzzef, Bdonckel, ChuispastonBot, Ericjster, Jalsck, ClueBot NG, Swarming, Anne Koziolek, Helpful Pixie Bot, Younessabdussalam, BG19bot, Rijinatwiki, Eugenecheung, ChrisGualtieri, Σμήνος, رضا, Garyatvocal, Me, Myself, and I are Here, Mark viking, Sharkyangliu916, Rcrumpf, Mehr86, Monkbot, Ephramac, Rezabny, HelpUsStopSpam, Waterfall mg, MatDian, Amer mohey, Nawabalam.iitr, Hossein6785, Joeyjoon, Hugo Amorim Neto, KolbertBot and Anonymous: 174

- **Swarm behaviour** Source: https://en.wikipedia.org/wiki/Swarm_behaviour?oldid=800732915 Contributors: Tarquin, Lexor, Dante Alighieri, Yann, Stan Shebs, Ronz, Kils, BAxelrod, José Ph, Itai, Omegatron, Phoebe, Pollinator, Phil Boswell, Robbot, Chris 73, Vespristiano, Altenmann, Lowellian, Rholton, Robinh, Gtrmp, MSGJ, Patrick-br, Andycjp, LucasVB, Onco p53, Pgreenfinch, Neves2882, Joyous!, Rich Farmbrough, Supercoop, Night Gyr, Bender235, Kross, Tom, Remember, Fir0002, Alansohn, JYolkowski, Jwinius, Shoefly, Svend17, PoccilScript, Uncle G, Ch'marr, Winged-stone, Kbdank71, Ketiltrout, Rjwilmsi, Eyu100, FlaBot, Mitsukai, YurikBot, RobotE, Gaius Cornelius, Epipelagic, Whitejay251, Arthur Rubin, Curpsbot-unicodify, SmackBot, McGeddon, Macintosh User, RDBrown, Dr. Dan, Slogan621, Wikiklaas, NessieVL, Zearin, Disavian, The Man in Question, SmokeyJoe, Renebeto, Iridescent, Kaarel, Jason7825, MGlosenger, Tawkerbot2, Connection, Omicronpersei8, Dyanega, Thijs!bot, Camelcast, Headbomb, Sobreira, Marek69, Frici, Ozzieboy, Magioladitis, Fabrictramp, Dan Pelleg, MartinBot, XRiffRaffx, R'n'B, CommonsDelinker, Nono64, Mausy5043, Mkruijtf, 1000Faces, Tarotcards, S (usurped also), Squids and Chips, Idioma-bot, Fences and windows, WOSlinker, LuckyInWaco, Nonstandard, Van Parunak, Mbz1, Cactus26, Jojalozzo, Blueclaw, Snigbrook, Fox, Silent Key, SuperHamster, CohesionBot, Jusdafax, SchreiberBike, Carriearchdale, AP500, Phynicen, Bilsonius, Danielscarvalho, Fyrael, Ezmindegy, MrOllie, Redheylin, Squandermania, Carlos Rosa PT, Tide rolls, Jarble, Yobot, AnomieBOT, ThaddeusB, Citation bot, Maiella, LilHelpa, Sionus, Capricorn42, Omnipaedista, Zohan140, Aaron Kauppi, Some standardized rigour, FrescoBot, Citation bot 1, ExcessPhase, Pinethicket, Halfway to never, Spidey104, Tom.Reding, Beardmites, Юрий Зорѧ, Phrontis, Full-date unlinking bot, Bolufe, Trappist the monk, Oxfordmathematician, Theo10011, Dgreenheck, Dcirovic, MacDaid, Maschen, ClueBot NG, Tanbircdq, DustinIngram, MerlIwBot, Helpful Pixie Bot, Bibcode Bot, Plantdrew, BG19bot, Snow Rise, BattyBot, RobertInoyes, Cyberbot II, ChrisGualtieri, Khazar2, Mogism, Pintoch, Federicoaolivieri, YiFeiBot, Jianhui67, Lizia7, Jdhale, Fixuture, Monkbot, Rhiever, Katieortman, Krausea, Amanda.Kalupa, Annamargit, Asuscreative, Ira Leviton, Eggcram3, Helenaxeros, Rasikareddy1019, Adaerr, InternetArchiveBot, Bender the Bot, PrimeBOT, Anonymousme24, Walkerka, KolbertBot and Anonymous: 83

- **Network science** Source: https://en.wikipedia.org/wiki/Network_science?oldid=802931970 Contributors: Edward, Michael Hardy, Kku, IceKarma, Klamma, RayBirks, D6, Giraffedata, Mdd, Jérôme, Oleg Alexandrov, Imersion, Rjwilmsi, Salix alba, Nihiltres, Gcalda, Madcoverboy, Cedar101, SmackBot, Chris the speller, BullRangifer, Loodog, Kvng, Mdanziger, Quaeler, Twas Now, Connection, CmdrObot, Sprhodes, Headbomb, Labongo, Douglas R. White, Yaron K., R'n'B, AgarwalSumeet, Skullers, Geekdiva, Yecril, JohnDoe0007, Dggreen, Anna Lincoln, WereSpielChequers, Iamthedeus, Jojalozzo, Antonio Lopez, ClueBot, Antipopxx, Sun Creator, Arjayay, Addbot, JBsupreme, Skynet1, MrOllie, Download, Chicagoian, Smoke73, Netzwerkerin, NYNetwork, Luckas-bot, Yobot, 2themoon, Jean.julius, Vkchoudhary, AnomieBOT, Andrewrp, Materialscientist, GaborPete, Citation bot, Aquacool91, Quebec99, Xqbot, TheAMmollusc, Anna Frodesiak, Crzer07, FrescoBot, Mentatseb, Sidna, Plan92bsure, Asaalt, Jonesey95, Tom.Reding, Beteltreuse, Trappist the monk, Worldcontrol, Brentd25, Onel5969, John of Reading, Karayang, Dewritech, Dcirovic, Mambonotive, Djdjdjdjb, Traxs7, Sharkface1, Anselrill, Akseli.palen, Zdorovo, Pribs, Gary Dee, ClueBot NG, Satellizer, Gavin.perch, Helpful Pixie Bot, Bibcode Bot, Whatsamattau, Jmcatania, BG19bot, Bereziny, 507WVS, Alchames, Ettypaldos, Admatkin2, Sharkbite3, TigerDover, 1hipster, Blah314, The Vintage Feminist, Daviscientist, Nrifel, Carloseu, Mhbeals, LieutenantLatvia, Cypherquest, Yeda123, Snoodlebug, Monkbot, Sefer12, Srijankedia, Sidharth10, Pcurley, Thebucketmanfromhades, GlennLawyer, Narky Blert, What4uknow, BenjaminDHorne, Shikang Liu, Edric Dayne, Puccio.b, Bhatia.u, Nicholasbarry, Datzr, BrunoCoutinho, BrunoGCoutinho, Aimlessslyknows, Kent Krupa, Calvinius, Mil686, Mpritham, Dedalus1234, RobbieIanMorrison, SlvrKy, Raydyday, Networkscientist, Quitute, Khassan du, Alongapo, PrimeBOT, Liu.zhen337, John46666, Nkoub and Anonymous: 73

- **Scale-free network** Source: https://en.wikipedia.org/wiki/Scale-free_network?oldid=803073643 Contributors: DavidLevinson, Heron, Dwheeler, Edward, Michael Hardy, Kku, Meekohi, CesarB, Ronz, Yaronf, Cryoboy, Charles Matthews, Nickg, Peak, DavidCary, Harp, WiseWoman, Lethe, Anville, Dratman, Kmote, SanderSpek-enwiki, ChaTo, TedPavlic, JimR, Longhair, Htmlism, 3mta3, Mdd, Jérôme, Cheezycrust, PAR, Andreala, Caesura, Dirac1933, Oleg Alexandrov, Mindmatrix, Dandv, Marudubshinki, Jshadias, MarkHudson, Rjwilmsi, Gareth McCaughan, JFromm, Mathbot, GünniX, Gcalda, Madcoverboy, Conradl, Zwobot, FlyingPenguins, Gadget850, Fmccown, SmackBot, RDBrown, Gragus, Derlikous, Dreftymac, Danlev, Dlohcierekim, CmdrObot, Agathman, Gritzko, Myasuda, Ventania-enwiki, Cosmi, Headbomb, AntiVandalBot, Joe Schmedley, Dougher, JAnDbot, Douglas R. White, Typochimp, Sangak, David Eppstein, Onlynone, Gmagkots, Atddta, Darwin-Peacock, Chiswick Chap, Pphaneuf, BarroColorado, Sandman2007, Econterms, PaulTanenbaum, Hannes Röst, Dirkbb, SieBot, Kl4m, DFRussia, Infoeco, Alexbot, SilvonenBot, Addbot, DOI bot, TutterMouse, Download, SamatBot, Echinoidea, Luckas-bot, Yobot, Nog33, Citation bot, Control.valve, Miym, JonDePlume, Dkasthurirathna, Metasoarous, BenzolBot, RjwilmsiBot, Brteag00, WikitanvirBot, Massimo.franceschet, Dcirovic, Smiling1126, ZéroBot, Snehalshekatkar, Bobsponj, Octochimps, Mouse20080706, Ugronugron, Helpful Pixie Bot, T214OU, Bibcode Bot, BG19bot, Tsndiffopera, Deltasun, CitationCleanerBot, BattyBot, Alexandersaschawolff, ChrisGualtieri, Ophthalmol, Dexbot, Jaspermogg, 14GTR, FrigidNinja, Stamptrader, Monkbot, Malexmave, Hou710, Pariah24, Anarchyte, Nurbudapest, Shifra987, Mil686, Slaplagne, Khassan du, Abhijitjimmi2, AlienGod, JCW-CleanerBot, Evanusam and Anonymous: 114

- **Social network analysis** Source: https://en.wikipedia.org/wiki/Social_network_analysis?oldid=800194314 Contributors: The Anome, Enchanter, SimonP, DavidLevinson, AdamRetchless, R Lowry, Edward, D. Boud, Michael Hardy, Vaughan, Isomorphic, Kku, Lquilter, Karada, Mmorabito67, CesarB, Radicalsubversiv, Ronz, Jonathan Chang, Glenn, Ciphergoth, Nikai, BAxelrod, Mxn, CAkira, Mydogategodshat, Timwi, Nohat, Andrevan, Wikiborg, Kierant, Buridan, Ebricca, The lorax, Robbot, R3m0t, Vespristiano, Chocolateboy, ZimZalaBim, Romanm, Postdlf, Davodd, Sunray, Ramir, Adam78, David Gerard, Ancheta Wis, Cokoli, Joeboy, Everyking, Niteowlneils, Khalid hassani, Pne, Alan Au, Edcolins, JRR Trollkien, Stevietheman, Antandrus, Dasch, Superborsuk, Piotrus, ShakataGaNai, Ot, Rdsmith4, Khaydarian, Sam Hocevar, Bbpen, Dhn, Robin klein, Gerrit, Kousu, Jwdietrich2, Pasquale, Discospinster, Luisrull, Rhobite, Westendgirl, Goochelaar, Bender235, ZeroOne, Brian0918, Rgdboer, Alex Kosorukoff, RoyBoy, Chriscf, Perfecto, Renice, Grick, A Wikipedia user from Minnesota, TACD, Wrs1864, Nsaa, Mdd, Goodoldpolonius, Danski14, Alansohn, Gary, Kurieeto, Goodoldpolonius2, Malo, DreamGuy, RussAbbott, Wtmitchell, Velella, Pixel-enwiki, Versageek, Sether, Recury, SteveWilhelm, Philthecow, 2004-12-29T22:45Z, Mindmatrix, Wnjr, Berti, Zenkat, Faznar, WelshAspie, JeremyA, Wikiklrsc, Eras-mus, Waldir, Imersion, FreplySpang, RadioActive-enwiki, Effeietsanders, Rjwilmsi, Philcomputing-enwiki, Somesh, Vegaswikian, Arbor, ElKevbo, Jehochman, Bhadani, DoubleBlue, Aapo Laitinen, Christion123, Nihiltres, Crazycomputers, Anurag Garg, Jsheehy, Jrtayloriv, Psantora, Chobot, DVdm, UkPaolo, YurikBot, Wavelength, Pip2andahalf, RussBot, Petiatil, Slproxy, Briaboru, Bhny, Mark Ironie, Stephenb, Gaius Cornelius, Ksyrie, Shaddack, Artgirl88, Morphh, NawlinWiki, Nowa, Martin Fasani, Madcoverboy, Koskimaki, Grafen, Drderail, Korny O'Near, Cleared as filed, Jpbowen, Mikeblas, Syrthiss, Hakeem.gadi, Caseyhelbling, Wknight94, PGPirate, Unforgiven24, Tetracube, Getcrunk, Mamawrites, Zzuuzz, 2bar, Raul Lapeira, Kazazz, Josh3580, GraemeL, Nae'blis, Houdani, GrinBot-enwiki, Oldhamlet, Veinor, Joshbuddy, Mosiah-enwiki, JJL, SmackBot, Deborah909, RedHouse18, Reedy, KnowledgeOfSelf, Deon Steyn, Leki, LaurensvanLieshout, Jtneill, Typewriter, Jonobennett, Uttoransen, Gilliam, Ohnoitsjamie, Hmains, Betacommand, Qtoktok, Wykis, Uba33, Sparsefarce, Gragus, OneVeryBadMan, Onorem, Sftone, Azumanga1, CNHolbein, Anthon.Eff, Memming, EPM, Pdixon, Dreadstar, Bigmantonyd,

47.10. TEXT AND IMAGE SOURCES, CONTRIBUTORS, AND LICENSES

RJBurkhart, Marktan, Sadi Carnot, Kukini, Clicketyclack, SashatoBot, Lambiam, Mokshjuneja, JzG, Kuru, Rigadoun, Ptroxler, Timclare, Beetstra, Treyp, Jcbutler, Hu12, DabMachine, Mashable, SimonD, Levineps, Susko, Kencf0618, IvanLanin, Telrod, Danlev, Tawkerbot2, Dave Runger, George100, Dan1679, Dimz793, FatalError, Freud's Genius, JForget, Mikiemike, Rizen, Phauly, DeLarge, Bill.albing, Thomasmeeks, Phnk, Tychay, Myasuda, Rdepontb, Cydebot, JohnUpp, Abeg92, Mato, Tseeker, Gogo Dodo, Solidcore, JFreeman, Pascal.Tesson, Dancter, NMChico24, BetacommandBot, Mattisse, Epbr123, Barticus88, Jheuristic, Pajz, N5iln, ClosedEyesSeeing, Headbomb, Maxelrod, EmilyChew, Markusvinzent, SusanLesch, Alessandriana, Prcoulson, Hmrox, AntiVandalBot, RobotG, Rprout520, Seaphoto, Myrockstar, Emulsionla, Nithin-Bekal, Alessandro De Rossi~enwiki, Alphachimpbot, VictorAnyakin, Danellicus, Sluzzelin, JAnDbot, Tdunvan, Athkalani~enwiki, Barek, MER-C, Dsp13, Tech2blog, Phidman, Rubyji, Dmcgrew1, Lvsubram, Douglas R. White, Korotayev, .anacondabot, Meeples, Magioladitis, Bongwarrior, G S P. Ramanbasu, Avicennasis, Engineman, Isepellissery, Ferran.cabrer, Matt B., Ankitdoshi1, MartinBot, Toracle~enwiki, R'n'B, Mike Restivo, Cnawan, Emily GABLE, RockMFR, J.delanoy, Trusilver, Bellagio99, Maurice Carbonaro, FreshBreeze, Ginsengbomb, McSly, DarwinPeacock, Punkandbarbies, Jazzguy3333, E31029, Cg2916, Bonadea, Shergul, JohnDoe0007, Funandtrvl, Abeusher, Pleasantville, HughD, MultimediaGuru, Sudipdasin, Sdsds, Refsworldlee, Pealmasa, Oliver.perry, Tomsega, Wikidemon, Terryoleary1981, Einphrey, Id4abel, Qxz, Mobnews, Starmuscle, SirReid, LeaveSleaves, Boomgate, UnitedStatesian, Nigelcopley, Tesfatsion, Mitsubishi Zero, STOwiki, Wikigregor~enwiki, Mind123, Lova Falk, Buburuza, Chicago andrew, Socialgovernance, Maallyis, NeniPogarcic, Mnontala, Theredmonkey, Acm1989, Isis07, Mark.Crosby, JulieWohlberg, Kmouly, SieBot, Tresiden, Rlendog, Jauerback, Legion fi, Joshua.hammond, Keilana, Tedstanton, Bentogoa, MaynardClark, Pm master, Oxymoron83, Elsesteban, Joetroll, DancingPhilosopher, Diego Grez-Cañete, Ayumijomori, Sean.hoyland, Hariva, WikiLaurent, Cms3rd, Eredux, Linforest, Shelliwright, Ricklaman, MenoBot, ClueBot, Kl4m, Kai-Hendrik, Jackollie, Rgerstley, Fox, Jthewombat, Redskinsfan1, Jan1nad, Tmp373, Franamax, EddieMo, Jcoplien, Wikijens, Jelena filipov, Darcfudg, EnriqueMurillo, Viper80, Rebecita.angle, Excirial, Nero8858, Nathanbanton, Ashesofwind, Gobeshock Gobochondro Gyanotirtho, Igorberger, Tjwood26, Smundy, Bluemosquito, Skyebend, Jezramsforth, Flower Priest, Apparition11, Peasantwarrior, XLinkBot, Ianmw, Rankiri, Cathyatscholar360, Slifty, Yuwiepal, AndreNatas, NellieBly, Terrillfrantz, Alexius08, MystBot, Camoz87, JanaDiesner, Airplaneman, Haganrich, Kbdankbot, Addbot, Proofreader77, DOI bot, Blethering Scot, Ronhjones, Fieldday-sunday, Heisss, Angrysusan, USchick, Polmorry, MrOllie, LaaknorBot, CarsracBot, JACKAL-XXX, Favonian, Getmoreatp, Cdagnino, WgMp, NYNetwork, Justin534, Rimapacowa, Numbo3-bot, Principe massimo, Tide rolls, Lightbot, Antiselfpromotion, Matěj Grabovský, Zorrobot, Jarble, Billhansen5, Kurtis, Dillardjj, Yobot, Themfromspace, Sogge~enwiki, TaBOT-zerem, Goapsy, Shrikrishnabhardwaj, Stephroj, Eduen, Eric-Wester, Tizzey, AnomieBOT, Pelayo el Sabio, Enisten, Mbiama Assogo Roger, Coethnic~enwiki, Flewis, Dan53, 90 Auto, The High Fin Sperm Whale, Citation bot, Taeshadow, Jtamad, Tezpur4u, SeventhHell, Ppg90828302187, Neurolysis, LilHelpa, EagleWang, Xqbot, Golubchikav~enwiki, Capricorn42, Fabio007fedel, Dickreuter, SocialNetworking, Harvey the rabbit, Quietquite1234, Tdoug870, XEsup, Jmundo, Anna Frodesiak, Habit 247, JoshXF, Shirik, Tore.opsahl, Kevinshroff, Shadowjams, Kwinkunks, Robykiwi~enwiki, JVans, Mmmalexand, Sayyid Muhammad Emadi, Logos.undp, FrescoBot, Vicvicvicvic, Sky Attacker, Jc3s5h, Alxeedo, Phanhaitrieu, IllinoisTeacher, QuéSéYo2, Doitinpublicpr, Jimernst, Louperibot, Citation bot 1, Nnehaa, Cansem, Kiefer.Wolfowitz, Jonesey95, Reconsider the static, Sroel, Trappist the monk, Buddy23Lee, Lotje, LilyKitty, Shavonbradshaw, Cirmion, Antipastor, Hak686, SkiLLru, Reach Out to the Truth, Sharon08tam, DARTH SIDIOUS 2, Pboodark, RjwilmsiBot, Wassermann7, Realglobalist, Me1krishna, Salvio giuliano, Per7, Djjr, EmausBot, Lordknox, WikitanvirBot, Immunize, GUIDIUnknown, Deep1256, Dewritech, Going-Batty, RenamedUser01302013, My-dfp, L235, Itzcyndi, Corrossive, Fæ, Jbtrout, Qiob, Winquan1, Rosti99, Jpvandijk, Richie.lukas, Gz33, Wayne Slam, Erianna, Calenet, M0joer1013, Tectonicura, IGeMiNix, Gray eyes, Coasterlover1994, L Kensington, Hodzha, Socialnetworks, Logos.noreply, Shi Hou, Federalostt, Hamadashow, ClueBot NG, Psorakis, Cocal2, Histree, Quiname, Mesoderm, Gwenchlan~enwiki, MerllwBot, Kataksham100, Helpful Pixie Bot, Vrlab, BG19bot, Bmusician, Tvmlawstudents, Simonrodan, Solar Police, Williamkurman, BMT2627, Hallows AG, Paganinip, Marcocapelle, FaFlo, Usjiechen, TheDarkKnight107, Westminsterx, Meclee, Jacklee2013, Caypartisbot, Racheycocoleman, RroccoMaroc, AustenHead, BobDohse, ChrisGualtieri, Khazar2, Skatalites, Robbord, KLongAus, Gabelglesia, Readyheavygo, Me, Myself, and I are Here, Dianep2013, Miriamdavidadca, Bodzsoo, Ldemat, Mohamed-Ahmed-FG, Ulfaslak90, MikeWilcox417, Fixuture, Sfeverton18, Idanoyes, Eva-schiffer, Pagina100, Nkhemani, Sark7, Pritam.Gundecha, Googleguacamole, Jnce13, Mjguo, Gdrarshs, SlvrKy, Heysarahfu, Babajido, CarleneRenee, Bender the Bot, Enaul, KolbertBot and Anonymous: 812

- **Small-world network** Source: https://en.wikipedia.org/wiki/Small-world_network?oldid=800185761 Contributors: William Avery, Heron, Edward, Michael Hardy, Kku, Meekohi, Karada, Ronz, AugPi, Jitse Niesen, Bevo, Markus Krötzsch, Ds13, Dratman, Mboverload, Discospinster, Lejean2000, Paul August, Fenice, Davidgothberg, Tgr, Danski14, Cheyinka, Linas, Mihai Damian, Kzollman, Stochata, XaosBits, BD2412, Rjwilmsi, NatusRoma, Mbutts, JFromm, Gurch, Debivort, Dbagnall, Wavelength, H005, JabberWok, Madcoverboy, Slarson, Daniel Mietchen, Nethgirb, Cconnett, That Guy, From That Show!, SmackBot, Golbeck, Commander Keane bot, Drknexus, Chris the speller, Oli Filth, Uthbrian, Rludlow, Meson537, Ligulembot, Dankonikolic, Loodog, JHunterJ, Kvng, Dragon guy, Myasuda, Gregbard, Headbomb, KConWiki, David Eppstein, User A1, PaulBHartzog, G.A.S, Urrameu, Joelmiller, Eloz002, AKA MBG, DarwinPeacock, Fheyligh, KylieTastic, Pleasantville, Dggreen, Mkcmkc, Econterms, Wiae, Joetroll, Admiral Norton, Alexbot, Dwiddows, Terrillfrantz, Addbot, DOI bot, Obobskivich, EconoPhysicist, Netzwerkerin, SamatBot, Lightbot, AnomieBOT, Flopsy Mopsy and Cottonmouth, Citation bot, Xqbot, TRNiekras, Anna Frodesiak, Shadowjams, Urgos, Rafaelgoogle, ΙωάννηςΚαραμήτρος, Citation bot 1, Trappist the monk, Wolongzhiyong, Geostudent, Dangling Reference, Dcirovic, Dennishouston, Helpful Pixie Bot, Bibcode Bot, BG19bot, Panchobook, Bereziny, CitationCleanerBot, Zujua, Nevik.R, BattyBot, WH98, ChrisGualtieri, DamascusGirlGeek, MidnightRequestLine, YiFeiBot, Schulllz, Lagoset, Monkbot, ClassicOnAStick, BxtrsChin, Shmall;D, Shifra987, Bsenek, Atashsiah, KolbertBot and Anonymous: 64

- **Centrality** Source: https://en.wikipedia.org/wiki/Centrality?oldid=797812714 Contributors: Michael Hardy, Kku, Cherkash, BAxelrod, Dbabbitt, Giftlite, Sepreece, Piotrus, Icairns, Tgr, Diego Moya, Bkkbrad, Ruud Koot, Stochata, BD2412, Rjwilmsi, Hathawayc, Wavelength, Mgwalker, SmackBot, RedHouse18, Reedy, Kazastankas, Flyingspuds, Michael Rogers, Vigna, Loodog, Eassin, Vanisaac, CmdrObot, Vyznev Xnebara, RafG, Alaibot, Headbomb, Edchi, Utopiah, Douglas R. White, RogierBrussee, JJ Harrison, David Eppstein, Zuludogm, J.delanoy, Rocchini, DarwinPeacock, Mstuomel, Nburden, Borgatts, Sandal bandit, Trumpsternator, ClueBot, Syhon, Sevenp, Frozen4322, Addbot, Fgnievinski, Mrocklin, Netzwerkerin, Upulcranga, Yobot, AnomieBOT, Citation bot, Thegzak, Harthur, Tore.opsahl, FrescoBot, Orubt, Citation bot 1, Fabercap, Jonesey95, Tom.Reding, WaddSpoiley, Mdvs, Ergotius, EbrushERB, Seanandjason, Minimac, RjwilmsiBot, Chmod644, Robd75, Maddendalybrokaw, Shil88, Psorakis, ChristophE, BG19bot, Paolo Lipparini, Matthieu Vergne, Sundirac, Compsim, Kroq-gar78, Marlon'n'marion, Mogism, 7804j, Carlwitt, Cypherquest, Aftersox, Yeda123, Morgoth106, Monkbot, Dvoina13, Spariggio82, Jmagdanz, Magriteappleface, GlennLawyer, Loraof, Trantd.vn, Jean-Pierre de la Croix, Zachwf, HelpUsStopSpam, Tapiocozzo, Ajalvare, InternetArchiveBot, GreenC bot, Hyperbolick and Anonymous: 90

- **Network motif** Source: https://en.wikipedia.org/wiki/Network_motif?oldid=801767018 Contributors: Rich Farmbrough, Btyner, Rjwilmsi,

Jknabe, Madcoverboy, Opethmetal-enwiki, Mebden, SmackBot, RDBrown, Tdmg, CmdrObot, Headbomb, Nono64, J.delanoy, Metophile, Duncan.Hull, Marashie, FghIJklm, Download, Netzwerkerin, Yobot, Citation bot, LilHelpa, 564dude, Jesse V., Reverus, EmausBot, Dcirovic, Puppetmaster87, Ryanalane, A930913, Wmayner, Boris Breuer, ChristophE, BG19bot, GKFX, Radu.zabet, Khazar2, Dan traveller, Illia Connell, Skhakabi, Yaarasegre, Jodosma, Luismeira, Tchanders, Anrmusna, Monkbot, Srijankedia, Jorge Guerra Pires, RolandKluge, Dmuhs, Stormlovetao and Anonymous: 32

- **Graph theory** *Source:* https://en.wikipedia.org/wiki/Graph_theory?oldid=803057103 *Contributors:* AxelBoldt, Kpjas, LC-enwiki, Robert Merkel, Zundark, Taw, Jeronimo, BlckKnght, Dze27, Oskar Flordal, Andre Engels, Karl E. V. Palmen, Shd-enwiki, XJaM, JeLuF, Arvindn, Gianfranco, Matusz, PierreAbbat, Miguel-enwiki, Boleslav Bobcik, FvdP, Camembert, Hirzel, Tomo, Patrick, Chas zzz brown, Michael Hardy, Wshun, Chris-martin, Glinos, Meekohi, Jakob Voss, TakuyaMurata, GTBacchus, Grog-enwiki, Pcb21, Dgrant, CesarB, Looxix-enwiki, Ellywa, Ams80, Ronz, Nanshu, Gyan, Nichtich-enwiki, Mark Foskey, Александър, Poor Yorick, Caramdir-enwiki, Mxn, Charles Matthews, Berteun, Almi, Hbruhn, Dysprosia, Daniel Quinlan, Gutza, Doradus, Zoicon5, Roachmeister, Populus, Zero0000, Doctorbozzball, McKay, Shizhao, Optim, Robbot, Brent Gulanowski, Fredrik, Altenmann, Dittaeva, Gandalf61, MathMartin, Sverdrup, Puckly, KellyCoinGuy, Thesilverbail, Bkell, Paul Murray, Fuelbottle, ElBenevolente, Aknxy, Dina, Tea2min, Giftlite, Dbenbenn, Thv, The Cave Troll, Elf, Lupin, Brona, Pashute, Duncharris, Andris, Jorge Stolfi, Tyir, Sundar, GGordonWorleyIII, Alan Au, Bact, Knutux, APH, Tomruen, Tyler McHenry, Naerbnic, Peter Kwok, Robin klein, Ratiocinate, Andreas Kaufmann, Chmod007, Madewokherd, Discospinster, Solitude, Guanabot, Qutezuce, Mani1, Paul August, Bender235, Zaslav, Tompw, Diego UFCG-enwiki, Chalst, Shanes, Renice, C S, Csl77, Jojit fb, Photonique, Jonsafari, Obradovic Goran, Tsirel, Jumbuck, Msh210, Alansohn, Liao, Mailer diablo, Marianocecowski, Aquae, Blair Azzopardi, Oleg Alexandrov, Youngster68, Linas, LOL, Ruud Koot, Tckma, Astrophil, Davidfstr, GregorB, SCEhardt, Stochata, Xiong, Graham87, Magister Mathematicae, BD2412, SixWingedSeraph, Rjwilmsi, Gmelli, George Burgess, Eugeneiim, Arbor, Kalogeropoulos, Fred Bradstadt, FayssalF, FlaBot, PaulHoadley, RexNL, Vonkje, Chobot, Jinma, YurikBot, Wavelength, Michael Slone, Gaius Cornelius, Alex Bakharev, Morphh, SEWilcoxBot, Jaxl, Ino5hiro, Xdenizen, Daniel Mietchen, Shepazu, Voidxor, Rev3nant, Lt-wiki-bot, Jwissick, Arthur Rubin, Netrapt, LeonardoRob0t, Agro1986, Eric.weigle, Allens, Sardanaphalus, Melchoir, Brick Thrower, Ohnoitsjamie, Oli Filth, OrangeDog, Taxipom, Esokullu, DHN-bot-enwiki, Tsca.bot, Onorem, GraphTheoryPwns, Lpgeffen, Jon Awbrey, Henning Makholm, Mlpkr, SashatoBot, Whyfish, Disavian, MynameisJayden, Idiosyncratic-bumblebee, Dicklyon, Quaeler, Lanem, Tawkerbot2, Ylloh, Mahlerite, CRGreathouse, Dycedarg, CBM, Requestion, Bumbulski, Myasuda, RUVARD, The Isiah, Ntsimp, Abeg92, Corpx, DumbBOT, Anthonynow12, Thijs!bot, Jheuristic, King Bee, Pstanton, Hazmat2, Mojo Hand, Headbomb, Marek69, Eleuther, AntiVandalBot, Whiteknox, Hannes Eder, Spacefarer, Myanw, JAnDbot, MER-C, The Transhumanist, Igodard, Restname, Sangak, Tmusgrove, Feeeshboy, Usien6, Ldecola, David Eppstein, Kope, DerHexer, Oroso, MartinBot, R'n'B, Uncle Dick, Joespiff, Ignatzmice, Shikhar1986, Tarotcards, Policron, XxjwuxX, Yecril, JohnBlackburne, Dggreen, Oshwah, Anonymous Dissident, Alcidesfonseca, Anna Lincoln, Ocolon, Magmi, PaulTanenbaum, Geometry guy, Fivelittlemonkeys, Sacredmint, Spitfire8520, Radagast3, SieBot, Dawn Bard, Toddst1, Jon har, Bananastalktome, Titanic4000, Beda42, Maxime.Debosschere, Aechase1, Damien Karras, ClueBot, DFRussia, PipepBot, Justin W Smith, Vacio, Wraithful, Garyzx, Mild Bill Hiccup, DragonBot, Fchristo, Hans Adler, Dafyddg, Razorflame, Rmiesen, Kruusamägi, Pugget, Darkicebot, XLinkBot, BodhisattvaBot, Dekart, Tangi-tamma, Addbot, Dr.S.Ramachandran, Cerber, DOI bot, Ronhjones, Low-frequency internal, CanadianLinuxUser, MrOllie, Protonk, LaaknorBot, Smoke73, Delaszk, Favonian, Maurobio, Lightbot, Jarble, Ettrig, Luckas-bot, Yobot, Kilom691, Trinitrix, Jean.julius, AnomieBOT, Womiller99, Sonia, Jim1138, Piano non troppo, Gragragra, RandomAct, Citation bot, Ayda D, Xqbot, Jerome zhu, Capricorn42, Nasnema, Miym, GiveAFishABone, RibotBOT, Jalpar75, Aaditya 7, Ankitbhatt, MultiPoly, FrescoBot, Mark Renier, SlumdogAramis, Citation bot 1, Launchballer, Sibian, Maggyero, Pinethicket, RobinK, Wsu-dm-jb, D75304, Wsu-f, Xnn, Obankston, Andrea105, RjwilmsiBot, TjBot, Powerthirst123, Aaronzat, EmausBot, Domesticenginerd, EleferenBot, Jmencisom, Dcirovic, Slawekb, Akutagawa10, D.Lazard, Netha Hussain, Tolly4bolly, ChuispastonBot, EdoBot, ClueBot NG, Wcherowi, Watersmeetfreak, Matthiaspaul, MelbourneStar, Outraged duck, OMurgo, Bazuz, Aks1521, Masssly, Joel B. Lewis, Johnsopc, HMSSolent, 4368a, BG19bot, Ajweinstein, Канеюку, MusikAnimal, AvocatoBot, Bereziny, Brad7777, Sofia karampataki, ChrisGualtieri, GoShow, SuperbowserX, Dexbot, DaltonCastle, Cerabot-enwiki, Omgigotanaccount, Me, Myself, and I are Here, Wikiisgreat123, Faizan, Maxwell bernard, Bg9989, Zsoftua, SakeUPenn, Yloreander, StaticElectricity, Gold4444, Cyborgbadger, Monkbot, Zachwaltman, Gr pbi, Tourorist, Jdcomix, KasparBot, Lr0^^k, PA Math Prof, Meachamus.Prime, Baking Soda, KGirlTrucker81, GreenC bot, Ywang416, Anish karimaloor, Jetroberts, Mitsou-dewiki, Disuja1975, JCW-CleanerBot and Anonymous: 414

- **Scalability** *Source:* https://en.wikipedia.org/wiki/Scalability?oldid=799109920 *Contributors:* Kpjas, The Anome, Awaterl, Matusz, Michael Hardy, Kku, TakuyaMurata, Jjshapiro, Bearcat, Sander123, Jondel, Dbroadwell, SpellBott, Lysy, Javidjamae, Leonard G., Stevietheman, Gdr, Beland, Urhixidur, Hugh Mason, Ferrans, Shiftchange, FT2, Mazi, Dtremenak, Liao, Calton, Pion, Suruena, Kusma, Mattbrundage, Tyz, Undefined-enwiki, BD2412, Rjwilmsi, Quiddity, Williamborg, Fred Bradstadt, Aapo Laitinen, FlaBot, Intgr, Dalef, Agil-enwiki, YurikBot, Whoisjohngalt, NTBot-enwiki, Michael Slone, Bovineone, Moe Epsilon, Leotohill, .marc., Xpclient, LeonardoRob0t, Stumps, SmackBot, Irnavash, KelleyCook, Ohnoitsjamie, Thumperward, Jammus, Javalenok, Ascentury, Frap, JonHarder, Cyhatch, BWDuncan, Andrei Stroe, Harryboyles, Writtenonsand, 16@r, Swartik, Hu12, UncleDouggie, Tawkerbot2, FatalError, CBM, Cydebot, Thijs!bot, Uiteoi, Marokwitz, Kdakin, JAnDbot, NapoliRoma, Crazytonyi, Shar1R, SunSw0rd, Raffen, Joshua Davis, FienX, RockMFR, Shellwood, Auroramatt, 1000Faces, NewEnglandYankee, Doria, Jottinger, Izno, VolkovBot, TXiKiBoT, CHaoTiCa, Falcon8765, Suction Man, Bpringlemeir, Paladin1979, StAnselm, DigitalDave42, JCLately, Luciole2013, Gp5588, Sanya3, Dangelow, Nvrijn, Elnon, Tearaway, Mild Bill Hiccup, Saruvu2k, M4gnum0n, Friendlydata, Shiro jdn, MPH007, XLinkBot, Philippe Giabbanelli, Avoided, Klungel, MystBot, Dsimic, Addbot, Jncraton, Tonkie67, Fluffernutter, MrOllie, Latiligence, Kiril Simeonovski, Teles, Luckas-bot, Yobot, Davew haverford, Terrifictriffid, AnomieBOT, Materialscientist, Obersachsebot, Xqbot, Miym, GrouchoBot, W Nowicki, Sae1962, MastiBot, Jandalhandler, Akolyth, Jesse V., Mean as custard, Gf uip, EmausBot, John of Reading, Anirudh Emani, Josve05a, Cosmoskramer, AManWithNoPlan, Music Sorter, Tsipi, ChuispastonBot, ClueBot NG, Widr, Daniel Minor, Meniv, Helpful Pixie Bot, MarkusWinand, Electriccatfish2, BG19bot, ElphiBot, Wikicadger, Anbu121, Srenniw, BattyBot, Cyberbot II, CGBoas, K0zka, Mwaci11, Paul2520, Shahbazali101, Slashdottir, Igorghisi, Vieque, KN2731, Verbal.noun, Mantraman701, KasparBot, The Quixotic Potato, KealanJH, InternetArchiveBot, Bender the Bot, DFinlaysonMD, Zhanjiaxin, Ramesh Namachivayam, Biografer and Anonymous: 134

- **Robustness (computer science)** *Source:* https://en.wikipedia.org/wiki/Robustness_(computer_science)?oldid=803397853 *Contributors:* Kku, Marteau, Silverfish, Tea2min, Wiki Wikardo, Beland, Shiftchange, Bgwhite, Gilliam, Derek R Bullamore, Wizard191, Avigreen18, Magioladitis, Swpb, Chrisportelli, Sun Creator, Addbot, Yobot, AnomieBOT, Materialscientist, Makeswell, FrescoBot, Pepper, RedBot, EmausBot, WikitanvirBot, Cat4567nip, ChuispastonBot, ClueBot NG, Trunks ishida, BG19bot, Juro2351, ElectricUvula, Yegor256, Kpalaniswamy, Alpha Kand, YFdyh-bot, BethNaught, Dtgee and Anonymous: 24

47.10. TEXT AND IMAGE SOURCES, CONTRIBUTORS, AND LICENSES

- **Systems biology** *Source:* https://en.wikipedia.org/wiki/Systems_biology?oldid=801391167 *Contributors:* Edward, Michael Hardy, Lexor, Kku, Netsnipe, Zoicon5, Steinsky, Fredrik, Stewartadcock, Jondel, Fuelbottle, Alan Liefting, Giftlite, Dmb000006, Waltpohl, Tagishsimon, Quarl, APH, Thorwald, D6, Jwdietrich2, Discospinster, Rich Farmbrough, Bender235, Ceolas, Fenice, Syp, DanielNuyu, Zlite, Lenov, Rajah, Srlasky, Mdd, Arthena, Ombudsman, GJeffery, Danhash, Bobrayner, Rwcitek, CharonZ, Ggonnell, Halx, BD2412, Grammarbot, Kane5187, Rjwilmsi, Edaddison, Mohawkjohn, Jongbhak, Nihiltres, NavarroJ, Vonkje, Chobot, Whosasking, Electric sheep~enwiki, YurikBot, Wavelength, Neilbeach, PaulGarner, Aeusoes1, Cquan, Jpbowen, Kkmurray, HexiToFor~enwiki, Zargulon, Arthur Rubin, SmackBot, Eskimbot, Ohnoitsjamie, Betacommand, David Ludwig, Jethero, RDBrown, Miguel Andrade, DHN-bot~enwiki, Can't sleep, clown will eat me, Sholto Maud, Karthik.raman, Kcordina, SeanAhern, Kleuske, O RLY?, Mkotl, Nick Green, Ben Moore, Ckatz, Satish.vamm, Pkahlem, Lkathmann, Bad Cat, Andreas td, FLeader, Thomas81, Patho~enwiki, CRGreathouse, CmdrObot, N2e, Pgr94, CX, Klipkow, Skittleys, Blueleezard, Nbaliga, Narayanese, Michael Fourman, Letranova, Thijs!bot, Droyarzun, Crodriguel, Second Quantization, Lauranrg, Vangos, Mbadri, Opertinicy, Natelewis, Asadrahman, NBeale, Gem, Freshacconci, Magioladitis, VoABot II, MikeHucka, BatteryIncluded, Charlenelieu, JaGa, WLU, Synthetic Biologist, Molelect, Pvosta, Yobol, Garychurchill, Urselius, Nono64, Erkan Yilmaz, Robnpov, Senu, Unauthorised Immunophysicist, Boku wa kage, Gwolfe, Erick.Antezana, Eveillar, Pleasantville, AlnoktaBOT, Fences and windows, Triamus, Benjamin Barenblat, Rvencio, SeeGee, JRSocInterface, IPSOS, Amaher, Meters, ColinGillespie, Biochaos, Amandadawnbesemer, Flyer22 Reborn, JulioVeraGon, Strife911, Ngriffeth, Svick, Lilia Alberghina, Massbiotech, Drgarden, Heisner, Touchstone42, ClueBot, DFRussia, Trivialist, NIH Media, ChandlerMapBot, Aua, Klenod, Calimo, Insouciantfiend, Agor153, Mobashirgenome, Dubitzky, Versus22, Cantor, Djhbrown, DumZiBoT, MystBot, Addbot, DOI bot, Captain-tucker, Seuss01, Daehlerr, MrOllie, Reggiebird, Favonian, Foggy29, Duelist135, جر, Luckas-bot, Yobot, Ptbotgourou, Dirk Hans, Mmxx, AnomieBOT, Zinovyev, Citation bot, Bci2, Biophysik, LilHelpa, AlirezaShaneh, Amirsnk, Bamess, Xqbot, Mkuiper, Bio-ITWorld, Oddleik, Xeaa, Dr Oldekop, RibotBOT, Jdegreef, Linkman21, Claronow, Fletcher04, FrescoBot, KirbyRandolf, Gdrahnier, GabEuro, Haeinous, Citation bot 1, PointOfPresence, Nemenman, DrilBot, OrcaMorgan, Trappist the monk, Kieran Mace, Amkilpatrick, GlenBrydon, Gauravsjbrana, Slon02, EmausBot, Smythph, Clayrat, Listmeister, ZéroBot, U+003F, Sunur7, EdoBot, TheoThompson, Giovannistefano35, Helpful Pixie Bot, Curb Chain, Bibcode Bot, BG19bot, Stevetihi, Fred Stupor Mundi, Benedict Pope, Zuck3434, Snow Rise, Hequba, DGRichard, The Bald Russian, Avimaayan2012, Dexbot, Joeinwiki, Mark viking, Randykitty, Hsyoo, Aszarsha, A1078558, Andy Quarry, Evolution and evolvability, G13140, Éthèdaligrh, H4n9t3nn, Ethically Yours, Nekokoe, Monkbot, TVleck1971, Jorge Guerra Pires, Loraof, Malloy.65, Bhishek, Shikang Liu, Puccio.b, KasparBot, CAPTAIN RAJU, Ramblingsteve, Erictleung, Quinton Feldberg, Dockabo, PLAKK, Sincosx, Timm6205, KolbertBot and Anonymous: 262

- **Dynamic network analysis** *Source:* https://en.wikipedia.org/wiki/Dynamic_network_analysis?oldid=801558266 *Contributors:* Michael Hardy, Imersion, SmackBot, Bluebot, JonHarder, Beetstra, Ace Frahm, Dan1679, Argon233, Diabloblue, Mattdereno, Porqin, MarshBot, Douglas R. White, SiobhanHansa, AZK, Erkan Yilmaz, Supertabular, Supernet, Wikidemon, Mind123, Seinecle, Melcombe, Terrillfrantz, MrOllie, Delaszk, Drpickem, Lauyukpui, 477TataB, Anna Frodesiak, Yassens, Nameless23, LittleWink, Deirovic, Robbiemorrison, Payyan-2000, Ricardohz, BG19bot, Meclee, Wade.kodrin, Mboydos, Me, Myself, and I are Here, Faizan, Totobalito, Monkbot, Pcurley, CAPTAIN RAJU, Liu.zhen337, JCW-CleanerBot and Anonymous: 23

- **Complex adaptive system** *Source:* https://en.wikipedia.org/wiki/Complex_adaptive_system?oldid=798802726 *Contributors:* Michael Hardy, Lexor, Kku, Ronz, Goethean, Blainster, Giftlite, Robin klein, RevRagnarok, LeeHunter, Fenice, Viriditas, Flammifer, Mdd, Gwendal (usurped), Acadac, BDD, Nightstallion, Ceyockey, Firien, BD2412, Imersion, Rjwilmsi, JFromm, Bgwhite, FrankTobia, YurikBot, Bhny, Arthur Rubin, Garion96, SmackBot, Betacommand, Tyciol, RDBrown, RandyBurge, John D. Croft, Jon Awbrey, Dankonikolic, Sina2, Beetstra, Dl2000, Mmwaldrop, Mr3641, George100, N2e, AndrewHowse, Cydebot, Thijs!bot, Headbomb, Pbramer, TimVickers, Eurobas, Magioladitis, Snowded, Filll, Maurice Carbonaro, Fheyligh, STBotD, Kenneth M Burke, Korotkikh, Slowhand181, Elizabeth McMillan, Jeff G., Dggreen, Lordvolton, IPSOS, Mfmoore, Tesfatsion, Jamelan, Kilmer-san, Slowwriter, Scarian, Dr Paul Thomas, LSmok3, Escape Orbit, ShelleyAdams, Kai-Hendrik, R000t, Spirilis, Alexbot, Sun Creator, Chris4uk, Bcastel3, XLinkBot, NICO-CANet, Rholladay1, Addbot, DOI bot, Montgomery '39, Garrycl, MrOllie, LaaknorBot, Meamus, Моржан, Luckas-bot, Yobot, DrPTThomas, Jean.julius, Cmbarton54, AnomieBOT, Steamturn, Ivythorn, Pandroozie, Citation bot, ArthurBot, GrouchoBot, 9645a9645, FrescoBot, Citation bot 1, RedBot, Niazim1, Slatteryz, Jarpup, Rayman60, Dalssoft, GoingBatty, Deirovic, To0808, ClueBot NG, Georgehobbs, Torbrax, Helpful Pixie Bot, Bibcode Bot, BG19bot, Marcocapelle, Pvnanini, ChrisGualtieri, IjonTichyIjonTichy, Peterknows, Isarra (HG), Me, Myself, and I are Here, Mark viking, LogertGilby, Biogeographist, Jodosma, Ginsuloft, Joe.w.graham, NobreSoldado, Rubbish computer, Cruella wiki, InternetArchiveBot, Molisi, Dadif'89, Driverin, JerseyWill, LPonyets and Anonymous: 96

- **Evolution** *Source:* https://en.wikipedia.org/wiki/Evolution?oldid=802993154 *Contributors:* AxelBoldt, Paul Drye, Lee Daniel Crocker, Eloquence, Vicki Rosenzweig, Mav, Wesley, Bryan Derksen, Robert Merkel, Zundark, The Anome, Taw, Slrubenstein, David Merrill, Ed Poor, RK, Alex.tan, LA2, Josh Grosse, Danny, XJaM, Rgamble, Fredbauder, Christian List, Matusz, PierreAbbat, Fubar Obfusco, M~enwiki, Hannes Hirzel, Zoe, AdamRetchless, Graft, Heron, Camemberт, Ryguasu, Jaknouse, Tijmz, Gog, Q~enwiki, Emmett~enwiki, Bth, Netesq, R Lowry, Fxmastermind, Tbarron, Someone else, Leandrod, Stevertigo, Edward, Dwmyers, Chas zzz brown, Smelialichu, JohnOwens, Michael Hardy, Alan Peakall, Ezra Wax, Zashaw, Fred Bauder, Lexor, DopefishJustin, Dante Alighieri, Vacilandois, Shyamal, MartinHarper, Gabbe, Jketola, Tannin, Ixfd64, Dcljr, Cyde, Shoaler, Delirium, Alfio, Kosebamse, Goatasaur, 168..., MartinSpamer, Ahoerstemeier, Cyp, Jimfbleak, JWSchmidt, Bueller 007, Darkwind, Randywombat, Andrewa, Michael, Kevin Baas, Julesd, Amcaja, Bogdangiusca, Nikai, Evercat, TonyClarke, KayEss, Mxn, Quizkajer, Valluvan~enwiki, Ec5618, Vanished user 5zariu3jisj0j4irj, Timwi, Mkrose, Dcoetzee, Andrevan, Lfh, Lou Sander, Fuzheado, Rednblu, Greenrd, Wik, Steinsky, Quux, DJ Clayworth, Haukurth, Markhurd, Tpbradbury, Marshman, Maximus Rex, Hyacinth, Azra99, Saltine, Kaal, Ed g2s, Philopp, Tom Allen, Samsara, Thue, Bevo, Jecar, Rls, Fvw, Wilke, Raul654, Power~enwiki, Wetman, Johnleemk, Flockmeal, DLR (usurped), David.Monniaux, Frazzydee, JackH, AnthonyQBachler, Pollinator, Shantavira, Jason Potter, Rossnixon, Frank A. Pigsonthewing, Fredrik, Daelin, Moondyne, Goethean, ZimZalaBim, Altenmann, Nurg, Ppe42, Naddy, Modulatum, Sam Spade, Lowellian, Mirv, Postdlf, Academic Challenger, Texture, Premeditated Chaos, Auric, Blainster, Hippietrail, Davodd, Hadal, UtherSRG, Johnstone, MykReeve, Raeky, Anthony, HaeB, Arm, Pengo, Dina, Pablo-flores, Alan Liefting, Enochlau, Mor~enwiki, Stirling Newberry, Ancheta Wis, Michael2, Connelly, Centrx, Giftlite, Christopher Parham, MPF, Recentchanges, Polsmeth, Pretzelpaws, Kim Bruning, Wolfkeeper, Lethe, Fastfission, MSGJ, Peruvianllama, Wwoods, Chardon, Everyking, No Guru, Curps, Michael Devore, FeloniousMonk, Scott Noyes, Maver1ck, Faradn, Thetorpedodog, Duncharris, H-2-O, Iota, Ptk~enwiki, Sundar, Siroxo, Gzornenplatz, Chameleon, SWAdair, Kandar, Christopherlin, Wmahan, MSTCrow, Reilly, Adenosine, ChicXulub, Mporch, StuartH, Gadfium, Tpduden, Utcursch, Pgan002, Dullhunk, Alexf, Lst27, Traumerei, Nova77, Geni, Popefauvexxiii, CryptoDerk, Vanished user svinet8j3ogifm98wjfgoi3tjosfg, Slowking Man, GeneralPatton, Antandrus, Beland, Onco p53, GD~enwiki, Thomas Veil, MisfitToys, Piotrus, Elembis, Vanished user 1234567890, PDH, Saucepan, PhDP, Jossi, Gmlk, Rd-smith4, LeperColony, JimWae, Tomandlu, Ricimer, Tatarize, RetiredUser2, Tothebarricades.tk, Bodnotbod, Icairns, Tail, Vasile, Timothy Usher, J0m1eisler, Kmweber, Cynical, Gscshoyru, ArcticFrog, Jmeppley, Asbestos, Neutrality, Sam, Joyous!, Irpen, JohnArmagh, Kelsey Francis, Syvanen, Muijz, MattDal, Barnaby dawson, Adashiel, Trevor MacInnis, Randwicked, Grunt, Eisnel, Canterbury Tail, Mike Jones, Kate, DavidL (usurped), Mike Rosoft, Rfl, Freakofnurture, Duja, DanielCD, Jiy, Mindspillage, Noisy, Discospinster, ElTyrant, Rich Farmbrough, KillerChihuahua, Rhobite, Leibniz, FT2, Oliver Lineham, Regebro, Paulr-enwiki, Vsmith, Jpk, Silence, Dave souza, Fourier-enwiki, David Schaich, Ascánder, Samboy, Nondescript, Grutter, Jimaltieri, Paul August, MarkS, Robertbowerman, SpookyMulder, Bender235, ESkog, Android79, Jaberwocky6669, Kaisershatner, Ignignot, Brian0918, BlueNight, Frenchgeek, CanisRufus, Aecis, Mr. Billion, El C, Mwanner, Chairboy, Aude, Shanes, El-Spectre, Art LaPella, RoyBoy, Keno, Maqsarian, Adambro, Guettarda, Causa sui, Bobo192, Cretog8, O18, NetBot, Adraeus, Walkiped, BrokenSegue, StoatBringer, Wisdom89, .:Ajvol:., Naoh-enwiki, Adrian-enwiki, 9SGjOSfyHJaQVsEmy9NS, Slicky, La goutte de pluie, Malcolm rowe, Hob Gadling, Athf1234, Numerousfalx, Cinnamon-enwiki, Pschemp, Ardric47, Apostrophe, Vanished user 19794758563875, John Fader, Haham hanuka, Pharos, Crust, Fox1, Alimustafakhan, Matthewcieplak, Orangemarlin, Swaggart, GK, Ranveig, Süm, Alansohn, JYolkowski, Sully, Etxrge, Theuser, Duffman-enwiki, Hackwrench, Polarscribe, Jordan117, Jamyskis, 119, Borisblue, Uogl, Mattpickman, Rd232, Andrewpmk, Craigy144, Raymond, Plumbago, Robin Johnson, CheeseDreams, Wikidea, Demi, Jnothman, Aza-Toth, Yamla, Fraslet, Lightdarkness, InShaneee, Dark Shikari, Warpsmith, BG-enwiki, SMesser, Spangineer, Hu, Malo, Titanium Dragon, Bart133, Caesura, DreamGuy, Schaefer, Gilgameshfuel, Isaac, Binabik80, ClockworkSoul, Gdavidp, KingTT, Rebroad, Mcy jerry, Almafeta, Knowledge Seeker, Aaarrrggh, Harej, Tony Sidaway, RJFJR, Mikeo, Mcmillin24, LFaraone, Ndteegarden, Gortu, Itsmine, Versageek, Axeman89, Oking83, HenryLi, Bookandcoffee, Ceyockey, Harvestdancer, Angelo, Tariqabjotu, Tom.k, A D Monroe III, Siafu, StuTheSheep, Hojimachong, DarTar, Stemonitis, BerndH, Gmaxwell, Googleaseerch, Weyes, Boothy443, Richard Arthur Norton (1958-), Kelly Martin, Firsfron, Jeffrey O. Gustafson, Nicklott, OwenX, Woohookitty, Garylhewitt, Mindmatrix, TigerShark, Anilocra, StradivariusTV, Uncle G, Benhocking, TomTheHand, Poiuyt Man, Mazca, BlankVerse, Veratien, Urod, WadeSimMiser, JeremyA, MONGO, Alexwebb2, Jok2000, Yqbd, Trevor Andersen, Bbatsell, Thesquire, Striver, Sengkang, GregorB, Maartenvdbent, Andromeda321, Macaddct1984, Karmosin, Waldir, Mary Calm, Zzyzx11, TheAlphaWolf, GalaazV, Wayward, Joke137, Eriathwen, MarcoTolo, Raybechard, Temtem, Boffey, LexCorp, Paxsimius, GSlicer, Sin-man, RichardWeiss, Rnt20, Ashmoo, Graham87, Marskell, WBardwin, Alienus, Deltabeignet, Magister Mathematicae, Jan van Male, Ilya, Galwhaa, Teflon Don, Chun-hian, FreplySpang, RxS, Jclemens, Bikeable, Dpr, Vanderdecken, Tsackton, Whoutz, Sjö, Drbogdan, Rjwilmsi, Shauri, Koavf, Kugamazog, Fred Hsu, Panoptical, JedRothwell, Vary, Ikh, MarSch, Bill37212, Thomas Arelatensis, MZMcBride, Scott1328, Tawker, Skywriting, SpNeo, Tixity, You wouldnt dare block me.., Abuzin-enwiki, ZA-enwiki, Jlefler, Oblivious, Ligulem, Fisher-enwiki, Jehochman, RE, Bubba73, Durin, Brighterorange, The wub, TheIncredibleEdibleOompaLoompa, Sarg, FlavrSavr, MikeJ9919, Wdavies, Sango123, Jesus Is Love, Yamamoto Ichiro, Dracontes, Mayumi, Leithp, Gringer, Scorpionman, Dionyseus, Titoxd, JeffStickney, Ian Pitchford, RobertG, Mishuletz, Salva31, Winhunter, Nihiltres, JiFish, Gary Cziko, GünniX, Nivix, Chanting Fox, Pathoschild, RexNL, Gurch, Verybigfish86, Theforester, Jumbo Whales-enwiki, NoSeptember, Jrtayloriv, TeaDrinker, Skiddum, Wikipedia Administration, Alphachimp, Diza, Andriesb, Tysto, Daycd, BradBeattie, TheSun, Butros, WhyBeNormal, Vind r, CJLL Wright, Nicholasink, Kazuba, Parallel or Together?, Bornhj, DVdm, Mhking, Digitalme, Gwernol, Algebraist, Jimbobsween, Uriah923, The Rambling Man, Wavelength, John Stumbles, TheTrueSora, Deeptrivia, Tznkai, Arof, JustSomeKid, John Callender, RussBot, Crazytales, PWhittle, Pippo2001, WAvegetarian, Loom91, Splash, Mark Ironie, Chris Capoccia, Marcperkel, Ytrottier, SpuriousQ, Wavesmikey, Van der Hoorn, Stephenb, Lord Voldemort, Rintrah, Flo98, Gaius Cornelius, Mike Young, Quadraxis, CambridgeBayWeather, Kyorosuke, Chaos, Owendude1210, Wimt, Ritchy, Ugur Basak, Pftaylor, Shanel, NawlinWiki, Nowa, Dysmorodrepanis-enwiki, Wiki alf, Spike Wilbury, ChadThomson, Waldow, Neural, Wtfmate, Rjoebrandon, Nick Thorne, NickBush24, Raymondofrish, Tailpig, Chunky Rice, Taco325i, Uberjivy, Joelr31, Stephen e nelson, Arker, Sitearm, BirgitteSB, The Obfuscator, Thevenerablez, Apokryltaros, Irishguy, Nick, Matnkat, Kdbuffalo, Jpowell, Banes, Daniel Mietchen, PhilipO, Rmky87, Pyroclastic, Wap, Sfnhltb, Stevenwmccrary58, Misza13, Killdevil, Iamnotanorange-enwiki, Tony1, Epipelagic, Bucketsofg, Rich333, JNeal, Aaron Schulz, Nethgirb, DeadEyeArrow, Psy guy, Pierpontpaul2351, Sleepnomore, Elkman, Haemo, Maunus, Unclekirk, Nick123, Joshurtree, Wknight94, Ms2ger, Igiffin, WAS 4.250, FF2010, Illuminatiscott, Calaschysm, Enormousdude, Elijya, Theodolite, Zzuuzz, Andrew Lancaster, Encephalon, NAZISMISNTCOOL, RDF, Cynicism addict, Nikkimaria, Rhynchosaur, Closedmouth, Fang Aili, Davril2020, Th1rt3en, Varith, Noodleman, Petri Krohn, GraemeL, JoanneB, TBadger, Waterspyder, Alias Flood, Delphinian, Karatenerd, Emc2, Willtron, Spliffy, ArielGold, Phsource, X-mass, Skittle, Staxringold, Sancassania, Kungfuadam, RG2, Sciolus, NeilN, Shadowin, Dabs, DVD R W, CIreland, Eenu, Tobyk777, AndrewWTaylor, Luk, Snalwibma, FieryPhoenix, Tynews2001, Tuatara, A bit iffy, Remiel, Laurence Boyce, SmackBot, YellowMonkey, Teenwriter, ManaUser, PiCo, Ashenai, Jclerman, Haymaker, David Kernow, Jedi Davideus, R0M4NC1NG H3LL, Prodego, KnowledgeOfSelf, Olorin28, McGeddon, David.Mestel, Pgk, C.Fred, Jim62sch, Ramdrake, Rokfaith, Phaldo, Jacek Kendysz, Jagged 85, Nickst, EncycloPetey, Jrockley, Eaglizard, Frymaster, Dafergu3, DLH, David Fuchs, Tim@, Xaosflux, Portillo, Watercactus3210, Jdfoote, Ohnoitsjamie, Betacommand, Chaojoker, ERcheck, Squiddy, Qtoktok, Saros136, Chris the speller, Master Jay, Kurykh, Audacity, SlimJim, Persian Poet Gal, RDBrown, Justforasecond, Jdhunt, Raymond arritt, Kitzke, Ediblehuman, Silly rabbit, Muza47, SchfiftyThree, Joel.Gilmore, The Rogue Penguin, Afasmit, Deli nk, Dustimagic, J. Spencer, Jfsamper, Kungming2, Delta Tango, CMacMillan, Pasado, Cassan, Gracenotes, Mikker, Clean Sweep, Zachorious, DocJohnny, Simpsons contributor, Zsinj, Can't sleep, clown will eat me, Briguy, Jeffhre, AVengel, Smallbones, Danielkueh, Mourn, Max David, The Placebo Effect, Awon, Onorem, Valich, Vanished User 0001, Nixeagle, Sommers, Snowmanradio, Avb, MHall, H-J-Niemann, Darthgriz98, Korinkami, Xiner, Justbob, Homestarmy, Jesus Loves You, Flubbit, Edivorce, Midnightcomm, Meepster, Phaedriel, Jedi of redwall, Stevenmitchell, Tyro55555, Aldaron, PrometheusX303, Smooth O, PopiethePopester, Mojosam, CanDo, Cybercobra, Khukri, Digresser, Nakon, Savidan, TedE, Rossman123, VegaDark, Ravenswood Media, Minasbeede, RJN, John D. Croft, MichaelBillington, Andymarczak, Ebush, Whoistheroach, Fgleb, Richard001, Paul H., MrMorgan, PTHS, Nrcprm2026, Squeakytoad, Mirlen, MBCF, Learscad, BlueGoose, Agentmoose, PatrickA, Mister Five, Andrew c, Wisco, Fagstein, DMacks, Just plain Bill, DragonFlySpirit, Kotjze, N Shar, Jitterro, Hrimfaxi, Where, Evlekis, Alan G. Archer, Candorwien, Rock nj, Sadi Carnot, Ck lostsword, Igilli, Pilotguy, Leon..., Kukini, StN, Ged UK, Byelf2007, Madeleine Price Ball, VictorAu, Danielrcote, Joe11-enwiki, Mchavez, Nishkid64, Salanus, Arnoutf, Hanksname, Rory096, Bcasterline, Swatjester, Sophia, AlanD, SuperTycoon, Dbtfz, Soap, Kuru, John, Joe.aston, AmiDaniel, Lakinekaki, Scientizzle, Unlocked, Dougp59, J 1982, Asemoasyourmom, 8r13n, Heimstern, Joelsmades6, Webbrg, Aron-Ra-enwiki, Svartkell, Mat8989, Gleng, Tktktk, ConservativeChristian, This user has left wikipedia, JoshuaZ, Edwy, Minna Sora no Shita, Mgiganteus1, Zarniwoot, Celador, Seb951, Osbus, Ocatecir, PseudoSudo, Misteror, Dipset1991, GodsWarrior, Ian Dalziel, Soulkeeper, Tarcieri, JHunterJ, Smith609, Chris Melton, Crazytail2, Kirbytime, Munita Prasad, Ems2, Nsheetz, Kondspi, Mr Stephen, Rickert, Brazucs, Fbartolom, Bendzh, Micahbode, SandyGeorgia, Mathsci, Intranetusa, Ryulong, Halaqah, Hari Seldon (usurped), Noleander, LaMenta3, Liquid2ice, Olav L, Grapplequip, Sheep81, Autonova, Sifaka, Axa4975, Amitch, Saltforkgunman, CerealKiller-enwiki, Ginkgo100, Levineps, W00tboy, BranStark, Vanished user, Friedgreenkillertomatoes, HisSpaceResearch, Roland Deschain, Iridescent, K, Axel147, Astrobayes, Spark, CuteWombat, Jason.grossman, NativeForeigner, Sander Säde, Kinst, J Di, Jlrobertson, Hipporoo,

47.10. TEXT AND IMAGE SOURCES, CONTRIBUTORS, AND LICENSES

KsprayDad, Rubisco-enwiki, Manph, CapitalR, Bharatveer, Az1568, Adambiswanger1, Achoo5000, Dontlookatme, Ande B., 2357, Vlipvlop, Markbassett, Billy Hathorn, Tawkerbot2, Dlohcierekim, Daniel5127, Theoldhenk, Yosef_52, Chris55, DKqwerty, Iankeir, Tooth Fairy, If humans came from apes, why are there still apes?, IronChris, Biglonstud, CalebNoble, Silvstridr, Fvasconcellos, Mnptl, Jjb123, BrianG5, AH9, Brainstormer1980, Levi P., Sleeping123, JF Mephisto, CmdrObot, Ale jrb, Mattbr, Wafulz, Hdhest95, Satyrium, Kaboomywhackwhack, Agathman, SupaStarGirl, BeenAroundAWhile, John Wilkins, ClarenceCM3, MFlet1, GHe, KnightLago, Kylu, Cwilldagangsta, Toothturtle, Dgw, Questioning, ShelfSkewed, Scruffy1, FlyingToaster, MarsRover, Ballista, Standonbible, Moreschi, SpeakEasier, FlyingOrca, Cold Sandwhich, Tim1988, Nnp, Memills, No1lakerstan, VirtualEye, MrFish, Besidesamiracle, BigGoose2006, Tylertreso, Bowslayer, Gregbard, AaronFX, Equendil, Rudjek, Logicus, GeoMor, Refusedalways, Cydebot, Fariduddien, Pce3@ij.net, Nbound, Ryan, Tntnnbltn, Reywas92, Ag545, Pendragonneopets, WillowW, Steel, SpaceTycoon, Michaelas10, Clayoquot, Gogo Dodo, Colin Keigher, Flowerpotman, Frosty0814snowman, Lscminn, Ttiotsw, A Softer Answer, Chasingsol, Eu.stefan, Wildnox, Retired user 0002, Miguel de Servet, BillySharps, Raphus Cucculatus, B, Tawkerbot4, Doug Weller, DumbBOT, Chrislk02, Caasiopia68, Apostlealex, CommonJoe, Narayanese, Optimist on the run, Gnfnrf, Adz71, Garik, Abtract, Chicken Soup, TxMCJ, Omicronpersei8, Greeneto, ForbiddenWord, Landroo, Konjikiashisogijizo, UberScienceNerd, Nol888, Gimmetrow, Dw4372, Michael Johnson, Elamere, Daa89563, Coppit, Wikid77, Dubc0724, Opabinia regalis, Neohart, G-123, Ultimus, Kajisol, Nowimnthing, Ryankhart, Kablammo, Ucanlookitup, Geothermal, Mpallen, Sting, PrivateRyan, Mungomba, Bobby Mason, Headbomb, Moulder, John254, Kangaru99, Woody, Cutsman-enwiki, Edwardmayhem, Picus viridis, Z10x, Joymmart, Davidhorman, Renamed user 5197261az5af96as6aa, Scfencer, Rosencrantz1, Infophile, EdJohnston, Zachary, Worship Him!, BlytheG, Sikkema, Moon&Nature, SusanLesch, Escarbot, BlueFireIce, Ju6613r, Thadius856, David D., AntiVandalBot, Ggbroad, AlanHarmony, Luna Santin, Turbotong, Lostcaesar, Dbrodbeck, Lyotchyk, Proopnarine, Outoftuneviolin, Prolog, Gnixon, Doc Tropics, ASDFGHJKL, AaronY, Myrridias, Fic-in, NCartmell, KeloheDeb, TimVickers, Mdotley, Tmopkisn, Zappernapper, MarcIam, Sillygrin, Semantics-enwiki, Maork, Dashmast3r, Hanoachide, Zero g, Yarovit, MECU, Shlomi Hillel, Wing Nut, Unclebulgaria, Gdo01, Dancingspring, Dinaayoub, Bio-queen, Odinbolt, OGGVOB, Isac-enwiki, Pizzaghost, Canadian-Bacon, SamuelGrauer, Nate Slayer0, Sluzzelin, JAnDbot, Narssarssuaq, NBeale, GromXXVII, Inthebeginning, Superior IQ Genus, Bakasuprman, MER-C, Plantsurfer, Iamme06, The Transhumanist, Matthew Fennell, Arch dude, Mikldy Mad, Adamsiepel, Ricardo sandoval, Welsh Crazy Frog, BenB4, Ikanreed, Hut 8.5, 100110100, Nate slayer0, MarineHebert, Rorrenig, Tstrobaugh, Coopercmu, Savant13, Jrmccall, GN-enwiki, Truth777, Dar book, Vituperex, Jsjacobs, Y2kcrazyjoker4, Insearchoftruth, Acroterion, Wasell, Bencherlite, Coffee2theorems, Casmith_789, Magioladitis, Dr. Boom, Creationlaw, Cougarkid, Daniel Cordoba-Bahle, Bongwarrior, VoABot II, Back23, SDas, Ajf99, Jack Walsh, Sushant gupta, AuburnPilot, Outoftuneviola, Outoftuneviola, Boomcoach, JNW, Yandman, Agrabiea4, JamesBWatson, Jiejunkong, McBeardo, BlackMaria, Swpb, سلام, Bwhack, ProfessorRight, Thorht, Feeeshboy, Outoftuneguitar, Trishm, Depolarizer, Outoftunetrumpet, Pixel :-), Psychohistorian, OutOfTuneViolin, Operation V.I.O.L.I.N., Foxxmuuldr-enwiki, Roadsoap, Outoftuneviolin 3, Outoftuneviolin 4, Outoftuneclarinet, SparrowsWing, Avicennasis, Calamine Lotion, Outoftunebassguitar, RebekahThorn, Outoftunebassoon, Bubba hotep, Outoftuneviolin 9, Theroadislong, Tune......., Illspirit, Benzocane, Outoftuneviolin and his sockpuppets, Lailablossom, Teh one who blocks me is an evolutionist, Rusty Cashman, Animum, Sugarcaddy, Mooselower801, Niaga Niloivenutfotuo, Ciar, Another sockpuppet of Outoftuneviolin, BatteryIncluded, Lenin13, ArthurWeasley, Torchiest, REF4, O. O.Teavee, Adrian J. Hunter, Terjen, LookingGlass, !!!niloivenutfotuO, NiLoIvEnUtFoTuO, Emw, Cpl Syx, Fang 23, Badreligion, Vssun, Outoftuneviolin's son, JoergenB, Aldenrw, THobern, GetAgrippa, OFVWT, Glen, DerHexer, Superdudefreak, Khalid Mahmood, Die4christ, Johnbrownsbody, Wi-king, Baristarim, Urco, Saganaki-, Martycota, PieCam, Malvaro, DGG, NatureA16, Wassupwestcoast, Stephenchou0722, Mickchaaya, FisherQueen, Sonikkua, Hdt83, MartinBot, Crazytail3, PAK Man, Arjun01, Sjsitler92, Christian41691, Smilingsuzy, Skamel85, Mike6271, Grendel's mother, Keith D, WatchingYouLikeAHawk, Fastman99, CommonsDelinker, AlexiusHoratius, Tweetbird, Vox Rationis, LightningPower, VirtualDelight, Joshjoshjosh1, Papapryor, Hairchrm, Beltho, Wiki Raja, Those Debate Studs, Danerz34, Mausy5043, Thirdright, Bam2014, Wlodzimierz, AlphaEta, J.delanoy, Filll, Raiph, DrKay, CFCF, Danimoth-enwiki, Sikatriz, Gurvindert, Inic, Philchea, Ali, Bogey97, UBeR, Poopmaster1234567, Blue401, Delarge3, Tony 1212, Monkey4160, Radred, TomS TDotO, Ian.thomson, Magician that makes things white, Tom Schmal, Eviom, Tdadamemd, OctopusHands, Jaydge, Staticmemory, Icseaturtles, Jokerst44, Gzkn, Acalamari, Mcapplbee, Textangel, Pbarnes, MirDoc, Random Replicator, FrummerThanThou, Katalaveno, Smeira, McSly, Da Gingerbread Man, Cards44izzy, Renamed user 5417514488, P4k, Ooogage-enwiki, Number 04, Mikael Häggström, Hillock65, Memestream, Vainolo, MAXimum Xtreme, Evolutionswrong, Linecircle, Tarinth, Zeromegamanx, Hm john morse, HiLo48, Chiswick Chap, Margareta, Rappadizydizzy, Knight of BAAWA, Ludachris78641, Matt poo watt, Davud363000, Richard D. LeCour, NewEnglandYankee, DanTheMan2, Nwbeeson, Philipholden, WeddingCrashError, SmilesALot, DEFRT, Bollyx, Joka1991, SJP, Newtman, MKoltnow, Burtonsarpa, Macdonwald, Ly4321, Jhouser24, Nathanielsmithisaloser, Lickarice, Izztek, Wfward, Marc210, Mkmori, Cometstyles, Dongless, Unhinged1122, Jamesofur, Footighter20x, Tygrrr, Messenger777, AzureCitizen, Treisijs, Tsusurfanami, Redrocket, Samgoody777, HiEv, Mattunseal, JosephPayseur, Rugby471, ComputerPlace, Country050, Sparrows point, Bkl3x, DSG2, Amarilloarmadillo, Beezhive, Furrypig, Tfietkau, Yorek, Mastahcheeph, Onlyme31, Realknowledge, Xiahou, CardinalDan, Idioma-bot, Spellcast, Snakey Jake, Sawahlstrom, AndrewTJ31, Livingboner, Jenequa, Abi Don, Genius523, Black Kite, Lights, SuperLexicon, 123heyho, Malik Shabazz, Deor, Hawknel, Kay of Kintyre, VolkovBot, Danielparker, DrMicro, Camrn86, Ljbhebjbb, Bassbreaker, The Duke of Waltham, Jeff G., Heyits5, Nilli, Butwhatdoiknow, Ymous, Billystut, Brent0270, StudyAndBeWise, Dggreen, Vlmastra, Breed3011, Ryan032, NuttyProSci-Fi3000, QuackGuru, Grandpa2390, Drpsarge, PNG crusade bot, Wild Deuce, Hy Brasil, Dash92, GimmeBot, Cosmic Latte, 99DBSIMLR, Cheshire Boy, I.scheme.a.lot, GOC-in-C, Electric free, Red lorry, Red Act, LifeScience, Vipinhari, A4bot, River flowing, Boring bus, Busy ironing, Laughing tyre, Copper canned, Vain vase, Nexon123, Walor, Cloud silver, Shed 8*6, Nxavar, Soundofmusicals, Lawn 42, Z.E.R.O., Sankalpdravid, Hensa, Charlesdrakew, Shayne T. Thiessen, Someguy1221, Caltechdoc, James.Spudeman, Warrush, Cherhillsnow, 0nlyth3truth, Siddiq ahmed, The pupils president, Unflappable, Abdullais4u, Ripepette, Kljenni, Harehawk, KC Panchal, Noformation, Driski555, Kilda, Fen zero, Tile join, Dial lean, Plaster of, Skid slid, Boo wall, Andrewrost3241981, Jvbishop, Eel eat, Crab or, Robert1947, Sodicadl, Dg10050, Barbary lion, Golden two, Dfarrar, Mgs 90, Conspiracyfactory, Bow bowl, Dachshundboy25, Dreary101, Wolfrock, Dressagebea, Peace237, Liz de Quebec, Enviroboy, Broalka, CephasE, Screen five, Iamdadawg1, Judgeking, Northfox, Zx-man, B89smith, K. Aainsqatsi, Peter borger, Esseh, Gilawson, Meson man, Hrafn, Canavalia, Breathe ing, Wool sixty, Doggie Bark, Prom ten, Table pot, Macdonald-ross, Dan ald, Strum enge, Maz haz, Eath Br, Oboeboy, GoonerDP, SieBot, Mikemoral, Weightofair, ShiftFn, Fm.illuminatus, Caulde, Invmog, DDek, Su huynh, Dawn Bard, Viskonsas, ConfuciusOrnis, God Emperor, Nutella002, TheSlowLife, Ncaroe, Pine fame, Ethancole117, Fingers80, Africangenesis, Zbvhs, CouldOughta, Arbor to SJ, Wilson44691, Yerpo, Mimihitam, VVtam, Oxymoron83, Artoasis, Faradayplank, Lancer873, Pmrich, RyanParis, Algorithms, Sunrise, Five part, EP8841, Maelgwnbot, Tar concr, CharlesGillingham, LonelyMarble, Pjotr Morgen, Akarkera, NameThatWorks, Gamall Wednesday Ida, Alpha166, Sean.hoyland, Mygerardromance, Randomblue, Oasrocks, BlueYellowRed, Spotty11222, Wikiskimmer, Loom peat, West store, Felizdenovo, Yongqunh, Eve oft, Lukasz Lukomski, Vquirk, Furry great, Scen heal, Raz grime, Kanonkas, Khirurg, The sunder king, SallyForth123, Leranedo, Mx. Granger, YSSYguy, Separa, Ray vivid, Payp two, ClueBot, Loom yellow, Wildie, Deviator13,

Tmol42, Artichoker, Gits (Neo), Snigbrook, Gorillasapiens, Fixtgear, Ear Sid, Poise too, The Thing That Should Not Be, Yournoangei, TCMagic87, Taffboyz, Jan1nad, TychaBrahe, Nsk92, Ajgisme, Herakles01, Gregcaletta, Sting au, Kniknight, Drmies, Jumacdon, Jeremypyle, Mild Bill Hiccup, Grooy, Staam, Natman43, Stuthomas4, Here4thefood, Tlame, Niceguyedc, Baegis, Blanchardb, Leadwind, Grandmastergrawp, Peteruetz, Flaiw, Sratt, Ploum-enwiki, Freestyle10evan, Ballin789, Hmpxrii, Bauxzaux, AlienHook, Laorv, Aartp, Zarru, Jurto, Dhomm, Lakle, Thuom, Wacxi, Thinradred9, Aua, Dr. B. R. Lang, Qomee, Aarnu, Juamh, Flimg, Laegg, Noivb, Verwoerd, Javascap, Abrech, Tgees, Ghaor, Mseas, Lartoven, Salmanmdkhan, Rhododendrites, Aks818guy, Sally1309, Supercrazy617, Skaterboy, NuclearWarfare, Arjayay, Hubcapwiki, Njardarlogar, Cmayy, Betbs, Ocmec, DeltaQuad, TomShmells247, Jackrm, Hans Adler, Navme, Basil45, Icthus123, Dekisugi, Amused Usher, ChrisHodgesUK, Uuger-enwiki, La Pianista, Thingg, Dangee63, Vegetator, Aitias, ForestDim, Wildfirejmj, Ra2007, Dana boomer, Adriansrfr, Amaltheus, Tezero, AC+79 3888, Johnuniq, Kwion, Lremo, Gcior, Rveek, Mlaat, Editor2020, Franz lino, Crazy Boris with a red beard, Heironymous Rowe, Mjharrison, Samsara noadmin, RightGot, Spitfire, Nathan Johnson, Maky, PervyPirate, Stickee, Steht, Zyasu, IanCheesman, JimmyButler, UTvolfan711, Nicolae Coman, Rcuub, Vduer, Bdaay, Hriow, Pydan, Sweetpoet, JinJian, Aunt Entropy, Dcooh, Wennj, Pleuu, Tteil, Pyiid, Mikearion, Yuodd, Dyoes, Kteey, Sharifneuro, Xeatc, Tocir, Ynist, Hollywoodgenes, D.M. from Ukraine, Pbuee, Dveim, Compute14, Hbeop, Addbot, Xp54321, Twilh, Otterathome, TheNightRyder, Plow Col, May Ced, Five edges, Manuel Trujillo Berges, Donhoraldo, Knuv moor, Two teem, 12 Cove, Pluthra, Broabey, Moss oul, Glomove, Ystram, Srivange, Laik G, Vewar, DOI bot, Jojhutton, Zadeh79, Vuri, Pond sew, Inn par, Fott leigh, Betterusername, Montgomery '39, Sarrh, Wdraa, Lif 91, Jos Jio, Seven rawt, Stone Oak Drive, Draco 2k, Fieldday-sunday, Zarcadia, CanadianLinuxUser, Dsmith77, Ekologkonsult, Proxima Centauri, Yday, Giuegg, Ptaat, 6 Four, Red went, Type how, Bernstein0275, AndersBot, Debresser, CUSENZA Mario, Favonian, LemmeyBOT, LinkFA-Bot, Quietmarc, Phaup, Nvav, Politoed666, Guoed, Wteaw, Leeg 7, Six moss, Soom zo, Tide rolls, Thermalimage, Joe1978, MuZemike, Wpaa, Dinaey, Aggn, Xted, Daeoque, Pseiir, Temmor, Skilltim, Ettrig, Geddone, Volnav, 3 Lane, Matekm, Krukouski, Vsast, It da, GloriousLeader, Legobot, Luckas-bot, 7 Fej, Yobot, JohnnyCalifornia, TaBOT-zerem, Louisstar, Rsquire3, Adi, Mgegh, Kev 63, Finn up, Kdeo, U mene, Ydonne 2, Boe syl, Mal hen, 489thCorsica, Becky Sayles, Intercalate, KamikazeBot, Byrll, 4 trin, 6 y go, Scipain, Qcoo, Yiwb, 9 noon, Azcolvin429, Lazza 99, Jak Tac, Melono V, Farsight001, Armchair info guy, Backslash Forwardslash, ElkeK, AnomieBOT, Donnellyj, Jim1138, Galoubet, Uesonne, Paw yanne, Craigmac41, Hadrian89, Lobe lob, Wawmaw, Cyr egi, Gooago, Cen Upp, Ykse, Goxxen, Trabucogold, Sfvace, Jo3sampl, Nacre 10, NickK, Mann jess, Materialscientist, Senortypant, Danno uk, Citation bot, Myoe83, Qymme, 4 Gnon, Fr8we, Yar owe, ArthurBot, Lapabe, Donner meat123, Acsparkman, Sumurai8, Shglien, Xqbot, Wyher, Timir2, Swazamee, Drilnoth, GeometryGirl, Spotfixer, A455bcd9, Oontoo, RoNeunzig, Songwriter17, Onecatowner, St3p5, JCrue, Gap9551, 40 tune, Dr Oldekop, Montana's Defender, Triqqi, Marr Plv, Abce2, Ten tinted, Strreeh, ProtectionTaggingBot, StealthCopyEditor, Omnipaedista, Thosjleep, Imag Qu, Mixmoney, RibotBOT, Rab Eyye, Thompsma, Wegge 9, Hubrid Noxx, Doulos Christos, Tiptophiphop, Aurush kazemini, Kanfood, Moxy, Shadowjams, Methcub, Howdy541, Sbarry12, Erik9, 11cookeaw1, Binarypascal, Doll ten, FrescoBot, Paleoderek, Doalla 8, Twci, Sandgem Addict, Mark Renier, IzzyReal, KMFDM Fan, Non believer evolution, Arlen22, Comder, Keet Onn, VI, HJ Mitchell, Solaricon, When nine, Matthew Ackerman, Citation bot 1, Igor233, Sopher99, Winterst, SixPurpleFish, Gaba p, Pinethicket, Base tonne, Vicenarian, HRoestBot, Alphazeta33, Hard Sin, Tom.Reding, A8UDI, Lars Washington, HXJ, Axxel Sel, Barras, David Z. Smith, Dude1818, Sobored1, J.O.94, Christpower79, IVAN3MAN, Lhasanimir, Risker (Anne Criske) is old & has gray hair & wears glasses. 16, FoxBot, Thrissel, TobeBot, Arkatox, Fama Clamosa, Barny Fife, Lotje, Parrikk, Sviev, Fen Fin, J4V4, Tie Bon, Sted Go, Wuob, LilyKitty, Sagi Nahor, Neptunerover, Silicon-28, Visite fortuitement prolongée, GGT, Chronulator, Techyactor15, Zyyx, Tbhotch, Jesse V., RjwilmsiBot, Phe enn, Jedwina, Venn 87, Rammadd, NameIsRon, PPdd, Bhawani Gautam, Thiridaz, Minia23, Dyac L, NerdyScienceDude, Burmiester, Plommespiser, Nath Hebb, Aircorn, Salvio giuliano, Billare, Juffe, Mivmirre, Mukogodo, EmausBot, Anglyn, Orphan Wiki, Surlyduff50, Gfoley4, Dominus Vobisdu, Najeeb1010, Faceless Enemy, Redioz, Rarevogel, Jmv2009, Bull Market, Maxozo, Tommy2010, Kiran Gopi, Yaqix, Lisasilzaz, StacyMJC, Plaonui, ZéroBot, Bwatb, Lateg, Alpha Quadrant, AvicAWB, H3llBot, Gniniv, Sepia officinalis, SporkBot, Wayne Slam, Ocaasi, Lime Ore, Resprinter123, Ventus55, Rcores, L Kensington, Ludovica1, Donner60, Ollyoxenfree, Ems2715, Snubcube, Joannamasel, NorCal764, Z1nemo, Sciencenews, Dylan Flaherty, Wellwicked94, KimaniJamaica, Herk1955, Mjbmrbot, Vindicated vigilanty, Woodsrock, ClueBot NG, Pandelver, Horoporo, Zytigon, Shorrec, Roccw, Somedifferentstuff, Jacksoncw, LeftCoastMan, Hoothe, Rainbowwrasse, Nazzare, Nezzere, Darkjedi10, Jj1236, Deivu, Croacka, Kjspring1, Frietjes, StKyrie, Rerrenn, Widr, JoetheMoe25, Ryan Vesey, Helpful Pixie Bot, KieranKiwiNinja, Bibcode Bot, BG19bot, Wpete510, Areawa, Verum solum, Hurricanefan25, Zenkai251, Rorree, Salmonne, Rastapunk, Chris the Paleontologist, Mark Arsten, Umar farooq miana, Dkspartan1, Michael Barera, Cadiomals, YimmyZeee, Rawrisherz, Delay learn, V6mustang, OwnAllow, Othurba, Revolve Ion, Cowsgobob, Elronhir, Civcraz, Homur, Knoids, Tilevo, P'tit Pierre, Harizotoh9, Claviclehorn, TheProfessor, Dr. Coal, Dontreader, Hamish59, Oct13, Peru Serv, EvolveBuster, Rolandwilliamson, Compuh, Neeshi, Wernot, Willietell, Bonkers The Clown, BattyBot, SkepticalRaptor, Soulbust, Saedon, Illia Connell, JYBot, AthanasiusOfAlex, IjonTichyIjonTichy, Dexbot, Br'er Rabbit, Mogism, Jinx69, Alphama, TippyGoomba, Josephk, Lugia2453, ItsLuke, SFK2, Rul3rOfW1k1p3d1a, K.Hollingsworth486, Corinne, KidRocket71, Reatlas, Joeinwiki, NathanWubs, Apidium23, Royroydeb, CDH31211811, Quacod, DonaldKronos, Eyesnore, Havensfire, Raptormimus456, Everymorning, Hendrick 99, Splicevariant, Marchino61, Abdalla Dabdoub, NK2015, Comp.arch, Evolution and evolvability, Blackbombchu, Citytownhome, Finnusertop, Cyborg1981, Oranjelo100, Astredita, Gharris7, Rosemaryshanley, VAleles, Anrnusna, Meteor sandwich yum, Dodi 8238, Tjiolye, BillyTanjung, Sharif Uddin, 19grays, Ordessa, 22merlin, Wyatt ax 4, Racerdas7, Monkbot, Mama meta modal, VeniVidiVicipedia, Tjonile, CongoJumper, Chippewa69, Physicsandwhiskey, Kyle1009, Joinwit, Bodhisattwa, AsteriskStarSplat, WikiTrollTerminator, Gonzales John, Serevix, Aslamthelion, Sarr Cat, Pigmentkleur, Joseph2302, Isambard Kingdom, Caprockranger, TeirJ, Equivocasmannus, Gavinomelia, KasparBot, D.g.lab., Eulalefty, Magomahot, Dunkleosteus77, Versallies89, Xayaz, Agent of the nine, Sro23, Pixelgraph, Guitarguy84, Sleety Dribble, Mebennett49, Spreetycakes, Milku3459, Preciousjfm, InternetArchiveBot, Motivação, Holy Goo, Bender the Bot, Eearboy, Cheshirecat8021, Dockabo, Socceristhebest, PaleoNeonate, KolbertBot and Anonymous: 1542

- Adaptation *Source:* https://en.wikipedia.org/wiki/Adaptation?oldid=801843844 *Contributors:* Timo Honkasalo, Andre Engels, Stevertigo, Michael Hardy, Lexor, Gabbe, Habj, Big iron, Andres, Charles Matthews, Steinsky, DJ Clayworth, Samsara, Jerzy, Robbot, Sander123, Vespristiano, Tea2min, ZeroJanvier, Gadfium, Korou, Thincat, Discospinster, Rich Farmbrough, FT2, Vsmith, DerekLaw, HCA, Xezbeth, Bender235, Kbh3rd, MBisanz, Shanes, RoyBoy, Femto, Bobo192, Pdorrell, Vanished user sdfkjertiwoi1212u5mcake, Stesmo, Shenme, Townmouse, Dmanning, Orangemarlin, Danski14, Alansohn, Gary, Delmonte, PAR, Velella, Brookie, Stemonitis, Woohookitty, Cimex, Mandarax, Graham87, BD2412, Josh Parris, Sjö, Drbogdan, Rjwilmsi, Angusmclellan, Eptalon, Bhadani, Protez, Nihiltres, RexNL, Pete.Hurd, ImpalerBugz, Alex314058, Chobot, DVdm, Antiuser, Bgwhite, Roboto de Ajvol, YurikBot, Pip2andahalf, Hede2000, Mark Ironie, Leandro Palacios-enwiki, Gaius Cornelius, Rsrikanth05, Wimt, NawlinWiki, Dysmorodrepanis-enwiki, Waldow, Snek01, Deskana, Iridium ionizer, Nutmeg, Bucketsofg, Nethgirb, BOT-Superzerocool, DeadEyeArrow, Jwing79, Allens, NeilN, Elliskev, The Claw, SmackBot, Amcbride, Enlil Ninlil, ElectricRay, Gilliam, Skizzik, CanbekEsen, Chris the speller, Unint, Miquonranger03, Ed Hagen, Royboycrashfan, NYKevin, Egsan Bacon, Dodgydobson, Liontooth, Rrburke, Addshore, EPM, LazyBoi633, Ravenswood Media, Richard001, DMacks, Salamurai, Alan G. Archer,

GoldenTorc, SashatoBot, Mchavez, John, Heimstern, Gobonobo, Joelmills, Extremophile, Ckatz, Ex nihil, Werdan7, Munita Prasad, Muadd, Optakeover, Fangfufu, Waggers, Doczilla, BananaFiend, Electrified mocha chinchilla, JoeBot, Bobamnertiopsis, Ale jrb, Van helsing, 0zymandias, Memills, Gregbard, Fl, MC10, Steel, Gogo Dodo, Shirulashem, Omicronpersei8, Thijs!bot, Epbr123, Mojo Hand, Headbomb, Marek69, SomeStranger, James086, Z10x, Dmitri Lytov, Nick Number, Pie Man 360, Mentifisto, Porqin, AntiVandalBot, Yuanchosaan, Majorly, Gioto, Seaphoto, Doc Tropics, Smartse, Danger, Myanw, Sluzzelin, Tio1, JAnDbot, Nthep, Andonic, Kipholbeck, Bencherlite, Bongwarrior, Sushant gupta, StripeyBadger, Ghost2, ArchStanton69, 28421u2232nfenfcenc, DerHexer, JaGa, S3000, MartinBot, Joie de Vivre, EverSince, Thirdright, J.delanoy, Trusilver, Rgoodermote, Redwrathe, Numbo3, Uncle Dick, Mike.lifeguard, Deathgod1995, Dr Dima, AntiSpamBot, Chiswick Chap, Sewings, Juliancolton, SixteenBitJorge, Guyzero, Martial75, Idioma-bot, Jeff G., AlnoktaBOT, Pilotbob, Philip Trueman, TXiKiBoT, Oshwah, Vipinhari, Olly150, Vanished user ikijeirw34iuaeolaseriffic, Melsaran, JhsBot, Abdullais4u, Whatiguana, FinnWiki, Lova Falk, Kingjalis3, MCTales, Brianga, AlleborgoBot, Logan, PGWG, Fanatix, Macdonald-ross, Calliopejen1, Tiddly Tom, DGCollard, Yintan, Keilana, Happysailor, Flyer22 Reborn, Paps 70, Fimbriata, Steven Crossin, Tombomp, Pmrich, Mercenario97, Svick, Addaick, Arnold90, WikipedianMarlith, ClueBot, The Thing That Should Not Be, Gunnar Mikalsen Kvifte-enwiki, Lawrence Cohen, Drmies, Mild Bill Hiccup, Vangoghhasamachette, Kalem, Harland1, DragonBot, Takeaway, Excirial, Lindenburg, Gtstricky, YDaniel7, ZuluPapa5, NuclearWarfare, Razorflame, Muro Bot, Aitias, Johnuniq, SoxBot III, Egmontaz, XLinkBot, Dxballer08, WikiDao, Aunt Entropy, On the other side, Habbilow, Addbot, Polinizador, Fyrael, Friginator, Fgnievinski, Mrsssgt, Golden.josh, Ronhjones, Flufternutter, OliverTwisted, LaaknorBot, RTG, 10outof10die, Glane23, Favonian, Tassedethe, Numbo3-bot, Tide rolls, Trak Nar, Mx740, Jackelfive, Luckas-bot, Yobot, WikiDan61, 2D, Amirobot, Azcolvin429, Eric-Wester, InfoCan, Pyrrhus16, Bluerasberry, Materialscientist, Citation bot, Xqbot, Plumpurple, Gatorgirl7563, Dr Oldekop, J04n, Jobin104, SassoBot, Thompsma, Brambleshire, Mathonius, Amaury, Bigger digger, Methcub, WaysToEscape, JayJay, Jisily, Rang3rofskil, Finalius, Citation bot 1, Soccerdu82, Intelligentsium, DrilBot, Pinethicket, Jonesey95, Xfact, Jandalhandler, Jujutacular, Reconsider the static, TobeBot, Trappist the monk, PiRSquared17, Animalparty, Callanecc, Seahorseruler, Weedwhacker128, PleaseStand, Nascar1996, Minimac, JordanZimme, Hornlitz, DARTH SIDIOUS 2, Onel5969, RjwilmsiBot, Bhawani Gautam, Chrisbkooho, Salvio giuliano, DASHBot, WikitanvirBot, Faceless Enemy, Vbhubeny, Tommy2010, Uleli, Wikipelli, Dcirovic, K6ka, Antigrandiose, MigueldelosSantos, Cvata, The Nut, Wayne Slam, Ocaasi, Openstrings, L Kensington, Joannamasel, Mc4th, Sven Manguard, ClueBot NG, Rich Smith, Jack Greenmaven, CocuBot, MelbourneStar, This lousy T-shirt, Satellizer, Darwin's Teapot, Thejavadrinker, Widr, Sariovski, Excerpted31, LaLaLanderMan, BG19bot, Tgbyhymhhtg, PTJoshua, PhnomPencil, Umar farooq miana, Altair, CitationCleanerBot, Totedati, Rhinomantis88, 220 of Borg, Deedee318, Kgundle, Prof. Squirrel, E prosser, Sminthopsis84, FlashierHornet5, Lugia2453, Mark Bao, David Tar, Fmoreno212, Cadillac000, Hillbillyholiday, Epicgenius, FallingGravity, Lollipopfudge, Notebook600, Soham, Dogmanmister, NottNott, Ginsuloft, Jacko39067, Quenhitran, AddWittyNameHere, Stomperinky, Anrnusna, Stamptrader, Golden Bosnian Lily, JaconaFrere, Juliansilvestre, ThatRusskiiGuy, Kkk690, Mr. Smart LION, John Santana, Isabellelouise, Dattatreya101, Hanna.225, Nandini Batta, Kenarose 1095, Hvgvfcdvddddde, LarryBoy79, Somenoisyperson, Yonzhouisyourlord, AmazedLogic, Merritt1234567777777, Anpanman, Jfrab, Jack1610, Peterbelle, Kyle123783, Sanket Edits Wiki, Simplexity22, MrDelta292, PigeonOfTheNight, Entranced98, Helocke, ✍, Exoloverforever, Nalak Dey, Youtubebobrime and Anonymous: 653

- **Artificial neural network** *Source:* https://en.wikipedia.org/wiki/Artificial_neural_network?oldid=802763737 *Contributors:* Magnus Manske, Ed Poor, Iwnbap, PierreAbbat, Youandme, Susano, Hfastedge, Mrwojo, Nealmcb, Michael Hardy, Erik Zachte, Oliver Pereira, Kku, Bobby D. Bryant, Zeno Gantner, Tgeorgescu, Parmentier-enwiki, Delirium, Pieter Suurmond, (, Alfio, 168..., Ellywa, Ronz, Snoyes, Den fjättrade ankan-enwiki, Cgs, Glenn, Cyan, Hike395, Hashar, Novum, Charles Matthews, Guaka, Timwi, Reddi, Andrewman327, Munford, Furrykef, Bevo, Fvw, Raul654, Nyxos, Unknown, Pakcw, Robbot, Chopchopwhitey, Bkell, Hadal, Wikibot, Diberri, Xanzzibar, Wile E. Heresiarch, Connelly, Giftlite, Rs2, Markus Krotzsch, Spazzm, Seabhcan, BenFrantzDale, Zigger, Everyking, Rpyle731, Wikiwikifast, Foobar, Edrex, Jabowery, Wildt-enwiki, Wmahan, Neilc, Pgan002, Quadell, Beland, Onco p53, Lylum, Gene s, Sbledsoe, Mozzerati, Karl-Henner, Jmeppley, Asbestos, Fintor, Asqueella, AAAAA, Splatty, Rich Farmbrough, Pak21, NeuronExMachina, Michal Jurosz, Pjacobi, Mecanismo, Zarutian, Dbachmann, Bender235, ZeroOne, Ben Standeven, Violetriga, Mavhc, One-dimensional Tangent, Gyll, Stephane.magnenat, Mysteronald, .:Ajvol:., Fotinakis, Nk, Tritium6, M0rph, JesseHogan, Mdd, Passw0rd, Zachlipton, Alansohn, Jhertel, Anthony Appleyard, MrTree, Denoir, Arthena, Fritz Saalfeld, Sp00n17, Rickyp, Hu, Tyrell turing, Cburnett, Notjim, Drbreznjev, Forderud, Oleg Alexandrov, Mogigoma, Madmardigan53, Justinlebar, Olethros, Ylem, Ruud Koot, Dr.U, Gengiskanhg, Male1979, BarOn, Waldir, Eslip17, Yoghurt, Ashmoo, Graham87, BD2412, Qwertyus, Imersion, Grammarbot, Rjwilmsi, Jeema, Venullian, SpNeo, Intgr, Predictor, Kri, BradBeattie, Plarroy, Windharp, Mehran.asadi, AllyD, Commander Nemet, Wavelength, Borgx, IanManka, Rsrikanth05, Philopedia, Ritchy, David R. Ingham, Grafen, Nrets, Exir Kamalabadi, Deodar-enwiki, Mosquitopsu, Jpbowen, Dennis!, JulesH, Moe Epsilon, Supten, Dbtirs, DeadEyeArrow, Eclipsed, SamuelRiv, Tribaal, Chase me ladies, I'm the Cavalry, CWenger, Donhalcon, Banus, Shepard, John Broughton, Attilios, A13ean, SmackBot, PinstripeMonkey, McGeddon, CommodiCast, Jfmiller28, Stimpy, Commander Keane bot, Feshmania, Yamaguchi???, ToddDeLuca, Diegotorquemada, Patrickdepinguin, KYN, Gilliam, Bluebot, Oli Filth, Gardoma, Complexica, Nbarth, Nossac, Hongooi, Pdtl, Izhikevich, Trifon Triantafillidis, SeanAhern, Neshatian, Daniel.Cardenas, Vernedj, Dankonikolic, Chymicus, Rory096, Sina2, SS2005, Kuru, Plison, Lakinekaki, Bjankuloski06en-enwiki, IronGargoyle, WMod-NS, Mbisgaier, Dicklyon, Citicat, StanfordProgrammer, Dl2000, Ojan, Chi3x10, Aeternus, CapitalR, Atreys, George100, Gveret Tered, Devourer09, SkyWalker, CmdrObot, Leonoel, CBM, Mcstrother, MarsRover, CX, Arauzo, Peterdjones, Josephorourke, Kozuch, ClydeC, NotQuiteEXPComplete, Irigi, Mbell, Oldiowl, Tolstoy the Cat, Headbomb, Mitchell.E.Timin, Davidhorman, Sbandrews, KrakatoaKatie, QuiteUnusual, Prolog, AnAj, LinaMishima, Whenning, Lfstevens, Hamaryns, Daytona2, JAnDbot, Oxinabox, MER-C, Dcooper, Extropian314, Magioladitis, VoABot II, JamesBWatson, Amitant, Jimjamjak, SSZ, Robotman1974, David Eppstein, User A1, Martynas Patasius, Pmbhagat, JaGa, Tuhinsubhrakonar, SoyYo, Nikoladie-enwiki, R'n'B, Maproom, Rod57, Nemo bis, K.menin, Gill110951, Tarotcards, Plasticup, Margareta, Paskari, Jamesontai, Kiran uvpce, Llorenzi, Malik Shabazz, Jamiejoseph, Error9312, Jlaramee, Jeff G., Jwilleke, MrRK, Like.liberation, Antoni Barau, A4bot, Singleheart, Ebbedc, Lordvolton, Ask123, CanOfWorms, Mundhenk, Wiae, M karamanov, Softtest123, Enkya, Moltean, Blumenkraft, Twikir, Mikemoral, Oldag07, Smsarmad, Flyer22 Reborn, MaynardClark, Janopus, Bwieliczko, Dhatfield, F.j.gaze, Mark Lewis Epstein, S2000magician, WingedPig, PuercoPop, Martarius, ClueBot, Ignacio Javier Igjav, Ahyeek, The Thing That Should Not Be, Fadesga, Zybler, Plastikspork, Midiangr, Epsilon60198, Thomas Tvileren, Wduch, Excirial, Txbangert, Three-quarter-ten, J+, Skbkekas, Chaosdruid, Aprock, Certes, Qwfp, Jean-claude perez, Achler, XLinkBot, AgnosticPreachersKid, BodhisattvaBot, Stickee, Cmr08, Porphyro, Fippy Darkpaw, Addbot, DOI bot, AndrewHZ, Thomblake, Techjerry, Looie496, MrOllie, Transmobilator, Jarble, TundraGreen, Ettrig, Yobot, Blm19732008, Nguyengiap84-enwiki, SparkOfCreation, AnomieBOT, DemocraticLuntz, Tryptofish, Trevithj, Jim1138, Durran65, MockDuck, JonathanWilliford, Materialscientist, Citation bot, Brightgalrs, Eumolpo, Twri, NFD9001, Isheden, J04n, Omnipaedista, Mark Schierbecker, RibotBOT, RoodyBeep, Gunjan verma81, FrescoBot, X7q, Ömer Cengiz Çelebi, Outback the koala, ChrstphrChvz, Citation bot 1, Tylor.Sampson, Jonesey95, Tom.Reding, Calmer Waters, Skyerise, Dividingbyzerofordummies, Trappist the monk, Krassotkin, Cjlim, Fox Wilson, The Strategist, LilyKitty, Eparo, בוצ׳י, Horcrux92, Jfmantis, Onel5969, Mehdiabbasi, VernoWhitney, Wiknn,

BertSeghers, DASHBot, Steve03Mills, EmausBot, Johncasey, Nacopt, Dzkd, Lepsyleon, Dewritech, Racerx11, Primefac, Japs 88, GoingBatty, RaoInWiki, Roposeidon, Epsiloner, Stheodor, Benlansdell, Radshashi, Dcirovic, K6ka, Jasonanaggie, D'oh!, THNTK, Thisisentchris87, Aavindraa, Chire, SporkBot, Glosser.ca, IGeMiNix, Kapil.xerox, Donner60, Yoshua.Bengio, Orange Suede Sofa, Shinosin, Venkatarun95, Dschotze, ChuckNorrisPwnedYou, Petrb, ClueBot NG, Vladpaln, Raghith, Robiminer, Snotbot, Tideflat, Frietjes, Gms3591, Ryansandersuk, Widr, MerllwBot, Helpful Pixie Bot, Trepier, Wbm1058, BG19bot, Thwien, Adams7, TillF, Joeykai, Rahil2000, Chafe66, Michaelmalak, Compfreak7, Kirananils, Altaïr, Medende, BattyBot, J.Davis314, Attleboro, Pratyya Ghosh, JoshuSasori, Ferrarisailor, Eugenecheung, Mtschida, ChrisGualtieri, Dave2k6inthemix, Whebzy, EyedMoon, Enterprisey, Raymond1922A, Dexbot, JurgenNL, Oritnk, Stevebillings, Djfrost711, BillTonnies, Sa publishers, ⌧, Me, Myself, and I are Here, Mark viking, Phamnhatkhanh, Markus.harz, Deeper Learning, Vinchaud20, Soueumxm, Iztwoz, Toritris, Vbrm, Comp.arch, Evolution and evolvability, Sboddhu, Sharva029, YiFeiBot, Weiping.thu, Quenhitran, Oranjelo100, ParliamentFunk, Paheld, Putting things straight, Frosztbyte, *thing goes, Rosario Berganza, Monkbot, CaraniS, Zidong Du, Buggiehuggie, Santoshwriter, Likerhayter, Joma.huguet, Bclark401, Mvdyck, Alinowe, Rahulpratapsingh06, Donkeychee, Michaelwine, Xsantostill, Jorge Guerra Pires, Narky Blert, Sometree, Loraof, Alexander.yermakov.1990, Wfwhitney, Sayan98, HelpUsStopSpam, Miraclexix, Loïc Bourgois, I enjoy sandwiches, KasparBot, RippleSax, Stümper, Akritasa, Ghost30, Dubhcloch, Lr0^^k, The Quixotic Potato, Vishnuanand.ckdy, AlF6Na3, K.ghiasi, Bhattasamuel, PcPrincipal, GSS-1987, Lasgbjames, Waterfall mg, Mercyse, Awesomemeeos, ColeDeanShepherd, Fmadd, Frugal fisherman, MMC's, Don hayler, Juillermo, Mani monisha, Alex4532, Bender the Bot, 72, JoMM, IAmPyroglyph, BronHiggs, Selichm, Anximander, Renamed user BM, Alisohani, L8 ManeValidus, MfortyoneA, Memento Mori 4, Botatao, Magic links bot, Optate823, IAFSK, KolbertBot and Anonymous: 597

- **Evolutionary computation** *Source:* https://en.wikipedia.org/wiki/Evolutionary_computation?oldid=803272623 *Contributors:* Michael Hardy, Togelius, Lexor, Karada, Ronz, Samsara, Power~enwiki, Jiy, Alex Kosorukoff, Jjmerelo~enwiki, Gary, MikiWiki, Babajobu, Woohookitty, Ruud Koot, BD2412, Rjwilmsi, Joe Decker, Chobot, Bgwhite, Adoniscik, YurikBot, Wavelength, Ritchy, RKUrsem, Tony1, Epipelagic, SmackBot, Moxon, Mneser, Inego~enwiki, Oli Filth, JonHarder, Luís Felipe Braga, FerzenR, Antonielly, Bilby, Mulder416sBot, George100, Horsman, Juckele, Peterdjones, Mercere99, Biblbroks, Daniel, Pruetboonma~enwiki, Headbomb, TimVickers, Ehasl, JAnDbot, Jwojt, Mdorigo, Calltech, Obscurans, Salih, Fheyligh, Pdcook, JohnBlackburne, Dggreen, Tense, Rei-bot, Lordvolton, Epktsang, Michael Allan, Zwgeem, Yintan, Jdaloner, CharlesGillingham, Sean.hoyland, Zilupe, Bahriyebasturk, Addbot, Ettrig, Fryed-peach, Luckas-bot, Yobot, AnomieBOT, Paskornc, HRV, LilHelpa, TheAMmollusc, Kamitsaha, FrescoBot, Mohdavary, Jlaire, Trappist the monk, Lotje, Sergey539, Vrenator, Duoduoduo, Wo.luren, Rayman60, Dzkd, Primefac, JustinTime55, Arkenflame, Midas02, ChuispastonBot, ClueBot NG, Baldy Bill, Zach Winkler, Helpful Pixie Bot, Jr271, Solomon7968, Intelcompu, Harizotoh9, MBazzy, TheProfessor, Илтıорел, ChrisGualtieri, Oritnk, Me, Myself, and I are Here, Mark viking, Phamnhatkhanh, Xzesey, Anrnusna, Ccottap, Marco castellani 1965, Gusz Eiben, Boky90, Muaafa, Yafrani, RichardF77, Emir of Wikipedia, Jeanpmartins, Theelepeltje, Ernesto.benini, RSRV0306, Magic links bot, Trbiyanto, KolbertBot and Anonymous: 119

- **Genetic algorithm** *Source:* https://en.wikipedia.org/wiki/Genetic_algorithm?oldid=803284550 *Contributors:* Magnus Manske, The Epopt, Taw, Piotr Gasiorowski, Vignaux, William Avery, Waveguy, Edward, Simonham, Michael Hardy, Soegoe, Kwertii, Felsenst, Lexor, David Martland, AdamRaizen, Dan Koehl, Shyamal, Kku, Bobby D. Bryant, Ahoerstemeier, Ronz, Ijon, Qed, Marco Krohn, Kyokpae~enwiki, Nikai, Evercat, BAxelrod, Smack, Hike395, Disdero, Timwi, Jitse Niesen, Greenrd, Steinsky, Furrykef, Grendelkhan, AnthonyQBachler, Aliekens, Robbot, Chocolateboy, Arkuat, Chopchopwhitey, Stewartadcock, Texture, Hippietrail, Mark Krueger, Unyounyo, Centrx, Giftlite, Sankar netsoft, DavidCary, Jyril, Lee J Haywood, Stefano KALB~enwiki, Curps, Duncharris, Utcursch, KaHa242, Antandrus, Oneiros, Aabs, Asbestos, Hellisp, Goobergunch, SamuelScarano, Andreas Kaufmann, RevRagnarok, Jwdietrich2, AAAAA, Rfl, Vincom2, 2fargon, Yinon, Cap'n Refsmmat, DerrickCheng, Alex Kosorukoff, Spoon!, .:Ajvol:., Malafaya, Giraffedata, Cpcjr, Mdd, Phyzome, Larham, Alansohn, Silver hr, Crispin Cooper, BryanD, Hu, Brinkost, CloudNine, Artur adib, Zawersh, Oleg Alexandrov, Postrach, Stuartyeates, Ruud Koot, Jeff3000, GregorB, Dionyziz, Male1979, VsevolodSipakov, MassGalactusUniversum, Qwertyus, Terryn3, Avinesh (usurped), Tailboom22, CoderGnome, Kane5187, Rjwilmsi, Strait, Loudenvier, MikeMayer, Arbor, Mpo~enwiki, Vietbio~enwiki, Mathbot, Ewlyahoocom, Diza, Kri, Chobot, Bgwhite, Manu3d, Dúnadan, YurikBot, Wavelength, RussBot, Freiberg, Arado, Bhny, Chris Capoccia, Stephenb, Gaius Cornelius, Bovineone, Madcoverboy, LMSchmitt, RKUrsem, Toncek, Mikeblas, Brat32, Bota47, Pegship, Tribaal, Twelvethirteen, Open2universe, Lt-wiki-bot, Arthur Rubin, Janto, GraemeL, CWenger, QmunkE, Wjousts, Allens, A.Nath, Xiaojeng~enwiki, Edin1, Pecorajr~enwiki, SmackBot, Tarret, InverseHypercube, Brick Thrower, Eskimbot, Mikołaj Koziarkiewicz, Sundaryourfriend, Oli Filth, Conway71, MalafayaBot, Nosophorus, DHNbot~enwiki, Mohan1986, Ludvig von Hamburger, Simpsons contributor, Jasonb05, Gragus, Chlewbot, MattOates, Parent5446, Spacecaldwell, Wen D House, Radagast83, "alyosha", Acdx, Bidabadi~enwiki, Kuzaar, Thomas weise, Antonielly, Jcmiras, Beetstra, Negrulio, DabMachine, Pelotas, Poweron, Tawkerbot2, George100, Purplesword, Pgr94, Bumbulski, Grein, Simeon, Gpel461, VladB, Cydebot, Carl Turner, ST47, Tawkerbot4, Roberta F., Scarpy, Omicronpersei8, Klausikm~enwiki, Lawrenceb, Techna1, Bockbockchicken, Pruetboonma~enwiki, Massimo Macconi, Tapan bagchi, KrakatoaKatie, AntiVandalBot, Baguio, TimVickers, Temporary-login, North Shoreman, Kdakin, JAnDbot, MER-C, CosineKitty, TAnthony, Raduberinde, SDas, Xn4, PeterStJohn, Eleschinski2000, Destynova, WhatamIdoing, SSZ, David Eppstein, User A1, Jetxee, Diroth, Francob, A. S. Aulakh, Projectstann, Sm8900, Glrx, Tulkolahten, J.delanoy, Stochastics, TempestCA, DeKXer, Rod57, Silas S. Brown, K.menin, Edrucker, Plasticup, Fheyligh, DavidCBryant, Riccardopoli, Ratfox, Useight, CardinalDan, Marksale, VolkowBot, JohnBlackburne, AlnoktaBOT, Kyle the bot, TXiKiBoT, Esotericengineer, Tameeria, Rei-bot, TyrantX, Kjells, Keburjor, Jasper53, Broadbot, Darioizzo, Andy Dingley, Kindyroot, AlterMind, Cnilep, Thric3, Moltean, SieBot, Bjtaylor01, Zwgeem, ToePeu.bot, Kaell, Yuanwang200409, Antzervos, Raulcleary, Algorithms, CharlesGillingham, Anchor Link Bot, DixonD, Novablogger, Guang2500, CShistory, AussieScribe, Cngoulimis, ImageRemovalBot, Djhache, Euhapt1, SlackerMom, ClueBot, Toshke~enwiki, Justin W Smith, Ahyeek, Swarmcode, Adrianwn, Razimantv, Mild Bill Hiccup, Bradka, Dylan620, Otolemur crassicaudatus, Jkolom, DragonBot, SteelSoul, Calimo, Leonard^Bloom, Arjayay, Chaosdruid, Unixcrab, Frongle, Josilber, LieAfterLie, Johnuniq, DumZiBoT, XLinkBot, Avoided, Addbot, DOI bot, Download, ChenzwBot, SamatBot, Tide rolls, Chipchap, Angrysockhop, Legobot, Luckas-bot, Gth-au, Yobot, Twexcom, Armchair info guy, AnomieBOT, Paskornc, Jim1138, No1sundevil, Citation bot, ArthurBot, Breeder8128, Xqbot, TheAMmollusc, Jeffrey Mall, DSisyphBot, GrouchoBot, Tarantulae, BulldogBeing, FrescoBot, Mark Renier, Mohdavary, Citation bot 1, Keladosi, Biker Biker, Tom.Reding, RedBot, Kon michael, Orenburg1, Docurbs, TobeBot, Trappist the monk, Gatator, Jonkerz, Rdelcueto, YouAndMeBabyAintNothingButCamels, RjwilmsiBot, Justinaction, VernoWhitney, BertSeghers, Lbartnik, Kencflewo, Dzkd, Sigurdurbj, Yuejiao Gong, Sunandwind, Gretchen Hea, Wikipelli, Daryakav, Optimering, Nentrex, H3llBot, SporkBot, AManWithNoPlan, Kcwong5, RockMagnetist, Tezzet, Sednodna, ClueBot NG, Mathstat, Wcherowi, Mctechuciztecatl, Helpful Pixie Bot, Jr271, J.Dong820, Lugel74, BG19bot, Papadim.G, Qx2020, Metricopolus, Anubhab91, Compfreak7, Darkeffy, BattyBot, Ferrarisailor, ChrisGualtieri, Mediran, AusCanBri, Oritnk, Mogism, Kub1x, Mrzazz001, Thomas Jeal, Me, Myself, and I are Here, Mark viking, Selvi muthukumar, Epicgenius, Bhrnjica, Jramsden271, Brzydalski, David.conradie, Kks11deq, DGent, Monkbot, Btomoiaga, Sofia Koutsou-

47.10. TEXT AND IMAGE SOURCES, CONTRIBUTORS, AND LICENSES

veli, Moshaydi, DAColey, Pierre-delmoral, Vitor, Olosko, Boky90, SharifCS, Muaafa, KasparBot, Cruella wiki, ArguMentor, Bender the Bot, Jeanpmartins, LordBali, Libraryyyyy, Magic links bot, Johnhiltonw, KolbertBot and Anonymous: 486

- **Genetic programming** *Source:* https://en.wikipedia.org/wiki/Genetic_programming?oldid=797732393 *Contributors:* Bryan Derksen, Hari, Nealmcb, Lexor, David Martland, Cyde, Minesweeper, Ahoerstemeier, Ronz, BAxelrod, Guaka, Timwi, Furrykef, Stewartadcock, Tualha, TittoAssini, Jleedev, Pengo, DavidCary, Duncharris, Vanished user wdjklasdjskla, Pgan002, Klemen Kocjancic, Andreas Kaufmann, Michal Jurosz, Wrp103, Aris Katsaris, Xgenei, Nuwanda, BrokenSegue, Mdd, Liao, Terrycojones, Crispin Cooper, Diego Moya, Mr-enwiki, Artur adib, Linas, Ruud Koot, Waldir, Qwertyus, Rjwilmsi, Mahlon, Lmatt, Kri, PowerMacX, Chobot, YurikBot, Wavelength, Squillero, Joel7687, D jinn, Bob0the0mighty, Brat32, Bota47, MartijnBodewes, Allens, SmackBot, Gwax, Gragus, Tamfang, Firegnome, Thomas weise, Chris55, Ceran, CmdrObot, Chuffy, DeLarge, Cydebot, Simenheg, Klausikm-enwiki, Mentifisto, JAnDbot, Barek, Magioladitis, Nyq, Botfob, Ianboggs, Uncoolbob, Riccardopoli, Especialist, EminNew, Knomegnome, Rogerfgay, Don4of4, Wingedsubmariner, Cmbay, SQL, Micklin, Moltean, Karlyoxall, Soler97, Algorithms, Mrberryman, CharlesGillingham, Capitalismojo, Classicalecon, Explicit, ClueBot, Ggiavelli, Biggerj1, Jamesmichaelmcdermott, ChrisKalt, DumZiBoT, NicMcPhee, XLinkBot, Farthur2, Addbot, MrOllie, Feijai, Yobot, Themfromspace, YetAnotherMatt, Jorge.maturana-enwiki, Thattommyguy, Chchen, Tarantulae, RibotBOT, Halhen, FrescoBot, Mark Renier, Mohdavary, ParadoxGreen, Tudlio, RozanovFF, Roboo.jack, Sergey539, EmausBot, Arkenflame, Newpotty, Nextrex, Josve05a, Alaa safeef, Snowscorpio, ClueBot NG, Teja.Nanduri, BG19bot, Ilent2, CitationCleanerBot, Golmschenk, Harsh4101991, SpoonUnit, Oritnk, Me, Myself, and I are Here, Bhrnjica, Davidgumberg, Robevans123, Dodi 8238, Superploro, Pblowry, SylerHaker, Rizzank, HelpUsStopSpam, Nichaelcramer, Latosh Boris, W102102, RobbieIanMorrison, GreenC bot, Jeanpmartins, Alisohani, KolbertBot and Anonymous: 191

- **Artificial life** *Source:* https://en.wikipedia.org/wiki/Artificial_life?oldid=801378440 *Contributors:* Brion VIBBER, Mav, Bryan Derksen, The Anome, Pietro speroni, Heron, Lexor, Bobby D. Bryant, Nine Tail Fox, Ronz, Extro, CatherineMunro, Kimiko, Kwekubo, Emperorbma, Guaka, Timwi, David Latapie, Slark, Wik, Samsara, Wilke, Hajor, Robbot, Stewartadcock, Aniu-enwiki, Wikibot, Tea2min, Ancheta Wis, Psb777, Matthew Stannard, DavidCary, Peruvianllama, Ds13, Dratman, Zumbo, Iota, Macrakis, Zeimusu, Bcameron54, MisfitToys, Sam Hocevar, Sam, MakeRocketGoNow, GreenReaper, EagleOne, Jwdietrich2, AAAAA, Rfl, Discospinster, Michal Jurosz, Pjacobi, Dbachmann, Bender235, Kbh3rd, El C, RoyBoy, Erauch, Cmdrjameson, Jjmerelo-enwiki, Ferkel, Mdd, Markus.Waibel, Pinar, Cough, TheCoffee, Marasmusine, JimmyShelter-enwiki, SDC, Jackobogger, Graham87, Rjwilmsi, Zbxgscqf, Ligulem, Husky, Erikwithaknotac, Margosbot-enwiki, JiFish, Quuxplusone, -ts-, Predictor, Chobot, Gaius Cornelius, Chaos, Draeco, William Caputo, Trovatore, Joel7687, Tailpig, Thsgrn, Nick, William R. Buckley, CLW, Saulkaiserman, SaTaMaS, Arthur Rubin, Levil, JLaTondre, JDspeeder1, NeilN, SmackBot, Numsgil, Smec, Chris the speller, OrphanBot, Onorem, MattOates, Richard001, Newsmare, Jon Awbrey, Barbalet, Sina2, Khazar, Snleo, Ckatz, Tarcieri, Loadmaster, Beetstra, BenRayfield, Skinsmoke, MrDolomite, Chris55, James pic, Liam Skoda, Mercere99, Kenrinaldo, Seth Manapio, Headbomb, Davidhorman, Oddity-, MattBan, The Transhumanist, Oliviermichel, RainbowCrane, BrotherE, Magioladitis, VoABot II, N16HTM4R3, BatteryIncluded, David Eppstein, DerHexer, Squidonius, Stephenchou0722, MartinBot, Francis Tyers, Mikael Häggström, Fheyligh, Bofoc Tagar, Steel1943, RashmiPatel, Dggreen, Myles325a, Lordvolton, Atomcoeur, Tesfatsion, Milkbreath, Michafaw, SpikeZOM, Hingfat, Spacemonster, Omermar, Why Not A Duck, Kenstauffer, Dan Polansky, Ivan Štambuk, Spamburgler, Jdaloner, CharlesGillingham, SallyForth123, Martarius, Ahyeek, Franksbnetwork, ForestDim, DerBorg, Rankiri, Tautrimas, Eric Catoire, Truthnlove, Svea Kollavainen, Addbot, MrOllie, Lightbot, Zeugma fr, Drpickem, Yobot, Themfromspace, Taxisfolder, AnomieBOT, DemocraticLuntz, Eumolpo, BloodGrapefruit, GliderMaven, Mauriceling, Timawesomeness, DrilBot, Cedric71, MKFI, LilyKitty, Diannaa, Rzuwig, Jfmantis, Cdocrun, Adhemarius, In ictu oculi, DASHBot, Primefac, Tommy2010, TuHan-Bot, Phoenixthebird, H3llBot, Demomoer, Ventus55, ClueBot NG, Payyan-2000, Frietjes, Zach Winkler, Davidcofer73, MerIlwBot, Helpful Pixie Bot, Wbm1058, TillF, ArmorShieldA99, Larryy, JYBot, Robingras, Me, Myself, and I are Here, Phamnhatkhanh, Jamie Hlusko, FockeWulf FW 190, Herve Noel, Wallnut tree, Lev Kalmykov, Fixuture, Monkbot, KreittonVictor, KasparBot, AmitMakwana008, Linguist111, InternetArchiveBot, GreenC bot, Shaofengmo, Magic links bot, KolbertBot and Anonymous: 169

- **Machine learning** *Source:* https://en.wikipedia.org/wiki/Machine_learning?oldid=802701162 *Contributors:* Arvindn, ChangChienFu, Michael Hardy, Kku, Delirium, Ahoerstemeier, Ronz, BenKovitz, Mxn, Hike395, Silvonen, Furrykef, Buridan, Jmartinezot, Phoebe, Shizhao, Topbanana, Robbot, Plehn, KellyCoinGuy, Ancheta Wis, Fabiform, Centrx, Giftlite, Seabhcan, Levin, Dratman, Jason Quinn, Khalid hassani, Utcursch, Beland, APH, Gene s, Paulscrawl, Neutrality, Clemwang, Tordek ar, Nowozin, Silence, Bender235, ZeroOne, Superbacana, Aaronbrick, Jojit fb, Nk, Rajah, Tritium6, Haham hanuka, Mdd, HasharBot-enwiki, Vilapi, Arcenciel, Denoir, Diego Moya, Wjbean, Stephen Turner, LearnMore, Danhash, Rrenaud, Leondz, Soultaco, Ruud Koot, BlaiseFEgan, JimmyShelter-enwiki, Essjay, Joerg Kurt Wegner, Adiel, RichardWeiss, BD2412, Qwertyus, Imersion, Rjwilmsi, Emrysk, Feelmybrainwaves, VKokielov, Eubot, Celendin, Intgr, Predictor, Kri, BMF81, Irregulargalaxies, Chobot, Bobdc, Bgwhite, Adoniscik, YurikBot, Misterwindupbird, Trondtr, Nesbit, CambridgeBayWeather, Grafen, Gareth Jones, Srinivasasha, Raikkonen, Crasshopper, Tony1, DaveWF, Masatran, Cedar101, SMcCandlish, CWenger, Fram, KnightRider-enwiki, SmackBot, Mneser, InverseHypercube, CommodiCast, Jyoshimi, Mcld, KYN, Gilliam, Ohnoitsjamie, Chris the speller, FidesLT, Nbarth, Cfallin, Moorejh, JonHarder, Baguasquirrel, Krexer, Shadow1, Philpraxis-enwiki, Daniel.Cardenas, Sina2, Kuru, ChaoticLogic, NongBot-enwiki, RexSurvey, Beetstra, WMod-NS, Julthep, Dsilver-enwiki, Dicklyon, Vsweiner, Optakeover, Ctacmo, MTSbot-enwiki, Ralf Klinkenberg, Dave Runger, Chris55, GerryWolff, Doceddi, Scigrex14, Pgr94, Innohead, Bumbulski, Peterdjones, Sonda, Sytelus, Dancter, Msnicki, Quintopia, Thijs!bot, Mereda, Perrygogas, Djbwiki, GordonRoss, Kinimod-enwiki, Damienfrancois, Natalie Erin, Seaphoto, AnAj, Ninjakannon, Kimptoc, Penguinbroker, The Transhumanist, Jrennie, Hut 8.5, Kyhui, Magioladitis, Ryszard Michalski, Jwojt, Transcendence, Tedickey, Pebkac, Robotman1974, Jroudh, Businessman332211, Pmbhagat, Calltech, NatGertler, Carlosayam, STBot, Keith D, Glrx, Nickvence, Gem-fanat, Salih, AntiSpamBot, Gombang, Chriblo, Mxwsn, Dana2020, DavidCBryant, Bonadea, WinterSpw, Llorenzi, RJASE1, Funandtrvl, James Kidd, LokiClock, Redgecko, Markcsg, Jrljrl, Like.liberation, A4bot, Daniel347x, Joel181, Wikidemon, Lordvolton, Defza, Chrisoneall, Wingedsubmariner, Spiral5800, Kesshaka, Cvdwalt, Why Not A Duck, Sebastjanmm, LittleBenW, Gal chechik, Amir Rahat, Biochaos, Cmbishop, Jbmurray, IradBG, Smsarmad, Flyer22 Reborn, Scorpion451, Kumioko (renamed), CharlesGillingham, StaticGull, CultureDrone, Anchor Link Bot, ImageRemovalBot, ClueBot, GorillaWarfare, Acfleiss, Ahyeek, Sonu mangla, Ggia, Debejyo, D.scain.farenzena, He7d3r, Magdon~enwiki, WilliamSewell, Jim15936, Vanished user uih38riiw4hjlsd, Evansad, Roxy the dog, Jytdog, PseudoOne, André P Ricardo, Agamemnonc, Mudfud85, MystBot, Dsimic, YrPolishUncle, MTJM, Addbot, Mortense, Fyrael, Aceituno, MrOllie, LaaknorBot, Quercus solaris, Jarble, Movado73, Luckas-bot, QuickUkie, Yobot, NotARusski, Genius002, Examtester, AnomieBOT, Erel Segal, Piano non troppo, Materialscientist, Clickey, Devantheryv, Vivohobson, ArthurBot, Quebec99, LilHelpa, Xqbot, Happyrabbit, Gtfjbl, Kithira, J04n, SinPantuflas, Addingrefs, Webidiap, Shirik, Joehms22, Aaron Kauppi, Velblod, Prari, FrescoBot, Jdizzle123, Olexa Riznyk, Featherard, WhatWasDone, Siculars, Proffviktor, Cnnsbs, Boxplot, Swordsmankirby, I dream of horses, Wikinacious, Skyerise, Mostafa mahdieh, Lars Washington, Serols, AvnishIT, Docurbs, TobeBot, OnceAlpha, AXRL, Mean as custard, Њь Љю Ха, BertSeghers, Edouard.darchimbaud, Winnerdy, Zosoin, Helwr, EmausBot, John of Reading, Johncasey, Dzkd, Dewritech, Primefac, MartinThoma, Jasonanaggie, MarsTrombone, Wht43, Chire, Bilbo571, GZ-Bot, Jcautilli, Jorjulio, AManWithNoPlan, Staszek Lem, Pintaio, L Kensington, Ataulf, Zfeinst, Yoshua.Bengio, Casia wyq, Ego White Tray, Blaz.zupan, Shinosin, Randallbritten, Marius.andreiana, Lovok Sovok, Bobbybobbie, ManU0710, Graytay, Liuyipei, ClueBot NG, Tillander, Keefaas, Lawrence87, Aiwing, Pranjic973, Candace Gillhoolley, Robiminer, Leonardo61, Wrdieter, Arrandale, O.Koslowski, MathKeduor7, Joel B. Lewis, WikiMSL, Helpful Pixie Bot, RobertPollak, BG19bot, Smorsy, Mohamed CJ, Lisasolomonsalford, Anubhab91, Chafe66, Solomon7968, Ishq2011, Autologin, Brooksrichardbrown, DasAllFolks, Billhodak, Debora.riu, Ohandyya, Davidmetcalfe, David.moreno72, Mdann52, JoshuSasori, Ulugen, IjonTichyIjonTichy, Keshav.dhandhania, Dexbot, Mogism, Djfrost711, Bkuhlman80, DougBoost, Frosty, Jamesx12345, Shubhi choudhary, Jochen Burghardt, Joeinwiki, Aomatveev, Brettrmurphy, Phamnhatkhanh, Ppilotte, Delafé, InnocuousPilcrow, Kittensareawesome, Statpumpkin, Neo Poz, Master of Time, TJLaher123, Ankit.ufl, Francisbach, Aleks-ger, MarinMersenne, Weiping.thu, LokeshRavindranathan, Tonyszedlak, Proneat123, GrowthRate, Kojikawano, Sami Abu-El-Haija, Mpgoldhirsh, Work Shop Corpse, Superploro, Riceissa, Dawolakamp, Waggie, VeniVidiVicipedia, Justincahoon, Gufosowa, Jorge Guerra Pires, Hm1235, Velvel2, Cowprophet, Vidhul sikka, Ian (Wiki Ed), Erik Itter, Annaelison, Tgriffin9, Chazdywaters, Rmashrmash, Komselvam, Nbro, Robbybluedogs, HelpUsStopSpam, EricVSiegel, KenTancwell, Kaartic, Justinqnabel, Rusky.ai, Datapablo, Aetilley, JenniferTheEmpress0, Dsysko, Ghost30, Haodong123, Lr0^^k, BNoack, NightOwl15, Latosh Boris, Dorianinou, User8178, Thejavis86, Faustino.josemar, Muratovst, Pinsi281, ArguMentor, Datakeeper, Nesny, Doctasarge, Espyromi, Kailey 2001, WunderStahl, Natenatenatenate, Fmadd, Vladiatorr, Ayasdi, ChillyBlue, Hyksandra, Famousceleb, Hexacta BA, AllenAkhaumere, Bender the Bot, Ajav13, Aios3837, An Inconvenient Truth, PouyaRZ, PrimeBOT, Caoanexo, GeneSobolev, BunnyShark, SecurityPanther, Wang Zichen, Alex-jprof, DrStrauss, Clipname, Johnnyballs454545, Noah Kastin, Schellluri, Peterwittek, CobraL0rd, Emoody17, Shubh24aug, Jacob.stein, Blankshot, TuCo, Erukhhirhsr318, Ml 771, KolbertBot, Clarities, RachelTan, Emily ly and Anonymous: 497

- **Evolutionary developmental biology** *Source*: https://en.wikipedia.org/wiki/Evolutionary_developmental_biology?oldid=802839663 *Contributors*: Mav, Zundark, Slrubenstein, PierreAbbat, DavidLevinson, Adam Retchless, Stevertigo, Edward, Lexor, Shyamal, Gabbe, Deljr, JWSchmidt, Charles Matthews, Samsara, Robbot, Naddy, Cholling, Nmg20, NiteowInneils, Alteripse, APH, Jania902, Rich Farmbrough, Dave souza, LemRobotry, JackWasey, Cohesion, Katefan0, Woohookitty, Mindmatrix, Karmosin, Wdanwatts, RichardWeiss, BD2412, Drbogdan, Rjwilmsi, Jivecat, Vary, Ian Pitchford, Bgwhite, Adonisick, YurikBot, Jknabe, Tavilis, Spike Wilbury, Yahya Abdal-Aziz, Sitearm, William R. Buckley, Ospalh, Maunus, Culmination, Leptictidium, Mike Dillon, Wikipeditor, SmackBot, Bluebot, Ed Hagen, Vanished User 0001, Vina-iwbot~enwiki, StN, Clicketyclack, Mat8989, Olin, AdultSwim, Roland Deschain, Bobamnertiopsis, Courcelles, CmdrObot, Outriggr (2006-2009), Kupirijo, Narayanese, Pjvpjv, Lauranrg, Magioladitis, Bookuser, Theroadislong, GetAgrippa, Miaers, Pekaje, Stan J Klimas, Memestream, SteveChervitzTrutane, Chiswick Chap, LittleHow, Ale2006, VolkovBot, WOSlinker, Philip Trueman, Antoni Barau, Insanity Incarnate, SieBot, Lightmouse, Jacek FH, Ecobion, Gauravm1312, SchreiberBike, Trigley, Alamowit, Johnuniq, Roosme, Frostus, DOI bot, Moosepuggle, Fgnievinski, 10outof10die, Zorrobot, Ettrig, Legobot, Luckas-bot, Yobot, WikiDan61, Azcolvin429, CHW100, AnomieBOT, Floquenbeam, Citation bot, ArthurBot, Xqbot, FrescoBot, Citation bot 1, Winterst, Tom.Reding, MondalorBot, Pbsouthwood, Trappist the monk, Wotnow, Fama Clamosa, RjwilmsiBot, Deirovic, ZéroBot, Rmarquezp, Ventus55, Ernestfax, Gongoozler123, Joannamasel, Woodsrock, ClueBot NG, Helpful Pixie Bot, Bibcode Bot, Plantdrew, BG19bot, CitationCleanerBot, Harizotoh9, SaudiPseudonym, Lirusaito, Sminthopsis84, Isarra (HG), Corinne, Joeinwiki, CalaClii, Hawkeye-gnome, Iztwoz, Charvetej, Anrnusna, FSProctor, Dodi 8238, Monkbot, Mama meta modal, Shlomi85, Oyeleke2, Ian (Wiki Ed), Kashish Arora, KasparBot, A little angry, Rtrust and Anonymous: 61

- **Evolutionary robotics** *Source*: https://en.wikipedia.org/wiki/Evolutionary_robotics?oldid=802408752 *Contributors*: Togelius, Ronz, BAxelrod, AlainV, Marcika, NiteowIneils, Gene s, Juxi, Andreas Kaufmann, Stephane.magnenat, Markus.Waibel, Triddle, Bluemoose, CharlesC, Bgwhite, The Rambling Man, Searchme, Masatran, Open2universe, SmackBot, Moxon, Reedy, Thunder8, MungoZero, Chris the speller, Breno, Stefano.nolfi, Androidchild, Aboeing, IDSIAupdate, Kenrinaldo, Fayenatic london, TAnthony, Ben Ram, Jedbailey, Jim.henderson, Jiuguang Wang, Jkaplan, Fheyligh, Rogerfgay, Tikuko, Altermike, Sfan00 IMG, ClueBot, Bwfrank, Chaosdruid, XLinkBot, Addbot, Download, Luckas-bot, Grantmidnight, FrescoBot, DrilBot, Nekonaute, Robertalbertallen, EmausBot, S doncieux, Aurelien.desbrieres, ClueBot NG, Davimaro74, Danim, Coenig, Me, Myself, and I are Here, DavidLeighEllis, Sboddhu, Dtm22, Mohamed-Ahmed-FG, GuidoAI, InternetArchiveBot, Aiesparcia, Evert.haasdijk, Magic links bot and Anonymous: 53

- **Pattern formation** *Source*: https://en.wikipedia.org/wiki/Pattern_formation?oldid=802563648 *Contributors*: Kku, Nina, Pharos, Wtmitchell, Oleg Alexandrov, BD2412, Josh Parris, Rjwilmsi, DonSiano, JarrahTree, Daniel Mietchen, Reyk, JesseStone, Berland, LeoNomis, Wizard191, Miketwardos, Narayanese, Wikid77, Utopiah, Ste4k, Lopkiol, JaGa, R'n'B, Dr d12, Chiswick Chap, The enemies of god, Danwills, LucDecker, Paradoctor, Forluvoft, SchreiberBike, DumZiBoT, Addbot, Rascallgh, Yobot, FrescoBot, Reirobros, Algorithmgeek, Blueshifting, GoingBatty, WarEqualsPeace, Deirovic, Danielse, Gongoozler123, Vaulttech, BattyBot, ChrisGualtieri, Cerabot~enwiki, Mk85 2, Monkbot, CStreiss, Bender the Bot, Cyrus noto3at bulaga and Anonymous: 23

- **Reaction–diffusion system** *Source*: https://en.wikipedia.org/wiki/Reaction%E2%80%93diffusion_system?oldid=800691531 *Contributors*: Michael Hardy, Jitse Niesen, Markhurd, Aetheling, Giftlite, Rich Farmbrough, Lenov, Oleg Alexandrov, BD2412, Spencerk, Borgx, Welsh, Allens, SmackBot, Complexica, CBM, Neelix, Kupirijo, Oink54321, Headbomb, R'n'B, Chiswick Chap, Danwills, Jmath666, Arcfrk, Tiddly Tom, Huboedeker, Vagogan, XLinkBot, Addbot, Mjl2008, MrOllie, Jockelinde, بن سعد, Yobot, Ptbotgourou, KamikazeBot, Geek1337~enwiki, Omnipaedista, Rhodydog, FrescoBot, Petrelharp, TharsHammar, D'ohBot, Jonesey95, Wrettyfugi, EdoBot, Xanchester, Vaulttech, Schenad, Kryomaxim, Bhavya1333, Jasherratt, Awliehr, Andy Quarry, Latinoo.west, Bender the Bot and Anonymous: 45

- **Partial differential equation** *Source*: https://en.wikipedia.org/wiki/Partial_differential_equation?oldid=796675578 *Contributors*: AxelBoldt, CYD, The Anome, Tarquin, Roadrunner, Nealmcb, Patrick, Michael Hardy, Oliver Pereira, Ixfd64, Ahoerstemeier, Stevenj, Silverfish, Charles Matthews, Timwi, Bemoeial, Dysprosia, Jitse Niesen, Gutza, Tpbradbury, Topbanana, Robbot, MathMartin, Moink, Cbm, Robinh, Filemon, Tosha, Giftlite, Jyril, BenFrantzDale, Michael Devore, Kmote, Waltpohl, Rpchase, PeR, Fintor, PhotoBox, L-H, Mazi, ArnoldReinhold, Paul August, Bender235, Djordjes, Ub3rm4th, Pt, El C, David Crawshaw, Chbarts, Photonique, Crust, Msh210, Arthena, EmmetCaulfield, Jon Cates, Kbolino, Oleg Alexandrov, Tbsmith, Linas, WadeSimMiser, Mpatel, MFH, Isnow, Mandarax, Rnt20, Magister Mathematicae, BD2412, Rjwilmsi, Bertik, Andrei Polyanin, Nneonneo, R.e.b., Gseryakov, Arnero, Mathbot, Gurch, Srleffler, Chobot, Eienmaru, YurikBot, Wavelength, Borgx, Pacaro, Jschlosser, Wavesmeister, Prime Entelechy, Ojcit, FF2010, Zzuuzz, DrHok, Brian Tvedt, GraemeL, Eigenlambda, Bjorn.sjodin, NSiDms, DStoykov, RayAYang, Nbarth, Giese, DHN-bot~enwiki, Hongooi, D.328, Yaje, Mhym, Germandemat, Donludwig, Ckatz, CyrilB, Dicklyon, Stizz, Paul Matthews, Richard77, CmdrObot, Myasuda, AndrewHowse, Cydebot, Quibik, Dharma6662000, Thijs!bot, Egriffin, Epbr123, Headbomb, Marek69, JustAGal, Cj67, Ben pcc, Mhaitham.shammaa, Sbarnard, Erxnmedia, The Transhumanist, Thenub314, Hut 8.5, Burga, VoABot II, JNW, David Eppstein, User A1, JaGa, Coffeepusher, Marupio, R'n'B, DominiqueNC, Manticore, Maurice Carbonaro,

Foober, Salih, Belovedfreak, JonMcLoone, Policron, Izno, Cuzkatzimhut, Affluent Rider, Wtt, Jmath666, Dirkbb, Wolfrock, SieBot, Tiddly Tom, Roesser, Lagrange613, OKBot, Spartan-James, Shooke, Yhkhoo, Jonathanstray, LikeFunYouAre, ClueBot, Rememberlands, Excirial, Winston365, Evankeane, Muro Bot, Crowsnest, Siegmaralber, Forbes72, Jss214322, PL290, Super Cleverly, Addbot, AkhtaBot, Download, Pranagailu1436, Jasper Deng, Jarble, Luckas-bot, Yobot, OrgasGirl, Fraggle81, THEN WHO WAS PHONE?, SwisterTwister, Wsulli74, Xqbot, Unigfjkl, Frosted14, Oscarjquintana, Amaury, FrescoBot, Aliotra, D'ohBot, Sławomir Biały, Kwiki, Gaj0129, Pinethicket, I dream of horses, Tom.Reding, Wikiain, Iwfyita, Rausch, Orenburg1, SobakaKachalova, Onel5969, Pokespa, Mandolinface, Chris in denmark, EmausBot, Dcirovic, Bkocsis, Mathsfreak, Druzhnik, Gerasime, Alpha Quadrant (alt), Donner60, ChuispastonBot, ClueBot NG, ChristophE, HappySophie, Frietjes, MathKeduor7, Widr, Rafnuss, Helpful Pixie Bot, J824h, Noam Duke, Solomon7968, F=q(E+v^B), Beckman16, NotWith, Fuse809, Vanished user lt94ma34le12, YFdyh-bot, Illia Connell, BASANTDUBE, Me, Myself, and I are Here, Hillbillyholiday, Jamesmcmahon0, Mathmensch, Leanedjuneon, JaconaFrere, Mgkrupa, Sanoonan, Pandaboy117, Treasuredwealth, KasparBot, Sonicrs, ProboscideaRubber15, Lahwaacz, Leschnei, Heluos, GreenC bot, Fmadd, Bear-rings, Manishkrisna108, DrStrauss and Anonymous: 247

- **Dissipative system** Source: https://en.wikipedia.org/wiki/Dissipative_system?oldid=800263194 Contributors: Mav, Michael Hardy, Lexor, Stevenj, BillBell, Stone, Laussy, Jtvisona, Phys, Omegatron, Lumos3, Robbot, Goethean, Dratman, Pearbonn, Fanghong~enwiki, Neurophyre, Mdd, Linas, CLW, SmackBot, WebDrake, Vald, Cazort, Chris the speller, Seberia Eagle, Headbomb, WinBot, Narssarssuaq, Tigga, James-BWatson, Pere prlpz, Mårten Berglund, Tomasao, Cyprium29, Ordermaven, EverGreg, Wolfch, Addbot, BartNotelaers, Luckas-bot, Yobot, Ptbotgourou, KamikazeBot, AnomieBOT, Daniele Pugliesi, Omnipaedista, The Wiki ghost, Chjoaygame, Taweetham, RedBot, EmausBot, WikitanvirBot, Bibcode Bot, BG19bot, Teika kazura, Cyberbot II, Khazar2, Garuda0001, Andy Quarry, Sdurietz, Prestigiouzman, Beanstash, DoubleDr, Ssmmachen, GreenC bot, Mmartre88, Molisi, Murakamifan93, Driverin, KolbertBot and Anonymous: 32

- **Percolation** Source: https://en.wikipedia.org/wiki/Percolation?oldid=787495084 Contributors: DavidLevinson, RTC, Michael Hardy, Lexor, Delirium, Dgrant, Docu, Snoyes, Doradus, Steinsky, Robbot, Giftlite, Legarcia, Pgreenfinch, Antaeus Feldspar, Brian0918, Mdd, Guy Harris, Melaen, Rafti Institute, Uffish, Oleg Alexandrov, Linas, Ruud Koot, BD2412, Rjwilmsi, Windchaser, Vonkje, YurikBot, TexasAndroid, MadeYourReadThis, Shell Kinney, NawlinWiki, Welsh, BazookaJoe, CedricVonck, 2over0, SmackBot, Stepa, Gilliam, Mhym, Mwtoews, BranStark, Iridescent, JMK, PetaRZ, Tawkerbot2, Nunquam Dormio, Neelix, Edchi, Escarbot, F.Shelley, VoABot II, The Big Man, Seba5618, J.delanoy, Idunno271828, Mikael Häggström, Bonadea, Signalhead, Lights, SCriBu, Biscuittin, THEMONKEYZIP, Anton Petrov, Flyer22 Reborn, Oda Mari, Bakashi10, Svick, Water and Land, JohnnyMrNinja, Rziff, Tanvir Ahmmed, ClueBot, Wikeepedian, SilvonenBot, Addbot, Jacopo Werther, Quercus solaris, Jarble, Yobot, E mraedarab, SwisterTwister, AnomieBOT, Kingpin13, Materialscientist, Gggmaster, Dr. Perfessor, Omnipaedista, Mmru, Pinethicket, Tom.Reding, Felix0411, Heracles31, Erianna, Sitic, Berberisb, ClueBot NG, Frietjes, Bibcode Bot, BG19bot, BendelacBOT, JMtB03, BattyBot, AustenHead, ChrisGualtieri, CarrieVS, Compsim, Nunoaraujo, Derosigur, KasparBot, Shifra987, Boomer Vial, Weiner18, PrimeBOT, Rkorstanje and Anonymous: 100

- **Cellular automaton** Source: https://en.wikipedia.org/wiki/Cellular_automaton?oldid=800940171 Contributors: AxelBoldt, LC~enwiki, Bryan Derksen, XJaM, PierreAbbat, Ezubaric, Hephaestos, Jose Icaza, Hfastedge, Michael Hardy, Jdandr2, Kku, Eric119, Angela, Error, AugPi, Samw, Schneelocke, Iseeaboar, Dysprosia, Jogloran, Wik, Zoicon5, Selket, Saltine, Bevo, Kizor, Kyber~enwiki, Chopchopwhitey, P0lyglut, TittoAssini, Tea2min, Giftlite, Smjg, Curps, MingMecca, Guanaco, Jasper Chua, LucasVB, Watcher, Joseph Myers, Elektron, Robin klein, Chmod007, Yonkeltron, Balsarxml, Imroy, Peak Freak, On you again, ZeroOne, Ben Standeven, Kb, Edward Z. Yang, Dalf, Erauch, Christian Kreibich, ACW, Photonique, Ferkel, Perceval, Keenan Pepper, Benjah-bmm27, Kotasik, Alexwg, Samohyl Jan, LukeSurl, Oleg Alexandrov, Marasmusine, Zorbid, Woohookitty, Mihai Damian, Kzollman, Lgallindo, -Ril-, GregorB, CharlesC, Graham87, Deltabeignet, Rjwilmsi, MarSch, Sbp, Pygy, FlaBot, Mathbot, RexNL, Orborde, Quuxplusone, Mahlon, Srleffler, Kri, Ahunt, Visor, DVdm, Hmonroe, Wavelength, Karlscherer3, RussBot, Allister MacLeod, Xihr, ENeville, Welsh, R.e.s., Lpdurocher, JocK, Chakazul, Pcorteen, Raven4x4x, Scs, William R. Buckley, Iztok.jeras~enwiki, Silverhill, Ninly, Bhumiya, MaNeMeBasat, Curpsbot-unicodify, Ilmari Karonen, Tropylium, Banus, Nekura, Mosiah~enwiki, SmackBot, RDBury, PEHowland, InverseHypercube, K-UNIT, Axd, Hmains, Chris the speller, Dra, Zom-B, Felicity Knife, Froese, Gragus, Crazilla, Phaedriel, Radagast83, Spectrogram, Nakon, Jon Awbrey, Sadi Carnot, Joeyramoney, Sam Tobar, SashatoBot, Metric, Acidburn24m, AnonEMouse, Mgiganteus1, Alpha Omicron, Ckatz, 524, Dl2000, The Temple Of Chuck Norris, Warrado, EmreDuran, Iridescent, JoeBot, Mudd1, Xerophytes, Argon233, Requestion, Cydebot, DumbBOT, Scolobb, Mattisse, Headbomb, Oubiwann, I do not exist, Ideogram, Dawnseeker2000, Navigatr85, Geneffects, AntiVandalBot, Dhushara, Gioto, Widefox, Caileagleisg, Hannes Eder, Pixelface, Kaini, JAnDbot, Asmeurer, Avaya1, Hillgentleman, Baccyak4H, EagleFan, Torchiest, David Eppstein, Cypherzero0, JaGa, Hiner, NAHID, J.delanoy, Marcus Wilkinson, Chiswick Chap, DadaNeem, B.huseini, Kneb~enwiki, Mydogtrouble, Dcornforth, Torcini, Pleasantville, TXiKiBoT, Yugsdrawkcabeht, Mbaudier, Calwiki, Chuckwolber, RyanB88, Anonymous Dissident, Akramm1, Bearian, Lamro, SQL, Dmcq, AlleborgoBot, EmxBot, AdRock, SieBot, Setoodehs, BotMultichill, Gerakibot, Pi is 3.14159, Lightmouse, Sph110, JL-Bot, FerrenMacI, Beeblebrox, ClueBot, NickCT, Justin W Smith, HairyFotr, Sun Creator, Gleishma, Olivierse2, Cstheoryguy, Versus22, DumZiBoT, Gthen, XLinkBot, Pichpich, Jytdog, Mandalaschmandala, Dekart, Bprentice, Shoemaker's Holiday, Addbot, Wli625, MrOllie, LinkFA-Bot, Hope09, ScAvenger, Jarble, Yobot, AnomieBOT, JackieBot, Flewis, JuliusCarver, Materialscientist, Citation bot, Watertree, Loveless, Nippashish, Artem M. Pelenitsyn, Sharkyangliu, OgreBot, Citation bot 1, Kiefer.Wolfowitz, MondalorBot, Dave Feldman, Throwaway85, Jonkerz, BorysB, RjwilmsiBot, Bento00, Laesod, EmausBot, Johnhwynne, Svrist, Dcirovic, Sumanafsu, GSM83~enwiki, Agora2010, Grondilu, ZéroBot, PBS-AWB, Ὁ οἶστρος, Beddowve, Tijfo098, ChuispastonBot, Wolfpax50, Mishaell, ClueBot NG, Deer*lake, LunchboxGuy, Helpful Pixie Bot, Jlopez1967, Bibcode Bot, Alvyray, BG19bot, Pasicles, Tklauser, Dexbot, Ashleydan, Luanoz, Frizzil, 420mysteryman69, Mark viking, Captain Ford, PierreYvesLouis, Pdecalculus, Genaro.juarez_martinez, Andy Quarry, Sharkyangliu916, Nigellwh, Ginsuloft, JeremyLThompson, Prestigiouzman, Lev Kalmykov, Todd Rowland, Monktues, Monkbot, Joseph2302, Nbro, KasparBot, Mhanga parto, Nicole tylor, Fahkr smith, Socaacos, InternetArchiveBot, Gulumeemee, TracieBurns, Bender the Bot, Ariefyudho, Hocimi, Magic links bot, WikiSquirrel42, Zozo1220, Asontha, KolbertBot and Anonymous: 286

- **Spatial ecology** Source: https://en.wikipedia.org/wiki/Spatial_ecology?oldid=772348046 Contributors: Kku, Bovlb, Guettarda, Woohookitty, Rjwilmsi, Jrbouldin, Epipelagic, Reyk, SmackBot, EncycloPetey, Chris the speller, Bluebot, Davinaq, Funandtrvl, DrMicro, AlkaIIn, Addbot, Yobot, AnomieBOT, Minnecologies, J04n, I dream of horses, Trappist the monk, RjwilmsiBot, Look2See1, Liuqx315, Helpful Pixie Bot, SilverEyedLionQueen, Stefanocasalegno, Monkbot, Oleandir, Jmcgnh and Anonymous: 10

- **Self-replication** Source: https://en.wikipedia.org/wiki/Self-replication?oldid=796081190 Contributors: Damian Yerrick, LC~enwiki, Lee Daniel Crocker, Mav, Bryan Derksen, Andre Engels, Ray Van De Walker, AdamRetchless, AdSR, Lexor, Dominus, Kku, BigFatBuddha, Kimiko, Gamma~enwiki, Timwi, Hyacinth, Omegatron, Finlay McWalter, Robbot, Altenmann, Hadal, Tea2min, Enochlau, DavidCary, Alexf, Spottedowl, TheObtuseAngleOfDoom, DanielCristofani, Eric Shalov, RJHall, Liberatus, Sietse Snel, Rebroad, Woohookitty, David Haslam, Barrylb,

Apokrif, Rjwilmsi, Zbxgscqf, RexNL, Pete.Hurd, Diza, SteveBaker, Gwernol, RussBot, Gaius Cornelius, Grafen, BirgitteSB, Mosquitopsu, William R. Buckley, Mais oui!, Curpsbot-unicodify, Kiv, SmackBot, Bbewsdirector, DCDuring, Chris the speller, D-Rock, Dacoutts, Sagaciousuk, Stannered, Gioto, Just Chilling, MER-C, Avaya1, VoABot II, BatteryIncluded, David Eppstein, Gwern, R'n'B, Tarotcards, Chiswick Chap, Dxhtml, Moshesipper, Jamelan, Omermar, Sunrise, JL-Bot, Arnos78, Tweetlebeetle367, DumZiBoT, XLinkBot, Ost316, Addbot, Hakan Kayı, CarsracBot, Lightbot, Guyonthesubway, Yobot, PMLawrence, 1exec1, ChildofMidnight, Plasmon1248, J04n, Solphusion~enwiki, Brunonar, Miyagawa, CES1596, FrescoBot, Foobarnix, RjwilmsiBot, Mehdiabbasi, Lopifalko, Racerx11, Faolin42, Starcheerspeaksnewslostwars, Mussermaster, Ontyx, Mikhail Ryazanov, BG19bot, Knowledge Examiner, OakRunner, Me, Myself, and I are Here, Comp.arch, Fixuture, R1D1 and Anonymous: 73

- **Geomorphology** Source: https://en.wikipedia.org/wiki/Geomorphology?oldid=802876054 Contributors: Mav, Heron, Seav, Ellywa, Ahoerstemeier, Stan Shebs, Cferrero, Crissov, Steinsky, SEWilco, Lumos3, RedWolf, Altenmann, Thomas Wozniak, Alan Liefting, Giftlite, Legarcia, Zigger, Tagishsimon, LiDaobing, Adamsan, APH, GeoGreg, Zfr, Shiftchange, Geof, Vsmith, Mani1, Dmr2, CanisRufus, Aude, Octavius~enwiki, Maureen, Kentin, Siim, Andrewpmk, Paleorthid, Pauldavidgill, Denniss, BlueCanoe, Woohookitty, Matijap, Duncan.france, DanHobley, Palica, BD2412, FreplySpang, Sjö, Rjwilmsi, Danielcollins1, Daniel Collins, Bhadani, FlaBot, Margosbot~enwiki, Nihiltres, RexNL, Bgwhite, YurikBot, RobotE, Sceptre, Chris Capoccia, Wbfergus, MadeYouReadThis, Joelr31, Epipelagic, Tachs, Stefeyboy, Ratagonia, MaNeMeBasat, Anclation~enwiki, AlexD, Mmcannis, GrinBot~enwiki, Samuel Blanning, SmackBot, Thunderboltz, Gilliam, Hmains, Chris the speller, TimBentley, MK8, Mdwh, Robth, Darth Panda, Diyako, Rohanuk, Bud0011, Hoof Hearted, Mwtoews, Epf, Alþykkr, SashatoBot, Gobonobo, JorisvS, Peterlewis, IronGargoyle, Dicklyon, Geologyguy, Novangelis, Keith-264, Iridescent, Finn Krogstad, Eassin, Courcelles, MarylandArtLover, Postmodern Beatnik, Jordanotto, WeggeBot, Bumbulski, Myasuda, Cydebot, Tawkerbot4, Naudefj, Laramide, Mattisse, Thijs!bot, Headbomb, John254, RickinBaltimore, Divercol, Tocharianne, Dawnseeker2000, Just Chilling, Mack2, Mikenorton, Dentren, Gomm, NatureA16, MartinBot, BetBot~enwiki, Anaxial, CommonsDelinker, J.delanoy, Numbo3, Leaflet, Yonidebot, MrBell, PhoenixBlitzkrieg, GeoWriter, Nwbeeson, Hydroli, Cometstyles, Brianwhalley, Geekdiva, DorganBot, Idioma-bot, Funandtrvl, Deor, Macedonian, Gaianauta, TXiKiBoT, Geohumphrey, JhsBot, Andrewaskew, AlleborgoBot, Logan, PericlesofAthens, SieBot, Flyer22 Reborn, Geomorphologe, Pkumar.iitkgp, Oxymoron83, Ioverka, Walbe470, Drgarden, ClueBot, IceUnshattered, Rjd0060, Fossiliferous, Mild Bill Hiccup, Melizg, Masterpiece2000, Awickert, NuclearWarfare, BodhisattvaBot, Qfl247, Pee Tern, Addbot, Willking1979, Paris 16, Kevmus, Tide rolls, Zorrobot, Legobot, Luckasbot, Xeliff, Yobot, Themfromspace, AnomieBOT, Piano non troppo, Materialscientist, ArthurBot, LilHelpa, Ejsamoht, Hi878, GrouchoBot, Omnipaedista, GhalyBot, Shadowjams, LucienBOT, Recognizance, Micromesistius, Jonesey95, Tom.Reding, BRUTE, Serols, Jandalhandler, FoxBot, Trappist the monk, Callanecc, TjBot, EmausBot, WikitanvirBot, Nick Moyes, Winner 42, Wikipelli, Chiton magnificus, ZéroBot, Allforrous, Thine Antique Pen, Staszek Lem, RockMagnetist, ClueBot NG, Stromattolite, PaleCloudedWhite, HMSSolent, Gob Lofa, Bibcode Bot, KawaiiNippon, Otogi, Gorthian, Harizotoh9, Zedshort, ChrisGualtieri, Shyncat, EuroCarGT, Dexbot, Caroline1981, Me, Myself, and I are Here, Mark viking, Awesome monk, Star767, Prokaryotes, NottNott, Lizia7, Ebgoldstein, TheEpTic, Monkbot, Jens.turowski, GuyEmerson, Crystallizedcarbon, Mtcrsx28, Izkala, Tetra quark, MicroPaLeo, Isambard Kingdom, Supdiop, KasparBot, Breadedchicken, Sro23, Lappspira, 1416domination, Magic links bot and Anonymous: 158

47.10.2 Images

- **File:(a)Original_Image_(b)Image_Generated_using_equation(1)_(c)Image_generated_using_equation(2)_(d)_Image_generate).jpg** Source: https://upload.wikimedia.org/wikipedia/commons/e/ee/%28a%29Original_Image_%28b%29Image_Generated_using_equation%281%29_%28c%29Image_generated_using_equation%282%29_%28d%29_Image_generated_using_equation%283%29_%28e%29Image_generated_using_equation%284%29.jpg License: CC BY-SA 4.0 Contributors: Own work Original artist: MaryThomas29

- **File:6_centrality_measures.png** Source: https://upload.wikimedia.org/wikipedia/commons/1/11/6_centrality_measures.png License: CC BY-SA 4.0 Contributors: Own work Original artist: Tapiocozzo

- **File:6n-graf.svg** Source: https://upload.wikimedia.org/wikipedia/commons/5/5b/6n-graf.svg License: Public domain Contributors: Image: 6n-graf.png simlar input data Original artist: User:AzaToth

- **File:ACEGESGUI.png** Source: https://upload.wikimedia.org/wikipedia/commons/4/4a/ACEGESGUI.png License: GFDL Contributors: http://www.aceges.org/ Original artist: Dr. Vlasios Voudouris

- **File:ADN_static.png** Source: https://upload.wikimedia.org/wikipedia/commons/c/c2/ADN_static.png License: Public domain Contributors: Single frame of Image:ADN animation.gif, created with data from Edwards K, Brown D, Spink N, Skelly J, Neidle S (1992). "Molecular structure of the B-DNA dodecamer d(CGCAAATTTGCG)2. An examination of propeller twist and minor-groove water structure at 2.2 A resolution". J. Mol. Biol. **226** (4): 1161–73. PMID 1518049. Original artist: Brian0918

- **File:ANT_Antenna_1.jpg** Source: https://upload.wikimedia.org/wikipedia/commons/b/bd/ANT_Antenna_1.jpg License: CC0 Contributors: Own work Original artist: Swadim

- **File:ANT_antenna_2.jpg** Source: https://upload.wikimedia.org/wikipedia/commons/7/75/ANT_antenna_2.jpg License: CC0 Contributors: Own work Original artist: Swadim

- **File:A_Swarm_of_Ancient_Stars_-_GPN-2000-000930.jpg** Source: https://upload.wikimedia.org/wikipedia/commons/6/6a/A_Swarm_of_Ancient_Stars_-_GPN-2000-000930.jpg License: Public domain Contributors: Great Images in NASA Description Original artist: NASA, The Hubble Heritage Team, STScI, AURA

- **File:A_Trajectory_Through_Phase_Space_in_a_Lorenz_Attractor.gif** Source: https://upload.wikimedia.org/wikipedia/commons/1/13/A_Trajectory_Through_Phase_Space_in_a_Lorenz_Attractor.gif License: CC BY-SA 3.0 Contributors: Own work Original artist: Dan Quinn

- **File:A_rep-tile-based_setiset_of_order_4.png** Source: https://upload.wikimedia.org/wikipedia/commons/4/45/A_rep-tile-based_setiset_of_order_4.png License: CC BY-SA 3.0 Contributors: http://www.leesallows.com/files/reptile_demo1a.png Original artist: Lee Sallows

- **File:Aco_TSP.svg** Source: https://upload.wikimedia.org/wikipedia/commons/2/2a/Aco_TSP.svg License: CC-BY-SA-3.0 Contributors: Own work Original artist: Nojhan

47.10. TEXT AND IMAGE SOURCES, CONTRIBUTORS, AND LICENSES

- **File:Aco_shortpath.svg** *Source:* https://upload.wikimedia.org/wikipedia/commons/1/17/Aco_shortpath.svg *License:* CC-BY-SA-3.0 *Contributors:* Own work *Original artist:* Nojhan
- **File:Alfred-Russel-Wallace-c1895.jpg** *Source:* https://upload.wikimedia.org/wikipedia/commons/d/d4/Alfred-Russel-Wallace-c1895.jpg *License:* Public domain *Contributors:* First published in Borderland Magazine, April 1896 *Original artist:* London Stereoscopic & Photographic Company (active 1855-1922)
- **File:Allele-frequency.png** *Source:* https://upload.wikimedia.org/wikipedia/commons/4/46/Allele-frequency.png *License:* CC-BY-SA-3.0 *Contributors:* Transferred from en.wikipedia to Commons by Leptictidium using CommonsHelper. *Original artist:* The original uploader was Esurnir at English Wikipedia
- **File:Allosaurus_Jardin_des_Plantes.png** *Source:* https://upload.wikimedia.org/wikipedia/commons/1/1b/Allosaurus_Jardin_des_Plantes.png *License:* CC BY-SA 1.0 *Contributors:* Own work, formerly uploaded on WP-fr *Original artist:* Eric Gaba (Sting - fr:Sting)
- **File:Ambox_important.svg** *Source:* https://upload.wikimedia.org/wikipedia/commons/b/b4/Ambox_important.svg *License:* Public domain *Contributors:* Own work based on: Ambox scales.svg *Original artist:* Dsmurat, penubag
- **File:Animation2.gif** *Source:* https://upload.wikimedia.org/wikipedia/commons/c/c0/Animation2.gif *License:* CC-BY-SA-3.0 *Contributors:* Own work *Original artist:* MG (talk · contribs)
- **File:Ann_dependency_(graph).svg** *Source:* https://upload.wikimedia.org/wikipedia/commons/d/dd/Ann_dependency_%28graph%29.svg *License:* CC BY-SA 3.0 *Contributors:* Vector version of File:Ann dependency graph.png *Original artist:* Glosser.ca
- **File:Ants_eating_fruit.jpg** *Source:* https://upload.wikimedia.org/wikipedia/commons/c/ca/Ants_eating_fruit.jpg *License:* CC BY 3.0 *Contributors:* Own work *Original artist:* Zainichi Gaikokujin until nobody found original :D (Kidding, don't mind)
- **File:Ape_skeletons.png** *Source:* https://upload.wikimedia.org/wikipedia/commons/4/49/Ape_skeletons.png *License:* Public domain *Contributors:* Transferred from en.wikipedia to Commons. *Original artist:* The original uploader was TimVickers at English Wikipedia
- **File:Application_of_collective_intelligence_in_Millennium_Project.png** *Source:* https://upload.wikimedia.org/wikipedia/commons/3/3d/Application_of_collective_intelligence_in_Millennium_Project.png *License:* CC BY-SA 4.0 *Contributors:* http://www.millennium-project.org/millennium/GFIS.html *Original artist:* Jerome Glenn
- **File:Argon_ice_1.jpg** *Source:* https://upload.wikimedia.org/wikipedia/commons/0/0d/Argon_ice_1.jpg *License:* CC-BY-SA-3.0 *Contributors:* No machine-readable source provided. Own work assumed (based on copyright claims). *Original artist:* No machine-readable author provided. Deglr6328~commonswiki assumed (based on copyright claims).
- **File:Artificial_neural_network.svg** *Source:* https://upload.wikimedia.org/wikipedia/commons/e/e4/Artificial_neural_network.svg *License:* CC-BY-SA-3.0 *Contributors:* This vector image was created with Inkscape. *Original artist:* en:User:Cburnett
- **File:Auklet_flock_Shumagins_1986.jpg** *Source:* https://upload.wikimedia.org/wikipedia/commons/5/5e/Auklet_flock_Shumagins_1986.jpg *License:* Public domain *Contributors:* images.fws.gov ([1]) *Original artist:* D. Dibenski
- **File:Automorphisms_of_a_subgraph.jpg** *Source:* https://upload.wikimedia.org/wikipedia/en/c/ca/Automorphisms_of_a_subgraph.jpg *License:* Cc-by-sa-3.0 *Contributors:*
Own work
Original artist:
Skhakabi (talk) (Uploads)
- **File:Autoregulation_motif.png** *Source:* https://upload.wikimedia.org/wikipedia/commons/1/1e/Autoregulation_motif.png *License:* CC BY-SA 3.0 *Contributors:* Wikipedia *Original artist:* en:User:Marashie
- **File:Badlands_at_the_Blue_Gate,_Utah.JPG** *Source:* https://upload.wikimedia.org/wikipedia/commons/5/54/Badlands_at_the_Blue_Gate%2C_Utah.JPG *License:* CC BY-SA 4.0 *Contributors:* Own work *Original artist:* DanHobley
- **File:Bangkok_skytrain_sunset.jpg** *Source:* https://upload.wikimedia.org/wikipedia/commons/f/f6/Bangkok_skytrain_sunset.jpg *License:* CC-BY-SA-3.0 *Contributors:* Own work *Original artist:* User:Diliff
- **File:Barabasi-albert_model_degree_distribution.svg** *Source:* https://upload.wikimedia.org/wikipedia/commons/a/a8/Barabasi-albert_model_degree_distribution.svg *License:* CC BY-SA 3.0 *Contributors:* Created by the NetworkX module of the Python *Original artist:* Arpad Horvath
- **File:Beaver_dam_in_Tierra_del_Fuego.jpg** *Source:* https://upload.wikimedia.org/wikipedia/commons/c/c2/Beaver_dam_in_Tierra_del_Fuego.jpg *License:* CC BY-SA 1.0 *Contributors:* Own work *Original artist:* User:IlyaHaykinson
- **File:Bee_swarm_feb08.jpg** *Source:* https://upload.wikimedia.org/wikipedia/commons/b/b4/Bee_swarm_feb08.jpg *License:* GFDL 1.2 *Contributors:* Own work *Original artist:*
fir0002 | flagstaffotos.com.au
- **File:Bienenschwarm_17c.jpg** *Source:* https://upload.wikimedia.org/wikipedia/commons/1/17/Bienenschwarm_17c.jpg *License:* CC-BY-SA-3.0 *Contributors:* Own work *Original artist:* Waugsberg
- **File:Biston.betularia.7200.jpg** *Source:* https://upload.wikimedia.org/wikipedia/commons/6/6c/Biston.betularia.7200.jpg *License:* CC-BY-SA-3.0 *Contributors:* ? *Original artist:* ?
- **File:Biston.betularia.f.carbonaria.7209.jpg** *Source:* https://upload.wikimedia.org/wikipedia/commons/d/db/Biston.betularia.f.carbonaria.7209.jpg *License:* CC-BY-SA-3.0 *Contributors:* ? *Original artist:* ?
- **File:Border_Collie_sheepdog_trial.jpg** *Source:* https://upload.wikimedia.org/wikipedia/commons/5/52/Border_Collie_sheepdog_trial.jpg *License:* CC BY-SA 2.0 *Contributors:* Flickr *Original artist:* Scot Campbell

- **File:Braitenberg_vehicle_(simulation_made_with_breve).jpg** *Source:* https://upload.wikimedia.org/wikipedia/commons/c/c5/Braitenberg_vehicle_%28simulation_made_with_breve%29.jpg *License:* CC-BY-SA-3.0 *Contributors:* Transferred from en.wikipedia to Commons by Roberta F. using CommonsHelper. *Original artist:* The original uploader was Rxke at English Wikipedia
- **File:CA-Moore.png** *Source:* https://upload.wikimedia.org/wikipedia/en/d/d2/CA-Moore.png *License:* CC0 *Contributors:* Own work
  Original artist:
  Torchiest (talk) (Uploads)
- **File:CA-von-Neumann.png** *Source:* https://upload.wikimedia.org/wikipedia/en/5/56/CA-von-Neumann.png *License:* CC0 *Contributors:* Own work
  Original artist:
  Torchiest (talk) (Uploads)
- **File:CA_rule110s.png** *Source:* https://upload.wikimedia.org/wikipedia/commons/f/fa/CA_rule110s.png *License:* CC0 *Contributors:* Own work by the original uploader *Original artist:* Grondilu (talk) (Uploads)
- **File:CA_rule30s.png** *Source:* https://upload.wikimedia.org/wikipedia/commons/9/9d/CA_rule30s.png *License:* CC-BY-SA-3.0 *Contributors:* ? *Original artist:* ?
- **File:CI_types1s_2.jpg** *Source:* https://upload.wikimedia.org/wikipedia/commons/3/3e/CI_types1s_2.jpg *License:* CC BY 2.5 *Contributors:* Transferred from en.wikipedia to Commons. *Original artist:* Olga Generozova
- **File:Causes_for_c.png** *Source:* https://upload.wikimedia.org/wikipedia/commons/9/98/Causes_for_c.png *License:* CC BY-SA 4.0 *Contributors:* Own work *Original artist:* PaDBu
- **File:Causeway-code_poet-4.jpg** *Source:* https://upload.wikimedia.org/wikipedia/commons/c/c0/Causeway-code_poet-4.jpg *License:* CC BY-SA 2.0 *Contributors:* http://www.flickr.com/photos/alphageek/20005235/ *Original artist:* code poet on flickr.
- **File:Charles_Darwin_aged_51.jpg** *Source:* https://upload.wikimedia.org/wikipedia/commons/4/42/Charles_Darwin_aged_51.jpg *License:* Public domain *Contributors:* ? *Original artist:* ?
- **File:Collaborative_tagging.png** *Source:* https://upload.wikimedia.org/wikipedia/commons/f/f1/Collaborative_tagging.png *License:* CC BY-SA 4.0 *Contributors:* https://www.researchgate.net/figure/271658600_fig1_Fig-1-Parenting-social-network-and-collaborative-tagging-as-pillars-for-automatic-IPTV *Original artist:* Ana Fernández-Vilas Rebeca P. Díaz-Redondo Rebeca P. Díaz-Redondo Sandra Servia-Rodríguez Sandra Servia-Rodríguez
- **File:Colored_neural_network.svg** *Source:* https://upload.wikimedia.org/wikipedia/commons/4/46/Colored_neural_network.svg *License:* CC BY-SA 3.0 *Contributors:* Own work, Derivative of File:Artificial neural network.svg *Original artist:* Glosser.ca
- **File:Commons-logo.svg** *Source:* https://upload.wikimedia.org/wikipedia/en/4/4a/Commons-logo.svg *License:* PD *Contributors:* ? *Original artist:* ?
- **File:Comparison_carbon_dioxide_water_phase_diagrams.svg** *Source:* https://upload.wikimedia.org/wikipedia/commons/4/40/Comparison_carbon_dioxide_water_phase_diagrams.svg *License:* CC BY-SA 3.0 *Contributors:* Own work *Original artist:* Cmglee
- **File:Comparison_of_Three_Invertebrate_Chordates.svg** *Source:* https://upload.wikimedia.org/wikipedia/commons/4/44/Comparison_of_Three_Invertebrate_Chordates.svg *License:* CC BY-SA 4.0 *Contributors:* Own work *Original artist:* Basketball1713
- **File:Complex-adaptive-system.jpg** *Source:* https://upload.wikimedia.org/wikipedia/commons/0/00/Complex-adaptive-system.jpg *License:* Public domain *Contributors:* Own work by Acadac : Taken from en.wikipedia.org, where Acadac was inspired to create this graphic after reading: *Original artist:* Acadac
- **File:Complex_adaptive_system.gif** *Source:* https://upload.wikimedia.org/wikipedia/commons/8/81/Complex_adaptive_system.gif *License:* CC BY-SA 4.0 *Contributors:* http://integral-options.blogspot.com/2012/06/what-im-reading-part-one-complex.html *Original artist:* William Harryman
- **File:Complex_network_degree_distribution_of_random_and_scale-free.png** *Source:* https://upload.wikimedia.org/wikipedia/commons/3/39/Complex_network_degree_distribution_of_random_and_scale-free.png *License:* Public domain *Contributors:* Drawn by the author *Original artist:* user:Sazaedo (ja:user:?????)
- **File:Complex_systems_organizational_map.jpg** *Source:* https://upload.wikimedia.org/wikipedia/commons/d/de/Complex_systems_organizational_map.jpg *License:* CC BY-SA 3.0 *Contributors:* Created by Hiroki Sayama, D.Sc., Collective Dynamics of Complex Systems (CoCo) Research Group at Binghamton University, State University of New York *Original artist:* Hiroki Sayama, D.Sc.
- **File:Complexity_Map.svg** *Source:* https://upload.wikimedia.org/wikipedia/commons/8/8b/Complexity_Map.svg *License:* CC BY-SA 3.0 *Contributors:* Own work *Original artist:* Brian Castellani
- **File:Computational_collective_intelligence.jpg** *Source:* https://upload.wikimedia.org/wikipedia/commons/8/88/Computational_collective_intelligence.jpg *License:* CC BY-SA 4.0 *Contributors:* https://www.amazon.com/Computational-Collective-Intelligence-Tadeusz-Szuba/dp/0471349666 *Original artist:* Tadeusz M. Szuba
- **File:Cono_de_Arita,_Salar_de_Arizaro_(Argentina).jpg** *Source:* https://upload.wikimedia.org/wikipedia/commons/9/9e/Cono_de_Arita%2C_Salar_de_Arizaro_%28Argentina%29.jpg *License:* CC BY 2.0 *Contributors:* http://www.flickr.com/photos/96935551@N02/8943680306/in/photostream/ *Original artist:* Ben Stubbs
- **File:ConvectionCells.svg** *Source:* https://upload.wikimedia.org/wikipedia/commons/f/f5/ConvectionCells.svg *License:* CC-BY-SA-3.0 *Contributors:* Own work *Original artist:* Eyrian
  Con-struct

47.10. TEXT AND IMAGE SOURCES, CONTRIBUTORS, AND LICENSES

- **File:Copepodkils.jpg** *Source:* https://upload.wikimedia.org/wikipedia/commons/2/28/Copepodkils.jpg *License:* CC-BY-SA-3.0 *Contributors:* ? *Original artist:* ?
- **File:Crowd_04378.JPG** *Source:* https://upload.wikimedia.org/wikipedia/commons/5/54/Crowd_04378.JPG *License:* CC-BY-SA-3.0 *Contributors:* Own work *Original artist:* Nevit Dilmen
- **File:DNA_chemical_structure.svg** *Source:* https://upload.wikimedia.org/wikipedia/commons/e/e4/DNA_chemical_structure.svg *License:* CC-BY-SA-3.0 *Contributors:* <a href='//validator.w3.org/' data-x-rel='nofollow'><img alt='W3C' src='https://upload.wikimedia.org/wikipedia/commons/thumb/1/1a/Valid_SVG_1.1_%28green%29.svg/88px-Valid_SVG_1.1_%28green%29.svg.png' width='88' height='30' style='vertical-align: top' srcset='https://upload.wikimedia.org/wikipedia/commons/thumb/1/1a/Valid_SVG_1.1_%28green%29.svg/132px-Valid_SVG_1.1_%28green%29.svg.png 1.5x, https://upload.wikimedia.org/wikipedia/commons/thumb/1/1a/Valid_SVG_1.1_%28green%29.svg/176px-Valid_SVG_1.1_%28green%29.svg.png 2x' data-file-width='91' data-file-height='31' /></a>The source code of this SVG is <a data-x-rel='nofollow' class='external text' href='//validator.w3.org/check?uri=https%3A%2F%2Fcommons.wikimedia.org%2Fwiki%2FSpecial%3AFilepath%2FDNA_chemical_structure.svg..&..ss=1#source'>valid</a>. *Original artist:* Madprime (talk · contribs)
- **File:DNA_nanostructures.png** *Source:* https://upload.wikimedia.org/wikipedia/commons/5/55/DNA_nanostructures.png *License:* CC BY 2.5 *Contributors:* Strong M: *Protein Nanomachines.* PLoS Biol 2/3/2004: e73. http://dx.doi.org/10.1371/journal.pbio.0020073 *Original artist:* (Images were kindly provided by Thomas H. LaBean and Hao Yan.)
- **File:Darwin'(}s_finches.jpeg** *Source:* https://upload.wikimedia.org/wikipedia/commons/9/97/Darwin%27s_finches.jpeg *License:* Public domain *Contributors:* From "Voyage of the Beagle" as found on [1] and [2] *Original artist:* John Gould (14.Sep.1804 - 3.Feb.1881)
- **File:Desktop_computer_clipart_-_Yellow_theme.svg** *Source:* https://upload.wikimedia.org/wikipedia/commons/d/d7/Desktop_computer_clipart_-_Yellow_theme.svg *License:* CC0 *Contributors:* https://openclipart.org/detail/17924/computer *Original artist:* AJ from openclipart.org
- **File:Different_occurrences_of_a_sub-graph_in_a_graph.jpg** *Source:* https://upload.wikimedia.org/wikipedia/en/2/27/Different_occurrences_of_a_sub-graph_in_a_graph.jpg *License:* Cc-by-sa-3.0 *Contributors:* ? *Original artist:* ?
- **File:Drosophila_early_embryo_protein_gradients.svg** *Source:* https://upload.wikimedia.org/wikipedia/commons/d/db/Drosophila_early_embryo_protein_gradients.svg *License:* GFDL *Contributors:* <a href='//commons.wikimedia.org/wiki/File:Drosophila_early_embryo_protein_gradients.png' class='image'><img alt='Drosophila early embryo protein gradients.png' src='https://upload.wikimedia.org/wikipedia/commons/thumb/7/7a/Drosophila_early_embryo_protein_gradients.png/100px-Drosophila_early_embryo_protein_gradients.png' width='100' height='65' srcset='https://upload.wikimedia.org/wikipedia/commons/thumb/7/7a/Drosophila_early_embryo_protein_gradients.png/150px-Drosophila_early_embryo_protein_gradients.png 1.5x, https://upload.wikimedia.org/wikipedia/commons/thumb/7/7a/Drosophila_early_embryo_protein_gradients.png/200px-Drosophila_early_embryo_protein_gradients.png 2x' data-file-width='353' data-file-height='230' /></a> *Original artist:* Fred the Oyster
- **File:Drugroutemap.gif** *Source:* https://upload.wikimedia.org/wikipedia/commons/6/64/Drugroutemap.gif *License:* Public domain *Contributors:* CIA Employee *Original artist:* CIA Employee
- **File:DynamicNetworkAnalysisExample.jpg** *Source:* https://upload.wikimedia.org/wikipedia/en/2/25/DynamicNetworkAnalysisExample.jpg *License:* PD *Contributors:* ? *Original artist:* ?
- **File:ER_model.png** *Source:* https://upload.wikimedia.org/wikipedia/commons/a/a2/ER_model.png *License:* CC BY-SA 3.0 *Contributors:* Own work *Original artist:* Jmcatania
- **File:ESU-Tree.jpg** *Source:* https://upload.wikimedia.org/wikipedia/en/7/7f/ESU-Tree.jpg *License:* Cc-by-sa-3.0 *Contributors:* ? *Original artist:* ?
- **File:Earth_surface_NGDC_2000.jpg** *Source:* https://upload.wikimedia.org/wikipedia/commons/1/1c/Earth_surface_NGDC_2000.jpg *License:* Public domain *Contributors:* ? *Original artist:* ?
- **File:Edit-clear.svg** *Source:* https://upload.wikimedia.org/wikipedia/en/f/f2/Edit-clear.svg *License:* Public domain *Contributors:* The Tango! Desktop Project. *Original artist:*

  The people from the Tango! project. And according to the meta-data in the file, specifically: "Andreas Nilsson, and Jakub Steiner (although minimally)."
- **File:Editorial_cartoon_depicting_Charles_Darwin_as_an_ape_(1871).jpg** *Source:* https://upload.wikimedia.org/wikipedia/commons/6/6f/Editorial_cartoon_depicting_Charles_Darwin_as_an_ape_%281871%29.jpg *License:* Public domain *Contributors:* Originally published in *The Hornet* magazine; this image is available on University College London Digital Collections (18886) *Original artist:* Unknown<a href='https://www.wikidata.org/wiki/Q4233718' title='wikidata:Q4233718'><img alt='wikidata:Q4233718' src='https://upload.wikimedia.org/wikipedia/commons/thumb/f/ff/Wikidata-logo.svg/20px-Wikidata-logo.svg.png' width='20' height='11' srcset='https://upload.wikimedia.org/wikipedia/commons/thumb/f/ff/Wikidata-logo.svg/30px-Wikidata-logo.svg.png 1.5x, https://upload.wikimedia.org/wikipedia/commons/thumb/f/ff/Wikidata-logo.svg/40px-Wikidata-logo.svg.png 2x' data-file-width='1050' data-file-height='590' /></a>
- **File:Elmer-pump-heatequation.png** *Source:* https://upload.wikimedia.org/wikipedia/commons/c/cd/Elmer-pump-heatequation.png *License:* CC BY-SA 3.0 *Contributors:* ? *Original artist:* ?
- **File:Emile_Durkheim.jpg** *Source:* https://upload.wikimedia.org/wikipedia/commons/2/24/Emile_Durkheim.jpg *License:* Public domain *Contributors:* http://www.marxists.org/glossary/people/d/pics/durkheim.jpg *Original artist:* Unknown<a href='https://www.wikidata.org/wiki/Q4233718' title='wikidata:Q4233718'><img alt='wikidata:Q4233718' src='https://upload.wikimedia.org/wikipedia/commons/thumb/f/ff/Wikidata-logo.svg/20px-Wikidata-logo.svg.png' width='20' height='11' srcset='https://upload.wikimedia.org/wikipedia/commons/thumb/f/ff/Wikidata-logo.svg/30px-Wikidata-logo.svg.png 1.5x, https://upload.wikimedia.org/wikipedia/commons/thumb/f/ff/Wikidata-logo.svg/40px-Wikidata-logo.svg.png 2x' data-file-width='1050' data-file-height='590' /></a>

- **File:Eroding_Mesas_Forming_Seif_and_Barchan_Dunes_in_Hellespontus_region.jpg** *Source:* https://upload.wikimedia.org/wikipedia/commons/f/ff/Eroding_Mesas_Forming_Seif_and_Barchan_Dunes_in_Hellespontus_region.jpg *License:* Public domain *Contributors:* HiRISE webpage, description, file *Original artist:* NASA/JPL/University of Arizona
- **File:Evolsex-dia1a.png** *Source:* https://upload.wikimedia.org/wikipedia/commons/f/fc/Evolsex-dia1a.png *License:* CC-BY-SA-3.0 *Contributors:* ? *Original artist:* ?
- **File:Evolution_of_complexity.svg** *Source:* https://upload.wikimedia.org/wikipedia/commons/3/3b/Evolution_of_complexity.svg *License:* Public domain *Contributors:* Vector version of en:Image:Evolution of complexity.png *Original artist:* Fvasconcellos, original by Tim Vickers
- **File:Expansion_Tree.jpg** *Source:* https://upload.wikimedia.org/wikipedia/en/8/8c/Expansion_Tree.jpg *License:* Cc-by-sa-3.0 *Contributors:* ? *Original artist:* ?
- **File:Feed-forward_motif.GIF** *Source:* https://upload.wikimedia.org/wikipedia/en/f/f2/Feed-forward_motif.GIF *License:* Cc-by-sa-3.0 *Contributors:* ? *Original artist:* ?
- **File:Ferguson-slide.jpg** *Source:* https://upload.wikimedia.org/wikipedia/commons/7/75/Ferguson-slide.jpg *License:* CC BY 3.0 *Contributors:* Own work *Original artist:* Eeekster
- **File:Firefighters_in_Parade.jpg** *Source:* https://upload.wikimedia.org/wikipedia/commons/d/d8/Firefighters_in_Parade.jpg *License:* CC0 *Contributors:* http://digital.lib.uh.edu/cdm4/item_viewer.php?CISOROOT=/p15195coll32&CISOPTR=68&DMSCALE=12.5&DMWIDTH=600&DMHEIGHT=600&DMMODE=viewer&DMTEXT=&REC=10&DMTHUMB=1&DMROTATE=0 *Original artist:* Unknown<a href='https://www.wikidata.org/wiki/Q4233718' title='wikidata:Q4233718'><img alt='wikidata:Q4233718' src='https://upload.wikimedia.org/wikipedia/commons/thumb/f/ff/Wikidata-logo.svg/20px-Wikidata-logo.svg.png' width='20' height='11' srcset='https://upload.wikimedia.org/wikipedia/commons/thumb/f/ff/Wikidata-logo.svg/30px-Wikidata-logo.svg.png 1.5x, https://upload.wikimedia.org/wikipedia/commons/thumb/f/ff/Wikidata-logo.svg/40px-Wikidata-logo.svg.png 2x' data-file-width='1050' data-file-height='590' /></a>
- **File:Fitness-landscape-cartoon.png** *Source:* https://upload.wikimedia.org/wikipedia/commons/6/67/Fitness-landscape-cartoon.png *License:* Public domain *Contributors:* ? *Original artist:* ?
- **File:Folder_Hexagonal_Icon.svg** *Source:* https://upload.wikimedia.org/wikipedia/en/4/48/Folder_Hexagonal_Icon.svg *License:* Cc-by-sa-3.0 *Contributors:* ? *Original artist:* ?
- **File:Fugle,_ornsø_073.jpg** *Source:* https://upload.wikimedia.org/wikipedia/commons/d/d6/Fugle%2C_%C3%B8rns%C3%B8_073.jpg *License:* Public domain *Contributors:* Own work *Original artist:* Christoffer A Rasmussen (Rasmussen29892 at da.wikipedia)
- **File:Gap_gene_expression.svg** *Source:* https://upload.wikimedia.org/wikipedia/commons/7/73/Gap_gene_expression.svg *License:* GFDL *Contributors:* <a href='//commons.wikimedia.org/wiki/File:Gap_gene_expression.png' class='image'><img alt='Gap gene expression.png' src='https://upload.wikimedia.org/wikipedia/commons/thumb/5/53/Gap_gene_expression.png/100px-Gap_gene_expression.png' width='100' height='103' srcset='https://upload.wikimedia.org/wikipedia/commons/thumb/5/53/Gap_gene_expression.png/150px-Gap_gene_expression.png 1.5x, https://upload.wikimedia.org/wikipedia/commons/5/53/Gap_gene_expression.png 2x' data-file-width='200' data-file-height='205' /></a> *Original artist:* Fred the Oyster
- **File:Gene-duplication.svg** *Source:* https://upload.wikimedia.org/wikipedia/commons/5/5d/Gene-duplication.svg *License:* Public domain *Contributors:* Own work *Original artist:* K. Aainsqatsi
- **File:Gene_Regulatory_Network.jpg** *Source:* https://upload.wikimedia.org/wikipedia/commons/c/c4/Gene_Regulatory_Network.jpg *License:* Public domain *Contributors:* ? *Original artist:* ?
- **File:Genes_hox.jpeg** *Source:* https://upload.wikimedia.org/wikipedia/commons/0/0c/Genes_hox.jpeg *License:* CC0 *Contributors:* Imagem original da revista Nature, disponível em < http://cienciaxreligiao.blogspot.com.br/2009_01_01_archive.html> *Original artist:* Nature
- **File:Genetic_Distribution.svg** *Source:* https://upload.wikimedia.org/wikipedia/commons/6/62/Genetic_Distribution.svg *License:* CC BY-SA 4.0 *Contributors:* Own work *Original artist:* Ealbert17
- **File:Genetic_Program_Tree.png** *Source:* https://upload.wikimedia.org/wikipedia/commons/7/77/Genetic_Program_Tree.png *License:* Public domain *Contributors:* Transferred from en.wikipedia to Commons. *Original artist:* BAxelrod at English Wikipedia
- **File:Genomics_GTL_Pictorial_Program.jpg** *Source:* https://upload.wikimedia.org/wikipedia/commons/0/01/Genomics_GTL_Pictorial_Program.jpg *License:* Public domain *Contributors:* ? *Original artist:* ?
- **File:Glacial_landscape_LMB.png** *Source:* https://upload.wikimedia.org/wikipedia/commons/4/4b/Glacial_landscape_LMB.png *License:* Public domain *Contributors:* ? *Original artist:* ?
- **File:Global_brain.jpg** *Source:* https://upload.wikimedia.org/wikipedia/commons/5/5d/Global_brain.jpg *License:* CC BY-SA 4.0 *Contributors:* http://nexusilluminati.blogspot.com/2011/04/emerging-global-brain.html *Original artist:* Ben Goertzel
- **File:Gospers_glider_gun.gif** *Source:* https://upload.wikimedia.org/wikipedia/commons/e/e5/Gospers_glider_gun.gif *License:* CC-BY-SA-3.0 *Contributors:* Own work *Original artist:* Kieff
- **File:Graph_betweenness.svg** *Source:* https://upload.wikimedia.org/wikipedia/commons/6/60/Graph_betweenness.svg *License:* CC BY 2.5 *Contributors:* Own work *Original artist:* Claudio Rocchini
- **File:Haeckel_drawings.jpg** *Source:* https://upload.wikimedia.org/wikipedia/commons/0/08/Haeckel_drawings.jpg *License:* Public domain *Contributors:* Romanes, G. J. (1892). Darwin and After Darwin. Open Court, Chicago. *Original artist:* Romanes, G. J.; uploaded to Wikipedia by en:User:Phlebas; authors of the description page: en:User:Phlebas, en:User:SeventyThree
- **File:Heat_eqn.gif** *Source:* https://upload.wikimedia.org/wikipedia/commons/a/a9/Heat_eqn.gif *License:* Public domain *Contributors:* This graphic was created with MATLAB. *Original artist:* Oleg Alexandrov

47.10. TEXT AND IMAGE SOURCES, CONTRIBUTORS, AND LICENSES

- **File:Heliconius_erato_Richard_Bartz.jpg** *Source:* https://upload.wikimedia.org/wikipedia/commons/e/ea/Heliconius_erato_Richard_Bartz.jpg *License:* CC BY-SA 2.5 *Contributors:* Own work *Original artist:* Richard Bartz, Munich aka Makro Freak <a href='//commons.wikimedia.org/wiki/File:Makro_Freak_bar.jpg' class='image'><img alt='Makro Freak bar.jpg' src='https://upload.wikimedia.org/wikipedia/commons/8/8b/Makro_Freak_bar.jpg' width='80' height='15' data-file-width='80' data-file-height='15' /></a>
- **File:Heliconius_melpomene_2b_Richard_Bartz.jpg** *Source:* https://upload.wikimedia.org/wikipedia/commons/f/f6/Heliconius_melpomene_2b_Richard_Bartz.jpg *License:* CC BY-SA 2.5 *Contributors:* Own work *Original artist:* Richard Bartz, Munich aka Makro Freak <a href='//commons.wikimedia.org/wiki/File:Makro_Freak_bar.jpg' class='image'><img alt='Makro Freak bar.jpg' src='https://upload.wikimedia.org/wikipedia/commons/8/8b/Makro_Freak_bar.jpg' width='80' height='15' data-file-width='80' data-file-height='15' /></a>
- **File:Heringsschwarm.gif** *Source:* https://upload.wikimedia.org/wikipedia/commons/4/4d/Heringsschwarm.gif *License:* CC-BY-SA-3.0 *Contributors:* Transferred from en.wikipedia. *Original artist:* Kils at en.wikipedia
- **File:Herringramkils.jpg** *Source:* https://upload.wikimedia.org/wikipedia/commons/9/98/Herringramkils.jpg *License:* CC-BY-SA-3.0 *Contributors:* Transferred from en.wikipedia to Commons by Sreejithk2000 using CommonsHelper. *Original artist:* The original uploader was Kils at English Wikipedia
- **File:Homebrew_reaction_diffusion_example_512iter.jpg** *Source:* https://upload.wikimedia.org/wikipedia/en/6/67/Homebrew_reaction_diffusion_example_512iter.jpg *License:* PD *Contributors:* ? *Original artist:* ?
- **File:Homology_vertebrates-en.svg** *Source:* https://upload.wikimedia.org/wikipedia/commons/5/5e/Homology_vertebrates-en.svg *License:* CC BY-SA 4.0 *Contributors:* Own work *Original artist:* Волков Владислав Петрович
- **File:Hoxgenesoffruitfly.svg** *Source:* https://upload.wikimedia.org/wikipedia/commons/d/da/Hoxgenesoffruitfly.svg *License:* Public domain *Contributors:* self-made, base on Hoxgenesoffruitfly.png *Original artist:* PhiLiP
- **File:Improvisational_actors.jpg** *Source:* https://upload.wikimedia.org/wikipedia/commons/3/32/Improvisational_actors.jpg *License:* CC BY-SA 4.0 *Contributors:* http://www.hideouttheatre.com/about/what-is-improv *Original artist:* Hideouttheatre
- **File:Internet_map_1024.jpg** *Source:* https://upload.wikimedia.org/wikipedia/commons/d/d2/Internet_map_1024.jpg *License:* CC BY 2.5 *Contributors:* Originally from the English Wikipedia; description page is/was here. *Original artist:* The Opte Project
- **File:Issoria_lathonia.jpg** *Source:* https://upload.wikimedia.org/wikipedia/commons/2/2d/Issoria_lathonia.jpg *License:* CC-BY-SA-3.0 *Contributors:* ? *Original artist:* ?
- **File:Jelly_cc4.jpg** *Source:* https://upload.wikimedia.org/wikipedia/commons/b/ba/Jelly_cc4.jpg *License:* CC BY-SA 2.0 *Contributors:* ? *Original artist:* ?
- **File:John_von_Neumann_ID_badge.png** *Source:* https://upload.wikimedia.org/wikipedia/commons/d/d9/John_von_Neumann_ID_badge.png *License:* Public domain *Contributors:* ? *Original artist:* ?
- **File:Kencf0618FacebookNetwork.jpg** *Source:* https://upload.wikimedia.org/wikipedia/commons/9/90/Kencf0618FacebookNetwork.jpg *License:* CC BY-SA 3.0 *Contributors:* Own work *Original artist:* Kencf0618
- **File:Kilobot_robot_swarm.JPG** *Source:* https://upload.wikimedia.org/wikipedia/commons/d/d9/Kilobot_robot_swarm.JPG *License:* CC BY-SA 4.0 *Contributors:* Own work *Original artist:* asuscreative
- **File:Knapsack_ants.svg** *Source:* https://upload.wikimedia.org/wikipedia/commons/e/ec/Knapsack_ants.svg *License:* CC BY-SA 2.5 *Contributors:* ? *Original artist:* ?
- **File:Konigsberg_bridges.png** *Source:* https://upload.wikimedia.org/wikipedia/commons/5/5d/Konigsberg_bridges.png *License:* CC-BY-SA-3.0 *Contributors:* Public domain (PD), based on the image
  - <a href='//commons.wikimedia.org/wiki/File:Image-Koenigsberg,_Map_by_Merian-Erben_1652.jpg' class='image'><img alt='Image-Koenigsberg, Map by Merian-Erben 1652.jpg' src='https://upload.wikimedia.org/wikipedia/commons/thumb/1/15/Image-Koenigsberg%2C_Map_by_Merian-Erben_1652.jpg/120px-Image-Koenigsberg%2C_Map_by_Merian-Erben_1652.jpg' width='120' height='84' srcset='https://upload.wikimedia.org/wikipedia/commons/thumb/1/15/Image-Koenigsberg%2C_Map_by_Merian-Erben_1652.jpg/180px-Image-Koenigsberg%2C_Map_by_Merian-Erben_1652.jpg 1.5x, https://upload.wikimedia.org/wikipedia/commons/thumb/1/15/Image-Koenigsberg%2C_Map_by_Merian-Erben_1652.jpg/240px-Image-Koenigsberg%2C_Map_by_Merian-Erben_1652.jpg 2x' data-file-width='628' data-file-height='437' /></a>

Original artist: Bogdan Giuşcă

- **File:Krill_swarm.jpg** *Source:* https://upload.wikimedia.org/wikipedia/commons/1/1f/Krill_swarm.jpg *License:* Public domain *Contributors:* NOAA *Original artist:* Jamie Hall
- **File:Lac_Operon.svg** *Source:* https://upload.wikimedia.org/wikipedia/commons/2/22/Lac_Operon.svg *License:* CC BY-SA 3.0 *Contributors:* Own work *Original artist:* T A RAJU
- **File:Land_ocean_ice_cloud_hires.jpg** *Source:* https://upload.wikimedia.org/wikipedia/commons/6/6b/Land_ocean_ice_cloud_hires.jpg *License:* Public domain *Contributors:* ? *Original artist:* ?
- **File:Lock-green.svg** *Source:* https://upload.wikimedia.org/wikipedia/commons/6/65/Lock-green.svg *License:* CC0 *Contributors:* en:File:Free-to-read_lock_75.svg *Original artist:* User:Trappist the monk
- **File:Logical_deterministic_individual-based_cellular_automata_model_of_interspecific_competition_for_a_single_limited_resource.gif** *Source:* https://upload.wikimedia.org/wikipedia/commons/1/10/Logical_deterministic_individual-based_cellular_automata_model_of_interspecific_competition_for_a_single_limited_resource.gif *License:* CC BY-SA 4.0 *Contributors:* Own work *Original artist:* Lev Kalmykov

- **File:Logical_deterministic_individual-based_cellular_automata_model_of_single_species_population_growth.gif** *Source:* https://upload.wikimedia.org/wikipedia/commons/b/bf/Logical_deterministic_individual-based_cellular_automata_model_of_single_species_population_growth.gif *License:* CC BY-SA 4.0 *Contributors:* Own work *Original artist:* Lev Kalmykov
- **File:Logo_sociology.svg** *Source:* https://upload.wikimedia.org/wikipedia/commons/a/a6/Logo_sociology.svg *License:* Public domain *Contributors:* Own work *Original artist:* Tomeq183
- **File:Lorenz_attractor_yb.svg** *Source:* https://upload.wikimedia.org/wikipedia/commons/5/5b/Lorenz_attractor_yb.svg *License:* CC-BY-SA-3.0 *Contributors:* Own work based on images Image:Lorenz system r28 s10 b2-6666.png by User:Wikimol and Image:Lorenz attractor.svg by User:Dschwen *Original artist:* User:Wikimol, User:Dschwen
- **File:Lucretius_Rome.jpg** *Source:* https://upload.wikimedia.org/wikipedia/commons/b/bd/Lucretius_Rome.jpg *License:* Public domain *Contributors:* Own work, photo by Colle Pincio, digital new versione *Original artist:* Photo: StefanoRR; Sculpture: Unknown Italian artist
- **File:Manual_coffee_preperation.jpg** *Source:* https://upload.wikimedia.org/wikipedia/commons/9/92/Manual_coffee_preperation.jpg *License:* CC BY-SA 2.0 *Contributors:* Flickr *Original artist:* miheco from California, USA
- **File:Mariehønseår.jpg** *Source:* https://upload.wikimedia.org/wikipedia/commons/8/8c/Marieh%C3%B8nse%C3%A5r.jpg *License:* CC BY-SA 2.5 dk *Contributors:* Own work *Original artist:* Hdalgaard
- **File:Mass_collaboration.jpg** *Source:* https://upload.wikimedia.org/wikipedia/commons/5/5d/Mass_collaboration.jpg *License:* CC BY-SA 4.0 *Contributors:* https://markfoden.com/2011/09/toogoodtowaste/ *Original artist:* Mark Elliot
- **File:Mathematical_models_for_complex_systems.jpg** *Source:* https://upload.wikimedia.org/wikipedia/commons/f/f3/Mathematical_models_for_complex_systems.jpg *License:* CC BY-SA 4.0 *Contributors:* Own work *Original artist:* Lev Kalmykov
- **File:Merge-arrows.svg** *Source:* https://upload.wikimedia.org/wikipedia/commons/5/52/Merge-arrows.svg *License:* Public domain *Contributors:* ? *Original artist:* ?
- **File:Metric_vs_topological_distance_in_schools_of_fish.png** *Source:* https://upload.wikimedia.org/wikipedia/commons/b/b8/Metric_vs_topological_distance_in_schools_of_fish.png *License:* Public domain *Contributors:* Derivative of File:Metric vs topological distance for animal aggregations.png *Original artist:* User:Murphd84
- **File:MexicaliEarthquakeSwarm.gif** *Source:* https://upload.wikimedia.org/wikipedia/commons/a/a4/MexicaliEarthquakeSwarm.gif *License:* Public domain *Contributors:* From the federal U.S. Geological Survey website: data.scec.org *Original artist:* Unknown<a href='https://www.wikidata.org/wiki/Q4233718' title='wikidata:Q4233718'><img alt='wikidata:Q4233718' src='https://upload.wikimedia.org/wikipedia/commons/thumb/f/ff/Wikidata-logo.svg/20px-Wikidata-logo.svg.png' width='20' height='11' srcset='https://upload.wikimedia.org/wikipedia/commons/thumb/f/ff/Wikidata-logo.svg/30px-Wikidata-logo.svg.png 1.5x, https://upload.wikimedia.org/wikipedia/commons/thumb/f/ff/Wikidata-logo.svg/40px-Wikidata-logo.svg.png 2x' data-file-width='1050' data-file-height='590' /></a>
- **File:MoabAlcove.JPG** *Source:* https://upload.wikimedia.org/wikipedia/commons/7/73/MoabAlcove.JPG *License:* CC BY-SA 3.0 *Contributors:* I (Qfl247 (talk)) created this work entirely by myself. (Original uploaded on en.wikipedia) *Original artist:* Qfl247 (talk) (Transferred by Citypeek/Original uploaded by Qfl247)
- **File:MontreGousset001.jpg** *Source:* https://upload.wikimedia.org/wikipedia/commons/4/45/MontreGousset001.jpg *License:* CC-BY-SA-3.0 *Contributors:* Self-published work by ZA *Original artist:* Isabelle Grosjean ZA
- **File:Moofushi_Kandu_fish.jpg** *Source:* https://upload.wikimedia.org/wikipedia/commons/3/32/Moofushi_Kandu_fish.jpg *License:* CC BY-SA 2.5 it *Contributors:* Own work *Original artist:* Bruno de Giusti
- **File:Moreno_Sociogram_1st_Grade.png** *Source:* https://upload.wikimedia.org/wikipedia/commons/4/4b/Moreno_Sociogram_1st_Grade.png *License:* CC BY-SA 4.0 *Contributors:* Own work *Original artist:* Martin Grandjean
- **File:Moreno_Sociogram_2nd_Grade.png** *Source:* https://upload.wikimedia.org/wikipedia/commons/b/b6/Moreno_Sociogram_2nd_Grade.png *License:* CC BY-SA 4.0 *Contributors:* Own work *Original artist:* Martin Grandjean
- **File:Mutation_and_selection_diagram.svg** *Source:* https://upload.wikimedia.org/wikipedia/commons/f/f3/Mutation_and_selection_diagram.svg *License:* CC-BY-SA-3.0 *Contributors:* GPL image Image:643px-Explanation of Evolution v2.1.PNG *Original artist:* Elembis
- **File:Nanga_Parbat_Indus_Gorge.jpg** *Source:* https://upload.wikimedia.org/wikipedia/commons/7/7f/Nanga_Parbat_Indus_Gorge.jpg *License:* CC BY-SA 3.0 *Contributors:* Own work *Original artist:* Heavyrunner
- **File:Nb3O7(OH)_self-organization2.jpg** *Source:* https://upload.wikimedia.org/wikipedia/commons/3/3f/Nb3O7%28OH%29_self-organization2.jpg *License:* CC BY 3.0 *Contributors:* http://pubs.rsc.org/en/content/articlehtml/2014/ta/c4ta02202e *Original artist:* Sophia B. Betzler et al.
- **File:Nuvola_apps_kaboodle.svg** *Source:* https://upload.wikimedia.org/wikipedia/commons/1/1b/Nuvola_apps_kaboodle.svg *License:* LGPL *Contributors:* http://ftp.gnome.org/pub/GNOME/sources/gnome-themes-extras/0.9/gnome-themes-extras-0.9.0.tar.gz *Original artist:* David Vignoni / ICON KING
- **File:Office-book.svg** *Source:* https://upload.wikimedia.org/wikipedia/commons/a/a8/Office-book.svg *License:* Public domain *Contributors:* This and myself. *Original artist:* Chris Down/Tango project
- **File:OpenSystemRepresentation.svg** *Source:* https://upload.wikimedia.org/wikipedia/commons/7/77/OpenSystemRepresentation.svg *License:* CC BY-SA 4.0 *Contributors:* Own work *Original artist:* Krauss
- **File:Open_Access_logo_PLoS_transparent.svg** *Source:* https://upload.wikimedia.org/wikipedia/commons/7/77/Open_Access_logo_PLoS_transparent.svg *License:* CC0 *Contributors:* http://www.plos.org/ *Original artist:* art designer at PLoS, modified by Wikipedia users Nina, Beao, and JakobVoss
- **File:Open_book_nae_02.svg** *Source:* https://upload.wikimedia.org/wikipedia/commons/9/92/Open_book_nae_02.svg *License:* CC0 *Contributors:* OpenClipart *Original artist:* nae

- **File:Oscillator.gif** *Source:* https://upload.wikimedia.org/wikipedia/commons/8/86/Oscillator.gif *License:* CC-BY-SA-3.0 *Contributors:* Transferred from en.wikipedia to Commons. Self-made with Java program. *Original artist:* Grontesca at English Wikipedia
- **File:PAX6_Phenotypes_Washington_etal_PLoSBiol_e1000247.png** *Source:* https://upload.wikimedia.org/wikipedia/commons/1/14/PAX6_Phenotypes_Washington_etal_PLoSBiol_e1000247.png *License:* CC BY 2.5 *Contributors:* Figure 1 of Washington et al.: <a data-x-rel='nofollow' class='external text' href='http://www.plosbiology.org/article/info%3Adoi%2F10.1371%2Fjournal.pbio.1000247'>"Linking Human Diseases to Animal Models Using Ontology-Based Phenotype Annotation." PLoS Biol 7(11): e1000247. doi:10.1371/journal.pbio.1000247</a> (PDF) *Original artist:* Washington NL, Haendel MA, Mungall CJ, Ashburner M, Westerfield M, Lewis SE.
- **File:PSM_V84_D217_2_Flocking_habit_of_migratory_birds_fig5.jpg** *Source:* https://upload.wikimedia.org/wikipedia/commons/f/f5/PSM_V84_D217_2_Flocking_habit_of_migratory_birds_fig5.jpg *License:* Public domain *Contributors:* C. C. Trowbridge: *On the origin of the flocking habit of migratory birds*. The Popular science monthly, Volume 84, p213. New York, Popular Science Pub. Co., March 1914. Online: archive.org. *Original artist:* C. C. Trowbridge
- **File:PSO_Meta-Fitness_Landscape_(12_benchmark_problems).JPG** *Source:* https://upload.wikimedia.org/wikipedia/commons/a/ac/PSO_Meta-Fitness_Landscape_%2812_benchmark_problems%29.JPG *License:* Public domain *Contributors:* Own work *Original artist:* Pedersen, M.E.H., Tuning & Simplifying Heuristical Optimization, PhD Thesis, 2010, University of Southampton, School of Engineering Sciences, Computational Engineering and Design Group.
- **File:Palais_de_la_Decouverte_Tyrannosaurus_rex_p1050042.jpg** *Source:* https://upload.wikimedia.org/wikipedia/commons/a/ab/Palais_de_la_Decouverte_Tyrannosaurus_rex_p1050042.jpg *License:* CC-BY-SA-3.0 *Contributors:* Own work *Original artist:* Copyright © 2005 David Monniaux
- **File:ParticleSwarmArrowsAnimation.gif** *Source:* https://upload.wikimedia.org/wikipedia/commons/e/ec/ParticleSwarmArrowsAnimation.gif *License:* CC BY-SA 4.0 *Contributors:* Own work *Original artist:* Ephramac
- **File:People_icon.svg** *Source:* https://upload.wikimedia.org/wikipedia/commons/3/37/People_icon.svg *License:* CC0 *Contributors:* OpenClipart *Original artist:* OpenClipart
- **File:Pfau_imponierend.jpg** *Source:* https://upload.wikimedia.org/wikipedia/commons/1/1c/Pfau_imponierend.jpg *License:* CC-BY-SA-3.0 *Contributors:* Photo taken by user BS Thurner Hof *Original artist:* BS Thurner Hof
- **File:Pfeil_SO.svg** *Source:* https://upload.wikimedia.org/wikipedia/commons/a/a1/Pfeil_SO.svg *License:* Public domain *Contributors:* made by me (Inkscape or Corel-Draw or Flash) *Original artist:* user:Mjchael
- **File:Phase-diag2.svg** *Source:* https://upload.wikimedia.org/wikipedia/commons/3/34/Phase-diag2.svg *License:* CC-BY-SA-3.0 *Contributors:* SVG conversion from raster image Image:Phase-diag.png; some additions from Image:Phase diagram.png *Original artist:* me
- **File:Phase_change_-_en.svg** *Source:* https://upload.wikimedia.org/wikipedia/commons/0/0b/Phase_change_-_en.svg *License:* Public domain *Contributors:* Own work *Original artist:* F l a n k e r, penubag
- **File:Plumpollen0060.jpg** *Source:* https://upload.wikimedia.org/wikipedia/commons/3/39/Plumpollen0060.jpg *License:* CC-BY-SA-3.0 *Contributors:* ? *Original artist:* ?
- **File:Police_protect_Nick_Altrock_from_adoring_crowd,_1906_World_Series.jpg** *Source:* https://upload.wikimedia.org/wikipedia/commons/2/23/Police_protect_Nick_Altrock_from_adoring_crowd%2C_1906_World_Series.jpg *License:* CC BY 2.0 *Contributors:* originally posted to Flickr as Police protect Nick Altrock from adoring crowd, 1906 World Series *Original artist:* BPL
- **File:Portal-puzzle.svg** *Source:* https://upload.wikimedia.org/wikipedia/en/f/fd/Portal-puzzle.svg *License:* Public domain *Contributors:* ? *Original artist:* ?
- **File:Question_book-new.svg** *Source:* https://upload.wikimedia.org/wikipedia/en/9/99/Question_book-new.svg *License:* Cc-by-sa-3.0 *Contributors:*
  Created from scratch in Adobe Illustrator. Based on Image:Question book.png created by User:Equazcion *Original artist:*
  Tkgd2007
- **File:Rail_Bridge_Swarm_of_Starlings._-_geograph.org.uk_-_124591.jpg** *Source:* https://upload.wikimedia.org/wikipedia/commons/1/1b/Rail_Bridge_Swarm_of_Starlings._-_geograph.org.uk_-_124591.jpg *License:* CC BY-SA 2.0 *Contributors:* From geograph.org.uk *Original artist:* John Holmes
- **File:Random_graph_gephi.png** *Source:* https://upload.wikimedia.org/wikipedia/commons/a/ac/Random_graph_gephi.png *License:* CC BY 3.0 *Contributors:* Own work *Original artist:* Schulllz
- **File:Reaction_diffusion_spiral.gif** *Source:* https://upload.wikimedia.org/wikipedia/commons/c/c5/Reaction_diffusion_spiral.gif *License:* CC-BY-SA-3.0 *Contributors:* Transferred from en.wikipedia to Commons. *Original artist:* Huboedeker at English Wikipedia
- **File:Reaction_diffusion_stationary_ds.gif** *Source:* https://upload.wikimedia.org/wikipedia/commons/4/42/Reaction_diffusion_stationary_ds.gif *License:* CC-BY-SA-3.0 *Contributors:* Transferred from en.wikipedia to Commons. *Original artist:* Huboedeker at English Wikipedia
- **File:Reaction_diffusion_target.gif** *Source:* https://upload.wikimedia.org/wikipedia/commons/2/28/Reaction_diffusion_target.gif *License:* CC-BY-SA-3.0 *Contributors:* Transferred from en.wikipedia to Commons. *Original artist:* Huboedeker at English Wikipedia
- **File:Recurrent_ann_dependency_graph.png** *Source:* https://upload.wikimedia.org/wikipedia/commons/7/79/Recurrent_ann_dependency_graph.png *License:* CC-BY-SA-3.0 *Contributors:* ? *Original artist:* ?
- **File:Restricted_Boltzmann_machine.svg** *Source:* https://upload.wikimedia.org/wikipedia/commons/e/e8/Restricted_Boltzmann_machine.svg *License:* CC BY-SA 3.0 *Contributors:* Own work *Original artist:* Qwertyus
- **File:Robot-army.png** *Source:* https://upload.wikimedia.org/wikipedia/commons/8/8e/Robot-army.png *License:* GPL *Contributors:* [1], from the Swarmbot.org project *Original artist:* Serg (Sergey Kornienko (?)), cropped by Zanaq

- **File:Runtimes_of_algorithms.jpg** *Source:* https://upload.wikimedia.org/wikipedia/en/3/32/Runtimes_of_algorithms.jpg *License:* Cc-by-sa-3.0 *Contributors:* ? *Original artist:* ?
- **File:SWARM-vs-guerilla.png** *Source:* https://upload.wikimedia.org/wikipedia/commons/1/10/SWARM-vs-guerilla.png *License:* Public domain *Contributors:* US Army National Ground Intelligence Center *Original artist:* Sean J.A. Edwards
- **File:Safari_ants.jpg** *Source:* https://upload.wikimedia.org/wikipedia/commons/3/34/Safari_ants.jpg *License:* CC-BY-SA-3.0 *Contributors:* Own work *Original artist:* Mehmet Karatay
- **File:Salp.jpg** *Source:* https://upload.wikimedia.org/wikipedia/commons/6/63/Salp.jpg *License:* CC BY-SA 2.0 *Contributors:* Salp, influencers on the planet's climate - P1040203 *Original artist:* Lars Plougmann from London, United Kingdom
- **File:Sand_dune_ripples.jpg** *Source:* https://upload.wikimedia.org/wikipedia/commons/c/cd/Sand_dune_ripples.jpg *License:* CC BY-SA 2.0 *Contributors:* http://www.flickr.com/photos/shirazc/3387882509/ *Original artist:* Shiraz Chakera http://www.flickr.com/photos/shirazc/
- **File:Scale-free_network_sample.png** *Source:* https://upload.wikimedia.org/wikipedia/commons/7/77/Scale-free_network_sample.png *License:* CC-BY-SA-3.0 *Contributors:* ? *Original artist:* ?
- **File:Schwarm_Wanderheuschrecke.jpg** *Source:* https://upload.wikimedia.org/wikipedia/commons/a/a5/Schwarm_Wanderheuschrecke.jpg *License:* Public domain *Contributors:* Brehms Thierleben. Allgemeine Kunde des Thierreichs, Neunter Band, Vierte Abtheilung: Wirbellose Thiere, Zweiter Band: Die Niederen Thiere. Leipzig: Verlag des Bibliographischen Instituts, 1887. http://www.zeno.org/Naturwissenschaften/I/bt09550a.jpg *Original artist:* ?
- **File:Scree_plot_showing_percent_of_explained_variance_for_the_first_five_factors_in_Woolley_et_al.'s_(2010)_two_original_studies_as_well_as_the_individual_intelligence_test_for_all_participants_(assessed_with_Wonderlic_Personnel_Test).png** *Source:* https://upload.wikimedia.org/wikipedia/commons/8/85/Scree_plot_showing_percent_of_explained_variance_for_the_first_five_factors_in_Woolley_et_al.%27s_%282010%29_two_original_studies_as_well_as_the_individual_intelligence_test_for_all_participants_%28assessed_with_Wonderlic_Personnel_Test%29.png *License:* CC BY-SA 4.0 *Contributors:* Own work *Original artist:* PaDBu
- **File:Searchtool.svg** *Source:* https://upload.wikimedia.org/wikipedia/en/6/61/Searchtool.svg *License:* ? *Contributors:* ? *Original artist:* ?
- **File:Self-replication_of_sphynx_hexidiamonds.svg** *Source:* https://upload.wikimedia.org/wikipedia/commons/f/fa/Self-replication_of_sphynx_hexidiamonds.svg *License:* Public domain *Contributors:* en:Image:Sphnxhex.png *Original artist:* en:User:Spottedowl, User:Stannered
- **File:Shadow_Hand_Bulb_large.jpg** *Source:* https://upload.wikimedia.org/wikipedia/commons/c/c5/Shadow_Hand_Bulb_large.jpg *License:* CC-BY-SA-3.0 *Contributors:* http://www.shadowrobot.com/media/pictures.shtml *Original artist:* Richard Greenhill and Hugo Elias (myself) of the Shadow Robot Company
- **File:Signal_transduction_pathways.svg** *Source:* https://upload.wikimedia.org/wikipedia/commons/b/b0/Signal_transduction_pathways.svg *License:* CC BY-SA 3.0 *Contributors:* This file was derived from: Signal transduction v1.png
 Original artist: cybertory
- **File:Single-layer_feedforward_artificial_neural_network.png** *Source:* https://upload.wikimedia.org/wikipedia/commons/3/32/Single-layer_feedforward_artificial_neural_network.png *License:* CC BY-SA 4.0 *Contributors:* Own work *Original artist:* Akritasa
- **File:Single_layer_ann.svg** *Source:* https://upload.wikimedia.org/wikipedia/commons/b/be/Single_layer_ann.svg *License:* CC BY 3.0 *Contributors:* Own work *Original artist:* Mcstrother
- **File:Sinosauropteryxfossil.jpg** *Source:* https://upload.wikimedia.org/wikipedia/commons/c/c5/Sinosauropteryxfossil.jpg *License:* CC BY-SA 2.0 *Contributors:* Dinosaurs! *Original artist:* Sam / Olai Ose / Skjaervoy from Zhangjiagang, China
- **File:Small-world-network-example.png** *Source:* https://upload.wikimedia.org/wikipedia/commons/3/37/Small-world-network-example.png *License:* CC BY-SA 3.0 *Contributors:* Own work *Original artist:* Schulllz
- **File:Snapshot_of_weighted_stochastic_lattice.jpg** *Source:* https://upload.wikimedia.org/wikipedia/commons/a/a7/Snapshot_of_weighted_stochastic_lattice.jpg *License:* CC BY-SA 4.0 *Contributors:* https://arxiv.org/abs/1409.7928 *Original artist:* M. K. Hassan
- **File:SnowflakesWilsonBentley.jpg** *Source:* https://upload.wikimedia.org/wikipedia/commons/c/c2/SnowflakesWilsonBentley.jpg *License:* Public domain *Contributors:* Plate XIX of "Studies among the Snow Crystals ... " by Wilson Bentley, "The Snowflake Man." From Annual Summary of the "Monthly Weather Review" for 1902. *Original artist:* Wilson Bentley
- **File:Sound-icon.svg** *Source:* https://upload.wikimedia.org/wikipedia/commons/4/47/Sound-icon.svg *License:* LGPL *Contributors:* Derivative work from Silsor's versio *Original artist:* Crystal SVG icon set
- **File:South_Africa-Mpumalanga-Gods_Window002.jpg** *Source:* https://upload.wikimedia.org/wikipedia/commons/a/ac/South_Africa-Mpumalanga-Gods_Window002.jpg *License:* CC BY-SA 3.0 *Contributors:* Own work *Original artist:* NJR ZA
- **File:Speciation_modes_edit.svg** *Source:* https://upload.wikimedia.org/wikipedia/commons/d/d3/Speciation_modes_edit.svg *License:* Public domain *Contributors:* ? *Original artist:* ?
- **File:Srep17095-f1.jpg** *Source:* https://upload.wikimedia.org/wikipedia/commons/4/45/Srep17095-f1.jpg *License:* CC BY-SA 4.0 *Contributors:* Own work *Original artist:* Ajalvare
- **File:St_5-xband-antenna.jpg** *Source:* https://upload.wikimedia.org/wikipedia/commons/f/ff/St_5-xband-antenna.jpg *License:* Public domain *Contributors:* ? *Original artist:* ?
- **File:Standardized_Regression_Coefficients.png** *Source:* https://upload.wikimedia.org/wikipedia/commons/4/41/Standardized_Regression_Coefficients.png *License:* CC BY-SA 4.0 *Contributors:* Own work *Original artist:* PaDBu
- **File:Svm_max_sep_hyperplane_with_margin.png** *Source:* https://upload.wikimedia.org/wikipedia/commons/2/2a/Svm_max_sep_hyperplane_with_margin.png *License:* Public domain *Contributors:* Own work *Original artist:* Cyc
- **File:Symbol_book_class2.svg** *Source:* https://upload.wikimedia.org/wikipedia/commons/8/89/Symbol_book_class2.svg *License:* CC BY-SA 2.5 *Contributors:* Mad by Lokal_Profil by combining: *Original artist:* Lokal_Profil

47.10. TEXT AND IMAGE SOURCES, CONTRIBUTORS, AND LICENSES

- **File:Synapse_deployment.jpg** *Source:* https://upload.wikimedia.org/wikipedia/en/2/22/Synapse_deployment.jpg *License:* CC-BY-SA-2.5 *Contributors:* ? *Original artist:* ?
- **File:Synchropredation.gif** *Source:* https://upload.wikimedia.org/wikipedia/commons/5/59/Synchropredation.gif *License:* CC-BY-SA-3.0 *Contributors:* http://en.wikipedia.org/wiki/Image:Synchropredation.gif *Original artist:* Mr. Kils 320×200 (17,314 bytes) (synchropredation animation by uwe kils gfdl self)
- **File:TPI1_structure.png** *Source:* https://upload.wikimedia.org/wikipedia/commons/1/1c/TPI1_structure.png *License:* Public domain *Contributors:* based on 1wyi (http://www.pdb.org/pdb/explore/explore.do?structureId=1WYI), made in pymol *Original artist:* —<a href='//commons.wikimedia.org/wiki/User:A₂₃T_oth' title='User:A₂₃T_oth'>A₂₃</a><a href='//commons.wikimedia.org/wiki/User_talk:A₂₃T_oth' title='User talk:A₂₃T_oth'>T_oth</a>
- **File:TalusConesIsfjorden.jpg** *Source:* https://upload.wikimedia.org/wikipedia/commons/d/d5/TalusConesIsfjorden.jpg *License:* Public domain *Contributors:* Own work *Original artist:* Wilson44691
- **File:Tampering_W_Nature_Guacharacas.jpg** *Source:* https://upload.wikimedia.org/wikipedia/commons/2/28/Tampering_W_Nature_Guacharacas.jpg *License:* CC BY-SA 4.0 *Contributors:* Own work *Original artist:* Roberto Galindo Deshays
- **File:Termite_Cathedral_DSC03570.jpg** *Source:* https://upload.wikimedia.org/wikipedia/commons/7/73/Termite_Cathedral_DSC03570.jpg *License:* CC-BY-SA-3.0 *Contributors:* [1] *Original artist:* taken by w:User:Yewenyi
- **File:Text_document_with_red_question_mark.svg** *Source:* https://upload.wikimedia.org/wikipedia/commons/a/a4/Text_document_with_red_question_mark.svg *License:* Public domain *Contributors:* Created by bdesham with Inkscape; based upon Text-x-generic.svg from the Tango project. *Original artist:* Benjamin D. Esham (bdesham)
- **File:Textile_cone.JPG** *Source:* https://upload.wikimedia.org/wikipedia/commons/7/7d/Textile_cone.JPG *License:* CC-BY-SA-3.0 *Contributors:* Own work; Location: Cod Hole, Great Barrier Reef, Australia *Original artist:* Richard Ling <wikipedia@rling.com>
- **File:ThaiFledermaus.gif** *Source:* https://upload.wikimedia.org/wikipedia/commons/2/28/ThaiFledermaus.gif *License:* CC-BY-SA-3.0 *Contributors:* aufgenommen von Paul Lenz *Original artist:* Paul Lenz, Plenz at German Wikipedia
- **File:Thamnophis_sirtalis_sirtalis_Wooster.jpg** *Source:* https://upload.wikimedia.org/wikipedia/commons/f/f7/Thamnophis_sirtalis_sirtalis_Wooster.jpg *License:* Public domain *Contributors:* Own work *Original artist:* Wilson44691
- **File:The_pattern_tree_in_FPF_algorithm.jpg** *Source:* https://upload.wikimedia.org/wikipedia/en/4/41/The_pattern_tree_in_FPF_algorithm.jpg *License:* Cc-by-sa-3.0 *Contributors:* ? *Original artist:* ?
- **File:Thomas_Robert_Malthus_Wellcome_L0069037_-crop.jpg** *Source:* https://upload.wikimedia.org/wikipedia/commons/d/d5/Thomas_Robert_Malthus_Wellcome_L0069037_-crop.jpg *License:* CC BY 4.0 *Contributors:*
  http://wellcomeimages.org/indexplus/obf_images/fa/25/d2c7707f809bd259eb86d61d1cc5.jpg
  Original artist: John Linnell
- **File:Tiger_Bush_Niger_Corona_1965-12-31.jpg** *Source:* https://upload.wikimedia.org/wikipedia/commons/1/1b/Tiger_Bush_Niger_Corona_1965-12-31.jpg *License:* Public domain *Contributors:* Data available from the U.S. Geological Survey *Original artist:* US Agency
- **File:Torus.png** *Source:* https://upload.wikimedia.org/wikipedia/commons/1/17/Torus.png *License:* Public domain *Contributors:* This image was created with POV-Ray *Original artist:* LucasVB
- **File:Travelling_wave_for_Fisher_equation.svg** *Source:* https://upload.wikimedia.org/wikipedia/commons/7/78/Travelling_wave_for_Fisher_equation.svg *License:* Public domain *Contributors:* Own work, created using Matlab *Original artist:* Jitse Niesen
- **File:Tree_of_life.svg** *Source:* https://upload.wikimedia.org/wikipedia/commons/0/09/Tree_of_life.svg *License:* CC-BY-SA-3.0 *Contributors:* No machine-readable source provided. Own work assumed (based on copyright claims). *Original artist:* No machine-readable author provided. Vanished user fijtji34toksdcknqrjn54yoimascj assumed (based on copyright claims).
- **File:Tree_of_life_by_Haeckel.jpg** *Source:* https://upload.wikimedia.org/wikipedia/commons/d/de/Tree_of_life_by_Haeckel.jpg *License:* Public domain *Contributors:* First version from en.wikipedia; description page was here. Later versions derived from this scan, from the American Philosophical Society Museum. *Original artist:* Ernst Haeckel
- **File:Tripletsnew2012.png** *Source:* https://upload.wikimedia.org/wikipedia/commons/4/43/Tripletsnew2012.png *License:* CC BY 4.0 *Contributors:* Own work *Original artist:* Thinkbig-project
- **File:Turing_bifurcation_1.gif** *Source:* https://upload.wikimedia.org/wikipedia/commons/1/1d/Turing_bifurcation_1.gif *License:* CC-BY-SA-3.0 *Contributors:* Transferred from en.wikipedia to Commons. *Original artist:* Huboedeker at English Wikipedia
- **File:Turing_bifurcation_2.gif** *Source:* https://upload.wikimedia.org/wikipedia/commons/1/1d/Turing_bifurcation_2.gif *License:* CC-BY-SA-3.0 *Contributors:* Transferred from en.wikipedia to Commons. *Original artist:* Huboedeker at English Wikipedia
- **File:Turing_bifurcation_3.gif** *Source:* https://upload.wikimedia.org/wikipedia/commons/6/68/Turing_bifurcation_3.gif *License:* CC-BY-SA-3.0 *Contributors:* Transferred from en.wikipedia to Commons. *Original artist:* Huboedeker at English Wikipedia
- **File:Two-layer_feedforward_artificial_neural_network.png** *Source:* https://upload.wikimedia.org/wikipedia/commons/5/58/Two-layer_feedforward_artificial_neural_network.png *License:* CC BY-SA 4.0 *Contributors:* Own work *Original artist:* Akritasa
- **File:Two_layer_ann.svg** *Source:* https://upload.wikimedia.org/wikipedia/commons/7/7f/Two_layer_ann.svg *License:* CC BY 3.0 *Contributors:* Own work *Original artist:* Mestrother
- **File:U.S._states_(and_territories)_by_election_methods,_2016.svg** *Source:* https://upload.wikimedia.org/wikipedia/commons/3/31/U.S._states_%28and_territories%29_by_election_methods%2C_2016.svg *License:* CC BY 4.0 *Contributors:* Own work; Map is based on here. *Original artist:* Ali Zifan
- **File:UNU_Predicts.gif** *Source:* https://upload.wikimedia.org/wikipedia/commons/0/06/UNU_Predicts.gif *License:* CC BY-SA 4.0 *Contributors:* Unanimous AI *Original artist:* UNU

- **File:US_Navy_100531-N-7676W-075_Visitors_interact_with_the_mobile,_dexterous,_social_(MDS)_robot_Octavia_at_the_Office_of_Naval_Research_(ONR)_exhibit_during_Fleet_Week_New_York_2010.jpg** *Source:* https://upload.wikimedia.org/wikipedia/commons/5/53/US_Navy_100531-N-7676W-075_Visitors_interact_with_the_mobile%2C_dexterous%2C_social_%28MDS%29_robot_Octavia_at_the_Office_of_Naval_Research_%28ONR%29_exhibit_during_Fleet_Week_New_York_2010.jpg *License:* Public domain *Contributors:*
  This Image was released by the United States Navy with the ID 100531-N-7676W-075 <a class='external text' href='//commons.wikimedia.org/w/index.php?title=Category:Files_created_by_the_United_States_Navy_with_known_IDs,,&,,filefrom=100531-N-7676W-075#mw-category-media'>(next)</a>.
  This tag does not indicate the copyright status of the attached work. A normal copyright tag is still required. See Commons:Licensing for more information.
  Original artist: U.S. Navy photo by John F. Williams
- **File:VU0K1843_(39985550).jpg** *Source:* https://upload.wikimedia.org/wikipedia/commons/f/fa/VU0K1843_%2839985550%29.jpg *License:* CC BY 2.0 *Contributors:* VU0K1843.jpg *Original artist:* Christopher Michel from San Francisco, USA
- **File:Velke_Hincovo_pleso.jpg** *Source:* https://upload.wikimedia.org/wikipedia/commons/4/49/Velke_Hincovo_pleso.jpg *License:* CC-BY-SA-3.0 *Contributors:* Own work *Original artist:* Kristo
- **File:Wallula-Gap-the-sisters.JPG** *Source:* https://upload.wikimedia.org/wikipedia/commons/6/62/Wallula-Gap-the-sisters.JPG *License:* CC-BY-SA-3.0 *Contributors:* ? *Original artist:* ?
- **File:Wasp_mimicry.jpg** *Source:* https://upload.wikimedia.org/wikipedia/commons/b/b5/Wasp_mimicry.jpg *License:* CC BY 2.5 *Contributors:* PLoS *Original artist:* Image Credit: (A, C, E, and F) by Rob Knell; (B and D) by Tom Ings
- **File:Water_Crystals_on_Mercury_20Feb2010_CU1.jpg** *Source:* https://upload.wikimedia.org/wikipedia/commons/7/77/Water_Crystals_on_Mercury_20Feb2010_CU1.jpg *License:* CC BY-SA 3.0 *Contributors:* I photographed a car window with my Kodak digital camera *Previously published:* Published on Wikipedia, deleted by a vandal, unfortunately *Original artist:* Rusfuture
- **File:Watts-Strogatz-rewire.png** *Source:* https://upload.wikimedia.org/wikipedia/commons/e/e1/Watts-Strogatz-rewire.png *License:* CC BY-SA 3.0 *Contributors:* Own work *Original artist:* Jmcatania
- **File:Whale_skeleton.png** *Source:* https://upload.wikimedia.org/wikipedia/commons/1/16/Whale_skeleton.png *License:* Public domain *Contributors:* Meyers Konversionlexikon 1888 *Original artist:* Meyers Konversionlexikon
- **File:Wiki_letter_w_cropped.svg** *Source:* https://upload.wikimedia.org/wikipedia/commons/1/1c/Wiki_letter_w_cropped.svg *License:* CC-BY-SA-3.0 *Contributors:* This file was derived from Wiki letter w.svg: <a href='//commons.wikimedia.org/wiki/File:Wiki_letter_w.svg' class='image'><img alt='Wiki letter w.svg' src='https://upload.wikimedia.org/wikipedia/commons/thumb/6/6c/Wiki_letter_w.svg/50px-Wiki_letter_w.svg.png' width='50' height='50' srcset='https://upload.wikimedia.org/wikipedia/commons/thumb/6/6c/Wiki_letter_w.svg/75px-Wiki_letter_w.svg.png 1.5x, https://upload.wikimedia.org/wikipedia/commons/thumb/6/6c/Wiki_letter_w.svg/100px-Wiki_letter_w.svg.png 2x' data-file-width='44' data-file-height='44' /></a>
  Original artist: Derivative work by Thumperward
- **File:Wikibooks-logo-en-noslogan.svg** *Source:* https://upload.wikimedia.org/wikipedia/commons/d/df/Wikibooks-logo-en-noslogan.svg *License:* CC BY-SA 3.0 *Contributors:* Own work *Original artist:* User:Bastique, User:Ramac et al.
- **File:Wikibooks-logo.svg** *Source:* https://upload.wikimedia.org/wikipedia/commons/f/fa/Wikibooks-logo.svg *License:* CC BY-SA 3.0 *Contributors:* Own work *Original artist:* User:Bastique, User:Ramac et al.
- **File:Wikinews-logo.svg** *Source:* https://upload.wikimedia.org/wikipedia/commons/2/24/Wikinews-logo.svg *License:* CC BY-SA 3.0 *Contributors:* This is a cropped version of Image:Wikinews-logo-en.png. *Original artist:* Vectorized by Simon 01:05, 2 August 2006 (UTC) Updated by Time3000 17 April 2007 to use official Wikinews colours and appear correctly on dark backgrounds. Originally uploaded by Simon.
- **File:Wikipedia_multilingual_network_graph_July_2013.svg** *Source:* https://upload.wikimedia.org/wikipedia/commons/5/5b/Wikipedia_multilingual_network_graph_July_2013.svg *License:* CC BY-SA 3.0 *Contributors:* Own work *Original artist:* Computermacgyver
- **File:Wikiquote-logo.svg** *Source:* https://upload.wikimedia.org/wikipedia/commons/f/fa/Wikiquote-logo.svg *License:* Public domain *Contributors:* Own work *Original artist:* Rei-artur
- **File:Wikisource-logo.svg** *Source:* https://upload.wikimedia.org/wikipedia/commons/4/4c/Wikisource-logo.svg *License:* CC BY-SA 3.0 *Contributors:* Rei-artur *Original artist:* Nicholas Moreau
- **File:Wikiversity-logo-Snorky.svg** *Source:* https://upload.wikimedia.org/wikipedia/commons/1/1b/Wikiversity-logo-en.svg *License:* CC BY-SA 3.0 *Contributors:* Own work *Original artist:* Snorky
- **File:Wiktionary-logo-en-v2.svg** *Source:* https://upload.wikimedia.org/wikipedia/commons/9/99/Wiktionary-logo-en-v2.svg *License:* CC-BY-SA-3.0 *Contributors:* ? *Original artist:* ?
- **File:Wiktionary-logo-v2.svg** *Source:* https://upload.wikimedia.org/wikipedia/en/0/06/Wiktionary-logo-v2.svg *License:* CC-BY-SA-3.0 *Contributors:* ? *Original artist:* ?
- **File:World_Brain_HG_Wells_1938.jpg** *Source:* https://upload.wikimedia.org/wikipedia/commons/9/9e/World_Brain_HG_Wells_1938.jpg *License:* Public domain *Contributors:* http://www.abebooks.co.uk/servlet/BookDetailsPL?bi=11792769744 *Original artist:* H. G. Wells
- **File:Yemen_Chameleon_(cropped).jpg** *Source:* https://upload.wikimedia.org/wikipedia/commons/f/fc/Yemen_Chameleon_%28cropped%29.jpg *License:* CC BY-SA 3.0 *Contributors:* Own work *Original artist:* Chiswick Chap

47.10.3 Content license

- Creative Commons Attribution-Share Alike 3.0

www.ingramcontent.com/pod-product-compliance
Lightning Source LLC
Chambersburg PA
CBHW082319220526
45470CB00008B/2359